Equipment Commonly Used in the Organic Chemistry Laboratory

Filter flask

Büchner funnel

Hirsch funnel

Thiele tube

Pasteur or micro pipets

Separatory funnel (Squibb type)

Separatory funnel with ground-glass joints

Bent adapter

Clamp holder

Three-pronged clamp

Vacuum adapter with ground-glass joints

West condenser

Hempel fractionating column

Round-bottom flask

Claisen connecting tube

Still head

Thermometer adapter with neoprene fitting

Rubber septum

Experimental Organic Chemistry

A Miniscale Approach

Experimental Organic Chemistry

A Miniscale Approach

Royston M. Roberts
The University of Texas at Austin

John C. Gilbert
The University of Texas at Austin

Stephen F. Martin
The University of Texas at Austin

Saunders Golden Sunburst Series
Saunders College Publishing
Harcourt Brace College Publishers

Fort Worth • Philadelphia • San Diego • New York • Orlando • Austin
San Antonio • Toronto • Montreal • London • Sydney • Tokyo

Text Typeface: Times Roman
Compositor: Monotype Composition Company, Inc.
Publisher: John Vondeling
Developmental Editor: Beth Rosato
Managing Editor: Carol Field
Project Editor: Janet Nuciforo
Copy Editor: Andrew Potter
Manager of Art and Design: Carol Bleistine
Art Directors: Anne Muldrow, Susan Blaker
Text Designer: Irwin Hahn
Cover Designer: Lou Fuiano
Text Artwork: Rolin Graphics
Director of EDP: Tim Frelick
Production Manager: Charlene Squibb
Marketing Manager: Majorie Waldron

Cover Photo: Malonic acid recrystallized from the melt. Photographed with a Nikon polarizing microscope, crossed polars and filtered with a quarter-wave plate. © Arthur L. Meyer.

Printed in the United States of America

Experimental Organic Chemistry: A Miniscale Approach

ISBN: 0-03-029008-2

Library of Congress Catalog Card Number: 93-086830

3456   32   987654321

This book was printed on paper made from waste paper, containing 10% post-consumer waste and 40% pre-consumer waste, measured as a percentage of total fiber weight content.

PREFACE

The laboratory course in organic chemistry has undergone significant changes in the past decade. The motivating forces behind these changes are mainly associated with concerns about the increasing costs of chemicals and their disposal. In response to these concerns, teachers have turned to microscale organic chemistry, a trend that we believe is pedagogically unwise. The financial benefits associated with this option are far outweighed by the tedium that attends working with only milligrams of material, and the failure to observe any product. More than once we have heard students say that the only way they knew a product had been formed was by smelling it! This type of laboratory experience does not reflect the real world of experimental organic chemistry.

We wrote this textbook with the goal of balancing the financial and environmental considerations that preclude performing experiments on a multigram scale and the educational benefits associated with doing reactions on a scale that allows isolation and characterization of products without using exotic equipment and tedious experimental techniques. We believe we achieved that goal through development of experimental procedures that generally require less than 3 grams of individual starting materials. Working at this "miniscale" level still allows use of the type of apparatus normally found in organic laboratories but at a lower expense for the purchase and disposal of chemicals.

The philosophy we adopted in developing the discussions preceding the Experimental Procedures was to provide in-depth analysis of the theoretical and technical considerations so the textbook is self-contained. To some extent, these descriptions may duplicate those in your lecture textbook. However, the coverage of the particular topics related to our experiments varies greatly from one textbook to another, and this variation fully justifies the possibility of some duplication. The focus provided by our own discussions is intended to give the experimentalist the basic information needed for complete understanding of the mechanistic and practical aspects of the procedures being performed.

The experiments we selected are intended not only to reinforce concepts given in the lecture component of organic chemistry but also to provide the student with a sense of the techniques that organic chemists use to test mechanistic hypotheses and to prepare a wide variety of compounds. Thus, experiments that illustrate concepts such as kinetic and thermodynamic control of reactions (Chapter 13), generation, reactions, and rearrangements of carbocations (Chapters 10 and 15), electrophilic aromatic and nucleophilic substitution processes (Chapters 15 and 14, respectively), selectivity of free-radical substitution (Chapter 9), and the stereochemistry and regiochemistry of addition reactions (Chapters 10, 11, and 17) are designed to augment the student's knowledge of these important subjects. Other experiments provide experience in fundamental laboratory techniques such as various forms of distillation, liquid-liquid and liquid-solid extraction, thin-layer and column chromatography, the basic principles for which are described in Chapters 4, 5, and 6, respectively. The value of enzymes for effecting enantioselective reactions is illustrated in Chapter 17, and the *in situ* generation of unstable or potentially hazardous reagents is presented in Chapters 10 and 12. Examples of multistep synthesis are contained in Chapters 15, 18, and 20, among others, and a rationale approach to solving the structures of unknowns with and without the aid of spectroscopic data is given in Chapter 23. In summary, the experimental procedures cover a broad range of topics and provide a sufficient number of options so that the instructor may focus the course in whatever manner is desired.

With each procedure we incorporate sections entitled **"Safety Alert"** and **"Finishing Touches."** These entries alert the instructors and students to possible hazards associated with the experimental operations being performed and to the methods for minimization and proper disposal of the chemical by-products of the procedures. Careful adherence to the safety recommendations we provided will make the laboratory a safer workplace, and the use of the recommended methods for handling chemical by-products will protect the environment and lessen the costs of the necessary disposal.

The importance of spectroscopy in modern experimental organic chemistry cannot be overemphasized. Consequently, a thorough discussion of the theory and practical techniques for infrared, nuclear magnetic resonance, and ultraviolet spectroscopy is presented in Chapter 8. Moreover, the infrared and nuclear magnetic spectra of essentially all of the organic starting materials and products and the ultraviolet spectra of a few compounds are provided in this textbook, and they offer an excellent opportunity to learn how to interpret spectral data.

Several "Historical Highlights" are included as part of the chapters to which they are related. These brief essays are intended to provide you with a sense of the excitement and insight on the part of the individuals whose observations evolved into the experiments that you perform in the laboratory. As you will see, luck has played a major role in some very important scientific discoveries!

Pre-Lab and Post-Lab Exercises

Each experiment in this textbook is accompanied by a set of questions. The "Pre-Lab Exercises" are intended to be worked as part of the preparation for the laboratory and are contained in a separate ancillary that accompanies the book. These exercises are designed to test your understanding of the basic concepts underlying the experiments that are being performed and the techniques required to complete the procedures in a *safe* and successful manner. Your instructor may ask that answers to these questions be submitted before you may begin work in the laboratory. Even if submission of answers is not required, working the exercises will help you derive the greatest educational benefit from the laboratory experience.

The "Post-Lab Exercises" are found under the heading, "Exercises," at the end of each Experimental Procedure. These questions are written to reinforce the principles that are illustrated by the experiments and to determine whether you understand the observations you made or the particular laboratory operations you performed.

Instructor's Manual

Teachers who adopt this textbook may obtain an *Instructor's Manual*. Its contents include listings of the laboratory equipment and chemicals needed and comments on the individual "Experimental Procedures." The manual also contains answers to all the pre- and post-lab exercises, as well as assignments of the ^{13}C spectra provided in this book.

Acknowledgments

A number of individuals made important contributions to this textbook. These include Cassandra Hutson, Lilly Yung Chih Hsu, Kevin Wheeler, Janet Macdonald and Teresa Winterringer (The University of Texas at Austin), and James Garrett (Stephen F. Austin University). In addition, we thank the many individuals at Saunders College Publishing for cajoling and encouraging us throughout the course of this project.

Royston M. Roberts
John C. Gilbert
Stephen F. Martin
November 1993

CONTENTS OVERVIEW

CONTENTS

IR, NMR, and UV Spectra

Compound	IR	^{1}H NMR	^{13}C NMR	UV
p-Acetamidobenzenesulfonamide	20.7	20.8a	20.8b	
p-Acetamidobenzenesulfonyl chloride	20.5	20.6a	20.6b	
Acetanilide	3.3	3.4a	3.4b	
Adduct, maleic anhydride and dienes	12.16	12.17a	12.17b	
Adipic acid (*See* Hexanedioic acid)				
Aniline	20.3	20.4a	20.4b	
trans-p-Anisalacetophenone	18.12	18.13a	18.13b	18.14
trans-p-Anisalacetophenone dibromide	18.19	18.20a	18.20b	18.21
p-Anisaldehyde	18.10	18.11a	18.11b	
Anisole	15.24	15.25a	15.25b	
Anthracene	5.6	5.7a	5.7b	
Benzaldehyde	16.9	16.10a	16.10b	
Benzoic acid	3.1	3.2a	3.2b	
Benzoic anhydride	8.17			
Benzonitrile	8.19			
Benzoyl chloride	8.16			
Benzyl alcohol	8.11	16.11a	16.11b	
Benzyl chloride	18.1	18.2a	18.2b	
Bromobenzene	15.8	15.9a	15.9b	
1-Bromobutane	14.3	14.4a	14.4b	
1-Bromohexane	10.30	10.31a	10.31b	
2-Bromohexane	10.32	10.33a	10.33b	
o-Bromonitrobenzene	15.12	15.13a	15.13b	
p-Bromonitrobenzene	15.10	8.35	15.11	
p-Bromophenol	15.26	15.27a	15.27b	
Butanal (*n*-butyraldehyde)		8.47		
1-Butanol	14.1	8.33	14.2	
1-Butanol (in DMSO-d_6)		8.34		
2-Butanone (methyl ethyl ketone)		8.43	8.51	

(Continued)

Compound	IR	^{1}H NMR	^{13}C NMR	UV
2-Butanone ([1]H-decoupled)			8.52	
Butanoic acid (butyric acid)		8.44		
3-Buten-2-one (methyl vinyl ketone)	18.32	18.33a	18.33b	
n-Butyl acetate (butyl ethanoate)	4.6	4.7a	4.7b	
	8.14			
t-Butylbenzene	9.16	9.17a	9.17b	
Butyl ethanoate (*See* n-Butyl acetate)				
n-Butyraldehyde (*See* Butanal)				
Butyric acid (*See* Butanoic acid)				
Carvone	7.15ab	7.16a	7.16b	
1-Chlorobutane	9.8	9.9a	9.9b	
Chlorocyclohexame	9.6	9.7a	9.7b	
2-Chloro-2-methylbutane	10.2	10.3a	10.3b	
1-Chloro-2-methylpropane		8.40		
Cinnamaldehyde				
N-Cinnamylidene-m-nitroaniline	17.9	17.10a	17.10b	
N-Cinnamyl-m-nitroaniline	17.11	17.12a	17.12b	
Citral	4.12	4.13a	4.13b	8.69
Cumene (*See* Isopropylbenzene)				
Cyclobutane-1,1-dicarboxylic acid	8.9a,b			
Cyclododecanol	16.5	16.6a	16.6b	
Cyclododecanone	16.7	16.8a	16.8b	
Cyclohexane	9.4	9.5a	9.5b	
Cyclohexane-cis-1,2-dicarboxylic acid	17.3	17.4a	17.4b	
cis-1,2-Cyclohexanediol	7.3	7.4a	7.4b	
trans-1,2-Cyclohexanediol	7.1	7.2a	7.2b	
Cyclohexanol	10.23	10.24a	10.24b	
Cyclohexanone	13.3	13.4a	13.4b	
Cyclohexanone semicarbazone	13.8	13.9a	13.9b	
Cyclohexene	10.25	10.26a	10.26b	
4-Cyclohexene-cis-1,2-dicarboxylic acid	17.1	17.2a	17.2b	
4-Cyclohexene-cis-1,2-dicarboxylic anhydride	12.8	12.9a	12.9b	
1,3-Cyclopentadiene	12.2	12.3a	12.3b	
1,x-Dichlorobutanes, mixture		9.10		
Decanedioyl dichloride	21.5	21.6a	21.6b	
Diethyl ether	8.18			
2,5-Dihydrothiophene-1,1-dioxide (*See* 3-Sulfolene)				
N,N-Dimethylacetamide	8.15			

(Continued)

Compound	IR	^{1}H NMR	^{13}C NMR	UV
1,3-Dimethylanthracene	15.22	15.23a	15.23b	
2,4-Dimethylanthraquinone	15.18	15.19a	15.19b	
Dimethylanthrones	15.20	15.21a	15.21b	
2-(2',4'-dimethylbenzoyl)benzoic acid	15.16	15.17a	15.17b	
2,3-Dimethylbutane	8.12a			
4,4-Dimethyl-2-cyclohexen-1-one	18.34	18.35a	18.35b	18.36
Dimethyl fumarate	7.8	7.9a	7.9b	
Dimethyl heptane-1,1-dicarboxylate	18.26	18.27a	18.27b	
Dimethyl maleate	7.6		7.7	
Dimethyl maleate and dimethyl fumarate, mixture		7.5		
Dimethyl malonate	18.24	18.25a	18.25b	
trans, trans-Diphenyl-1,3-butadiene	18.8	18.9a	18.9b	
Diphenyl ether	15.28	15.29a	15.29b	
Ethyl acetate (ethyl ethanoate)	4.4	4.5a	4.5b	
		8.46		
N-Ethylaniline		8.45		
Ethylbenzene	9.12	9.13a	9.13b	
Ethyl ethanoate (*See* Ethyl acetate)				
Ethyl phenyl ether (phenetole)		8.42		
Fluorene	6.4	6.5a	6.5b	
Fluorenol	17.13	17.14a	17.14b	
Fluorenone	6.6	6.7a	6.7b	
2-Furaldehyde	13.5	13.6a	13.6b	13.7
2-Furaldehyde semicarbazone	13.10	13.11a	13.11b	
3-(2-Furyl)-1-(3-nitrophenyl)propenone	18.17	18.18a	18.18b	
Heptane-1,1-dicarboxylic acid	18.28	18.29a	18.29b	
1,6-Hexanediamine	21.7	21.8a	21.8b	
Hexanedioic acid (adipic acid)	16.3	16.4a	16.4b	
1-Hexene	10.28	10.29a	10.29b	
Hydrocinnamic acid (*See* 3-Phenylpropanic acid)				
Isobutyl chloride (*See* 1-Chloro-2-methyl-propane)				
Isobutyraldehyde (*See* 2-Methylpropanal)				
(-)-Isopinocampheol	10.40	10.41a	10.41b	
Isopropylbenzene (cumene)	9.14	9.15a	9.15b	
Isopropyl-*p*-xylene	15.5b	15.6b		
Maleic anhydride	12.4	12.5a	12.5b	
2-Methoxyethyl acetate	23.2a	23.2b		

(Continued)

Compound	IR	^{1}H NMR	^{13}C NMR	UV
Methyl acetoacetate	17.16	17.17a	17.17b	
N-Methylaniline	8.20			
Methyl benzoate	19.1	19.2	8.53	
2-Methylbutane	8.12b			
2-Methyl-2-butanol (t-pentyl alcohol)	14.3	14.4a	14.4b	
3-Methyl-2-butanone			8.54	
2-Methyl-1-butene	10.6	10.7a	10.7b	
2-Methyl-2-butene	10.8	10.9a	10.9b	
Methylbutanes, mixtures from dehalogenation of 2-chloro-2-methylbutane		10.4 10.5		
2-Methyl-3-butyn-2-ol	11.1	11.2a	11.2b	
3-Methyl-4-cyclohexene-cis-1,2-dicarboxylic anhydride	12.16	12.17a	12.17b	
3-Methyl-2-cyclohexenone	8.3			
Methylcyclopentane		8.36		
Methyl ethyl ketone (See 2-Butanone)				
2-Methyl-3-heptanol	19.5	19.6a	19.6b	
3-Methyl-3-hydroxy-2-butanone	11.3	11.4a	11.4b	
Methyl 3(S)-hydroxybutyrate	17.18	17.19a	17.19b	
Methyl 3(S/R)-hydroxybutyrate/Eu(hfc)$_3$ complex (racemic)		17.20		
Methyl 3(S/R)-hdroxybutyrate/Eu(hfc)$_3$ complex (S enriched)		17.21		
4-Methyl-2-pentanol	10.11	10.12a	10.12b	
2-Methyl-1-pentene	10.19	10.20a	10.20b	
2-Methyl-2-pentene	10.21	10.22a	10.22b	
4-Methyl-1-pentene	10.13	8.30 10.14a	10.14	
cis-4-Methyl-2-pentene	10.17	10.18a	10.18b	
trans-4-Methyl-2-pentene	10.15	10.16a	10.16b	
4-Methyl-3-penten-2-one				8.68
2-Methylpropanal (isobutyraldehyde)	16.1	16.2a	16.2b	
2-Methyl-1-propanol (isobutyl alcohol)	8.55a	8.55b	8.55c	
Methyl vinyl ketone (See 3-Buten-2-one)				
Mineral Oil	8.8			
Naphthalene	3.5	3.6a	3.6b	
α-Naphthol	15.30	15.31a	15.31b	
m-Nitroacetophenone	18.15	18.16a	18.16b	
m-Nitroaniline	17.7	17.8a	17.8b	
p-Nitroaniline	5.4	5.5a	5.5b	

(Continued)

Compound	IR	¹H NMR	¹³C NMR	UV
Nitrobenzene	20.1	20.2a	20.2b	
o-Nitrobromobenzene (See *o*-Bromonitro-benzene)				
p-Nitrobromobenzene (See *p*-Bromonitro-benzene)	15.10	8.35	15.11	
1-Nitropropane		8.26		
Norbornene	10.34	10.35a	10.35b	
exo-Norbornanol	10.36	10.37a	10.37b	
Octanoic acid	18.30	18.31a	18.31b	
1,3-Pentadiene, commercial mixture of isomers	12.10	12.11		
cis-1,3-Pentadiene	12.14	12.15a	12.15b	
trans-1,3-Pentadiene	12.12	12.13a	12.13b	
Pentane		8.39		
t-Pentyl alcohol (*See* 2-Methyl-2-butanol)				
Perfluorokerosene	8.8			
Phenetole (*See* ethyl phenyl ether)				
Phenol	15.32	15.32a	15.32b	
Phenylacetylene (*See* Phenylethyne)				
1-Phenylethanamine	7.20	7.21a	7.21b	
Phenylethyne (phenyl acetylene)	8.13			
Phenylmethanol (*See* Benzyl alcohol)				
Phthalic anhydride	15.14	15.15a	15.15b	
(+)-α-Pinene	10.38	10.39a	10.39b	
Polystyrene	8.10			
n-Propyl-*p*-xylene	15.5a	15.6a		
cis-Stilbene	18.3	18.4a	18.4b	18.7
trans-Stilbene	18.5	18.6a	18.6b	18.7
Styrene	21.1	21.2a	21.2b	
Sulfanilamide	20.9	20.10a	20.10b	
3-Sulfolene (2,5-dihydrothiophene-1,1-dioxide)	12.6	12.7a	12.7b	
Tartaric acid	7.18	7.19a in D_2O	7.19b	
Trimyristin	5.1	5.2a	5.2b	
Triphenylmethanol	19.3	19.4a	19.4b	
Triptycene	8.7			
m-Xylene	15.12	15.13a	15.13b	
p-Xylene	15.3	15.4a	15.4b	

EXPERIMENTAL PROCEDURES

Experimental Organic Chemistry

A Miniscale Approach

INTRODUCTION, RECORD KEEPING, AND LABORATORY SAFETY

1.1 INTRODUCTION

The laboratory component of a course in organic chemistry has an important role in developing and augmenting your understanding of the subject matter. The theoretical concepts, functional groups, and reactions presented in the lecture part of the course may be abstract, but should become more understandable as a result of the experiments you perform. The "hands-on" experience gained in the laboratory, as you gather and interpret data from a variety of reactions, provides a sense of organic chemistry that is nearly impossible to communicate in lectures. For example, it is one thing to be told that the addition of bromine (Br_2) across the π-bond of most alkenes is a rapid process at room temperature but quite another to personally observe the *immediate* decoloration of a reddish solution of bromine in carbon tetrachloride (Br_2/CCl_4) as a few drops of it are added to cyclohexene. In this and many other cases you will encounter in the laboratory, the principles developed in the lectures will help you to predict what reaction(s) should occur when various reagents are combined and to understand the mechanistic course of the process(es). Performing reactions allows you to verify what you have been told in lecture.

Of course, the laboratory experience in organic chemistry has another important function beyond reinforcing the concepts presented in lecture—and that is to introduce you to the techniques and procedures that are of great importance to the successful practice of experimental organic chemistry. You will learn how to handle a variety of chemicals safely and how to manipulate apparatus properly, talents that are critical to your success as a student of the chemical sciences. Along with becoming more skilled in the technical aspects of laboratory work, you should also develop a proper scientific approach to the execution and interpretation of experiments. By reading and understanding this chapter, you will be better able to achieve these valuable goals.

1.1.1 PREPARING FOR THE LABORATORY

One of the most common misconceptions students have about performing experiments is that it is much like cooking; that is, one merely follows the directions given—the "recipe"—and the desired product or data will result. Such students enter the laboratory prepared to follow the provided experimental procedure in a more or less rote manner. This is an unfortunate attitude that can lead to inefficiencies, accidents, and minimal educational benefit and enjoyment from the laboratory experience.

The correct approach to being successful in the laboratory is *never* to begin any experiment until you understand its overall purpose and the reasons for each operation that you are to do. This means that you must *study*, not *just read*, the entire experiment *prior* to arriving at the laboratory. Rarely, if ever, can you complete the necessary preparation in five or ten minutes, which means that you should not wait until just before the laboratory period begins to do the study, thinking, and writing that are required. *Planning* how to spend your time in the laboratory is the key to efficient completion of the required experiments. Your performance in the laboratory will benefit enormously with proper advance work and so will your grade!

The specific details of what you should do prior to coming to the laboratory will be provided by your instructor. However, to encourage you to prepare in advance, we have developed a set of "Pre-Lab exercises" for each of the experimental procedures in this textbook. Your instructor may require you to submit the answers to the exercises for approval before authorizing you to do the assigned experiments. Even if you are not asked to complete the exercises, you will find that working them *prior* to the laboratory period will be a valuable educational tool to self-assess your understanding of the experiments to be done.

You will undoubtedly be required to maintain a laboratory notebook, which will serve as a complete, accurate, and neat record of the experimental work that you do. Once more, your instructor will provide an outline of what specific information should appear in this notebook, but part of what is prescribed will probably necessitate advance preparation, which will further enhance your ability to complete the experiments successfully. The laboratory notebook is a *permanent record* of your accomplishments in the course, and you should take pride in the quality and completeness of its contents!

1.1.2   WORKING IN THE LABORATORY

You should be aware that experimental organic chemistry is *potentially* dangerous, since many of the chemicals used are toxic and/or highly flammable, and most of the procedures require the use of glassware that is easily broken. Careless handling of these chemicals and sloppy assembly of apparatus are sources of danger not only to yourself but also to those working near you. You should *not* be afraid of the chemicals and equipment that you will be using, but you *should* treat them with the respect and care associated with safe experimental practice. To facilitate this, there is an emphasis on the proper handling of chemicals and apparatus throughout the textbook, and the importance of paying particular attention to these subjects *cannot* be overemphasized. In a sense, laboratory safety is analogous to a chain that is only as strong as its weakest link: the possibility that an accident will occur is only as great as the extent to which unsafe practices that are followed. In other words, if you and your labmates adhere to proper laboratory procedures, the risk of an accident will be minimized.

It is important that you closely follow the experimental procedures in this textbook. There is a good reason why each operation should be performed as it is described, although that reason may not be immediately obvious to you. Suffice it to say that it is *dangerous* for a beginning experimentalist to be "creative" when it comes to modifying the steps and approaches that we have specified! As you become more experienced in the organic laboratory, you may wish to develop alternate procedures for performing a reaction or purifying a desired product, but *always* check with your instructor *before* trying any modifications.

Note that rather detailed experimental procedures are given early in the textbook, whereas somewhat less detailed instructions are provided later on. This is because many of the basic laboratory operations will have become familiar to you and need not be spelled out. It is hoped that this approach to the design of procedures will decrease your tendency to think that you are essentially following a recipe in a cookbook. Moreover, many of the experimental procedures given in the literature of organic chemistry are relatively brief and require the chemist to "fill in the blanks," so it is valuable to gain some initial experience in figuring out some details on your own.

Most of the previous experience you have had in a chemistry laboratory has probably required that you measure quantities fairly precisely, using analytical balances, burets, and the like. Indeed, if you have done any quantitative inorganic analysis, you know that it is often necessary to measure weights to the third or fourth decimal place and volumes to at least the first. Experiments in organic chemistry normally do not demand such precision, so weighing of reagents to the nearest tenth of a gram is usually satisfactory, as is measuring out liquids in graduated cylinders, which are only accurate to $\pm$ 10%. For example, if you are directed to add 30 mL of diethyl ether to serve as a solvent for a reaction, this need not be 30.0 mL. In fact, it probably will make little difference to the success of the process whether anywhere from 25 to 35 mL of the solvent is added. This is not to say that care should not be taken in measuring out the amounts of materials that you use, but rather that valuable time need not be wasted in making these measurements highly precise.

Markers, in the form of **stars** (*), have been inserted within many of the experimental procedures in this textbook. These indicate places where the procedure can be interrupted without affecting the final outcome of the experiment. These markers are designed to help you make the most efficient use of the time you have in the laboratory. For example, you may be able to start a procedure at a point in the period when there is insufficient time to complete it, but enough to be able to work through to the location of a star (*); you can then safely store the reaction mixture and finish the sequence during the next laboratory period. Stars have *not* been inserted at every possible stopping point, only at those where it is not necessarily obvious that interruption of the procedure will have no effect on the experimental results. Consult your instructor if in doubt about whether a proper stopping point has been reached.

As noted above, a *carefully* written **notebook** and *proper* **safety procedures** are important components of an experimental laboratory course. These aspects are further discussed in the following two sections.

1.2   THE LABORATORY NOTEBOOK

One of the most important characteristics of successful scientists is the habit of keeping a complete and understandable record of the experimental work that has been done. Did a precipitate form? Was there a color change during the course of the reaction? What was the temperature at which the reaction was performed and for how long did the reaction proceed? Was the reaction mixture homogeneous or heterogeneous? These are observations and data that may seem insignificant at the time but may later prove critical to the interpretation of an experimental result or to the ability of another person to reproduce your work. All of them belong in a properly kept laboratory notebook. Suggestions for such a document follow. Your instructor may specify other items to be included, but the list we give is representative of a good notebook.

1.2.1   PROTOCOL FOR THE LABORATORY NOTEBOOK

1. Use a *bound* notebook for your permanent laboratory record to minimize the possibility that pages will be lost. If a number has not been printed on each page, do so manually. Some laboratory notebooks are designed with pairs of identically numbered pages so that a carbon copy of all entries can be made. The duplicate page can then be removed and turned in to your instructor or put in a separate place for safekeeping. Many professional scientists use this type of notebook.

2. Reserve the first page of the notebook for use as a title page, and leave several additional pages blank for a Table of Contents.

3. As the main criterion for what should be entered in the notebook, adopt the rule that the record should be sufficiently complete so that anyone who reads it will know exactly what you did and will be able to repeat the work in precisely the same way as it was done originally.

4. Record all experimental observations and data in the notebook as they are obtained. Include the *date* and, if appropriate, the time when you did the work. In a legal sense, the information entered into the notebook *at the time of performance* constitutes the primary record of the work, and it is important for you to follow this principle. Many patent cases have been determined on the basis of dates and times recorded in a laboratory notebook.

5. Make all entries in ink, and *do not delete anything you have written* in the notebook. If you make a mistake, cross it out and record the correct information. Do not scribble notes on odd bits of paper with the intention of later recording the information in your notebook. Such bad habits only lead to problems, since the scraps of paper are easily lost or mixed up. They are also inefficient, since transcribing the information to your notebook means that you must write it a second time. This procedure can also result in errors if you miscopy the data. Finally, do not trust your memory with respect to observations that you have made. When the time comes to write down the information, you may have forgotten a key observation that is critical to the success of the experiment.

6. Unless instructed to do otherwise, do not copy *detailed* experimental procedures provided elsewhere in your notebook; this consumes valuable time. Rather, provide a specific reference to the source of the detailed procedure and enter a *synopsis* of the written procedure that contains enough information that (1) you need not refer to the source while performing the procedure and (2) another chemist will be able to *duplicate* what you did. For example, when performing an experiment from this textbook, a reference should be given to the page number on which the procedure appears, and any *variations* made in the procedure should be detailed along with the reason(s) for doing so.

7. Provide a title for and start the description of each experiment on a new page. The recording of data and observations from several different procedures on the same page can lead to confusion, both for yourself and for others who may read your notebook.

1.2.2   Types of Organic Experiments and Notebook Formats

This textbook contains two general classes of experiments: **investigative experiments** and **preparative experiments**. Investigative experiments normally involve making observations and learning techniques that are common to laboratory work in organic chemistry but do not entail conversion of one compound into another. Some examples are solubility tests, distillation, recrystallization, and qualitative organic analysis. Preparative experiments, in contrast, involve interconversion of different compounds. Most of the procedures described in this textbook fall into the latter category.

The format of the laboratory notebook is usually different for these two types of experiments. Once again, your instructor may have a particular style that is recommended, but we provide suggested formats in the sections that follow.

Notebook Format for Investigative Experiments

1. *Introduction.* Give a brief introduction to the experiment in which you clearly state the purpose(s) of the procedure. This should require no more than one-fourth of a page.

2. *Experiments and Results.* Enter a one- or two-line statement for each part of an experiment. Reserve sufficient room to record results as they are obtained. As noted in Part (6) of "Protocol for the Laboratory Notebook," do *not* copy the experimental procedure from the textbook, but provide a synopsis of it.

Much of this section of the write-up can be completed before coming to the laboratory to ensure that you understand the experiment and that you will perform all parts of it.

3. *Conclusions.* Record the conclusions that can be reached, based on the results you have obtained in the experiment. If the procedure has involved identification of an unknown compound, summarize your findings in this section.

4. *Answers to Exercises.* Enter answers to any exercises for the experiment that have been assigned from the textbook.

A sample write-up of an investigative experiment is given in Figure 1.1.

FIGURE 1.1 Sample notebook format for investigative experiments.

(1) **Your Name**
Date

Separation of Green Leaf Pigments by TLC

Reference: *Experimental Organic Chemistry: A Miniscale Approach,*
by Gilbert, Martin, and Roberts, Section 6.3.

(2) 1. INTRODUCTION

The pigments in green leaves are to be extracted into an organic solvent, and the extract is to be analyzed by thin-layer chromatography (TLC). The presence of multiple spots on the developed TLC plate will indicate that more than a single pigment is contained in the leaves.

(3) 2. SYNOPSIS OF AND NOTES ON EXPERIMENTAL PROCEDURE—RESULTS

Procedure: Grind five spinach leaves in mortar and pestle with 5 mL of 2:1 pet. ether and EtOH. Swirl soln. with 3×2-mL portions H_2O in sep. funnel; dry org. soln. for few min over anhyd. Na_2SO_4 in Erlenmeyer. Decant and concentrate soln. if not dark-colored. Spot 10-cm $\times$ 2-cm TLC plate about 1.5 mm from end with dried extract; spot should be less that 2-mm diam. Develop plate with $CHCl_3$.
Variances and observations: Procedure followed exactly as described in reference. Org. soln. was dark green in color; aq. extracts were yellowish. Half of org. layer lost. TLC plate had five spots having colors and R_f-values shown on the drawing below.

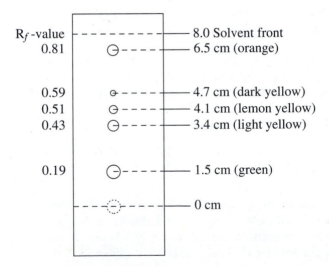

(4) 3. CONCLUSIONS

Based on TLC analysis, the procedure used allows the extraction of at least five different pigments from the spinach leaves. Judging from colors, one of these is a carotene, three are xanthophylls, and the last is chlorophyll *b*.

(5) 4. ANSWERS TO EXERCISES

(*Answers omitted intentionally*)

Notebook Format for Preparative Experiments

1. *Introduction.* Give a brief introduction to the experiment in which you clearly state the purpose(s) of the procedure. This should require no more than one-fourth of a page.

2. *Main Reaction(s) and Mechanism(s).* Write *balanced* equations giving the main reaction(s) for conversion of starting material(s) to product(s). The reason for balancing the equations is discussed in Part (4) below. Whenever possible, include the detailed mechanisms for the reactions that you have written.

3. *Table of Reactants and Products.* Set up a Table of Reactants and Products as an aid in summarizing the amounts and properties of reagents and catalysts being used and the product(s) being formed. Only those reactants, catalysts, and products that appear in the main reaction(s) should be listed in the table; many other reagents may be used in the work-up and purification of the reaction mixture, but these should *not* be entered in the table.

Your instructor will have specific recommendations about what should appear in the table, but the following items are illustrative.

a. The name and/or structure of each reactant, catalyst, and product.

b. The molecular weight of each compound.

c. The weight used, in grams, of each reactant and the volume of any liquid reactant. We recommend that the weight and/or volume of any catalysts used be entered for purposes of completeness.

d. The molar amount of each reactant used; this can be calculated from the data in Parts (b) and (c).

e. The theoretical mole ratio, expressed in whole numbers, for the reactants and products; this ratio is determined by the *balanced* equation for the reaction, as given in Part (2).

f. Physical properties of the reactants and products; this entry might include data such as boiling and/or melting point, density, solubility, color, and odor.

4. *Yield Data.* Compute the maximum possible amount of product that can be formed, called the **theoretical yield.** This can easily be calculated from the data in the Table of Reactants and Products as follows. First determine which of the reactants corresponds to the **limiting reagent.** This is the reagent that is used in the least molar amount relative to what is required theoretically. In other words, the reaction will stop once this reactant is consumed, so its molar quantity will define the maximum quantity of product that can be produced. From the number of moles of limiting reagent involved and the balanced equation for the reaction, determine the theoretical yield, in moles (mol), of product. This value can then be converted into the theoretical yield in grams, based on the molecular weight of the product.

Once the isolation of the desired product(s) has been completed, you should also calculate the **percent yield,** which is a convenient way to express the overall efficiency of the reaction. This is done by obtaining the **actual yield** of product(s) in grams, and then applying the expression in Equation (1.1).

$$\text{Percent yield} = \frac{\text{actual yield (g)}}{\text{theoretical yield (g)}} \times 100 \qquad \textbf{(1.1)}$$

Generally, the calculated value of percent yield is rounded to the nearest whole number. As points of reference, most organic chemists consider yields of 90% or greater as being "excellent," and those below 20% as "poor."

5. *Synopsis of and Notes on Experimental Procedure.* Provide an outline of the experimental procedure that contains enough detail so that you do not have to refer to the textbook repeatedly while performing the experiment. Note any variations that you use, as compared to the referenced procedure, and observations that you make while carrying out the formation and isolation of the product(s).

6. *Observed Properties of Product.* Record the physical properties of the product that you have isolated in the experiment. Appropriate data under this heading might include boiling and/or melting point, odor, color, and crystalline form, if the product is a solid. Compare your observations with those available on the compound in various reference books (for example, the *CRC Handbook of Chemistry and Physics* or Lange's *Handbook of Chemistry*).

7. *Side Reactions.* List possible **side reactions;** those reactions leading to undesired products, that are likely to occur in the experiment. It is important to consider such processes because the by-products that are formed must be removed by the process used to purify the desired product. You may need to consult your lecture notes and textbook in order to predict what side reactions might be occurring.

8. *Other Methods of Preparation.* If instructed to do so, suggest alternate methods for preparing the desired compound. Such methods may involve using entirely different reagents and reaction conditions. Your lecture notes and textbook can serve as valuable resources for providing possible entries for this section.

9. *Method of Purification.* Develop a flow chart that summarizes the sequence of operations that will be used to purify the desired product. The chart will show at what stages of the work-up procedure unchanged starting materials and unwanted by-products are removed. By understanding the logic of the purification process, you will know why each of the various operations specified in the purification process is performed.

The purification of the final product of a reaction can be the most challenging part of an experimental procedure. Professional organic chemists are constantly required to develop work-up sequences that allow isolation of a pure product, free from starting materials and other contaminants. They do this by considering the chemical and physical properties of both the desired and undesired substances, and it is important for you to gain experience in devising such schemes as well.

10. *Answers to Exercises.* Enter answers to any exercises for the experiment that have been assigned from the textbook.

A detailed example of the write-up for a preparative experiment involving the dehydration of cyclohexanol (see Sec. 10.2) is given in Figure 1.2. You may not actually perform this reaction; nevertheless, you should carefully study the example in order to see how to prepare specific entries for the first eight items listed. The various entries in Figure 1.2 are labeled with circled, **boldface** numbers, and are discussed further in the following paragraphs. It is assumed for illustrative purposes that an actual yield of 2.7 g is obtained.

FIGURE 1.2   Sample notebook format for preparative experiments.

---

①   **Your Name**
                                                           Date

Dehydration of Cyclohexanol

Reference: *Experimental Organic Chemistry: A Miniscale Approach,*
by Gilbert, Martin, and Roberts, Section 9.2.

② 1. **INTRODUCTION**

    Cyclohexene is to be prepared by the acid-catalyzed dehydration of cyclohexanol.

③ 2. **MAIN REACTION(S) AND MECHANISM(S)**

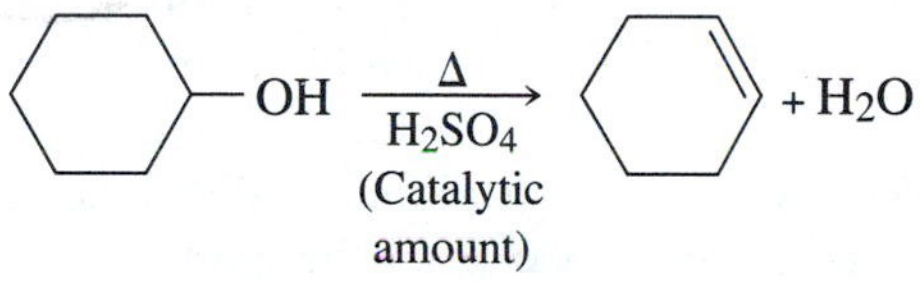

 $-$OH $\xrightarrow[\text{H}_2\text{SO}_4]{\Delta}$ $+ \text{H}_2\text{O}$
(Catalytic
amount)

        1   **2**

(Mechanism omitted intentionally)

④ 3. **TABLE OF REACTANTS AND PRODUCTS**

⑤   ⑥   ⑦a   ⑦b   ⑧   ⑨   ⑩

Compound	M.W.	Volume used (mL)	Weight used (g)	Moles used	Moles required	Other data
Cyclohexanol	100.2	5.2	5	0.05	1	bp 161 °C (760 torr), mp 25.1 °C, d 0.962 g/mL, colorless
Sulfuric Acid (9 *M*)	98.1	2.5	#	#	0	d 1.84 (18 *M* H$_2$SO$_4$)
Cyclohexene	82.2	*	*	*	1	bp 83 °C (760 torr), mp – 103.5 °C, d 0.810, colorless

#Entry left blank because this row is for the catalyst.

*Entry left blank because this row is for the product.

⑪ Limiting Reagent: **CYCLOHEXANOL**

⑫ 4. **YIELD DATA**

    Theoretical yield of cyclohexene = moles of limiting reagent (cyclohexanol) $\times$ M.W. of cyclohexene

                        = 0.05 mol $\times$ 82.2 g/mol

                        = 4.1 g

    Actual yield = 2.7 g   *(Continued)*

FIGURE 1.2   (Continued)

Percent yield = [Actual yield (g)/theoretical yield (g)] × 100

= [2.7/4.1] × 100 = 66%

⑬ 5.   SYNOPSIS OF AND NOTES ON EXPERIMENTAL PROCEDURE—RESULTS

Procedure: Put alcohol in 25-mL rb flask and add H_2SO_4. Mix, add boiling chips, attach to fractional dist. apparatus. Heat with oil bath; heating rate such that head temp. stays below 90 °C. Stop when 2.5 mL remain in rxn. flask. Put distillate in 25-mL Erlenmeyer and add 1–2 g K_2CO_3.✳ Occasionally swirl mix. for 15 min and transfer liquid to 10-mL rb by decantation or pipet. Add boiling stones and do simple distillation (no flames!); receiver must be close to drip tip of adapter to minimize losses by evaporation. Collect product at 80–85 °C (760 torr). *Variances and observations:* Procedure followed exactly as described in reference. Distillate cloudy throughout dehydration step; formed two layers in receiver. Head temperature never exceeded 77 °C. Liquid in stillpot darkened as reaction proceeded. Addition of carbonate (1 g) to distillate caused evolution of a few bubbles of gas (CO_2?) Had to add about 0.5 g more of carbonate to get rid of cloudiness. Left solution over drying agent for one week (next lab period). Used pipet to transfer dried liquid to distillation flask. Collected cyclohexene in ice-cooled 10-mL rb flask attached to vacuum adapter protected with $CaCl_2$ tube. Stopped distillation when about 1 mL of yellowish liquid remained in stillpot.

⑭ 6.   OBSERVED PROPERTIES OF PRODUCT

bp 80–84 °C (760 torr); colorless liquid; insoluble in water; decolorizes Br_2/CCl_4 solution and produces brown precipitate upon treatment with $KMnO_4/H_2O$.

⑮ 7.   SIDE REACTIONS

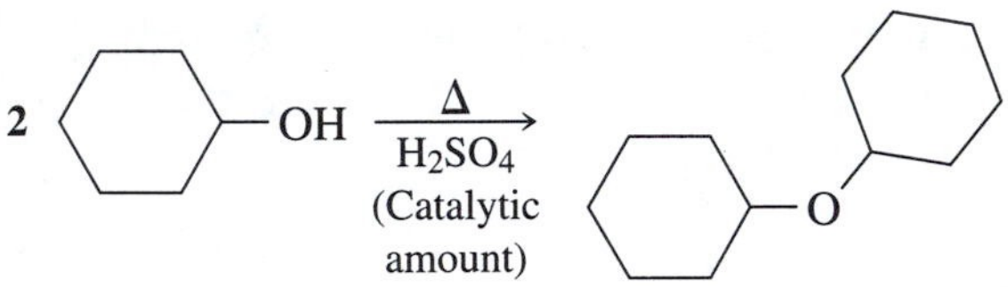

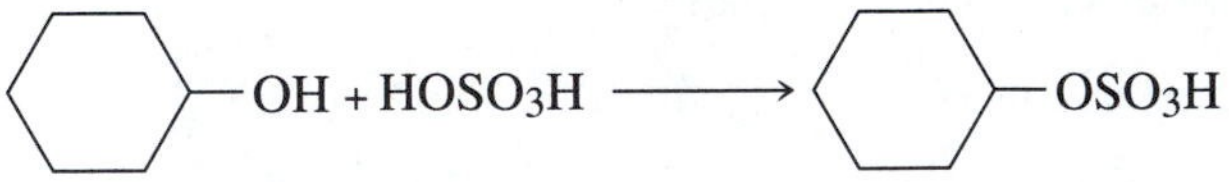

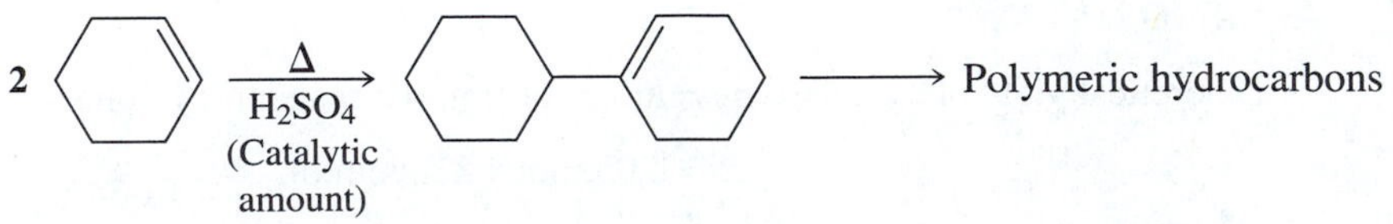

(Continued)

(16)   **7. OTHER METHODS OF PREPARATION**

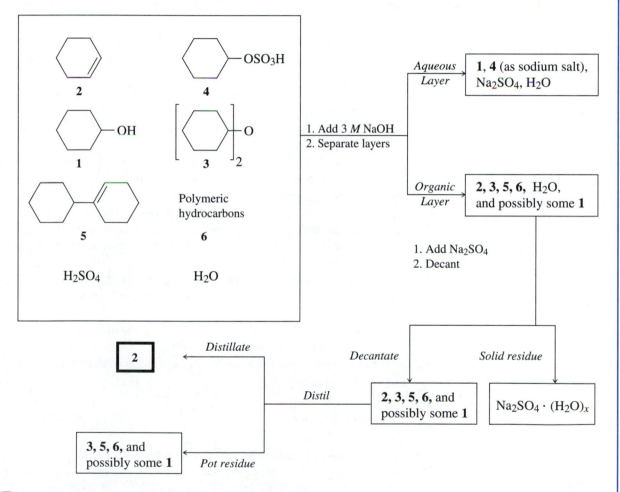

(X = halogen)

(17)   **8. FLOW CHART FOR PURIFICATION**

(1) Use a new page of the notebook to start the entries for the experiment. Provide information that includes your name, the date, the title of the experiment, and a reference to the place in the laboratory textbook or other source where the procedure can be found.

(2) Self-explanatory.

(3) There is only a single reaction in our example, but in many cases more than one step is involved; write equations for *all* of the main reactions. A mechanism for the reaction is intentionally omitted in our example.

(4) Use the illustrated format for the Table of Reactants and Products unless instructed to do otherwise.

(5) Enter the name or structure of each reactant, catalyst, if any, and desired product.

(6) Record the molecular weight (MW) of each reactant, and desired product. For completeness, make an entry for any catalyst used, although this may be optional.

(7a) Give the volume, in milliliters (mL), of each *liquid* reactant and catalyst.

(7b) Record the weight, in grams (g), of each reactant. This entry is optional for liquid catalysts, but should be provided for reference purposes.

(8) Calculate the moles (mol) used of each reactant. For completeness, a value for the catalyst is computed in our example.

(9) Obtain the **theoretical ratio** for reactant(s) and product(s) by referring to the *balanced* main equation(s) for the reaction.

(10) List selected physical properties of reactant(s) and product(s). The information needed is generally available in reference books.

(11) Determine the **limiting reagent** in the following way. Compare the actual ratio of reactants used to that theoretically required. The reagent that is used in the least molar amount, relative to the theoretical amount, is the limiting reagent. In our example, there is only a single reactant, cyclohexanol, so it obviously must be the limiting reagent.

(12) Calculate the *theoretical yield* of the desired product both in moles and in grams. Knowing in our case that the limiting reagent is cyclohexanol and, from the main equation, that 1 mol of alcohol yields 1 mol of cyclohexene, it is clear that no more than 0.05 mol of the alkene can be formed.

Assuming that you were able to isolate 2.7 g of pure cyclohexene in the experiment, the **percent yield** would be calculated according to Equation (1.1).

(13) Self-explanatory.

(14) Self-explanatory.

(15) Self-explanatory.

(16) Self-explanatory.

(17) Develop this diagram by considering what components, in addition to the desired product, may be present in the reaction mixture *after* the main reaction is complete. The chart shows how and where each of the inorganic and organic contaminants of the product is removed by the various steps of the work-up procedure. Ideally, pure cyclohexene results.

(18) Self-explanatory.

1.2.3   SAMPLE CALCULATIONS

Students frequently have difficulty in setting up Tables of Reactants and Products and calculating theoretical yields, so two hypothetical examples are provided for your reference.

Example 1

Problem. Consider the reaction shown in Equation (1.2). Assume that you are to use 5 g of the alkene and 25 mL of concentrated HBr solution. Prepare a Table of Reactants and Products, determine the limiting reagent, and calculate the theoretical yield for the reaction.

$$CH_2 = CHCH_2CH_2CH_3 + HBr \longrightarrow CH_3CH(Br)CH_2CH_2CH_3 \qquad \textbf{(1.2)}$$

Answer. First of all note that the equation is balanced, since the "1" that signifies that 1 mol of each reactant will react to produce 1 mol of product is omitted by convention. Since an aqueous solution of HBr, rather than the pure acid, is being used, the amount of HBr present must be determined. Concentrated HBr is 47% by weight in the acid, and its density is 1.49 g/mL, a value that would be recorded in the column headed "Other Data." Consequently, 25 mL of this solution contains 17.5 g of HBr (25 mL × 1.49 g/mL × 0.47). The needed data can then be entered into Table 1.1.

The limiting reagent is seen to be the 1-pentene, since theory requires that it and HBr react in a 1:1 molar ratio, yet they have been used in a ratio of 0.07:0.22. This means that no more than 0.14 mole of product can be formed, since theory dictates that the ratio between 2-bromopentane and 1-pentene also be 1:1. The calculation of the theoretical yield is then straightforward.

TABLE 1.1   Table of Reactants and Products for Preparation of 2-Bromopentane

Compound	M.W.	Volume Used (mL)	Weight Used (g)	Moles Used	Moles Required	Other Data
1-Pentene	70.14	—	5	0.07	1	*
HBr	80.91	25	17.5	0.22	1	*
2-Bromopentane	151.05	†	†	†	1	*

* These entries have been intentionally omitted in this example.

† These entries are left blank because this line is for the product.

Limiting reagent: 1-Pentene

Theoretical yield: 151.05 g/mol × 0.07 mol = 10.5 g

TABLE 1.2   Table of Reactants and Products for Preparation of Diethyl Ether

Compound	M.W.	Volume Used (mL)	Weight Used (g)	Moles Used	Moles Required	Other Data
Ethanol	46.07	7	5.5	0.12	2	*
Sulfuric Acid	†	0.1	†	†	†	*
Diethyl ether	74.12	‡	‡	‡	1	*

* These entries have been intentionally omitted in this example.

† These entries are left blank for reactants that serve only as catalysts

‡ These entries are left blank because this line is for the product.

  Limiting reagent: Ethanol

  Theoretical yield: 74.12 g/mol × 0.12 mol · 0.5 = 4.4 g

Example 2

Problem. Now consider the transformation illustrated in Equation (1.3). Assume that you are to use 7 mL of ethanol and 0.1 mL of concentrated H_2SO_4 as the catalyst. Prepare a Table of Reactants and Products, determine the limiting reagent, and calculate the theoretical yield for the reaction.

$$2 \ CH_3CH_2OH \xrightarrow[\text{(Catalytic amount)}]{H_2SO_4} CH_3CH_2OCH_2CH_3 + H_2O \tag{1.3}$$

Answer. As in the previous example, a volumetric measurement must first be converted to a weight. The density of ethanol is 0.789 g/mL, information that would be entered under the column headed "Other Data," so that means that 5.5 g is being used. Table 1.2 can then be completed. Note that the catalyst, although recorded in the table, is not used in any of the calculations because, by definition, it is not consumed during the reaction. Its inclusion should help remind the experimentalist that it is indeed required to make the reaction work!

Calculation of the theoretical yield is performed as in Example 1, with the important exception that a factor of 0.5 is incorporated to adjust for the fact that only one-half mole of diethyl ether would be produced for each mole of ethanol that is used.

1.3 SAFE LABORATORY PRACTICE

There is little question that one of the most important abilities that you, the aspiring organic chemist, can bring to the laboratory is a sound knowledge of how to perform experimental work in a safe manner. But just knowing *how* to work safely is insufficient! You must also make a *serious* commitment to follow standard safety protocols. In other words, having the knowledge about safety is useless if you do not put that knowledge into practice. What you actually do in the laboratory will determine whether you and your labmates are working in a safe environment.

Chemistry laboratories are potentially dangerous because they commonly house flammable liquids, fragile glassware, toxic chemicals, and equipment that may be under vacuum or at pressures above atmospheric. They may also contain gas cylinders that are under high pressure. The gases themselves are hazardous—for example, nitrogen is not whereas hydrogen certainly is—but the fact that their containers are under pressure makes them so. Imagine what might happen if a cylinder of nitrogen fell and ruptured: You could have a veritable rocket on your hands, and, as in the case of hydrogen, the "rocket" might even come equipped with a fiery tail! This is another way of saying *all* substances are hazardous under specified conditions.

Fortunately the laboratory need be no more dangerous than a kitchen or bathroom, but this depends on you *and* your labmates practicing safety as you work. Should you observe someone else doing anything that is unsafe, let them know about it in a friendly manner. Everyone will benefit from your action. We shall be alerting you repeatedly to the possible dangers associated with the chemicals and apparatus that you will use so that you can become well-trained in safe laboratory practice. Mastery of the proper procedures is just as important in the course as obtaining high yields of pure products, and careful reading of our suggestions will assist you in this goal. Some safety information will be contained in the text describing a particular experiment or in the experimental procedure itself. It will also appear in highlighted sections titled "Safety Alert." These are designed to draw your special attention to aspects of safety that are of particular importance. Read these sections carefully and follow the guidelines in them carefully.

1.3.1 Safety: General Discussion

We highlight here, in the form of a Safety Alert, some general aspects regarding safe practices in the laboratory.

Safety Alert

Personal Attire
1. *Do not wear shorts or sandals in the laboratory.* Proper clothing gives protection against chemicals that may be spilled accidentally. It is advisable to wear a laboratory coat, but in any case, the more skin that is protected by clothing the better.
2. *Always wear safety glasses or goggles in the laboratory.* This applies even when you are writing in your laboratory notebook or washing glassware, since workers nearby may have an accident. Unless your laboratory instructor says otherwise, do *not* wear contact lenses in the laboratory. Even if you are wearing eye protection, chemicals may get into your eyes, and you may not be able to get the contact lenses out

(Continued)

before damage has occurred. Should you have to wear corrective glasses while working in the laboratory, make certain that the lenses are shatter-proof. Wearing goggles over such glasses is recommended because the goggles give additional protection from chemicals entering your eyes from the side of the lenses.

General Considerations

1. *Become familiar with the layout of the laboratory room.* Locate the exits from the room and the fire extinguishers, fire blankets, eye-wash fountains, safety showers, and first-aid kits in and near your workspace. Consult with your instructor regarding the operation and purpose of each of the safety-related devices.
2. *Find the nearest exits from your laboratory room to the outside of the building.* Should evacuation of the building be necessary, use stairways rather than elevators to exit. Remain calm during the evacuation, and walk rather than run to the exit.
3. *Become knowledgeable about basic first-aid procedures.* The damage from accidents will be minimized if first aid is applied promptly. Read the section, "First Aid in Case of An Accident" on the inside front cover of this book.
4. *Never work alone in the laboratory.* In the event of an accident, you may need the immediate help of a co-worker. Should you have to work in the laboratory outside of the regularly scheduled periods, do so only with the express permission of your instructor and in the presence of at least one other person.

Apparatus and Chemicals

1. *Always check carefully for imperfections in the glassware that you will be using.* This should be done not only when checking into the laboratory for the first time but also when setting up the apparatus needed for each experiment. Look for cracks, chips, or other imperfections in the glass that weaken it. Use care in storing your glassware so that it is not damaged upon opening or closing of your locker or drawer.

   Pay particular attention to the condition of round-bottom flasks and condensers. The flasks often have "star" cracks (multiple cracks emanating from a central point) as a result of being banged against a hard surface. Heating or cooling a flask having this type of flaw may cause the flask to rupture with loss of its contents. This could result in a serious fire, not just loss of the desired product. To detect such cracks, hold the flask up to the light and look at all its surfaces closely. With respect to condensers, their most vulnerable points are the ring seals (the points where the inner tube and the water jacket of the condenser are joined). Special care must be paid to examining these seals for defects because if cracks are present, water might leak into your apparatus and cause violent reactions.

   If you detect imperfections in your glassware, consult with your teacher immediately regarding replacement. Cracked or seriously

(Continued)

chipped apparatus should always be replaced, but glassware with slight chips may still be safe to use.

2. *Dispose of glassware properly.* The laboratory should be equipped with a properly labeled special container for broken glassware and disposable glass items such as Pasteur pipets and melting point capillaries. **It is not appropriate to throw such items in the regular trash containers because maintenance personnel may injure themselves while removing the trash.** Broken thermometers are a special problem because they usually contain residual mercury, which is toxic and relatively volatile. There should be a separate, closed container for disposal of thermometers. If mercury has spilled as a result of the breakage, it should be cleaned up immediately. Consult with your instructor about appropriate procedures for doing so.

3. *Know the properties of the chemicals used in the experiments.* Understanding the properties of the chemicals that you will be using helps you to take the proper precautions when handling them and to minimize danger in case of an accident. **Handle all chemicals with care.**

   Refer to Material Safety Data Sheets (see Sec. 1.3.2) to learn about toxicity and other potential hazards associated with the chemicals you use. Most chemicals are at least slightly toxic and many are *very* toxic and irritating if inhaled or allowed to come in contact with the skin. It is a good laboratory practice to wear plastic or rubber gloves when handling chemicals, and there may be times when it is imperative to do so. Your instructor will advise you on the need for gloves.

   Should chemicals come in contact with your skin, they can usually be removed by a thorough and *immediate* washing of the affected area with soap and water. Do *not* use organic solvents like ethanol or acetone to rinse chemicals from your skin, as these solvents may actually assist the absorption of the substances into your skin.

4. *Avoid the use of flames as much as possible.* Most organic substances are flammable, and some are highly volatile as well, which increases their potential for being ignited accidentally. Examples of these are diethyl ether, commonly used as a solvent in the organic laboratory, and acetone. Occasionally, open flames, such as that from a Bunsen burner, must be used for flame-drying an apparatus or distilling a high-boiling liquid, for example. In such cases, a **"SAFETY ALERT"** section will give special precautions for their use. Some general guidelines follow.

   a. *Never use an open flame without the permission of your instructor.*

   b. *Never utilize a flame to heat a flammable liquid in an open container.* Use a steam bath, hot plate, or similar electrical heating device instead. If a flammable liquid is to be heated with an open flame, equip the container holding the liquid with a *tightly* fitting reflux condenser.

(Continued)

Information about the flammability of many commonly used organic solvents is provided in Table 3.1. Do *not* assume, however, that a solvent is not flammable just because it is not listed in the table. In such cases, refer to the Material Safety Data Sheets (see discussion in Sec. 1.3.2) or other sources to determine flammability.

c. *Do not pour flammable liquids when there are open flames within several feet.* The act of transferring the liquid from one container to another will release its vapors into the laboratory, and these could be ignited by a flame some distance away.

d. *Do not pour flammable water-insoluble organic solvents into drains or sinks.* First of all, this is an environmentally unsound way to dispose of waste solvents, and second, the solvents may be carried to locations where there are open flames which could ignite them. Water-soluble solvents can be flushed down the drain if local regulations permit; consult your instructor about this.

5. *Avoid inhaling vapors of organic and inorganic compounds.* Although many of the odors you encounter in everyday life are organic in nature, it is prudent not to expose yourself to such vapors in the laboratory. Work at a fume hood when handling particularly noxious chemicals, such as bromine or acetic anhydride, and, if possible, when performing reactions that produce toxic gases.

6. *Never taste anything in the laboratory unless specifically instructed to do so.* You should also never eat or drink in the laboratory, as your food may become contaminated by the chemicals that are being used.

7. *Minimize the amounts of chemicals you use and dispose of chemicals properly.* This aspect of laboratory practice is so important that we have devoted a portion of Section 1.3.2 to it. Read the relevant paragraphs *carefully* and consult with your instructor if there are any questions about the procedures.

1.3.2 SAFETY: MATERIAL SAFETY DATA SHEETS

The variety and potential danger of chemicals used in the organic chemistry laboratory probably exceed that of any laboratory course you have had. It is imperative to understand the nature of the substances with which you are working. Fortunately, the increased emphasis on the proper handling of chemicals has led to a number of different types of publications containing key information about the chemical, physical, and toxicological properties of the majority of organic and inorganic compounds used in the experiments in this textbook. The most comprehensive of the sources is *The Sigma-Aldrich Library of Chemical Safety Data* (see Reference 8), and it or similar compilations should be available in your library or some other central location. Alternatively, the information may be available to you on a computer-accessible database; consult with your instructor about this possibility. The data provided by such references are basically a summary of the information contained in the Material Safety Data Sheets (MSDSs) published by the supplier of the chemical of interest. Your instructor may also be able to provide these sheets because, by Federal regulation, an MSDS must be delivered to the buyer each time a chemical is purchased.

The information contained in an MSDS can be overwhelming to the student. For example, the official MSDS for sodium bicarbonate is currently *seven* pages although it was originally only two pages length! The data provided in most compilations, such as *The Sigma-Aldrich Library of Chemical Safety Data*, are summaries of the complete documents and contain the type of information shown in Figure 1.3; the specific data given are those for diethyl ether. Entries regarding the structure and physical properties of the compound, including melting point ("MP"), boiling point ("BP"), and flash point ("FP"), are included

FIGURE 1.3   Summary of MSDS for diethyl ether.

Name	Ether	Reviews and Standards	OSHA standard-air: TWA 400 ppm
Other Names	Diethyl ether	Health Hazards	May be harmful by inhalation, ingestion, or skin absorption. Vapor or mist is irritating to the eyes, mucous membranes, and upper respiratory tract. Causes skin irritation. Exposure can cause coughing, chest pains, difficulty in breathing and nausea, headache, and vomiting.
CAS Registry No.	60-29-7	First Aid	In case of contact, immediately flush eyes or skin with copious amounts of water for at least 15 min while removing contaminated clothing and shoes. If inhaled remove to fresh air. If not breathing give artificial respiration; if breathing is difficult, give oxygen. If ingested, wash out mouth with water. Call a physician.
Structure	$(CH_3CH_2)_2O$	Incompatibilities	Oxidizing agents and heat.
MP	$-116°C$	Extinguishing Media	Carbon dioxide, dry chemical powder, or alcohol or polymer foam
BP	$34.6\,°C$ (760 torr)	Decomposition Products	Toxic fumes of carbon monoxide, carbon dioxide
FP	$-40\,°C$	Handling and Storage	Wear appropriate respirator, chemical-resistant gloves, safety goggles, other protective clothing. Safety shower and eye bath. Do not breathe vapor. Avoid contact with eyes, skin, and clothing. Wash thoroughly after handling. Irritant. Keep tightly closed. Keep away from heat, sparks, and open flame. Forms explosive peroxides on prolonged storage. Refrigerate. Extremely flammable. Vapor may travel considerable distance to source ignition. Container explosion may occur under fire conditions. *Danger:* tends to form explosive peroxides especially when anhydrous. Inhibited with 0.0001% BHT
Appearance	Colorless liquid	Spillage	Shut off all sources of ignition. Cover with activated carbon adsorbent, place in closed containers, and take outdoors.
Irritation Data	Human eye 100 ppm	Disposal	Store in clearly labeled containers until container is given to approved contractor for disposal in accordance with local regulations.
Toxicity Data	Man, oral LDL_0 260 mg/kg		

along with its CAS (Chemical Abstracts Service) Registry number, which is unique for each different chemical substance (see Chap. 24 for a further discussion of CAS registry numbers), and RTECS (Registry of Toxic Effects of Chemical Substances) number. Further data are provided concerning its toxicity, the permissible levels set by OSHA for exposure to it in the air you breathe (time-weighted average of 400 ppm), and possible health consequences resulting from contact with the compound. For diethyl ether, the entry for "Toxicity Data" represents the *lowest* recorded *lethal* concentration for ingestion of the chemical. Valuable information is also given regarding first-aid procedures, classes of substances with which diethyl ether reacts, and thus is "incompatible," products of its decomposition, and materials suitable for extinguishing fires involving ether. Finally, protocols for safe handling and storage are included, along with procedures for disposal of diethyl ether and cleaning up spills of it.

Remember that the information in an MSDS is targeted toward dealing with much larger quantities of the chemical than would be used in the laboratory. For example, it is unnecessary to wear a respirator when working with the amounts of diethyl ether normally encountered in a teaching or research laboratory! Moreover, some statements describe needs that are standard to the properly equipped laboratory (for example, safety goggles, shower, and eye-wash fountain) and to routine laboratory practice—such as not breathing the vapor and avoiding contact with skin and clothing. However, some comments deserve particular attention—such as those regarding the flammability of diethyl ether and its vapors, and the tendency of this substance to form explosive peroxides upon prolonged storage.

You may be required to show your knowledge of safety procedures associated with handling chemicals by incorporating of some of the MSDS data into your laboratory notebook. One possible format for doing so is shown in Figure 1.4. The first time that a particular chemical is used in the course, data like those shown can be recorded at the end of your laboratory notebook on pages reserved for such information. Whenever this same substance is used in later experiments, a reference to the page of the notebook containing these data is all that would be needed. *However,* you should reread the information that has originally been recorded so that you can continue to handle the chemical properly.

To summarize, you may think that reading about and recording data like those contained in Figures 1.3 and 1.4 are not a good investment in time. This is

FIGURE 1.4 Abstract of MSDS for diethyl ether.

Compound	Health hazards, First Aid, Incompatibilities, Extinguishing Media, and Handling
Diethyl Ether	May be harmful by inhalation, ingestion, or skin absorption. Avoid contact with eyes, skin, and clothing. In case of contact, immediately flush eyes or skin with copious amounts of water. Incompatible with oxidizing agents and heat. Keep tightly closed when not in use. Keep away from heat, sparks, and open flames. Extremely flammable. Vapor may travel considerable distance to source of ignition. If spilled, shut off all sources of ignition. Extinguish fire with carbon dioxide or dry chemical extinguisher.

absolutely *wrong*! By knowing more about the chemicals that are used in the laboratory, you will be able to work safely and to deal with accidents, should they occur. The end result will be that you should accomplish a greater amount of laboratory work and have a more valuable educational experience.

Safety: Disposal of Chemicals.   The proper disposal of inorganic and organic chemicals is one of the biggest responsibilities that you have in the organic laboratory. Your actions, and those of your labmates, can minimize the environmental impact and even financial cost to your school of handling the waste chemicals that are necessarily produced in the experiments you do.

The experimental procedures in this textbook have been designed at a scale that should allow you to isolate an amount of product sufficient to see and manipulate, but they also involve the use of minimal quantities of reactants, solvents, and drying agents. Bear in mind, however, that the minimization of amounts of chemicals that are used is only the *first* part of an experimental design that results in the production of the least possible quantity of waste. The *second* part is to reduce the amounts of materials that you, the experimentalist, *defines* as waste, thereby making the material subject to regulations for its disposal. From a legal standpoint, the laboratory worker is empowered to declare material as waste; that is, unneeded materials are not waste until you say they are! Consequently, a part of most of the experimental procedures in this textbook is reduction of the quantity of residual material that eventually must be consigned to waste. This means some additional time will be required for completion of the experiment, but the benefits, educational, environmental, and economic in nature, fully justify your efforts. The recommended procedures that should be followed are described under the heading, **"Finishing Touches."**

How do you properly dispose of spent chemicals at the end of an experiment? In some cases this involves simply flushing chemicals down the drain with the aid of large volumes of water. As an example, solutions of sulfuric acid can be neutralized with base, and the aqueous sodium sulfate that is formed can safely be washed into the sanitary sewer system. However, the environmental regulations that apply in your particular community may require use of alternative procedures. *Be certain to check with your instructor before flushing any chemicals down the drain!*

For water-insoluble substances, and even for certain water-soluble ones, this option is not permissible under *any* circumstances, and other procedures must be followed. The laboratory should be equipped with various containers for disposal of both liquid and solid chemicals; the latter should not be thrown in a trash can because (1) this exposes maintenance personnel to potential danger, and (2) it is environmentally unsound. The containers must be properly labeled as to what can be put in them, because it is very important for safety and environmental reasons that different categories of spent chemicals be segregated from one another. Thus, you are likely to find the following types of containers in the organic laboratory: hazardous solids, nonhazardous solids, halogenated organic liquids, hydrocarbons, and oxygenated organic liquids. Each student must assume the responsibility for seeing that his or her spent chemicals go into the appropriate container; otherwise dangerous combinations of chemicals might result and/or a much more expensive method of disposal be required.

REFERENCES

1. Lunn, G.; Sansone, E.B. *Destruction of Hazardous Chemicals in the Laboratory,* John Wiley & Sons, New York, 1990. A handbook providing procedures for decomposition of 44 materials or classes of materials commonly used in the laboratory.

2. Committee on Hazardous Substances in the Laboratory. *Prudent Practices for Disposal of Chemicals from Laboratories,* National Academy Press, Washington, D.C., 1983. An excellent reference containing information for the minimization of waste generated in the laboratory and for the proper handling and disposal of waste chemicals, both organic and inorganic.

3. Young, J.A., ed. *Improving Safety in the Chemical Laboratory: A Practical Guide,* 2nd ed., John Wiley & Sons, New York, 1991. A book containing thorough discussions of the full range of safe practices in the laboratory.

4. Mahn, W.J. *Fundamentals of Laboratory Safety. Physical Hazards in the Academic Laboratory,* Van Nostrand Reinhold, New York, 1991.

5. Lide, D.A., ed. *CRC Handbook of Chemistry and Physics,* annual editions, CRC Press, Boca Raton, FL.

6. Dean, J.A., ed. *Lange's Handbook of Chemistry,* 14th ed., McGraw-Hill, New York, 1992.

7. Bretherich, L., ed. *Hazards in the Chemical Laboratory,* 4th ed., The Royal Society of Chemistry, London, 1986.

8. Lenga, R.E., ed. *The Sigma-Aldrich Library of Chemical Safety Data,* 2nd ed., Sigma-Aldrich, Milwaukee, WI, 1988.

9. *Merck Index of Chemicals and Drugs,* 11th ed., Merck and Co., Rahway, N.J., 1989.

TECHNIQUES AND APPARATUS

This chapter introduces the basic experimental techniques and associated glassware and apparatus that are commonly used in the organic chemistry laboratory. The theory of some of the techniques is discussed in later chapters. In many of the experiments that you will perform, specific reference is made to a technique and/or a piece of apparatus that is described in this chapter. Thus, you will have occasion to refer to this chapter frequently. Be certain to familiarize yourself with these techniques *prior* to conducting the assigned experiments.

2.1 GLASSWARE: PRECAUTIONS AND CLEANING

Laboratory experiments in organic chemistry are commonly conducted in specialized glassware, which is usually expensive. Since you are responsible for the maintenance of your glassware, you should follow proper procedures for safely handling and cleaning it. Failure to do so is likely to result in injury to yourself and broken or dirty glassware that is difficult to clean.

The cardinal rule in handling and using laboratory glassware is *never apply undue pressure or strain to any piece of glassware*. Strained glassware may break at the moment the strain is induced, when it is heated, or even upon standing for a period of time. When setting up a glassware apparatus for a particular experiment, be sure that the glassware is properly positioned and supported so that strain does not develop.

Sometimes it is necessary to insert thermometers or glass tubes into rubber or cork stoppers or rubber tubing. *If you have to force it, do not do it!* Either make the hole slightly larger or use a smaller piece of glass. A useful aid for minimizing the hazard of this procedure is to lubricate the glass tube with a little water containing soap or glycerol prior to insertion into stoppers or tubing. *Always* grasp the glass piece as close as possible to the rubber or cork part when trying to insert it. It is also wise to wrap a towel around the glass tube and the rubber or cork stopper while inserting the tube. This usually prevents a serious cut in the event the glass happens to break.

Glassware should be thoroughly cleaned *immediately* after use. Residues from chemical reactions may attack the surface of the glass, and cleaning becomes more difficult the longer you wait. Before washing glassware, it is good practice to wipe off any lubricant or grease from standard-taper ground-glass joints (see Sec. 2.2) with a towel or tissue moistened with a solvent such as acetone or methylene chloride. This prevents the grease from being transferred during washing to inner surfaces of the glassware, where it may be difficult to remove. Most chemical residues can be removed by washing the glassware using a brush, special laboratory soap and water. Acetone, which is miscible with water, dissolves most organic residues and thus is commonly used to clean glassware; use as little solvent as possible to do the job. Acetone should not be used to clean equipment that contains residual amounts of bromine, however, since a powerful lachrymator, bromoacetone, may form. A **lachrymator** is a chemical that adversely affects the eyes and causes crying, and may also affect the lungs and produce a burning sensation; tear gas is a lachrymator. After use, the spent wash acetone should be transferred to an appropriately marked container for disposal.

Stubborn residues may sometimes remain in your glassware. Often these may be removed by carefully scraping the glassware with a bent spatula in the presence of soap and water or acetone. If this technique fails, more powerful cleaning solutions may be required, but these must be used with great care as they are highly corrosive. *Do not allow these solutions to come into contact with your skin or clothing; they will cause severe burns and produce holes in your clothing.* Chromic acid, which is made from concentrated sulfuric acid and chromic anhydride or potassium dichromate, is sometimes an effective cleaning agent, but since it is a strong oxidizing acid, it must be used with great care. When handling chromic acid, always wear rubber gloves and pour it *carefully* into the glassware to be cleaned. After the glassware is clean, pour the chromic acid solution into a specially designated bottle, *not* into the sink. Another powerful cleaning solution is alcoholic potassium hydroxide. This is most conveniently prepared by putting some ethanol into the flask, adding a few pellets of solid potassium hydroxide, and warming the solution gently while swirling it around inside the flask. When the glassware is clean, pour the solution into a specially designated bottle, and rinse the glassware thoroughly with soap and water to complete the process. *Before using any cleaning solutions other than soap and water, consult your instructor for permission and directions concerning their safe handling and disposal.*

Brown stains of manganese dioxide that are left in the glassware can generally be removed by rinsing the apparatus with a 30% (4 M) aqueous solution of sodium bisulfite, $NaHSO_3$. If this fails, wash the equipment with water and then add a small amount of 6 M HCl. This must be done in a good hood, since chlorine gas is evolved; try this only after obtaining permission from your instructor.

The fastest method for drying your glassware is to rinse the residual droplets of water from the flask with a small volume (< 5 mL) of acetone; this acetone may be recovered as wash acetone to remove organic residues in the future. Flasks and beakers should be inverted to allow the last traces of solvent to drain. Final drying may also be accomplished by directing a *gentle* stream of dry compressed air into the flask or beaker or by placing the glassware in a drying oven.

2.2 STANDARD-TAPER GLASSWARE

Your laboratory kit probably contains **standard-taper glassware** with **ground-glass joints;** this equipment is safe and convenient to use. Before the advent of standard-taper glassware, chemists had to bore corks or rubber stoppers to fit

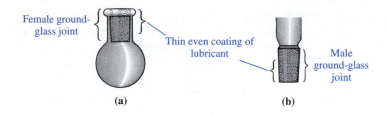

Female ground-glass joint

Thin even coating of lubricant

Male ground-glass joint

(a)   (b)

FIGURE 2.1 Standard-taper (ᵀ§) joints: (a) female, (b) male.

each piece of equipment, and the glass apparatus was assembled by connecting the individual pieces of equipment with glass tubing. This was not only time-consuming, but many chemicals reacted with and/or were absorbed by the cork or rubber stoppers. It was necessary to use new stoppers for each different reaction. Moreover, cuts were a common occurrence for inexperienced chemists. Although standard-taper glassware has greatly simplified the task of assembling the glass apparatus required for numerous routine laboratory operations, it is expensive, so handle it carefully.

A pair of standard-taper joints is depicted in Figure 2.1. Regardless of the manufacturer, a given size of a male standard-taper joint will fit a female joint of the same size. The joints are tapered to ensure a snug fit and a tight seal. Standard-taper joints come in a number of sizes, and are designated by the symbol ᵀ§ followed by two sets of numbers separated by a slash as 14/20, 19/22, 24/40. The first number is the diameter of the joint in millimeters at its widest point, and the second number is the length of the joint in millimeters. A standard-taper joint that is designated as ᵀ§ 19/22 therefore has a widest diameter of 19 mm and a length of 22 mm.

When using glassware with standard-taper ground-glass joints, you must be sure that the joints are properly lubricated so that they do not freeze and become difficult, if not impossible, to separate. Lubrication is accomplished by spreading a *thin* layer of joint grease around the outside of the upper half of the male joint, mating the two joints, and then rotating them gently together to cover the surfaces of the joints with a thin coating of the lubricant. Applying the correct amount of grease to the joints is important. If you use too much, the contents of the flask, including your product, may become contaminated; if too little lubricant is used, the joints may freeze. As soon as you have completed the experiment, disassemble the glassware to lessen the likelihood that the ground-glass joints will stick. If the pieces do not separate easily, the best way to pull them apart is to grasp the two pieces as close to the joint as possible and try to loosen the joint with a *slight* twisting motion.

Sometimes the pieces of glass will still not separate. In these cases there are a few other tricks that can be tried. These include the following options: (1) Tap the joint *gently* with the wooden handle of a spatula, and then try pulling the joint apart as described earlier. (2) Heat the joint in hot water or a steam bath before attempting to separate the pieces. (3) As a last resort, heat the joint *gently* in the yellow portion of the flame of a Bunsen burner. Heat the outer part of the joint slowly and carefully until the outer joint breaks away from the inner section. Wrap a cloth towel around the hot joint to avoid burning yourself, and pull the joint apart as described earlier. Consult your instructor before attempting this final option, as it is tricky.

Apparatus that contains standard-taper glassware is illustrated throughout this text; common articles of glassware are pictured on the endsheets of this laboratory book. When you set up your equipment according to these diagrams, be careful that no strain is put on the joints such that breakage occurs; support the apparatus using clamps as shown in the illustrations. If standard-taper glassware is not available in your laboratory, your instructor will demonstrate the proper techniques for setting up the apparatus using glassware fitted with bored corks or stoppers.

2.3 MELTING-POINT METHODS AND APPARATUS

The laboratory practices and apparatus that are used to determine **melting points** of solids are discussed in this section, and the theory and use of melting points are described in Section 3.3. The task of determining a melting point of a compound simply involves heating a small amount of a solid and determining the temperature at which it melts. Many different types of heating devices can be used, but most equipment utilizes a capillary tube to contain the sample so that only a small amount of the sample is required.

2.3.1 CAPILLARY TUBES AND SAMPLE PREPARATION

The first step in determining a melting point is transferring the sample into a capillary tube. Capillary tubes that have a sealed and an open end are commercially available. The proper method for loading the sample into the capillary tube is as follows: Place a small amount of the solid whose melting point you wish to determine on a clean watchglass and tap the open end of the capillary tube on the solid on the glass to force a small amount of solid about 2–3 mm into the tube (Fig. 2.2a). To move the solid to the closed end of the tube, take a piece of 6–8 mm glass tubing about 1 m long, place this tube vertically on a hard surface (bench top or floor), and drop the capillary tube through the large tube several times with its sealed end *down* (Fig. 2.2b). This operation packs the solid sample at the closed end of the capillary tube.

2.3.2 MELTING-POINT DETERMINATION

The melting point of the solid is then determined by heating the packed capillary tube until the solid melts. Some of the appropriate apparatus for this procedure are presented in Section 2.3.3. The most reproducible and accurate results are obtained by heating the sample at the rate of about 1–2 °C/min to ensure that heat is transferred to the sample at the same rate as the temperature increases and that the mercury in the thermometer and the sample in the capillary tube are in thermal equilibrium. Many organic compounds undergo a change in crystalline structure just before melting, perhaps as a consequence of release of the solvent of crystallization. The solid takes on a softer, "wet" appearance, which may also be accompanied by shrinkage of the sample in the capillary tube. These changes in the sample should *not* be interpreted as the beginning of the melting process. Wait for the first tiny drop of liquid to appear. Melting invariably occurs over a

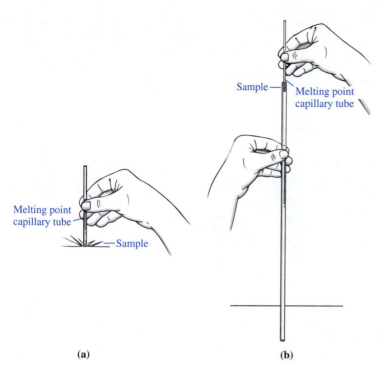

FIGURE 2.2   (a) Filling a capillary melting-point tube. (b) Packing the sample at the bottom of the capillary tube.

range of several degrees, and the **melting-point range** is defined as the temperature at which the first tiny drop of liquid appears and extends to and includes the temperature at which the solid has completely melted.

2.3.3   MELTING-POINT APPARATUS

A simple type of melting-point apparatus is the **Thiele tube**, which is shown in Figure 2.3(a). This tube is shaped such that the heat applied to a heating liquid in the sidearm by a burner is distributed evenly to all parts of the vessel by convection currents, so stirring is not required. Temperature control is accomplished by adjusting the flame produced by the microburner; this may seem difficult at first but can be mastered with practice.

Proper use of the Thiele tube is required to obtain reliable melting points. Secure the capillary tube to the thermometer at the position indicated in Figure 2.3(b) using either a rubber band or a small segment of rubber tubing. Be sure that the band holding the capillary tube on the thermometer is as close to the top of the tube as possible. Now support the thermometer and the attached capillary tube containing the sample in the apparatus either with a cork, as shown in Figure 2.3(a), or by carefully clamping the thermometer so that it is immersed in the oil. The thermometer and capillary tube must *not* contact the glass of the Thiele tube. Since the oil will expand on heating, make sure that the height of the heating fluid is approximately at the level indicated in Figure 2.3(a) and that the rubber band is in the position indicated. Otherwise, the hot oil will come in contact with the rubber, causing the band to expand and loosen; the sample tube may then fall into the oil. Heat the Thiele tube at the rate of 1–2 °C min in order to determine the melting point. The maximum temperature to which the appara-

FIGURE 2.3 (a) Thiele melting-point apparatus. (b) Arrangement of sample and thermometer for determining melting point.

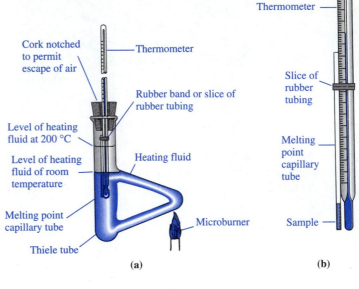

Cork notched to permit escape of air

Thermometer

Rubber band or slice of rubber tubing

Level of heating fluid at 200 °C

Heating fluid

Level of heating fluid of room temperature

Melting point capillary tube

Microburner

Thiele tube

(a)

Thermometer

Slice of rubber tubing

Melting point capillary tube

Sample

(b)

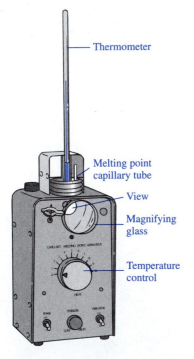

Thermometer

Melting point capillary tube

View

Magnifying glass

Temperature control

FIGURE 2.4 Thomas-Hoover melting-point apparatus. (Courtesy of Arthur H. Thomas Company.)

tus can be heated is dictated by the nature of the heating fluid, a topic that is discussed in Section 2.5.2.

The Thiele tube has been replaced in many laboratories by various **electric melting point devices,** which are more convenient to use. One common type of electric melting-point apparatus is the Thomas-Hoover melting-point unit, which is shown in Figure 2.4. This particular unit has a built-in vibrating device to pack the sample in the capillary tube, and it also allows for the determination of the melting points of up to five samples simultaneously. The oil bath in this unit is electrically heated and stirred. An electrical resistance heater is immersed in a container of silicone oil. The voltage across the heating element is varied by turning the large knob in the front of the apparatus so that the oil is heated at a slow, controlled rate. A motor drives a stirrer in the oil bath to ensure even heating; the rate of stirring is controlled by a knob at the bottom of the unit. Some models are equipped with a movable magnifying lens system that gives the user a better view of the thermometer and the sample in the capillary tube. The capillary tube containing the sample is inserted into the apparatus as illustrated in Figure 2.4.

The Mel-Temp® apparatus shown in Figure 2.5 is another electrical unit that utilizes a heated metal block rather than a liquid for transferring the heat to the capillary tube. A thermometer is inserted into a hole bored into the block, and the thermometer gives the temperature of the block and the capillary tube. Heating is accomplished by controlling the voltage applied to the heating element contained within the block.

2.4 BOILING-POINT APPARATUS

Techniques have also been developed for determining the boiling points of small amounts of liquids, and a simple **micro boiling-point** apparatus may be easily constructed. First, prepare a capillary ebullition tube by taking a standard melting-point capillary tube, which is already sealed at one end, and make a seal in it about 1 cm from the open end, using a hot flame (Fig. 2.6a). Alternatively, two melting-point capillary tubes can be joined by heating the closed ends in a hot flame; a clean cut is then made about 1 cm below the point where the tubes have been joined (Fig. 2.6b). Second, seal a piece of 4–6-mm glass tubing at one end using a hot flame and cut it to a length about 1 cm longer than the capillary ebullition tube. These tubes are prepared most easily by using a glassblowing torch, and they may be provided by your instructor.

Attach the 4–6-mm tube to a thermometer with a rubber ring near the top of the tube. The bottom of the tube should be even with the mercury bulb of the thermometer. Put the capillary ebullition tube into the larger glass tube, and with a Pasteur pipet add the liquid whose boiling point you wish to determine until the level of the liquid is about 2 mm above the seal of the capillary tube (Fig. 2.6c).

Immerse the thermometer and the attached tubes in a heating bath. A Thiele tube (Fig. 2.3) is convenient for this purpose, but other heating baths can be used. *Be sure that the rubber ring is well above the level of the oil in the heating bath.* Heat the oil bath at the rate of about 5 °C/min until a *rapid* and *continuous* stream of bubbles comes out of the capillary ebullition tube. Before this occurs, you may see some bubbles form in an erratic fashion. This is due to expansion of air trapped in the capillary tube. You should see a marked change from the slow evolution of air bubbles to the rapid evolution of bubbles resulting from the boiling action of the liquid as the boiling point is reached. *However, this is **not** the boiling point!* Remove the heating source and allow the bath to cool slowly. As the liquid starts to rise into the capillary tube, note the temperature measured by the thermometer; *this is the boiling point of the liquid.* If the liquid rises sufficiently slowly into the capillary tube, note the temperatures at which the liquid starts to rise and at which the capillary tube is full. This will be the **boiling-point range** of the liquid. Remove the capillary ebullition tube and expel the liquid from the small end by gently shaking the tube. Replace it in the sample tube and repeat the determination of the boiling point by heating the oil bath at the rate of 1–2 °C/min when you are within 10–15 °C of the approximate boiling

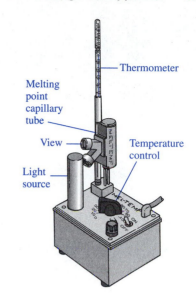

FIGURE 2.5   Mel-Temp melting-point apparatus. (Courtesy of Laboratory Devices.)

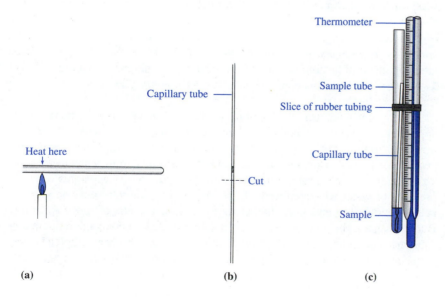

(a)   **(b)**   **(c)**

FIGURE 2.6   Micro boiling-point apparatus: (a) using a single capillary tube; (b) joining capillary tubes and cutting one end off; (c) assembly of micro boiling-point apparatus, showing correct placement of ebullition tube, sample tube, and thermometer.

point as determined in the previous experiment. Observed boiling points may be reproduced to within 1 or 2 °C.

The physical basis of this technique is interesting. Before the liquid is heated, the capillary tube is filled with air. As the bath is heated, the air in the capillary tube is driven out and replaced with the vapor of the liquid. When the apparatus is heated until vigorous boiling of the liquid is observed, the actual boiling point of the liquid has been exceeded, and the air in the capillary tube has been completely replaced by the vapor of the liquid. On cooling, the vapor pressure of the liquid becomes equal to the external pressure, thus allowing the liquid to rise into the capillary tube. The temperature at which this occurs is, by definition, the boiling temperature of the liquid (see Sec. 4.1).

2.5 HEATING METHODS

Heating is an important laboratory technique that serves a variety of functions. Many chemical reactions must be heated to proceed at a reasonable rate. Heating is used to evaporate excess solvents during the work-up of a reaction; it is also used to purify liquids by distillation and to dissolve solids in suitable solvents when purifying solid products by recrystallization. Remember that to minimize the chance of fire and to avoid filling the room with solvent vapors, *the safest way to heat organic solvents is in a hood.* Some common heating techniques along with the advantages and disadvantages of each are discussed in the following paragraphs.

2.5.1 BURNERS

Most chemistry laboratories are supplied with natural gas for use with various types of burners. A burner provides the convenience of a rapid and reasonably inexpensive source of heat. Many organic substances, however, especially solvents such as ether and hexane, are highly flammable, and good judgment must always be exercised when considering the use of a burner for heating volatile organic compounds. *Before using a burner to heat anything in the laboratory, consult with your instructor for the proper precautions and directions.* If an alternate mode of heating is available, choose it in preference to a burner. ***Never use a burner to heat flammable materials in open containers, such as beakers or Erlenmeyer flasks.***

There are a number of situations in the laboratory where burners are an appropriate heat source. They may be used to heat Thiele tubes in the determination of melting or boiling points of organic substances. Burners may be used to heat a water bath to obtain and maintain temperatures from ambient to about 90 °C. Burners may be safely substituted for heat guns to dry apparatus when the presence of water will interfere with the desired reaction. On occasion you will need a burner to bend glass tubing or to fashion a piece of glass apparatus. Burners can also be used to heat aqueous solutions that do not contain flammable substances or to heat higher-boiling flammable liquids that are completely contained in round-bottom flasks either fitted with a reflux condenser (see Sec. 2.19)

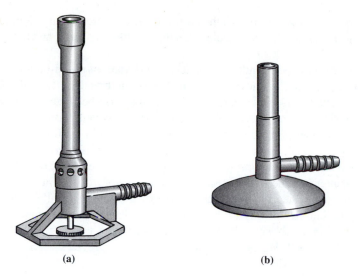

FIGURE 2.7 Laboratory burners: (a) Bunsen burner; (b) microburner.

(a)   (b)

or equipped for distillation as discussed in Sections 2.7 and 2.8. In these instances, it is important to lubricate the joints of the apparatus with a hydrocarbon or silicone grease to avoid freezing of the points or the leakage of vapors through them.

You must be aware of the type of work others are doing in the laboratory. Although you might be using a burner to perform a completely safe operation, someone else nearby may be working with a very volatile, flammable solvent, some of which can creep along the bench top for several feet! These vapors or those in the room may be ignited explosively by an open flame.

Two common types of laboratory burners are pictured in Figure 2.7. The classic **Bunsen burner**, named after its inventor, is shown in Figure 2.7(a). The needle valve at the bottom of the burner serves as a fine adjustment of the gas flow, and turning the barrel of the burner regulates the air flow; adjustment of gas and air flow provides control of the flame. In a **microburner** (Fig. 2.7b), the air flow is adjusted at the baffle at the bottom of the burner and the gas flow is adjusted at the gas valve on the laboratory bench.

Heating a flask with a burner may produce "hot spots" if most of the heat is supplied to a small area on the bottom of the flask. Hot spots can lead to severe bumping, since the heat must be dispersed throughout the liquid by convection or by means of the turbulence caused by boiling. Hot spots can be avoided by holding the burner and slowly moving the flame over the bottom of the flask. Alternatively, a piece of wire gauze, which diffuses the heat reaching the flask, may be placed between the flame and the flask; the gauze is supported with an iron ring.

2.5.2 ELECTRICAL HEATING

Heating with electrical devices is generally the method of choice in the organic laboratory, since it is much safer than using open flames. Electrical heating devices have two essential components: (1) a resistance element in which electrical energy is converted into thermal energy; and (2) a variable AC transformer to control the voltage reaching the resistance element. A variable transformer is common to all electrical heating apparatus, but there are several possibilities for the heating element itself. These include heating mantles, oil baths, and hot plates.

Heating Mantles. A widely used electrical device for round-bottomed flasks is a **heating mantle**, as shown in Figure 2.8(a); the mantle is typically fashioned from a woven blanket of spun fiberglass but may also be ceramic. These mantles have an electrical resistance coil embedded in the woven-glass fabric or ceramic core, and consequently care must be exercised to avoid spilling any type of liquid on them. Since heating mantles are constructed of nonferrous material, they can be used in conjunction with magnetic stirring (see Sec. 2.6.2), thus permitting simultaneous heating and stirring of the contents of the flask. Most heating mantles have a hemispherical cavity so that a different mantle is required for each different size of flask; other mantles are shaped to accommodate different styles of flasks. A special cord is used to connect the heating mantle to the Variac (Fig. 2.8b); *heating mantles must **never** be plugged directly into the wall outlet.* The other wires that come out of a heating mantle are connected to a thermocouple inside the mantle so that the internal temperature of the mantle can be measured, but this is seldom done in undergraduate laboratories.

There are some drawbacks to using heating mantles. They heat up rather slowly, and it is difficult to obtain a given temperature or maintain a constant temperature. Heating mantles have a high heat capacity, so if it becomes necessary to discontinue heating suddenly (for example, if a reaction begins to get out of control), *the heating mantle must be removed immediately from below the flask* to allow the flask to cool, either on its own or by means of a cooling bath; it is not sufficient simply to lower the voltage or turn off the electricity. After the mantle is removed, the electricity should be turned off at the transformer. To facilitate quick removal of the heating mantle or any other heat source, the apparatus should be assembled so that the mantle or heat source is supported above the bench by an iron ring or a laboratory jack (Fig. 2.8b); lowering the ring or jack then provides a simple means of removing the heat source from the flask.

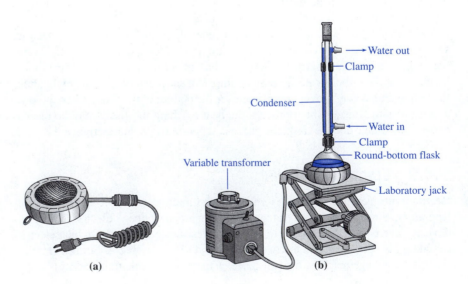

FIGURE 2.8 (a) Woven-glass heating mantle. (b) Use of heating mantle and variable transformer to heat a reaction mixture under reflux.

Water out
Clamp
Condenser
Water in
Clamp
Round-bottom flask
Variable transformer
Laboratory jack

(a)

(b)

The amount of heat supplied by a heating mantle is moderated by the boiling liquid contained in the flask because the hot vapors of the liquid transfer heat away from the mantle. If the flask becomes dry or nearly so during a distillation, the mantle can become sufficiently hot to melt the resistance wire inside, thus causing the mantle to "burn out." *Do not heat an empty flask with a heating mantle.* Moreover, most mantles are marked with a maximum voltage to be supplied, and this should not be exceeded.

Oil Baths. Electrically heated **oil baths,** which typically contain either mineral oil or silicone oil, are commonly employed in the laboratory. These baths are heated either by placing the bath container on a hot plate or by inserting a coil of Nichrome resistance wire in the bath. In the latter case, the resistance wire, which serves as the heating element, is attached to a transformer with an electrical cord and plug (Fig. 2.9). Thermal equilibration is best achieved by placing a paper clip in the oil bath and using a magnetic stirrer (see Sec. 2.6.2) to spin the paper clip to stir the oil.

Heating baths offer several important advantages over heating mantles. The temperature of the bath can easily be determined by inserting a thermometer in the liquid, and a given bath temperature may be obtained and accurately maintained by careful adjustment of the variable transformer. Although heat is transferred uniformly to the surface of the flask in the bath and there are no hot spots, there typically will be a temperature gradient of about 10 °C between the bath and the contents of the flask.

Some inconveniences are also encountered using heating baths. If the volume of heating liquid is fairly large, it may take a while to reach the desired bath temperature. The maximum temperature that may be safely attained in an oil bath is limited by the type of heating liquid being used. Silicone oils are more expensive but are generally preferable to mineral oils because they can be heated to 200 – 275 °C without reaching the **flash point**, the temperature at which a liquid can burst into flame, and without thickening through decomposition. Mineral oil should *not* be heated above about 200 °C because it will begin to smoke, and there is the potential danger of flash ignition of the vapors. Water must not be present in mineral and silicone oils, since at temperatures of about 100 °C, the water will boil, spattering hot oil. If water drops are present in the oil, the heating fluid should be changed, and the container should be cleaned and dried before refilling. Like heating mantles, oil baths have a relatively high heat

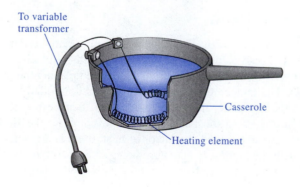

To variable transformer

Casserole

Heating element

FIGURE 2.9 An electrically heated oil bath.

FIGURE 2.10 (a) Hot plate. (b) Stirring hot plate.

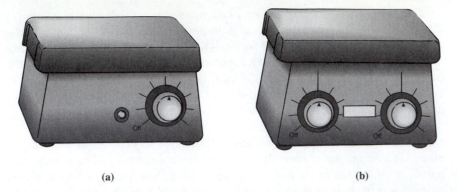

(a) (b)

capacity, and a reaction flask or a distillation pot must be removed completely from the bath to discontinue heating. This is most conveniently accomplished by supporting the oil bath on a laboratory jack in a fashion similar to that illustrated in Figure 2.8(b).

A minor nuisance associated with oil baths is removing the film of mineral or silicone oils, both of which are water-insoluble, from the outer surface of the flask. This is best done by wiping the flask using a *small* amount of hexane or methylene chloride on a paper towel prior to washing the flask with soap and water.

Hot Plates. When flat-bottom containers such as beakers or Erlenmeyer flasks must be heated, **hot plates** and **stirring hot plates,** which are hot plates with built-in magnetic stirrers (Fig. 2.10), are convenient heat sources; round-bottom flasks cannot be heated effectively with hot plates. The flat upper surface of the hot plate is heated by electrical resistance coils to a temperature that is controlled by a built-in voltage regulator, which is varied by turning a knob on the front of the unit. A hot plate generally should be limited to heating liquids such as water, mineral or silicone oil, and nonflammable organic solvents such as chloroform and carbon tetrachloride. *Under no circumstances should a hot plate be used to boil highly flammable organic solvents,* since the vapors may ignite as they billow onto the surface of the hot plate or the electrical resistance coils inside the hot plate. Furthermore, many hot plates use a relay that turns the electricity on and off to maintain the desired temperature, and these relays are often not explosion-proof and may produce sparks that can ignite fires.

2.5.3 SAND BATHS

Sand baths are convenient devices for heating small volumes of material in small flasks. These baths may be easily prepared by placing about 1–3 cm of sand in a Pyrex crystallizing or Petri dish and then placing the dish on a hot plate (Fig. 2.11). The temperature in the sand bath is controlled by varying the heat setting on the hot plate; the temperature may then be monitored with a thermometer inserted into the bath to the same depth as the flask being heated. Because sand is not a good conductor of heat, there is a temperature gradient in the bath, with the highest temperature being closest to the hot plate. This gradient can be exploited: The flask can be deeply immersed in the sand bath for

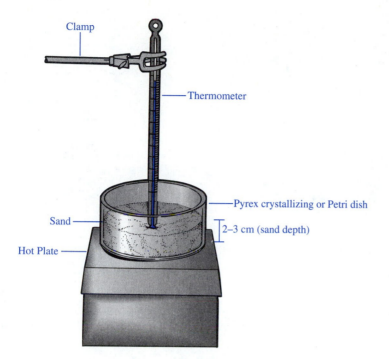

Clamp

Thermometer

Pyrex crystallizing or Petri dish

Sand

2–3 cm (sand depth)

Hot Plate

FIGURE 2.11   Sand bath, with hot plate and thermometer.

rapid heating; and once the mixture has begun to boil, the flask can be raised to slow the rate of reflux or boiling.

Although sand is cleaner than mineral and silicone oil, there are some limitations to using sand baths. For example, sand baths are not normally used to heat flasks to temperatures higher than about 200 °C, since the glass dish might break. Since sand is a poor heat conductor and it is not possible to stir the bath, precise temperature control is difficult. Furthermore, different hot plates may heat the sand bath to a different temperature gradient, so you must calibrate each hot plate/sand bath combination for reproducible results.

2.5.4   STEAM HEATING

Steam provides a useful and safe source of heat when temperatures only up to 100 °C are required. The steam outlet is connected to either a **steam bath** or a **steam cone** (Fig. 2.12), both of which have an outlet at the bottom to drain the condensed water. The tops of steam baths and cones are typically fitted with a series of overlapping concentric rings that may be removed in succession to provide openings of variable size. For example, if a rapid rate of heat transfer to a flask is desired, the rings are removed until up to one-half of the surface of the flask is immersed in the steam (Fig. 2.12a); cloth towels may be wrapped around the flask to facilitate heat transfer. If a slower rate of heating is desired, the opening may be adjusted so less of the flask is in direct contact with the steam. To heat beakers and Erlenmeyer flasks, the opening should be small enough so that the container sits directly on top of the steam bath with only its lower surface exposed to steam.

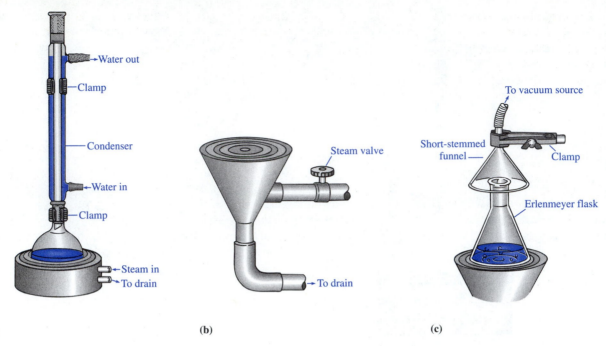

FIGURE 2.12   (a) Steam bath being used to heat a reaction mixture. (b) Steam cone. (c) Heating an Erlenmeyer flask on a steam cone, with an inverted funnel attached to a vacuum source to remove solvent vapors and keep them out of the laboratory.

When the steam valve is first opened, several minutes are usually required for the condensed water to drain out of the steam lines. Once the steam is issuing smoothly, the steam valve should be adjusted to provide a *slow,* steady flow of steam. There is no benefit to having a fast flow of steam, other than to fog your safety glasses and give everyone in the lab a sauna bath! Regardless of how fast the steam is flowing, the temperature will never exceed 100 °C. Another disadvantage of a fast flow of steam is that water will condense on your equipment and may even find its way into the reaction flask.

Steam baths are especially useful heat sources for the safe evaporation of volatile solvents during the course of working up a reaction and removing the solvent used for the reaction or extraction (see Sec. 2.25) or when purifying a solid product by recrystallization (see Sec. 3.1). As an alternative to working at a hood, an inverted funnel attached to a vacuum source may be placed over the top of the container (Fig. 2.12c) to remove vapors and keep them from entering the room.

2.6   STIRRING METHODS

Heterogeneous reaction mixtures must be stirred to distribute the reactants uniformly and facilitate the chemical reactions. Whenever the contents of a flask are being heated or cooled, stirring also ensures thermal equilibration. If a mixture is boiling, the associated turbulence is usually sufficient to provide reasonable mixing; however, stirring a boiling mixture is an alternative to using boiling chips to maintain smooth boiling action and avoid bumping. Stirring is most effectively achieved using mechanical or magnetic stirring devices, but often swirling is sufficient.

2.6.1 SWIRLING

The simplest means of mixing the contents of a flask is **swirling,** which is accomplished by manually rocking the flask with a circular motion. If a reaction mixture must be swirled, carefully loosen the clamp(s) that support the flask and attached apparatus, and swirl the contents periodically during the course of the reaction. If the entire apparatus is supported by clamps attached to a single ring stand, the clamp(s) attached to the flask do not have to be loosened. Make sure all the clamps are tight, pick up the ring stand, and gently move the entire assembly in a circular motion to swirl the contents of the flask.

2.6.2 MAGNETIC STIRRING

Magnetic stirring equipment consists of a **magnetic stirrer,** which houses a large bar magnet that is rotated by a variable-speed electric motor, and a **magnetic stirring bar,** which is placed in the flask whose contents are to be stirred. The stirring bar, which is coated with a chemically inert substance such as Teflon® or glass, should be placed *gently* in the flask before addition of any solvent or reagents. This operation is best performed by tilting the flask and letting the stirring bar slide down the side of the flask; dropping the stirring bar in directly is likely to crack or break the flask.

A flat-bottomed container such as a beaker or Erlenmeyer flask may be placed directly on top of the stirrer (Fig. 2.13a), whereas a round-bottom flask must be clamped directly above the center of the stirrer (Fig. 2.13b). The flask containing the magnetic stirring bar should be *centered* on the magnetic stirrer so that the stirring bar rotates smoothly and does not wobble. As the motor of the magnetic stirrer turns, the magnetic stirring bar rotates in phase with the motor-driven magnet, thereby stirring the contents of the flask. The stirring rate may be adjusted using the knob on the front of the magnetic stirring motor to control the spin-rate of the bar, but excessive speed often causes the stirring bar to wobble rather than to rotate smoothly.

The use of magnetic stirring in conjunction with heating is illustrated in Figure 2.13(b), which depicts an example in which the contents of a flask are being heated with an oil bath while being stirred magnetically. In this case, the contents of the flask *and* the heating bath are stirred simultaneously with a single magnetic stirrer; a large stirring bar or paper clip is used in the bath. Stirring the heating bath maintains a homogeneous temperature throughout the heating fluid. A heating mantle can be used in place of the heating bath, but it is then not easy to monitor the temperature of the heating source.

Magnetic stirring is probably the most common technique used in the laboratory for stirring mixtures. However, it is not effective for stirring *viscous* liquids or reaction mixtures, and mechanical stirring must be employed instead.

FIGURE 2.13 (a) Magnetic stirring of the contents of a beaker. (b) Use of an oil bath with magnetic stirring of a reaction mixture.

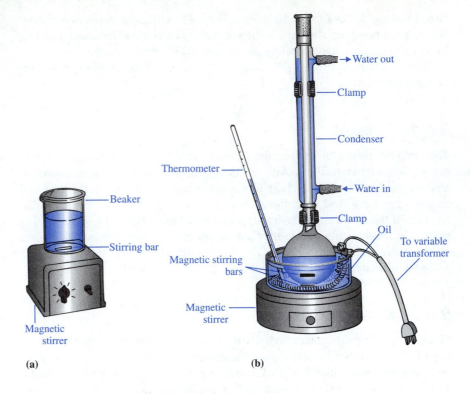

(a)

(b)

2.6.3 Mechanical Stirring

Thick mixtures and large volumes of fluids are most efficiently mixed using a mechanical stirrer; a typical set-up is depicted in Figure 2.14. A variable-speed, explosion-proof, electric motor drives a stirring shaft and paddle that extend into the flask containing the mixture to be stirred. The motor should have high torque, so that it has sufficient power to turn the shaft and stir highly viscous mixtures. The stirrer shaft is usually constructed of glass, and the paddle, that agitates the contents of the flask is constructed of an inert material such as stainless steel, Teflon, or glass. A glass paddle must be used to stir reaction mixtures containing active metals such as sodium or potassium. The paddle is easily removed from the shaft to facilitate cleaning, and different-sized paddles can be used according to the size of the flask. The glass shaft and the inner bore of the standard-taper bearing are ground to fit each other precisely. A cup at the top of the bearing is used to hold a few drops of silicone or mineral oil, which lubricates the shaft and provides an effective seal.

The stirrer shaft is connected to the motor with a short length of heavy-walled rubber tubing that is secured with twisted copper wire or a hose clamp. The motor and shaft *must* be carefully aligned to avoid wear on the glass surfaces of the shaft and bearing and to minimize vibration of the apparatus that could result in breakage. The bearing is held in place in the flask with either a rubber band or a clamp so that it does not work loose while the motor is running. The rate of stirring is controlled by varying the speed of the motor with either a built-in or separate variable transformer.

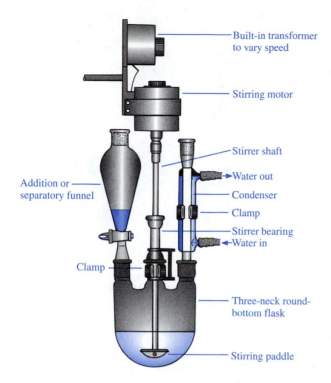

Built-in transformer to vary speed

Stirring motor

Stirrer shaft

Water out

Condenser

Clamp

Stirrer bearing

Water in

Addition or separatory funnel

Clamp

Three-neck round-bottom flask

Stirring paddle

Various operations can be performed while using mechanical stirring. For example, the flask in Figure 2.14 is a three-neck, standard-taper, round-bottom flask that is equipped with an addition funnel and a condenser. This apparatus could be used in cases where dropwise addition of a reagent to a stirred and heated reaction mixture is required.

2.7   SIMPLE DISTILLATION

Simple distillation is a useful method for isolating a pure liquid from other substances that are not volatile. The experimental aspects of this technique are described in this section, whereas the theory and application of the method are discussed in detail in Section 4.2.1.

Typical examples of laboratory apparatus for performing simple distillations are shown in Figure 2.15. The operation entails heating the liquid contained in the **stillpot,** also called the **distillation flask,** to its **boiling point,** which is defined as the temperature at which the total vapor pressure of the liquid is equal to the external pressure. The vapors then pass from the pot into a water- or air-cooled **condenser** where they condense to form the liquid phase that is collected in the **receiver.** The thermometer measures the temperature of the vapors. However, in order to obtain an accurate temperature reading, the thermometer must be carefully positioned so that the top of the mercury bulb is approximately even with the bottom of the sidearm outlet as indicated in the inset in Figure

FIGURE 2.15 (a) Typical apparatus for simple distillation at atmospheric pressure or under vacuum. Inset shows correct placement of thermometer in stillhead. (b) Typical apparatus for semi-micro distillation with ice bath to collect condensate in receiver.

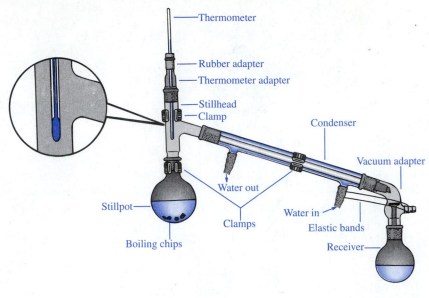

(a)

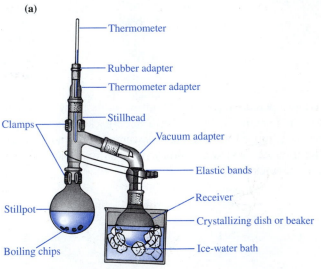

(b)

2.15. To ensure an accurate temperature reading, a drop of condensate *must* adhere to the bottom of the mercury bulb. If a pure liquid is being distilled, the temperature read on the thermometer, which is termed the **head temperature,** will be identical to the temperature of the boiling liquid, which is called the **pot temperature,** provided that the liquid is not superheated. The head temperature corresponds to the boiling point of the liquid, and will remain constant throughout the distillation. If a mixture of volatile compounds is being distilled, the head and pot temperatures will be different (see Sec. 4.2.2).

To avoid possible breakage and spillage, certain techniques must be followed when assembling the apparatus shown in Figure 2.15. The apparatus in Figure 2.15(a) is typically used for volumes of distillate in excess of about 10 mL. If a condenser is not necessary, the modified apparatus shown in Figure 2.15(b) comprised of a distillation head and a bent vacuum adapter, is well suited for distilling quantities in the range of 1–10 mL; the shorter path length in this set-up minimizes material losses. *You should thoroughly familiarize yourself with the following general guidelines before attempting to set up your apparatus.*

1. Assemble the equipment so that the distillation flask is elevated 15 cm (6 in.) or so above the bench to allow placement of a suitable heat source.

2. Review the various heating techniques discussed in Section 2.5. Heating mantles may be used to heat the stillpot, but an oil heating bath is generally the best way to heat the flask, since the temperature of the bath, and hence the stillpot, is easier to control. The level of the liquid in the distillation flask should be below the level of the oil in the heating bath to minimize the risk of bumping. *Do not use a flame unless directed to do so by your instructor.* The flask should *never* be heated *directly* with a flame because this may produce hot spots and cause bumping; the flask may only be heated with a flame *if* wire gauze is placed under the flask to diffuse the heat.

3. When distilling materials that have boiling points in excess of 125 °C, wrap the stillhead with glasswool and then aluminum foil to prevent excessive heat loss; failure to do so usually means that it will be necessary to heat the distillation flask to higher temperature to drive over the last portion of product.

4. Place several boiling stones in the distillation flask along with the liquid to be distilled *before* attaching the rest of the glassware; the flask should not be more than half-filled. *Start assembling the equipment by clamping the stillpot in position,* and attach the rest of the glassware by next putting the stillhead in place, then the condenser, and finally the vacuum adapter at the end of the condenser. Place a thin film of lubricant (see Sec. 2.2) on each of the standard-taper glass joints before mating them. Joints must fit snugly so that flammable vapors do not leak from the apparatus into the room.

5. Note the location of the clamps in Figure 2.15; *do not "overclamp" the apparatus.* An apparatus that is too rigidly clamped may be so stressed that it breaks during its assembly or the distillation. Align the jaws of the clamp *parallel* to the piece of glass being clamped so that the clamp may be tightened without twisting or torquing the glass and either breaking it or pulling a joint loose. Tighten the clamps only enough to hold each piece securely in place. *Do not tighten the clamp unless the piece of glassware is correctly positioned and the clamp is properly aligned.* Always remember the rule: *Do not apply undue pressure to the glassware.*

6. Use a strong rubber band to hold the vacuum adapter in place. Since rubber has a tendency to deteriorate in the presence of organic vapors, use only rubber bands that are uncracked and have sufficient elasticity and strength. Secure the receiving flask to the vacuum adapter by twisting a short piece of stiff wire about the neck of the flask and fastening the flask to the vacuum adapter by a rubber band. *Do not depend on rubber bands to support relatively heavy weights* however. Additional support should be provided for flasks of 100-mL capacity or larger; smaller flasks may also require additional support if they become more than half-full during a distillation. Clamp the receiver. An iron ring holding a piece of wire gauze or a cork ring may be used to provide additional support underneath the receiving flask.

7. Note the location of the "water-in" and "water-out" nipples on the condenser. The tube carrying the incoming water is *always* attached to the lower point, which ensures that the condenser is filled with water at all times.

8. Adjust the water flow through the condenser to a modest flow rate. There is no benefit to a fast flow, and the increased pressure in the apparatus may cause a piece of rubber tubing to pop off, spraying water everywhere. Showers in the laboratory should be restricted to the emergency shower! It is good practice to *wire* the hoses to the condenser and to the water faucet to minimize the danger that they will break loose.

2.8 FRACTIONAL DISTILLATION

The technique of **fractional distillation** is a useful method for isolating the individual pure liquid components from a mixture containing two or more volatile substances. This technique is described in this section, whereas the theory and application of this method are discussed in detail in Section 4.2.2. An apparatus for fractional distillation at atmospheric pressure or vacuum is shown in Figure 2.16. The principal difference between an apparatus for fractional and simple distillation is the presence of a **fractional distillation column** (Fig. 2.17) between the stillpot and the stillhead. This column is similar to a condenser, the

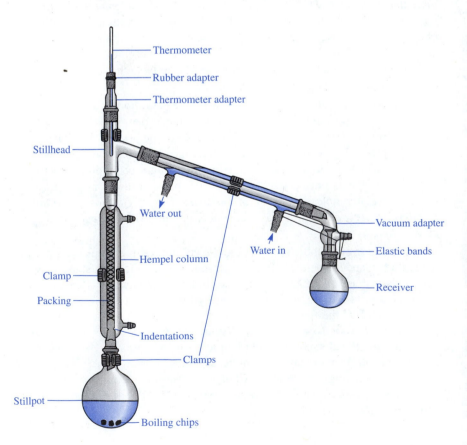

FIGURE 2.16 Typical apparatus for fractional distillation, either at atmospheric pressure or under vacuum.

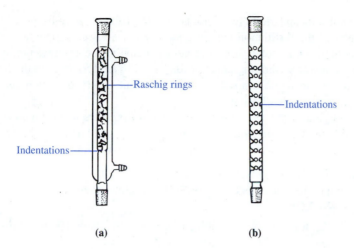

FIGURE 2.17 Fractional distillation columns. (a) Hempel column filled with Raschig rings (or other type of packing materials). (b) Vigreux column.

Raschig rings

Indentations

Indentations

(a)   (b)

major difference being that a fractionating column has a large outside jacket and some indentations at the male end to hold the packing in place. Unpacked distillation columns such as the Hempel column shown in Figure 2.17(a) can also be used as condensers, but condensers lack the indentations and cannot serve as packed distillation columns.

Before assembling the apparatus, clean and dry the inner tube of the distillation column; it is not necessary to dry the water jacket, since traces of water in it will not affect the distillation. Pack the fractional distillation column by adding the desired column packing, a small quantity at a time, through the top of the column while holding it vertical. The column packing is an inert material such as glass, ceramic, or metal pieces in a variety of shapes (helices, saddles, woven mesh, and so on); common packings include glass tubing sections, glass beads, glass helices, or a stainless steel sponge. Some column packings are sufficiently large that they will stop at the indentations. If the packing falls through the column, put a small piece of glasswool, wire sponge, or wire screen into the column just above the indentations by pushing it down the column with a wooden dowel or a piece of glass tubing. The column packing should extend to the top of the water jacket but should not be packed too tightly.

Assemble the apparatus using the general guidelines for simple distillation apparatus (see Sec. 2.7). *Read those instructions carefully before proceeding with setting up the apparatus in Figure 2.17.* Start by putting the liquid to be distilled, along with several boiling chips, in the distillation flask, then clamp the flask in place. Do not fill the flask more than half-full. Attach the distillation column and make sure that it is *vertical*. Lubricate and tighten all the joints after the stillhead and condenser are in place. Clamp only the distillation flask and condenser. *Do not run water through the jacket of the distillation column.*

2.9 VACUUM DISTILLATION

It is most convenient to distil liquids at atmospheric pressure (760 torr), but compounds that have high molecular weights or numerous functional groups may decompose, oxidize, or undergo molecular rearrangement at temperatures below their atmospheric boiling points. These problems may frequently be circumvented if the distillation can be conducted at a lower temperature. Since a liquid boils when its total vapor pressure is equal to the external pressure, it is possible to lower the boiling point of the liquid by performing the distillation at a pressure *less* than one atmosphere. The technique involved in such distillations is termed **vacuum distillation.**

Although accurate estimates of the effect of pressure upon the boiling point of a liquid may be made by use of charts or a nomograph (Fig. 2.18); two useful *approximations* of the effect of lowered pressure on boiling points are:

1. Reduction from atmospheric pressure to 25 torr lowers the boiling point of a compound boiling at 250–300 °C at atmospheric pressure by about 100–125 °C.

2. Below 25 torr, the boiling point is lowered by about 10 °C each time the pressure is reduced by one-half.

Reduced pressures may be obtained by connecting a **water aspirator pump** or a mechanical vacuum pump to a vacuum adapter that is fitted between the condenser and the receiving flask. The vacuum produced by a water aspirator is limited by the vapor pressure of the water at the ambient temperature and the condition of the aspirator pump. Pressures as low as 8–10 torr may be obtained from a water aspirator with cold water, but pressures in the range of 15–25 torr are more common. A good **mechanical vacuum pump** can evacuate the apparatus to less than 0.01 torr; it is important to clean the oil periodically and maintain tight connections in the distillation apparatus to achieve the lowest possible pressures for a particular pump. Some laboratories are equipped with "house" vacuum lines that are connected to a large central vacuum pump, but such vac-

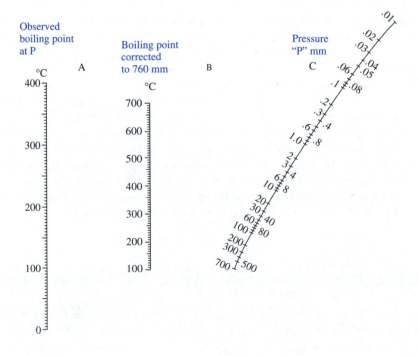

FIGURE 2.18   Pressure–temperature alignment nomograph. How to use the nomograph: Assume a reported boiling point of 120 °C at 2 torr. To determine the boiling point at 20 torr, connect 120 °C (column A) to 2 torr (column C) with a transparent plastic rule and observe where this line intersects column B (about 295 °C). This value would correspond to the normal boiling point. Next, connect 295 °C (column B) with 20 torr (column C) and observe where this intersects column A (160 °C). The approximate boiling point will be 160 °C at 20 torr.

uum systems seldom reduce the pressure to less than 50 torr. The pressure is measured by a device called a **manometer.**

Typical examples of laboratory apparatus for executing vacuum distillations are shown in Figure 2.19. The apparatus depicted in Figure 2.19(a) is typically used for volumes of distillate in excess of about 10 mL. If a condenser is not necessary, the modified apparatus, shown in Figure 2.19(b) and simply constituted of a distillation head and a bent vacuum adapter, is well suited for distilling smaller quantities in the range of 1–10 mL; the shorter path length in this set-up minimizes material losses. There are several important differences between the apparatus shown in Figure 2.15 and those in Figure 2.19. A special connecting tube, a **Claisen head,** is placed between the distillation flask and the stillhead so that a capillary tube may be inserted into the distillation flask, and the vacuum adapter is connected to either a vacuum or an aspirator pump. If a water aspirator is used, the vacuum adapter is connected to a safety flask, which serves as a trap to prevent the backup of water into the apparatus. This may occur if there is a sudden decrease in the water pressure in the aspirator. When properly equipped with a stopcock release valve, the safety flask provides a means of connecting the apparatus and the manometer to the source of the vacuum and of releasing the vacuum when desired.

The apparatus for a vacuum distillation (Fig. 2.19) is assembled according to the same general guidelines that were used to set up a simple distillation apparatus (see Sec. 2.7). *Read those instructions carefully before proceeding with setting up the apparatus in Figure 2.19.* Start by putting the liquid to be distilled in the stillpot. *Do not fill the flask more than half-full.* If magnetic stirring is available, tilt the flask to one side and add a stirbar to its contents. Then clamp the flask in place. After lubricating all of the joints, attach the Claisen head, the stillhead, and the thermometer adapters fitted accordingly with a capillary tube or a thermometer. Continue with the set-up as described in Section 2.7, clamping the condenser, if one is used, and the receiver. Make sure all of the joints are snug. Connect the apparatus to the vacuum source and the manometer.

The volume and density of the vapor formed by volatilization of a given amount of liquid are highly pressure-dependent. For example, the volume of vapor formed from vaporization of a drop of liquid will be about 20 times as great at 38 torr as it would be at 760 torr. During a vacuum distillation, large quantities of vapor are produced in the distillation flask, and these vapors enter the condenser with high velocity, whereupon they re-form a drop of liquid. Since the density of the vapor is much lower at reduced pressure than at atmospheric pressure, it may be difficult to control the rate of vaporization and thereby minimize the difference in pressure between the distillation flask and the manometer where the pressure must be accurately measured.

The increased rate of vaporization at reduced pressures causes the formation of large bubbles of vapor. When these escape from the liquid, they may cause bumping or vigorous splashing and splattering. The insertion of a Claisen connecting tube between the distillation flask and the stillhead (Fig. 2.19) will partially solve this problem by preventing the liquid in the distillation flask from splashing into the sidearm leading to the condenser. To ensure optimal control of the rate of distillation, however, it is better to control the ebullition of the vapors from the liquid. Boiling chips are generally ineffective. Magnetic stirring (see Sec. 2.6.2) is the most convenient means of obtaining an even rate of vapor-

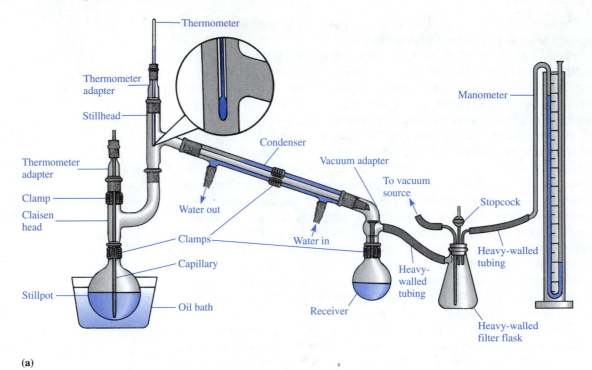

(a)

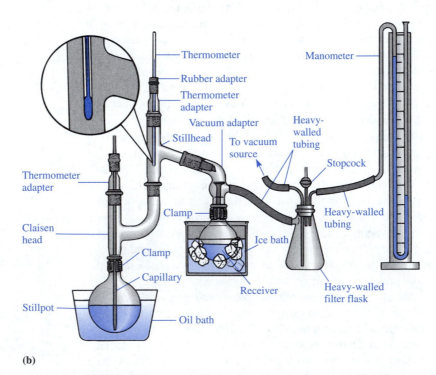

(b)

FIGURE 2.19 (a) Typical apparatus for vacuum distillation. (b) Modified apparatus for semi-micro vacuum distillation.

ization and should be used if possible. If magnetic stirrers are not available, then a *very thin,* flexible capillary tube that is extended nearly to the bottom of the distillation flask introduces a fine stream of air bubbles that provide nuclei for regular vaporization. The volume of air so introduced is small compared with the evacuating capacity of a water or oil pump, so the "leak" has no significant effect on the vacuum in the system. The capillary is drawn from a piece of 6-mm glass tubing and should be thin enough to allow only a slow stream of fine bubbles to emerge from the tip. The capillary tube may be tested by blowing air through the tube with its capillary end inserted into a test tube containing acetone. A thin wooden applicator stick is sometimes used as a boiling stick in place of a capillary. These do not always work well, especially if the vacuum must be released during the distillation, in which case the used stick should be replaced with a new one.

The manometer measures the pressure at which the distillation is being conducted; this value is important and is reported with the boiling point. For example, benzaldehyde boils at 180 °C at atmospheric pressure and at 87 °C at 35 torr, and these two boiling points are reported in the format: bp 180 °C (760 torr) and 87 °C (35 torr). To obtain an accurate measure of the pressure, however, you must carefully *control the rate of distillation.* The problem arises because a drop of condensate forms from a larger volume of vapor as the pressure is lowered, thereby causing higher vapor velocities to enter the condenser. This creates a back pressure so that the pressure in the distillation flask and stillhead is higher than that measured on the manometer, which is located beyond the condenser and receiver and is therefore insensitive to back pressure. The difference between the actual and measured pressure can be minimized by distilling the liquid at a slow but steady rate. Superheating the vapor can be avoided by maintaining the oil bath at a temperature no more than 25–35 °C higher than the head temperature.

The following paragraphs provide a general procedure for performing a vacuum distillation using the apparatus shown in Figure 2.19(a). *Read these instructions carefully prior to executing a vacuum distillation.*

1. *Never use glassware with cracks, thin walls, or flat bottoms, such as Erlenmeyer flasks, in vacuum distillations.* Hundreds of pounds of pressure may be exerted on the total exterior surfaces of systems under reduced pressure, even if only a water aspirator pump is used. Weak or cracked glassware may *implode,* and the air rushing into the apparatus will shatter the glassware violently in a manner little different from that of an explosion. Additional dangers that could arise from imploding glassware are burns from the hot oil of the bath and the possibility of starting a serious fire. *Examine the glassware carefully and always wear safety glasses when doing a vacuum distillation.*

2. Lubricate and seal all glass joints carefully during assembly of the apparatus; this will help avoid air leaks and provide lower pressure. The rubber fittings holding the thermometer and capillary in place must be tight. The neoprene fittings normally used with the thermometer adapters may be replaced, if necessary, with short pieces of heavy-walled tubing. Thermometers and capillary tubes that contain standard-taper joints may be available in your laboratory; these are preferable to using rubber connections. Do not use rubber stoppers elsewhere in the apparatus, since direct contact between the rubber and the hot vapors during distillation may cause contamination. The three-holed rubber stopper in the safety trap should fit snugly to the flask and the pieces of glass tubing. The safety trap must be made of heavy-walled glass and wrapped with electrical tape as protection in case it implodes. Heavy-walled vacuum tubing must be used for all vacuum connections. *Check the completely assembled apparatus to make sure that all joints and connections are tight.*

3. Now turn on the vacuum; if you are using a water aspirator pump, you should *fully open the water faucet.* The release valve on the safety flask should be *open. Slowly* close the release valve, but be prepared to reopen it if necessary. If the liquid contains small quantities of low-boiling solvents, as is often the case, foaming and bumping are likely to occur in the distillation flask. If this occurs, adjust the release valve until the foaming abates. This may have to be done several times until the solvent has been removed completely. When the surface of the liquid in the flask is relatively quiet, fully evacuate the system to the desired pressure. The release valve may have to be opened slightly until the desired pressure is obtained; a needle valve from the base of a Bunsen burner may be used for fine control of the pressure. Check the manometer for constancy of pressure. Note and record the pressure at which the distillation is being performed. This pressure should be monitored throughout the course of the distillation, even when fractions are not being collected. Begin heating the flask, and maintain the temperature of the bath so that the distillate is produced at the rate of 3–4 drops per minute. *Do not begin heating the flask until the system is fully evacuated and the vacuum is stable.* The best way to heat the stillpot is with an oil bath, since it is easier to control the temperature in the bath and distillation flask; however, other techniques (see Sec. 2.5) may be used.

4. If it is necessary to use multiple receivers to collect fractions of different boiling ranges, the distillation must be interrupted to change flasks. Remove the heating source with caution and allow the stillpot to cool somewhat. *Slowly* open the vacuum release valve to re-admit air to the system. When atmospheric pressure is attained, change receivers, close the release valve, re-evacuate the system to the same pressure as previously, reapply heat, and continue distilling. The operation may result in a different pressure in the fully evacuated system. Periodically monitor and record the head temperature and the pressure, particularly just before and after changing receivers.

5. After the distillation is complete, discontinue heating, allow the pot to cool somewhat, slowly release the vacuum, and turn off the source of the vacuum.

One of the most inconvenient aspects of this procedure is the disruption of the distillation in order to change receivers. In order to eliminate this problem, you may wish to use a "cow" receiver, which may accommodate three or four round-bottom flasks (Fig. 2.20). These flasks are successively used as receivers by rotating them into the receiving position.

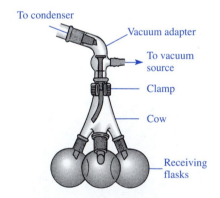

To condenser

Vacuum adapter

To vacuum source

Clamp

Cow

Receiving flasks

FIGURE 2.20 Multiple-flasked receiver for vacuum distillation.

2.10 STEAM DISTILLATION

Steam distillation is a useful and mild method for separating and purifying volatile liquid or solid organic compounds that are immiscible or insoluble in water. This technique is not applicable to substances that react with water, decompose on prolonged contact with steam or hot water, or have a vapor pressure of less than about 5 torr at 100 °C. The practical features and the apparatus of this technique are described in this section; its theory and applications are discussed in detail in Section 4.2.5.

The two basic techniques commonly used to carry out a steam distillation in the laboratory are differentiated based on whether the steam is introduced from an external source or generated internally. For larger scale reactions, the most common and most efficient method for conducting a steam distillation involves placing the organic compound(s) to be distilled in a round-bottom flask equipped with a Claisen head, a stillhead, and a water-cooled condenser, as depicted in Figure 2.21. The combination of introducing steam from an external source into the distillation flask via the inlet tube and of the turbulence associated with the boiling action tends to cause occasional violent splashing, and the Claisen head is necessary to prevent the mixture from splattering into the condenser. Steam may be produced externally in a generator as shown in Figure 2.22, but it is more conveniently obtained from a laboratory steam line. A trap (Fig. 2.23) is then placed between the steam line and the distillation flask to permit removal of any water and/or impurities present in the steam. During the course of the distillation, water may condense in the distillation flask and fill it to undesirable levels. This problem can be avoided by gently heating the flask with a Bunsen burner or a heating mantle.

If only a small amount of steam is necessary to separate a mixture completely, a simplified method involving *internal steam generation* may be employed. Water is added directly to the distillation flask together with the organic compounds to be separated. The flask is equipped for steam distillation by setting up the apparatus as shown in Figure 2.21 *except* that the steam inlet tube is replaced with a stopper. The flask is then heated with a Bunsen burner or heating mantle, and the distillate is collected. This technique is generally not applicable for distillations that require large amounts of steam, since it would be necessary to use an inappropriately large flask or to replenish the water in the flask frequently by using an addition funnel. However, for most miniscale reactions, this procedure is satisfactory and convenient.

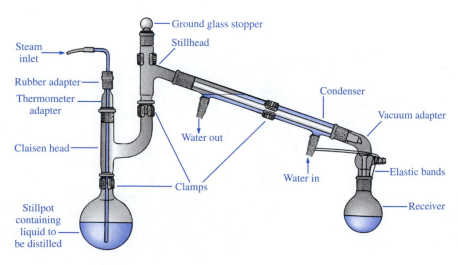

Steam inlet

Rubber adapter

Thermometer adapter

Claisen head

Stillpot containing liquid to be distilled

Ground glass stopper

Stillhead

Water out

Clamps

Condenser

Water in

Vacuum adapter

Elastic bands

Receiver

FIGURE 2.21 Apparatus for steam distillation. The steam tube is replaced by a stopper if steam is generated by direct heating.

FIGURE 2.22 Steam generator. The round-bottom flask is initially half-filled with water, and boiling chips are added before heating. The safety tube serves to relieve internal pressure if steam is generated at too rapid a rate.

Safety tube (35–50 cm long, i.d. ≥ 8mm)

Steam to apparatus

Cork or rubber stopper

Clamp

Round-bottom flask

Water

Wire gauze

Ring

Bunsen burner

Ring stand

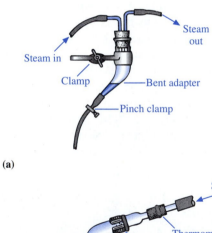

Steam out

Steam in

Clamp

Bent adapter

Pinch clamp

(a)

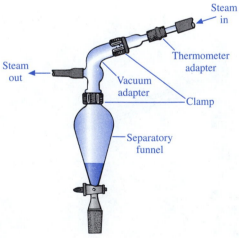

Steam in

Thermometer adapter

Steam out

Vacuum adapter

Clamp

Separatory funnel

FIGURE 2.23 Water traps for use in steam distillations.

(b)

2.11   FLUTED FILTER PAPER

"Fluted" filter paper is used in many filtering operations to increase the surface area and thus the rate and ease of filtration. Although fluted filters are commercially available, they may be quickly prepared using several methods, one of which is shown in Figure 2.24. Fold the paper in half, and then into quarters. Fold edge 2 onto 3 to form edge 4, and then 1 onto 3 to form 5 (Fig. 2.24a). Now fold edge 2 onto 5 to form 6, and 1 onto 4 to form 7 (Fig. 2.24b). Continue by folding edge 2 onto 4 to form 8, and 1 onto 5 to form 9 (Fig. 2.24c). The paper now appears as shown in Figure 2.24(d). All folds thus far have been in the same direction. Do not crease the folds tightly at the center, because this might weaken the paper and cause it to tear during filtration. Now make folds in the *opposite* direction between edges 1 and 9, 9 and 5, 5 and 7, and so on, to produce the fanlike appearance shown in Figure 2.24(e). Open the paper (Fig. 2.24f) and fold each of the sections 1 and 2 in half with reverse folds to form paper that is ready to use (Fig. 2.24g).

2.12   GRAVITY FILTRATION

Gravity filtration is a technique that is commonly used to remove solids such as impurities, drying agents (see Sec. 2.22), or decolorizing carbon (see Sec. 2.14) from liquids prior to evaporation or distillation. The folded or fluted filter paper is first placed in a funnel and *wetted* with the appropriate solvent. The solution containing the solid is then poured onto the filter paper or fluted filter paper (see Sec. 2.11) with the aid of a glass stirring rod to minimize dripping down the sides of the flask. As the solution passes through the paper, the solid remains on the filter paper. A typical experimental set-up is shown in Figure 2.25. If you do not use a ring stand to support the funnel, you should insert a paper clip or a bent

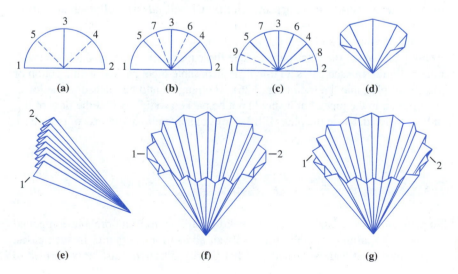

FIGURE 2.24   Folding filter paper to produce fluted filter paper.

FIGURE 2.25 Gravity filtration.

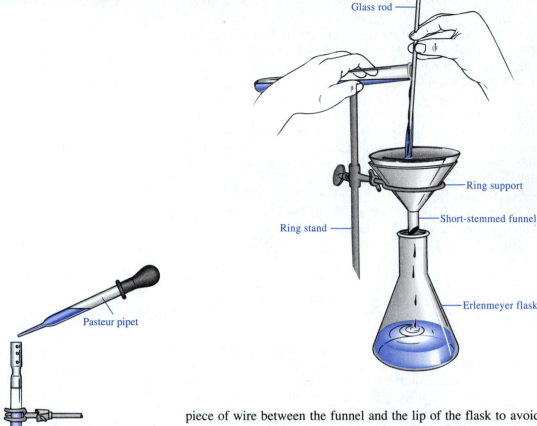

FIGURE 2.26 Use of a filtering pipet.

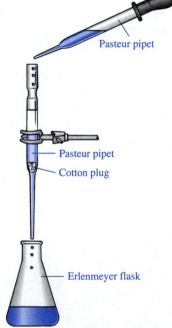

piece of wire between the funnel and the lip of the flask to avoid the formation of a solvent seal between the flask and the funnel. Such a seal will interfere with the filtration, since there will be no vent for the air displaced by the filtrate.

On occasion, it may be necessary to remove a finely divided or colloidal solid that passes through the filter paper. In such cases, 1 to 2 of a **filter-aid,** such as Celite®, is added to the solution before filtration. A filter-aid is a finely divided, inert material, usually diatomaceous earth, that will *adsorb* colloidal substances and prevent them from passing through the filter paper or clogging its pores.

When volumes less than about 5 mL must be filtered, there is a danger of losing material on the filter paper, since the paper will absorb a significant volume of liquid. In such cases a Pasteur or disposable pipet packed with a cotton or glasswool plug may be used (Fig. 2.26). The plug should be pushed to the lower constriction in the pipet, but it should not be packed so tightly that the flow of liquid is restricted. Rinse the plug with 1–2 mL of solvent to avoid loss of material.

2.13 HOT GRAVITY FILTRATION

Sometimes it is necessary to remove solid impurities from an organic compound that is only sparingly soluble in the solvent at room temperature. In such cases, the mixture must then be heated, and hot gravity filtration must be performed to

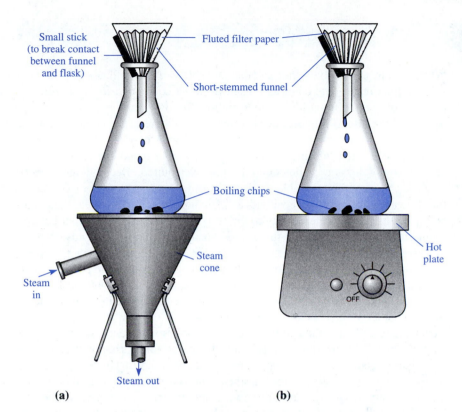

remove the insoluble solids without leaving the desired material in the funnel, too. Such a situation commonly arises during purification of an organic solid by recrystallization when decolorizing carbon (see Sec. 2.14) or other insoluble residues must be removed; a filter-aid (see Sec. 2.12) may be required for a successful filtration.

The apparatus for performing a hot gravity filtration is shown in Figure 2.27. A small volume of the *pure* solvent is first placed in an Erlenmeyer flask containing a boiling stick or several boiling chips, and a short-stemmed or stemless filter funnel containing a piece of fluted filter paper is placed in the mouth of the flask. A small stick or piece of thick paper is inserted between the flask and the funnel to allow the vapors of solvent to rise around the outside of the funnel. The flask is heated with either a steam bath or hot plate until the solvent boils and the vapors rise up the sides of the flask and around the funnel; the solvent vapors heat the filter funnel to minimize problems of crystallization of the desired material in the funnel. A steam bath is preferred for flammable solvents that boil below 90 °C, whereas a hot plate can be used for all nonflammable solvents and, with proper care and in a hood, for flammable solvents boiling above 90 °C.

When the flask and funnel are hot, the hot solution is poured onto the filter paper. After all the hot solution has been poured onto the filter paper, several milliliters of hot solvent are added to the flask that contained the original recrystallization solute and solvent. The additional hot solvent is poured onto the filter paper to ensure the complete transfer of material to the flask containing the filtered solution.

2.14   DECOLORIZING CARBON

When a solid is dissolved in a suitable solvent for recrystallization, a colored solution may result. The color can result from the compound itself or from the presence of colored impurities in the solution. Similarly, an organic liquid may be colored because of impurities or because of the nature of the compound itself. If you are in doubt about the cause of the color, you may try to remove it by adding activated **decolorizing carbon** to the solution or liquid and then removing the carbon by filtration; sometimes heating is required. Decolorizing carbon, which is commercially available under trade names such as Norite® or Darco®, is finely divided, activated charcoal with an extremely large surface area. Polar, colored impurities bind preferentially to the surface of decolorizing carbon, but you should be aware that some of the desired compound may also be adsorbed, so excessive quantities of decolorizing carbon should *not* be used. After the decolorizing carbon is removed by filtration, the liquid or solution should be free of colored impurities; any remaining color is probably due to the compound itself.

In a typical procedure, the liquid or solution, contained in an Erlenmeyer flask, is first brought to a temperature 10–20 °C below its boiling point. Decolorizing carbon is then added, and the hot solution is gently swirled. There are no firm rules concerning the amount of decolorizing carbon that should be used, but a good first approximation is to add about 0.5 to 1 g per 100 mL of solution. The decolorizing carbon is then removed by gravity filtration or hot gravity filtration, not vacuum filtration, using fluted filter paper; a filter-aid (Sec. 2.12) will facilitate complete removal of all of the carbon. Hot gravity filtration must be used whenever an organic solid is dissolved in a hot solvent, such as during a recrystallization. The solution or liquid must be refiltered if it contains any small black specks of carbon or any other solid material. Normally this extra step can be avoided if care is exercised when filtering the mixture and a filter-aid is used. For example, the filter paper should not contain any tears or holes, and the mixture should be carefully poured onto the filter paper so that none of it runs down the side of the funnel. If the compound is known to be colorless and if the decoloration process does not remove the color, the procedure should be repeated.

2.15   DECANTING SOLUTIONS

Small amounts of dense solids are sometimes removed by simple **decantation.** This technique is a viable alternative to filtration for removing solid drying agents (see Sec. 2.20), but it cannot be used to remove finely dispersed solids such as decolorizing carbon. To decant a liquid from a solid, the solid should first be allowed to settle to the bottom of the container. The container is then *carefully* tilted, and the liquid is *slowly* poured into a clean container, possibly with the aid of a glass stirring rod to minimize dripping down the sides of the flask (see Fig. 2.25); the solid should remain in the original container. Decantation is preferable to gravity filtration when working with very volatile organic liquids, since filtration is likely to result in considerable evaporation and loss of material. When decanting from an Erlenmeyer flask, a *loosely packed* ball of glasswool can be put in the neck of the flask to help keep the solid in the flask.

One major disadvantage of decantation is that some liquid will remain in the flask; this problem may be minimized by using a micropipet to transfer the last few milliliters and rinsing the residue with several milliliters of pure solvent that are also transferred with a micropipet.

2.16 VACUUM FILTRATION

Vacuum filtration is a technique that is used to collect crystalline solids, usually products, from solvents after recrystallization or precipitation. A typical apparatus is shown in Figure 2.28. Either a **Büchner** or a smaller **Hirsch funnel** is fitted to a vacuum filter flask using a neoprene adapter or a rubber stopper. The sidearm of the filter flask is connected to an aspirator or house vacuum line through a **trap** using heavy-walled rubber tubing. The trap prevents water from the aspirator from backing up into the filter flask when there is a loss of water pressure. The trap should be a *heavy-walled* bottle or a second vacuum filter flask wrapped with electrical or duct tape. If a filter flask is used for the trap, it should be equipped with a two-holed stopper with its sidearm attached to the first filter flask with heavy-walled tubing. The glass tube that extends to within approximately 2–3 cm of the bottom of the filter trap is connected with heavy-walled tubing to the water aspirator pump so that any water that collects slowly in the trap is evacuated back through the aspirator pump and into the drain. If a sudden back-up occurs, the vacuum should be released immediately so that the trap does not fill completely with water. Any filtrate that may overflow the filter flask will also be collected in the trap. *The filter flask should be cleaned before doing the filtration,* since it may be necessary to save the filtrate.

A piece of filter paper is placed in a Büchner or Hirsch funnel and should lie flat on the funnel plate, covering all the small holes in the funnel. It should *not* extend up the sides of the funnel. A vacuum is applied to the system, and the filter paper is "wetted" with a small amount of pure solvent in order to form a seal with the funnel so that crystals do not pass around the edges of the filter

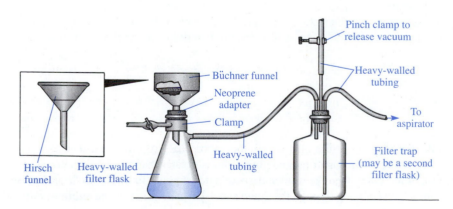

FIGURE 2.28 Apparatus for vacuum filtration of solids; inset shows Hirsch funnel, which may be used for small quantities of solid.

paper and through the holes in the filter. The solution containing the crystals is then transferred to the funnel. The flask containing the crystals is swirled to suspend the crystals in the solvent, and the solution containing the crystals is poured slowly onto the funnel. A stirring rod or spatula may be used to aid the transfer. The last of the crystals may be transferred to the funnel by washing them from the flask with some of the filtrate, which is called the **mother liquor.** When all the solution has passed through the filter, the vacuum is released slowly by opening the screw clamp or stopcock on the trap. The crystals are washed to remove the mother liquor, which contains impurities, by adding a small amount of *cold, pure solvent* to the funnel to just cover the crystals. Vacuum is reapplied to remove the wash solvent, and the crystals are pressed as dry as possible with a clean spatula or a cork while the funnel is under vacuum.

Most of the solvent may be evaporated from the crystals by allowing the vacuum to pull air through the crystals on the funnel for a few minutes. A clean spatula is used to scrape the crystals gently from the filter paper and then to transfer them to a clean watchglass; care should be exercised to avoid contaminating the crystals with torn bits of filter paper. The crystals may be dried completely by allowing them to air-dry for a few hours or by leaving them loosely spread on a filter paper in a locker until the next class. The drying process may be accelerated by placing the crystals in an oven, but the temperature of the oven should be at least 20–30 °C below the melting point of the crystals. A vacuum desiccator also can be used to hasten the drying process. The vacuum desiccator should either be wrapped with electrical or duct tape or placed in a metal cage as protection should the desiccator implode. Specially designed desiccators permit heating samples under vacuum, but the temperature must be kept about 20–30 °C below the melting point of the solid. Heat or vacuum desiccators should not be used to dry crystals of compounds that sublime readily (see Sec. 2.17).

Vacuum filtration also can be used to remove undesired solids from a solution. If the solid is finely divided or colloidal, a pad of a filter-aid such as Celite (see Sec. 2.12) can be used to ensure complete removal of the solid. A pad of filter-aid can be formed on the filter funnel by first making a slurry of 0.5 to 1 g of filter-aid in a few milliliters of the solvent being used. The slurry is poured onto a filter paper in a Büchner or Hirsch funnel that is attached to a *clean* filter flask; vacuum is then slowly applied to draw the solvent through the filter paper, leaving a thin, even pad of the filter-aid; the pad should be about 2–3 mm thick. The solution containing the solid is then filtered as described above. Obviously this technique is useful for removing solid impurities *only* when the *solution* contains the substance of interest; a solid product is not collected in this manner as it would be necessary to separate the desired material from the filter-aid.

2.17 SUBLIMATION

Like the vapor pressure of a liquid, the vapor pressure of a solid increases with temperature. If the vapor pressure of a solid is greater than the ambient pressure at its melting point, then the solid undergoes a direct phase transition to the gas phase without first passing through the liquid state. This process is called **sublimation.**

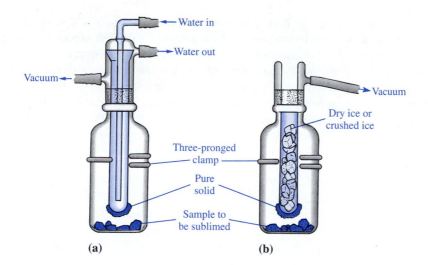

FIGURE 2.29   Sublimation apparatus: (a) using water as the coolant; (b) using low-temperature coolants.

The ease with which a molecule may escape from the solid to the vapor phase is determined by the strength of the intermolecular attractive forces between the molecules of the solid. Symmetrical structures have relatively uniform distributions of electron density, and they have small dipole moments compared to less symmetrical molecules. Since electrostatic interactions are the strongest intermolecular forces in the crystal lattice, molecules with smaller dipole moments will have higher vapor pressures. Sublimation is thus generally a property of relatively nonpolar substances having fairly symmetrical structures. Van der Waals attractive forces are also important but less so than electrostatic ones. Van der Waals forces increase in magnitude with increasing molecular weight, and large molecules, even if symmetrical, tend not to sublime.

In order to purify a compound by sublimation, it must have a relatively high vapor pressure, and the impurities must have vapor pressures significantly lower than the compound being purified. If the impurities in a compound have similar vapor pressures, recrystallization (see Sec. 3.2) or column chromatography (see Sec. 6.2) may be used to purify the compound. Since few organic solids exhibit vapor pressures high enough to sublime at atmospheric pressure, most sublimations are performed at reduced pressure; this is analogous to using vacuum distillation for high-boiling liquids.

Two common types of sublimation apparatus are shown in Figure 2.29. In each case there is a chamber that may be evacuated using a water aspirator, house vacuum, or vacuum pump and a cold finger in the center of the vacuum chamber to provide a surface upon which the sublimed crystals may form. The cold finger is cooled by water (Fig. 2.29a) or another cooling medium such as ice/water or Dry Ice/acetone (Fig. 2.29b). A simple sublimation apparatus may be assembled inexpensively from two test tubes, one having a side arm and rubber stoppers (Fig. 2.30a). In place of the test tube with a sidearm, a small vacuum filtration flask may also be used (Fig. 2.30b). The cold finger may be cooled either with running water or with ice chips. If the cold finger is cooled with running water, the water hoses must be securely attached to the inlet and outlet of the condenser by means of a piece of twisted copper wire or a hose clamp.

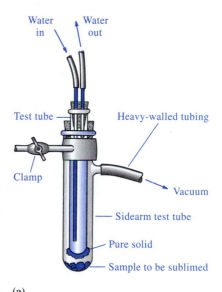

(a)

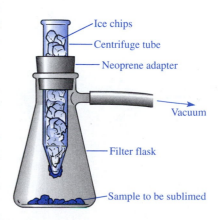

(b)

FIGURE 2.30   Simple sublimation apparatus: (a) test tube sublimator; (b) filter-flask sublimator.

In order to purify an impure substance by sublimation, it is first placed at the bottom of the sublimation chamber. The sample is then heated under reduced pressure, using an oil or sand bath or a small flame, to a temperature *below the melting point* of the solid. The solid will be vaporized and transferred via the vapor phase to the surface of the cold finger, where it condenses directly as a pure material to the solid state. After the sublimation is complete, the pressure must be released carefully to avoid dislodging the crystals from the cold finger with a surge of air. For similar reasons, care must be exercised when removing the cold finger from the sublimation apparatus. The pure crystals are scraped from the cold finger with a spatula.

2.18 SEPARATORY FUNNELS AND THEIR USE

Separatory funnels are used to separate two immiscible liquid phases. These funnels are available in many different shapes, ranging from almost spherical to an elongated pear shape (Fig. 2.31); the more elongated the funnel, the longer the time required for the two liquid phases to separate after the funnel is shaken. Although separatory funnels may be equipped with either a glass or Teflon stopcock, the latter which does not require lubrication is the preferred type, because the solutions being separated do not get contaminated with stopcock grease. Separatory funnels are most commonly used during work-up procedures after completion of a chemical reaction. For example, they are used for extracting the desired product from one immiscible liquid phase into another (see Secs. 5.2 and 5.4) and for "washing" organic layers to remove undesired substances such as acids or bases from the desired organic compound.

There are a number of general guidelines for using separatory funnels that merit discussion:

1. *Filling Separatory Funnels.* The stopcock should be closed and a clean beaker placed under the funnel before any liquids are added to the funnel in case the stopcock leaks or is not completely closed. A separatory funnel should never be more than three-quarters full, especially when doing an extraction. The upper opening of the funnel is then stoppered either with a ground-glass, plastic, or rubber stopper; most separatory funnels are now fitted with a ground-glass or plastic stopper.

2. *Holding and Using Separatory Funnels.* If the contents of the funnel are to be shaken, it is held in a specific manner. If the user is right-handed, the stopper should be placed against the base of the index finger of the left hand and the funnel grasped with the first two fingers and the thumb. The thumb and the first two fingers of the right hand can then be curled around the stopcock (Fig. 2.32a). Holding the funnel in this manner permits the stopper and the stopcock to be held tightly in place during the shaking process. A left-handed person might find it easier to use the opposite hand for each position.

3. *Shaking Separatory Funnels.* A separatory funnel and its contents should be shaken as shown in Figure 2.32(b) to mix the immiscible liquids as intimately

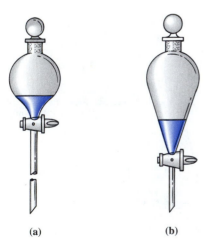

(a)   (b)

FIGURE 2.31 Separatory funnels: (a) conical; (b) pear-shaped.

as possible. The shaking process increases the surface area of contact between the immiscible liquids so that the equilibrium distribution of the solute between the two layers will be attained quickly; however, overly vigorous or lengthy shaking may produce undesired **emulsions** (discussed below). *The funnel must be **vented** every few seconds to avoid the buildup of pressure within the funnel.* Venting is accomplished by *inverting the funnel with the stopcock pointing upward and away from you and your neighbors* and slowly opening it to release any pressure (Fig. 2.32c). If the funnel is not carefully vented, liquid may be violently expelled, covering you and your laboratory partners with the contents. Venting is particularly important when using volatile, low-boiling solvents such as diethyl ether or methylene chloride; it is also necessary whenever an acid is neutralized with either sodium carbonate or sodium bicarbonate, since CO_2 is produced. If the funnel is not vented frequently, the stopper may be accidentally blown out; under extreme circumstances the funnel might blow up. Any sudden release of pressure is likely to result in the contents of the flask being lost and spattered on you and your co-workers. The funnel may be vented simply by holding it as described previously and opening the stopcock by twisting the fingers curled around it without readjusting your grip on the funnel. At the end of the period of shaking of (1–2 min are usually sufficient if the shaking is vigorous), the funnel is vented a final time. It is then supported on an iron ring (Fig. 2.33), and the layers are allowed to separate. (When an iron ring is used to support a separatory funnel, the ring should be covered with a length of rubber tubing to prevent breakage. This may be accomplished by slicing the tubing along its side and slipping it over the ring.) Copper wire may be used to fix the tubing permanently in place. The lower layer in the separatory funnel is then carefully drawn into a flask through the stopcock while the interface between the two layers is watched. If small quantities of insoluble material collect at the boundary between the layers and make it difficult to see this interface, it is best to remove these solids with the *undesired* liquid layer; a small amount of the desired layer is inevitably lost by this procedure. An alternative procedure is to remove the solids by gravity or vacuum filtration before separating the layers.

   4. *Layer Identification.* It is important to ascertain which of the two layers in a separatory funnel is the aqueous layer and which is the organic; this may be easily accomplished with a little care and thought. Since the layers will usually separate so that the denser solvent is on the bottom, knowledge of the densities of the liquids being separated provides an important clue for this identification; the densities of several water-insoluble solvents are found in Table 3.1. This generalization is not foolproof, however, because a high concentration of a solute in one layer may reverse the relative densities of the two liquids. You must not confuse the identity of the two layers in the funnel and then discard the layer containing your product. *Both layers should always be saved until there is no doubt about the identity of each.*

   Since one of the layers is usually aqueous and the other is organic, there is a simple and foolproof method to identify the two layers. Withdraw a few drops of the upper layer with a pipet and add these drops to about 0.5 mL of water in a test tube. If the upper layer is aqueous, these drops will be miscible with the water in the test tube and will dissolve, but if the upper layer is organic, the droplets will not dissolve and will remain visible.

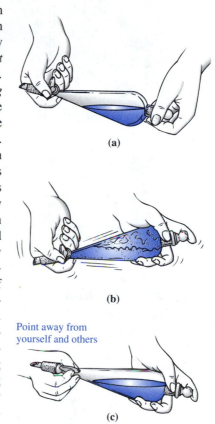

(a)

(b)

Point away from
yourself and others

(c)

FIGURE 2.32   (a) Efficient method of holding a separatory funnel; (b) shaking a separatory funnel; (c) venting a separatory funnel.

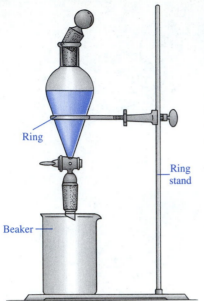

FIGURE 2.33 Separatory funnel positioned on iron ring with a beaker underneath to collect the contents in case of funnel malfunction.

5. *Emulsions.* Occasionally the two immiscible liquids will not separate cleanly into two distinct layers after shaking because of an **emulsion** that results from a colloidal mixture of the two layers. If prior experience leads you to believe that an emulsion might form, you should avoid shaking the funnel too vigorously; instead, swirl it gently to mix the layers. Encountering an emulsion during an experiment can be extraordinarily frustrating, because there are no infallible convenient procedures for breaking up emulsions. An emulsion left unattended for an extended period of time sometimes separates. However, it is usually more expedient to attempt one or more of the following remedies:

a. Add a few milliliters of a *saturated* solution of aqueous sodium chloride, commonly called brine, to the funnel and gently reshake the contents. This increases the ionic strength of the water layer, which helps force the organic material into the organic layer. This process can be repeated, but if it does not work the second time, other measures must be taken.

b. Filter the heterogeneous mixture by vacuum filtration through a thin pad of a filter-aid (see Sec. 2.12), and return the filtrate to the separatory funnel; if a filter-aid is not used, the pores of the filter paper may become clogged and the filtration will be slow. Sometimes emulsions are caused by small amounts of gummy organic materials whose removal will often remedy the problem.

c. Add a *small* quantity of water-soluble detergent to the mixture and reshake the mixture. This method is not as desirable as the first two techniques, particularly if the desired compounds are in the water layer, because the detergent adds an impurity that must be removed later.

d. Sometimes intractable emulsions that appear to be stabilized by small trapped air bubbles are encountered during the work-up of phase-transfer reactions. If the separatory funnel is thick-walled, apply a gentle vacuum with a water aspirator to speed the separation of the phases.

e. If all these procedures fail, it may be necessary to select a different extraction solvent.

2.19 HEATING UNDER REFLUX

The term **heating under reflux** means that a reaction mixture is heated at its boiling point in a flask equipped with a **reflux condenser** to allow continuous return of the volatile materials to the flask; in this manner no solvent or reactant is removed or lost. Since the reaction may be conducted at higher temperatures using this technique, less time is required for its completion.

A typical apparatus used for heating under reflux is shown in Figure 2.34a. The solvent, if any, and reactants are placed in a boiling flask, together with several boiling chips or a magnetic stirring bar; this flask should be set up about 15 cm (6 in.) or more from the bench top to allow for fast removal of the heating source (see Sec. 2.5) if necessary. The flask is then fitted with a reflux condenser with water running slowly through it; the water hoses should be secured to the inlet and outlet of the condenser with copper wire or a hose clamp. *Do not stopper the top of the condenser, since a closed system should never be heated.* The heating source is then placed under the flask, and the contents are slowly heated

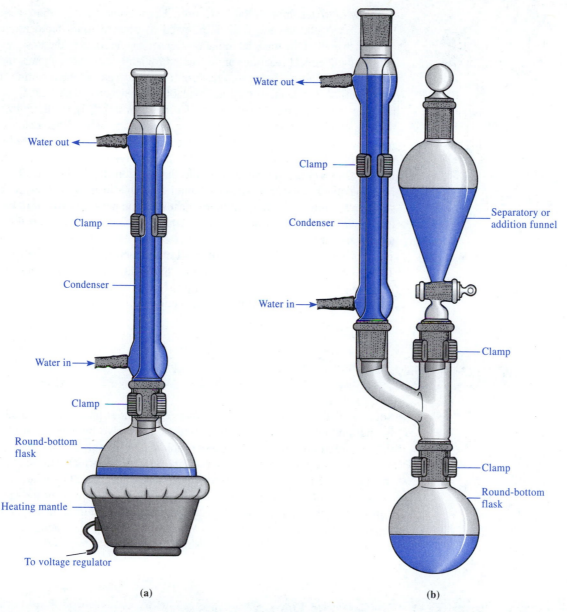

FIGURE 2.34   (a) Apparatus for heating a reaction mixture under reflux; (b) apparatus for adding liquid reagent or solution by means of an addition funnel to a reaction flask equipped with condenser.

to the boiling point of the mixture; slow heating makes it possible to control any sudden exothermicity more readily. The volatile components of the mixture vaporize and reliquefy in the condenser, and the condensate returns to the boiling flask, where the process continues. The vapors should rise 3 to 6 cm (1 to 2 in.) above the bottom of the condenser. If they rise to the top of the condenser or are "seen" being emitted into the room, then too much heat is being supplied to the flask, the flow of water through the condenser is insufficient, or the condenser may be too small to permit adequate cooling of the vapors. The problem may usually be corrected by lowering the temperature of the heat source or adjusting

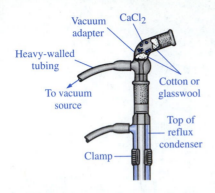

FIGURE 2.35 Gas trap for removal of gases from a reaction mixture.

the flow of water through the condenser. Upon completion of the reaction, the heat is removed, and the flask is allowed to cool to room temperature. This ensures that all of the material in the condenser has returned to the flask.

Many chemical reactions are executed by adding all the reagents, including solvent and catalyst, if any, to a reaction flask and heating the resulting mixture under reflux; however, sometimes one of the reactants must be added to the reaction during reflux. For example, if a reaction is highly exothermic, its rate may be controlled by adding one of the reagents slowly. Other reactions may require that one of the reagents be present in high dilution to minimize the formation of side products. Figure 2.34(b) shows a typical apparatus that allows a liquid reagent or solution to be added to the reaction flask using an addition funnel. The standard-taper separatory funnel available in many glassware kits can serve as an addition funnel, but there must be some provision for air to be admitted above the surface of the liquid to equalize the pressure inside; this can be accomplished by *not* placing the stopper on top of the funnel or by inserting a small piece of filter paper or copper wire between the stopper and the funnel. The flask can be heated to reflux during the addition, or the temperature in the flask can be maintained at room temperature or below with an ice-water bath and heated under reflux following completion of the addition.

2.20 GAS TRAPS

Some organic reactions release noxious gases, whose presence in the laboratory is undesirable. In some cases, these reactions involve materials that react with the moisture in the air. Both of these problems can be solved with moderate success by utilizing the **gas trap** pictured in Figure 2.35, which can be attached to the top of a reflux condenser or directly to a reaction flask. Granular, anhydrous calcium chloride in the vacuum adapter removes moisture from the air as it is pulled through the adapter by *gently* applying a vacuum from either a water aspirator or house vacuum system. Any gas formed by the reaction is removed by vacuum caused by the dry air passing through the gas trap.

2.21 DRYING AGENTS, DESICCANTS, AND DRYING LIQUIDS

Drying a reagent, solvent, or product will be encountered at some stage of nearly every reaction performed in the organic chemistry laboratory. The techniques of drying solids and liquids are described in this and the following sections.

2.21.1 DRYING AGENTS AND DESICCANTS

Most organic liquids are distilled at the end of the purification process, and any residual moisture that is present may react with the compound during the distillation; water may also co- or steam-distil with the liquid and contaminate the distillate. In order to remove these small traces of moisture before distillation, **drying agents,** sometimes called **desiccants,** are used. There are two general requirements for a drying agent: (1) neither it nor its hydrolysis product may react chemically with the organic liquid being dried, and (2) it must be *completely* and *easily* removed from the dry liquid. A drying agent should also be efficient so that the water is removed by the desiccant in a reasonably short period of time.

Some commonly used drying agents and their properties are listed in Table 2.1. These desiccants function in one of two ways: (1) the drying agent interacts *reversibly* with water by the process of adsorption or absorption (Eq. 2.1); or (2) it reacts irreversibly with water by serving as an acid or a base.

TABLE 2.1   Table of Common Drying Agents, Their Properties, and Uses

Drying Agent	Acid-Base Properties	Comments[1]
$CaCl_2$	Neutral	High capacity and fast action with reasonable efficiency; good preliminary drying agent; readily separated from dried solution because $CaCl_2$ is available as large granules; cannot be used to dry either alcohols and amines (because of compound formation) or phenols, esters, and acids (because drying agent contains some $Ca(OH)_2$).
Na_2SO_4	Neutral	Inexpensive, high capacity; relatively slow action and low efficiency; good general preliminary drying agent; preferred physical form is that of small granules, which may be easily separated from the dry solution by decantation or filtration.
$MgSO_4$	Weakly acidic	Inexpensive, high capacity, rapid drying agent with moderate efficiency; excellent preliminary drying agent; requires filtration to remove drying agent from solution.
H_2SO_4	Acidic	Good for alkyl halides and aliphatic hydrocarbons; cannot be used with even such weak bases as alkenes and ethers; high efficiency.
P_2O_5	Acidic	See comments under H_2SO_4; also good for ethers, aryl halides, and aromatic hydrocarbons; generally high efficiency; preliminary drying of solution recommended; dried solution can be distilled from drying agent.
CaH_2	Basic	High efficiency with both polar and nonpolar solvents, although inexplica-

TABLE 2.1 *(Continued)*

Drying Agent	Acid-Base Properties	Comments[1]
		bly it fails with acetonitrile; somewhat slow action; good for basic, neutral, or *weakly* acidic compounds; cannot be used for base-sensitive substances; preliminary drying of solution is recommended; dried solution can be distilled from drying agent. *Caution:* Hydrogen gas is evolved with this drying agent.
Na or K	Basic	Good efficiency but slow action; cannot be used on compounds sensitive to alkali metals or to base; care must be exercised in destroying excess drying agent; preliminary drying *required;* dried solution can be distilled from drying agent. *Caution:* Hydrogen gas is evolved with this drying agent.
BaO or CaO	Basic	Slow action but high efficiency; good for alcohols and amines; cannot be used with compounds sensitive to base; dried solution can be distilled from drying agent.
Molecular Sieve #3A or #4A[2]	Neutral	Rapid and highly efficient; preliminary drying recommended; dried solution can be distilled from drying agent if desired. *Molecular sieves* are aluminosilicates, whose crystal structure contains a network of pores of uniform diameter; the pore sizes of sieves #3A and #4A are such that only water and other small molecules such as ammonia can pass into the sieve; water is strongly adsorbed as water of hydration; hydrated sieves can be reactivated by heating at 300–320 °C under vacuum or at atmospheric pressure.

[1] Capacity, as used in this table, refers to the amount of water that can be removed by a given weight of drying agent; efficiency refers to the amount of water, if any, in equilibrium with the hydrated desiccant.

[2] The numbers refer to the nominal pore size, in Ångström units, of the sieve.

With drying agents that function by reversible hydration, a certain amount of water will remain in the organic liquid in equilibrium with the hydrated drying agent. The lesser the amount of water left at equilibrium, the greater the efficiency of the desiccant. A drying agent that forms a hydrate (Eq. 2.1) must be *completely* removed by gravity filtration (see Sec. 2.12) or by decantation (see

Sec. 2.15) *before* the dried liquid is distilled, since most hydrates decompose with loss of water at temperatures above 30–40 °C.

$$\text{Drying agent (solid)} + x\,H_2O\ \text{(liquid)} \rightleftharpoons \text{Drying agent} \cdot x\,H_2O\ \text{(solid)} \quad \textbf{(2.1)}$$

Anhydrous   Hydrate

Drying agents that remove water by an irreversible chemical reaction are very efficient, but they are generally more expensive than other types of drying agents. Such drying agents are sometimes more difficult to handle and are normally used to remove *small* quantities of water from reagents or solvents prior to a chemical reaction. For example, phosphorus pentoxide, P_2O_5, removes water by reacting vigorously with it to form phosphoric acid (Eq. 2.2). Desiccants such as calcium hydride (CaH_2) and sodium (Na) metal also react vigorously with water. When CaH_2 or Na metal is used as a drying agent, hydrogen gas is evolved, and appropriate precautions must be taken to vent the hydrogen and prevent buildup of this highly flammable gas.

$$P_2O_5\ \text{(solid)} + 3\,H_2O\ \text{(liquid)} \longrightarrow 2\,H_3PO_4 \quad \textbf{(2.2)}$$

Of the drying agents listed in Table 2.1, *anhydrous* calcium chloride, sodium sulfate, and magnesium sulfate will generally suffice for the needs of this introductory laboratory course. Both sodium sulfate and magnesium sulfate have high capacities and absorb a large amount of water, but magnesium sulfate dries a solution more completely. Calcium chloride has a low capacity, but it is a more efficient drying agent than magnesium sulfate. Do *not* use an unnecessarily large quantity of drying agent when drying a liquid, since the desiccant may adsorb or absorb the desired organic product along with the water. Mechanical losses on filtration or decantation of the dried solution may also become significant. The amount of drying agent required depends upon the quantity of water present, the capacity of the drying agent, and the amount of liquid to be dried.

Some organic solvents may be dried by a process called **azeotropic distillation** (see Sec. 4.2.3) in which the water codistils with the solvent in the first fraction of the distillation. For example, benzene or toluene can be dried reasonably well in this manner. Water can also be removed from 95% ethanol by adding benzene and distilling the ternary azeotropic mixture of benzene, water, and ethanol; the pure ethanol may then be collected by distillation.

2.21.2 EXPERIMENTAL PROCEDURE FOR DRYING LIQUIDS

The organic liquid to be dried should be placed in an Erlenmeyer flask of suitable size so that it will be no more than half-filled with liquid. In general, a portion of drying agent that covers the bottom of the flask should be sufficient. Start by adding a small amount of drying agent and swirling the flask gently. Swirling the flask of liquid and desiccant enhances the rate of drying with most desiccants. This is because swirling increases the surface area for contact between the solid and liquid phases. If the *liquid* still appears cloudy after the solid has settled to the bottom of the flask, add more drying agent and swirl. Repeat this process until the liquid appears clear. Remove the drying agent by gravity filtration (see Sec. 2.12) if the liquid is moderately high-boiling and nonvolatile or by decantation (see Sec. 2.15) if it is low-boiling and volatile.

2.22   DRYING SOLIDS

Solid organic compounds must be dried because the presence of water or organic solvents will affect their weight, melting points, quantitative elemental analysis, and spectra. Since proton sources must be excluded from some reactions, it is also necessary to remove all traces of moisture or protic solvents from a solid prior to performing such a reaction.

A solid that has been recrystallized from a volatile organic solvent can usually be dried satisfactorily by allowing it to air-dry at room temperature, provided it is not **hygroscopic** and thus absorbs moisture from the air. After the solid is collected on a Büchner or Hirsch funnel fixed on a filter flask, it is first pressed as dry as possible with a clean spatula or cork while air continues to be pulled through the funnel and solid by use of a water aspirator or house vacuum (see Sec. 2.16). The solid is then spread on a piece of filter paper, which absorbs the excess solvent, or on a clean watch glass and allowed to stand overnight or longer.

Water is sometimes removed from a solid by dissolving the solid in a suitable organic solvent such as chloroform or toluene, removing any water by azeotropic distillation, and then recovering the solid by removal of the solvent. However, water is more commonly removed from organic solids using desiccators containing desiccants such as silica gel, phosphorus pentoxide, calcium chloride, or calcium sulfate. The desiccator may be used at atmospheric pressure or under a vacuum; however, if a vacuum is applied, the desiccator must be enclosed in a metal safety cage or wrapped with electrical or duct tape. Desiccators or tightly stoppered bottles containing one of these desiccants may also be used to store dry solids contained in small vials.

If the sample is hygroscopic or if it has been recrystallized from water or a high-boiling solvent, it must be dried in an oven operating at a temperature *below* the melting or decomposition point of the sample. The oven-drying process can be performed at atmospheric pressure or under vacuum. Air-sensitive solids must be dried either in an inert atmosphere, such as in nitrogen or helium, or under vacuum. Samples to be submitted for quantitative elemental analysis are normally dried to constant weight by heating them under vacuum.

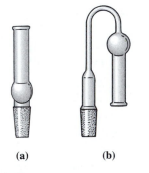

(a)   **(b)**

FIGURE 2.36   Drying tubes: (a) straight; (b) bent or U-shaped.

2.23   DRYING TUBES

Frequently it is necessary to protect a reaction mixture from moisture. There are advanced techniques for doing this that involve performing the reaction under an inert, dry atmosphere of nitrogen or argon gas, but a simple and moderately effective procedure simply utilizes a drying tube containing a drying agent. Two types of drying tubes are shown in Figure 2.36; the straight tube is more common in the introductory organic laboratory. A drying tube is prepared by placing a loose plug of glasswool at the bottom of the tube and filling the tube with a *granular* drying agent such as anhydrous calcium chloride or calcium sulfate. The drying agent commonly contains a blue indicator that turns pink when the drying agent is no longer effective. Although it is not necessary to cap the straight tube with glasswool, a loose plug of glasswool must also be placed in the other end of the bent drying tube to prevent the drying agent from falling out. Neither the desiccant nor the glasswool plug should be packed too tightly.

2.24 DRYING OF APPARATUS

Before a reaction that must be carried out under anhydrous conditions is started, and prior to introducing any reagents, as much moisture as possible must be removed from the apparatus. The apparatus, which must have one opening to the atmosphere, is gently warmed with a Bunsen burner or microburner, a process called **flaming.** The apparatus is heated first at the point most remote from the opening to the atmosphere, and the flame is gradually worked toward the opening. The moisture in the apparatus will be driven through the opening. If a condenser is part of the apparatus, no water should be in or flowing through it. When the apparatus is warm, a filled drying tube (see Sec. 2.23) is inserted into the opening, and the system is allowed to cool to room temperature. As cooling occurs, air from the atmosphere is drawn into the apparatus through the drying agent, which removes the moisture, and the dry apparatus is ready to be used.

2.25 EVAPORATION OF SOLVENTS

For many experiments and laboratory operations, it is necessary to remove the excess solvents to recover the product. Although an Erlenmeyer flask or beaker containing the solution could simply be left unstoppered in the hood until the solvent has evaporated, this is impractical. Most of the solvent may be removed by simple distillation (see Sec. 2.7) with heating being discontinued when only a small amount of solvent remains. The stillpot should be cooled to room temperature and then attached to a vacuum source, such as a water aspirator, for a few minutes to remove the last traces of solvent. Alternatively, several boiling chips or a wooden boiling stick may be added to the flask and the flask heated on a steam bath or hot plate or in a sand or oil bath so that the solvent boils gently and evaporates in a short time. The evaporation rate can be increased by directing a stream of air or nitrogen into the flask (Fig. 2.37a), or a vacuum can be applied over the flask to help blow away solvent vapors (Fig. 2.37b). The solution also could be placed in a filter flask or a test tube with a sidearm and a wooden boiling stick; after the flask is subjected to a gentle vacuum, it is swirled over the heat source to facilitate smooth evaporation and reduce the possibility of bumping (Fig. 2.37c).

FIGURE 2.37   Typical techniques for evaporating solvents.

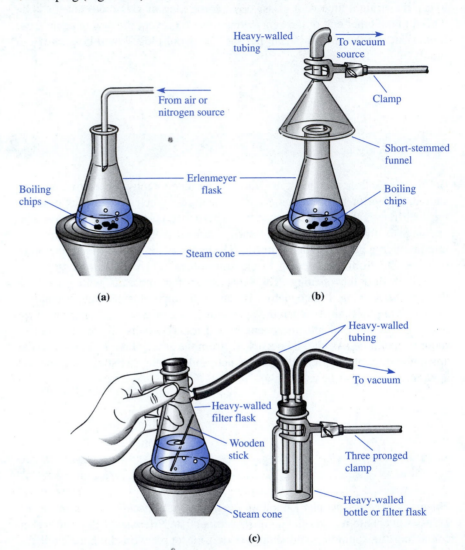

(a)

(b)

(c)

SOLIDS

RECRYSTALLIZATION AND MELTING POINTS

3.1   INTRODUCTION

The organic chemist usually works with substances that are in the liquid or solid state. The purpose of this and the succeeding two chapters is to present an introduction to the theory and practice of the most important methods for separation and purification of mixtures of organic compounds in these two physical states. As you read these chapters, you should also refer to the corresponding sections of Chapter 2 that describe the apparatus used for the various techniques. References to the appropriate sections have been provided for your convenience.

A pure compound is a homogeneous sample consisting only of molecules having the same structure. However, substances believed to be pure on the basis of certain criteria may actually contain small amounts of contaminants. Indeed, the presence of impurities in what were believed to be pure compounds has sometimes led to incorrect structural assignments and scientific conclusions. The possibility of making such errors was of particular concern prior to the advent of the powerful chromatographic (see Chap. 6) and spectral (see Chap. 8) techniques that have been developed since the 1950s. It is now relatively easy for the chemist to purify solids and liquids and to demonstrate their purity.

A compound formed in a chemical reaction or extracted from some natural source is rarely pure when initially isolated. For example, a chemical transformation intended to produce a single product almost invariably yields a reaction mixture containing a number of contaminants. These may include the products of side reactions proceeding concurrently with the main reaction, unchanged starting materials, inorganic materials, and solvents. Unfortunately, even chemicals purchased commercially are not always pure, owing to the expense of the needed purification process or to decomposition that may occur during storage.

Organic chemists devote considerable effort to the isolation of pure products, the ultimate goal being to obtain a substance that cannot be purified further by any known experimental techniques. This chapter focuses on the purification of solids by recrystallization and their characterization by the physical property of melting points.

3.2 RECRYSTALLIZATION

Recrystallization of solids is a valuable technique that should be mastered because it is one of the methods used most often for purification of solids. Other techniques for purifying solids include **sublimation** (see Sec. 2.17), **extraction** (see Chap. 5) and **chromatography** (see Chap. 6). Nevertheless, even when one of these alternative methods of purification has been used, the solid material thus isolated may still be recrystallized to achieve the highest possible state of purity.

The process of recrystallization involves dissolution of the solid in an appropriate solvent at an elevated temperature and the subsequent re-formation of the crystals upon cooling, so that any impurities remain in solution. This technique, called **solution recrystallization,** is discussed here. An alternative approach involves melting the solid in the absence of solvent and then allowing the crystals to re-form so that impurities are left in the melt. This method is seldom used in the organic laboratory because the crystals often form out of a viscous oil that contains the impurities and from which it is difficult to separate the desired pure solid. It is interesting to note, however, that this is the technique used to prepare the high-purity single crystals of silicon used in computer chips.

Almost all solids are *more* soluble in a *hot* than in a *cold* solvent, and solution crystallization takes advantage of this fact. Thus, if a solid is first dissolved in an amount of hot solvent insufficient to dissolve it when cold, crystals should form when the hot solution is allowed to cool. The extent of precipitation of the solid depends on the difference in its solubility in the particular solvent at temperatures between the extremes used. The upper extreme is determined by the boiling point of the solvent, whereas the lower limit is usually dictated by experimental convenience. For example, an ice-water bath is often used to cool the solution to 0° C, whereas ice-salt and Dry Ice–acetone baths are commonly used to cool solutions to –20° C and –78° C, respectively. The solid should be recovered with greater efficiency at these temperatures, provided the solvent itself does not freeze.

If the impurities present in the original solid mixture have dissolved and *remain* dissolved after the solution is cooled, isolation of the crystals that have formed should *ideally* provide pure material. Alternatively, the impurities may not dissolve at all in the hot solution and may be removed by filtration *before* the solution is cooled. The crystals that subsequently form should be purer than the original solid mixture. Solution recrystallization is seldom quite so simple in practice, but these two idealized generalizations do outline the basic principles of the technique.

Even after a solid has been recrystallized, it may still not be pure. Thus, it is important to determine the purity of the sample, and one of the easiest methods to do this is by determining the melting point of the solid. This technique is described in Section 3.3.

The technique of solution recrystallization involves the following steps:

1. **selection** of an appropriate solvent

2. **dissolution** of the solid to be purified in the solvent near or at its boiling point

3. **decoloration** with an activated form of carbon, if necessary, to remove colored impurities and **filtration** of the hot solution to remove insoluble impurities and the decolorizing carbon

4. **formation** of crystalline solid from the solution as it cools

5. **isolation** of the purified solid by filtration

6. **drying** the crystals

Each step of the sequence is discussed in the following subsections and representative experimental procedures are presented at the end of the discussion.

3.2.1 SELECTION OF SOLVENT

The choice of solvent is perhaps the most critical step in the process of recrystallization since the correct solvent must be selected to form a product of high purity and good recovery or yield. Consequently a solvent must satisfy certain criteria for use in recrystallization: (**a**) The desired compound should be reasonably soluble in the *hot* solvent, about 1 g/20 mL being satisfactory, and **insoluble** or *nearly* **insoluble** in the *cold* solvent. Note that the reference temperature for determination of the solubility in "cold" solvent is often taken to be room temperature. This combination of solute and solvent will allow dissolution to occur in an amount of solvent that is not unduly large and will also permit recovery of the purified product in high yield. A solvent having this type of solubility properties as a function of temperature would be said to have a favorable **temperature coefficient** for the desired solute. (**b**) Conversely, the impurities should either be insoluble in the solvent at all temperatures *or* must remain at least moderately soluble in the cold solvent. In other words, if the impurities are soluble, the temperature coefficient for them must be *unfavorable*; otherwise the desired product *and* the impurities would both crystallize simultaneously from solution. (**c**) The boiling point of the solvent should be low enough so that it can readily be removed from the crystals. (**d**) The boiling point of the solvent should generally be lower than the melting point of the solid being purified. (**e**) The solvent should not react chemically with the substance being purified.

The chemical literature is a valuable source of information about solvents suitable for recrystallizing known compounds. If the compound has *not* been prepared before, it is necessary to resort to trial-and-error techniques to find an appropriate solvent for recrystallization. The process of selection can be aided by consideration of some generalizations about solubility characteristics for classes of solutes. Polar compounds are normally soluble in polar solvents and insoluble in nonpolar solvents, for example, whereas nonpolar compounds are more soluble in nonpolar solvents. Such characteristics are summarized by the adage, *"like dissolves like."* Of course, although a highly polar compound is unlikely to be soluble in a hot, nonpolar solvent, it *may* be very soluble in a cold, very polar solvent. In this case, a solvent of intermediate polarity *may* be the choice for a satisfactory recrystallization.

The solvents commonly used in recrystallizations range widely in polarity, a property measured by the **dielectric constants** (ϵ) listed in Table 3.1. Those solvents with dielectric constants in the range of 2–3 are considered **nonpolar,** and those with constants above 10 as **polar.** Solvents in the 3–10 range are of intermediate polarity. Of the various solvents listed, *petroleum ether* deserves special mention because of its confusing common name. This solvent does *not* contain the *ether* functional group at all; rather it is a mixture of volatile aliphatic hydrocarbons obtained from the refining of petroleum. The composition and boiling point of the mixture depends on the particular distillation "cut" obtained. Thus, the boiling range of this type of solvent is usually given, as in the description, "petroleum ether, bp 60–80 °C (760 torr)."

Occasionally a mixture of solvents is required for satisfactory recrystallization of a solute. The mixture is usually comprised of only two solvents; one of these dissolves the solute even when cold and the other one does not. The logic of this will become clear when reading about this technique in Section 3.2.2, "Dissolution."

Chemical Structures of Common Recrystallization Solvents Found in Table 3.1

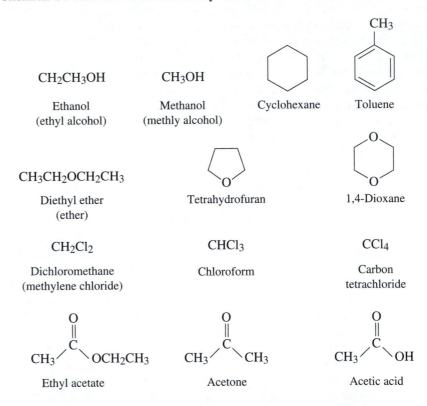

| CH$_2$CH$_3$OH | CH$_3$OH | Cyclohexane | Toluene |
| Ethanol (ethyl alcohol) | Methanol (methly alcohol) | | |

CH$_3$CH$_2$OCH$_2$CH$_3$ Tetrahydrofuran   1,4-Dioxane
Diethyl ether (ether)

CH$_2$Cl$_2$ CHCl$_3$ CCl$_4$
Dichloromethane (methylene chloride)   Chloroform   Carbon tetrachloride

Ethyl acetate   Acetone   Acetic acid

TABLE 3.1   Common Solvents for Recrystallization[a]

Solvent	Boiling Point °C (760 torr)	Freezing Point, °C[b]	Water Soluble	Dielectric Constant (ϵ)	Flammable
Petroleum ether	Variable		No	1.9	Yes
Diethyl ether	35		Slightly	4.34	Yes
Dichloromethane[c]	41		No	9.08	No
Acetone	56		Yes	20.7	Yes
Chloroform[c]	61		No	4.81	No
Methanol	65		Yes	32.63	Yes
Tetrahydrofuran	65		Yes	7.58	Yes

TABLE 3.1   *(Continued)*

Solvent	Boiling Point °C (760 torr)	Freezing Point, °C[b]	Water Soluble	Dielectric Constant (ϵ)	Flammable
Carbon tetrachloride[c]	77		No	2.23	No
Ethyl acetate	77		Yes	6.02	Yes
Ethanol (95%)	78		Yes	24.6	Yes
Cyclohexane	81	6	No	1.9	Yes
Water	100	0	N/A	78.54	No
1,4-Dioxane	101	11	Yes	2.21	Yes
Toluene	111		No	2.38	Yes
Acetic acid	118	17	Yes	6.15	Yes

[a] Benzene has been omitted from this list, owing to its toxicity. Cyclohexane can usually be substituted for it.

[b] Freezing points not listed are below 0 °C.

[c] As a general rule, avoid the use of chlorocarbon solvents *if another equally good solvent is available.* The toxicity of such solvents is greater and disposal of them is more difficult than for other types of solvents.

3.2.2   DISSOLUTION

This step may involve the handling of relatively large volumes of volatile solvents. Although most solvents used in the organic laboratory are of relatively low toxicity, it is prudent to avoid inhaling their vapors. For this reason, the following operations are best performed in a hood. If this is not possible, clamping an inverted funnel over the recrystallization flask and connecting the stem of the funnel to a source of a vacuum (see Fig. 2.37b) will help to minimize exposure. Another precaution is to choose the source of heat for the recrystallization carefully. As noted in Table 3.1, many solvents are highly flammable and should *never* be heated with an open flame; rather, steam or oil baths or hot plates should be used in such cases.

The solid to be purified is weighed and placed in an appropriately sized **Erlenmeyer flask.** A beaker is an inappropriate container for recrystallization because it has a relatively large surface that provides a greater area for adherence of recrystallized product; the efficiency of recovery will suffer as a result. Some crystals of the impure solid should always be retained, for they may be needed as "seeds" to induce crystallization. A few milliliters of solvent are added to the flask, and the mixture is then heated to the boiling point. More solvent is added to the boiling mixture in *small* portions by pipet until the solid just dissolves; enough time should be allowed for the solution to boil again after each addition. After all of the solid has dissolved, it is generally wise to add 2 to 5% of additional solvent to prevent premature crystallization of the solute during hot filtration, if this step is necessary. A large excess of solvent should be avoided, as this will decrease the recovery of solute. Sometimes, addition of solvent fails to cause dissolution of more of the solid, *particularly when only a relatively small quantity of solid remains.* The insoluble materials are likely to be impurities, which will be removed in the filtration step. If the recovery of the purified product is poor, too much solvent has probably been used.

The same general approach is followed when mixed solvents are employed. However, there are two variations for effecting dissolution once the solvents have been selected. In one option, the solid to be purified is first dissolved in a *minimum* volume of the hot solvent in which it is soluble; the second solvent is then added to the *boiling* solution until it turns cloudy. The cloudiness signals initial formation of crystals, caused by the fact that addition of the second solvent results in a solvent mixture in which the solute is less soluble. Finally, more of the first solvent is added *dropwise* until the solution clears. Two further aspects of this variant should be noted. First, the solution must be cooled slightly below the lower boiling point before the second solvent is added *if* this solvent has a boiling point *lower* than the first; otherwise the addition of this solvent could cause sudden and vigorous boiling of the mixture and hot solvent might spew from the apparatus. Second, hot filtration should be performed if needed *before* addition of the second solvent; this will prevent crystallization during the filtration step. A potential disadvantage of this method for mixed solvent recrystallization is that unduly large volumes of the second solvent may be required if excessive amounts of the initial solvent have been used.

In the second variant, the solute is added to the solvent in which it is insoluble, and the mixture is heated near the boiling point of the solvent; the second solvent is then added in small portions until the solid just dissolves. As with recrystallization from a single solvent, it is generally wise to add 2 to 5% of additional solvent to prevent premature crystallization of the solute during hot filtration, if this step is necessary. The use of this approach to mixed solvent recrystallization also has the disadvantage that using too much of the first solvent will require the addition of undesirably large volumes of the second solvent.

3.2.3 DECOLORATION AND HOT FILTRATION

After dissolution of the solid mixture, the solution may be colored. This signals the presence of impurities if the desired compound is colorless. If the compound is colored, contaminants may alter the color of the solution; for example, impurities should be suspected if the substance is yellow but the solution is green. Of course, if the solution is colorless, the decoloration step is unnecessary.

Colored impurities may often be removed by adding a small amount of **decolorizing carbon** (see Sec. 2.14) to the hot, *but not boiling*, solution. Decolorizing carbon is a finely divided, activated form of carbon with a high surface area, and adding it to a boiling solution is likely to cause the liquid to froth over the top of the flask which can result in the loss of product and possible injury. The amount of decolorizing carbon required is usually small; in fact, the quantity that can be loaded on the tip of a small spatula will normally suffice. After this addition, the solution is heated to boiling for a few minutes while being continuously stirred or swirled to prevent bumping of the boiling mixture.

The complete removal of the carbon by **hot filtration** (see Sec. 2.13) from the decolorized solution may be difficult. The decolorizing carbon is so finely divided that it may pass through the filter paper, as evidenced by the presence of a dark tint in the filtrate. Consequently, the following steps are recommended to alleviate this problem. The solution is allowed to cool slightly below the boiling point, a small amount of **filter-aid** is added to *adsorb* the decolorizing carbon, the mixture is briefly reheated to boiling and is then subjected to hot filtration. It may be necessary to repeat this procedure on the filtrate if some of the carbon manages to pass through the filter paper despite the use of a filter-aid.

This technique for decolorizing solutions works because colored impurities as well as the compound being purified are adsorbed on the surface of the carbon particles. For electronic reasons, the colored substances adsorb more strongly to the surface, and this factor, combined with the fact that the impurities are normally present in minor amounts, results in complete removal of the colored contaminants. Of course, because of its adsorption on the charcoal, less of the desired product will be recovered if too much carbon is added.

Insoluble impurities, including dust and decolorizing carbon, if used, are removed by **gravity filtration** of the hot solution (see Sec. 2.13); this step is not necessary if the hot solution is clear and homogeneous. Gravity filtration is normally preferred to **vacuum filtration** because the latter technique may cause cooling and concentration of the solution, owing to evaporation of the solvent, and this may result in premature crystallization. A short-stemmed or stemless glass funnel should be used to minimize crystallization in the funnel, and use of a **fluted filter paper** (see Sec. 2.11) will minimize crystallization on the filter. To keep liquid from flowing over the top of the funnel, the top of the paper should not extend above the funnel.

3.2.4 CRYSTALLIZATION

The hot solution of solute is allowed to cool slowly to room temperature, and crystallization should occur. Rapid cooling by immersion of the flask in water or an ice-water bath is undesirable because the crystals formed tend to be very small, and their resulting large surface area may foster adsorption of impurities from solution; in this sense, the crystals are functioning like decolorizing carbon! Generally the solution should not be disturbed as it cools, since this also leads to production of small crystals. The formation of crystals larger than about 2 mm should be avoided because some of the solution may become occluded or trapped *within* the crystals. The drying of such crystals is more difficult, and impurities may be left in them. Should overly large crystals begin to form, brief, gentle agitation of the solution normally induces production of smaller crystals.

Failure of crystallization to occur after the solution has cooled somewhat usually means that either too much solvent has been used or that the solution is supersaturated. A supersaturated solution can usually be made to produce crystals by **seeding.** A crystal of the original solid is added to the solution to induce crystallization, which may then be quite rapid. Another method to initiate formation of crystals involves using a glass rod to scratch the inside surface of the flask *at or just above the air-solution interface.* This should be done with care, as excessive force could result in a broken rod or flask.

Occasionally the solute will separate from solution as an oil rather than a solid. This type of separation, which is sometimes called "oiling out," is undesirable for purification of solutes because the oils usually contain significant amounts of impurities. Two general approaches are helpful in solving this problem: (1) Oils may persist on cooling with no evidence of crystallization. These may often be induced to crystallize by scratching the oil against the side of the flask with a glass stirring rod at the interface of the oil and the solution. If this fails, several small seed crystals of the original solid may be added to the oil, and the mixture allowed to stand for a period of time. Failure of these alternatives may necessitate separation of the oil from the solution and crystallization of it from another solvent. (2) Oils may form from the hot solution and then solidify to an amorphous mass at lower temperatures; in the meantime, crystals of the solute may precipitate from the mother liquor. Because the oil is not a pure liquid the solid mass produced from it will be impure as noted earlier. In a case such as this, the usual remedy is to reheat the entire mixture to effect dissolution, to add a few milliliters of additional pure solvent, and to allow the resulting solution to cool.

3.2.5 FILTRATION AND SOLVENT REMOVAL

The crystalline product is isolated by **vacuum filtration** using a Büchner or Hirsch funnel and a clean, dry filter flask (see Sec. 2.16). The crystals normally are washed with a small amount of *pure, cold* solvent, with the vacuum shut off; the vacuum is then reapplied to remove as much solvent as possible from the filter cake. Care must be taken in this step to ensure that the filter paper is not lifted off the bed of the filter while the vacuum is off; this could result in loss of product when vacuum is reapplied to remove the washes.

Further cooling of the filtrate, sometimes called the **mother liquors,** in an ice-water or ice-salt bath may allow isolation of a second crop of crystals. The filtrate can also be concentrated by evaporating part of the solvent and cooling the residual solution. The crystals isolated as a second or even a third crop are likely to be less pure than those in the first. Consequently, the various crops should *not* be combined until their purity has been assessed by comparison of their melting points as described in Section 3.3 for example.

3.2.6 DRYING THE CRYSTALS

Residual solvent is removed by transferring the crystals from the filter paper to a watchglass or vial and either air- or oven-drying them (see Sec. 2.22). With the latter option, the temperature of the oven must be below the melting point of the solid. Solids can also be dried on fresh pieces of filter paper. This is a less desirable option, however, because fibers from the paper may become mixed with the product when it is transferred to a vial for submission to your instructor.

Pre-Lab exercises for Section 3.2, "Recrystallization," are found on page PL. 1.

EXPERIMENTAL PROCEDURE

SAFETY ALERT

1. Do not use a burner in these procedures unless instructed to do so. Most solvents used for recrystallization are flammable (see Table 3.1).

2. When using a hot plate, do not set it at its highest value. A moderate setting will prevent overheating and the resultant bumping and splashing of materials from the flask. Do not employ hot plates for heating volatile or flammable solvents; rather use a steam bath.

3. Avoid excessive inhalation of solvent vapors. If a hood is not available to you, clamp an inverted funnel just above the Erlenmeyer flask in which you will be heating solvents. Attach this funnel to a source of vacuum by means of rubber tubing (see Fig. 2.37b).

4. When pouring or transferring solutions, either wear latex gloves or avoid getting these solutions on the skin. Organic compounds are much more rapidly absorbed through the skin when they are in solution, particularly in water-soluble solvents such as acetone and ethanol. For this reason do not rinse organic materials off your skin with solvents such as acetone; wash your hands thoroughly with soap and hot water instead.

5. When using decolorizing carbon, *never add it to a boiling solution;* to do so may cause the solution to boil out of the flask. Add the carbon only when the temperature of the solvent is *below* the boiling temperature. This same precaution applies when using a filter-aid to assist in the removal of the carbon during the hot filtration step.

A. SOLVENT SELECTION

Heat a beaker of water to a gentle boil on a hot plate and use the hot water to serve as a heating medium; replenish the water as necessary. Alternatively, a steam bath may be used (see Sec. 2.5.4).

steam bath in hood

Several options are available for selecting the solvent for a recrystallization, and these may be applied to known compounds or to an unknown compound. Your instructor will tell you which option to use. Different criteria are used for defining solubility. For purposes of this experiment, use the following definitions: (a) *soluble*—20 mg of solute will dissolve in 0.5 mL of solvent; (b) *slightly soluble*—some but not all of the 20 mg of solute will dissolve in 0.5 mL of solvent; (c) *insoluble*—none of the solute appears to dissolve in 0.5 ml of solvent.

1. Known Compounds. The procedures given here involve using several common organic crystalline compounds: resorcinol (**1**), naphthalene (**2**), benzoic acid (**3**), and acetanilide (**4**). Be certain to record all observations made regarding solubilities in your notebook.

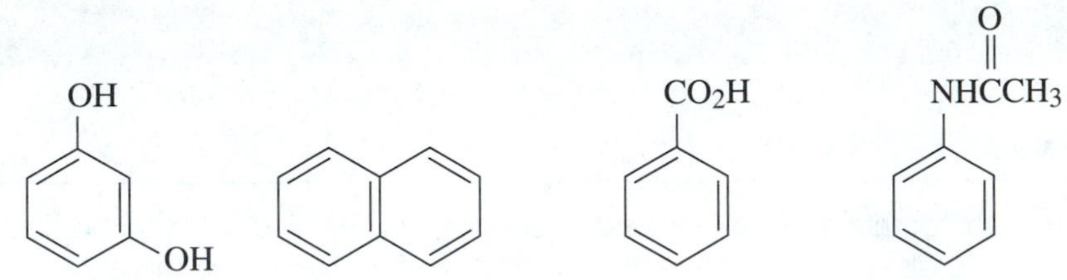

| **1** | **2** | **3** | **4** |
| Resorcinol | Naphthalene | Benzoic Acid | Acetanilide |

(handwritten margin notes, left side):
solvents
solutes
Water
Ethanol
Petroleum Ether
(mixture of low-
boiling hydrocarbons)

find an appropriate
solvent for solute

20 drops = 1 mL

if it recrystallizes it's a
good candidate
soluble cold — don't
waste time

Place about 20 mg (a small spatula-tip full) of finely crushed resorcinol in a small test tube and add about 0.5 mL of water to the tube. Stir the mixture with a glass rod and determine whether resorcinol is soluble in water at room temperature. If the resorcinol is not completely soluble at room temperature, place the test tube in the hot-water bath, and with stirring or swirling of the tube determine whether resorcinol is soluble in hot water.

Repeat the solubility test for resorcinol using 95% ethanol and then petroleum ether (bp 60–80 °C, 760 torr). If additional practice is desired, similarly determine the solubility properties of naphthalene, benzoic acid, and acetanilide in water, in 95% ethanol, and in petroleum ether.

If any of these solutes is soluble in the hot solvent but only slightly soluble or insoluble in the cold solvent, allow the hot solution to cool slowly to room temperature and compare the quantity, size, color, and form of the resulting crystals with the original solid material. Note which solvent you would consider best suited for recrystallization of each of the solutes.

2. Unknown Compounds. If you are given an unknown solid to recrystallize, first determine its solubility in various common solvents. This should be done on a *small* scale, using about 20 mg of solute and 0.5 mL of solvent. Once the most suitable solvent has been selected, the process can be repeated on a larger scale, as described in Part B of this experiment. Table 3.1 lists the properties of various solvents that are generally useful for recrystallization of organic solids. It should not be necessary to test all the solvents, but you should consider trying those solvents that are denoted with an asterisk in the table. Your instructor may also indicate which solvents to try.

A *systematic* approach to testing the solubility of an unknown compound is important. A protocol for doing so follows. Clean and dry enough small test tubes so that one can be used for each solvent to be tested. Place about 20 mg (a small spatula-tip full) of the finely crushed unknown in each test tube and add about 0.5 mL of a solvent to a tube containing the solid. Stir each mixture with a glass rod and determine the solubility of the unknown in each solvent at room temperature. Use the definitions of *soluble, slightly soluble,* or *insoluble* given earlier. If the unknown is insoluble in a particular solvent, place the test tube in the hot-water bath. Stir or swirl the contents of the tube and note whether the unknown is soluble in hot solvent. If the solid is soluble in the hot solvent but only slightly soluble or insoluble at room temperature, allow the hot solution to cool to room temperature slowly. If crystals form in the cool solution, compare their quantity, size, color, and form with the original solid material and with the other samples. Construct a table containing the solubility data and use it to select the solvent that appears best suited for recrystallization. It is a good idea to test the solubility in a variety of solvents. Even though nice crystals may form in the first solvent you try, another solvent might prove better if it provides either more or better crystals.

If these solubility tests produce no clear choice for the solvent, mixed solvents might be considered. Review the discussion presented earlier in this section for the procedure for using a mixture of two solvents. Before trying any combinations of solvent pairs, take about 0.2 mL of each *pure* solvent being considered and mix them to ensure that they are miscible in one another. If they are not, that particular combination *cannot* be used.

B. RECRYSTALLIZATION OF IMPURE SOLIDS

Carefully read the discussion given previously in this section so that you understand the techniques involved in the following procedure. This information will help you avoid or overcome most of the problems encountered when purifying impure solids by recrystallization. Only abbreviated directions are provided in the experiments that follow.

Obtain one or more samples of impure solids for recrystallization. Among those that might be assigned are naphthalene (**2**), benzoic acid (**3**), acetanilide (**4**), or an unknown. If assigned an unknown, determine the appropriate solvent(s) to use, as described in Part A, and then proceed by following the general instructions given in the discussion associated with this section.

For each compound that is recrystallized, determine the melting point, weight, and percent recovery of the pure crystals that are obtained. Also measure the melting point of the impure solid, for purposes of comparison. Melting points and the procedure for determining them are discussed in Sections 2.3 and 3.3. Percent recovery may be calculated by knowing the weight of the impure solid and that of the recrystallized material, as shown in Equation (3.1). As an example, isolation of 0.3 g of recrystallized product from 0.5 g of impure solid would correspond to a percent recovery of (0.3 g/0.5 g) × 100 = 60%.

$$\text{Percent recovery} = \frac{\text{weight of pure crystals recovered}}{\text{weight of original sample}} \times 100 \qquad (3.1)$$

no

1. Benzoic Acid. Place 0.5 g of impure benzoic acid in a clean 25-mL Erlenmeyer flask. Measure 12 mL of water in a graduated cylinder and add a 5-mL portion of it to the benzoic acid. Heat the mixture to a gentle boil with a microburner or hot plate. Add water in 0.5-mL portions until no more solid appears to dissolve in the boiling solution. Record the total volume of water used. No more than 10 mL should be required.

Pure benzoic acid is colorless, so a colored solution indicates that treatment with decolorizing carbon (see Sec. 2.14) is necessary. *(Caution:* Do not add decolorizing carbon to a *boiling* solution!) Cool the solution slightly, add approximately 0.05 g of carbon, and reheat to boiling for a few minutes. To aid in the removal of the finely divided carbon by filtration, allow the solution to cool slightly, add about 0.2 g of a filter-aid, and reheat.

Perform a hot filtration according to the directions provided in Section 2.13. Rinse the empty flask with about 1 mL of *hot* water and filter this rinse into the main solution. If the filtered solution remains colored, repeat the treatment with decolorizing carbon. Cover the flask with a watchglass or inverted beaker, and allow the filtrate to stand undisturbed until it has cooled to room temperature and no more crystals form.✱ To complete the crystallization, place the flask in an ice-water bath for at least 15 min.

Collect the white crystals on a Büchner or Hirsch funnel by vacuum filtration (see Sec. 2.16) and wash the filter cake with two small portions of *cold* water. Press the crystals as dry as possible on the funnel with a clean cork or spatula. Spread the crystals on a watchglass and allow them to air-dry completely. Determine the weight and melting point of the purified product. Calculate the percent recovery.

2. Acetanilide.   Place 1 g of impure acetanilide in a 50-mL Erlenmeyer flask. Measure 20 mL of water into a graduated cylinder and add a 10-mL portion of it to the crude acetanilide. Boil the mixture gently with the aid of a burner or hot plate.

A layer of oil should form when the stated amount of water is added. (If you have not done so already, review the discussion of "Crystallization," in this section with emphasis on how to crystallize compounds that form oils.) This layer consists of a solution of water *in* acetanilide. More water must be added to effect complete solution of the acetanilide *in* water (see below). However, even if a homogeneous solution is produced at the boiling point of the mixture, an oil may separate from it as cooling begins. The formation of this second liquid phase is known to occur only under specific conditions: the acetanilide/water mixture must have a composition that is between 5.2% and 87% in acetanilide and be at a temperature *above* 80 °C. Because the solubility of acetanilide in water at temperatures near 100 °C exceeds 5.2%, a homogeneous solution formed by using the *minimum* quantity of water meets these criteria. Such a solution will yield an *oil* on cooling to about 83 °C; *solid* begins to form below this temperature.

Continue adding 3–5 mL of water in 0.5-mL portions to the boiling solution until the oil has completely dissolved. Any solid present at this point must consist of insoluble impurities. Once the acetanilide has just dissolved, add an additional 1 mL of water to prevent formation of oil during the crystallization step. If oil forms at this time, reheat the solution and add a little more water. Record the total volume of water used.

If the solution is colored, allow it to cool below boiling and add about 0.05 g of decolorizing carbon; with stirring, gently boil the solution for a few minutes. Allow the solution to cool somewhat, and then add about 0.2 g of a filter-aid, stirring the mixture thoroughly. Reheat to boiling, and perform a hot filtration according to the directions given in Section 2.13.

Cover the flask with a watchglass or an inverted beaker, and allow the filtrate to stand undisturbed until it has cooled to room temperature and no more crystals form.✱ To complete the crystallization, place the flask in an ice-water bath for at least 15 min.

Collect the crystals on a Büchner or Hirsch funnel by vacuum filtration (see Sec. 2.16) and wash the filter cake with two small portions of *cold* water. Press the crystals as dry as possible on the funnel with a clean cork or spatula. Spread the crystals on a watchglass and allow them to air-dry completely. Determine the weight and melting point of the purified acetanilide. Calculate the percent recovery.

3. Naphthalene.   Naphthalene may be conveniently recrystallized from methanol, 95% ethanol, or 2-propanol. Because these solvents are somewhat toxic and/or flammable, proper precautions should be taken. The sequence of steps up through the hot filtration should be performed at a hood if possible. If a hood is not available, position an inverted funnel connected to a vacuum source above the mouth of the flask being used for recrystallization (see Fig. 2.37b).

Place 1 g of impure naphthalene in a 50-mL Erlenmeyer flask and dissolve it in the minimum required amount of boiling alcohol. (*Caution:* Use a steam, oil, or hot-water bath for heating; do *not* use a burner.) Add 0.5 mL of additional solvent. If the solution is colored, treat it with decolorizing carbon (see Sec. 2.14) and a filter-aid according to the procedure given in Part 2 for acetanilide. Perform a hot filtration (see Sec. 2.13), cover the flask containing the hot filtrate with either a watchglass or an inverted beaker, and allow the filtrate to stand undisturbed until it has cooled to room temperature and no more crystals form.✳ To complete the crystallization, place the flask in an ice-water bath for at least 15 min.

Collect the crystals on a Büchner or Hirsch funnel by vacuum filtration (see Sec. 2.16) and wash them with two small portions of *cold* solvent. Press the crystals as dry as possible on the funnel with a clean cork or spatula. Transfer the crystals to a watchglass and allow them to air-dry. Determine the weight and melting point of the purified naphthalene. Calculate the percent recovery.

4. Unknown Compound.   Accurately weigh about 1 g of the unknown compound and place it in a 50-mL Erlenmeyer flask. Measure about 10 mL of the solvent you have selected for the recrystallization and place it in the flask. Bring the mixture to a gentle boil with a steam bath or, *if* water is the solvent, a flame, add a 1-mL portion of the solvent, and again boil the solution. Continue adding 3-mL portions of solvent, one portion at a time, until the solid has completely dissolved. Bring the solution to boiling after adding each portion of solvent. Record the total volume of solvent that is added.

If the solution is colored, treat it with decolorizing carbon (see Sec. 2.14) and a filter-aid according to the procedure given in Part 1 for benzoic acid. Perform a hot filtration by gravity according to the directions of Section 2.13. If the solution is colorless but contains small particles of insoluble material, omit the treatment with decolorizing carbon but perform the hot filtration. In either case, rinse the empty flask with about 1 mL of *hot* solvent, pour the rinsings through the filter, and collect them in the flask containing the original filtrate. If the filtered solution remains colored or still contains insoluble particles, repeat the appropriate steps to eliminate these impurities. Cover the flask containing the filtrate with a clean watchglass or an inverted beaker, and allow the filtrate to stand undisturbed until it has cooled to room temperature and no more crystals form.✳ Complete the crystallization by placing the flask in an ice-water bath for at least 15 min.

Collect the crystals by vacuum filtration using the technique described in Section 2.16. Wash the crystals that have been collected on the filter with two 1- to 3-mL portions of *cold, pure* solvent, and allow them to dry as thoroughly as possible on the filter funnel. Dry the crystals as described in Section 2.22.

Determine the weight and melting point of the purified product and calculate the percent recovery.

Finishing Touches   Flush any *aqueous filtrates* or *solutions* down the drain. With the advice of your instructor, do the same with the *filtrates* derived from use of *alcohols, acetone,* or other *water-soluble solvents.* Use the appropriate containers for the *filtrates* containing *halogenated solvents* or *hydrocarbon solvents.* Put *filter papers* in the container for nontoxic waste, unless instructed to do otherwise.

SPECTRA OF STARTING MATERIALS

FIGURE 3.1 IR spectrum of benzoic acid (KBr pellet).

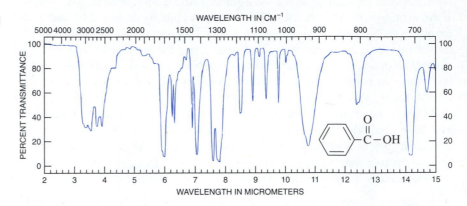

(a) ^{1}H NMR spectrum (60 MHz).

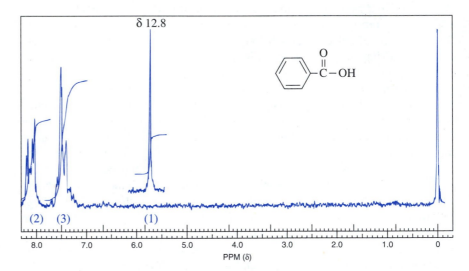

FIGURE 3.2 NMR data for benzoic acid.

(b) ^{13}C NMR data.
Chemical shifts: δ 128.4, 129.5, 130.3, 133.8, 172.7.

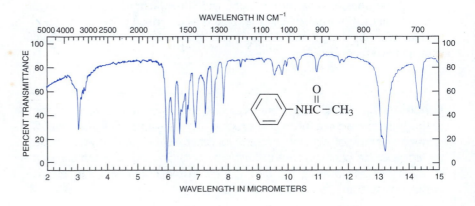

FIGURE 3.3 IR spectrum of acetanilide (KBr pellet).

(a) ^{1}H NMR spectrum (60 MHz).

FIGURE 3.4   NMR data for acetanilide.

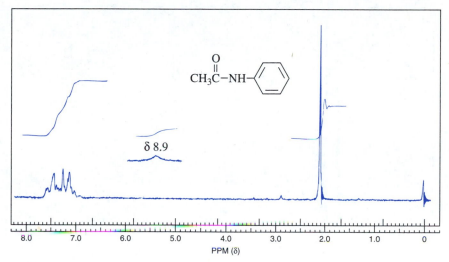

$$CH_3C-NH-\phi$$

δ 8.9

PPM (δ)

(b) ^{13}C NMR data.

Chemical shifts: δ 24.0, 119.7, 123.3, 128.5, 139.2, 168.7.

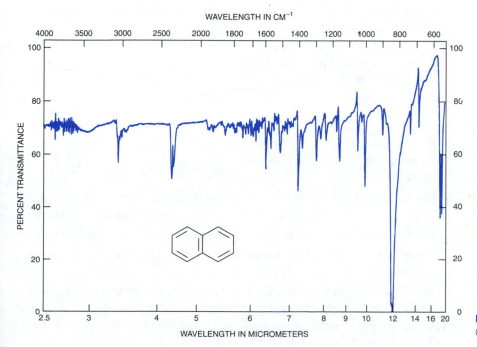

WAVELENGTH IN CM^{-1}

PERCENT TRANSMITTANCE

WAVELENGTH IN MICROMETERS

FIGURE 3.5   IR spectrum of naphthalene (KBr pellet).

FIGURE 3.6   NMR data for naphthalene.   **(a)** ^{1}H NMR spectrum (90 MHz).

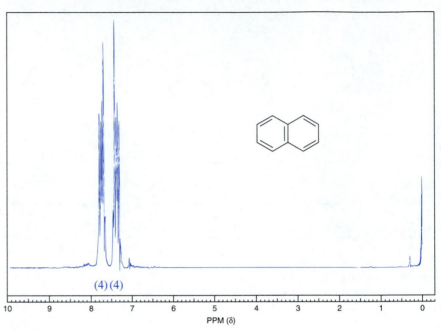

(b) ^{13}C NMR data.
   Chemical shifts: δ 126.0, 128.1, 133.7.

EXERCISES

1. Define or describe each of the following terms as applied to recrystallization:

   a. solution recrystallization
   b. temperature coefficient of a solvent
   c. the relationship between dielectric
      constant and polarity of a solvent

d. petroleum ether	j. fluted filter paper	p. seeding
e. mixed solvents	k. mother liquors	q. washing crystals
f. solvent selection	l. filtrate	r. drying crystals
g. decoloration	m. solute	s. percent recovery
h. hot gravity filtration	n. solvent	
i. vacuum filtration	o. occlusion	

2. Briefly explain how a colored solution may be decolorized.

3. Briefly explain how insoluble particles can be removed from a hot solution.

4. List five criteria that should be used in selecting a solvent for a recrystallization.

5. When is hot gravity filtration used and when is vacuum filtration used in recrystallization? What is the purpose of each type of filtration?

6. When might Celite® (filter-aid) be used in recrystallization, and why is it used?

7. In hot gravity filtration, what might happen if the filter funnel is not pre-heated before the solution is poured through it?

8. Why should the filter flask not be connected *directly* to a water aspirator pump when performing a vacuum filtration?

9. Briefly describe how a mixture of sand and benzoic acid, which is soluble in hot water, might be separated to provide pure benzoic acid.

10. Describe each of the following pieces of equipment and its use:
   a. filter flask
   b. filter trap
   c. Büchner funnel.

11. The following solvent selection data were collected for two different impure solids:

Solid A

Solvent	Solubility at Room Temperature	Solubility When Heated	Crystals Formed When Cooled
Methanol	Insoluble	Insoluble	—
Chloroform	Insoluble	Soluble	Very few
Cyclohexane	Insoluble	Soluble	Many
Toluene	Insoluble	Soluble	Very few

Solid B

Solvent	Solubility at Room Temperature	Solubility When Heated	Crystals Formed When Cooled
Water	Soluble	—	—
Ethanol	Soluble	—	—
Carbon tetrachloride	Insoluble	Insoluble	—
Petroleum ether	Insoluble	Insoluble	—
Toluene	Insoluble	Insoluble	—

Based on these results, what solvents or mixture of solvents might you consider using to recrystallize solids A and B?

12. List each of the steps in the systematic procedure for recrystallization. Indicate briefly the functional purpose of each of these steps in accomplishing the purification of the originally impure solid.

13. The goal of the recrystallization procedure is to obtain *purified* material with a *maximized* recovery. For each of the items listed, explain why this goal would be adversely affected.
   a. In the solution step, an unnecessarily large volume of solvent is used.
   b. The crystals obtained by vacuum filtration are *not* washed with fresh cold solvent before drying.

    c. The crystals referred to in (b) are washed with fresh *hot* solvent.

    d. A large quantity of decolorizing carbon is used.

    e. Crystals are obtained by breaking up the solidified mass of an oil that originally separated from the hot solution.

    f. Crystallization is accelerated by immediately placing the flask of hot solution in an ice-water bath.

14. A second crop of crystals may be obtained by concentrating the vacuum filtrate and cooling. Why is this crop of crystals probably less pure than the first crop?

15. Explain why the rate of dissolution of a crystalline substance may depend on the *size* of its crystals.

16. The solubility of benzoic acid at 0 °C is 0.02 g per 100 mL of water, and that of acetanilide is 0.53 g per 100 mL of water. If you performed either of these recrystallizations, calculate, with reference to the total volume of water you used in preparing the hot solution, the amount of material in your experiment that was unrecoverable by virtue of its solubility at 0 °C.

17. Assuming that either solvent is otherwise acceptable in a given instance, what advantages does ethanol have over 1-octanol as a crystallization solvent? hexane over pentane? water over methanol?

18. Look up the solubility of benzoic acid in hot water. According to the published solubility, what is the minimum amount of water in which 1 g of benzoic acid can be dissolved?

3.3 PHYSICAL CONSTANTS: MELTING POINTS

3.3.1 PHYSICAL CONSTANTS

Physical constants of compounds are numerical values associated with measurable properties of these substances. These properties are *invariant* and are useful in the identification and characterization of substances encountered in the laboratory so long as accurate measurements are made under specified conditions such as temperature and pressure. Physical constants are useful only in the identification of *previously known* compounds, however, because it is not possible to predict accurately the values of such properties. Among the more frequently measured physical properties of organic compounds are **melting point (mp), boiling point (bp), index of refraction (n), density (d), specific rotation ([α]),** and **solubility.** Melting points, discussed below, boiling points, described in Section 4.1, and solubilities, outlined in Section 3.2, are the properties most commonly encountered. Index of refraction and density are mentioned in Chapter 23. Specific rotation is discussed in Chapters 7 and 22 but applies only to molecules that are **optically active.** Whether the substance is known or unknown, such values along with other properties like color, odor, and crystal form should be recorded in the laboratory notebook.

    The values of one or two of the common physical properties *may* be identical for more than one compound, but it is most unlikely that values of several such properties will be the same for two different compounds. Consequently, a list of physical constants is a highly useful way to characterize a substance. Extensive compilations of the physical constants are available (see Chap. 24). One of the most convenient is the *CRC Handbook of Chemistry and Physics,* which contains a tabulation of the physical constants and properties of a large number of inorganic and organic compounds. The *Handbook of Tables for Organic Compounds* is especially useful for organic compounds. Neither of these books is comprehensive; rather, they contain entries for only the more common organic and inorganic substances. So many compounds are known that multivolume sets of books are required to list their physical properties (see Chap. 24).

3.3.2   MELTING POINT OF A PURE SUBSTANCE

The **melting point** of a substance is defined as the temperature at which the liquid and solid phases exist in equilibrium with one another without change of temperature. Ideally, addition of heat to a mixture of the solid and liquid phases of a pure substance at the melting point will cause no rise in temperature until all the solid has melted. Conversely, removal of heat from the equilibrium mixture will produce no decrease in temperature until all the liquid solidifies. This means that the melting and freezing points of a pure substance are identical.

The relationship between phase composition, total heat supplied, and temperature for a pure compound is plotted in Figure 3.7. It is *assumed* that heat is being provided to the compound at a *slow* and *constant* rate, and that the elapsed time of heating is a cumulative measure of the heat supplied. At temperatures below the melting point, the compound exists in the solid phase, and the addition of heat causes the temperature of the solid to rise. As the melting point is reached, the first small amount of liquid appears, and equilibrium is established between the solid and liquid phases. The temperature of the mixture does not change as more heat is supplied, because the additional heat causes the solid to be converted to liquid, with both phases remaining in equilibrium. At this point in the melting process, the energy being added to the system in the form of heat is being used to break down the forces that hold the molecules together in the solid state, the so-called **crystal lattice energy;** this accounts for the constancy

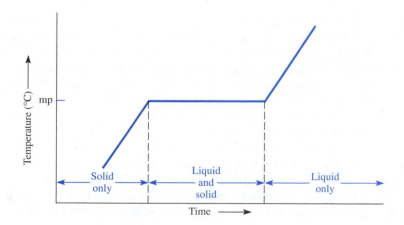

FIGURE 3.7   Phase changes with time and temperature.

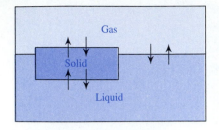

FIGURE 3.8 *Phase equilibria between the solid, liquid, and gas phases.*

of the temperature at this stage. When the last of the solid melts, the heat subsequently added causes the temperature to rise linearly at a rate that depends on the rate of heating.

The interconversion of the liquid and solid phases also may be described in terms of their respective **vapor pressures,** which are directly related to the rate of passage of the molecules from one phase to the other. In terms of melting points, only the transfer of molecules between the solid and liquid phases is considered. Vapor pressure may appear to be a misnomer here, but the same "escaping tendency" that produces the measurable equilibrium vapor pressure between liquid and gas phases is responsible for the transfer of molecules between the solid and liquid phases. Figure 3.8 represents the equilibria existing between the solid, liquid, and gas phases. At the melting point, the vapor pressures of the solid and liquid phases are equal, and there is no net transfer of molecules from one phase to the other *unless* there is a change in the amount of heat supplied to the system.

The equilibria illustrated in Figure 3.8 exist for a particular compound at all times, but the most important equilibria depend on the nature of the substance being studied. With respect to melting points, the important equilibrium is the one between the solid and liquid phases, since the vapor pressure of the liquid is *very* low. For boiling points (see Sec. 4.1), on the other hand, the equilibrium between liquid and gas phases is of prime importance because no solid is present. There are a few substances in which the dominant equilibrium at atmospheric pressure is between the solid and gas phases, with the liquid phase being essentially absent. Common examples include solid and gaseous carbon dioxide (Dry Ice), and solid and gaseous iodine. The phenomenon of *direct* conversion of the solid phase into the gaseous phase *without* the intervention of the liquid phase is called **sublimation** (see Sec. 2.17).

The preceding discussion of the theory of melting points assumes that the experimental apparatus for melting-point determinations allows measurement of *extremely* small changes in the heat supplied to a solid sample. This is difficult and very time-consuming to do in practice. Moreover, most melting-point apparatus for the organic laboratory is designed for ease of use and does not permit the determination of the *exact* temperature at which a compound melts, that is, the true melting point. This term is somewhat misleading because what is actually measured is a **melting range,** although the two terms are used interchangeably. If a solid substance is *pure,* it should melt over a very *narrow* or sharp range, which should normally be no more than 1 °C if the melting point is determined carefully. The melting ranges reported for many pure compounds may be greater than 1 °C because the particular compound was not quite pure or the melting point was not measured properly.

The melting point is expressed as the temperature range over which the solid *starts* to melt and then is completely converted to liquid. Although the process of melting may actually begin by "softening," as evidenced by an apparent shrinking of the solid, such softening is difficult to observe. Thus, for our purposes the start of melting is defined as the temperature at which the first droplet of *liquid* can be detected. Note that it is improper and inexact to report a single temperature, such as 118 °C for a melting point; rather, a range of 117–119 °C or 117.5–118.0 °C, for example, should be recorded.

A narrow melting-point range has the important implication that the compound is highly *pure.* For example, the previous section contains procedures for purifying solids by recrystallization. A convenient and relatively rapid method for ascertaining purity of the solid is the measurement of its melting point, a procedure that is given at the end of this section. If the observed melting point is sharp, the tentative conclusion is that the recrystallization was successful and that the substance is pure. This is not always the case, however, since **eutectic mixtures** are impure solids that give a narrow melting-point range.

3.3.3 Effect of Impurities on Melting Points; Mixtures

Many solid substances prepared in the organic laboratory are initially impure, so the effect of impurities on melting-point ranges deserves further discussion. Although this topic is discussed in freshman chemistry textbooks, a brief review of its basic principles is given here.

The presence of an impurity generally *decreases* the melting point of a pure solid. This is shown graphically by the melting point–composition diagram of Figure 3.9, in which points *a* and *b* represent the melting points of pure *A* and *B*, respectively. Point *E* is called the **eutectic point** and is determined by the equilibrium composition at which *A* and *B* melt in constant ratio. In Figure 3.9, this ratio is 60 mol % *A* and 40 mol % *B*; an impure solid comprised of *A* and *B* in this ratio would be called a **eutectic mixture.** The temperature at the eutectic point is designated by *e*.

Now consider the result of heating a solid mixture composed of 80 mol % *A* and 20 mol % *B*, a sample that might be considered as "impure *A*." As heat is applied to the solid, its temperature will rise. When the temperature reaches *e*, *A* and *B* will both begin to melt in the constant ratio defined by the composition at the eutectic point. Once all of the "impurity" *B* has melted, only solid *A* will be left in equilibrium with the melt. The remaining solid *A* will continue to melt as additional heat is supplied, and the percentage of *A* in the melt will increase, changing the composition of the melt from that of the eutectic mixture. This increases the vapor pressure of *A* in the solution according to Raoult's law (Eq. 4.2) and raises the temperature at which solid *A* is in equilibrium with the molten

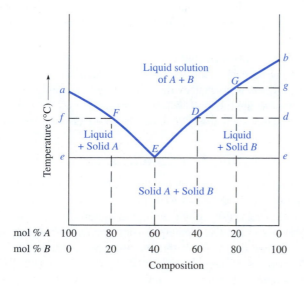

FIGURE 3.9 Melting-point–composition diagram for two hypothetical solids, *A* and *B*.

solution. The relationship between the equilibrium temperature and the composition of the molten solution is then represented by curve *EF* in Figure 3.9. When the temperature reaches *f,* no solid *A* will remain and melting of the sample will be complete. The impure sample *A* exhibits a melting "point" that extends over the relatively broad temperature range *e–f.* Because melting both begins and ends below the melting point of pure *A,* the melting point of *A* is said to be *depressed.*

The foregoing analysis is easily extended to the case in which substance *B* contains *A* as an impurity. In Figure 3.9, this simply means that the composition of the solid mixture is to the right of point *E.* The temperature during the melting process would follow curve *ED* or *EG,* and the melting range would now be *e–d* or *e–g.*

A sample whose composition is exactly that of the eutectic mixture (point *E,* Fig. 3.9) will exhibit a *sharp* melting point at the eutectic temperature. This means a eutectic mixture can be mistaken for a pure compound, since both have a sharp melting point.

It may be very difficult to observe the initial melting point of solid mixtures, particularly with the capillary tube melting-point technique used in the Experimental Procedure that follows. This is because the presence of only a minor amount of impurity means that only a tiny amount of liquid is formed in the stage of melting that occurs at the eutectic temperature. In contrast, the temperature at which the last of the solid melts (points *d* and *g,* Fig. 3.9) can be determined accurately. Consequently, a mixture containing smaller amounts of impurities will generally have both a higher final melting point and a narrower *observed* melting-point range than one that is less pure.

In summary, the melting-point range of a pure compound or a eutectic mixture is *sharp* and that of a noneutectic mixture is *broad.* The experimental observation of narrow melting-point ranges almost always indicates that the substance is pure, since the probability of having a eutectic mixture is low.

The broadening of the melting-point range that occurs upon introduction of an impurity into a pure compound may be used to advantage for the structural identification of a pure substance. The technique is known as a **mixture melting-point** determination, or less properly as a "mixed melting-point" determination, and is perhaps best illustrated by the following example. Assume that an unknown compound *X* melts at 134–135 °C, and you suspect it is either urea, H_2NCONH_2, or *trans*-cinnamic acid, $C_6H_5CH=CHCO_2H$, both of which melt in this range. If *X* is mixed intimately with urea and the melting point of this mixture is found to be lower than that of the pure compound and pure urea, then urea is acting as an impurity, and the compound *cannot* be urea. If the mixture melting point is identical to that of the pure compound and of urea, the compound is identified as urea. Unfortunately, this procedure is useful in identifying compounds only when authentic samples of the likely possibilities are available.

3.3.4 MICRO MELTING-POINT METHODS

The determination of highly accurate melting points of organic compounds can be time-consuming. Fortunately, micro methods are available that are convenient, require negligible amounts of sample, and give melting-point data that are satisfactory for most purposes. The technique using the capillary-tube melting-point procedure is described below and is the one used most commonly in the organic laboratory. Description of alternative methods, including the use of automated, microprocessor-controlled apparatus, is beyond the scope of this discussion.

There are practical considerations in determining melting points, and some of them are briefly noted here. First, the observed melting-point range depends on several factors, including the *quantity* of sample, its *state* of subdivision, the *rate* of heating during the determination, and the *purity* and *chemical characteristics* of the sample. The first three factors can cause the observed melting-point range to differ from the true value because of the time lag for transfer of heat from the heating medium to the sample and for conduction of heat within the sample. For example, if the sample is too large, the distribution of heat within it may not be uniform, and inaccurate melting ranges will result. A similar problem of non-uniform heat distribution is associated with use of crystals of large size: It will be difficult to pack the sample tightly in the capillary melting tube, and the resulting air-space results in poor conduction of heat. If the rate of heating is too fast, the thermometer reading will lag behind the actual temperature of the heating medium and produce measurements that are *low*. The chemical characteristics of the sample may be important if the compound tends to decompose upon melting. When this occurs, discoloration of the sample is usually evident, and may be accompanied by gas evolution. The decomposition products constitute impurities in the sample, and the true melting point is lowered as a result. The reporting of melting points for compounds that melt with decomposition should reflect this, as in "mp 195 °C (dec)."

The accuracy of any type of temperature measurement ultimately depends on the quality and calibration of the thermometer. A particular thermometer may provide accurate readings in some temperature ranges, but may be off by a degree or two in others. Melting points that have been determined using a *calibrated* thermometer may be reported in the form, "mp 101–102 °C (corr.)," where "corr." is the abbreviation for "*corrected*"; the corresponding abbreviation for values obtained with an uncalibrated thermometer is "uncorr." for "*uncorrected.*" The following Experimental Procedure contains a description of how to calibrate your thermometer.

Pre-Lab exercises for Section 3.3, "Melting Points," are found on page PL. 3.

EXPERIMENTAL PROCEDURE

SAFETY ALERT

1. If a burner is used in this experiment, be sure that no flammable solvents are nearby. Keep the rubber tubing leading to the burner away from the flame. Turn off the burner when it is not being used.

2. Some kinds of melting-point apparatus, such as the Thiele tube, use mineral or silicone oils as the heat transfer medium. These oils may *not* be heated safely if they are contaminated with even a few drops of water. Heating these oils above 100 °C may produce splattering of hot oil as a result of formation of steam from water. Fire can also result if splattered oil comes in contact with open flames. Examine your Thiele tube for evidence of water droplets in the oil. If any are observed, either change the oil or exchange tubes. Give the contaminated tube to your instructor.

(Continued)

3. Mineral oil is a mixture of high-boiling hydrocarbons and should *not* be heated above 200 °C because of the possibility of spontaneous ignition, particularly when a burner is used for heating. Silicone oils may be heated to about 300 °C without danger (*see* Sec. 2.5).
4. Be careful to avoid contact of chemicals with your skin. Clean up any spilled chemicals immediately with a brush or paper towel.
5. If you use a Thiele tube, handle it carefully when you are through using it. The tube cools slowly; to avoid burns, take care when removing it from its support.

A.   CALIBRATION OF THERMOMETER

Your instructor will advise you if calibration of the thermometer is required. The calibration involves the use of standard substances for the measurement of the temperature at a series of known points within the range of the thermometer and the comparison of the observed readings with the true temperatures. The difference between the observed and the true temperature measurement provides a correction that must be applied to the observed reading. Calibration over a range of temperatures is necessary because the error is likely to vary at different temperatures. The corrections can be plotted in your notebook as deviations from zero versus the temperature over the range encompassed by the thermometer. You will then be able to tell, for example, that at about 130 °C the thermometer gives readings that are 2 °C too low, or that at 190 °C the readings are about 1.5 °C too high. These values should then be applied to correct all temperature measurements taken. The corrections you obtain are valid only for the thermometer used in the calibration; if it is broken, the calibration must be repeated with a new one.

To calibrate a thermometer, *carefully* determine the capillary melting points (see Sec. 2.3) of a series of standard substances. A list of suitable standards is provided in Table 3.2. Your instructor may suggest other or additional standards to be used. The temperatures given in Table 3.2 correspond to the upper *limit* of the melting-point range for pure samples of these standards.

B.   DETERMINATION OF CAPILLARY-TUBE MELTING POINTS

1. *General Procedure.* The techniques for determination of a capillary-tube melting point are presented in Section 2.3. Read this section carefully before undertaking any experimental work.

2. *Mixture Melting Points.* To prepare a sample for mixture melting-point determination, the two components *must* be thoroughly and intimately mixed. This is best accomplished by grinding them together with a small mortar and pestle. If these are not available, a small, clean watchglass and glass stirring rod or metal spatula may be used. Be careful not to apply too much pressure to the glass rod, however, because it is more fragile than a pestle and may break. Do *not* perform this operation on a piece of filter paper because fibers from the paper may contaminate the sample.

TABLE 3.2   Standards for Thermometer Calibration

Compound	Melting Point (°C)
Ice water	0
3-Phenylpropanoic acid	48.6
Acetamide	82.3
Acetanilide	114
Benzamide	133
Salicylic acid	159
4-Chloroacetanilide	179
3,5-Dinitrobenzoic acid	205

3. *Experiments.* You may be assigned one or more of the following melting-point experiments, which should be performed using the general procedure given earlier. Your instructor will advise you about the type of apparatus to use.

In determining the melting point of a compound, *much* valuable time can be wasted waiting for melting to occur if the proper slow rate of heating is being used on a sample whose melting point is unknown to you. It is considerably more efficient to prepare *two* capillary tubes containing the compound being studied and determine the *approximate* melting point by rapidly heating one of them. Then the heating fluid is allowed to cool 10–15 °C below this approximate melting point, and an accurate melting point is obtained with the second sample. In case Thiele tubes and electric melting-point apparatus are both available, it may be convenient to determine the approximate melting point using a Thiele tube and the accurate melting-point range with the electric equipment.

Known Compounds.   Select one or two compounds from a list of available compounds of *known* melting point. Determine the melting-point ranges for each of these substances, using the capillary melting-point procedure. Repeat as necessary until you obtain accurate results and are confident with the procedure.

Mixture Melting Points.   Introduce 5–10% of a second substance as an impurity into a sample of one of the compounds whose melting range was determined in Part A. Thoroughly mix the two components and determine the melting range in order to study the effect of impurities on the melting range of a previously pure compound.

Unknown Compounds.   Accurately determine the melting range of an unknown pure compound supplied by your instructor. As noted earlier, it is more efficient to determine the approximate melting point by rapid heating before obtaining an accurate melting range.

Unknown Compound Obtained from Recrystallization.   In the event that you were given an impure mixture to purify by recrystallization (see Sec. 3.2), accurately determine the melting range of the recrystallized product. As noted earlier, it is more efficient to determine the approximate melting point by rapid heating before obtaining an accurate melting range.

If a broad melting range is obtained, it may result from contamination of the sample with solvent; additional drying should eliminate this problem. Another possible explanation is that the recrystallization was not completely successful in removing impurities. In this case the solid should be recrystallized using the same solvent, but if this fails to narrow the melting range, recrystallization with a different solvent should be attempted.

Finishing Touches   Return any *unused samples* to your instructor or dispose of them in the appropriate container for nonhazardous organic solids. Discard the used *capillary tubes* in a container for broken glass; do *not* leave them in the area of the melting-point apparatus or throw them in wastepaper baskets.

EXERCISES

1. Describe errors in procedure that may cause an observed capillary melting point of a pure compound
   a. to be *lower* than the correct melting point
   b. to be *higher* than the correct melting point
   c. to be *broad* in range (over several degrees)

2. Briefly define the following terms:
   a. vapor pressure as applied to melting
   b. melting point or melting-point range
   c. mixture or mixed melting point
   d. eutectic point
   e. eutectic mixture

3. Filter paper is usually a poor material on which to powder a solid sample before introducing it into a capillary melting-point tube because small particles of paper may end up in the tube along with the sample. Why is this undesirable, and how might the presence of paper in the sample make the melting-point determination difficult?

4. Criticize the following statements by indicating whether each is *true* or *false,* and if false, explain why.
   a. An impurity always lowers the melting point of an organic compound.
   b. A sharp melting point for a crystalline organic substance always indicates a pure single compound.
   c. If the addition of a sample of compound *A* to compound *X* does not lower the melting point of *X*, *X* must be identical to *A*.
   d. If the addition of a sample of compound *A* lowers the melting point of compound *X*, *X* and *A* cannot be identical.

5. The melting points of pure benzoic acid and pure 2-naphthol are 122.5 °C and 123 °C, respectively. Given a pure sample that is known to be either pure benzoic acid or 2-naphthol, describe a procedure you might use to determine the identity of the sample.

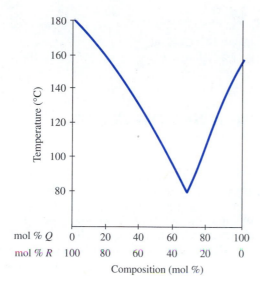

FIGURE 3.10   Melting-point–composition diagram for Exercise 7.

6. A student used the Thiele micro melting-point technique to determine the melting point of an unknown and reported it to be 182 °C. Is this value believable? Explain why or why not.

7. The melting-point–composition diagram for two substances, Q and R, is provided in Figure 3.10, which should be used to answer the following questions.

   a. What are the melting points of pure Q and R?
   b. What are the melting point and the composition of the eutectic mixture?
   c. Would a mixture of 20 mol % Q and 80 mol % R melt if heated to 120 °C? to 160 °C? to 75 °C?
   d. A mixture of Q and R was observed to melt at 105–110 °C. What can be said about the composition of this mixture? Explain briefly.

8. For the following melting points, indicate what might be concluded regarding the purity of the sample:

   a. 120–122 °C   **c.** 147 °C (dec)
   b. 46–60 °C   **d.** 162.5–163.5 °C

LIQUIDS

DISTILLATION AND BOILING POINTS

The purification of solids by recrystallization and the use of melting points as a criterion of their purity are discussed in Chapter 3. The techniques used for the purification of liquids, namely, **simple, fractional, steam,** and **vacuum distillation,** are topics of this chapter. Boiling points are also discussed as a physical property that can be used as one means of determining the purity and identity of liquids.

4.1 BOILING POINTS OF PURE LIQUIDS

The molecules of a liquid are in constant motion, and those at the surface are able to escape into the vapor phase. The consequences of vaporization of a liquid contained in a closed, *evacuated* system are considered first, and then the situation in which the system is open to the atmosphere is discussed.

In a closed evacuated system, the number of molecules in the gas phase will initially increase until the rate at which they reenter the liquid becomes equal to the rate at which they escape from it. At this point, no further *net* change is observed, and the system is said to be in a state of **dynamic equilibrium.** The molecules in the gas phase are in rapid motion and continually collide with the walls of the vessel, which results in the exertion of pressure against the walls. The magnitude of this vapor pressure at a given temperature is called the **equilibrium vapor pressure** of the liquid at that temperature. Vapor pressure is temperature dependent, as shown in Figure 4.1, and can be understood in terms of the tendency of the molecules to escape from the liquid. As the temperature of the system rises, the average kinetic energy of the liquid molecules increases, thus facilitating their escape into the gas phase. The rate of reentry of gaseous molecules into the liquid phase also increases, because this rate is proportional to the number of molecules in the gas phase and equilibrium is reestablished at the higher temperature. However, there are now more molecules in the gas phase than there were at the lower temperature, so the vapor pressure of the system is greater.

You should note that there is a very important **safety rule** based on the fact that the pressure in a closed system *increases* as the temperature of the system rises. Such a system should *not* be heated unless apparatus designed to withstand the pressure is used; otherwise, an *explosion* will result. For purposes of the first laboratory course in organic chemistry, the rule is simply, "*Never* heat a closed system!"

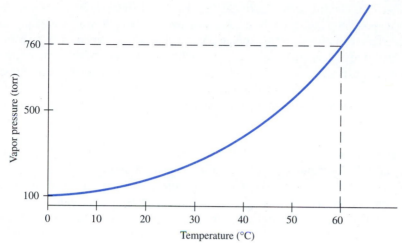

FIGURE 4.1 Graph of the dependence of vapor pressure on temperature for a liquid.

Now suppose that a liquid sample at a particular temperature is placed in an *open* container so that the molecules of the vapor over the liquid are mixed with air. The *total* or *external* pressure above the liquid is defined by **Dalton's law of partial pressures** (Eq. 4.1) and is equal to the sum of the partial pressures of the sample and of air. The partial pressure of the sample is equal to its **equilibrium vapor pressure** at the given temperature. When the temperature of the liquid is raised, the equilibrium vapor pressure of the sample will rise, and the number of gas molecules that have escaped from the liquid by the process of evaporation will increase in the space above the liquid. This will have the net effect of displacing some of the air. At the higher temperature, the partial pressure of the sample will be a *larger* percentage of the *total* pressure. This trend will continue as the temperature of the liquid is further increased, until the equilibrium vapor pressure of the sample equals the *total* pressure. At this point all of the air will have been displaced from the vessel containing the liquid. Entry of additional molecules from the liquid into the gas phase will only have the effect of further displacing those already in that phase; the partial pressure of the molecules of the sample will no longer increase.

$$P_{total} = P_{sample} + P_{air} \qquad (4.1)$$

Considering these facts prompts the conclusion that the equilibrium vapor pressure of the sample has an upper limit that is dictated by the total pressure. When the temperature of the liquid is such that the equilibrium vapor pressure of the sample equals the total pressure, the rate of evaporation increases dramatically, and bubbles are seen to form in the liquid. This is the boiling process, and the temperature associated with it is the **boiling point** of the liquid. Since the boiling point is dependent upon the total pressure, that pressure must be specified when boiling points are reported, for example, "bp 132 °C (748 torr)." The boiling point of water is 100 °C *only* when the external pressure is 760 torr, or 1 atm. The **normal boiling point** of a liquid is measured at 760 torr (1 atm) and is shown by the dashed lines in Figure 4.1 to be 60 °C. This figure also allows determination of what the boiling point of the liquid would be at various total pressures. For example, at 500 torr the boiling point will be about 50 °C; the decrease in temperature is accounted for by the fact that the total pressure in the system has decreased.

The dependence of vapor pressure upon total pressure, as reflected by the boiling point of a liquid, can be used to advantage in the following way. Suppose that a certain liquid has a normal boiling point such that it decomposes appreciably when heated to this temperature. Reduction in the total pressure of the system reduces the boiling point of the sample to a temperature at which it no longer decomposes. This technique, called **vacuum distillation,** is discussed in Section 2.9.

There are several commonly encountered examples of the effect of external pressure upon boiling point. For instance, it takes longer to cook a "3-minute egg" at high altitudes. The reason for this is that atmospheric pressure decreases at elevations above sea level, thereby lowering the boiling point of water below 100 °C. Consequently, the rates of the chemical and physical processes associated with the process of cooking are decreased. In contrast, foods cook faster in a pressure cooker because the internal pressure is greater than one atmosphere, and this makes the boiling point of water *higher*. The steam inside the cooker is in equilibrium with the boiling water and at a temperature higher than it would be at atmospheric pressure.

The discussions so far have considered the boiling points of pure liquids and the effect of pressure upon them. The presence of volatile and nonvolatile impurities in a liquid also affects its boiling point and this subject is discussed in Section 4.2.1.

Boiling points are useful for identification of pure liquids and some low-melting solids. In general, a *pure* liquid will boil at a constant temperature or a narrow temperature range, provided the total pressure in the system remains constant. On the other hand, most *mixtures* of liquids boil over a fairly wide temperature range at constant pressure. This effect is described in Section 4.2.2.

Determination of Micro Boiling Points

Pre-Lab exercises for Section 4.1, "Micro Boiling Points of Pure Liquids," are found on page PL. 5.

EXPERIMENTAL PROCEDURE

SAFETY ALERT

1. Burners are used in this experiment. Do not use them in the presence of volatile organic liquids. Review the material in Section 2.5.1 on the proper use of burners.
2. Check the heating fluid in the Thiele tube apparatus for the presence of moisture; see the discussion of oil baths in Section 2.5.2 for a description of the proper use of various types of heating fluids.
3. Use paper towels to clean up spilled liquids; discard the towels as directed by your instructor. Avoid contact of organic liquids with your skin; if this happens, wash the affected area thoroughly with soap and water.

Determine the boiling point of the liquid(s) assigned by your instructor. Follow the technique presented in Section 2.4 for using the micro boiling-point apparatus. Use a Bunsen burner or a microburner for heating. In the event you do not know the boiling point of liquid, *first* determine an approximate boiling-point range by heating the Thiele tube *fairly rapidly*. Repeat the measurement by heating the tube until the temperature is 20–30 °C *below* the approximate boiling point, and then heat the sample at a rate of 4–5 °C/min to obtain an accurate value. It may be desirable to repeat this procedure to obtain a more reliable boiling point.

EXERCISES

1. Refer to Figure 4.1 and answer the following:
   a. What total pressure would be required in the system in order for the liquid to boil at 30 °C?
   b. At what temperature would the liquid boil when the total pressure in the system is 100 torr?

2. Describe the relationship between escaping tendency of liquid molecules and vapor pressure.

3. Define the following terms:
   a. boiling point
   b. normal boiling point
   c. Dalton's law of partial pressure
   d. equilibrium vapor pressure

4. Explain why the boiling point at 760 torr of a solution of water, bp 100 °C (760 torr), and ethylene glycol ($HOCH_2$ CH_2OH), bp 196–198 °C (760 torr), exceeds 100 °C. For purposes of your answer, consider ethylene glycol as a nonvolatile compound.

5. Why should there be no droplets of water in the oil of a heating bath?

6. Why is the micro boiling-point technique not applicable for boiling points in excess of 200 °C if *mineral* oil rather than *silicone* oil is the heating fluid in the Thiele tube?

4.2   PURIFICATION OF LIQUIDS BY DISTILLATION

The most commonly used method for purifying *liquids* is distillation, a process that consists of evaporating a liquid by heating and condensing the vapor into a separate vessel to yield a **distillate.** After the distillation is complete, liquid remaining in the original stillpot or distillation flask is called the **pot residue.** The common types of distillation are **simple distillation** (see Sec. 2.7), **fractional distillation** (see Sec. 2.8), **vacuum distillation** (see Sec. 2.9), and **steam distillation** (see Sec. 2.10). The theory underlying each of these techniques will be considered in turn.

4.2.1 SIMPLE DISTILLATION

Simple distillation allows separation of distillates from less-volatile substances that remain as pot residue at the completion of the distillation. In the ideal case, only a single component of the mixture to be distilled will be volatile, so the distillate will be a pure compound. More commonly, however, several volatile components comprise the mixture. Simple distillation allows isolation of the various components of the mixture in acceptable purity if the *difference* between the boiling points of each pure substance is greater than 40–50 °C. For example, a mixture of diethyl ether, bp 35 °C (760 torr), and toluene, bp 111 °C (760 torr), could be separated by simple distillation, with the ether distilling first. Organic chemists frequently use this technique to separate a desired reaction product from the solvents used for the reaction or its work-up. The solvents are usually more volatile than the product and are readily removed from it by simple distillation.

To understand the principles of distillation, a review of the effect of impurities on the vapor pressure of a pure liquid is in order. The discussion starts with consideration of the consequences of having *nonvolatile* impurities present and then turns to the more common case of contamination of the liquid with other *volatile* substances.

Consider a homogeneous solution comprised of a nonvolatile impurity and a pure liquid; for the present purpose these are taken as sugar and water, respectively. The sugar reduces the vapor pressure of the water because this nonvolatile impurity lowers the concentration of the volatile constituent in the liquid phase. The consequence of this is shown graphically in Figure 4.2. In this figure, Curve 1 corresponds to the dependence of the temperature upon the vapor pressure of *pure* water and intersects the 760-torr line at 100 °C. Curve 2 is for a *solution* having a particular concentration of sugar in water. Note that the vapor pressure at any temperature is reduced by a constant amount by the presence of the nonvolatile impurity, in accord with Raoult's law as discussed later. The temperature at which this curve intersects the 760-torr line is higher because of the lower vapor pressure, and consequently the temperature of the boiling solution, 105 °C, is higher.

Despite the presence of the sugar in the solution, the **head temperature** (see Sec. 2.7) in the distillation will be the same as for the pure water, namely, 100 °C (760 torr), since the water condensing on the thermometer bulb is now

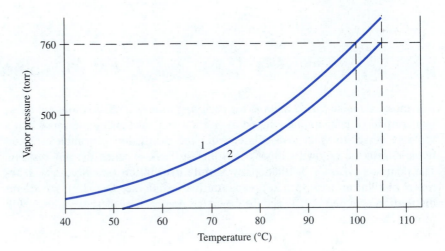

FIGURE 4.2 Diagram of the dependence of vapor pressure on temperature. Curve 1 represents pure water and curve 2 a pure water to which a sugar, a nonvolatile impurity, has been added.

uncontaminated by the impurity, sugar. The **pot temperature** will be *elevated,* however, owing to the decreased vapor pressure of the solution (Fig. 4.2). As water distils, the pot temperature will progressively rise because the concentration of the sugar in the stillpot increases, further lowering the vapor pressure of the water. Nevertheless, the head temperature will remain constant, just as though pure water were being distilled.

The quantitative relationship between vapor pressure and composition of homogeneous liquid mixtures is known as **Raoult's law** and may be expressed as shown in Equation 4.2. The factor P_X represents the partial pressure of component X, and is equal to the vapor pressure, P_X°, of pure X at a given temperature times the mole fraction N_X of X in the mixture. The **mole fraction** of X is defined as the fraction of *all* molecules present in the liquid mixture that are molecules of X. It is obtained by dividing the number of moles of X in a mixture by the sum of the number of moles of all components (Eq. 4.3). Raoult's law is strictly applicable only to **ideal solutions,** which are defined as those in which the interactions between *like* molecules are the same as those between *unlike* molecules. Fortunately, many organic solutions approximate the behavior of ideal solutions, so the following mathematical treatment applies to them as well.

$$P_X = P_X^\circ N_X \qquad \textbf{(4.2)}$$

$$N_X = \frac{nX}{nX + nY + nZ + \cdots} \qquad \textbf{(4.3)}$$

Note that the partial vapor pressure of X above an ideal solution depends *only* on its mole fraction in solution and is completely independent of the vapor pressures of the other volatile components of the solution. If all components other than X are nonvolatile, the total vapor pressure of the mixture will be equal to the partial pressure of X, since the vapor pressure of nonvolatile compounds may be taken as zero. Thus the distillate from such a mixture will always be pure R. This is the case for the distillation of a solution of sugar and water, as discussed earlier.

When a mixture contains two or more volatile components the *total* vapor pressure is equal to the *sum* of the partial vapor pressures of each such component. This is known as **Dalton's law** and is expressed by Equation 4.4, where P_X, P_Y, and P_Z refer to the vapor pressures of the volatile components. The process of distilling such a liquid mixture may be significantly different from that of simple distillation, because the vapors above the liquid phase will now contain some of each of the volatile components. Separation of the liquids in this case may require the use of fractional distillation, which is discussed in Section 4.2.2.

$$P_{\text{total}} = P_X + P_Y + P_Z + \cdots \qquad \textbf{(4.4)}$$

The experimental procedures and apparatus for simple distillation are given in Section 2.7. Experiments involving this technique are presented at the end of this section.

4.2.2 FRACTIONAL DISTILLATION OF IDEAL SOLUTIONS

Based on the discussion in Section 4.2.1, it clearly is easy to separate a volatile compound from a nonvolatile one by simple distillation. It is also possible to use this same technique to separate volatile compounds from one another if they differ in boiling point by at least 40–50 °C. If this is not the case, the technique of fractional distillation must normally be used to obtain each volatile component of a mixture in pure form. The theoretical basis of this technique is the subject of the following discussion.

Theory

For simplicity we shall consider only *ideal solutions* (see Sec. 4.2.1) consisting of two volatile components, designated X and Y. Solutions containing more than two such components are often encountered, but their behavior on distillation may be understood by extension of the principles developed here for a binary system.

The vapor pressure of a compound is a measure of the ease with which its molecules escape the surface of a liquid. When the liquid is comprised of two volatile components, in this case X and Y, the number of molecules of X and of Y in a given volume of the vapor above the mixture will be proportional to their respective partial vapor pressures. This relationship is expressed mathematically by Equation (4.5), where N_X'/N_Y' is the ratio of the mole fractions of X and Y in the *vapor* phase. The mole fraction of each component may be calculated from the equations $N_X' = P_X/(P_X + P_Y)$ and $N_Y' = P_Y/(P_X + P_Y)$. The partial vapor pressures, P_X and P_Y, are determined by the composition of the liquid solution according to Raoult's law (see Eq. 4.2). Since the solution boils when the sum of the partial vapor pressures of X and Y is equal to the external pressure, as expressed by Dalton's law (see Eq. 4.4), the boiling temperature of the solution is determined by its composition.

$$\frac{N_X'}{N_Y'} = \frac{P_X'}{P_Y'} = \frac{P_X^\circ N_X}{P_Y^\circ N_Y} \tag{4.5}$$

The relationship between temperature and the composition of the liquid and vapor phases of ideal binary solutions is illustrated by the diagram in Figure 4.3 for mixtures of benzene, bp 80 °C (760 torr), and toluene, bp 111 °C (760 torr). The lower curve is called the **liquid line** and gives the boiling points of all mixtures of these two compounds. The upper curve, the **vapor line,** is calculated using Raoult's law and defines the composition of the vapor phase in equilibrium with the boiling liquid phase at the same temperature. For example, a mixture whose composition is 58 mol % benzene and 42 mol % toluene will boil at 90 °C (760 torr), as shown by point A in Figure 4.3. The composition of the vapor in equilibrium with the solution when it *first* starts to boil can be determined by drawing a horizontal line from the *liquid line* to the *vapor line;* in this case, the vapor has the composition 78 mol % benzene and 22 mol % toluene, as shown by point B in Figure 4.3. This is a *key* point, for it means that at any given temperature the *vapor phase is richer in the more volatile component than is the boiling liquid with which the vapor is in equilibrium.* This phenomenon provides the basis of fractional distillation.

When the liquid mixture containing 58 mol % benzene and 42 mol % toluene is heated to its boiling point, which is 90 °C (760 torr), the vapor formed initially contains 78 mol % benzene and 22 mol % toluene. If this first vapor is condensed, the condensate would also have this composition and thus would be much richer in benzene than the original liquid mixture from which it was distilled. After this vapor is removed from the original mixture, the liquid remain-

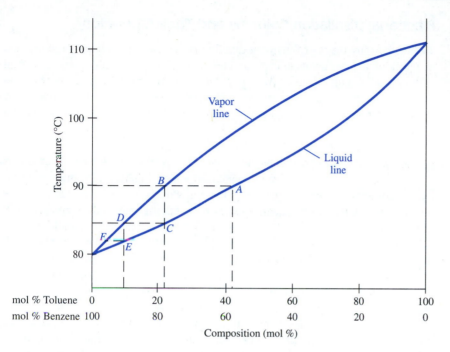

FIGURE 4.3   Temperature–composition diagram for a binary mixture of benzene and toluene.

ing in the stillpot will contain a smaller mole % of benzene and a greater mole % of toluene because more benzene than toluene was removed by vaporization. The boiling point of the liquid remaining in the distilling flask will rise as a result. As the distillation continues, the boiling point of the mixture will steadily rise until it approaches or reaches the boiling point of pure toluene. The composition of the distillate will change as well and will ultimately consist of "pure" toluene at the end of the distillation.

Now let us return to the first few drops of distillate that could be obtained by condensing the vapor initially formed from the original mixture. This condensate, as noted earlier, has a composition identical to that of the vapor from which it is produced. Were this liquid to be collected and then redistilled, its boiling point would be the temperature at point *C,* namely 85 °C; this boiling temperature is easily determined by drawing a vertical line from the vapor line at point *B* to the liquid line at point *C,* which corresponds to the composition of the distillate initially obtained. The first distillate obtained at this temperature would have the composition *D,* 90 mol % benzene and 10 mol % toluene; this composition is determined from the intersection with the vapor line of the horizontal line from point *C* on the liquid line. In theory, this process could be repeated again and again to give a very small amount of pure benzene. Similarly, collecting the *last* small fraction of each distillation and redistilling it in the same stepwise manner would yield a very small amount of pure toluene. If larger amounts of the initial and final distillates were collected, reasonable quantities of materials could be obtained, but a large number of individual simple distillations would be required. This process would be extremely tedious and time-consuming. Fortunately, the repeated distillation can be accomplished almost automatically in a single step by using a **fractional distillation column,** the theory and use of which are described next.

Fractional Distillation Columns and Their Operation

There are many types of fractional distillation columns, but all can be discussed in terms of a few fundamental characteristics. The column provides a vertical path through which the vapor must pass from the stillpot to the condenser before being collected in the receiver (see Fig. 2.16). This path is significantly longer than in a simple distillation apparatus. As the vapor from the stillpot rises up the column, some of it condenses *in the column* and returns to the distilling flask. *If the lower part of the distilling column is maintained at a higher temperature than the upper part of the column,* the condensate will be partially revaporized as it flows down the column. The uncondensed vapor, together with that produced by revaporization of the condensate in the column, rises higher and higher in the column and undergoes a repeated series of condensations and revaporizations. This repetitive process is equivalent to performing a number of simple distillations *within* the column, with the vapor phase produced in each step becoming increasingly richer in the *more* volatile component; the condensate that flows down the column correspondingly becomes richer in the *less* volatile component.

Each step along the path *A-B-C-D-E-F* of Figure 4.3 represents a *single* ideal distillation. One type of fractional distillation column, the bubble-plate column, was designed to effect one such step for each **plate** it contained. This led to the description of the efficiency of any fractional distillation column in terms of its equivalency to such a column in **theoretical plates.** Another index of the separating efficiency of a fractional distillation column is the **HETP,** which stands for *h*eight *e*quivalent to a *t*heoretical *p*late and is the vertical length of a column that is necessary to obtain a separation efficiency of one theoretical plate. For example, a column 60 cm long with an efficiency of 30 plates has an HETP value of 2 cm. Such a column would usually be better for research purposes than a 60-plate column that is 300 cm long (HETP = 5 cm) because of the small liquid capacity and **hold-up** of the shorter column. Hold-up refers to the condensate that remains in a column during and after distillation. When small amounts of material are to be distilled, a column must be chosen that has an efficiency, HETP, adequate for the desired separation and also a moderate to low hold-up.

As stated earlier, equilibrium between liquid and vapor phases must be established in a fractional distillation column so that the more volatile component is selectively carried to the top of the column and into the condenser, where the vapor condenses into the distillate. After all of the more volatile component is distilled, the less volatile one remains in the column and the stillpot; the heat supplied to the stillpot is then further increased in order to distil the second component. The most important requirements for performing a successful fractional distillation are (a) intimate and extensive contact between the liquid and the vapor phases in the column, (b) maintenance of the proper temperature gradient along the column, (c) sufficient length of the column, and (d) sufficient difference in the boiling points of the components of the liquid mixture. Each of these factors is considered here.

a. The desired contact between the liquid and vapor phases can be achieved by filling the column with an inert material having a large surface area. Examples of suitable packing materials include glass, ceramic, or metal pieces. Figure 2.17a shows a Hempel column packed with Raschig rings, which are pieces of glass tubing approximately 6 mm long. This type of column will have from two to four theoretical plates per 30 cm of length, if the distillation is carried out sufficiently slowly to maintain equilibrium conditions. Another type of fractional distillation column is the Vigreux column (see Fig. 2.17b), which is useful for small-scale distillations of liquid where low hold-up is of paramount importance. A 30-cm Vigreux column will only have about 1 or 2 theoretical plates and consequently will be less efficient than the corresponding Hempel column. The Vigreux column has the advantage of a holdup of less than 1 mL as compared to 2–3 mL for a Hempel column filled with Raschig rings.

A particularly effective type of fractional distillation apparatus is the **spinning-band column,** which contains a helical band of stainless steel or Teflon rotating at a high speed within the column. This band forces the condensed liquid *down* the column but leaves a very thin film of liquid on the inside surface of the column. This provides for *rapid* and *efficient* equilibration between the liquid and vapor. A well-insulated spinning-band column, 60 cm in length, may have a hold-up of only about 0.2 mL and a rating of 125 theoretical plates. It can readily separate the components having a difference in boiling points of only 2 °C.

b. **Temperature gradient** refers to the difference in temperature between the top and bottom of the column. The maintenance of the proper temperature gradient within the column is particularly important for an effective fractional distillation. Ideally, the temperature at the bottom of the column should be approximately equal to the boiling temperature of the solution in the stillpot, and it should decrease continually in the column until it reaches the boiling point of the more volatile component at the head of the column. The significance of the temperature gradient may be visualized by reference to Figure 4.3, where the boiling temperature of the distillate decreases with each succeeding step, for example, *A* (90 °C) to *C* (85 °C) to *E* (82 °C).

The necessary temperature gradient from stillpot to stillhead will, in most distillations, be established *automatically* by the condensing vapors *if* the rate of distillation is properly adjusted. Frequently, this gradient can be maintained only by insulating the column with a material such as glasswool or, most effectively, with a silver-coated vacuum jacket around the outside of the column. Insulation helps reduce heat losses from the column to the atmosphere. Even when the column is insulated, an insufficient amount of vapor may be produced to heat the column if the stillpot is heated too slowly, so that little or no condensate reaches the head. This rate must then be increased, but it must be kept below the point where the column is flooded. A flooded column is characterized by a column or "plug" of *liquid* that may be observed within the distillation column, often at the joint between it and the stillpot.

Factors directly affecting the temperature gradient in the column are the rate of heating of the stillpot and the rate at which vapor is removed at the stillhead. If the heating is too vigorous or the vapor is removed too rapidly, the entire column will heat up almost uniformly, and there will be no fractionation and thus no separation of the volatile components. On the other hand, if the stillpot is heated too strongly and if the vapor is removed too slowly at the top, the column will flood with returning condensate. Proper operation of a fractional distillation column thus requires *careful* control of the heat supplied to the stillpot and of the rate at which the distillate is removed at the stillhead. This rate should be *no more than* 1 drop every 2 to 3 sec.

The ratio of the amount of condensate returning to the stillpot and the amount of vapor removed as distillate per unit time is defined as the **reflux ratio.** A reflux ratio of 10:1, for example, means that 10 drops of condensate return to the stillpot for each drop of distillate that is obtained. In general, the higher the reflux ratio the more efficient is a fractional distillation.

c. Correct column length is difficult to determine in advance of performing a fractional distillation. The trial-and-error technique must normally be used, and if a particular column does not efficiently separate a certain mixture, a longer column or a different type of column or column packing must be selected.

d. The difference in boiling points between the two pure components of a mixture should be no less than 20–30 °C in order for a fractional distillation to be successful when a Hempel column packed with Raschig rings or a similar type of packing is used. As mentioned previously, modifications in column length and type may result in the successful separation of mixtures having smaller boiling point differences.

In summary, the most important variables that can be controlled experimentally in a fractional distillation are correct selection of the column and column packing, adequate insulation of the column, and careful control of the rate of heating so as to provide the proper reflux ratio and a favorable temperature gradient within the column.

4.2.3 EXPERIMENTS INVOLVING SIMPLE AND FRACTIONAL DISTILLATION

The following experiments have been designed to demonstrate the techniques of simple and fractional distillation and the relative efficiencies of the different types of apparatus that are used to separate mixtures of volatile components. Experimental procedures are given for the purification of ethyl acetate by simple distillation and the fractional separation of a mixture containing ethyl acetate and *n*-butyl acetate, but other pure compounds and mixtures may be assigned by your instructor.

In the simple distillation of a single volatile component, the head temperature should rise to a temperature that corresponds to the normal boiling point of the pure liquid and should remain constant until the distillation is complete. In the fractional distillation of a binary liquid mixture, the head temperature ideally should rise to the normal boiling point of the more volatile component and remain there until that component is completely removed. The head temperature may then drop somewhat, indicating that the more volatile component has been removed. If additional heat is provided, the less volatile component will begin to distil, and the now higher head temperature should remain constant until all of the second component has distilled. In other words, two different temperature **plateaus** should be observed. This idealized process will not be observed *unless* the amount of heat supplied to the distilling flask is *very carefully* adjusted, and this requires constant attention during the course of the distillation.

The fractional distillation of a binary mixture, such as one containing ethyl acetate and *n*-butyl acetate, raises the question of the effectiveness of the separation. In the ideal situation, only two distillation fractions should be obtained—one composed of pure ethyl acetate, the other of pure *n*-butyl acetate. The actual composition of the fractions can be conveniently determined by gas chromatography (GLC, see Sec. 6.4). This simple technique permits analysis of the components present in each distillation fraction and, if more than one is present, the percentage of each.

In preparation for performing either type of distillation, read the discussion of the techniques for simple and fractional distillation in Sections 2.7 and 2.8, respectively, *before* coming to the laboratory.

Pre-Lab exercises for Section 4.2, "Simple and Fractional Distillation," are found on page PL. 7.

EXPERIMENTAL PROCEDURE

SAFETY ALERT

1. Preferably, use a heating mantle or an oil bath as the heat source (see Sec. 2.5). If it is necessary to use burners, exercise *great* care: both ethyl acetate and *n*-butyl acetate are highly flammable.

2. Examine your glassware for cracks and other weaknesses before assembling the distillation apparatus. Look with particular care for "star" cracks in round-bottom flasks, because these can cause a flask to break upon heating.

3. Proper assembly of glassware is important in order to avoid possible breakage and spillage and to avoid the release of distillate vapors into the room. Be certain that all connections in the apparatus are tight *before* beginning the distillation. Have your instructor examine your set-up after it is assembled.

4. The apparatus used in these experiments *must* be open to the atmosphere at the receiving end of the condenser. *Never heat a closed system.* If heated, a closed system will build up pressure that may cause the apparatus to explode.

5. Be certain that the water hoses are securely fastened to your condensers so that they will not pop off and cause a flood. If heating mantles or oil baths are to be used for heating in this experiment, water hoses that come loose may cause water to spray onto electrical connections, into the heating mantle, or into *hot* oil, each of which is potentially dangerous to you and to those who work around you.

6. Avoid excessive inhalation of organic vapors at all times.

The experiments described here utilize ethyl acetate and *n*-butyl acetate. Other pure substances or liquid mixtures may be assigned by your instructor; if so, different boiling temperatures will be observed, and distillation cuts must be taken at temperatures different from those given in the procedure.

A. Simple Distillation of a Pure Liquid

Before starting experimental work, read Section 2.1 concerning the precautions and correct handling of glassware. Place 10 mL of ethyl acetate in a 25-mL round-bottom flask and add two or three boiling chips, or, if magnetic stirring is available, a stirring bar to ensure smooth boiling. Assemble the simple distillation apparatus described in Section 2.7 and shown in Figure 2.15a. The position of the thermometer in the stillhead is particularly important; the *top* of the mercury thermometer bulb should be level with the *bottom* of the sidearm of the distillation head. Collect the distillate in a dry Erlenmeyer flask or a dry graduated cylinder of appropriate volume. Have your instructor check your apparatus *before* you start heating the distilling flask. Use the method of heating specified by your instructor; details of various heating methods are discussed in Section 2.5, which should be consulted *before* you begin heating.

Begin heating the stillpot, and as soon as the liquid begins to boil *and the condensing vapors have reached the thermometer bulb,* regulate the heat supply so that distillation continues steadily at a rate of *2 to 4 drops per second;* if a drop of liquid cannot be seen suspended from the end of the thermometer, the rate of distillation is too *fast.* As soon as the distillation rate is adjusted and the head temperature is constant, note and record the temperature. Continue the distillation and periodically record the head temperature. Discontinue heating when only 2–3 mL of ethyl acetate remains in the distillation flask. Determine and record the volume of distilled ethyl acetate that you obtain.

Optional Procedure

You may be required to perform this distillation using the short-path apparatus discussed in Section 2.7 and illustrated in Figure 2.15b. After assembling the equipment, make certain that the *top* of the mercury in the thermometer bulb is level with the *bottom* of the sidearm of the distillation head. The preferred way to collect the distillate in this distillation is to attach a dry round-bottom flask to the vacuum adapter and put a drying tube containing $CaCl_2$ on the sidearm of the adapter to protect the distillate from moisture. The receiver should be cooled in an ice-water bath to prevent loss of product by evaporation and to ensure complete condensing of the distillate.

Finishing Touches: Unless directed otherwise, return the distilled and undistilled *ethyl acetate* to a bottle marked "Recovered Ethyl Acetate."

Spectra of Product

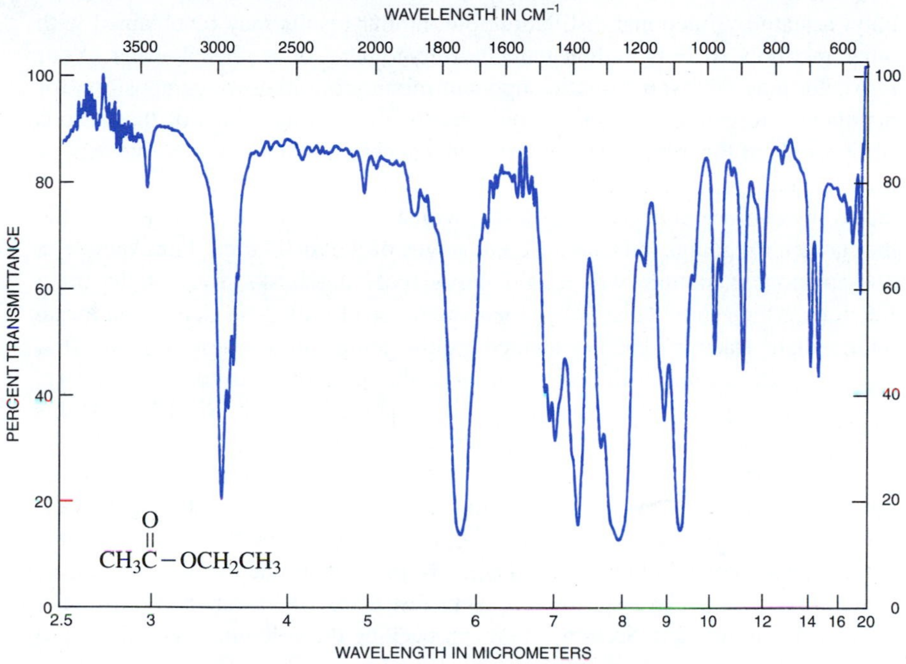

WAVELENGTH IN CM^{-1}

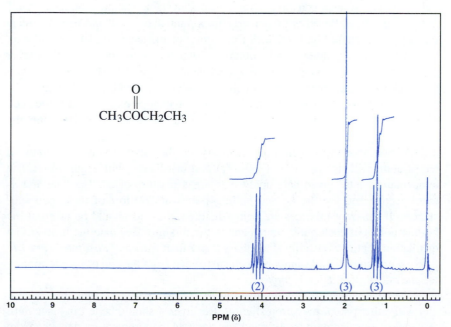

$$CH_3\overset{\displaystyle O}{\overset{\|}{C}} - OCH_2CH_3$$

WAVELENGTH IN MICROMETERS

FIGURE 4.4 IR spectrum of ethyl acetate.

(a) ^{1}H NMR spectrum (60 MHz).

$$CH_3\overset{\displaystyle O}{\overset{\|}{C}}OCH_2CH_3$$

PPM (δ)

(b) ^{13}C NMR data.
Chemical shifts: **δ** 14.3, 20.8, 60.4, 170.7.

FIGURE 4.5 NMR data for ethyl acetate.

B. Fractional Distillation of a Binary Mixture

This experiment provides for the separation of a mixture of ethyl acetate and *n*-butyl acetate by fractional distillation, but similar results may be obtained with other mixtures such as carbon tetrachloride–toluene or methanol–water. Your instructor may give you a multicomponent mixture of unknown composition for distillation. Regardless of your knowledge of the boiling points of the various components of the mixture to be separated, successful completion of this experiment depends on understanding that the fractions of distillate should be collected in *separate* dry receiving flasks, with flask changes occurring when the thermometer readings indicate different stages of the distillation. For example, a two-component mixture will require *three* receiving flasks, two for the main fractions, which should distil over a narrow range of only 2–4 °C, and one for an intermediate fraction that is obtained as the temperature rises from the first *plateau* to the second. If the composition of the mixture given to you is known, look up the respective boiling points in a suitable handbook. Read Section 2.8 for a discussion of the apparatus for fractional distillation.

Place 10 mL of ethyl acetate and 10 mL of *n*-butyl acetate in a 50-mL round-bottom flask, and add two or three boiling chips, or, if magnetic stirring is available, a stirring bar to ensure smooth boiling. Equip this flask for fractional distillation as shown in Figure 2.16. Pack a Hempel or similar distillation column, using the type of packing specified by your instructor. The technique for packing columns is discussed in Section 2.8. When packing the column, *be careful not to break off the glass support indentations at the base of the column.* Do not pack the column too tightly, for vapors cannot pass through a tightly packed column; heating a fractional distillation apparatus equipped with a column that is too tightly packed is analogous to heating a closed system. The position of the thermometer in the stillhead is particularly important; the *top* of the mercury thermometer bulb should be level with the *bottom* of the sidearm of the distillation head. Clean and dry three 25-mL containers, which may be bottles or Erlenmeyer flasks, for use as receiving flasks and label them *A, B,* and *C.* Have your instructor check your assembled apparatus before supplying heat to the distillation flask.

Place receiver *A* so that the tip of the vacuum adapter extends inside the neck of the container to minimize evaporation of the distillate. Begin heating the stillpot using the heating method specified by your instructor. See Section 2.5 for a discussion about various heating methods. As the mixture is heated, the head temperature will rise to 81 °C (760 torr) and distillation will begin. *Note:* The temperature "cuts" given here apply only for a mixture of ethyl acetate and *n*-butyl acetate and are the normal boiling points at 760 torr of these two substances. If different mixtures are being distilled, the cuts should be taken at the boiling points of each pure component of the mixture. Regulate the heat so that distillation continues steadily at a rate *no faster* than *1 drop of distillate every 1 to 2 sec;* if a drop of liquid cannot be seen suspended from the end of the thermometer, the rate of distillation is *too fast.*

The head temperature will remain at 81 °C for a period of time, but eventually will either rise or may drop slightly. Receiver *A* should be left in place until this increase or decrease is observed. As soon as the temperature deviates from 81 °C by more than ±3 °C, change to receiver *B* and increase the amount of heat supplied to the distillation flask. The temperature will again start to rise, and more liquid will distil. Leave receiver *B* in place until the temperature reaches 123 °C, and then change to receiver *C*. Continue the distillation until 1–2 mL of liquid remains in the pot, and then discontinue heating. Measure the volumes of the distillate collected in each receiver by means of a graduated cylinder, and record them. Allow the liquid in the column to drain into the distillation flask, then measure and record the volume of this pot residue.

   If instructed to do so, submit or save 0.2-mL samples of each fraction *A*, *B*, and *C* for GLC analysis (see Sec. 6.4). Columns containing either silicone gum rubber or SF-96 as the stationary phase give good separation of ethyl acetate and *n*-butyl acetate.

1. On the basis of the relative volumes of fractions *A*, *B*, and *C*, indicate what can be learned about the efficiency of the separation of the mixture.

2. If GLC analysis of the three fractions is available, calculate the percentage of ethyl acetate and *n*-butyl acetate in each fraction. What can be said regarding the purity of the three fractions?

3. Based on the volume and composition of the distillate collected in each fraction, does fractional distillation appear to provide a good method for separating ethyl acetate and *n*-butyl acetate?

> **Finishing Touches**   Unless directed to do otherwise, pour the *pot residue* into the container for nonhalogenated organic liquids and return the *distillation fractions* to a bottle marked "Recovered Ethyl and *n*-Butyl Acetate."

SPECTRA OF PRODUCTS

The IR and NMR spectra of ethyl acetate are presented in Figures 4.4 and 4.5, respectively.

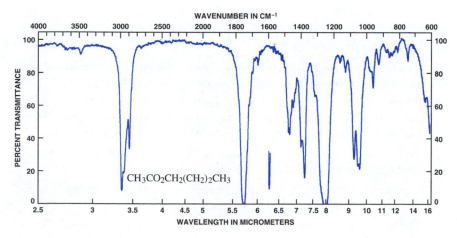

$CH_3CO_2CH_2(CH_2)_2CH_3$

FIGURE 4.6   IR spectrum of *n*-butyl acetate.

FIGURE 4.7 NMR data for *n*-butyl acetate.   **(a)** ^{1}H NMR spectrum (90 MHz).

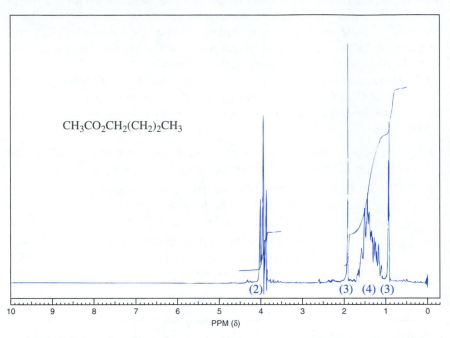

CH$_3$CO$_2$CH$_2$(CH$_2$)$_2$CH$_3$

(2)   (3) (4) (3)

PPM (δ)

(b) ^{13}C NMR data.
Chemical shifts: δ 13.8, 19.4, 20.8, 31.1, 64.2, 170.6.

EXERCISES

1. Define the following terms:

   a. simple distillation
   b. fractional distillation
   c. head temperature
   d. pot temperature
   e. Raoult's law
   f. ideal solution
   g. mole fraction
   h. height equivalent to a theoretical plate (HETP)
   i. temperature gradient
   j. Dalton's law
   k. reflux ratio

2. Sketch and completely label the apparatus required for (a) simple distillation and (b) fractional distillation.

3. Explain why a packed fractional distillation column is more efficient than an unpacked column for separating two closely boiling liquids.

4. If heat is supplied to the stillpot too rapidly, the ability to separate two liquids by fractional distillation may be drastically reduced. In terms of the theory of distillation presented in the discussion, explain why this is so.

5. Explain why the column of a fractional distillation apparatus should be aligned as near to the vertical as possible.

6. Explain the role of the boiling chips or magnetic stirring bar normally added to a liquid that is to be heated to boiling.

7. The top of the mercury bulb of the thermometer placed at the head of a distillation apparatus should be adjacent to the exit opening to the condenser.

Explain the effect on the observed temperature reading if the bulb is placed (a) below the opening to the condenser or (b) above the opening.

8. **a.** A mixture of 80 mol % *n*-propylcyclohexane and 20 mol % *n*-propylbenzene is distilled through a simple distillation apparatus; assume that no fractionation occurs during the distillation. The boiling temperature is found to be 157 °C (760 torr) as the first small amount of distillate is collected. The standard vapor pressures of *n*-propylcyclohexane and *n*-propylbenzene are known to be 769 torr and 725 torr, respectively, at 157.3 °C. Calculate the percentage of each of the two components in the first few drops of distillate.

   b. A mixture of 80 mol % benzene and 20 mol % toluene is distilled under exactly the same conditions as in Part a. Using Figure 4.3, determine the distillation temperature and the percentage composition of the first few drops of distillate.

   c. The normal boiling points of *n*-propylcyclohexane and *n*-propylbenzene are 156 °C and 159 °C, respectively. Compare the distillation results in Parts *a* and *b*. Which of the two mixtures would require the more efficient fractional distillation column for separation of the components? Why?

9. Examine the boiling-point–composition diagram for mixtures of toluene and benzene given in Figure 4.3.

   a. Assume you are given a mixture of these two liquids of composition 80 mol % toluene and 20 mol % benzene and that it is necessary to effect a fractional distillation that will afford at least some benzene of greater than 99% purity. What would be the *minimum* number of theoretical plates required in the fractional distillation column chosen to accomplish this separation?

   b. Assume that you are given a 20-cm Vigreux column having an HETP of 10 cm in order to distil a mixture of 58 mol % benzene and 42 mol % toluene. What would be the composition of the first small amount of distillate that you obtained?

10. At 100 °C, the vapor pressures for water, methanol, and ethanol are 760, 2625, and 1694 torr, respectively. Which compound has the highest normal boiling point and which the lowest?

11. At 50 °C, the vapor pressures for methanol and ethanol are 406 and 222 torr, respectively. Given a mixture at 50 °C that contains 0.1 mol of methanol and 0.2 mol of ethanol, compute the partial pressures of each liquid and the total pressure.

12. Figure 4.8 shows a temperature–composition diagram for a mixture of carbon tetrachloride and toluene. Answer the following questions, assuming that these substances form an ideal solution:

   a. What are the boiling points of pure carbon tetrachloride and toluene?

   b. Answer the following questions for a mixture that contains 50 mol % carbon tetrachloride and 50 mol % toluene:

   i. At what temperature will it boil?

   ii. What is the composition, in mole %, of the first small amount of vapor that forms?

   iii. If the vapor in (ii) is condensed and the resulting liquid is heated to its boiling point, what is the composition, in mole %, of the new vapor?

c. A mixture of unknown composition is heated until it just starts to boil, and the composition of the resulting vapor is found to be 50 mol % toluene and 50 mol % carbon tetrachloride. What is the percent composition of the original mixture, and at what temperature does the mixture start to boil?

d. Calculate the amount of carbon tetrachloride and toluene in the liquid and vapor phases in a boiling mixture of the two at 85 °C and at 100 °C. Express your answer both in mole percent and in mole fraction.

13. The data given in the table below were obtained from separate distillations on 22 mL of a mixture containing equal volumes of ethyl acetate and *n*-butyl acetate. The following types of distillation apparatus were used: simple distillation, fractional distillation with a packed column, and fractional distillation with an unpacked column.

a. Graph these data on the same piece of graph paper and plot cumulative volume (horizontally) vs. head temperature (vertically).

b. With reference to the graph, which distillation was the most effective in separating the mixture and which the least effective?

c. Suggest the type of apparatus that was used for distillations *A, B,* and *C.*

Distillation Data

Total Accumulated Volume (mL)	Head Temperatures (°C) for Distillation (760 torr)		
	A	*B*	*C*
1	80	80	80
2	82	82	82
3	83	83	83
4	83	85	84
5	83	86	85
6	83	88	86
7	83	89	87
8	83	92	89
9	85	95	91
10	95	98	95
11	110	104	103
12	122	112	110
13	123	117	115
14	123	122	119
15	123	122	119
16	123	123	122
17	123	123	123
18	123	123	123
19	123	123	123
20	123	123	123

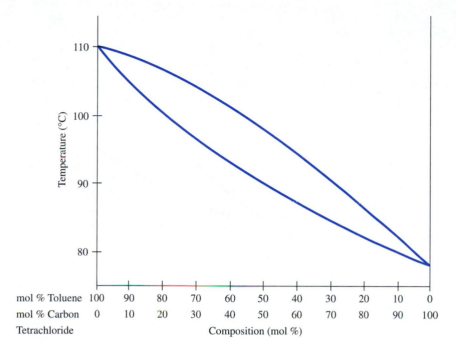

FIGURE 4.8 Temperature–composition diagram for Exercise 12.

4.2.4 FRACTIONAL DISTILLATION OF NONIDEAL SOLUTIONS

Most homogeneous organic liquid mixtures behave as ideal solutions, but examples are known in which the behavior is not ideal. In these solutions, the unlike molecules are not indifferent to the presence of one another. This results in deviations from Raoult's law in either of two directions: Some solutions display vapor pressures greater than expected and are said to exhibit **positive deviations;** others display vapor pressures lower than expected and are said to exhibit **negative deviations.** The discussions that follow are limited to binary mixtures, although some three-component systems exhibit similar deviations.

In the case of positive deviations, the forces of attraction between the molecules of the two components are *weaker* than those between the identical molecules of each component. The consequence of this is that over a certain range of compositions the combined vapor pressure of the two components is greater than the vapor pressure of the pure, more volatile component. Mixtures in this composition range, between *X* and *Y* in Figure 4.9, have boiling temperatures *below* that of either pure component. This mixture, which has composition *Z* in Figure 4.9, must be considered as if it were a third component. It has a constant boiling point because the vapor in equilibrium with the liquid has the same composition as the liquid itself. Consequently, this mixture is defined as a **minimum-boiling azeotropic mixture,** or **azeotrope.** Fractional distillation of such mixtures will *not* yield both of the components in pure form, only the azeotrope and the component present in excess of the azeotropic composition. For example, pure ethanol cannot be obtained by fractional distillation of aqueous solutions containing less than 95.57% ethanol, the azeotropic composition, even though the boiling point of this azeotrope is only 0.15 °C lower than the boiling point of pure ethanol. Since optimum fractional distillations of aqueous solutions containing less than 95.57% ethanol yield this azeotropic mixture, "95%

FIGURE 4.9 Temperature–composition diagram for a minimum-boiling azeotrope.

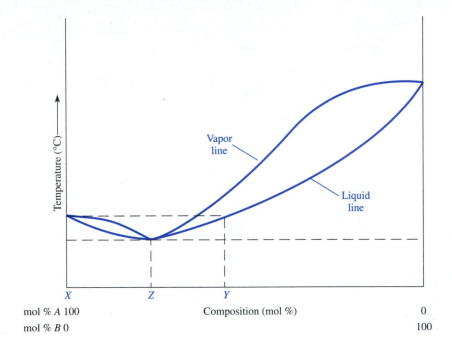

mol % A 100
mol % B 0

Composition (mol %)

0
100

ethyl alcohol" is the common composition of the commercial solvent ethanol. Pure or "absolute" ethanol can be obtained by chemically removing the water or by a distillation procedure that involves the use either of a ternary mixture of ethanol-water-benzene or of a vacuum (see below).

In the instance of a negative deviation from Raoult's law, the forces of attraction between the different molecules of the two components are *stronger* than those between identical molecules of each component. Here the result is that in a certain range of composition, the combined vapor pressure of the two components is less than the vapor pressure of the pure, less volatile component (Fig. 4.10). Thus, mixtures in the range between *X* and *Y* in Figure 4.10 boil at temperatures *higher* than the boiling temperature of *either* pure component. There is one particular composition, *Z* in Figure 4.10, that corresponds to a **maximum-boiling azeotrope.** Fractional distillation of mixtures of any composition other than that of the azeotrope will result in the distillation from the mixture of whichever of the two components is present in excess of the azeotropic composition *Z;* the composition of the pot residue will therefore approach that of *Z.*

As an illustration, formic acid and water form a maximum-boiling azeotrope having the composition 77.5% formic acid and 22.5% water. This azeotrope of pure formic acid, bp 100.7 °C (760 torr) and water, bp 100.0 °C (760 torr) boils at 107.3 °C (760 torr). The distillation of any mixture having other than the azeotropic composition will result in either water or formic acid, whichever is in excess, being distilled, and the azeotropic mixture will remain as the pot residue.

Data on other minimum- or maximum-boiling azeotropes may be found in the references at the end of the chapter, and extensive tables are given in the handbooks listed there. It is important to remember that azeotropes, like liquids, exhibit constant boiling points. As a result, they may be confused with pure compounds *if* boiling points are used as the sole criterion of purity. A similar source of confusion attends the use of melting points as the sole criterion of

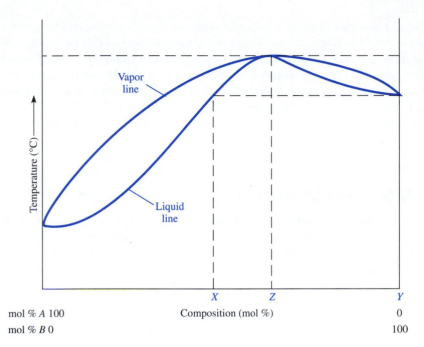

purity of solids; eutectic mixtures give sharp melting points, as do pure solids (see Secs. 3.3.2 and 3.3.3).

EXERCISES

1. Define the following terms:
   a. azeotrope
   b. minimum-boiling azeotrope
   c. maximum-boiling azeotrope
   d. positive deviation of vapor pressure
   e. negative deviation of vapor pressure

2. A certain liquid boils at a constant temperature, for example, 127 °C. Based on this boiling point alone, is it a pure liquid? Explain.

3. Suppose that equimolar amounts of water, bp 100 °C (760 torr), and ethanol, bp 78.32 °C (760 torr) are mixed and subjected to fractional distillation.
   a. The first liquid to distil boils at 78.17 °C (760 torr). Explain this result. What does this distillate contain?
   b. After all the material boiling at 78.17 °C (760 torr) has distilled, the next fraction to be collected boils at 100 °C (760 torr). No liquid boiling at 78.32 °C (760 torr) is collected. Explain this result.

4.2.5 VACUUM DISTILLATION

Both simple and fractional distillations are most conveniently performed at atmospheric pressure. As indicated in Section 4.1, however, the attempted distillation of some high-boiling liquids is often difficult or impossible, owing to thermal decomposition of the sample. This can usually be avoided by performing a simple or fractional distillation at reduced pressure so that the boiling temperature is reduced. The basic theory underlying this technique, termed *vacuum distillation,* is the same as that described earlier for simple and fractional distillation, with the exception that the "external pressure" is the pressure associated with the vacuum used, but in any case less than 760 torr. The effect of external pressure upon the boiling point of a pure liquid is discussed in Section 4.1. The experimental procedure for vacuum distillation is described in Section 2.9.

4.2.6 STEAM DISTILLATION

The separation and purification of *volatile* organic compounds that are immiscible or nearly immiscible with water can often be accomplished by steam distillation. The technique normally involves the codistillation of a mixture of organic liquids and water, although some organic solids can also be separated and purified by this means. Of the various distillation methods, steam distillation is utilized least frequently, owing to the rather stringent limitations on the types of substances for which it can be used. These limitations as well as the virtues of this technique are revealed by considering the physical principles underlying steam distillation.

Theory and Discussion

The partial pressure P_i of each component i of a mixture of *immiscible,* volatile substances at a given temperature is equal to the vapor pressure P_i° of the pure compound at the same temperature (Eq. 4.6) and does not depend on the mole fraction of the compound in the mixture; in other words, each component of the mixture vaporizes independently of the others. This behavior contrasts sharply with that exhibited by solutions of miscible liquids, for which the partial pressure of each constituent of the mixture depends on its mole fraction in the solution (Raoult's law, Eq. 4.2).

$$P_i = P_i^\circ \qquad (4.6)$$

Now recall that, according to Dalton's law (see Eq. 4.4), the total pressure, P_T, of a mixture of gases is equal to the sum of the partial pressures of the constituent gases, meaning that the total vapor pressure of a mixture of immiscible, volatile compounds is given by Equation 4.7. This expression indicates that the total vapor pressure of the mixture at any temperature is always greater than the vapor pressure of even the most volatile component at that temperature, owing to the contributions of the vapor pressures of the other constituents in the mixture. The boiling temperature of a mixture of immiscible compounds must then be *lower* than that of the lowest-boiling component.

$$P_T = P_a^\circ + P_b^\circ + \cdots + P_i^\circ \qquad (4.7)$$

Application of the principles just outlined is seen in an analysis of the steam distillation of an immiscible mixture of water, bp 100 °C (760 torr), and bromobenzene, bp 156 °C (760 torr). Figure 4.11 is a plot of the vapor pressure versus temperature for each pure substance and for a mixture of these compounds.

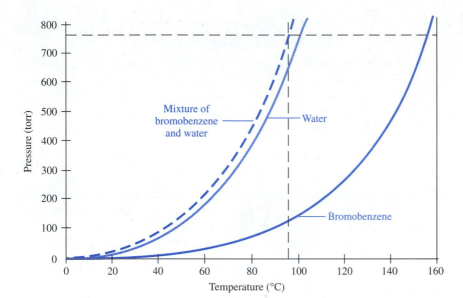

Analysis of this graph shows that the mixture should boil at about 95 °C (760 torr), the temperature at which the total vapor pressure equals standard atmospheric pressure. As theory predicts, this temperature is below the boiling point of water, which is the lowest-boiling component in this example.

The ability to distil a compound at the relatively low temperature of 100 °C or less by means of a steam distillation is often of great use, particularly in the purification of substances that are heat-sensitive and would therefore decompose at higher temperatures. It is useful also in the separation of compounds from reaction mixtures that contain large amounts of nonvolatile residues such as inorganic salts.

The composition of the condensate from a steam distillation depends upon the molecular weights of the compounds being distilled and upon their respective vapor pressures at the temperature at which the mixture steam-distils. To illustrate this, consider a mixture of two immiscible components, *A* and *B*. If the vapors of *A* and *B* behave as ideal gases, the ideal gas law can be applied and Equations (4.8a) and (4.8b) are obtained. In these two expressions, $P°$ is the vapor pressure

$$P_A°V_A = (g_A/M_A)(RT) \tag{4.8a}$$

$$P_B°V_B = (g_B/M_B)(RT) \tag{4.8b}$$

of the pure liquid, *V* is the volume in which the gas is contained, *g* is the weight in grams of the component in the gas phase, *M* is its molecular weight, *R* is the gas constant, and *T* is the absolute temperature in kelvins (K). Dividing Equation (4.8a) by Equation (4.8b) gives Equation (4.9).

$$\frac{P_A°V_A}{P_B°V_B} = \frac{g_A/M_A(RT)}{g_B/M_B(RT)} \tag{4.9}$$

Because the *RT* factors in the numerator and denominator are identical and because the volume in which the gases are contained is the same for both ($V_A = V_B$), these terms in Equation (4.9) cancel to yield Equation (4.10).

$$\frac{\text{grams of } A}{\text{grams of } B} = \frac{(P_A^\circ)(\text{molecular weight of } A)}{(P_B^\circ)(\text{molecular weight of } B)} \tag{4.10}$$

Now let the immiscible mixture of A and B consist of bromobenzene and water, whose molecular weights are 157 g/mol and 18 g/mol, respectively, and whose vapor pressures at 95 °C, as determined from Figure 4.11, are 120 torr and 640 torr, respectively. The composition of the distillate at this temperature can be calculated from Equation (4.10) as shown in Equation (4.11). This calculation

$$\frac{g_{\text{bromobenzene}}}{g_{\text{water}}} = \frac{(120)(157)}{(640)(18)} = \frac{1.64}{1} \tag{4.11}$$

indicates that on the basis of weight *more* bromobenzene than water is contained in the steam distillate, even though the vapor pressure of the bromobenzene is *much* lower at the temperature of the distillation.

Organic compounds generally have molecular weights much higher than that of water, so it is possible to steam-distil compounds having vapor pressures of only about 5 torr at 100 °C with a fair efficiency on a weight-to-weight basis. Thus, *solids* that have vapor pressures of at least this magnitude can even be purified by steam distillation. Examples are camphor, used in perfumes, and naphthalene, present in some brands of mothballs; the rather high vapor pressures of these solids is evidenced by the fact that their odors can be detected at room temperature.

In summary, steam distillation provides a method for separating and purifying moderately volatile liquid and solid organic compounds that are insoluble or nearly insoluble in water from nonvolatile compounds. Although relatively mild conditions are used in steam distillation, it cannot be used for substances that decompose on prolonged contact with steam or hot water, that react with water, or that have vapor pressures of 5 torr or less at 100 °C, all of which are significant limitations to the method. The technique for carrying out a steam distillation is presented in Section 2.10, and an experiment that uses this method is given in Section 4.2.7.

Steam distillation involves passing steam through a mixture of one or more volatile compounds that are immiscible with water and collecting the resulting distillate, which contains water and the desired compound(s). Some procedures involve the isolation of a desired product using a form of steam distillation in which the steam is *not* obtained from an external source, such as a steam line, but is generated internally. Several preparative experiments appearing later in this text utilize this form of steam distillation: 1-bromobutane (see Sec. 14.2), cyclohexene (see Sec. 10.2), and aniline (see Sec. 20.2). In each of these syntheses, a mixture containing water, the desired compound, and nonvolatile impurities is subjected to distillation, yielding a distillate that consists of water and the desired compound in moderately pure form.

4.2.7 NATURAL PRODUCTS; STEAM DISTILLATION OF CITRAL FROM LEMON GRASS OIL

The naturally occurring organic compounds found in and produced by living organisms have fascinated human beings for centuries. Through the years, so-called "natural products" have attracted attention for a variety of reasons, ranging from their practical applications in daily life to the scientific challenges they present. Such compounds have been used, for example, to alleviate pain, to cure diseases, to make colorful dyes, to flavor foods, and even to cause death. From the standpoint of scientific interest, natural products, because of their ready availability, provided chemists with many interesting experimental problems during the period when chemistry was emerging from alchemy into a more exact science. The isolation, identification, and synthesis of natural products remains a target of the research efforts of modern-day organic chemists.

Historically, most natural products have been extracted from plants rather than from animals. However, microorganisms are now assuming an increased importance as sources of natural products, particularly since the advent of molecular biology. The isolation of a natural product in pure form normally represents a significant undertaking for the experimentalist, mainly because even the simplest plants and microorganisms contain a multitude of organic compounds, often in only minute amounts.

The general approach used to isolate and purify a natural product may be summarized as follows: The plant or microorganism is ground or homogenized into fine particles, and the resulting material is then extracted with a solvent or mixture of solvents in which the desired natural product is expected to dissolve. An example is the extraction of trimyristin (**1**) from nutmeg using diethyl ether (see Sec. 5.3). Volatile natural products in the extract can be detected and possibly even isolated by gas chromatography. Often, though, the natural product is relatively nonvolatile, and removal of the solvent used for the extraction yields an oil or gum that requires further separation in order to obtain its various components in pure form.

$$
\begin{array}{l}
\mathrm{CH_2-O-\overset{\overset{\displaystyle O}{\|}}{C}-(CH_2)_{12}CH_3} \\[2pt]
\mathrm{\;|\qquad\qquad\;\; \overset{\overset{\displaystyle O}{\|}}{} } \\[2pt]
\mathrm{CH-O-\overset{\overset{\displaystyle O}{\|}}{C}-(CH_2)_{12}CH_3} \\[2pt]
\mathrm{\;|\qquad\qquad\;\; \overset{\overset{\displaystyle O}{\|}}{} } \\[2pt]
\mathrm{CH_2-O-\overset{\overset{\displaystyle O}{\|}}{C}-(CH_2)_{12}CH_3}
\end{array}
$$

1

Various chromatographic techniques are also used to isolate and purify natural products. These include thin- and thick-layer, liquid-liquid, and gas-liquid chromatography. Chapter 6 contains descriptions of many of these techniques and experiments illustrating their use are also found in that chapter.

The next stage facing the natural products chemist is determination of the structure of the isolated pure product. Traditional approaches to structural elucidation such as qualitative tests for various functional groups (see Chap. 23), and chemical degradations to known, simpler compounds, were and remain of great importance. More recently, spectroscopic techniques such as mass spectrometry, x-ray diffraction, and ultraviolet, infrared, and nuclear magnetic resonance spectroscopy have greatly facilitated the determination of structure. The theory and application of various spectroscopic methods are described in Chapter 8.

In many instances, the final goal of chemists working in this field is the total laboratory synthesis of a natural product. The synthesis of some natural products represents mainly an intellectual challenge and/or an opportunity to develop and demonstrate the utility of new synthetic techniques. In cases where the natural product has medicinal uses, the development of an efficient synthetic route may be of great importance, especially when the supply of the material from the natural source is severely limited or its isolation is difficult. Chapter 20 provides an example of a multistep synthesis of the medically important compound sulfanilamide (**2**), an antibiotic but not a natural product.

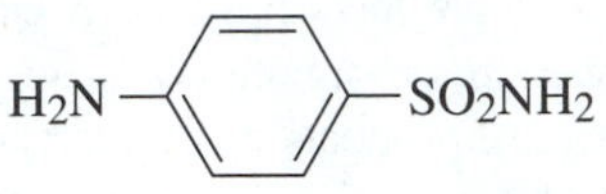

2

The experimental procedure of this section describes the isolation of citral (**3**), a natural product that possesses a pleasant lemonlike odor and taste. It is interesting that citral evokes pleasant odor and taste responses in humans, whereas it is less attractive to other forms of life. For example, certain insects such as ants are known to secrete citral in order to ward off potential predators; that is, **3** serves as a defense **pheromone.** Citral is of commercial importance as a constituent of perfumes in which a lemonlike essence is desired; it is also used as an intermediate for the synthesis of vitamin A (**4**).

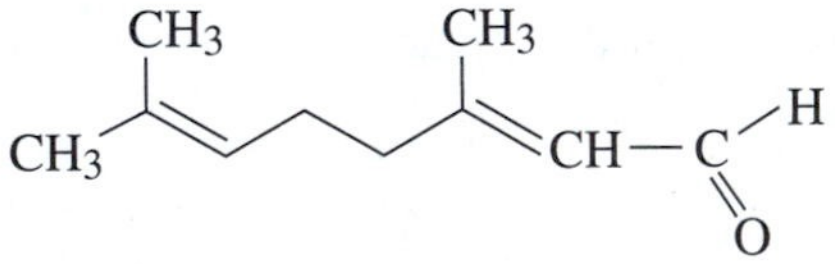

3

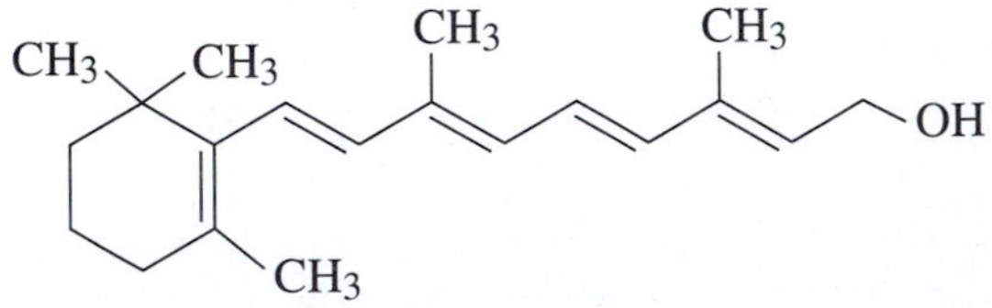

4

The commercial importance of citral has stimulated an extensive search for its presence in natural products. One source is the oil from the skins of lemons and oranges, although it is only a minor component of this oil. However, citral is the major constituent of the oil obtained from lemon grass, and in fact 75–85% of crude oil derived from pressing lemon grass is the desired natural product.

The structure of **3** contains two carbon-carbon double bonds, one of which is conjugated with the carbon-oxygen double bond of the functional group called an aldehyde. Among the chemical reactions of citral are the facile oxidation of its aldehyde group to a carboxylic acid, a reaction typical of aldehydes, and polymerization, resulting from the presence of the conjugated π-system in the molecule. Consequently, citral is a chemically labile substance under conditions that induce its oxidation and/or polymerization. Heat, light, and reagents such as acids, bases, and oxidizing agents promote these types of reactions. The isolation of pure citral therefore presents a challenge to the experimentalist.

This task is simplified, however, by the fact that citral, bp 229 °C, 760 torr, is relatively volatile and has a low solubility in water. These properties make it a suitable candidate for steam distillation, a technique that allows distillation of citral from crude lemon grass oil at a temperature below 100 °C, which is far below its normal boiling point. Neutral conditions are maintained in steam distillation as is the partial exclusion of oxygen, so the possibility of oxidation and/or polymerization of citral is minimal. This emphasizes the value of steam distillation, namely, that it is often the method of choice when reactive, volatile substances are to be separated from nonvolatile and/or water-soluble contaminants.

The citral isolated in this experiment will actually be a mixture of two **diastereomers,** the **geometric isomers** geranial (**5a**) and neral (**5b**). It is extremely difficult to separate these two isomers, and this will not be attempted in the present procedure.

that induce its oxidation and/or polymerization. Heat, light, and reagents such as

The steam distillation can be performed by two different methods, as described in Section 2.10. One involves the use of an external steam source, which may be either a laboratory steam line or a steam generator, and passage of the steam through a mixture of lemon grass oil and water. This method has some advantages but is experimentally more difficult than an alternate method, which involves heating a mixture of lemon grass oil and water and collecting the distillate. Although this latter procedure is simpler from the experimental standpoint, there are some limitations to its application. For example, in the steam distillation of only slightly volatile substances, a large initial volume of water will be required, or water must be added as the distillation proceeds, perhaps by means of an addition funnel. Procedures using both external and internal generation of steam are provided here.

After steam distillation of the lemon grass oil, the distillate contains a mixture of citral and water, from which the citral is isolated by **extraction** (see Chap. 5) with diethyl ether. The ethereal extracts are dried by a **drying agent,** anhydrous calcium chloride, and the diethyl ether is removed by evaporation under vacuum. Mild conditions are used for the extraction and evaporation in order to minimize the oxidation and/or polymerization of citral.

Structure Determination for Citral

Various spectroscopic and chemical analyses can be performed on the citral that is isolated from lemon grass oil. These analyses are mentioned briefly here, and references are given to other parts of the text where they are discussed in more detail.

That the compound you isolated from lemon grass oil is citral can be confirmed using infrared (IR), nuclear magnetic resonance (NMR), and/or ultraviolet (UV) spectroscopy. Its IR and H NMR spectra are given in Figures 4.12 and 4.13, respectively, and its UV spectrum is presented in Figure 8.69. Citral may also be analyzed using gas chromatography (see Sec. 6.4), which allows the assessment of the nature and purity of the product that has been isolated.

Various chemical methods can be used to characterize citral. For example, qualitative tests for the presence of carbon-carbon double bonds and for the aldehyde group can be performed, and solid derivatives can be prepared by utilizing reactions of the aldehyde group (see Chap. 23).

Pre-Lab exercises for Section 4.2.7, "Steam Distillation of Citral," are found on page PL. 9.

EXPERIMENTAL PROCEDURE

SAFETY ALERT

1. Steam distillation involves the use of glassware that becomes very hot. Exercise care when handling hot glassware.
2. If a house steam line is used as a source of external steam, allow the condensed water in the line to blow out before attaching the hose to the water trap (see Fig. 2.23). Do this by holding the end of the hose in the sink or drain and slowly opening the steam valve until no more water is emitted. Do *not* point the hose into the air; hot water and/or steam may issue from the hose and cause burns to yourself and your neighbors.
3. When passing steam from a house line through the mixture of lemon grass oil and water, control the flowrate of steam carefully so that pressure does not build up in the apparatus. This minimizes the potential for glass joints to separate and release the steam uncontrollably or for the apparatus to explode.
4. If the flow of steam from a house line is interrupted, a partial vacuum will be created in the line and water trap, and the mixture of water and lemon grass oil will be pulled into the trap. If this occurs, disassemble the apparatus and pour the mixture from the trap back into the distillation flask before resuming the steam distillation.
5. Upon completion of the steam distillation, open the drain from the water trap and remove the steam inlet tube from the distilling flask before turning off the source of external steam. Otherwise, the hot liquid in the flask will be pulled into the water trap.
6. Be certain that the steam distillate is cooled below 30 °C before extracting it with diethyl ether; otherwise excessive pressure may develop in the separatory funnel and may blow out the stopper.
7. Diethyl ether is *extremely* flammable. Be certain that there are no flames in the vicinity during its use and its removal from the citral.

Option 1: Distillation Using an External Steam Source

Determine the weight of 5 mL of lemon grass oil, and then place it and 50 mL of water in a 125-mL round-bottom flask. Equip this flask for steam distillation as described in Section 2.10 and as shown in Figure 2.21. Steam may be obtained from a steam line, if this is available in your laboratory, or may be generated using the apparatus shown in Figure 2.22. Regardless of the steam source, a water trap (see Fig. 2.23) must be placed *between* the steam source and the steam distillation apparatus. Steam-distil the mixture as rapidly as possible, and continue the distillation until droplets of oil no longer appear in the distillate; approximately 60–80 mL of distillate should be collected. Drain the condensed water from the water trap whenever necessary; do not allow the water trap to fill up.✳ Proceed to "Isolation of Citral," below.

Option 2: Steam Distillation Using Internal Source of Steam

Determine the weight of 5 mL of lemon grass oil, and then place it and 125 mL of water in a 250-mL round-bottom flask. Equip this flask for fractional distillation (see Sec. 2.8), but *do not pack the distillation column.* Heat the flask to boiling with a burner or heating mantle, as directed by your instructor. Adjust the heat source so that the distillation proceeds as rapidly as possible, but avoid applying excess heat to the distillation flask, especially when the volume of water in the flask has been reduced below about 50 mL. Continue the distillation until oil droplets no longer appear in the distillate, which should occur after about 60–80 mL of distillate has been collected. Allow the distillate to cool to room temperature or below, using an ice-water bath if necessary.✳ Proceed to "Isolation of Citral," below.

Isolation of Citral

Transfer the cooled distillate to a separatory funnel and extract it with 25 mL diethyl ether; the proper use of a separatory funnel is described in Section 2.18. The funnel should be vented *frequently* to avoid the buildup of pressure. Note that if the separatory funnel has a capacity of only 125 mL, it may be necessary to split the distillate into two or more portions in order to perform the extraction. The original portion of diethyl ether can be used to extract the entire amount of distillate, however.

Transfer the organic layer from the separatory funnel to a 50-mL Erlenmeyer flask, and add about 0.5 g of anhydrous calcium chloride. Allow the ethereal solution to remain in contact with the drying agent until the organic layer is dry, as evidenced by observing that it is completely clear. If the experiment is stopped at this point, loosely stopper the flask and store it in a hood; *never leave flasks containing diethyl ether in your locker drawer.*✳ Decant (see Sec. 2.15) the dried organic solution into a 125-mL round-bottom flask, and evaporate the solvent under vacuum, using a water aspirator or the house vacuum (see Sec. 2.25). When vacuum is applied, the diethyl ether is likely to start foaming; if this occurs, reduce the vacuum until foaming stops. It will be advantageous to place the flask in a pan of water *at room temperature* during the evaporation of the diethyl ether. After the ether is completely removed, the pot residue is crude citral (**3**). Determine its weight, and calculate the percentage recovery of citral, based on the weight of the original sample of lemon grass oil.

If requested to do so, save or submit a sample of citral for GLC (see Sec. 6.4) or IR, ^{1}H NMR, and/or UV spectral (see Chap. 8) analysis.

Chemical Characterization of Citral

You may be asked to perform one or more of the following tests on citral.

1. Test for carbon-carbon double bond. Perform the tests for unsaturation described in Section 23.5.2.
2. Test for an aldehyde. Perform the chromic acid test outlined in Section 23.5.1c.
3. Prepare solid derivatives. Prepare the 2,4-dinitrophenylhydrazone (see Sec. 23.5.1e) and the semicarbazone (see Sec. 23.5.1f) of citral. The melting points of these derivatives are given below.

Compound	Melting Point of 2,4-Dinitrophenylhydrazone	Melting Point of Semicarbazone
5a	134–135 °C	164–165 °C
5b	171–172 °C	125–126 °C

SPECTRA OF PRODUCT

The UV spectrum of citral is presented in Figure 8.69.

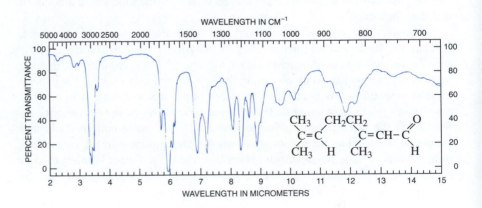

FIGURE 4.12   IR spectrum of citral.

(a) ^{1}H NMR spectrum (60 MHz).

FIGURE 4.13 NMR data for citral.

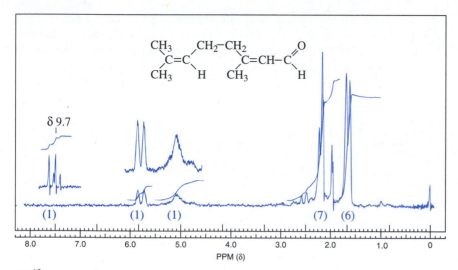

(b) ^{13}C NMR data.

Chemical shifts: δ 17.5, 17.7, 24.9, 25.6, 26.0, 27.2, 32.6, 40.6, 122.8, 123.0, 127.6, 128.8, 133.3, 163.0, 189.9, 190.5.

> **Finishing Touches** Flush the *aqueous solution* remaining in the distillation flask down the drain. Do the same with the *aqueous steam distillate* once you have completed the extraction. Pour the *solution from the test for unsaturation* into the container for halogen-containing liquids. Neutralize the *solution for the chromic acid test* and then pour it into the container for hazardous heavy metals.

EXERCISES

1. Define the term pheromone.

2. What type of product is expected from the reaction of citral (**3**) with Br$_2$/CCl$_4$? with chromic acid?

3. Provide structures for the semicarbazone and the 2,4-dinitrophenylhydrazone of citral.

4. What effect would there be on the melting point of a sample of the pure semicarbazone of **5a** if some of the semicarbazone from **5b** were added to the sample and the melting point of that mixture taken?

5. Why does the citral float on the surface of the aqueous distillate rather than sinking to the bottom?

6. Explain why the substitution of 1-propanol bp 97 °C (760 torr) for water in a steam distillation would not work.

7. Suppose that you are to steam-distil a sample of a natural product whose vapor pressure at 100 °C is known to be half that of citral. What conse-

quence would this have on the amount of distillate required per mole of the natural product present?

8. Both citral (**3**) and vitamin A (**4**) are members of the class of natural products called **terpenes.** This group of compounds have the common characteristic of being biosynthesized by linkage of the appropriate number of five-carbon units having the skeletal structure shown below. Determine the number of such units present in each of these terpenes and indicate the bonds linking the various individual units.

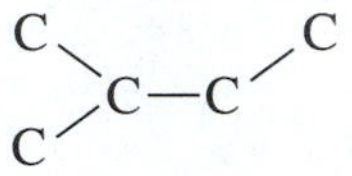

$$
\begin{array}{c}
\text{C} \\
\diagdown \\
\text{C}-\text{C} \\
\diagup \diagup \\
\text{C} \text{C}
\end{array}
$$

EXTRACTION

T he desired compound from a reaction is frequently part of a mixture, and
its isolation in pure form can be a significant experimental challenge. Two
of the more common methods for separating and purifying organic liquids and
solids are **recrystallization** and **distillation;** these procedures are discussed in
Chapters 3 and 4, respectively. Many other techniques are also available for
these purposes, two very important examples being **extraction** and **chromatography.** As you will see here and in Chapter 6, both of these methods involve
partitioning of compounds between two *immiscible* phases. This process is
termed **phase distribution** and can result in separation of compounds if they
distribute differently between phases.

Distribution of solutes between phases is the result of **partitioning** or
adsorption phenomena. Partitioning involves the difference in solubilities of a
substance in two immiscible solvents—in other words, *selective dissolution.*
Adsorption, on the other hand, is based on the selective attraction of a substance
in a liquid or gaseous mixture to the surface of a solid phase. The various chromatographic techniques depend on both of these processes, whereas the extraction process relies only on partitioning.

The process of extraction involves *selectively* removing one or more components of a solid, liquid, or gaseous mixture into a separate phase. The substance being extracted will distribute itself between the two immiscible phases
that are in contact. As would be expected, the ratio of its distribution between
the phases will depend on the relative solubility of the solute in each phase.

One of the most common methods for removing an organic compound from a
mixture is that of **liquid-liquid** extraction. This process is used by chemists not
only in the isolation of natural products but also in the isolation and purification
of products in virtually *every* organic synthesis.

Liquid-liquid extraction involves the distribution of a solute A between two
immiscible liquids, S_x, the **extracting phase,** and S_o the **original phase.** The
immiscible liquids normally encountered in the organic laboratory are water and
some organic solvent, such as diethyl ether, $(C_2H_5)_2O$, or dichloromethane,

CH_2Cl_2. At a given temperature the amount of A, in g/mL, in each phase is expressed *quantitatively* in terms of a constant, K, called the **distribution** or **partition coefficient** (Eq. 5.1). The process of liquid-liquid extraction can be considered a competition between two immiscible liquids for solute A, with solute A distributing between these two liquids when it is brought into intimate contact with them. The mathematical expression of Equation (5.1) shows that *at equilibrium* the ratio of concentrations of A in the two phases will always be *constant*. It is obvious from the equation that if A were completely dissolved in one of the solvents with none of A in the other, the value of K would be infinity or zero depending on the solvent in which the solute was dissolved. These limits are never obtained in practice.

$$K = \frac{[A] \text{ in } S_x}{[A] \text{ in } S_o} \qquad (5.1)$$

To simplify the following discussion, it will be assumed that solvent S_x is the solvent in which A has the *greater* solubility. The value of K, by definition, then will always be greater than 1.0.

A close approximation of the distribution coefficient K may be obtained by simply dividing the solubility of A in the extracting solvent by the solubility of A in the original solvent. A problem arises in doing this, however, because no two solvents are completely immiscible. The extent of their mutual solubilities alters their solvent characteristics and thus affects the experimental value of K. The less the mutual solubility of the two liquids, the closer the approximated value of K is to the actual one. Fortunately, even with solvents such as diethyl ether and water, whose mutual solubilities are significant (at 25 °C, the solubility of diethyl ether in water is 1.64 mol %, and that of water in diethyl ether is 5.15 mol %), the partition coefficients determined from solubility data of the solute in the pure solvents are usually satisfactory for most calculations.

Application of Equation (5.1) allows one to make a number of simple predictions. For example, given the assumption that $K > 1$, solute A will be present in greater quantity in the *extracting* solvent so long as the volume of $S_x \geq S_o$. The amount of solute remaining in S_o will depend on the value of K, of course. The equation can be recast into the form shown in Equation (5.2), since $[A] =$ grams of A/milliliter of solvent. This makes it obvious that when the volume of extracting solvent is equal to the volume of the original solvent, the value of K is simply the ratio of the number of grams of A in each of the solvents. It should be noted here that strictly speaking the *volumes* of *solution* should be used in the definition of $[A]$. However, if the solutions are dilute only slight errors result if *volumes* of *solvent* are used.

$$K = \frac{\text{grams of } A \text{ in } S_x}{\text{grams of } A \text{ in } S_o} \times \frac{\text{mL of } S_o}{\text{mL of } S_x} \qquad (5.2)$$

Since the product of the two ratios on the right side of Equation (5.2) *must* be constant for a given solvent pair, increasing or decreasing the value of one ratio will cause the opposite effect in the other. This means that increasing the amount of extracting solvent S_x will decrease the ratio, mL of S_o/mL of S_x, and have the net effect of increasing the amount of solute A in S_x.

It is clear from the preceding discussion that the more extracting solvent used, the more of the solute will be dissolved in that phase. Nonetheless, there are practical, economic, and environmental reasons that limit the amount of organic solvents that can realistically be utilized in extractions. The most efficient use of a given volume of extracting solvent thus becomes an important practical consideration when performing an extraction. The issue is basically whether it is better to carry out a *single* large extraction or *several* smaller ones. To answer this question, let us consider two scenarios in which it is assumed that an organic compound is dissolved in 10 mL of water and that there is a limit of 30 mL of diethyl ether allowed for the extraction. Two different values of the distribution coefficient *K* as defined in Equation (5.1) will be considered. In the first, *K* will be set at 2, in the second, at 5. Both values reflect a greater solubility of the compound in diethyl ether than in water, which is generally true for organic substances, but the smaller one would apply to relatively few cases because a fairly high solubility of the compound in water is required to give such a low equilibrium constant. Considering both scenarios, would it be more effective to do a *single* extraction or to carry out *three* successive extractions with 10-mL portions of ether?

Application of Equation (5.2) would provide the desired answer but would be tedious for the case of multiple extractions. Equation (5.3), which can be derived from Equation (5.2), accommodates multiple extractions in terms of the fraction of solute F_A remaining in the original solvent of volume V_o after *n* extractions, using volume V_x of an immiscible solvent for each extraction. In this expression, C_f is the *final* concentration of *A* in S_o, and C_i its *initial* concentration in S_o. Carrying out the calculations for the case of *K* = 2, we obtain 1/7 when a single extraction is performed, and 1/27 when three are done. This means that 14% of *A* remains in the aqueous phase when a *single* extraction is performed, whereas the value drops to only 4% in the case of *three* successive extractions with the same total volume of solvent. The amount of *A* that could be isolated is increased by some 10% with multiple extractions. Setting *K* at 5 gives corresponding values for F_A of 1/16 and 1/216, respectively, which translates to a 5.5% increase in the quantity of *A* that is removed from the aqueous layer with multiple extractions (see Exercise 1, Sec. 5.3).

$$F = \frac{C_f}{C_i} = \left(\frac{V_o}{V_o + KV_x}\right)^n \qquad \text{(5.3)}$$

From the standpoint of percentage recovery of product, it clearly is more efficient to perform successive extractions with smaller portions of extracting solvent than to do a single extraction using the same total amount of solvent, *regardless* of the value of *K*. The closer *K* is to 1, or in other words, the more water-soluble is the desired compound, the more advantageous multiple extractions become. From a practical standpoint, however, the relatively small increase in recovery may not justify the additional time required to do multiple extractions unless the product is of great value. The experimental procedures we use typically involve only one or two extractions, which reflects the fact that the partition coefficient *K* is generally large.

The selection of the appropriate extracting solvent is obviously a key to the success of this technique of isolating and purifying compounds, and important guidelines for making the choice are summarized here.

1. The extracting solvent *must not react* in a chemically irreversible way with the components of the mixture.

2. The extracting solvent *must be immiscible*, or nearly so, with the original solution.

3. The extracting solvent *must selectively remove* the desired component of the solution being extracted. That is, the distribution coefficient K of the component being removed must be *high* while the distribution coefficients of all other components should be *low*.

4. The extracting solvent *should be readily* separable from the solute. Use of a volatile solvent is advantageous because the solvent often is removed by simple distillation.

5.3 EXTRACTION OF NATURAL PRODUCTS

The historical importance of the isolation and characterization of natural products to the development of organic chemistry is discussed in connection with the technique of steam distillation (see Sec. 4.2.6), and it is shown how citral can be obtained from lemon grass oil by this technique (see Sec. 4.2.7). As noted in that section, solvent extraction is another valuable method for obtaining a desired substance from its natural source. A familiar example of this is the hot-water extraction from coffee beans and tea leaves of caffeine (**1**) and the various oils that constitute the flavor components of freshly brewed coffee and tea. Because most of the compounds we wish to isolate are *insoluble* in water, however, a variety of organic solvents such as diethyl ether, dichloromethane, ethanol, and acetone are used for the extraction.

1
Caffeine

As mentioned in Section 4.2.7, extraction of natural materials often produces complex mixtures of products, and additional procedures are required to separate and purify the individual components. These may involve techniques in which the components are separated on the basis of their acidic or basic properties (see Sec. 5.4) or by chromatographic methods (see Chap. 6). Thus, the isolation of pure natural products normally involves a series of complex and tedious operations. Fortunately there are some exceptions to this general rule. The isolation of trimyristin (**2**) from nutmeg is one of them.

Trimyristrin occurs in many vegetable fats and oils, most notably those from coconuts and nutmeg. In fact, it is the principal lipid that can be isolated from the latter source, a hard, aromatic-smelling seed of an East Indian tree (*Myristica fragrans*). Examination of the structure of this compound reveals it to be the triester of glycerol (**3**) and the fatty acid myristic acid (**4**); an alternate name of **2**, glyceryl trimyristate, reflects this combination of substituents. Naturally occurring fats and oils in general are esters of **3** and straight-chain carboxylic acids, the most common of which contain fourteen to twenty carbon atoms. Indeed, the distinction of whether such a material is called a fat or an oil simply depends on whether it is a solid or liquid at room temperature! The lack of any double bonds in the carbon chain of myristic acid (**4**) makes **2** a member of the family of saturated fats, which allegedly increase the risk of heart disease if present in excessive amounts in the diet.

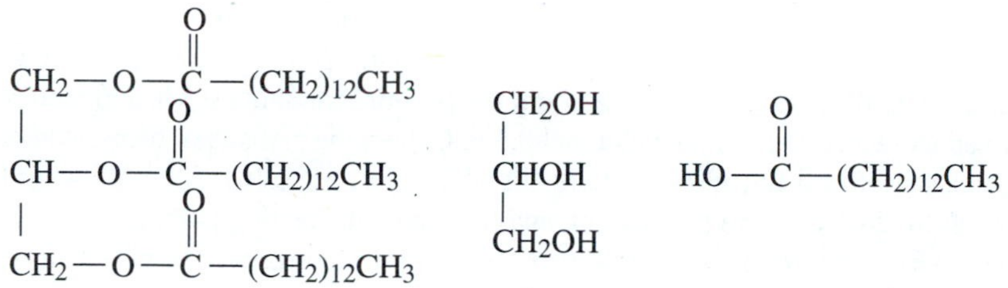

	2		3		4
	Trimyristin		Glycerol		Myristic acid

Nutmeg, the source of trimyristin in this experiment, has been a valued spice ever since its discovery in the Spice Islands of Indonesia by Portuguese sea captains over four centuries ago. The chemical makeup of nutmeg is somewhat unusual in that extraction of the ground seeds with diethyl ether yields **2** in high purity without contamination by other structurally related esters of glycerol and fatty acids. Thus, the procedure described here does *not* typify that required for isolation of most natural products, but its simplicity makes it suitable for demonstrating the general technique of extraction of biological materials from natural sources.

Pre-Lab exercises for Section 5.3, "Isolation Trimyristin from Nutmeg," are found on page PL. 11.

EXPERIMENTAL PROCEDURE

SAFETY ALERT

1. Diethyl ether and acetone are both highly volatile and flammable solvents. *Use no flames or electrical dryers in the laboratory during this experiment.*
2. Monitor the heating of diethyl ether under reflux so that the ring of condensate remains *in the lower third of the condenser.* This will ensure that no vapors escape from the condenser into the room. Very little heat will be required whether one is using steam, a hot-water or an oil bath to maintain reflux.

Place about 4 g of ground nutmeg in a 50-mL round-bottom flask containing a few boiling stones or a magnetic stirring bar and add 10 mL of diethyl ether. If the nutmeg is available only as the whole seed, the seed should first be ground with the aid of a mortar and pestle or a blender. Attach a water-cooled condenser to the flask and, using a *flameless* source of heating, heat the mixture under gentle reflux for approximately 0.5 h. Allow the contents of the flask to cool to room temperature. Filter the mixture through a fluted filter paper (see Sec. 2.11) into a 25-mL round-bottom flask, rinsing any residue remaining in the 50-mL flask onto the filter paper with an additional 2–4 mL of diethyl ether.

Equip the flask containing the filtrate for simple distillation, and remove the diethyl ether by distillation; again use *flameless* heating for this step! Dissolve the resulting yellow oil in 3–5 mL of acetone while warming the mixture over a steam or hot-water bath. Immediately pour the hot solution into a 25-mL Erlenmeyer flask. Allow the solution to stand at room temperature for about 0.5 h, and then cool it in an ice-water bath for an additional 0.5 h. Collect the crystals of trimyristin on a Büchner or Hirsch funnel by vacuum filtration (see Sec. 2.16). After allowing the crystals to air-dry, weigh the product, and determine its melting point (see Sec. 2.3). The reported melting point is 55–56 °C. Based on the original weight of nutmeg used, calculate the percent recovery. Submit the product to your instructor.

Finishing Touches   Once any residual diethyl ether has evaporated from the filter cake, the used *filter paper and its contents* should be placed in a container for nonhazardous solids. Any *diethyl ether* that has been isolated should be poured into a container for non-halogenated organic liquids. The *acetone-containing filtrate* from which the trimyristin is obtained can either be flushed down the drain or put in a container for nonhalogenated organic liquids.

SPECTRA OF PRODUCT

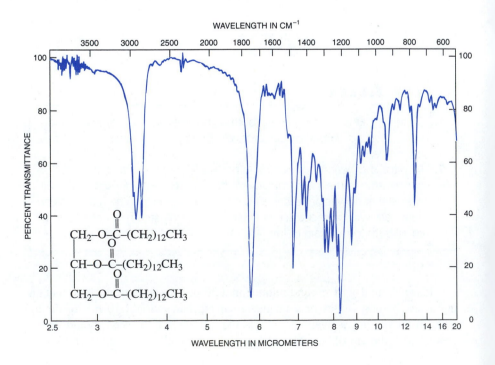

FIGURE 5.1   IR spectrum of trimyristin (KBr pellet).

(a) ^{1}H NMR spectrum (90 MHz).

FIGURE 5.2   NMR data for trimyristin.

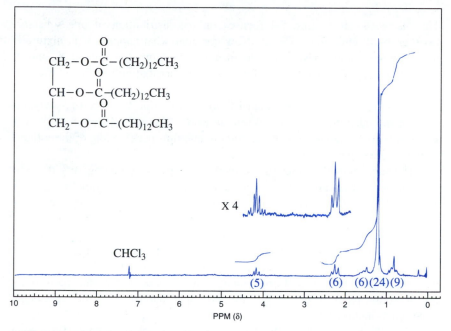

(b) ^{13}C NMR data.
Chemical shifts: δ 14.1, 22.7, 24.9, 29.1, 29.3, 29.4, 29.5, 29.6, 31.9, 34.0, 34.2, 62.1, 68.8, 172.8, 173.3.

EXERCISES

1. Confirm the statement made in the discussion of Equation (5.3) that a 5.5% increase in the recovery of solute A results from multiple extractions, rather than one extraction, when $K = 5$.

2. Why was diethyl ether rather than acetone chosen as the extraction solvent?

3. Would water be a good choice as an extracting solvent for trimyristin? Consider the anticipated solubility properties of trimyristin in providing an answer.

4. Why is gravity rather than vacuum filtration used to separate the ethereal extract from the residual nutmeg?

5. Why is fluted filter paper rather than plain filter paper used for the filtration procedure?

6. Judging from your examination of trimyristin, is this substance responsible for the odor of nutmeg?

7. Explain why it is necessary to crush the whole seeds if they are used as the source of nutmeg.

8. A certain plant material is known to contain mainly trimyristin and tripalmitin, in approximately equal amounts. Tripalmitin, the structure of which you should look up in a reference source, has mp 66–67 °C. Extraction of these two compounds from the plant material with diethyl ether gave an oil after removal of the solvent, and this oil was difficult to crystallize. Explain this result. (*Hint:* see Sec. 3.3.)

5.4 ACID AND BASE EXTRACTIONS

The discussion of Section 5.2 focuses on the distribution of *one* substance between two immiscible solvents. Now consider what happens if a mixture of *two or more* compounds is present in a given volume of solvent S_o and an extraction using a solvent S_x is performed. If the distribution coefficient K of one of the components, let us call it A, is *significantly* greater than 1.0 and if those of the others are *significantly* less than 1.0, the majority of A will be in S_x whereas most of the other compounds will remain in S_o. Physical separation of the two solvents will give at least partial separation, and thus purification, of the solute A from the other components of the mixture.

The success of this separation technique depends on having very different distribution coefficients for the solutes that are to be separated from one another. This is what would be expected if the components have widely different polarities. An extreme example would be the expected distribution of two organic solutes, one neutral and nonpolar, the other ionic and thus polar, between a nonpolar solvent and a polar solvent. The neutral component will be expected to have a distribution coefficient that will favor its presence in the nonpolar phase, whereas that of the charged constituent will promote preferential partitioning into the polar phase. It is precisely this principle that serves as the basis of the experiment in this section.

Consider the solubility characteristics of organic acids and bases. Carboxylic acids and phenols containing six or more carbons are generally either insoluble or only slightly soluble in water but are soluble in a number of different organic solvents, such as diethyl ether. These types of compounds usually dissolve in aqueous media, provided the pH is greater than 10. As shown in Equation (5.4), the

$$
\underset{\substack{\text{(water insoluble)}\\ K_{HA(H_2O/organic)} < 1}}{R-\overset{\overset{\displaystyle O}{\|}}{C}-O-H} \;+\; NaOH\,(aq) \;\longrightarrow\; \underset{\substack{\text{(water soluble)}\\ K_{A^-(H_2O/organic)} > 1}}{R-\overset{\overset{\displaystyle O}{\|}}{C}-O^-\,Na^+} \;+\; H_2O \qquad \textbf{(5.4)}
$$

reason for this is that an acidic organic compound HA will be deprotonated by the aqueous base to produce a salt A^-, which is the conjugate base of the original acid. The highly ionic character of the salt makes it soluble in water, a polar solvent. The distribution coefficient of the acid, K_{HA}, is such that HA partitions preferentially into the organic phase, whereas that of the conjugate base, K_{A^-}, favors partitioning of A^- into the aqueous phase. This means that acidic organic compounds in general can be rather efficiently separated from nonacidic compounds by extraction of a nonaqueous solution of the mixture with aqueous base. If the basic extract is then neutralized by addition of a mineral acid such as hydrochloric acid, the conjugate base will be *protonated* to regenerate the organic acid (Eq. 5.5). Because the acid is insoluble in the aqueous medium, it will appear as either a precipitate or a second layer, if it is a liquid. The desired organic acid can then be recovered by use of the appropriate technique, namely either filtration or separation of the layers.

$$\text{R} - \overset{\overset{\text{O}}{\|}}{\text{C}} - \text{O}^- \quad \text{Na}^+ + \text{H} - \text{Cl (aq)} \longrightarrow \text{R} - \overset{\overset{\text{O}}{\|}}{\text{C}} - \text{O} - \text{H} + \text{NaCl} \qquad \textbf{(5.5)}$$

(water soluble)   (water insoluble)

$K_{A^-(\text{H}_2\text{O/organic})} > 1$ $K_{\text{HA}(\text{H}_2\text{O/organic})} < 1$

In like fashion, organic bases, usually amines, that are insoluble in water will generally be soluble when converted to an ammonium salt, the conjugate acid of the amine, by treatment with an acidic solution whose pH is less than 4. The process is illustrated in Equation (5.6), in which HCl is used as the acid. Again, it is the enhanced ionic character of the salt that makes it water-soluble, $(K_{\text{BH}^+} > 1)$, whereas the parent base is not $(K_B < 1)$. Neutralization of the acidic solution by the addition of aqueous base fosters deprotonation of the ammonium ion to produce the original water-insoluble organic base (Eq. 5.7). It will now either precipitate from solution or form a separate layer. Recovery is accomplished by use of the appropriate method.

$$\text{R} - \overset{..}{\text{N}}\text{H}_2 \quad + \quad \text{H} - \text{Cl (aq)} \longrightarrow \text{R} - \overset{+}{\text{N}}\text{H}_3 \ \text{Cl}^- \qquad \textbf{(5.6)}$$

(water insoluble)   (water soluble)

$K_{A^-(\text{H}_2\text{O/organic})} < 1$ $K_{\text{HA}(\text{H}_2\text{O/organic})} > 1$

$$\text{R}\overset{+}{\text{N}}\text{H}_2\text{-H} \ \text{Cl}^- + \text{HO}^- \ \text{Na}^+ \text{(aq)} \longrightarrow \text{R} - \overset{..}{\text{N}}\text{H}_2 + \text{H}_2\text{O} + \text{NaCl} \qquad \textbf{(5.7)}$$

(water soluble)   (water insoluble)

$K_{\text{HA}(\text{H}_2\text{O/organic})} > 1$ $K_{A^-(\text{H}_2\text{O/organic})} < 1$

The pH-dependence of the solubility in water of various classes of organic compounds can thus serve as the basis for a powerful separation technique. Consider, for example, an ethereal solution of two water-insoluble compounds, one of which is acidic. Extraction of the solution with water at pH 7 would be fruitless because the distribution coefficient for both compounds disfavors their partitioning into the aqueous phase. However, if a basic aqueous solution is used for the extraction, the *acidic* organic component will be deprotonated, and the distribution coefficient for the resulting salt will be favorable for its partitioning into the aqueous layer. The result will be the desired separation of the organic compounds, since the nonacidic one can now be recovered from the organic phase by evaporation of solvent, and its acidic partner can be removed from the aqueous phase by neutralization followed by filtration or extraction. The overall process is summarized in Figure 5.3.

Using similar logic, it is possible to remove an organic base *selectively* from a mixture of water-insoluble compounds. Extracting the organic layer with aqueous acid will favor partitioning of the basic component into the aqueous phase, from which it can eventually be recovered by neutralization followed by filtration or extraction.

To illustrate these important principles of separations utilizing acid/base extractions, the separation of three organic compounds, benzoic acid (**5**), *p-*

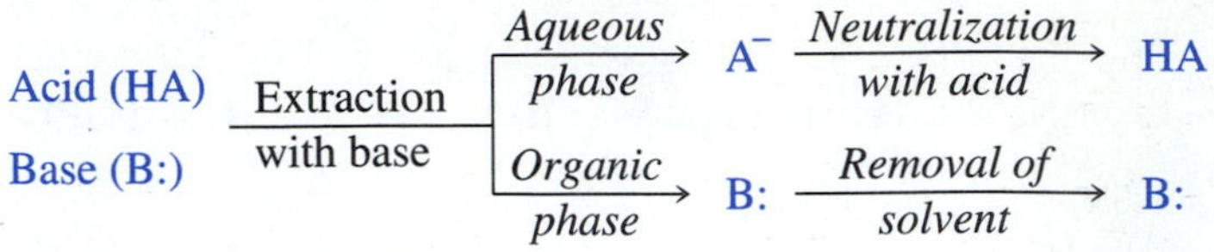

FIGURE 5.3 Separation of compounds based on their acidic and basic properties.

nitroaniline (**6**), and anthracene (**7**), is described in the following experimental procedure. Each of the compounds is soluble in dichloromethane, (CH_2Cl_2) but only slightly soluble in cold water.

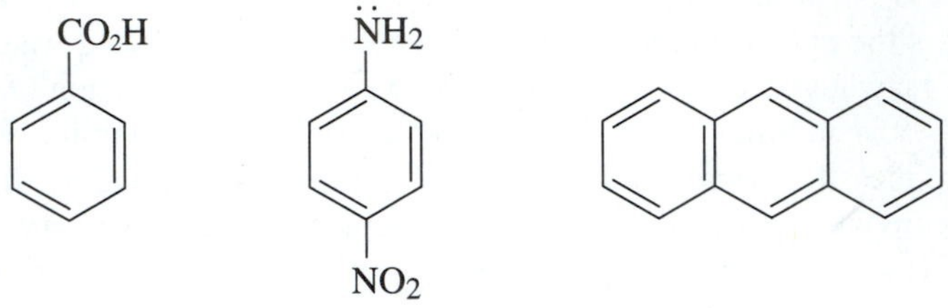

| **5** | **6** | **7** |
| Benzoic acid | *p*-Nitroaniline | Anthracene |

Pre-Lab exercises for Section 5.4, "Acid and Base Extraction," are found on page PL. 13.

EXPERIMENTAL PROCEDURE

SAFETY ALERT

1. Avoid letting any of the chemicals used in this experiment come in contact with your skin. If they do, *immediately* wash the affected areas with soap and water.
2. If you wear rubber gloves, use caution in handling wet glassware because it can easily slip out of your grasp.
3. Vent the separatory funnel *frequently* in order to avoid a build-up of pressure.

In preparation for this experiment, study the detailed instructions for the proper use of a separatory funnel as described in Section 2.18.

50 mL
~~125~~ mL
E Flask

Obtain a mixture of solids comprised of 0.5 g each of benzoic acid, *p*-nitroaniline, and anthracene. Dissolve the mixture in 40 mL of dichloromethane. If *small* quantities of solid do not dissolve, gravity-filter (see Sec. 2.25) the solution into a separatory funnel through a cotton plug or filter paper. Extract the organic solution twice using 25-mL portions of 6 *M* hydrochloric acid each time. Be certain that you have correctly identified the *aqueous* layer in the separatory funnel (see Section 2.18 for a way to do this). Combine the acidic extracts in an Erlenmeyer flask labeled "Acidic Extract."＊

remove stopper to drain

Extract the organic solution that remains in the separatory funnel twice using two 25-mL portions of 6 M aqueous sodium hydroxide. Place the combined basic extracts in a second Erlenmeyer flask labeled "Basic Extract."✳

Transfer the remaining organic solution into a third flask labeled "Neutral Fraction." Add 2–3 g of anhydrous sodium sulfate, and let the solution stand for about 15 min; swirl it occasionally to hasten the drying process.✳

Cool each of the flasks containing the aqueous extracts by placing them in an ice-water bath. Neutralize the acidic extract by careful addition of 6 M aqueous sodium hydroxide, using a slight excess in order to make the solution distinctly basic to litmus or pHydrion™ paper.✳ Similarly, carefully neutralize the contents of the flask labeled "Basic Extract" with 6 M hydrochloric acid, adding a slight excess of acid in order to make the solution distinctly acidic to litmus or pHydrion paper.✳ After carrying out the neutralizations, a precipitate should appear in each of the flasks.

Collect the precipitates separately by vacuum filtration (see Sec. 2.16) using a Büchner or Hirsch funnel. Wash each solid on the filter paper with *cold* distilled water. Separately transfer each of the solids to a labeled watchglass and allow the product to air-dry until the next laboratory period. As an alternative, place the samples in an oven having a temperature of 90–100 °C for about 1 h. After drying the solids, place them in dry, preweighed (tared) vials.

Separate the "Neutral Fraction" from the drying agent by gravity filtration through a cotton plug or by decantation. Remove the solvent by simple distillation. Discontinue heating when only a small amount of residue remains. Allow the stillpot to cool before attaching it to a vacuum source for a few minutes. This will remove essentially all of the residual dichloromethane. Transfer the contents of the flask to a watchglass, dry the solid either in the air or in the oven, and then place it in a third tared vial.

Reweigh each of the vials containing the dry products and determine the weights of the individual samples. Obtain melting points (see Sec. 2.3) for benzoic acid and *p*-nitroaniline. Do *not* attempt to measure the melting point of anthracene unless instructed to do so. The high melting point of this compound precludes using apparatus in which mineral oils are used as the heating medium. The melting point can be determined if silicone oil or a Mel-Temp apparatus is available. Your instructor may require that the isolated products be recrystallized to purify them further. Refer to Section 3.2 for the general procedure for recrystallization.

The reported melting points of the pure components of the mixture are as follows:

Benzoic acid 121–123 °C, *p*-Nitroaniline 149–151 °C, Anthracene 216–218 °C

Finishing Touches   Place the used *filter papers* in a container for non-toxic solid waste. Flush the *acidic* and *basic filtrates* down the drain. Pour any *dichloromethane* that has been isolated into a container for halogenated organic liquids.

Optional Experiment

You may be assigned an unknown mixture to be separated using the acid/base extraction procedure just described.

Spectra of Products

The IR and NMR spectra for benzoic acid are provided in Figures 3.1 and 3.2.

FIGURE 5.4 IR spectrum of *p*-nitroaniline (KBr pellet).

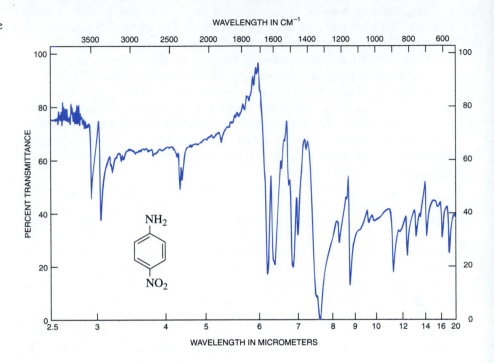

(a) ^{1}H NMR spectrum (90 MHz).

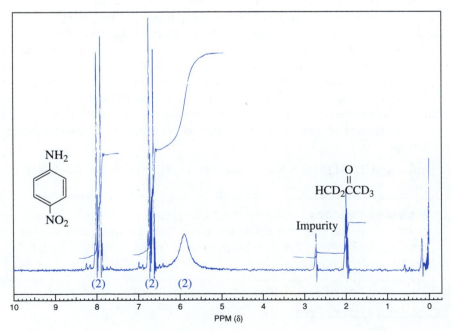

(b) ^{13}C NMR data.
Chemical shifts: δ 112.8, 126.3, 136.9, 155.1.

FIGURE 5.5 NMR data for *p*-nitroaniline.

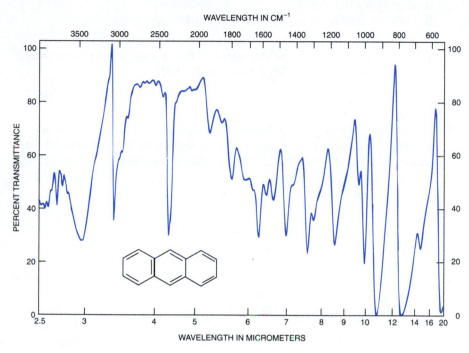

FIGURE 5.6   IR spectrum of anthracene (KBr pellet).

(a) ^{1}H NMR spectrum (90 MHz).

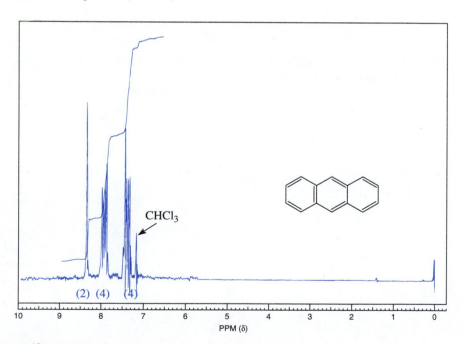

(b) ^{13}C NMR data.
    Chemical shifts: δ 125.1, 126.0, 127.9, 131.5.

FIGURE 5.7   NMR data for anthracene (KBr pellet).

Exercises

1. Define the following terms
   a. immiscible liquid phases
   b. partition coefficient
   c. adsorption
   d. absorption
   e. liquid-liquid extraction

2. Why is anhydrous sodium sulfate added to the organic solution remaining after the extractions with 6 M HCl and 6 M NaOH?

3. Explain why swirling of a solution and its drying agent hastens the drying process.

4. On the basis of what you learned in this experiment, plus the information that phenol is soluble in aqueous sodium hydroxide but not in aqueous sodium bicarbonate, Na HCO$_3$, solution, whereas benzoic acid is soluble in both solutions, show how you could separate a mixture containing phenol, benzoic acid, p-nitroaniline, and anthracene.

5. a. Given 400 mL of an aqueous solution containing 12 g of compound **A,** how many grams of **A** could be removed from the solution by a single extraction with 200 mL of diethyl ether?
   b. How many *total* grams of **A** can be removed with *three successive extractions* of 67 mL each? (Assume the distribution coefficient for **A** in diethyl ether/water is equal to 3.0.)

   Note: In solving problems of this type, one should recognize that Equation (5.3) applies to the situation pertaining *after* equilibrium has been reached. For example, a practical form of the equation is shown below, where a = grams of **A** originally present in water (S') and x = grams of **A** present in diethyl ether (S) after extraction.

$$K = \frac{x}{a-x} \times \frac{\text{mL of } S'}{\text{mL of } S}$$

6. a. The pKa of benzoic acid is 4.19. Show mathematically that this acid is 50% ionized at pH 4.19.
   b. Use the result of part (a) to explain why precipitation of benzoic acid is incomplete if the pH of an aqueous solution of benzoate ion is lowered only to pH 7 by addition of acid.

CHROMATOGRAPHY

T he common laboratory techniques of extraction, recrystallization, and distillation as methods for purifying organic compounds are discussed in Chapters 3 through 5. In many instances, however, the mixtures of products obtained from chemical reactions do not lend themselves to easy separation by any of these techniques because the appropriate physical properties of the individual components are too similar. In this chapter we learn some of the basic techniques that are used to separate mixtures by taking advantage of the differential distribution of the individual compounds in two immiscible phases, a principle that also forms the basis of extraction (see Chap. 5).

6.1 INTRODUCTION

The word **chromatography** was first used to describe the colored bands observed when a solution containing plant pigments is passed through a glass column containing an adsorbent packing material. From that origin, the term now encompasses a variety of separation techniques that are widely used for analytical and preparative purposes. All methods of chromatography operate on the principle that the components of a mixture will distribute unequally between two immiscible phases. The **mobile phase** is generally a liquid or a gas and flows continuously over the fixed **stationary phase,** which may be a solid or a liquid. The individual components of the mixture have different affinities for the mobile and stationary phases so a dynamic equilibrium is established in which each component is selectively, but temporarily, removed from the mobile phase by binding to the stationary phase. When the equilibrium concentration of that substance in the moving phase decreases, it is released from the stationary phase and the process continues. Since each component partitions between the two phases with a different equilibrium constant or **distribution coefficient,** the components separate into separate regions termed **migratory bands** (Fig. 6.1). *The component that interacts with or binds more strongly to the stationary phase moves more slowly in the direction of the flow of the mobile phase.* The attractive forces that are involved in this selective adsorption are the same forces that cause attractive interactions between any two molecules: electrostatic and dipole–dipole interactions, hydrogen bonding, complexation, and van der Waals forces.

The chromatographic methods used by modern chemists to identify and/or purify components of a mixture may be characterized by the nature of the mobile and stationary phases. For example, the techniques of **column, high-**

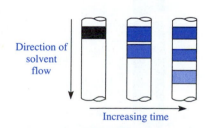

Direction of solvent flow

Increasing time

FIGURE 6.1 Separation of components by column chromatography.

pressure (or **high-performance**) **liquid** (HPLC), and **thin-layer chromatography** (TLC) each involve *liquid–solid* phase interactions. **Gas–liquid partition chromatography** (GLC), also known as **gas chromatography** (GC), involves distributions between a mobile *gas* phase and a stationary *liquid* phase coated on a solid support. These important techniques can be used as tools to analyze and identify the components in a mixture as well as to separate the mixture into its pure components for preparative purposes. Although there are other chromatographic techniques, such as ion exchange and paper chromatography, a review of those methods is beyond the scope of this discussion.

6.2   COLUMN CHROMATOGRAPHY

Column chromatography is a form of **solid–liquid adsorption** chromatography and is a powerful technique in preparative synthetic organic chemistry. In industry and research, it is frequently necessary to separate the components of a reaction mixture so that one of these may be used in a subsequent chemical reaction. Typically, a multicomponent mixture is dissolved in a small amount of an appropriate solvent and applied to the top of a packed column containing a finely divided, active solid **adsorbent** that serves as the stationary phase (Fig. 6.2). Next an **eluting solvent**, the mobile phase, is passed through the column. The individual components of the mixture, which were initially adsorbed on the stationary phase at the top of the column, begin to move downward with the eluting solvent. These components travel at different rates depending on their relative affinities for the packing material; a more weakly adsorbed compound is eluted faster from the column than is a more strongly adsorbed compound. As the individual components exit from the bottom of the column in bands or zones, they are collected in separate containers. The solvent is then removed from each fraction by evaporation to provide the pure components, which are characterized and identified by determining their physical constants (see Chaps. 3 and 4) and spectral properties (see Chap. 8).

When all of the components of a mixture are colored, their bands are easily observed. Many organic compounds are colorless, however, and other methods are required for detecting the bands as they elute from the column. For those organic compounds that absorb ultraviolet or visible light (see Sec. 8.4), electronic devices, called **detectors,** that measure differences in the absorption of light as the solvent exits the column are used to locate the bands of the individual components. Detectors that measure differences in the **refractive index** of the eluent are also used to identify the different bands; such detectors do not rely on absorption of light by the organic components.

If a detector is not available, the progress of the chromatographic separation can be conveniently followed using thin-layer chromatography (TLC) to analyze the eluent at regular intervals (see Sec. 6.3). Another, albeit more laborious, method involves collecting small, equal fractions of the eluent from the column in a series of flasks. The solvent in each flask is then evaporated, and the presence or absence of a solute in the residue then provides a means of locating the bands of each component. If there is adequate separation of the different bands, a given flask will often contain no more than one constituent of the original mixture. That component will normally appear in a number of consecutive fractions, however.

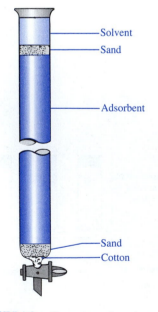

Solvent
Sand

Adsorbent

Sand
Cotton

FIGURE 6.2   Chromatography column.

6.2.1   ADSORBENTS

Selecting the proper adsorbent or stationary phase is one important decision that must be made before attempting a chromatographic separation. Solid *polar* adsorbents that are commonly used for the chromatographic separations of organic compounds include, in order of decreasing polarity, **alumina** (Al_2O_3), **silica gel** (silicic acid, $SiO_2 \cdot xH_2O$), and **Florisil**® (activated magnesium silicate). Glass beads coated with thin hydrocarbon films are used when a nonpolar adsorbent is required. Although Sephadex® and cellulose are frequently utilized as solid supports in biological applications, they are rarely encountered in routine separations of simple organic compounds.

Alumina, which is selected as the packing material in the experimental portion of this section, is a highly active and strongly adsorbing, polar substance. It is commercially available in three forms: neutral, acidic, and basic. Acidic and basic alumina can be used to separate basic and acidic compounds, respectively, but neutral alumina is frequently the packing of choice, especially for compounds sensitive to pH extremes. The presence of small amounts of water lowers the activity, or adsorptivity, of alumina, and activity grades of alumina are based on the weight percentage of water present.

Since the attraction of solute to adsorbent is a surface phenomenon, the most effective solid adsorbents have *uniform particle size* and *high specific area*. The higher the specific area, the faster the equilibrium of the solute between the mobile and solid phases is established and the narrower the bands. High specific areas on the order of several hundred m^2/g are common for good grades of alumina and silica gel.

The strength of the adsorption of an organic compound to the solid support depends not only on the polarity and nature of the adsorbent but also on the nature of the functional groups present in the molecule. In **normal-phase chromatography** the packing material for the stationary phase is a polar adsorbent such as alumina and silica gel, and organic solvents are used as the mobile phase. Under these conditions, compounds containing carboxyl groups and other polar functional groups are more tightly held than those containing halogen, as the **elutropic series** shown below indicates. More polar solvents must then be used as the mobile phase to elute highly polar molecules from the column.

In **reverse-phase chromatography,** the packing material for the stationary phase consists of glass beads coated with a nonpolar, hydrocarbon film, and mixtures of water and organic solvents are generally used as the eluting solvents. Under these conditions, nonpolar organic molecules are more strongly attracted to the nonpolar stationary phase, whereas polar solutes are more strongly attracted to the mobile phase. The order of elution is then the *reverse* of that shown earlier, with the more polar components of a mixture eluting more rapidly than the less polar ones. Reverse-phase chromatography may sometimes be used to separate mixtures that are inseparable by normal-phase chromatography.

6.2.2 SOLVENT

Selecting the eluting solvent or the mobile liquid phase is the other important decision that must be made in planning a chromatographic separation. The best solvent or combination of solvents for a given separation is determined by trial and error. Often the most efficient method to determine the optimal solvent system(s) for column chromatography is to perform a series of trial runs using TLC (see Sec. 6.3). These experiments may be performed quickly, and the amount of material required for the trials is small.

Certain criteria guide the selection of the solvent. For example, an effective eluting solvent must readily dissolve the solute but not compete with it for binding sites on the stationary phase. If the mixture to be separated is not soluble in the solvent, the individual components may remain permanently adsorbed on the stationary phase. Another criterion is that a solvent should not be too polar because it may bind strongly to the adsorbent and force the solute to remain in the mobile phase. In such circumstances, the components will move rapidly on the column, offering little opportunity to establish the equilibria required for separation. Consequently, the eluting solvent must be significantly *less* polar than various components of the mixture to obtain an effective separation. As a rule, the relative ability of different solvents to move a given substance down a column of alumina, silica gel, or Florisil is termed **eluting power** and is generally found to follow the order shown below. In reverse-phase chromatography the situation is just the opposite, with less-polar solvents having the greater eluting power.

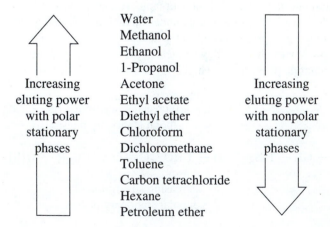

6.2.3 EXPERIMENTAL TECHNIQUE

With this general discussion as background, we now discuss some of the experimental aspects of performing column chromatography. The optimal adsorbent and eluting solvent(s) typically are first determined using TLC (see Sec. 6.3) and then the column is packed with the adsorbent. The amount of adsorbent used to prepare the column varies according to the differences in distribution coefficients and polarities of the individual components in the chromatographic system. For simple separations, it is possible to use as little as 10 g of adsorbent for 1 g of the mixture, but when the components of the mixture have similar polarities, this ratio must be increased to as much as 100–200:1; a ratio of about 25:1 is a convenient starting point. As a general rule of thumb, the height-to-diameter ratio for the packed column should be about 8:1.

The column is made of glass and is fitted with a Teflon stopcock, or a segment of polyethylene tubing with a screw clamp to control the flow of solvent through the column. Using a long glass rod or a piece of wire, a plug of cotton or glasswool is first inserted into the small end of the column to prevent the packing materials from washing from the bottom of the column. A layer of clean white sand, approximately 1 cm deep, is then added to provide an even bed for the adsorbent (see Fig. 6.1).

The manner in which a column is packed ultimately determines its separation efficiency. Proper packing of the column is vital to the success of column chromatography because this step ultimately determines the efficiency of separation. Two general protocols are followed for this important operation. The first of these, the **dry pack** method, involves pouring the solid dry adsorbent slowly into a vertical glass column that is approximately half-filled with the solvent that has been selected as the eluent. The other technique is the **wet pack** method, in which a slurry of the adsorbent in the eluting solvent is added to the column; this is the preferred procedure when silica gel is the adsorbent. With both methods, the column is constantly tapped as the solid settles through the liquid to ensure an even and firm packing of the adsorbent and to remove any trapped air bubbles. Some solvent may be drained from the column during this operation, but the liquid level in the column should *never* be allowed to fall below the top of the adsorbent. If this occurs, the air bubbles that form in the column will allow *channeling*, which results in poor separations because the components are eluted as ragged rather than sharp bands. *Uniform packing of the adsorbent is essential* so the solvent will move down the column with a horizontal front.

To complete packing of the column, a layer of sand is normally placed on top of adsorbent as shown in Figure 6.2. The purpose of the sand is twofold: (1) It allows the sample to flow evenly onto the surface of the adsorbent, and (2) it prevents disruption of the packing material as eluting solvent is added.

The mixture to be separated is typically dissolved in a *minimal* volume of a solvent and transferred to the top of the column; a liquid may be transferred directly to the column. It is important to distribute the sample evenly on the surface of the adsorbent and to use as little solvent as possible in loading the column. This procedure ensures that with the proper choice of eluting solvent, the bands that form during the development of the chromatogram will be narrow, thereby providing the best possible separation. If too much solvent is used to dissolve the sample, the initial band will be broad, and poor resolution of the mixture may result.

Once the sample has been loaded onto the column, there are several different techniques that may be used to elute its components. In a **simple elution** experiment, a *single* solvent is passed through the column during the entire course of the separation. This procedure works well for the separation of mixtures containing only two or three compounds having similar polarities. However, the more common chromatographic procedure is **stepwise** or **fractional elution.** In this technique, a series of increasingly more polar solvents is used to elute the mixture from the column. A nonpolar solvent such as petroleum ether, which is a mixture of low-boiling hydrocarbons, or hexane, is first used to move the least polar component of the mixture down the column while the others remain at or near the top of the column. After elution of the first band, the polarity of the eluent is gradually increased using mixed solvent combinations so that the bands are individually eluted from the column. Systematic increases in solvent polarity are essential so that individual bands remaining on the column separate and do not coelute. For example, after starting with pure hexane, a solvent mixture having the composition of 95% hexane and 5% chloroform could be used, followed by solvent mixtures containing 10, 15, 20, 40, and 80% chloroform. As the polarity of the solvent system increases, those components of the mixture that are more tightly adsorbed on the column will begin to move. As a rule of thumb, a volume of solvent approximately equal to three times the column volume is passed through the column prior to switching to a solvent of higher polarity.

The separation is monitored using one of a variety of methods. Unfortunately, most organic compounds are not highly colored, and sophisticated devices for their detection are rarely available in the undergraduate laboratory. The most effective technique for following the separation is to collect fractions of equal volume in tared flasks, to concentrate the solvent, and to reweigh the flasks. The fractions containing the different bands may then be easily identified by the relative amounts of solute in each flask. One may also use TLC (see Sec. 6.3) to monitor the separation.

An experimental variant of normal column chromatography has recently been developed in which slight positive pressure is applied to the head of the column to speed the flow of the solvent through the column. Since faster flow rates are typically used in this technique, it is necessary to use smaller particles of adsorbent to increase the surface area of the adsorbent. Otherwise, equilibration of the components of the mixture between solid and liquid phases does not occur, and incomplete separations may result.

The need for faster, higher-resolution separations led to the development of another improved form of column chromatography called **high-pressure,** or **high-performance, liquid chromatography,** usually abbreviated as HPLC. In HPLC, the liquid phase is pumped through small, efficient columns packed with uniformly sized particles of solid adsorbent that are as small as 6 microns. Using finely divided packing materials dramatically increases the surface area of the stationary phase. Because the more tightly packed adsorbent particles have considerably smaller interstitial volumes, equilibration of the components of the mixture between the solid and liquid phases occurs rapidly. The tighter packing of these adsorbent particles severely restricts the flow of solvent through the column. To overcome this problem, pressure must be applied evenly to the head of the column to obtain reasonable flow rates, and pumping systems have been developed that produce flow rates of 1–500 mL/min at head pressures ranging from 250 to 6000 psi. Although the liquid phase enters the column at high pressure, the effluent from it issues at atmospheric pressure. The components of the mixture in the eluting solvent are detected in a flow-through cell by continu-

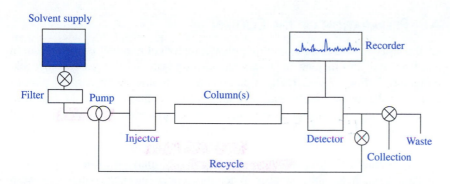

FIGURE 6.3   Block diagram of HPLC system.

ously monitoring the ultraviolet absorption or the refractive index of the efflu-ent, and the detector signal is fed to a chart recorder to provide a chromatogram. A typical experimental set-up is illustrated in Figure 6.3. Special HPLC devices are also available that allow the recycling of sample through the column if a sin-gle pass through the column does not give a satisfactory separation. Compounds can be identified by retention time and peak enhancement. Pure samples of the individual components may be obtained by collecting the separate fractions and evaporating the eluting solvent. The compounds may then be characterized by physical and chemical techniques.

In the experimental section, column chromatography will be used to sepa-rate fluorene (**1**) from an oxidation product, fluorenone (**2**). One of these com-pounds is white, the other is yellow. Consequently, the progress of the chromatography may be followed by evaporation of the solvent at periodic inter-vals as well as by visual observation of the slower-moving yellow band.

1
Fluorene

2
Fluorenone

Pre-Lab exercises for Section 6.2, "Column Chromatography," are found on page PL. 15.

EXPERIMENTAL PROCEDURE

SAFETY ALERT

Petroleum ether is a highly volatile and flammable mixture of low-boiling hydrocarbons. During the preparation and development of the chromato-graphic column, be certain that there are *no flames* in the vicinity.

A. PREPARATION OF THE COLUMN

Clamp a dry 50-mL glass buret or chromatography column that is about 1 cm in diameter and 25 cm long in a vertical position with its *ungreased* stopcock, preferably Teflon, *closed*. Using a piece of glass tubing, insert a small plug of cotton or glasswool loosely into the bottom of the column. Cover the cotton or glasswool plug with enough clean sand to form a layer about 1 cm thick, and add approximately 30 mL of petroleum ether, bp 30–60 °C (760 torr). Place a funnel on top of the column, and *slowly* add 5 g of dry alumina to the column while constantly tapping the buret. A rubber "tapping mallet" may be made by placing a pencil in a one-hole rubber stopper. When this process has been completed, wash the inner walls of the column with additional petroleum ether to remove any alumina that may adhere to the sides. Cover the alumina with a 1-cm layer of clean sand, and open the stopcock to allow the solvent to drain until its level reaches just to the top of the alumina. The column is now ready for the addition of your sample mixture.

B. SEPARATION AND PURIFICATION OF FLUORENE AND FLUORENONE

Obtain a sample of an approximately 1:1 mixture of fluorene (**1**) and fluorenone (**2**) and accurately determine its melting-point range. In a small test tube, dissolve about 0.1 g of this mixture in 1 mL of petroleum ether by warming the tube slightly to effect solution. Using a Pasteur pipet, carefully transfer this solution directly to the top of the column. Open the stopcock until the liquid level is at the top of the alumina. *Do not allow the solvent to drain below the level of the alumina*, as air bubbles and channels might develop in the solid support. Add approximately 1–2 mL of fresh petroleum ether to the top of the column, and again allow the liquid to drain to the top of alumina.

Fill the buret with approximately 20 mL of fresh petroleum ether, open the stopcock, and collect the eluant in a 50-mL Erlenmeyer flask. Follow the progress of the chromatography by collecting a drop or two of eluant on a watchglass with every 5 mL that elutes from the column. When the solvent evaporates, any white solid present will be visible on the watchglass. Using this visualization technique, you can determine when all of the white solid has been eluted. Your instructor might also direct you to follow the chromatography by TLC using 15% dichloromethane in petroleum ether as the developing solvent (see Sec. 6.3). Most of the white solid should elute in a volume of 15–20 mL of petroleum ether, and slow movement of a yellow band down the column should occur. Wash any of the white solid from the tip of the column into your collection flask with fresh petroleum ether.

When all of the white solid has eluted from the column, change the collection flask to another clean 50-mL Erlenmeyer flask. Elute the column with about 5 mL of petroleum ether, and then change the eluant to dichloromethane, a more polar solvent. Watch the progress of the yellow band as it now proceeds rapidly down the column. When this yellow band just reaches the bottom of the column, change to a third clean 50-mL Erlenmeyer flask. The intermediate fraction should not contain significant amounts of solid; verify this by evaporating a few drops on a watchglass.

Continue eluting the column with dichloromethane until the eluant is color-less; approximately 10 mL will be required. The progress of the chromatography may be monitored visually by the color of the eluant, but the watchglass tech-nique just described should be used to confirm these observations.

The first and third fractions should contain pure samples of fluorene and fluorenone. Most of the solvent in these fractions may be removed by simple distillation or according to one of the techniques outlined in Section 2.25. Attach the flask to a water aspirator or house vacuum to remove the last traces of sol-vent under reduced pressure. When the crystals of each of the purified com-pounds are completely dry, determine their melting points. Fluorene is reported to have a melting point of 114–116 °C, whereas fluorenone has a reported melt-ing point of 82–85 °C. Use your experimentally determined melting points to identify the order of elution of fluorene and fluorenone from the column. Record your observations and conclusions in your notebook.

Finishing Touches   Place recovered *petroleum ether* in the container for nonhalogenated organic liquids, but pour the recovered *dichloro-methane* into a container for halogenated organic liquids. Put the recovered *fluorene* and *fluorenone* either in suitably marked containers for reuse or in a container for nonhazardous organic solids. Spread out the alumina adsor-bent, which is wet with organic solvent, in a hood to dry and then place it in the nonhazardous solid waste container.

SPECTRA OF STARTING MATERIALS

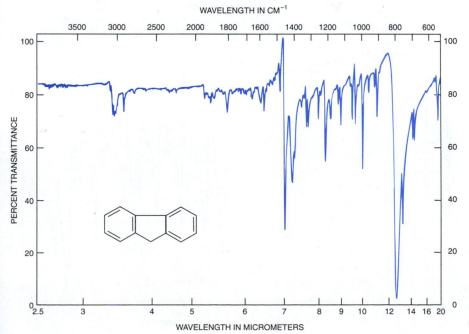

WAVELENGTH IN CM^{-1}

PERCENT TRANSMITTANCE

WAVELENGTH IN MICROMETERS

FIGURE 6.4   IR spectrum of fluorene (KBr pellet).

FIGURE 6.5 NMR data for fluorene.

(a) ^{1}H NMR spectrum (90 MHz).

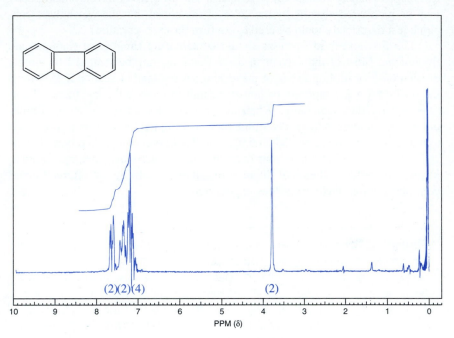

(2)(2)(4)   (2)

PPM (δ)

(b) ^{13}C NMR data.
Chemical shifts: δ 36.9, 119.8, 125.0, 126.7, 126.7, 141.7, 143.2.

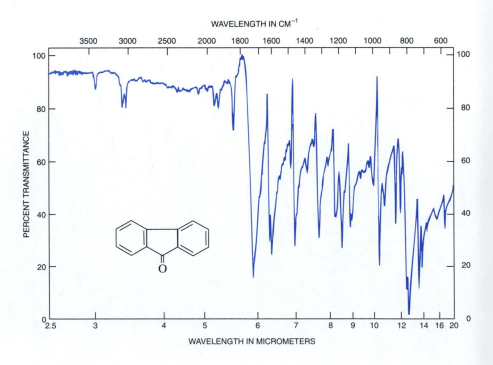

WAVELENGTH IN CM^{-1}

FIGURE 6.6 IR spectrum for fluorenone
(KBr pellet).

WAVELENGTH IN MICROMETERS

(a) ¹H NMR spectrum (90 MHz).

FIGURE 6.7   NMR data for fluorenone.

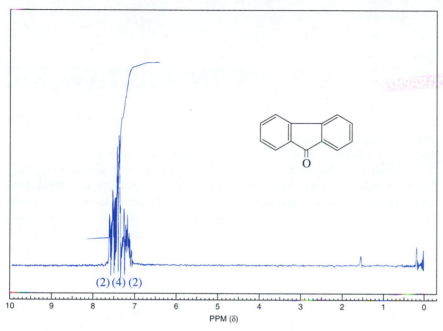

PPM (δ)

(2)(4)(2)

(b) ¹³C NMR data.
   Chemical shifts: δ 120.1, 123.8, 128.8, 133.9, 134.4, 144.1, 193.1.

EXERCISES

1. Why is it preferable to use a Teflon or an ungreased stopcock rather than a greased stopcock on a column used for column chromatography?

2. Why should care be exercised in the preparation of the column to prevent air bubbles from being trapped in the adsorbent?

3. Why is a layer of sand placed above the cotton plug prior to the addition of the column packing material?

4. Does fluorene or fluorenone move faster down the column when petroleum ether is used as the eluant? Why?

5. When separating a mixture by standard column chromatography, why is it better to change from a less-polar solvent to a more-polar solvent rather than the reverse?

6. Normally, when increasing the polarity of the eluant, the increase is made gradually; however, in this procedure, *pure* dichloromethane rather than intermediate mixtures of petroleum ether and dichloromethane was added to the column. Why is this variation from the usual technique acceptable in this case?

7. The observed melting point of the 1:1 mixture of fluorene and fluorenone should be relatively sharp, although lower than the melting point of either of the pure compounds. On the other hand, a 3:1 mixture has a broad melting-point range of about 60–90 °C. Explain these observations. (Hint: See Sec. 3.3.)

8. Define the following terms.
   a. eluate   c. absorbent
   b. eluent   d. adsorbent

6.3 THIN-LAYER CHROMATOGRAPHY

Thin-layer chromatography (TLC) is another form of **solid–liquid adsorption** chromatography and involves the same fundamental principles as column chromatography. The properties of the adsorbents and the relative eluting abilities of the solvents given for column chromatography (see Sec. 6.2) also apply to TLC. However, unlike column chromatography where the mobile phase moves down the column, the mobile phase in TLC *ascends* the thin layer of adsorbent. Another significant difference is that a TLC experiment may be performed more rapidly and requires much less sample, sometimes as little as 10^{-9} g. Thin-layer chromatography is an excellent analytical tool and is frequently used to determine the optimal combinations of solvent and adsorbent for preparative column chromatographic separations as well as to monitor the progress of a separation by column chromatography. Note that TLC is limited to relatively *nonvolatile* compounds such as solids and liquids whose boiling points are above 150 °C at atmospheric pressure (see Exercise 6).

6.3.1 EXPERIMENTAL TECHNIQUE

For TLC, the adsorbent, usually alumina or silica gel, is mixed with a small quantity of a **binder,** such as starch or calcium sulfate, and spread as a layer approximately 250 μ thick on either a glass or plastic plate. The binder is necessary for proper adhesion of the thin layer of adsorbent to the plate. Thin-layer chromatography plates should be dried in an oven for an hour or more at 110 °C prior to use to remove any adsorbed moisture. The presence of water molecules on the surface of the adsorbent decreases its activity and effectiveness in binding and separating the components of the mixture.

In a typical TLC analysis, a *small* drop of a solution of the mixture to be analyzed is carefully applied or spotted from a capillary to a small rectangular chromatographic plate near the edge of the narrow end as shown in Figure 6.8a. The plate is placed with spotted end down in a closed jar, called a **developing chamber,** containing a suitable eluting solvent, which may be either a pure solvent or a mixture of two or more solvents; the level of the eluent should be just below that of the spot (Fig. 6.8b). As the solvent moves up the plate, the components of the mixture are carried along at different rates to produce a series of spots on the plate as depicted in Figure 6.8b. When the solvent front has advanced nearly to the top of the TLC plate, the development of the chromatogram is complete, and the plate is withdrawn from the developing chamber.

The components of the mixture may be detected in a variety of ways. If the compounds being separated are colored, visual detection of the spots is easy.

FIGURE 6.8 Thin-layer chromatography. (a) Original plate. (b) Developed chromatogram.

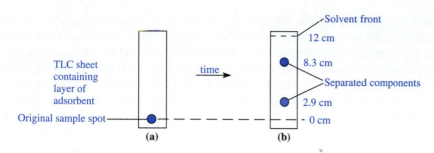

Many organic compounds are colorless, however, and a variety of methods have been developed to detect their presence on the plate:

1. Compounds that fluoresce may be located by placing the plate under an ultraviolet light. Since the spots disappear when the light is removed, it is necessary to circle the spots with a pencil in order to have a permanent record of the chromatogram. There are also commercially available plates that contain a fluorescent material as part of their coating; compounds that do not fluoresce but do absorb ultraviolet light then appear as dark spots under ultraviolet light.
2. The chromatographic plate may be sprayed with a variety of reagents such as sulfuric acid, potassium permanganate, phosphomolybdic acid, and ninhydrin; these reagents will react with the individual components to produce colored or dark spots.
3. The chromatographic plate may be exposed to iodine vapor by placing it in a closed chamber containing several crystals of iodine. As the iodine forms complexes with the various organic compounds, the spots become brown. Since the process is reversible and the spots fade, it is wise to circle the spots with a pencil in order to have a permanent record of the chromatogram.

Once the separation of the components of the mixture is complete and the individual spots have been detected, the **retention factor** (R_f) of each compound may be calculated as shown below for the chromatogram pictured in Figure 6.8b:

$$R_f = \frac{\text{distance traveled by substance}}{\text{distance traveled by solvent}}$$

$$R_f(\text{compound 1}) = \frac{2.9\ \text{cm}}{12\ \text{cm}} = 0.24$$

$$R_f(\text{compound 2}) = \frac{8.3\ \text{cm}}{12\ \text{cm}} = 0.69$$

The R_f value for a compound is a physical constant for a given set of chromatographic conditions, and consequently the adsorbent and the eluting solvent should be recorded together with the experimentally determined R_f values.

There are many important applications of TLC in modern organic chemistry. For example, TLC is commonly used to identify components of an unknown mixture by running chromatograms of the unknown sample side by side with known standards. Multiple aliquots of samples collected from chromatographic columns (see Sec. 6.2) may be analyzed by TLC to follow the chromatographic separation. Alternatively, it is possible to follow the course and progress of a reaction by TLC by monitoring the disappearance of starting material or the appearance of product. Samples are simply withdrawn from a reaction mixture and subjected to TLC analysis. Thin-layer chromatography may also be used for preparative purposes. Specially prepared TLC plates with thicker coats of adsorbent can accommodate samples as large as 5–10 mg. In such cases, recovery of the components can be achieved by scraping the individual spots from the plate and extracting them from the adsorbent with an appropriate solvent. It may be necessary to repeat the process a number of times in order to get an amount of compound sufficient for spectral analysis.

6.3.2 APPLICATIONS

Two experiments are presented here to demonstrate the TLC technique. The first involves the separation of the pigments present in spinach leaves. A variety of other sources including crushed tomato pulp or carrot scrapings as well as leaves from grasses, shrubs, and trees may be substituted; however, waxy leaves are not acceptable. A comparative study could be performed by having various members of the class analyze the composition of pigments of different leaves.

In the second experiment, the **geometric isomers** *syn*-azobenzene (**3**) and *anti*-azobenzene (**4**) are separated. Commercially available azobenzene consists predominantly of the more stable *anti-* form, but this isomer may be photochemically isomerized to the less stable *syn-* isomer by irradiation with ultraviolet or sunlight. Since the colors of the two isomers differ, they may be detected visually. The course of the reaction and the effectiveness of the irradiation is followed by placing spots of irradiated and nonirradiated samples side by side on the TLC strip.

3
Syn-azobenzene

4
Anti-azobenzene

Another application of TLC is demonstrated in Section 7.2, where the **diastereomers** of 1,2-hexanediol are separated. Although these diastereomers have similar physical properties and are difficult to separate by fractional crystallization, they are readily separable by chromatography. There is another opportunity to use TLC to analyze the products of the experiment in Section 15.3.

Pre-Lab exercises for Section 6.3, "Thin-Layer Chromatography," are found on page PL. 17.

SAFETY ALERT

Petroleum ether, ethanol, and acetone are highly volatile and flammable solvents. Be certain there are *no flames* in the vicinity during the extraction of the green pigments and during development of the chromatograms in Parts B and C of this experiment.

A. PREPARATION OF A DEVELOPING CHAMBER

Use a wide-mouth bottle with a tightly fitting screw-top cap or a gas collection bottle with a rubber stopper as a developing chamber; a beaker covered with a watchglass may also be used. As shown in Figure 6.9a, place a folded filter paper lengthwise in the developing chamber and add an amount of the eluting solvent so that it forms a 1-cm layer on the bottom of the vessel. Shake the container to saturate the atmosphere of the chamber with vapors of the solvent. Having a saturated atmosphere inhibits the evaporation of solvent from the plate during the development of the chromatogram (Fig. 6.9b); such evaporation would disturb the equilibration and partitioning of the components of the mixture between the mobile and solid phases, thereby adversely affecting the separation.

B. SEPARATION OF GREEN LEAF PIGMENTS BY TLC

Prepare a developing chamber as described in Part A using a 70:30 mixture of petroleum ether, bp 30–60 °C (760 torr) and acetone or chloroform as the developing solvent; *there should be no flames in the vicinity*. It is instructive to have half of the class use one developing solvent, the other half use the alternative; a comparison of the results will reveal which eluting solvent is superior for this separation.

Using a small mortar and pestle, thoroughly grind a spinach leaf in a mixture of 4 mL of petroleum ether and 2 mL of ethanol. Transfer the liquid extract

(a)　　　　　　　　(b)　　　　　　　　**FIGURE 6.9**　TLC chamber.

Adsorbent
Solvent front
Solvent
Spot

to a small separatory funnel or test tube using a Pasteur pipet and *swirl* the extract gently with an equal volume of water. Do not shake the funnel or test tube because emulsions are readily formed. Remove and discard the aqueous layer; if you do not know which layer is the aqueous one, perform the necessary test (see Sec. 2.18). Wash the organic layer with water two more times to remove the ethanol and any water-soluble materials that are present in the leaves. Transfer the petroleum ether extract to a 10-mL Erlenmeyer flask and add about 1 g of *anhydrous* sodium sulfate. After 5 to 10 min, decant the solution from the drying agent. If the solution is not deeply colored, concentrate it using a gentle stream of air or nitrogen to remove some of the solvent.

Obtain a 10-cm × 2-cm strip of silica gel chromatogram sheet without a fluorescent indicator. Handle the strip *only* by the sides to avoid contaminating the plate with oils from your hands. Place a pencil dot in the middle of the sheet about 1 cm from one end. Using a micro or a capillary pipet, apply a spot of pigment solution over the pencil dot by *lightly* and *briefly* applying the tip of the pipet to the surface of the plate; you may blow gently on the plate as the sample is applied. Do not allow the spot to diffuse to a diameter of more than 2 mm during application of the sample. Perform the spotting process an additional four or five times, allowing the solvent of each drop to evaporate before adding the next. When the spot has thoroughly dried, place the strip in the developing chamber, being careful not to splash solvent onto the plate. The spot must be *above* the solvent level. Allow the solvent front to move to within 2–3 mm of the top of the strip and then remove the strip. Mark the position of the solvent front with a pencil, and allow the plate to air-dry.

With a good separation, one may observe as many as eight colored spots. These are, in order of decreasing R_f values, the carotenes (two spots, orange), chlorophyll *a* (blue-green), the xanthophylls (four spots, yellow), and chlorophyll *b* (green).

Calculate the R_f values of all spots on your developed plate. In your notebook, include a picture of the developed plate drawn to scale as a permanent record.

Finishing Touches   Put the *petroleum ether* and *acetone* eluent in the container for nonhalogenated organic liquids and the *chloroform* in the container for halogenated organic liquids. Flush the saturated *salt solution* used in the extraction down the drain, and discard the dry chromatographic plates in the nonhazardous solid waste container.

C. SEPARATION OF *SYN-* AND *ANTI-*AZOBENZENES BY TLC

SAFETY ALERT

Since azobenzene is a *suspected carcinogen*, avoid contacting it with your skin or ingesting it.

Obtain a 10-cm × 3-cm strip of silica gel chromatogram sheet without a fluorescent indicator. Handle the strip *only* by the sides in order to avoid contaminating the plate with oils from your hands. Place one pencil dot about 1 cm from the left side and about 1 cm from one end of the sheet and another about 1 cm from the right side the same distance from the bottom as the first. Using a micro or capillary pipet, carefully apply a spot of a 10% solution of commercial azobenzene in toluene, which you should obtain from your instructor, over one of the pencil dots. Do not allow the spot to diffuse to a diameter of more than 1–2 mm during application of the sample. Allow the spot to dry and then expose the plate to sunlight for one to two hours (or a sun lamp for about 20 min).

While irradiating the TLC plate, prepare a developing chamber as described in Part A using a 9:1 mixture (by volume) of hexane and chloroform as the developing solvent; *there should be no flames in the vicinity.*

When the irradiation is complete, apply another spot of the *original* solution on the plate over the second pencil dot in the same manner as just described and allow the strip to dry. Place the strip in the developing chamber, being careful not to splash solvent onto the plate. Both spots must be *above* the solvent level. Allow the solvent to move to within approximately 2–3 mm of the top of the strip and then remove the strip. Mark the position of the solvent front with a pencil, and allow the plate to air-dry. Note the number of spots arising from each of the two original spots. Pay particular attention to the relative intensities of the two spots nearest the starting point in each of the samples; these are *syn*-azobenzene.

Calculate the R_f values of each of the spots on your developed plate. In your notebook include a picture of the developed plate drawn to scale as a permanent record.

> **Finishing Touches**   Pour the unused *mixture of hexane and chloroform* used as eluent into the container for halogenated solvents, and put the *dry chromatographic plates* in the hazardous solid waste container, since they contain small amounts of azobenzene.

EXERCISES

1. Which of the two diastereomers of azobenzene would you expect to be more thermodynamically stable? Why?

2. From the results of the TLC experiment with the azobenzenes, describe the role of sunlight.

3. In a TLC experiment why should the spot not be immersed in the solvent in the developing chamber? If you have time, see what happens when this is done.

4. Explain why the solvent must not be allowed to evaporate from the plate during development.

5. Explain why the spot should be no larger than 2 mm.

use small pipettes
initial after drying
and put under sunlamp
15-20 min.
Apply second spot
total 4 spots

6.4 GAS-LIQUID CHROMATOGRAPHY

Gas-liquid partition chromatography, (GLC), which is also called gas chromatography (GC), is a technique that may be used to separate mixtures of volatile compounds whose boiling points may differ by less than 0.5 °C. It can also be applied as an analytical tool to identify the components of a mixture or in preparative applications when quantities of the pure components are desired. Preparative GLC is far superior to fractional distillations (see Sec. 4.2), even those employing spinning-band columns, for carrying out separations and purifications of volatile substances on a small scale.

6.4.1 THEORY OF GAS-LIQUID CHROMATOGRAPHY

Gas-liquid chromatography operates on the principle of partitioning the components of a mixture between a mobile gaseous phase and a stationary liquid phase. In practice, a sample is injected into a heated chamber where it is immediately vaporized and carried through a column by a flowing inert gas such as helium or nitrogen, which is called the carrier gas. This gaseous mixture is the mobile phase. The column is packed with a finely divided solid support that has been coated with a viscous, high-boiling liquid, which serves as the stationary phase. As the mobile phase moves through the column, its components are continuously partitioned between the two phases. Those components that show a higher affinity for the mobile phase move through the column more quickly, whereas those with a stronger attraction to the stationary phase migrate more slowly, and separation occurs. As with fractional distillation, a given column may contain a large number of theoretical plates.

The retention time of a component is the elapsed time required for the compound to pass from the point of injection to the detector, and represents a physical property that may be used for purposes of identification. The retention time of a component is *independent* of the presence or absence of other components in the sample mixture. Four experimental factors influence a compound's retention time: (1) the *nature* of the stationary phase, (2) the *length* of the column, (3) the *temperature* of the column, and (4) the *rate of flow* of the inert carrier gas. Thus, for a particular column, temperature, and flowrate, the retention time will be the same for a specific compound.

6.4.2 INSTRUMENTATION

All commercially available gas chromatographs have a number of basic features in common. These are illustrated schematically in Figure 6.10. Parts 1–5 are associated with supplying the dry carrier gas, usually helium or nitrogen, and allowing an operator to control its flow. The mixture to be separated is injected using a special syringe through a rubber septum into the injection port (6), an individually heated chamber in which the sample is immediately vaporized. The sample then enters the flowing stream of carrier gas and is swept into and through the column (7), which is located in an oven (8) and consists of coiled aluminum, stainless steel, or glass tubing filled with an appropriate packing

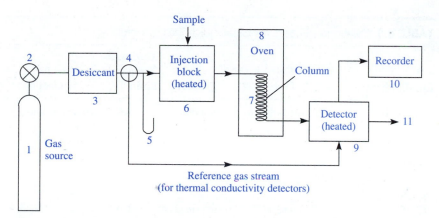

FIGURE 6.10 Schematic diagram of apparatus for gas-liquid chromatography. (1) Carrier gas supply. (2) Pressure-reducing valve. (3) Desiccant. (4) Fine-control valve. (5) Flowmeter. (6) Heated injection port. (7) Column. (8) Oven. (9) Detector. (10) Electronic recorder. (11) Exit port.

material. In the column, the individual components separate into bands that ultimately pass through a **detector** (9), which produces an electronic signal whose voltage is proportional to the amount of material other than carrier gas present in the mobile phase. One type of detector that is commonly used is the **thermal conductivity detector** (TCD), which operates on the basis of differences in the thermal conductivity of the mobile phase as a function of its composition. A **flame ionization detector** (FID) is much more sensitive and operates by detecting the number of ions produced by passing the mobile phase through a hydrogen flame. The **recorder** (10) plots the changes in voltage measured by the detector as a function of time to give the gas chromatogram. The vapors then pass from the detector into either the atmosphere or a collection device at the **exit port** (11).

6.4.3 STATIONARY PHASE

Although a large number of stationary liquid phases are available, only a few (Table 6.1) are widely used. Two criteria must be considered when selecting the stationary liquid phase that is best suited for separating a given mixture in a GLC experiment. First, *each liquid phase has a maximum temperature limit above which it cannot be used;* this temperature depends upon the stability and volatility of the liquid phase. At higher temperatures the liquid phase will vaporize and "bleed" from the column with the mobile phase. However, the most important factor in selecting the liquid phase is its ability to separate the components of the mixture.

The differences in the partition coefficients of the individual components of a mixture in GLC depend primarily upon the *differences in solubility of each of the components in the liquid phase*. The solubility of a gas in a liquid partly depends on its vapor pressure at the ambient temperature; the higher the vapor pressure of a gas, the lower its solubility in a liquid. Consequently, the components of a gaseous mixture tend to move through the column with the more volatile compounds eluting first. The solubility of a substance in a liquid is also influenced by the nature and magnitude of the interactions between the molecules of solute and solvent. In this regard, the well-established principle of solubility that *like dissolves like* serves as an excellent guide. Polar samples are most

TABLE 6.1   Stationary Phases for Gas-Liquid Chromatography

Liquid Phase	Type	Property	Maximum Temperature Limit, °C	Used for Separating
Apiezon-L	Hydrocarbon grease	Nonpolar	300	Hydrocarbons, general application
Carbowax 20M	Hydrocarbon wax	Polar	250	Alcohols, C_6–C_{18} aldehydes, sulfur compounds
DC-550	Silicone oil	Intermediate polarity	275	C_1–C_5 aldehydes, sulfur compounds
QF-1	Silicone (fluoro)	Intermediate polarity	250	Polyalcohols, alkaloids, halogen compounds, pesticides, steroids
SE-30	Silicone gum rubber	Nonpolar	375	C_5–C_{10} hydrocarbons, pesticides, steroids
Diethyleneglycol succinate (DEGS)	Polyester	Polar	190	Esters, fatty acids
Butanediol succinate	Polyester	Intermediate polarity	225	Esters, fatty acids

effectively separated by using a polar liquid phase, whereas nonpolar compounds are best separated using a nonpolar liquid phase.

The stationary liquid phase is normally coated as a thin film on an inert solid support. The support should be composed of small, uniformly meshed granules, so that a large surface area of the liquid phase is available for contact with the vapor phase. The high surface area is necessary to ensure efficient separation. Some common types of solid supports are given in Table 6.2. To coat the support with the mobile phase, a known weight of the solid is suspended in a solution of a low-boiling solvent containing the amount of the liquid phase required to give the desired loading, usually 2–25%. The low-boiling solvent is evaporated, leaving the solid granules evenly coated. The column is then filled with these granules. Columns are now commercially available with a wide variety of liquid phases on different solid supports.

TABLE 6.2   Solid Supports for Gas-Liquid Chromatography

Chromosorb P	Pink diatomaceous earth (surface area: 4–6 m²/g)
Chromosorb W	White diatomaceous earth (surface area: 1–3.5 m²/g)
Crushed Firebrick	40/60 mesh Teflon 6
Chromosorb T	

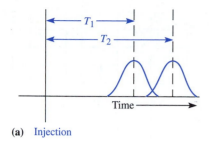

(a)   Injection

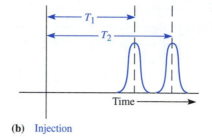

(b)   Injection

An alternative method of supporting the liquid phase is used in capillary columns of the Golay type. In these columns, which can be as long as 300 m and have diameters of only 0.1–0.2 mm, the liquid is coated directly onto the inner walls of the tubing. These columns are highly efficient but relatively expensive.

In general, the efficiency, or **resolution,** of a column increases with increasing length and decreasing diameter. Increasing the path length increases the difference in retention times between bands, whereas decreasing the diameter of the column gives rise to narrower bands. With a small band separation, as measured from the band centers, wide bands are more likely to overlap (Fig. 6.11a) than narrow bands (Fig. 6.11b).

Two other experimental factors may be varied and will affect the degree of separation of the bands: the *temperature* at which the column is maintained and the *flowrate* of the carrier gas. Increasing the temperature results in shorter retention times, because the solubility of gases in liquids decreases with increasing temperature. The partition coefficients are thus affected, and the bands move through the column at a faster rate. Higher flow rates also cause retention times to decrease. In spite of the decreased resolution and band separation obtained at higher temperatures and flowrates, these conditions are sometimes necessary for substances otherwise having very long retention times.

6.4.4   ANALYTICAL AND PREPARATIVE GAS-LIQUID CHROMATOGRAPHY

Gas-liquid chromatography may be used as an analytical tool for the qualitative and/or quantitative analysis of a sample or as a technique to separate, purify, and characterize the components of a mixture.

Qualitative Analysis.   The **retention time** of a pure compound is constant under a specified set of conditions, including the column, temperature, and flowrate used, as noted earlier. Consequently this property may be used as a first step in the identification of an unknown compound or the individual components in a mixture. In a typical experiment, an unknown compound or mixture is injected into the injection port of a GC, and the retention time(s) of the component(s) is (are) measured. A series of known samples are then injected under the same conditions. Comparison of the retention times of the standard samples with those of the unknown allows a preliminary identification of the component(s) of the unknown. However, observation of the same retention time for a known and an unknown substance is a *necessary but not sufficient* condition to establish identity, since it is possible for two different compounds to have the same retention

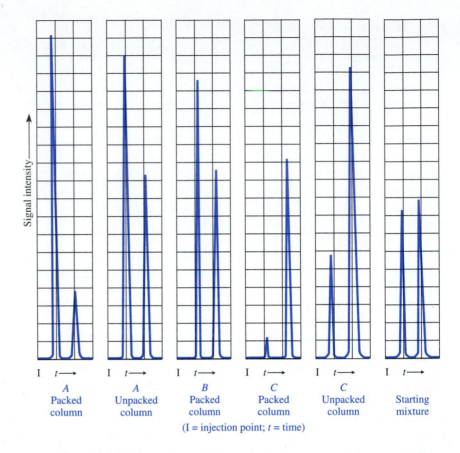

Signal intensity →

I t→	I t→	I t→	I t→	I t→	I t→
A	*A*	*B*	*C*	*C*	Starting
Packed	Unpacked	Packed	Packed	Unpacked	mixture
column	column	column	column	column	

(I = injection point; *t* = time)

time. Independent confirmation of the identity by spectral (see Chap. 8) or other means is imperative.

An example of the use of GLC as a qualitative, analytical tool is illustrated by examination of Figure 6.12. These sets of peaks represent a gas chromatographic separation of the distillation fractions of a mixture of ethyl acetate and *n*-butyl acetate similar to those obtained in the distillation experiment described in Section 4.2.3. The notations *A*, *B*, and *C* refer to the three fractions taken in that experiment. The individual peaks in the mixture may be identified by comparing their retention times with those of pure ethyl acetate and pure *n*-butyl acetate; the peak with the shorter retention time in the mixture is ethyl acetate, whereas the peak with the longer retention time is *n*-butyl acetate.

Quantitative Analysis. The voltage output of the detector is related to the mole fraction of the material being detected in the vapor, so there is a correlation between the *relative areas* under the peaks in the chromatogram and the *relative amounts* of each of the components in the mixture. The quantitative evaluation of the chromatogram thus requires reliable methods for determining these peak areas. An **electronic integrator,** which measures the intensity of detector output as a function of time, is the most accurate method for determining peak areas. However, since these devices are expensive, they are usually found only in research laboratories. A number that is proportional to the area of the peak may be obtained by manually tracing the perimeter of the peak with a device called a

planimeter. Since the thickness and density of chart paper is reasonably uniform, another means of determining the relative areas involves carefully cutting the peaks out with scissors; the peak areas are then assumed to be proportional to their weight, as measured on an analytical balance. The original chromatogram should be saved as a permanent record, so the peaks should be cut from a photocopy of the chromatogram, which must also be on uniformly thick and dense paper. If the peaks are symmetrical, as are those shown in Figure 6.12, the areas may be approximated by assuming them to be equilateral triangles. The area of a symmetrical peak is then determined by multiplying the width of the peak at its half-height times its height. The percentage of each component in the mixture may be computed as the area of the peak corresponding to that component, expressed as a percentage of the sum of the areas of all peaks in the chromatogram. A sample calculation of this type is shown in Figure 6.13.

Although the peak areas are related to the mole fraction of the component in the mobile phase, they are not *quantitatively* related, since detector response varies with the class of compound. Not all compounds have the same thermal conductivity (TCD), nor do they ionize in a hydrogen flame to form the same types or number of ions (FID). Thus, it is necessary to *correct* the measured areas in the chromatogram using the appropriate **response factor** to obtain an accurate quantitative analysis of the mixture. Although response factors for different compounds may be determined experimentally, approximate values may be obtained from monographs on gas chromatography. The response factors for thermal conductivity and flame ionization detectors for compounds that you may encounter in this experiment are given in Table 6.3. Notice that the correction factors vary more widely for flame ionization detectors than for thermal conduc-

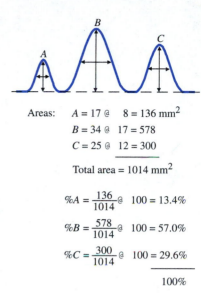

Areas: $A = 17$ @ $8 = 136\ mm^2$
$B = 34$ @ $17 = 578$
$C = 25$ @ $12 = 300$

Total area $= 1014\ mm^2$

$\%A = \dfrac{136}{1014}$ @ $100 = 13.4\%$

$\%B = \dfrac{578}{1014}$ @ $100 = 57.0\%$

$\%C = \dfrac{300}{1014}$ @ $100 = 29.6\%$

100%

FIGURE 6.13   Determination of percentage composition of a mixture by gas chromatography.

TABLE 6.3   Weight (W_f) and Molar (M_f) Correction Factors for Some Representative Substances[a]

Substance	Thermal Conductivity		Flame Ionization	
	W_f	M_f	W_f	M_f
Benzene	1.00	1.00	1.00	1.00
Toluene	1.02	0.86	1.01	0.86
Ethylbenzene	1.05	0.77	1.02	0.75
Isopropylbenzene	1.09	0.71	1.03	0.67
Ethyl acetate	1.01	0.895	1.69	1.50
n-Butyl acetate	1.10	0.74	1.48	0.995
Heptane	0.90	0.70	1.10	0.86
o-Xylene	1.08	0.79	1.02	0.75
m-Xylene	1.04	0.765	1.02	0.75
p-Xylene	1.04	0.765	1.02	0.75
Ethanol	0.82	1.39	1.77	3.00
Water	0.71	3.08	—	—

[a]McNair, H. M.; Bonelli, E. J. *Basic Gas Chromatography*, 3rd ed., Varian Aerograph, Walnut Creek, CA, 1967.

tivity detectors. In the experimental section, you may have the opportunity to calculate the response factor for an unknown compound.

To analyze a mixture of substances quantitatively when **weight factors** are known, the peak area for each component is simply *multiplied* by the weight factor for that particular compound. The resulting *corrected* areas are used to calculate the percentage composition of the mixture according to the procedure outlined in Figure 6.13. Note that using these factors provides the composition on a *weight percentage* basis.

Calculation of the composition of a mixture on a mole percentage basis requires the use of **mole factors,** M_f. These are obtained by dividing the weight factors by the molecular weights of each component of the standard solution and normalizing the resulting numbers. A sample calculation utilizing mole correction factors is provided below in the analysis of a mixture of ethanol, heptane, benzene, and ethyl acetate with a gas chromatograph equipped with a thermal conductivity detector. The last column shows the percentage composition calculated directly from the measured peak areas, without correction for detector response. The dramatic differences in the calculated composition with and without this correction, as noted in the last two columns, underscore the necessity of making this correction for quantitative analysis.

Compound	Area (A) (mm²)	M_f	$A \times M_f$	Mol % ($A \times M_f/194$ $\times$ **100**)	Uncorrected % ($A/207.1$ $\times$ **100**)
Ethanol	44.0	1.39	61.2	31.5	21.2
Heptane	78.0	0.70	54.6	28.1	37.7
Benzene	23.2	1.00	23.2	11.9	11.2
Ethyl acetate	61.9	0.895	55.4	28.5	29.9
Total	207.1		194.4	100	100

Let us now reexamine the chromatograms in Figure 6.12. A quick inspection of the chromatograms for fractions *A* and *C* reveals that the packing in the Hempel column has increased the number of theoretical plates in the column and increased the efficiency of the separation of ethyl acetate from *n*-butyl acetate. Although a semiquantitative evaluation of chromatograms is sufficient for some purposes, it is often necessary to establish the relative amounts of the different components of a mixture accurately. For example, the first step in determining the relative amounts of ethyl acetate and *n*-butyl acetate in fractions *A* and *C* involves measuring the relative areas for the two peaks in fractions *A* and *C*. These measured areas must then be corrected by multiplying the peak areas by the appropriate response factors, which vary with the type of detector used, for ethyl acetate and *n*-butyl acetate. For ethyl acetate and *n*-butyl acetate, the *weight correction factors* for a thermal conductivity detector are 1.01 and 1.10, whereas for a flame ionization detector the corresponding correction factors are 1.69 and 1.48. Using these corrected values for the peak areas, the composition of the mixture on a *weight percentage basis* may be accurately determined according to the procedure outlined in Figure 6.13.

Preparative Gas-Liquid Chromatography.   Frequently it is necessary to separate the components of a reaction mixture and isolate sufficient quantities of each for further characterization or use in a chemical reaction. The size of a sample that may be injected depends upon the diameter and length of the column and the percent loading of the liquid phase on the solid support. With the small columns typically used for analytical determinations, only small samples of approximately 0.1–5 μL may be injected. This is clearly inconvenient if significant quantities of material must be separated, so specially designed instruments having large columns up to 30 cm in diameter are used for preparative separations. Chromatographs fitted with thermal conductivity detectors are generally used in these applications because these detectors do not destroy the sample and there is no requirement for the high sensitivity that is characteristic of flame ionization detectors. The recorder indicates when the material is passing through the detector. As different peaks are observed, different collection vessels are used for each separate component.

Pre-Lab exercises for Section 6.4, "Gas-Liquid Chromatography," are found on page PL. 19.

EXPERIMENTAL PROCEDURE

SAFETY ALERT

> The solvents and other liquids that are used in sample mixtures for gas chromatography analysis are flammable and volatile. *No flames* should be used in the vicinity of these liquids.

A.   NOTES ON INJECTION OF SAMPLES INTO A GAS CHROMATOGRAPH

Liquid samples are injected into the heated injection port of a gas chromatograph using a syringe with a capacity of 1 to 10 μL for analytical work and 50 μL or more for preparative work. The sample is either injected neat or dissolved in a volatile liquid such as diethyl ether or pentane. The sample should not contain nonvolatile substances that may eventually clog the injection port or contaminate the stationary phase of the column.

Syringes are precision-made, gas-tight, and expensive. They should be handled with care and used according to the following procedure. To fill the syringe draw slightly more than the desired volume of the sample into the barrel by withdrawing the plunger as needed, point the syringe needle-up, push out the excess liquid, and then wipe the tip of the needle with a tissue. To inject the sample, insert the needle *straight* into the rubber injection septum as far as it will go, push the plunger all the way in with one quick motion, and remove the syringe from the septum while holding the plunger in place. If the sample is injected slowly, it will be spread along the column and the peaks will be broadened. Be careful to avoid bending the needle or the plunger during the course of injection. It is important to clean the syringe immediately after use. Rinse it with a volatile solvent such as acetone and dry it by drawing a stream of air through it.

Since there is considerable variation in the operating procedures of different commercial gas chromatographs, consult your instructor for specific directions to use the instrument in your laboratory.

B. IDENTIFICATION OF AN "UNKNOWN" COMPOUND BY GAS-LIQUID CHROMATOGRAPHY

Using a known mixture of compounds, obtain a gas chromatogram as directed by your instructor. A suitable mixture would be one containing some or all of the following compounds: ethyl acetate, toluene, *n*-butyl acetate, ethylbenzene, and isopropylbenzene.

Obtain a gas chromatogram of an authentic sample of one of the components of the mixture under the same instrumental conditions used for the known mixture. You will not know which of the components you have been given.

Identify the unknown compound as follows: (1) Compare the retention time of the unknown compound with the retention times of the known components of the mixture. (2) Prepare a sample mixture by adding about one volume of your unknown compound to two volumes of the known mixture, and prepare a gas chromatogram of this new sample mixture. (3) Confirm the identity of the unknown compound by observing which of the peaks of the known mixture has been amplified in the chromatogram of the sample mixture.

C. QUANTITATIVE ANALYSIS OF A MIXTURE BY GAS-LIQUID CHROMATOGRAPHY

Obtain a mixture of three or more of the compounds listed in Part A and prepare a gas chromatogram as directed by your instructor. Using one of the quantitative methods described in the Section 6.4.4, determine the relative amounts, in percentages, of the components of the mixture. Consult with your instructor to determine which, if any, correction factors to use.

D. DETERMINATION OF THE RESPONSE FACTOR FOR AN UNKNOWN

Prepare a standard solution containing known *weights* of the substances whose response factors are to be determined and a known weight of a standard substance, which may be any of the compounds given in Table 6.3. Determine the density of this solution so that you can calculate the exact weight of each component, including the standard, in the volume of sample you inject into the gas chromatograph. The size of the sample that you inject will depend on a variety of factors, and you should consult with your instructor for guidance on the exact amount to be used; a volume of 1–5 μL is normally suitable. Inject the mixture and obtain the chromatogram. You may need to obtain a chromatogram of each of the individual components so that you will know the retention time of each.

Determine the area of each of the various peaks, and divide each by the exact weight of that component in the sample. Normalize the *inverse* of each of the resulting numbers to the standard by dividing each by the number so obtained for the standard.

E. OTHER ANALYSES

Some of the mixtures encountered in other experiments in this text are appropriate for demonstrating gas chromatographic analysis and may be assigned by your instructor. These include:

1. The fractions collected from the fractional distillation of mixtures of ethyl acetate and *n*-butyl acetate in Section 4.2.2.

2. The mixture of chloroheptanes produced in the experiment of Section 9.2.

3. The mixture of alkenes produced by the dehydrochlorination of 2-chloro-2-methylbutane in the experiment in Section 10.2 or the dehydration of 4-methyl-2-pentanol in the experiment in Section 10.3.

4. The isomeric propyl-*p*-xylenes produced in the Friedel-Crafts alkylation experiment in Section 15.2.

5. The impure 2-methylpropanal prepared in the oxidation experiment in Section 16.2, which contains unchanged starting material.

Finishing Touches   Pour any unused *volatile organic compounds* into the appropriate container for nonhalogenated liquids or halogenated hydrocarbons; put any mixture that contains a halogenated hydrocarbon in the container for halogenated hydrocarbons.

EXERCISES

1. Define the following terms.
   a. stationary phase   d. retention time
   b. mobile phase   e. solid support
   c. carrier gas   f. thermal conductivity

2. Refer to the GLC traces given in Figure 6.12. These are analyses of the various fractions collected during the fractional distillation of the mixture of ethyl acetate and *n*-butyl acetate. Utilizing the GLC correction factors for a thermal conductivity detector provided in Table 6.3, determine accurately both the weight percent and the mole percent compositions of the fractions *A*, *B*, and *C* obtained using a packed distillation column.

3. Benzene (4 g, 0.05 mol) is subjected to Friedel-Crafts alkylation with 1-chloropropane (3.9 g, 0.05 mol) and $AlCl_3$. The product (4.8 g) is subjected to analysis on a gas chromatograph equipped with a thermal conductivity detector. The chromatogram shows two product peaks identified as *n*-propylbenzene (area = 60 mm^2; W_f = 1.06) and isopropylbenzene (area = 108 mm^2;

$W_f = 1.09$). Calculate the percent yield of each of the two isomeric products obtained in this reaction. Note that since each of the products has the same molecular weight of 120, the use of weight factors gives both weight and mole percent composition.

4. Explain how the response factor for a particular substance would have an effect on the sample size you use to determine the response factor of an unknown.

STEREOISOMERS

An important part of organic chemistry is knowledge of the three-dimensional characteristics of organic molecules. The experiments in this chapter are designed to provide experience with **stereoisomers.** Such isomers have molecular constitutions or molecular skeletons that are identical, but they differ in the three-dimensional orientation of their atoms in space. The two broad subclasses of stereoisomers that are of importance in organic chemistry are **conformational** isomers and **configurational** isomers.

Conformational isomers, as illustrated by two **Newman projections 1** and **2** for 1,2-dibromoethane, are stereoisomers that can be interconverted by rotation about single bonds. The interconversion of such isomers is usually a low-energy process and thus occurs rapidly at room temperature and below. Consequently, isolation of conformational isomers is rarely possible.

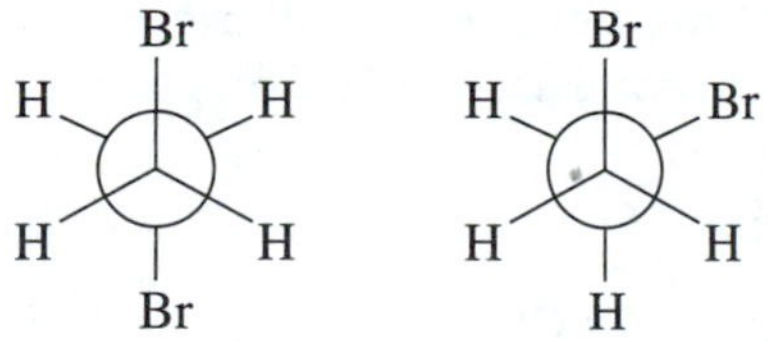

The second important class of stereoisomers is configurational isomers, whose interconversion would require breakage and re-formation of a chemical bond. An example of this type of stereoisomerism is seen with carvone, a natural product for which the two representations **3** and **4** can be written. These are related as **nonsuperimposable** mirror images, and by definition **3** and **4** are **enantiomers.** Molecules having a nonsuperimposable mirror image are said to be **chiral,** a term that is derived from the Greek word, *cheir,* meaning "hand." Because hands are perhaps the best-known example of nonsuperimposable mirror images, use of the word "chiral" to characterize molecules that have a "handedness" is certainly appropriate. A companion term, **achiral,** describes all molecules having **superimposable** mirror images. Under most circumstances, the chiral compounds that you will be preparing or using in the laboratory will be 50:50 mixtures of the two enantiomers, a composition referred to as the **racemate** or the **racemic mixture.** An equimolar mixture of **3** and **4** would thus be called a racemate.

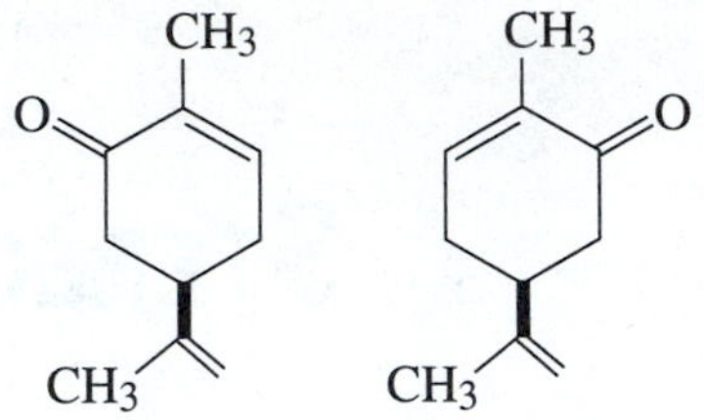

3   **4**

Because the interconversion of enantiomers involves bond cleavage and re-formation, it is generally a relatively high-energy process, so the separation of such stereoisomers from one another becomes possible. From the experimental standpoint, this separation, or **resolution,** as it is commonly called, can be diffi-cult because *enantiomers have identical physical and chemical properties in an achiral environment.* An important exception to this general statement is their effect on plane-polarized light: Enantiomers rotate the plane of such light an equal number of degrees but in *opposite* directions. For this reason they are sometimes called **optical isomers** and are said to possess **optical activity.** The effect of enantiomers on plane-polarized light is not useful in the resolution process, however, so various techniques have been developed to accomplish the resolution of enantiomers. These are described more fully in Section 7.6.

Even *achiral* molecules may exist as *separable* stereoisomers. Examples include *cis* and *trans*, or *E-* and *Z-*, alkenes such as **5** and **6,** and cyclic com-pounds such as **7** and **8.** For each of these isomeric pairs it can be seen that the molecules are nonsuperimposable, yet are not mirror images; that is, they are *not* enantiomers. Configurational isomers like these two pairs that are *not* enan-tiomers are designated **diastereomers,** and, as might be expected, such sub-stances have different chemical and physical properties.

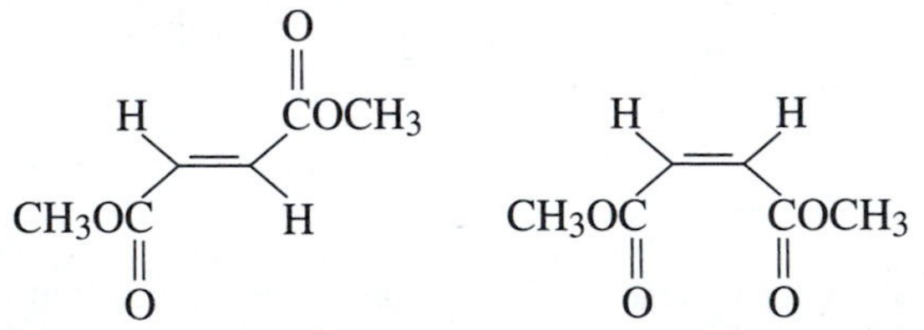

5   **6**

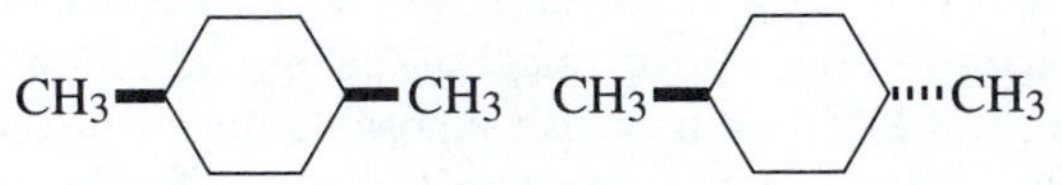

7   **8**

Chiral compounds can also exist in diastereomeric forms, although only if two or more **stereocenters** are present. In organic compounds a stereocenter is usually a carbon atom bearing four different substituents. Consider 1,2-cyclo-hexanediol, **9,** for example, which contains the two stereocenters marked by asterisks. Each center consists of a carbon atom bonded to (a) a hydroxyl group, (b) a hydrogen atom, (c) a methylene group, which is part of the ring, and (d) the carbon atom of the other stereocenter. Three configurational isomers, **10–12,** can then be written for **9.** It is clear that the *trans* isomer can exist as the enantiomers **10** and **11,** but that **12** is stereochemically different from both of them, being diastereomeric with each of the enantiomers **10** and **11.** Moreover, even though it contains two stereocenters, **12** has a *superimposable mirror image*. We could therefore variously describe this isomer as being achiral, optically inactive, or **meso.** The last term describes those configurational isomers that have two or more stereocenters *and* a superimposable mirror image. Although the identity of the physical and chemical properties of the enantiomers **10** and **11** might make their separation from one another difficult, the isolation of **12** free from contamination by **10** and **11** would be comparatively easy because diastereomers have different physical and chemical characteristics.

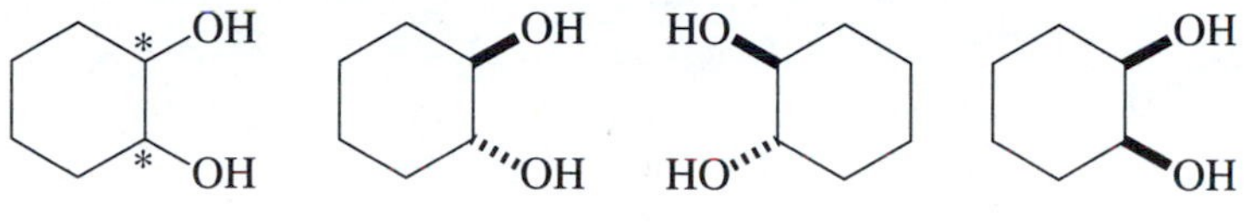

      9   **10**   **11**   **12**

The procedures in this chapter are chosen to illustrate: (a) separation of diastereomers by chromatography (Sec. 7.2); (b) conversion of one diastereomer into another (Sec. 7.3); (c) isolation of the enantiomers of carvone (Sec. 7.5); and (d) resolution of a racemate (Sec. 7.6). A description of the technique of polarimetry is given in Section 7.4 to support the last two studies.

7.2   SEPARATION OF DIASTEREOMERIC 1,2-CYCLOHEXANEDIOLS

The stereochemistry of cyclic 1,2-diols is a subject of interest because these compounds can be produced stereoselectively from cycloalkenes. For example, a racemic mixture of the enantiomeric *trans*-1,2-cyclohexanediols **10** and **11** is produced by reaction of a peracid with cyclohexene, followed by hydrolysis of the intermediate epoxide **13** (Eq. 7.1). Oxidation of this alkene by permanganate, on the other hand, gives *cis*-1,2-cyclohexanediol (**12**), as shown in Equation (7.2). The intermediate in this process is presumably the cyclic manganese-containing species **14,** which is not isolated.

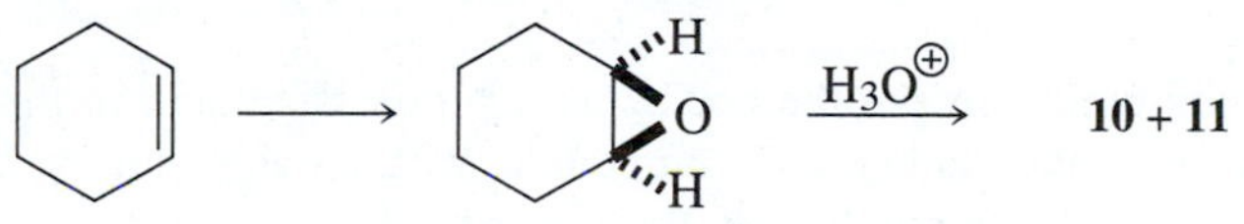

$$\qquad\qquad\qquad\qquad\qquad\qquad\qquad\qquad\qquad\qquad\qquad\qquad (7.1)$$

            13

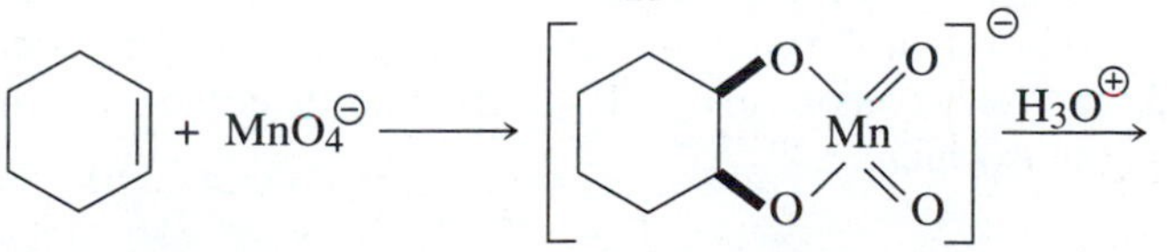

$$\qquad\qquad\qquad\qquad\qquad\qquad\qquad\qquad\qquad\qquad\qquad\qquad (7.2)$$

            14

In this experiment a commercial mixture containing the *racemate* **10, 11,** and the *meso* isomer **12** will be separated into its components; the mixture is produced by the catalytic hydrogenation of catechol (**15,** Eq. 7.3). Because the enantiomeric *trans*-diols **10** and **11** have identical physical properties, except for their effect on plane-polarized light, they will not be resolved from one another in the procedure of this section. The *meso*-isomer **12,** on the other hand, being diastereomeric to **10** and **11,** has physical and chemical properties that are sufficiently different that it can be separated from the other two stereoisomers.

$$\text{15} \quad + \ 3 \ \text{H}_2 \quad \xrightarrow[\text{Heat, pressure}]{\text{Catalyst}} \quad \textbf{10} + \textbf{11} + \textbf{12} \qquad \textbf{(7.3)}$$

The separation of diastereomers from one another may often be done solely on the basis of differing solubilities, but this is not the case with the 1,2-cyclohexanediols because they have similar solubilities. The desired separation is possible by chromatographic means, however, and the technique of TLC (see Sec. 6.3) is used here, although column (see Sec. 6.2) or gas chromatography (see Sec. 6.4) would also work.

A challenge to you, the experimentalist, is to identify which spot on the TLC plate corresponds to the racemic mixture of *trans* isomers and which to the *cis* diastereomer. In other words, it is one thing to separate different compounds and quite another to assign specific structures to them. This is a problem commonly faced by organic chemists and can be addressed in a variety of ways. For example, you might extract the portions of the TLC plate containing each of the isomers and determine the melting point or IR or NMR spectral characteristics of the solid left after removal of solvent. Spectral characterization is difficult to do in this case because only extremely small amounts of material would be obtained from the TLC plate. In addition, the melting points of the diastereomers are similar, being 103–104 °C for **9** and **10** and 100–101 °C for **11;** thus, traces of impurities in the *trans* isomer could easily lower its melting range to that of the *cis*, making unambiguous assignment of structure impossible.

The method used to make the needed structural assignment in this experiment is to compare the "unknown" compounds with an authentic specimen of "known" material, a technique frequently used by experimentalists for purposes of identification. Specifically, by spotting the TLC plate with authentic samples of either *cis*- or *trans*-1,2-cyclohexanediol as well as the commercial mixture, you are able to make a direct comparison of R_f values and determine which component of the mixture is which.

Pre-Lab exercises for Section 7.2, "Separation of Diastereomeric 1,2-Cyclohexanediols," are found on page PL. 21.

SAFETY ALERT

1. Acetone and petroleum ether are highly flammable; do not have any flames in the vicinity when they are being used.
2. Do not allow iodine to come in contact with your skin and do not inhale its vapor, as it is corrosive and toxic.

Consult Section 6.3.1 for directions for performing a TLC separation.

Obtain or prepare *ca.* 5% solutions in acetone of a commercial mixture of *cis*- and *trans*-1,2-cyclohexanediol and of pure *trans*-1,2-cyclohexanediol. Place spots of each solution side by side on a 10-cm × 3-cm silica gel TLC plate. Develop the chromatogram using a mixture of 75% petroleum ether, bp 60–80 °C (760 torr), and 25% 2-propanol by volume. Remove the plate from the solvent mixture when the solvent front approaches the top of the plate, mark the position reached on the plate by the solvent front, air-dry the plate for a few minutes, and place it in a closed container with a few iodine crystals to make the spots visible. Remove the plate and circle the location of any spots on it. Record the R_f-value of each spot.

Finishing Touches   Transfer any excess of the *eluting solvents* to a container for nonhalogenated organic liquids. Dispose of the used *TLC plates* in a container for nonhazardous materials.

SPECTRA OF STARTING MATERIALS

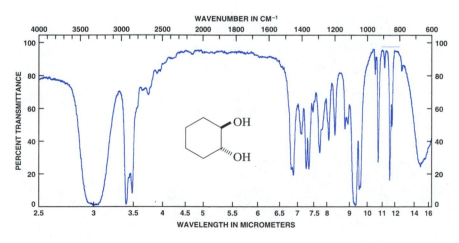

FIGURE 7.1 IR spectrum of *trans*-1,2-cyclohexanediol (KBr pellet).

FIGURE 7.2 NMR data for *trans*-1,2-cyclohexanediol.

(a) ^{1}H NMR spectrum (90 MHz).

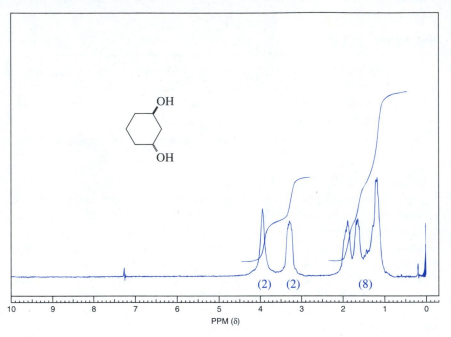

(b) ^{13}C NMR data.
Chemical shifts: δ 24.2, 32.8, 74.7.

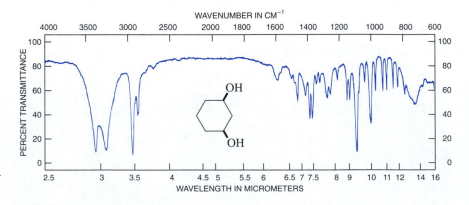

FIGURE 7.3 IR spectrum of *cis*-1,2-cyclohexanediol (KBr pellet).

(a) ¹H NMR spectrum (90 MHz).

FIGURE 7.4   NMR data for *cis*-1,2-cyclohexanediol.

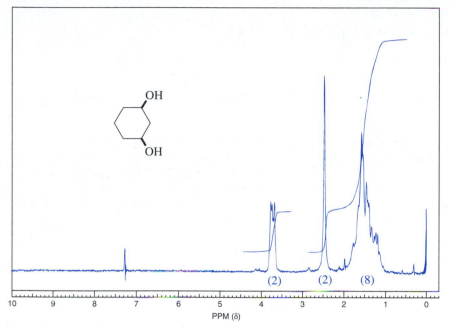

(b) ¹³C NMR data.
   Chemical shifts: δ 21.6, 30.0, 70.4.

EXERCISES

1. By comparing the R_f values of the separated spots from the mixture of the *cis* and *trans* diols with the R_f value of the *trans* isomer, decide which isomer is adsorbed more strongly by the silica gel. Give a reason for the difference in adsorptivity of the two isomers.

2. It may happen that the separation of the isomers by TLC is not complete—one of the spots may be found to be smeared out to some extent. Considering this, which one of the isomers could more easily be obtained pure by column chromatography?

3. What would be the consequence of using pure petroleum ether as the eluting solvent?

4. What would be the consequence of using pure 2-propanol as the eluting solvent?

5. Compare Figures 7.2 and 7.3. In what areas are the IR absorptions similar and in what areas are they different? Explain the similarities and the differences.

6. In Section 7.1 it was stated that interconversion of configurational isomers requires the cleavage and re-formation of a chemical bond. Indicate any such bonds in *cis*-1,2-cyclohexanediol (**12**) that upon being broken and then re-formed would afford the enantiomers **10** and **11**. Be sure to describe what must occur stereochemically during the process of bond cleavage and re-formation to effect the interconversion of the **12** to the other two diastereomers.

7. (a) Consider the bond dissociation energies of the various bonds in **12** and then predict which bond would be easiest to break for the conversion of **12** to **10** and **11**.

(b) Estimate the $\Delta H^{\ddagger}$ for this process.

7.3   ISOMERIZATION OF DIMETHYL MALEATE TO DIMETHYL FUMARATE

Dimethyl fumarate (**5**) and dimethyl maleate (**6**) are examples of *trans* and *cis* isomers, called **geometric isomers.** Their names do not reflect this relationship, however, because the parent diacids from which they can be prepared were discovered and named before this stereochemical fact was known. Being diastereomers, these diesters would be expected to have different physical and chemical properties, and indeed they do. The difference in their physical states at room temperature is the most dramatic: The fumarate **5** is a *solid* with a melting point of 103–104 °C, whereas **6** is a *liquid* with a freezing point of about –19 °C.

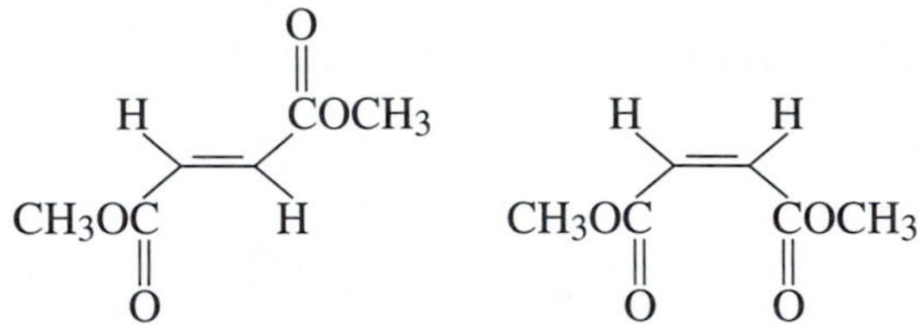

As discussed in Section 7.1, the interconversion of configurational isomers involves the cleavage and subsequent re-formation of at least one chemical bond. The geometric isomers **5** and **6,** for example, could be equilibrated by rupture of the π component of the double bond, rotation about the single bond that remains, and regeneration of the double bond. The bond cleavage necessary to initiate the isomerization can be promoted (a) thermally, (b) photochemically, or (c) chemically, by reversible addition of a reagent to the π-bond. Consequently, several mechanisms can be proposed for the interconversion of **5** and **6.** For illustrative purposes, one *possible* mechanism is given in Equation (7.4). In this particular example, addition of bromonium ion, Br^+, to the π-bond initiates the isomerization.

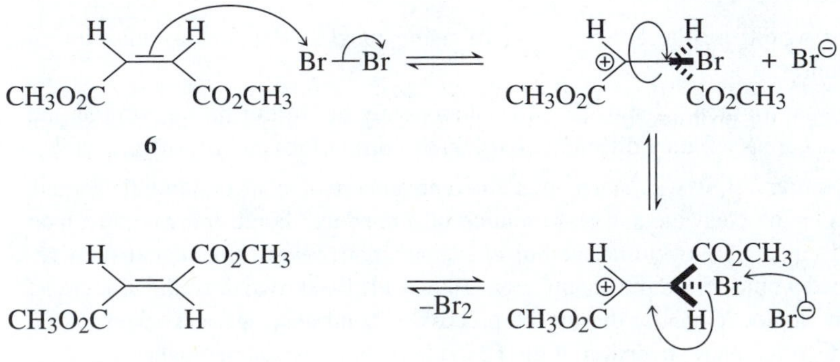

(7.4)

This experiment involves the conversion of dimethyl maleate (**6**) to dimethyl fumarate (**5**). Under the conditions used, a true equilibrium between these diastereomers is not established because **5** precipitates from solution as it is formed. Three different procedures are followed to study the reaction: exposure to light of two samples of **6,** one of which contains a catalytic amount of bromine, and treating a third sample with bromine but keeping the mixture in the dark. By thoughtful analysis of your results, it is possible to discover which of the three modes of breaking the π-bond, which were designated (a)–(c) in the previous paragraph, is responsible for the isomerization. The analysis also allows you to determine whether the mechanism involves polar intermediates or radical species (see Exercise 3).

Pre-Lab exercises for Section 7.3, "Isomerization of Dimethyl Maleate to Dimethyl Fumarate," are found on page PL. 23.

EXPERIMENTAL PROCEDURE

SAFETY ALERT

1. *Bromine is a hazardous chemical.* Do not breathe its vapors or allow it to come into contact with the skin because it may cause serious chemical burns. Perform all operations involving the transfer of solutions of bromine at a hood and wear rubber gloves when handling bromine or solutions of it. If you get bromine on your skin, immediately wash the affected area with soap and warm water and soak the skin in 0.6 *M* sodium thiosulfate solution, for up to 3 h if the burn is particularly serious.
2. Do *not* use your mouth to fill a pipet for transfer of the bromine/CCl$_4$ solution; use a dropper with a rubber bulb.
3. Bromine reacts with acetone to form α-bromoacetone, BrCH$_2$COCH$_3$, a powerful *lachrymator*. Do *not* rinse glassware that might contain residual bromine with acetone! Rather, follow the procedure described in Finishing Touches.

Place 0.5 mL of dimethyl maleate in each of three 100-mm test tubes and with a Pasteur pipet add enough of a 0.6 *M* solution of bromine in carbon tetrachloride to *two* of the tubes to give an orange solution. Add an equal volume of carbon tetrachloride to the third test tube. Stopper all the test tubes, place one of the tubes containing bromine in the dark, and expose the other two tubes to strong light. If decoloration of a solution should occur, add an additional portion of the bromine solution. After 30 min cool all three solutions in ice water, observe in which test tube(s) crystals appear, and isolate the precipitate by vacuum filtration (see Sec. 2.16). Wash the crystals free of bromine with a little *cold* carbon tetrachloride and press them as dry as possible on the filter disk with the aid of a clean cork or spatula. Recrystallize the product from ethanol (see Sec. 3.1) and determine its melting point and weight. The reported melting range for dimethyl fumarate is 101–102 °C.

Finishing Touches Decolorize any solutions or containers in which the color of *bromine* is visible by dropwise addition of cyclohexene, and then discard the resulting solutions in a container for halogenated organic liquids. Flush the *ethanolic filtrate* from the recrystallization process down the drain with water.

SPECTRA OF STARTING MATERIAL AND PRODUCT

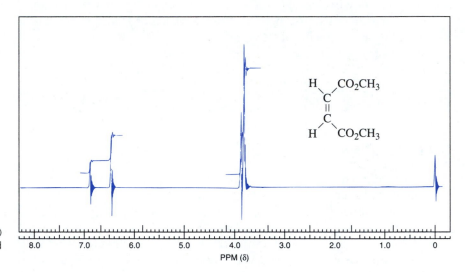

FIGURE 7.5 ^{1}H NMR spectrum (60 MHz) of a mixture of dimethyl fumarate and dimethyl maleate.

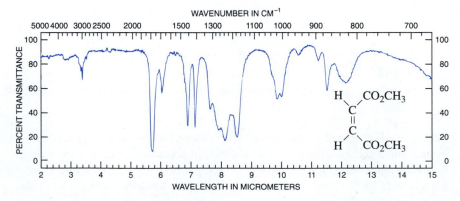

FIGURE 7.6 IR spectrum of dimethyl maleate.

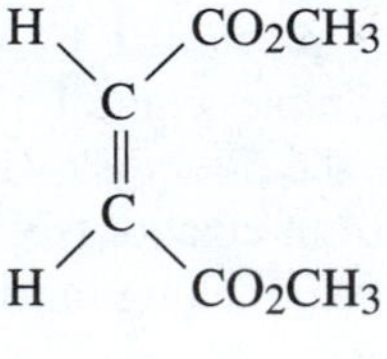

FIGURE 7.7 NMR data for dimethyl maleate.

Chemical shifts: δ 52.1, 130.1, 165.8.

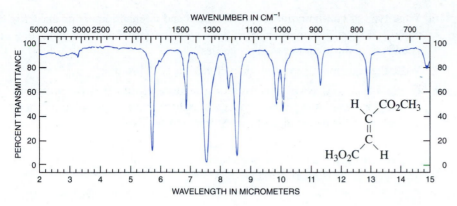

FIGURE 7.8 IR spectrum of dimethyl fumarate (KBr pellet).

(a) ^{1}H NMR Spectrum (60 MHz).

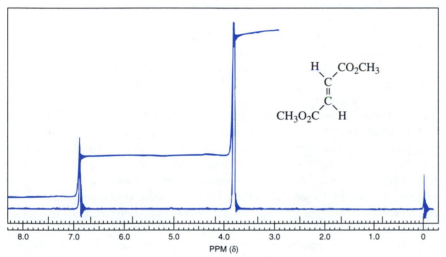

(b) ^{13}C NMR data.
Chemical shifts: δ 52.2, 133.5, 165.3.

FIGURE 7.9 NMR data for dimethyl fumarate.

EXERCISES

1. What observation allows elimination of thermal cleavage of the π-bond as the initiation step for isomerization of **6** to **5**?

2. What observation allows exclusion of direct photoexcitation of **6** as the means of initiating its conversion to **5**?

3. Write a stepwise mechanism for the isomerization of **6** to **5**. Be certain to account specifically for the roles, if any, that light and bromine play in promoting this process.

4. Explain why decoloration of a solution of bromine and dimethyl maleate is slow.

5. State which one of the two diesters **5** and **6** is predicted to be more stable thermodynamically and support your prediction.

6. What type of configurational isomers are **5** and **6,** enantiomers or diastereomers? How is your answer consistent with the fact that these two compounds have different physical properties?

7. Write the equation for reaction of cyclohexene with bromine.

8. Figure 7.5 is a ^{1}H NMR spectrum of a mixture of dimethyl fumarate and dimethyl maleate. Assign the resonances to the protons responsible for them and calculate the percentage of each isomer in the mixture.

9. The double bond of **5** and **6** is constituted of both a σ- and a π-bond. Explain why rupture of the π-bond rather than the σ-bond is the more probable event in the isomerization process.

7.4 POLARIMETRY

It was noted in Section 7.1 that chiral molecules are optically active; that is, they rotate the plane of polarized light as the light passes through the sample. An explanation of the physical basis for this phenomenon is beyond the scope of this discussion, but it is important to understand that achiral molecules do *not* exhibit this same property.

Of primary interest to organic chemists is the fact that each member of a pair of enantiomers rotates plane-polarized light by exactly the same amount but in *opposite* directions. This means that a solution containing equal amounts of two enantiomers will cause *no* net rotation of the light because the rotation in one direction by molecules of one enantiomer will be exactly offset with an opposite rotation by those of the other. Thus, a solution containing enantiomers can produce a rotation only if the amount of one enantiomer exceeds that of its mirror image.

The value **α** of an **observed rotation** depends on several factors: (a) the nature of the chiral compound; (b) its *concentration*, if in solution, or *density*, if a neat liquid; (c) the *pathlength* of the cell containing the sample; (d) the *temperature* at which the measurement is taken; (e) the nature of the *solvent*, if one is used; and (f) the *wavelength* of plane-polarized light used in the measurement. Factors (b) and (c) determine the average number of chiral molecules in the path of the light, and, in combination with the wavelength, have the greatest effect on the magnitude of the rotation of a sample.

The sign of α is defined by convention. When the observer views through the sample *toward* the source of the plane-polarized light, a clockwise rotation of the plane of the light is taken as positive or **dextrorotatory,** a counterclockwise rotation as negative or **levorotatory.**

Optical rotation represents a **physical constant** of a chiral compound *if* the variables listed above are considered. By specifying the temperature and wavelength at which the measurement is taken, and dividing the observed rotation by the factors that define the average number of molecules in the light path, a constant called the **specific rotation, [α]**, is obtained. This is expressed mathematically by Equation (7.5).

$$[\alpha]_{\lambda}^{T} = \frac{\alpha}{\ell \times c} \quad \text{or} \quad \frac{\alpha}{\ell \times d} \qquad (7.5)$$

where
$[\alpha]$ = specific rotation (degrees)

λ = wavelength (nanometers, nm)

T = temperature (°C)

α = observed rotation (degrees)

ℓ = path length of cell (decimeters, dm)

c = concentration (g/mL of solution)

d = density (g/mL, neat sample)

A specific rotation might be reported as $[\alpha]_{490}^{25}$ $-19.6°$ ($c = 1.8$, $CHCl_3$). Alternatively, the specific rotation might be written as $[\alpha]_{D}^{23}$ $+39.2°$ ($c = 12.1$, CH_3OH). The D refers to the fact that a sodium lamp emitting radiation at 589 nm, the D line of sodium, was used as the light source.

The basic components of all **polarimeters**, the instruments used for measuring optical rotations, are shown in Figure 7.10. Ordinary light first passes through a polarizer, such as a Nicol® prism or a sheet of Polaroid® film, and the plane-polarized light that emerges passes through the sample tube. If the tube contains an optically active substance, the plane of the light will be rotated either clockwise or counterclockwise. The degree of rotation can be determined by turning the analyzer, another Nicol prism or piece of Polaroid film, until the intensity of the light reaching the detector matches that observed in the absence of any of the optically active compound in the sample tube. The human eye or a photoelectric cell constitutes the detector. Finally, use of Equation (7.5) allows the conversion of the observed rotation α to the specific rotation $[\alpha]$.

REFERENCE

Shavitz, R. "An Easily Constructed Student Polarimeter," *Journal of Chemical Education*, **1978**, *55*, 682.

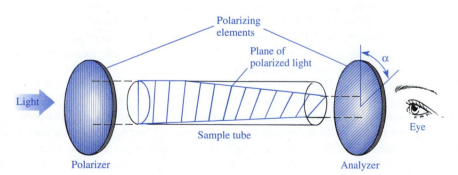

FIGURE 7.10 Schematic illustration of optical rotation (α = about $-45°$).

7.5 ISOLATION OF THE ENANTIOMERS OF CARVONE FROM ESSENTIAL OILS

It was noted in Section 7.1 that enantiomers have *identical* physical properties except with respect to their effect on plane-polarized light. Their chemical properties are also identical, with one *important* exception: One enantiomer undergoes a chemical change at a rate *different* from that of its mirror image *if* the reaction occurs in a stereochemically asymmetric environment, such as the active site of an enzyme or a biological receptor. This is because the active sites of enzymes and receptors are themselves chiral and, as a consequence, complex preferentially with one member of a pair of enantiomers. The complexed enantiomer will then undergo the enzyme-promoted reaction faster than will its mirror image.

A dramatic example of such differential complexation is the olfactory response to the two enantiomers of carvone. *R*-(–)-Carvone (**3**) smells like spearmint and is the principal component of spearmint oil, which also contains minor amounts of limonene (**16**) and α- and β-phellandrene (**17** and **18**). On the other hand *S*-(+)-carvone (**4**), along with limonene, is found in caraway seed and dillseed oils, and has been shown to be the compound largely responsible for the characteristic odor of these oils. This remarkable difference in how we sense these two enantiomers is because the odor receptors in our noses are chiral environments linked to the nervous system. Depending on which type of receptor-site complexes with a particular enantiomer will determine what odor is detected by the brain. Further information about the relationship between chirality and odor is found in the references at the end of this section.

It is unusual for both enantiomers of a compound to occur naturally and extremely rare for each of them to be available from readily accessible sources such as spearmint and caraway oils. This fortunate situation allows you to isolate **3** and **4** from these sources and, among other things, verify that these enantiomers have different odors. Bear in mind, however, that although most people can detect the difference, 8–10% cannot, and you may be in this group.

The isolation of the enantiomers of carvone requires their separation from **16–18**. All three of these isomers have boiling points in the range of 170–177 °C, whereas **3** and **4** boil at 231–233 °C (760 torr). This large difference in boiling points suggests that fractional distillation of the oils would be an effective way to achieve the separation, and that is the option that is followed. However, to minimize thermal decomposition of the oil, the distillation is performed under reduced pressure. This narrows the difference in boiling points somewhat, but not so much as to make the separation by fractional distillation difficult. For example, at 11 torr the boiling points of **16–18** are in the range of 57–63 °C, and **3** and **4** boil at 103–105 °C.

The pertinent tests that can be performed on the enantiomeric carvones will depend upon the time and facilities that are available. If gas chromatography is available, analyses should be performed on both the original oils and the isolated **3** and **4.** Typical chromatograms are given in Figures 7.11 through 7.14. Infrared and ^{1}H NMR spectra of the pure enantiomers of carvone should be identical, of course, but this should be confirmed if possible; spectra of the pure materials are provided in Figures 7.15 and 7.16.

Confirmation of the fact that **3** or **4** has been isolated is obtained by making the 2,4-dinitrophenylhydrazone, 2,4-DNP, of the product (see Sec. 23.5.1, Part A for the procedure) and comparing its melting point to that reported. By doing a mixture melting point of the 2,4-DNPs prepared from *both* **3** and **4,** you can determine whether one enantiomeric derivative depresses the melting point of the other. The ultimate test as to which enantiomer you have isolated depends on determination of the sign of its optical rotation.

Pre-Lab exercises for Section 7.5, "Isolation of the Enantiomers of Carvone from Essential Oils," are found on page PL. 25.

EXPERIMENTAL PROCEDURE

SAFETY ALERT

1. If an oil bath is used for heating, be sure that it is supported well, and do not handle it when it is hot.
2. Examine all glassware carefully, especially the stillpot, for stars and other cracks, which might lead to implosion during the vacuum distillation.

A. ISOLATION OF CARVONE BY VACUUM FRACTIONAL DISTILLATION OF SPEARMINT OIL OR CARAWAY SEED OIL

The most important aspect of this experiment is the proper assembly of the distillation apparatus so that a satisfactory vacuum can be maintained. The following notes duplicate parts of the discussion of this technique in Sections 2.9 and 4.2, but it will be helpful to repeat here some of the details that are most pertinent to this experiment.

A vacuum fractional distillation cannot be carried out efficiently with less than 15 mL of the essential oil. Therefore, in order to keep the amount of starting material required for a large class at a minimal level, two students should work together. The amount of carvone required for determination of its optical rotation in a simple student-type polarimeter may also be more than what will be obtained from 15 mL of the essential oil. Thus, it may be necessary for several students to combine their distillates for this determination as described in Section 7.5, Part B.

Assemble the apparatus for vacuum distillation using a 50-mL stillpot, a Claisen connecting tube, and a stillhead, as shown in Figure 2.19. Lubricate all the ground-glass joints lightly and make sure that all rubber stoppers fit well. Use rubber tubing that is thick-walled so that it will not collapse under vacuum.

TABLE 7.1   Boiling Points of Carvone and Limonene at Different Pressures

Pressure (torr)	Carvone bp, °C	Limonene bp, °C
760	231	177
100	157	110
40	132	87
30	125	79
20	115	71
10	100	55

Bore the holes in the rubber stopper used for the safety trap cleanly so that the glass tubes will seal well. Make sure the capillary ebullition tube is sufficiently fine that only a *very* small stream of bubbles is produced by blowing air through the large end of the tube with the capillary end immersed in acetone. Insulate the Claisen connecting tube and stillhead by wrapping them with glasswool and aluminum foil. If the stillpot is to be heated with an oil bath, use a liquid that has a flash point above 200 °C; silicone oil is ideal. The bath may be heated with a burner, but electrical heating of the bath is preferable. An electric heating mantle may also be used in conjunction with a variable transformer. An oil bath allows better temperature control because a thermometer suspended in the oil will give a direct measure of the heat supplied to the stillpot. Weigh the flasks that will be used for the stillpot and receivers before assembling the apparatus.

Place 15 mL of spearmint oil or caraway seed oil in the 50-mL stillpot, attach a 25-mL round-bottom flask as a receiver, make sure all connections are tight, and lower the pressure using an aspirator or whatever vacuum system is available in the laboratory. A pressure of about 25–30 torr is ideal for the distillation, and the following procedure is described for a distillation performed at this pressure. If you cannot achieve this pressure, use Table 7.1 or the monograph in Section 2.9 to interpolate the appropriate boiling temperatures corresponding to your pressure.

Begin the distillation by applying heat to the stillpot. If an oil bath is used, the first drops of distillate will appear when the oil bath is at about 150 °C; the stillhead temperature will be about 80 °C. Collect the fraction that distils in the range 80 °C to about 120 °C. In the case of caraway seed oil, the temperature may remain constant in the range 80–85 °C for a short time, since this oil contains a larger amount of limonene than does spearmint oil. In any event, when the stillhead temperature rises and then plateaus at about 120 °C, remove the heating device, open the stopcock that releases the vacuum, and remove the receiver containing the first distillation fraction. Attach another 10- or 25-mL round-bottom flask as a receiver, and close the stopcock at the trap to reduce the pressure again. When a stable vacuum has been reestablished, reheat the stillpot, and collect the distillate obtained over a boiling-point range of about 120–125 °C. During this distillation, the temperature of the oil bath may be raised to about 190 °C. When no more drops of distillate are obtained, discontinue the heating, release the vacuum, and remove the receiver containing the second fraction, which should be almost pure carvone. Weigh the receivers and record the weights of the two fractions. If time permits, allow the stillhead and Claisen connecting tube to drain into the stillpot for 10–15 min and then wipe the stillpot clean and weigh it to determine the amount of residual oil.

The amount of carvone obtained from 15 mL of the essential oils should be about 8 g from spearmint oil and about 6 g from caraway seed oil, but it may be more or less, depending on the source of the oils.

B. STUDIES OF THE ENANTIOMERS OF CARVONE FROM SPEARMINT OIL AND CARAWAY SEED OIL

1. Odor.   Compare the odor of the carvone obtained from spearmint oil with that of the carvone from caraway seed oil. Smell the first fraction obtained from each oil and compare its odor with that of the carvone fraction.

2. Test for Unsaturation.   Perform tests for unsaturation according to the procedures given in Section 23.5.2.

3. Gas Chromatography.   Obtain gas chromatograms of representative samples of carvone from spearmint oil and from caraway seed oil under identical conditions so that the retention times may be compared. Also obtain chromatograms of the original essential oils as well as one from the first distillation fraction, so an appraisal can be made of the success of the distillations. Suitable conditions for the analysis are use of a column with dimensions of 2-m × 3-mm and containing 15% QF-1 on Chromosorb W 80/100 support, and, if possible, a temperature program over the range of 100–220 °C at 10 °C/min; otherwise, the analysis should be performed at a column temperature of about 170 °C.

4. IR and ^{1}H NMR Spectra.   If possible, obtain an IR and a ^{1}H NMR spectrum from a sample of carvone from each of the essential oils, and compare the spectra with those of Figures 7.15 and 7.16.

5. Optical Rotation.   The amount of carvone obtained from 15 mL of the essential oils will probably not be enough to provide a 5-cm light path in a simple polarimeter such as that described in the reference provided in Section 7.4, so that the rotation of a neat sample can be determined. If insufficient carvone is obtained for determining its rotation, combine your distillate with that obtained by others from the same oil. Calculate the specific rotations from the observed rotations using the formula of Equation (7.5) and the description given in Section 7.4. The reported density of carvone is 0.960 at 20 °C, and it has specific rotation $[\alpha]_D^{20}$ of + or – 62° depending on its source.

6. 2,4-Dinitrophenylhydrazone of Carvone.   Prepare this derivative from a 0.5-g sample of your distilled carvone using the general procedure of Section 23.5.1, Part E. The reported melting point of the crystalline product is 193–194 °C. If recrystallization is necessary, a mixture of ethanol and ethyl acetate is a satisfactory solvent. Obtain a sample of a derivative from another student who used a carvone sample distilled from a different essential oil and make a 1:1 mixture with your own derivative. Determine the melting point of the mixture. You may wish to prepare mixtures of the derivatives in different ratios and determine their melting points as well.

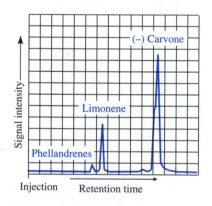

FIGURE   7.11   Gas chromatogram of spearmint oil.

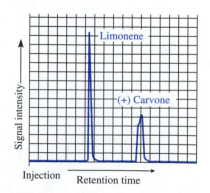

FIGURE 7.12   Gas chromatogram of caraway seed oil.

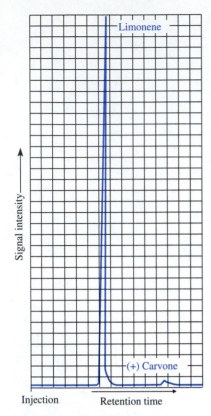

FIGURE 7.13 Gas chromatogram of Fraction 1 from distillation of caraway seed oil.

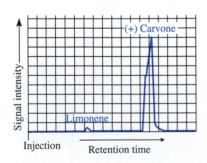

FIGURE 7.14 Gas chromatogram of Fraction 2 from distillation of caraway seed oil.

FIGURE 7.15 IR spectra of the enantiomeric carvones.

Finishing Touches Put the *pot residues* from the distillation of spearmint or caraway oil in a container for nonchlorinated organic waste. After the prescribed tests have been performed on your distillates, transfer the isolated *natural products* into appropriately labeled containers. Place the *carbon tetrachloride solution* from the test for unsaturation in a container for halogenated organic liquids; dispose of the *manganese dioxide*, from the test for unsaturation in a container for heavy metals.

SPECTRA OF PRODUCTS

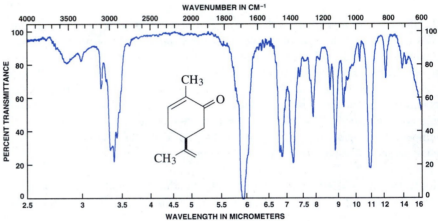

(a) *S*-(+)-Carvone from caraway seed oil.

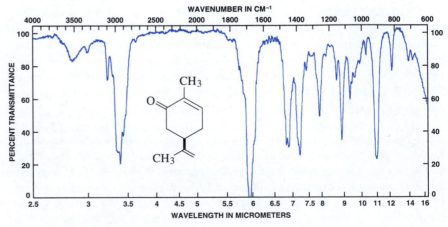

(b) *R*-(+)-Carvone from spearmint oil.

(a) ^{1}H NMR spectrum (60 MHz).

FIGURE 7.16   NMR data for enantiomeric carvones.

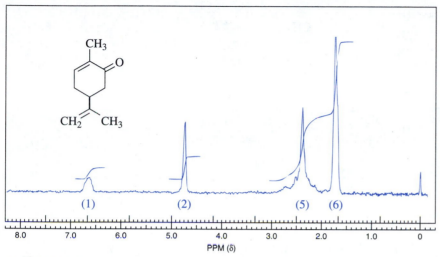

(b) ^{13}C NMR data.
Chemical shifts: δ 110.4, 135.4, 144.1, 146.7, 198.8.

EXERCISES

1. What difficulty would result from using an ebullition tube having a capillary bore that is too large? too small?

2. Why should a constant vacuum be established *before* heating is begun for the distillation?

3. If you observed a rotation of 180° for an optically active liquid in a simple polarimeter, how could you determine experimentally whether the sign of rotation is actually + or –?

4. What melting point should you expect from a mixture of the 2,4-dinitrophenylhydrazone from (+)-carvone and the 2,4-dinitrophenylhydrazone from (–)-carvone if both derivatives are pure?

5. Identify the chiral center in limonene (**16**) and draw the *R*-enantiomer of this molecule.

6. Determine whether the phellandrenes **17** and **18** are chiral.

7. In Section 7.1 it was stated that interconversion of configurational isomers involves the cleavage and re-formation of a chemical bond. Indicate the bonds in (+)-carvone that could be broken and then re-formed to produce (–)-carvone. Be sure to describe what must occur stereochemically during the process of bond cleavage/re-formation that effects the interconversion of one enantiomer to the other.

8. Suppose the following reactions were performed on the enantiomer of carvone that you isolated. Predict whether or not the product in each case would be optically active and support your predictions.

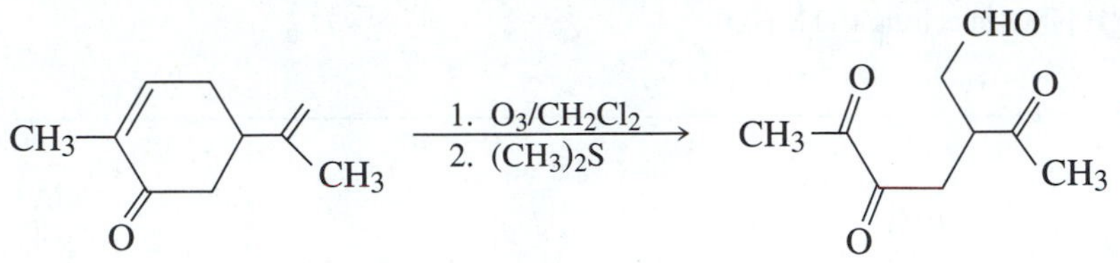

16

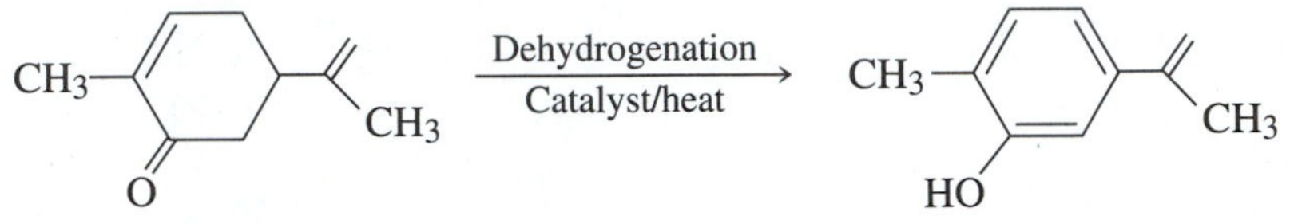

16

REFERENCES

1. Murov, S. L.; Pickering, M. *Journal of Chemical Education*, **1973**, *50*, 74.
2. Russell, G. F.; Hills, J. I. *Science*, **1971**, *172*, 1043.
3. Friedman, L.; Miller, J. G. *Science*, **1971**, *172*, 1044.

7.6 RESOLUTION OF RACEMIC 1-PHENYLETHANAMINE

Enantiomers cannot be physically separated because they have identical properties, except toward plane-polarized light (see Sec. 7.1). However, *diastereomers* do have different physical properties, such as solubility, melting and boiling points, and adsorptive characteristics in chromatography. Recognition of this characteristic of diastereomers has led to a general strategy for the separation of enantiomers that involves three steps: (1) conversion of the enantiomers to diastereomers by way of a chemical reaction; (2) separation of these diastereomers by any of a number of standard techniques, such as recrystallization and chromatography; and (3) regeneration of one or both of the pure enantiomers from the corresponding diastereomer. This is illustrated schematically in Equation (7.6), where R and S represent the enantiomers to be separated and R' stands for a single enantiomer of a compound, termed a **resolving agent,** with which both R and S can react. To convince yourself that RR' and SR' are actually *diastereomers*, you need only convince yourself that the *enantiomer* of RR', for instance, would be written as SS'.

$$R + S \xrightarrow{\;R'\;} RR' + SR' \xrightarrow[\text{technique}]{\text{Separation}} \begin{array}{l} RR' \xrightarrow{\text{Regeneration}} R + R' \\ SR' \xrightarrow{\text{Regeneration}} S + R' \end{array} \qquad (7.6)$$

| Racemic mixture | Diastereomeric mixture | | Separated diastereomers | | Pure Enantiomers |

Nature is the source of a number of different resolving agents that are used to prepare separable diastereomeric pairs from enantiomers. In this experiment, we use (+)-tartaric acid (**19**), which is produced from grapes in the production of wine, to resolve racemic 1-phenylethanamine (**20**), as shown in the resolution scheme outlined in Figure 7.17. The racemate of **20** may be prepared from the reaction of acetophenone (**21**) with ammonium formate (Eq. 7.7). Conversion of the enantiomers of **20** into separable diastereomers involves an acid-base reaction with **19**. The diastereomeric salts that result have different solubilities in methanol and can be separated by fractional crystallization. The individual enantiomers of **20** can then be regenerated by neutralization of the corresponding salts.

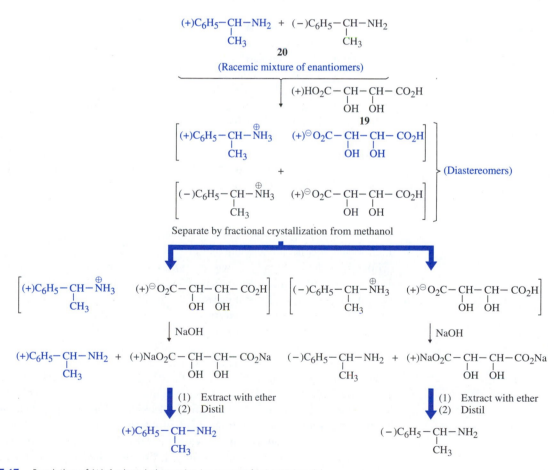

In principle, it would be possible to obtain both enantiomers of a chiral compound in high optical purity by this technique, but this generally would require an undesirably large number of crystallizations. Fortunately, one or two crystallizations usually suffice to allow isolation, in reasonable optical purity, of the enantiomer derived from the less soluble salt.

FIGURE 7.17 Resolution of (±)-1-phenylethanamine by means of (+)-tartaric acid.

Measurement of the optical rotation of the 1-phenylethanamine recovered in this experiment allows determination of whether it is the (+)- or (–)-enantiomer and of the extent of optical purification achieved. If you are to do this part of the experiment, consult with your instructor for special experimental directions.

Pre-Lab exercises for Section 7.6, "Resolution of Racemic 1-Phenylethanamine," are found on page PL. 27.

EXPERIMENTAL PROCEDURE

SAFETY ALERT

1. Methanol is flammable; do not heat it with a flame; use a steam or hot-water bath instead.
2. Avoid contact of 14 *M* NaOH with the skin as it is corrosive. If contact occurs, wash the area with copious amounts of water.
3. Diethyl ether is *extremely* volatile and flammable. When using it for extractions, be *certain* that there are *no flames* in the vicinity.

Dissolve an accurately weighed sample of approximately 5 g of racemic 1-phenylethanamine in 35 mL of methanol and determine its specific rotation [α] using a polarimeter and Equation (7.5). The amounts of 1-phenylethanamine and methanol specified here are those that give satisfactory results when a polarimeter sample tube with a light path of 2 dm or more and a volume of 35 mL or less is used. If a tube of shorter light path and/or larger volume is used, either the amount of 1-phenylethanamine used as starting material must be increased, or two or more students must combine their yields of resolved product to give a solution having sufficient concentration to allow accurate measurement of the optical rotation. Consult with your instructor about the type of polarimeter sample tube that is available.

Place 15.6 g of (+)-tartaric acid and 210 mL of methanol in a 500-mL Erlenmeyer flask and heat the mixture to boiling. To the stirred hot solution cautiously add the 35 mL of solution recovered from the polarimeter and enough additional racemic 1-phenylethanamine to make a total of 12.5 g of the amine. Allow the solution to cool slowly to room temperature and to stand undisturbed for 24 h or until the next laboratory period.✱ The amine hydrogen tartrate should separate in the form of white prismatic crystals. If the salt separates in the form of needlelike crystals, the mixture should be reheated until *all* the crystals have dissolved and then allowed to cool slowly. If any prismatic crystals of the salt are available, they should be used to seed the solution.

Collect the crystals of the amine hydrogen tartrate (8–10 g) on a filter, and wash them with a small volume of cold methanol. Do *not* discard the filtrate (see Finishing Touches). Dissolve the crystals in about four times their weight of water, and add 8 mL of 14 *M* sodium hydroxide solution. Extract the resulting mixture with four 35-mL portions of diethyl ether. Do *not* discard the aqueous solution (see Finishing Touches). Wash the combined ethereal extracts with 25 mL of saturated sodium chloride solution, dry the ethereal solution over anhydrous sodium sulfate, and then decant or filter it from the desiccant.✱

Remove the diethyl ether from the solution by simple distillation using a steam or oil bath as the heat source (*no flames!*),∗ and then distil the residue (1-phenylethanamine) under vacuum with a short-path apparatus (see Sec. 2.9 and the optional procedure of Sec. 4.2). It will be necessary to use a burner or an electric heater because the amine has a boiling point of 94–95 °C (28 torr). The yield of optically active 1-phenylethanamine should be 2–3 g.

Combine your product with that of another student to give about 5 g of 1-phenylethanamine. Weigh the combined product accurately, and then transfer it quantitatively into about 35 mL of methanol. Measure the volume of the methanol *solution* accurately. It is the weight, in grams, of the 1-phenylethanamine divided by the volume of the *solution*, in mL, that gives you the concentration *c* for Equation (7.5). Transfer the solution to the polarimeter sample tube, measure the observed rotation, and determine the specific rotation of the sample by using Equation (7.5). The reported specific rotation $[\alpha]_D^{25}$ of optically pure *R*-1-phenylethanamine is +40.1° (neat).

> **Finishing Touches:**   Transfer the recovered *diethyl ether* to a container for nonhalogenated organic liquids, and put the *methanolic filtrate* from isolation of the less-soluble amine tartrate either in this same container or in a special container from which the more-soluble amine tartrate could be recovered; consult your instructor for specific directions. Place the *methanolic solution of 1-phenylethanamine* either in a container so labeled or in the one labeled for nonhalogenated organic liquids. Transfer the *sodium hydroxide solution* into a container so labeled; the tartaric acid salt it contains can be recovered.

SPECTRA OF STARTING MATERIALS

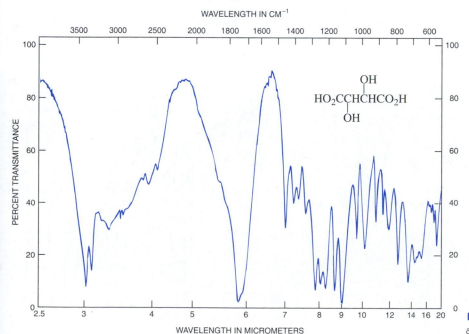

FIGURE 7.18   IR spectrum of (+)-tartaric acid (KBr pellet).

FIGURE 7.19   NMR data for (+)-tartaric acid.

(a) ^1H NMR spectrum (90 MHz).

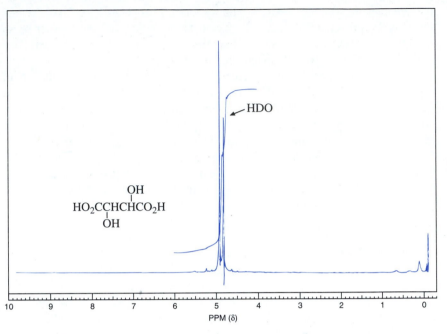

(b) ^{13}C NMR data.
   Chemical shifts: δ 72.7, 175.1.

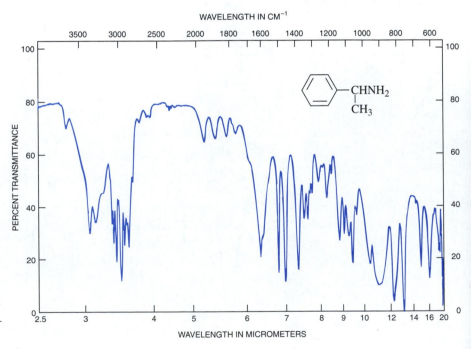

FIGURE 7.20   IR spectrum of racemic 1-phenylethanamine.

(a) ^{1}H NMR spectrum (90 MHz).

FIGURE 7.21 NMR data for racemic 1-phenylethanamine.

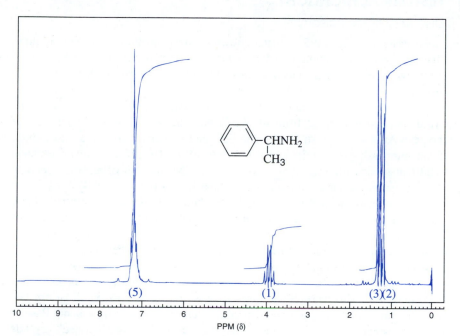

PPM (δ)

(b) ^{13}C NMR data.

Chemical shifts: δ 25.7, 51.2, 125.7, 126.6, 128.3, 148.0.

EXERCISES

1. When a label is prepared for the product that has been isolated, it should include the magnitude of the specific rotation as well as its sign. Suppose the observed rotation was found to be 180°. How could you determine whether the rotation was (+) or (–)?

2. Given the value for the specific rotation of optically pure 1-phenylethanamine, calculate the optical purity of the sample of resolved amine for which you obtained a rotation.

3. How might the optical purity of the product be increased?

4. Describe clearly the point in the experimental procedure at which the major part of the other enantiomer was removed from your product.

5. How could the (+)-tartaric acid be recovered so that it can be used over again to resolve additional racemic amine?

6. The absolute configuration of (+)-1-phenylethanamine has been shown to be *R*. Make a perspective drawing of this configuration.

7. Suppose you had prepared a racemic organic *acid* and it was necessary to resolve it. How could this be done?

8. Give a possible procedure for resolving a racemic *alcohol*.

9. Propose a stepwise reaction mechanism that accounts for the transformation shown in Equation (7.7).

HISTORICAL HIGHLIGHT

Discovery of Stereoisomers

Louis Pasteur (1822–1895) is better known for his contributions to microbiology than to chemistry, although he began his professional career as a chemist. In 1848, when Pasteur was only 24 years old and had just received his doctorate from the Sorbonne in Paris, he undertook the study of a salt of **racemic acid**, a substance deposited on wine casks during the fermentation of grape juice. (The word *racemic* is derived from the Latin word *racemus*, meaning a bunch of grapes.) Another chemist, E. Mitscherlich, had earlier reported that the sodium ammonium salt of racemic acid was identical in all ways to a salt of **tartaric acid** that is also found on wine casks, *except* that the salt of tartaric acid was optically active, whereas the salt of racemic acid was not.

Pasteur was puzzled by the fact that although the salts of tartaric acid and racemic acid were said to be identical in many ways, including chemical composition and even crystalline form, they had different effects on plane-polarized light. Specifically, the salt of racemic acid had *no* effect, whereas the salt of tartaric acid was dextrorotatory. When Pasteur examined the crystals of the salt of racemic acid produced according to Mitscherlich's description, he observed something that Mitscherlich had not. There were, in fact, *two* kinds of crystals present that were related to one another as a left hand is to a right hand. Pasteur carefully separated the "left-handed" crystals from the "right-handed" crystals using tweezers under a microscope. When he had a sufficient quantity of each type of crystal, he did something that arose from either a hunch or a flash of genius. He separately dissolved some of each kind of crystal in water and placed the solutions in turn in a polarimeter. He found that *both* solutions were optically active, the solution of the "left-handed" crystals rotating the polarized light counterclockwise, and the solution of the "right-handed" crystals rotating the polarized light clockwise. The author Vallery-Radot, in *The Life of Pasteur* (1902), reported that the young scientist was so excited by his discovery that he, "not unlike Archimedes, rushed out of the laboratory exclaiming 'I have it!'"

When Pasteur carefully measured the amounts of each kind of crystal for making the solutions, he found that equal amounts produced exactly the same degree of rotation, *but in opposite directions*. Moreover, the magnitude of rotation by the "right-handed" crystals was the same as that given by a similar solution of the salt of tartaric acid. Thus, Pasteur demonstrated that his "right-handed" racemic acid salt was actually identical with the dextrorotatory tartaric acid salt, and his "left-handed" racemic acid salt was a previously unknown mirror-image form of tartaric acid salt, as shown below. Finally, Pasteur made a mixture of equal amounts of the two kinds of crystals and found, as he expected, that a solution of this mixture was optically inactive.

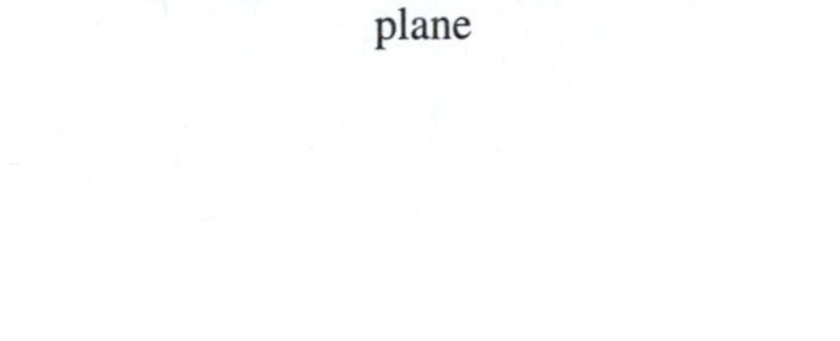

$$
\begin{array}{ccc}
& CO_2^{\ominus} & \\
H - & C - OH & \\
HO - & C - H & \\
& CO_2^{\ominus} &
\end{array}
\qquad
\begin{array}{c}
CO_2^{\ominus} \\
HO - C - H \\
H - C - OH \\
CO_2^{\ominus}
\end{array}
$$

(+)-Tartrate   Mirror   (–)-Tartrate
plane

By separating the two kinds of crystals of the salt of racemic acid, Pasteur accomplished the first and most famous example of what chemists now call **resolution** of a racemic mixture. The name of the specific acid, *racemic*, that Pasteur studied is now used generally to describe any equimolar mixture of **enantiomers** such as the "left-handed" and "right-handed" salts of Pasteur.

One is likely to think that the separation of the crystals of different shapes and the demonstration that their solutions rotate the plane of polarized light in opposite directions was a clever but rather trivial experiment. In a limited sense it was, but from a broader perspective the experiments had tremendous importance. This was the first time it had been shown that organic compounds exist in enantiomeric forms at the molecular level. Before Pasteur's experiments, the effect of quartz crystals on polarized light could be explained on the basis of the shape of the crystals themselves, because when the quartz crystals were melted the optical activity disappeared. Pasteur's work showed that the difference in the crystalline form of the salts of racemic acid reflected a difference in the three-dimensional shape of the molecules themselves. Thus, despite the fact that dissolving the salts of racemic acid in water destroyed their crystal structures just as that of quartz is destroyed by melting it, the resulting *solutions* of the two kinds of crystals of Pasteur's salt still exhibited optical activity. Pasteur later converted the separated salts of racemic acid into the corresponding acid forms and showed that there were indeed two isomeric forms, "left-handed" or (–)-tartaric acid and "right-handed" or (+)-tartaric acid.

Pasteur's observation of the different shapes of the crystals, his separation of them, and his deductions about the meaning of their opposite effects on polarized light were indeed acts of genius, but accidents played a large part in his discovery. There were two remarkable elements of chance that entered into the findings. The first was that the sodium ammonium salt of racemic acid, which was the one Pasteur examined, has been shown to be one of the only salts of this acid that crystallizes in mirror-image forms that are visually different and can be separated mechanically. Second, the crystallization in these two forms occurs only at temperatures below 26 °C (79 °F); above 26 °C the crystals that form are identical and have no optical activity. To obtain his crystals, Pasteur had placed the flask containing the solution of racemic acid on a cool window ledge in his laboratory in Paris and left it there until the next day for the crystallization to occur. Except for the combination of a fortuitous choice of the proper salt of racemic acid and the cool Parisian climate, Pasteur would not have made his important observations.

By his discovery, he recognized that a direct relationship exists between molecular geometry and optical activity. This led him to propose that molecules that rotate plane-polarized light in equal but opposite directions are related as an object and its mirror image. However, it remained for two other chemists, Van't Hoff and LeBel, to explain 25 years later exactly how the atoms could be assembled into such molecular structures. In the interim, Pasteur turned his attention with great success to the biological problems mentioned earlier, but it was his pioneering work on the resolution of racemic acid that led the way for other chemists to explain the relationship of chirality or "handedness" in molecular structure to biological activity, and this is the real significance of Pasteur's work.

The importance of the chirality of molecules is tragically illustrated by the case of thalidomide, shown below. In the 1950s and 1960s, this drug was prescribed to pregnant women as the racemate before it was known that only the

(+)-enantiomer of the molecule is safe and effective against morning sickness. Unfortunately, the (–)-enantiomer is a potent **teratogen,** an agent that causes fetal malformities. Numerous serious birth defects occurred before the problem was recognized and thalidomide was banned from the marketplace.

Thalidomide   Ibuprofen

Ibuprofen is an analgesic (pain-suppressing) agent that is a pharmacologically active component in over-the-counter products such as Nuprin® and Advil®. Like thalidomide, the molecule is chiral and only the S-enantiomer is effective for relief of pain. In this case, however, the R-isomer is slowly converted to the S-isomer in the body, so it too can serve as a therapeutically useful species. The commercially available drug is presently sold as the racemate.

That the different enantiomers may have vastly different pharmacological properties has led to great interest in developing methods for the synthesis of single enantiomers. A recent article describing the growing importance of chiral drugs in the development of new medicines is provided in *Chemical and Engineering News*, September 28, 1992, pages 46–78.

Relationship of the Historical Highlight to the Experiments

Enantiomers of chiral compounds can sometimes be distinguished not only by their equal but opposite effects on plane-polarized light, as first demonstrated by Pasteur, but also by their odors, if they are volatile enough. The enantiomers of carvone (see Sec. 7.5) fit this description, and their differing odors can be distinguished by most persons because the nasal receptors responsible for the detection of smells are chiral. The manner in which chiral odor receptors differentiate between enantiomers can be illustrated in the following way. Assume that your receptors are right-handed, symbolized as (+). Combining these with the (+)- and (–)-enantiomers of carvone gives complexes that can be represented as (+)/(–) for spearmint oil and (+)/(+) for caraway seed oil and dillseed oil, respectively. These complexes are diastereomeric, and they trigger unique nervous responses that we sense as different aromas. Thus, your nose is able to distinguish between diastereomers just as a chromatographic plate or column is able to differentiate between the diastereomers of 1,2-cyclohexanediol (see Sec. 7.2).

Pasteur's observation of the fortuitous resolution of the tartrates by crystallization of the racemate is an exception to the general methods required for separating enantiomers. The more usual case involves converting the racemic mixture into a mixture of diastereomers, which may then be separated on the basis of differing physical properties, such as solubility. The resolution of 1-phenylethanamine (see Sec. 7.6) is representative of this approach.

SPECTRAL METHODS

The major emphasis in the first laboratory course in organic chemistry is on the development of the skills and techniques for executing functional group transformations and isolating the products from such reactions in pure form. The starting materials and products that you normally use are "known" compounds because their molecular structures have been previously determined. Consequently, you are *not* usually faced with the necessity of determining the structures of the substances you produce. However, at times you may be presented with an "unknown" compound, one whose structure is known by your instructor but not by you. At such times you must undertake the challenge of determining the structure of the molecule. This is a challenge that most organic chemists face on a regular basis, whether the compounds are isolated from natural sources or from reactions that produce substances not characterized previously.

Determining the exact three-dimensional structures of organic compounds used to be a difficult and time-consuming undertaking because the "classical" method of characterizing a structure typically involved converting an unknown compound into a known substance (see Chap. 23). The advent of powerful spectroscopic methods for defining the structures of unknowns greatly simplified the task of structural elucidation and has been a catalyst for the rapid growth of organic chemistry. This chapter contains discussions of the application of various spectroscopic methods to the analysis of organic compounds.

The spectroscopic methods of greatest importance to the contemporary organic chemist are **infrared** (IR), **nuclear magnetic resonance** (NMR), **and ultraviolet** (UV) **spectroscopy, mass spectrometry** (MS) and **X-ray diffractometry.** The last two techniques are too specialized to merit discussion here. Of the remaining spectroscopic methods, we emphasize IR and NMR spectroscopy in this textbook because the necessary instrumentation is more commonly available to students in the organic laboratory course and because these techniques are the ones most frequently used by organic chemists.

With the notable exception of X-ray diffractometry, unequivocal definition of a structure is seldom possible by performing only a single type of spectroscopic analysis; rather, a combination of different analyses is generally required. This does not mean that the use of spectroscopic techniques has entirely eliminated the need to perform chemical transformations as part of the process of structure elucidation. This more classical approach, although less important than it once was, still has a role in the determination of structure, as discussed in Chapter 23.

The spectral methods described in this chapter depend on the absorption of radiation to produce an excitation that transforms the original molecule to a state of higher energy. The particular type of excitation that occurs is determined by the amount of energy associated with the radiation. The relationship between the energy of radiation (in kcal/mol) and its frequency ν, or wavelength, λ, is expressed in Equations (8.1a) and (8.1b), where N is Avogadro's number (6.023×10^{23}), h is Planck's constant (1.5825×10^{-37} kcal sec), c is the velocity of light (2.998×10^{14} μm/sec), and ν and λ are in Hertz (1.0 Hz = 1.0 cycle/sec) and micrometers (μm = 10^{-6} m), respectively. Because of the reciprocal relationship between ν and λ, the *higher* the frequency or the *lower* the wavelength, the *higher* is the energy associated with the radiation.

$$E = Nh\nu \tag{8.1a}$$

$$E = Nhc/\lambda \tag{8.1b}$$

The various spectroscopic techniques discussed in this chapter, the wavelengths and energies associated with them, and the type of molecular excitation that occurs in each are summarized in Table 8.1. A schematic diagram illustrating these excitations is presented in Figure 8.1. The highest energies, and thus the highest frequencies and shortest wavelengths of light, are associated with UV spectroscopy, which involves promoting an electron from a ground-state molecular orbital to an unfilled orbital to produce an **electronically excited** state. Such transitions, which require energies of 90 kcal/mol or more, are labeled E_1 and E_2 in Figure 8.1. In contrast, IR spectroscopy arises from transitions, among different molecular **vibrational-rotational levels** within the *same* electronic state, normally the ground state; these excitations require energies of 1–36 kcal/mol. The excitations labeled E_3 and E_4 in Figure 8.1 are examples.

TABLE 8.1 Relationships Between Wavelength, Energy, and Molecular Excitation

Wavelength (μm)	Energy (kcal/mol)	Type of Spectroscopy	Molecular Phenomenon
0.20 (200 nm)	143 ⎫		Excitation of valence electrons from filled to unfilled orbital, for example, $\pi \rightarrow \pi^*$ and $n \rightarrow \pi^*$
0.32 (320 nm)	89 ⎭	UV	
0.45 (450 nm)	71.5 ⎫		Same as above, except that absorption of light occurs in a region that is visible (colored) to the human eye
0.75 (750 nm)	38.1 ⎭	Visible	
0.8	36 ⎫		Stretching and bending of interatomic bonds
40	0.7 ⎭	IR	
5×10^4	5.7×10^{-4} ⎫		Realignment of nuclear spins in an applied magnetic field
2.1×10^9	1.4×10^{-8} ⎭	NMR	

The phenomenon associated with NMR spectroscopy is realignment of **nuclear spins** in a magnetic field and involves excitations of *very* low energy, less than 1 cal/mol; such transitions are not indicated in Figure 8.1. The types of spectroscopy listed in Table 8.1 are *nondestructive*—the sample is *unchanged* by the spectroscopic technique.

A more detailed discussion of each type of spectroscopy is given in the following sections. However, a summary of the structural information typically available from each spectral technique may be helpful at this point.

1. Infrared Spectroscopy (Sec. 8.2).   The data from IR spectroscopy are most useful for determining the presence or absence of **functional groups** in a molecule. For instance, examining the appropriate regions of an IR spectrum will show whether or not carbon-carbon multiple bonds, aromatic rings, carbonyl groups, or hydroxyl groups are present. This technique does not give quantitative information regarding the elemental composition of a compound, nor does it allow assignment of an exact structure to an unknown compound *unless* the IR spectrum of the unknown is shown to be identical to that of a known compound.

2. Nuclear Magnetic Resonance Spectroscopy (Sec. 8.3).   NMR spectroscopy provides information regarding the number and arrangement of various atoms in a molecule, although not all types of nuclei found in organic molecules can be detected by this method. Application of this technique to the analysis of hydrogen atoms in molecules, which is called ^{1}H NMR spectroscopy (Sec. 8.3.2), allows determination of the number of nearest neighbors to a hydrogen atom as well as the presence of certain functional groups, such as carbon-carbon multiple bands and carbonyl groups. A companion method involving carbon-13 atoms, and abbreviated as ^{13}C NMR spectroscopy (Sec. 8.3.4), permits analysis of the different types of carbon atoms present in a molecule and may be useful for detecting certain functional groups whose presence is not indicated in the ^{1}H NMR spectrum.

3. Ultraviolet Spectroscopy (Sec. 8.4).   This technique is most useful for detecting the presence of **conjugated** systems of π-bonds, as in 1,3-dienes, aromatic rings, and 1,3-enones. Because of its limited applicability, UV spectroscopy is considerably less important than the preceding two spectral methods.

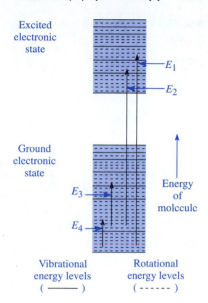

FIGURE 8.1   Rotational, vibrational, and electronic energy levels and transitions between them.

8.2 INFRARED (IR) SPECTROSCOPY

8.2.1 INTRODUCTION

The infrared portion of the electromagnetic spectrum spans from about 0.8 μm (4×10^{10} Hz) to 40 μm (7.5×10^8 Hz). Within this range, the region from 2.5 μm to 20 μm is of most interest to organic chemists because it is here that most of the common functional groups absorb IR radiation. Although wavelength, in μm, may be used to express the location of an IR absorption, current practice is to do so with another unit, **wavenumber** or **reciprocal centimeter,** written as $\tilde{\nu}$

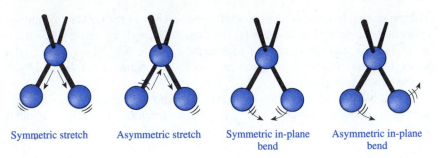

Symmetric stretch   Asymmetric stretch   Symmetric in-plane bend   Asymmetric in-plane bend

or cm^{-1}. The relationship between wavelength and wavenumber is expressed by Equation (8.2). Thus, the normal IR range of 4000 to 500 cm^{-1} corresponds to 2.5–20 μm. As with frequency v, the energy associated with radiation *increases* as $\tilde{v}$ increases. It is imprecise to call wavenumbers "frequencies," despite the mathematical relationship between the two (see Exercise 1 at the end of this section), although we do so for convenience.

$$\text{wavenumber (cm}^{-1}) = 10,000/\lambda \ (\mu\text{m}) \tag{8.2}$$

8.2.2 PRINCIPLES

The interaction of IR radiation with an organic molecule may be qualitatively understood by imagining that the covalent bonds between atoms are analogous to groups of molecular springs that are constantly undergoing **stretching, twisting**, and **bending** (Fig. 8.2). The characteristic frequencies, and thus the energies, of these vibrations depend on the masses of the atoms involved and the type of chemical bond joining the atoms. Indeed, a good approximation of the *stretching* frequency of bonds can be obtained by applying **Hooke's law** for simple harmonic oscillators (Eq. 8.3). The value of the **force constant** k, which essentially reflects the strength of the bond between the atoms A and B, is about 5×10^5 dyne cm^{-1} for single bonds and twice and three times this figure for double and triple bonds, respectively. The validity of Equation (8.3) for predicting stretching frequencies is illustrated by its use to give $\tilde{v}_{C-H}$ as 3040 cm^{-1}; the experimental value for this vibration is in the range of 2960–2850 cm^{-1}. Other consequences of Equation (8.3) are noteworthy. For instance, the wavenumber for stretching of a C–D bond is about 2150 cm^{-1}, a lower value than that for a C–H bond; this difference reflects the effect of atomic mass on the absorption frequency as accounted for by m^* in Equation (8.3) (see Exercise 2 at the end of this subsection). Similarly, the appearance of the stretching vibration of a carbon-carbon double bond at about 1620 cm^{-1} and that of a carbon-carbon triple bond at approximately 2200 cm^{-1} is because k (Eq. 8.3) approximately doubles in going from a double to a triple bond; this increases the energy required to stretch the bond between the atoms (see Exercise 3).

$$\tilde{v} = \frac{1}{2\pi c}\sqrt{\frac{k}{m^*}} \tag{8.3}$$

where $\tilde{v}$ = frequency of absorption in cm^{-1}

c = speed of light

k = force constant of bond

m^* = reduced mass of atoms joined by bond = $\dfrac{m_A m_B}{m_A + m_B}$

Because different energies are required to cause the molecular vibrations associated with various types of bonds, these functional groups can be excited sequentially and selectively by sweeping through the infrared range of frequencies. When the energy of the radiation corresponds to that required to produce a particular stretching or bending vibration, energy from the incident light will be absorbed. The efficiency of this absorption depends on several factors that are beyond the scope of the present discussion. Suffice it to say that the greater the change in the **dipole moment** associated with the particular vibration being excited, the more efficient is the transfer of energy to the molecule.

Most modern IR instruments are designed to measure the amount of light transmitted *through* the sample, the **transmittance**, rather than the **absorbance** of light *by* the sample. Consequently, the **infrared spectra** portrayed in this textbook are plots of transmittance, expressed in *percent* as defined in Equation

$$\%T = I/I_0 \times 100$$

I = intensity of radiation transmitted through sample at a particular wavelength

(8.4)

I_0 = intensity of radiation incident on sample at same wavelength

(8.4), versus the wavenumber or wavelength of the **incident radiation** (I_0) as seen in Figure 8.3. The lower the $\%T$ is, the greater is the amount of incident radiation being absorbed by the sample. For example, in the IR spectrum shown in Figure 8.3 the strongest absorption occurs at about 1675 cm^{-1} (5.97 μm). The energy required to produce the observed molecular excitations *decreases* in going from left to right in the spectrum.

It should be noted that the absorption of infrared radiation by a molecule increases its total energy (see Fig. 8.1), which results in an increase in **amplitude** of the particular molecular vibration that was responsible for absorption of the radiation. However, no irreversible change occurs in the molecule, as the excess energy is quickly dissipated in the form of heat. Obtaining an IR spectrum thus does *not* permanently change the structure of a molecule.

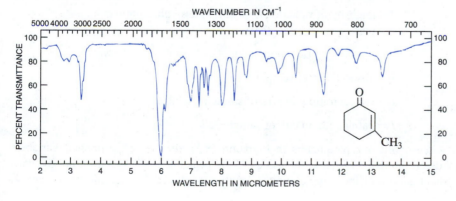

FIGURE 8.3 Infrared spectrum of 3-methyl-2-cyclohexenone.

8.2.3 PRACTICAL CONSIDERATIONS

Infrared spectra may be measured for liquids, solids, and gases, but only the experimental techniques for handling liquid and solid samples are described here. The apparatus required for trapping gaseous samples and transferring them to an IR cell is too specialized to be generally used in the first laboratory course in organic chemistry. Before discussing specific techniques, it is instructive to consider the criteria that define an acceptable IR spectrum. These criteria vary according to how the data are to be used, but as a general rule it is desirable to obtain a spectrum that shows detectable peaks for relatively weak absorptions but has no absorptions so strong that they go off-scale. Experimentally, a good operating principle is to modify the sample being analyzed so the most intense peak in its spectrum is close to 0 %*T*; this maximizes the possibility that even weak peaks will be observable. This strategy is seen in Figure 8.3, where the strongest absorption has a %*T* of nearly zero.

The factors that determine the intensity of an IR absorption are important because this intensity affects the appearance of the observed spectrum. The transmittance, *T,* of radiation at a given wavenumber is expressed by Equation (8.5), the **Beer-Lambert law**. The expression shows the *inverse* relationship between the amount of light transmitted through a sample and the factors *k*, *c*, and *l*. The absorptivity *k* is a constant that is characteristic of the sample, but its magnitude is wavenumber dependent. This variation in the value of *k* determines the *relative* intensities of the various peaks in the spectrum, with larger values of *k* providing relatively stronger absorptions, corresponding to lesser transmittance. The absorptivity is basically a measure of the efficiency with which radiation of a particular wavenumber is absorbed. As was noted previously, it has a higher value when a significant change in dipole moment is associated with the molecular excitation. This is why the intensity of the stretching mode of a carbonyl group, C=O, is almost invariably *greater* than that of a carbon-carbon double bond C=C: The dipole moment is little changed by lengthening the double bond of the latter functionality. This phenomenon is seen in Figure 8.3 by comparing the relative intensities of the peaks at about 1675 and 1630 cm^{-1} for the carbonyl and carbon-carbon double bond functions, respectively.

$$\log_{10}\left(\frac{1}{T}\right) = \log_{10} < \left(\frac{I_0}{I}\right) = kcl \qquad \text{(8.5)}$$

where *T* = transmittance

I = intensity of radiation transmitted through sample at a particular wavelength

I_0 = intensity of radiation incident on sample at same wavelength

k = absorptivity of sample

c = concentration, in g/L, of solute in solution

l = pathlength, in cm, of sample cell

Considering the parameters in Equation (8.5), the two experimental variables determining the *absolute* intensities of the peaks in an IR spectrum are the concentration *c* of the sample and the length *l* of the cell containing it. As with *k*,

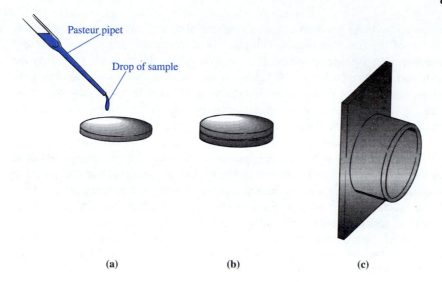

Pasteur pipet

Drop of sample

(a)   (b)   (c)

FIGURE 8.4   Simple demountable IR cell. (a) Salt plate for sample. (b) Assembled pair of salt plates. (c) Cell holder.

the magnitude of T or %T is *decreased* when the values of these variables are greater and the observed peak is more intense. In other words, the factors c and l define the *number* of molecules that are in the path of the radiation as it passes through the sample; the more molecules there are, the greater the absolute intensity of the absorption.

Now that the factors determining the transmittance of IR radiation have been considered, let us discuss the details of how a satisfactory IR spectrum is obtained.

1. Cells.   Two general types of cells are used to contain samples for analysis by IR spectroscopy: Cells in which the pathlength or thickness of the sample is fixed, and those in which the pathlength is variable. The latter type of cell can be very expensive, but a simple version of it is two transparent salt plates, or windows, between which the sample is sandwiched (Figs. 8.4 and 8.5). The cell is

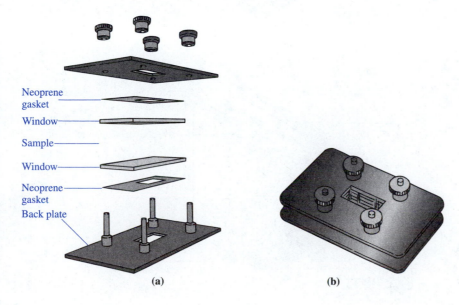

Neoprene gasket

Window

Sample

Window

Neoprene gasket

Back plate

(a)   (b)

FIGURE 8.5   A demountable IR cell. (a) Details showing correct assembly of cell. (b) Completely assembled cell.

prepared by putting a drop of sample on one plate and placing the second plate on top of the film that results. The assembled plates are then put into a holder that fits into the spectrometer. If a sample is too thick, some of it may be removed or the plates may be squeezed together, *carefully* to avoid breaking them, to produce a thinner film. However, if the sample is too thin, the plates can be separated and more material can be added. The plates can then be rejoined so as to leave the film of sample as thick as possible.

A **fixed-thickness cell** is constructed of two transparent salt plates separated by a plastic or metal gasket, which defines the thickness of the sample contained in the cell. One such cell is shown in Figure 8.6. A syringe is normally used to load the cell, which should be laid at an angle on top of a pen or pencil to keep an air bubble from forming within the cell as sample is introduced. If a bubble is observed, more sample must be added until the bubble disappears.

Proper care and handling of the windows used in IR cells are *very* important. The plates are generally a clear fused salt such as sodium chloride or potassium bromide, and contact with moisture must be minimized to keep the windows from becoming cloudy; this would decrease the amount of light trans-

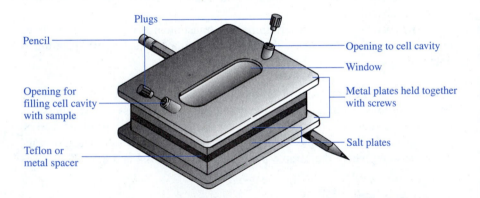

(a)

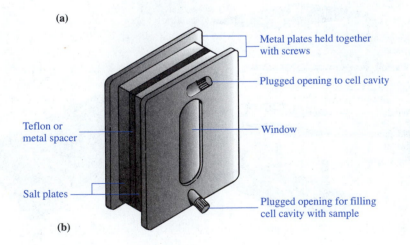

(b)

FIGURE 8.6 A fixed-thickness IR cell. (a) Cell with plugs removed for cleaning or filling. (b) Completely assembled cell.

mitted through the cell. Consequently, breathing on the windows or touching their faces must be avoided when handling assembled IR cells or the salt plates themselves. Moreover, any substances, including samples, that come in contact with the cells must be *dry*. This means that any materials used to clean an IR cell or salt plates must also be *dry*. This excludes 95% ethanol as a solvent for cleaning the windows, although *absolute* ethanol *can* be used. Other solvents suitable for cleaning cells are dichloromethane and chloroform. Permanent fixed-thickness cells are dried by passing a slow stream of *dry* nitrogen or air through them, whereas demountable cells and salt plates may also be dried by *gently* wiping or blotting them with soft laboratory tissue to avoid scratching the faces of the cells or plates. The clean, dry cells or plates are stored in a desiccator for protection from atmospheric moisture.

2. Liquids. The IR spectrum of a liquid sample is most conveniently obtained on the pure liquid, called a **neat sample.** Since concentration is not a variable in such samples, peak intensities may be modified only by changing the pathlength. This is done in one of two ways. One approach is to use a fixed-thickness cell with a pathlength of 0.020–0.030 mm. This short pathlength generally provides a spectrum in which strong absorptions, such as those associated with carbonyl groups, give peaks that remain on the chart paper, rather than going to 0 %*T*. If the spectrum is not suitable, however, different fixed thicknesses may be used. The other option is to use salt plates and to adjust the thickness of the film between them by gently pressing the plates together as described in Part (1). If the sample is volatile, having a boiling point of less than about 50 °C, salt plates would *not* be appropriate because some or all of the sample will evaporate as the spectrum is run, and inaccurate relative peak intensities will result. Alternatively, but less conveniently, the sample may be analyzed as a solution, just as a solid is. The technique for doing so is described in Part (3)(c).

3. Solids. There are several options for preparing a solid sample for IR spectroscopy. These include the preparation of **mulls, transparent films** and **KBr pellets,** and solutions. Each of these techniques is described in turn.

a. Mull. A mull is a suspension of the *finely divided* sample in a viscous carrier fluid such as mineral oil (Nujol®) or perfluorokerosene, which is a mixture of acyclic C_{12}–C_{18} linear hydrocarbons in which all of the hydrogen atoms have been replaced by fluorine atoms. The mulling fluid itself has absorptions in the IR region, as shown for mineral oil and perfluorokerosene in Figure 8.7. These two fluids absorb at different places in the IR spectrum, however, so preparing a mull in each of them will allow determination of the absorptions of the solid over the entire IR range. This is illustrated by the spectra in Figure 8.8 (p. 209).

   To prepare a mull, a few milligrams of the sample are finely ground in an agate mortar, several drops of the mulling fluid are added, and the mixture is ground further until a smooth paste is obtained. Failure to grind the sample finely enough results in an IR spectrum of poor quality because the solid particles disperse too much of the incident radiation, and little light is transmitted through the sample. As with a neat liquid, the paste is spread on a salt plate, a second plate is placed on top of the first, and the cell is put into a holder that fits into the IR spectrometer. The mull must be modified if the spectrum is too

FIGURE 8.7 Infrared spectra of the mulling fluids: (a) mineral oil (Nujol) and (b) perfluorokerosene.

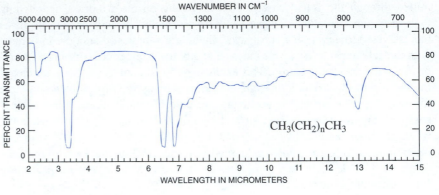

(a) Mineral oil (Nujol)

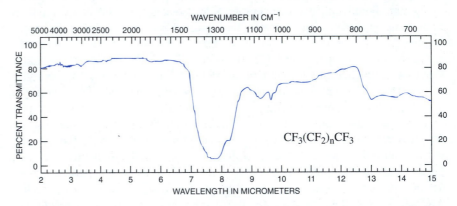

(b) Perfluorokerosene

strong or too weak. This is done by adding more mulling fluid if absorptions are too strong or more finely ground sample if they are too weak. The plates are cleaned by first removing most of the mull from them by wiping or blotting with a soft laboratory tissue and then rinsing them with an appropriate solvent, as discussed in Part (1).

b. Cast Film. It is sometimes possible to prepare *films* of solid samples that are sufficiently transparent for use in IR spectroscopy. The film is produced by putting a drop or two of a *solution* of the sample onto a salt plate and allowing the solvent to evaporate with the aid of mild heating or a gentle stream of *dry* nitrogen or air. The plate is then placed in a holder that fits into the spectrometer. The thickness of the film is controlled by the amount of solution put on the plate. If the solid produces a *crystalline* film, a spectrum of poor quality will likely result because of excessive scattering of the incident radiation.

c. Potassium Bromide (KBr) Pellet. Potassium bromide may be fused to produce a nearly transparent salt plate, and so an alternative to preparing a mull or a cast film is to form a KBr pellet. This is done by subjecting the KBr to high pressures in the range of 5–10 tons/in^2 to generate the heat required for the salt to become plastic and flow; release of the pressure then provides a clear plate. Thus,

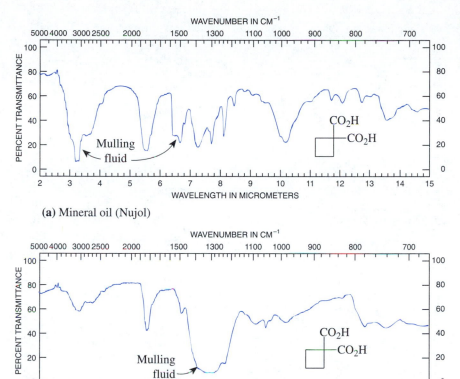

FIGURE 8.8 Infrared spectra of cyclobutane-1,1-dicarboxylic acid as mulls in (a) mineral oil (Nujol) and (b) perfluorokerosene.

(a) Mineral oil (Nujol)

(b) Perfluorokerosene

a KBr pellet containing the compound is prepared by intimately mixing about 1 mg of sample with 100 mg of *dry*, powdered KBr and subjecting the resulting mixture to high pressure in a die. Mixing is done with a mortar and pestle made of agate rather than ceramic materials, because ceramics may be abraded and contaminate the pellet with nontransparent particles. Once a pellet having satisfactory transparency has been produced, the IR spectrum is obtained. If the intensity of the bands in the spectrum is unsatisfactory, another pellet must be prepared in which the concentration of the solid sample has been modified accordingly.

Potassium bromide is extremely hygroscopic, so it should be stored in a tightly capped bottle in a desiccator except when needed for preparation of a pellet. Even with the greatest precautions, it is difficult to prepare a pellet that is completely free of water. Possible contamination of a KBr pellet with water may cause problems in interpreting IR absorptions in the range 3600–3200 cm^{-1} because the O–H stretching frequency occurs in this region. Consequently, moisture in the pellet may partially or completely obscure the absorptions due to the O–H and N–H stretching vibrations of alcohols and amines, which occur in this same region. Further, the appearance of an absorption in the aforementioned range may lead to an erroneous conclusion about the possible presence of O–H and N–H groups in the sample. The band at about 3600 cm^{-1} in Figure 8.9(a) might be mistaken for an O–H or N–H stretching absorption of the sample, for example. However, comparison of this spectrum to that of Figure 8.9(b), for

FIGURE 8.9 Infrared spectra of triptycene as a solution in chloroform and as a KBr pellet.

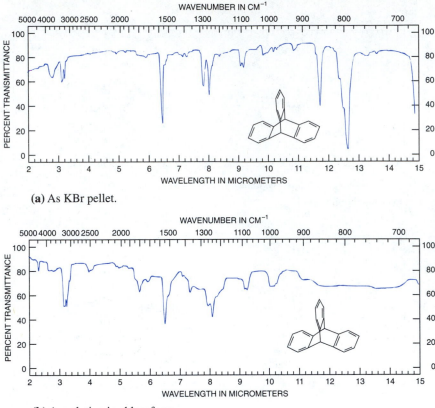

(a) As KBr pellet.

(b) As solution in chloroform.

which a solution of the sample was used, makes it clear that water in the pellet is responsible for this absorption.

The two main drawbacks to obtaining IR spectra of solids as KBr pellets are the difficulty in preparing the pellet and the confusion that may arise because of absorptions due to adventitious moisture. Moreover, this technique *cannot* be used with samples that decompose at the high temperatures reached when the pellet is being produced.

d. Solution. A fourth option for obtaining an IR spectrum of a solid is to analyze it in solution. The preferred solvents are carbon tetrachloride, chloroform, and carbon disulfide, in that order. Carbon tetrachloride is the first choice because it has relatively few absorptions higher than 1333 cm^{-1}. The solubility of the sample in the selected solvent must be in the range of 5–10 weight percent because solutions of such concentration generally provide suitable spectra in a fixed-thickness cell having a pathlength of 0.1 mm.

Two fixed-thickness cells having identical pathlengths, so called **matched cells,** are used if a *double beam* IR instrument is being used. One cell contains the solution, and is referred to as the **sample cell,** whereas the other contains pure solvent and is called a **reference cell.** The purpose of the reference cell is to allow cancellation of the absorption bands due to the solvent itself so that they do not appear in the spectrum. **Fourier transform** (FT) IR instrument is avail-

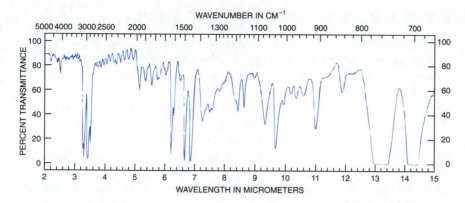

FIGURE 8.10 Infrared spectrum of polystyrene film.

able, the reference cell is not required provided the IR spectrum of the pure solvent has previously been stored in the memory of the spectrometer.

4. Spectral Calibration. It is good practice to calibrate the experimentally obtained IR spectra against a standard that has absorption bands at known positions. Calibration is necessary because problems such as misalignment of the chart paper in the instrument or mechanical slippage of gears may result in absorptions appearing at wavenumbers that are not correct. Polystyrene (Fig. 8.10) is a commonly used standard and is typically available as a thin, transparent film in a cardboard holder that fits into the spectrometer. After the spectrum of the sample is obtained, the standard is inserted in place of the sample cell, the reference cell, if used, is removed, and at least one band of the standard is recorded on the spectrum of the sample. In the case of polystyrene, the absorption at 1601 cm^{-1} is typically recorded (Fig. 8.11), although other bands may be used.

Once the reference peak of the standard is measured, the reported positions of the absorptions of the sample are adjusted accordingly. For example, if the 1601 cm^{-1} peak of polystyrene appears at 1595 cm^{-1} on the spectrum, the wavenumbers for the bands in the sample are corrected by *adding* 6 cm^{-1} to the values recorded on the chart paper. Because the source of the error may vary over the range, it is not strictly correct to use the same correction factor throughout the spectrum, but for most purposes this is a satisfactory protocol. For more precise work, calibration peaks should be recorded at several different locations in the IR spectrum.

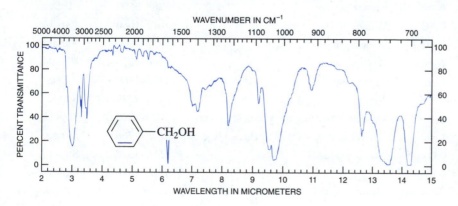

FIGURE 8.11 Infrared spectrum of benzyl alcohol with the 1601 cm^{-1} peak of polystyrene inserted for calibration.

8.2.4 USE AND INTERPRETATION OF IR SPECTRA

An IR spectrum of a compound is used to *identify* the *functional groups* present in a molecule, *evaluate* the *purity* of a known compound, and *determine* the *structure* of an unknown substance. These valuable applications are discussed in the following paragraphs.

Analysis of the IR spectrum of an unknown compound provides much useful information regarding the presence or absence of different functional groups. As a general rule, absorptions in the range of 4000–1250 cm^{-1} involve *vibrational* excitations that are associated with specific functional groups. Thus, the stretching mode of the carbonyl group of a ketone appears in the region of 1760–1675 cm^{-1} and that of a carbon-carbon double bond occurs in the vicinity of 1680–1610 cm^{-1} (see Fig. 8.3). A compilation of various functional groups and the region of the IR spectrum in which they absorb is provided in Table 8.2. Table 8.3 is a more detailed version of Table 8.2. Note that the *absence* of an

TABLE 8.2   Abbreviated Table of Infrared Absorption Ranges of Functional Groups

Bond	Type of Compound	Frequency Range, cm^{-1}	Intensity
C–H	Alkanes	2850–2970	Strong
		1340–1470	Strong
C–H	Alkenes 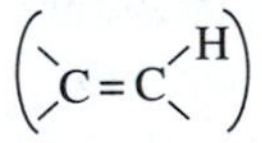 $\left(\begin{array}{c} \diagdown \\ \diagup \end{array} C = C \begin{array}{c} \diagup^H \\ \diagdown \end{array} \right)$	3010–3095	Medium
		675–995	Strong
C–H	Alkynes ($-C{\equiv}C-H$)	3300	Strong
C–H	Aromatic rings	3010–3100	Medium
		690–900	Strong
O–H	Monomeric alcohols, phenols	3590–3650	Variable
	Hydrogen-bonded alcohols, phenols	3200–3600	Variable, sometimes broad
	Monomeric carboxylic acids	3500–3650	Medium
	Hydrogen-bonded carboxylic acids	2500–2700	Broad
N–H	Amines, amides	3300–3500	Medium
C=C	Alkenes	1610–1680	Variable
C=C	Aromatic rings	1500–1600	Variable
C≡C	Alkynes	2100–2260	Variable
C–N	Amines, amides	1180–1360	Strong
C≡N	Nitriles	2210–2280	Strong
C–O	Alcohols, ethers, carboxylic acids, esters	1050–1300	Strong
C=O	Aldehydes, ketones, carboxylic acids, esters	1675–1760	Strong
NO$_2$	Nitro compounds	1500–1570	Strong
		1300–1370	Strong

TABLE 8.3 Detailed Table of Characteristic Infrared Absorption Frequencies

The hydrogen stretch region (3600–2500 cm^{-1}). Absorption in this region is associated with the stretching vibration of hydrogen atoms bonded to carbon, oxygen, and nitrogen. Care should be exercised in the interpretation of very weak bands because these may be overtones of strong bands occurring at frequencies one-half the value of the weak absorption, that is, 1800–1250 cm^{-1}. Overtones of bands near 1650 cm^{-1} are particularly common.

$\bar{\nu}$ (cm^{-1})	Functional Group	Comments
(1) 3600–3400	O–H stretching Intensity: variable	3600 cm^{-1} (sharp) unassociated O–H, 3400 cm^{-1} (broad) associated O–H; both bands frequently present in alcohol spectra; with strongly associated O–H (CO$_2$H or enolized β-dicarbonyl compound); band is very broad (about 500 cm^{-1} with its center at 2900–3000 cm^{-1}).
(2) 3400–3200	N–H stretching Intensity: medium	3400 cm^{-1} (sharp) unassociated N–H, 3200 cm^{-1} (broad) associated N–H; an NH$_2$ group usually appears as a doublet (separation about 50 cm^{-1}); the N–H of a secondary amine is often very weak.
(3) 3300	C–H stretching of an alkyne Intensity: strong	The *complete* absence of absorption in the region from 3300–3000 cm^{-1} indicates the absence of hydrogen atoms bonded to C=C or C≡C and *usually* signals the lack of unsaturation in the molecule.
(4) 3080–3010	C–H stretching of an alkene Intensity: strong to medium	Because this absorption may be very weak in large molecules, care should be exercised in this interpretation. In addition to the absorption at about 3050 cm^{-1}, aromatic compounds frequently show *sharp* bands of medium intensity at about 1500 *and* 1600 cm^{-1}.
(5) 3050	C–H stretching of an aromatic compound Intensity: variable; usually medium to weak	
(6) 3000–2600	OH strongly hydrogen-bonded Intensity: medium	A very broad band in this region superimposed on the C–H stretching frequencies is characteristic of carboxylic acids (see 1).
(7) 2980–2900	C–H stretching of an aliphatic compound Intensity: strong	As in previous C–H entries (3–5), *complete* absence of absorption in this region indicates absence of hydrogen atoms bound to tetravalent carbon atoms. The tertiary C–H absorption is weak.
(8) 2850–2760	C–H stretching of an aldehyde Intensity: weak	Either one or two bands *may* be found in this region for a single aldehyde function in the molecule.

(continued)

TABLE 8.3 *(continued)*

The triple-bond region (2300–2000 cm⁻¹). Absorption in this region is associated with the stretching vibration of triple bonds.

$\bar{\nu}$ (cm⁻¹)	Functional Group	Comments
(1) 2260–2215	C≡N Intensity: strong	Nitriles conjugated with double bonds absorb at *lower* end of frequency range; nonconjugated nitriles appear at *upper* end of range.
(2) 2150–2100	C≡C Intensity: strong in *terminal* alkynes, variable in others.	This band is absent if the alkyne is symmetrical, and will be very weak or absent if the alkyne is nearly symmetrical.

The double-bond region (1900–1550 cm⁻¹). Absorption in this region is *usually* associated with the stretching vibration of carbon-carbon, carbon-oxygen, and carbon-nitrogen double bonds.

$\bar{\nu}$ (cm⁻¹)	Functional Group	Comments
(1) 1815–1770	C=O stretching of an acid chloride Intensity: strong	Carbonyls conjugated with double bonds absorb at *lower* end of range; nonconjugated carbonyls appear at *upper* end of range.
(2) 1870–1800 and 1790–1740	C=O stretching of an acid anhydride Intensity: strong	*Both bands* are present; *each band* is altered by ring size and conjugation to approximately the same extent noted for ketones (see 4).
(3) 1750–1735	C=O stretching of an ester or lactone Intensity: very strong	This band is subject to all the structural effects discussed in 4; thus, a conjugated ester absorbs at about 1710 cm⁻¹ and a γ-lactone absorbs at about 1780 cm⁻¹.
(4) 1725–1705	C=O stretching of an aldehyde or ketone Intensity: very strong	This value refers to carbonyl absorption frequency of acyclic, nonconjugated aldehyde or ketone having no electronegative groups, for example halogens, near the carbonyl group; because this frequency is altered in a predictable way by structural alterations, the following generalizations are valid: (a) *Effect of conjugation:* Conjugation of carbonyl group with an aryl ring or carbon-carbon double or triple bond *lowers* the frequency by about 30 cm⁻¹. If the carbonyl group is part of cross-conjugated system (unsaturation on each side of the carbonyl group), the frequency is lowered by about 50 cm⁻¹.

$\bar{\nu}$ (cm−1)	Functional Group	Comments
		(b) *Effect of ring size:* Carbonyl groups in six-membered and larger rings exhibit approximately the same absorption as acyclic ketones; carbonyl groups in rings smaller than six absorb at *higher* frequencies: for example, a cyclopentanone absorbs at about 1745 cm^{-1} and a cyclobutanone at about 1780 cm^{-1}. The effects of conjugation and ring size are additive: A 2-cyclopentenone absorbs at about 1710 cm−1, for example.
		(c) *Effect of electronegative atoms:* An electronegative atom, especially oxygen or halogen, bound to the α-carbon atom of an aldehyde or ketone usually raises the position of the carbonyl absorption frequency by about 20 cm^{-1}.
(5) 1700	C=O stretching of an acid Intensity: strong	Conjugation *lowers* this absorption frequency, as noted under entry 4.
(6) 1690–1650	C=O stretching of an amide or lactam Intensity: strong	Conjugation *lowers* the frequency of this band by about 20 cm^{-1}. The frequency of the band is *raised* about 35 cm^{-1} in γ-lactams and 70 cm^{-1} in β-lactams.
(7) 1660–1600	C=C stretching of an alkene Intensity: variable	Conjugated alkenes appear at *lower* end of range, and absorptions are medium to strong; nonconjugated alkenes appear at *upper* end of range, and absorptions are usually weak. The absorption frequencies of these bands are raised by ring strain but to a lesser extent than noted with carbonyl functions (see 4).
(8) 1680–1640	C=N stretching Intensity: variable	This band is usually weak and difficult to assign.

The hydrogen bending region (1600–1250 cm^{-1}). Absorption in this region is commonly due to bending vibrations of hydrogen atoms attached to carbon and to nitrogen. These bands generally do not provide much useful structural information. In the listing, the bands that are most useful for structural assignment are marked with an asterisk.

$\bar{\nu}$ (cm−1)	Functional Group	Comments
(1) 1600	–NH$_2$ bending Intensity: strong to medium	In conjunction with bands in the 3300 cm^{-1} region, this band is often used to characterize primary amines and amides.

(continued)

TABLE 8.3   (continued)

$\bar{\nu}$ (cm⁻¹)	Functional Group	Comments
(2) 1540	–NH– bending Intensity: generally weak	In conjunction with bands in the 3300 cm⁻¹ region, this band is used to characterize secondary amines and amides. This band, like the N–H stretching band in the 3300 cm⁻¹ region, may be very weak in secondary amines.
(3) *1520 and 1350	NH₂ coupled stretching bands Intensity: strong	This pair of bands is usually very intense.
(4) 1465	–CH₂– bending Intensity: variable	Intensity of this band varies according to the number of methylene groups present; the more such groups, the more intense the absorption.
(5) 1410	–CH₂– bending of carbonyl-containing component Intensity: variable	This absorption is characteristic of methylene groups adjacent to carbonyl functions; its intensity depends on the number of such groups present in the molecule.
(6) *1450 and 1375	–CH₃ Intensity: strong	The band of lower frequency (1375 cm⁻¹) is usually used to characterize a methyl group. If two methyl groups are bound to one carbon atom, a characteristic doublet (1385 and 1365 cm⁻¹) is observed.
(7) 1325	–CH bending Intensity: weak	This band is weak and often unreliable.

The fingerprint region (1250–600 cm⁻¹). The fingerprint region of the spectrum is generally rich in detail, with many bands appearing. This region is particularly diagnostic for determining whether an unknown substance is identical to a known substance, the IR spectrum of which is available. It is not practical to make assignments to all these bands because many of them represent combinations of vibrational modes and therefore are very sensitive to the overall molecular structure; moreover, many single-bond stretching vibrations and a variety of bending vibrations also appear in this region. Suggested structural assignments in this region must be regarded as tentative and are generally taken as corroborative evidence in conjunction with assignments of bands at higher frequencies.

$\tilde{\nu}$ (cm⁻¹)	Functional Group	Comments
(1) 1200	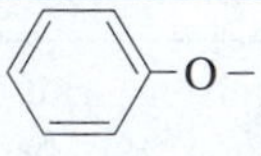 ⬡–O– Intensity: strong	It is not certain whether these strong bands arise from C–O bending or C–O stretching vibrations. One or more strong bands are found in this region in the spectra of alcohols, ethers, and esters. The relationship indicated between structure
(2) 1150	–C–O	and band location is only approximate,

$\bar{\nu}$ (cm^{-1})	Functional Group	Comments
(3) 1100	Intensity: strong $-CH-O$ Intensity: strong	and any structural assignment based on this relationship must be regarded as tentative. Esters often exhibit one or two strong bands between 1170 and 1270 cm^{-1}.
(4) 1050	$-CH_2-O-$ Intensity: strong	
(5) 985 and 910	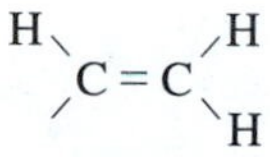 $\underset{H}{\overset{H}{\diagdown}}C=C\underset{H}{\overset{H}{\diagup}}$ C–H bending Intensity: strong	This pair of strong bands characterizes a terminal vinyl group.
(6) 965	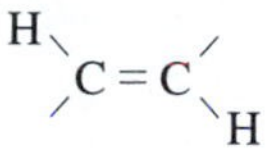 $\underset{}{\overset{H}{\diagdown}}C=C\underset{H}{\overset{}{\diagup}}$ C–H bending Intensity: strong	This strong band is present in the spectra of *trans*-1,2-disubstituted ethylenes.
(7) 890	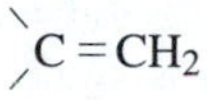 $\overset{\diagdown}{\underset{\diagup}{}}C=CH_2$ C–H bending Intensity: strong	This strong band characterizes a 1,1-disubstituted ethylene group and may be *raised* by 20–80 cm^{-1} if the methylene group is bound to an electronegative group or atom.
(8) 810–840	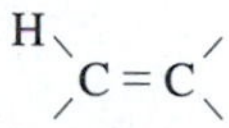 $\underset{}{\overset{H}{\diagdown}}C=C\underset{\diagdown}{\overset{\diagup}{}}$ Intensity: strong	Very unreliable; this band is not always present and frequently seems to be outside this range, since substituents are varied.
(9) 700	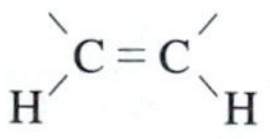 $\underset{H}{\overset{\diagdown}{\diagup}}C=C\underset{H}{\overset{\diagup}{\diagdown}}$ Intensity: variable	This band, attributable to a *cis*-1,2-disubstituted ethylene, is unreliable because it is frequently obscured by solvent absorption or other bands.
(10) 750 and 690	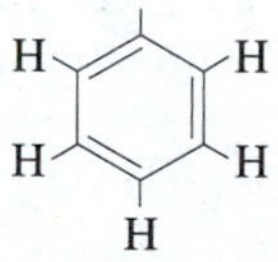 C–H bending Intensity: strong	These bands are of limited value because they are frequently obscured by solvent absorption or other bands. They are most useful when independent evidence leads to a structural assignment that is complete except for positioning of aromatic substituents.
(11) 750	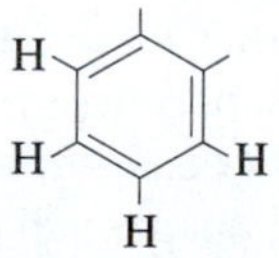 C–H bending Intensity: very strong	

(continued)

TABLE 8.3 *(continued)*

$\bar{\nu}$ (cm^{-1})	Functional Group	Comments
(12) 780 and 700	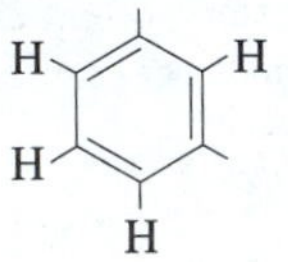 and 1, 2, 3, Intensity: very strong	
(13) 825	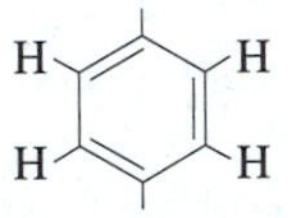 and 1, 2, 4, Intensity: very strong	
(14) 1400–1000	C–F Intensity: strong	The position of these bands is quite sensitive to structure. As a result, they are not particularly useful because the presence of halogen is more easily detected by chemical methods. The bands are usually strong.
(15) 800–600	C–Cl Intensity: strong	
(16) 700–500	C–Br Intensity: strong	
(17) 600–400	C–I Intensity: strong	

absorption may have as much significance as the *presence* of one. For example, if there is no strong band in the carbonyl region of the spectrum, this functionality is probably absent; this eliminates a number of functional groups such as aldehyde, ketone, and ester from consideration.

Assessing purity involves comparing the spectrum of the sample with that of the pure material; the observation of extra bands in the sample signals that contaminants are present. On the whole, the technique is *not* sensitive to low levels of impurities, however, since low levels of contamination in the range of 1–5% may easily go undetected.

Determining the structure of an unknown depends on the generally accepted premise that *if the IR spectra of two samples are superimposable, then the samples are identical*. The criterion of **superimposability** is a stiff requirement in the strictest sense because it means that the intensity, shape, and location of *every* absorption in the two spectra *must* be the same. This is difficult to achieve unless the *same* cells and instrument are used to obtain the IR spectra of *both* samples.

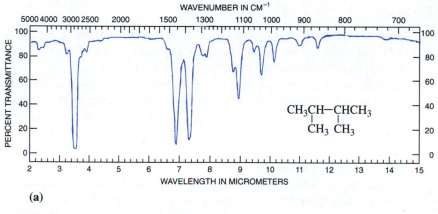

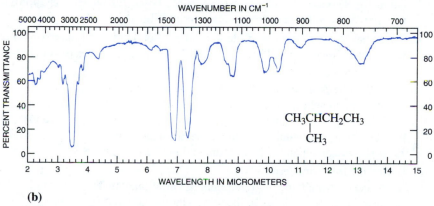

FIGURE 8.12 Infrared spectra of (a) 2,3-dimethylbutane and (b) 2-methylbutane.

Nevertheless, finding that the IR spectrum of an unknown material is very similar to the spectrum of a known material, even if the spectra are not obtained with the same instrument and cells, makes it probable that the structures of two samples are similar, if not identical. Alternative methods, such as NMR spectroscopy, preparing derivatives (see Sec. 23.5), and mixed melting points (see Sec. 3.3.3) are then used to determine whether or not the samples are indeed identical.

Absorptions that occur in the region from 1250–500 cm^{-1} usually arise from a complex *vibrational-rotational* excitation of the *entire* molecule. This portion of the spectrum typically is *unique* for a particular compound and is aptly described as the **fingerprint region.** Although the IR spectra of similar molecules may be comparable in the functional group region from 4000–1250 cm^{-1}, it is highly unlikely that this similarity will extend throughout the fingerprint region, as is illustrated in Figure 8.12.

It is important to learn to identify functional groups on the basis of the IR spectra of organic compounds. A useful way for the beginning student to do this is to examine the spectra of known compounds and correlate the key absorption bands with the functional groups responsible for them with the aid of Tables 8.2 and 8.3. Analysis of the IR spectra of the starting materials and products associated with the various experimental procedures in this textbook is an excellent way to sharpen your skills at interpreting such spectra.

EXERCISES

1. Derive the mathematical relationship between ν and $\tilde{\nu}$ by using Equations (8.1) and (8.2) and point out why it is imprecise to use the terms "frequently" and "wavenumber" interchangeably.

2. Compute the reduced mass, m^*, of a C–H and a C–D bond and then use Equation (8.3) to determine the wavenumber at which the C–D stretching vibration occurs, assuming the corresponding vibration for the C–H group is 3000 cm^{-1}.

3. Assume that the force constant k for a carbon-carbon double bond is twice that of a carbon-carbon single bond. Use Equation (8.3) to determine the wavenumber at which the C–C stretching vibration occurs, assuming the corresponding vibration for the C=C group is 1640 cm^{-1}. Repeat this calculation for the case of a carbon-carbon triple bond, for which k is approximately twice that for the carbon-carbon double bond.

4. Water has its O–H stretching mode at about 3600 cm^{-1}. At approximately what wavenumber would "heavy" water, D_2O, have the corresponding O–D stretch? Show your work.

5. **a.** If the strengths of a carbon-carbon triple bond and a carbon-nitrogen triple bond are assumed to be the same, as reflected in the force constant k (see Eq. 8.3), predict which of the two functionalities would appear at the higher wavenumber in the IR spectrum and support your prediction.
   b. Is this in agreement with experimental fact? (See Table 8.3.)

6. Define the following terms.
   a. percent transmittance   **f.** stretching vibrational mode
   b. neat liquid   **g.** bending vibrational mode
   c. hygroscopic   **h.** electronic excitation
   d. fixed-thickness cell   **i.** superimposability (as applied to IR spectra)
   e. functional group   **j.** mull

7. Convert the following values to micrometers, μm:
   a. 2250 cm^{-1} **b.** 1678 cm^{-1} **c.** 630 cm^{-1}

8. Transform the following values to wavenumbers, cm^{-1}:
   a. 6.05 μm **b.** 16.5 μm **c.** 3.75 μm

9. Compute the energy, in cal/mol, associated with radiation having a wavelength of 7.50 μm.

10. Explain why the windows of IR cells and KBr pellets should not be exposed to moisture.

11. Explain why the windows of IR cells should not be subjected to excessive pressure.

12. Why is an IR spectrum calibrated?

13. Name two solvents that are appropriate and two that are inappropriate for cleaning the faces of IR cells.

14. Suppose that the IR spectrum of a compound was obtained on a 10 mol % solution in CCl_4 contained in a cell having a pathlength of 0.25 mm. Assume that the transmittance, T, of a peak of interest in the solute was 0.50 at this concentration *and* that the absorptivity, k, (see Eq. 8.5) does not vary with concentration of the solution. Determine whether I will increase, decrease, or remain unchanged when the spectrum is measured under the following conditions, and briefly explain your answer.

  a. The concentration of the solution is decreased to 5 mol %, and the path-length of the cell remains at 0.25 mm.

  b. The concentration of the solution is increased to 15 mol %, and the path-length of the cell remains at 0.25 mm.

  c. The pathlength of the cell is increased to 0.30 mm, and the concentration of the solution remains at 10 mol %.

  d. The pathlength of the cell is decreased to 0.02 mm, and the concentration of the solution remains at 10 mol %.

15. What difficulties would be associated with obtaining a solution IR spectrum on a double beam spectrometer without using a reference cell containing the pure solvent?

16. What problems might be encountered in attempting to make a KBr pellet of an organic solid that has crystallized as a hydrate?

17. Figures 8.13–8.20 are the IR spectra of the compounds shown on the individual spectra. Assign as many of the major absorptions as possible in the spectrum of each compound, using Tables 8.2 and 8.3.

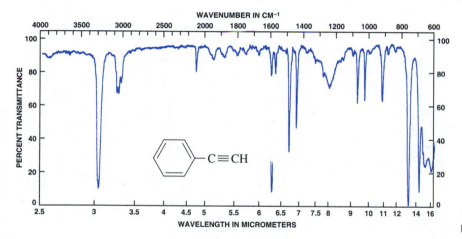

FIGURE 8.13   Infrared spectrum of phenylethyne (phenylacetylene).

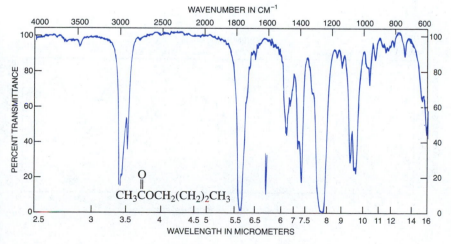

FIGURE 8.14   Infrared spectrum of *n*-butyl ethanoate (*n*-butyl acetate).

FIGURE 8.15 Infrared spectrum of *N,N*-dimethylacetamide.

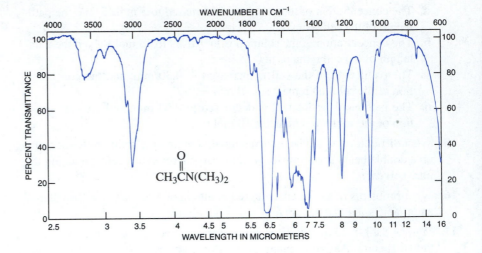

FIGURE 8.16 Infrared spectrum of benzoyl chloride.

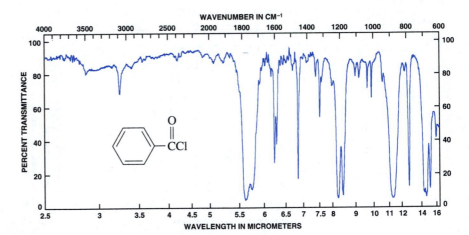

FIGURE 8.17 Infrared spectrum of benzoic anhydride (KBr pellet).

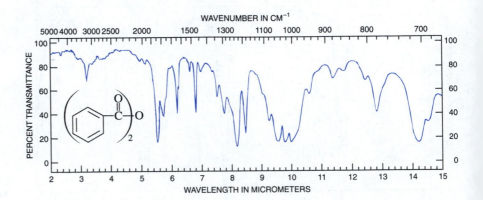

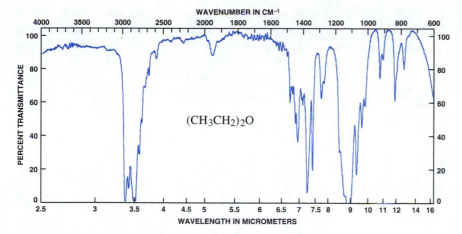

WAVENUMBER IN CM⁻¹

$(CH_3CH_2)_2O$

WAVELENGTH IN MICROMETERS

FIGURE 8.18 Infrared spectrum of diethyl ether.

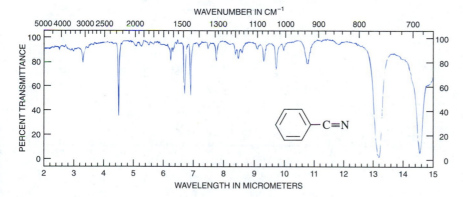

WAVENUMBER IN CM⁻¹

$C \equiv N$

WAVELENGTH IN MICROMETERS

FIGURE 8.19 Infrared spectrum of benzonitrile.

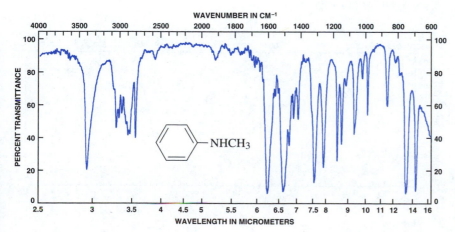

WAVENUMBER IN CM⁻¹

$NHCH_3$

WAVELENGTH IN MICROMETERS

FIGURE 8.20 Infrared spectrum of *N*-methylaniline.

FIGURE 8.21 Infrared spectra for Exercise 18.

18. Figure 8.21 contains the IR spectra of 4-methyl-2-pentanol and 4-methyl-1-pentene. Write the structure of each of these compounds and determine which spectrum is associated with each substance. Explain your assignment.

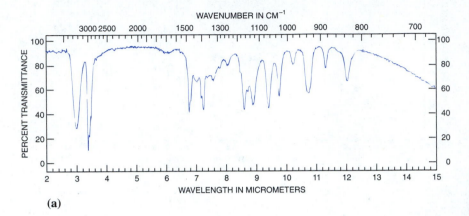

(a)

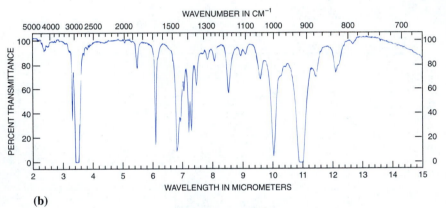

(b)

19. Figure 8.22 contains the IR spectra of dimethyl fumarate and *cis*-cyclohexane-1,2-dicarboxylic acid. Write the structure of each of these compounds and determine which spectrum is associated with each substance. Explain your assignment.

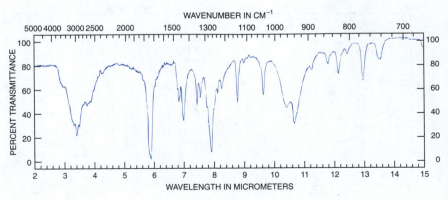

FIGURE 8.22 Infrared spectra for Exercise 19.

(a)

FIGURE 8.22 *(Continued)*

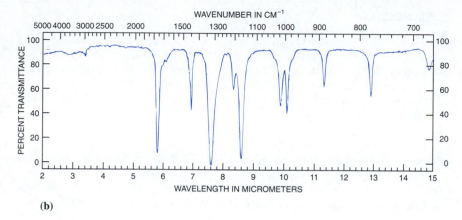

(b)

20. Figure 8.23 contains the IR spectra of acetanilide and styrene. Write the structure of each of these compounds and determine which spectrum is associated with each substance. Explain your assignment.

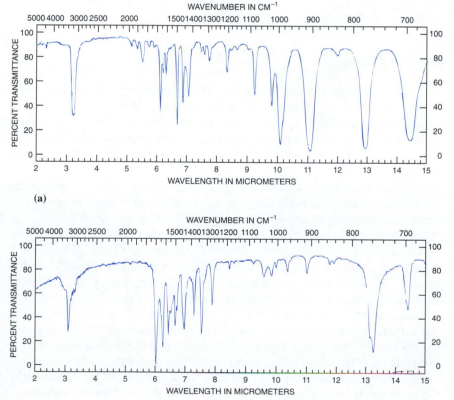

(a)

(b)

FIGURE 8.23 Infrared spectra for Exercise 20.

REFERENCES

1. Silverstein, R. M.; Bassler, G. C.; Morrill, T. C. *Spectrometric Identification of Organic Compounds*, 5th ed., John Wiley & Sons, New York, 1992, Chap. 3.
2. Colthup, N. B.; Daly, L. H.; Wiberley, S. E. *Introduction to Infrared and Raman Spectroscopy*, 3rd ed., Academic Press, New York, 1990.
3. Pouchert, C. J., ed. *Aldrich Library of Infrared Spectra*, vol. I–II, Aldrich Chemical Co., Milwaukee, WI, 1970.
4. Pouchert, C. J., ed. *Aldrich Library of FTIR Infrared Spectra*, vol. I–II, Aldrich Chemical Co., Milwaukee, WI, 1985.
5. *Sadtler Standard Infrared Prism Spectra*, vol. 1–107, Sadtler Research Laboratories, Philadelphia, PA. Compilation of 83,000 spectra, as of 1992.
6. *Sadtler Standard Infrared Grating Spectra*, vol. 1–107, Sadtler Research Laboratories, Philadelphia, PA. Compilation of 83,000 spectra, as of 1992.

8.3 NUCLEAR MAGNETIC RESONANCE (NMR) SPECTROSCOPY

Nuclear magnetic resonance (NMR) spectroscopy is probably the single most powerful spectroscopic method available to the modern organic chemist for analyzing compounds. This technique depends on the property of **nuclear spin** that is exhibited by certain nuclei when they are placed in a magnetic field. Some such nuclei commonly found in organic compounds are ^1H, ^2H, ^{19}F, ^{13}C, ^{15}N, and ^{31}P. Nuclei such as ^{12}C, ^{16}O, and ^{32}S do *not* have nuclear spin and thus cannot be studied by NMR spectroscopy.

8.3.1 PRINCIPLES

The theory underlying the technique of NMR spectroscopy is presented in any comprehensive lecture textbook of organic chemistry and is not discussed in detail here. It is important to recall, however, that the observation of absorptions in the NMR spectrum of a particular nucleus is due to *realignment of nuclear* spins in an **applied magnetic field,** H_0, as shown in Parts (b) and (c) of Figure 8.24. The energy associated with the realignment of spin or spin flip as it is often called, is expressed in Equation (8.6a) and depends on the strength of the applied field H_0 and the **magnetogyric ratio** γ that is characteristic of the particular nucleus being examined. In a modern NMR spectrometer operating at 90–300 MHz, the value of H_0 is in the range of 21,000 to 71,000 gauss (2.1–7.1 Tesla); this means that the energy required for the transition is less than 0.1 cal/mol for a nucleus such as ^1H. Another way of illustrating the relationship between the strength of the magnetic field H_0, and the energy difference between spin states is given in Figure 8.25, where it is seen that for a nucleus that has two spin states, labeled ±1/2, ΔE increases as H_0 increases. Energies of this sort are associated with radiation in the **radio-frequency** (rf) **range** of the electromagnetic spectrum; consequently, a radio-frequency oscillator is incorporated into the NMR spectrometer to provide the energy for the nuclear spin excitation. When the relationship between the magnetic field and the frequency of the oscillator is that defined by Equation (8.6b), the **resonance condition** is achieved, and the transition from one nuclear spin state to the other can occur.

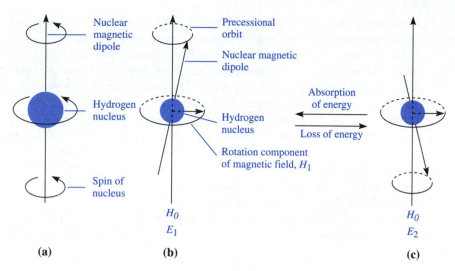

Nuclear magnetic dipole

Hydrogen nucleus

Spin of nucleus

(a)

Precessional orbit

Nuclear magnetic dipole

Hydrogen nucleus

Rotation component of magnetic field, H_1

Absorption of energy

Loss of energy

H_0
E_1

(b)

H_0
E_2

(c)

FIGURE 8.24 Spin properties of the hydrogen nucleus. (a) Rotation of hydrogen nucleus and its magnetic moment. (b) Magnetic moment of nucleus aligned with applied external magnetic field. (c) Magnetic moment of nucleus aligned against applied magnetic field.

$$\Delta E = h\gamma H_0/2\pi \qquad \textbf{(8.6a)}$$

$$h\nu = h\gamma H_0/2\pi \quad \text{or} \quad \nu = \gamma H_0/2\pi \qquad \textbf{(8.6b)}$$

where ΔE = energy difference between two spin states

h = Planck's constant

γ = magnetogyric ratio (a constant characteristic of a particular type of nucleus)

H_0 = strength of applied external magnetic field

ν = frequency of oscillator

To achieve the resonance condition of Equation (8.6b), H_0 could be held constant and ν varied, or ν could be kept constant and H_0 changed. In **continuous wave** (CW) spectrometers, the earliest form of commercial NMR instruments, the latter option is more common, and the spectrum is obtained by slowly sweeping through the range of field strengths required to produce resonance at a particular oscillator frequency. Modern instruments, called **Fourier transform** (FT) spectrometers, operate with a pulse technique in which all resonance frequencies are produced simultaneously while H_0 is held constant; this technique allows collection of spectral data in much less time than with a CW machine.

8.3.2 PROTON MAGNETIC RESONANCE (^{1}H NMR) SPECTROSCOPY

The NMR spectral technique involving the hydrogen nucleus, ^{1}H, which has a nuclear spin, I_z, of 1/2 is called **proton magnetic resonance** spectroscopy and is variously abbreviated as PMR or ^{1}H NMR; the second option is used here. Before discussing this type of NMR spectroscopy further, it is important to note that in the terminology of NMR spectroscopy a covalently bound hydrogen atom is often imprecisely called a proton.

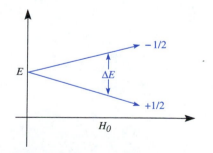

FIGURE 8.25 Energy difference between spin states as a function of strength of magnetic field, H_0.

Considering Equation (8.6) reveals that the *same* energy would be required to produce the resonance condition for *all* hydrogen nuclei in a molecule *if* all of these nuclei were magnetically identical. In other words, if the magnetic environment as defined by H_0 were identical for all of the hydrogen atoms, the same frequency ν would be needed for the spin flip. The consequence of this would be the appearance of a *single* absorption band in the ^{1}H NMR spectrum, a very uninformative result. Fortunately the nuclei are *not* all magnetically equivalent, because the three-dimensional electronic structure of a molecule produces variations in the magnetic environments of the hydrogen atoms in it (see **"Chemical Shift"** later in this section). Thus, achieving the resonance condition requires *varying* amounts of energy that results in a spectrum that contains valuable information for determining molecular structure.

Consider the ^{1}H NMR spectrum of 1-nitropropane (**1**), shown in Figure 8.26. The significant features illustrated in this spectrum are **chemical shift, spin-spin**

$$H_a\text{—}C\text{—}C\text{—}C\text{—}NO_2$$

(with H_a, H_a on first carbon; H_b, H_b on second carbon; H_c, H_c on third carbon)

1

1-Nitropropane

splitting, and **peak integration,** each of which is discussed in detail. First, however, a brief overview of the general format for presenting NMR data is necessary.

In Figure 8.26, the strength of the applied magnetic field H_0 is plotted along the horizontal axis and *increases* from left to right in the spectrum. This direction is called **upfield;** the opposite one is termed **downfield.** As an example, the group of peaks labeled B could be described as being upfield of the resonances labeled C. The *energy* needed to achieve resonance also *increases* in going from left to right. The intensity of the absorption band is plotted on the vertical axis and *increases* from the bottom, or baseline, of the plot.

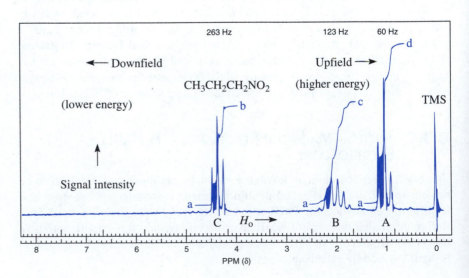

FIGURE 8.26 ^{1}H NMR spectrum of 1-nitropropane (60 MHz).

Depending on the type of spectrometer used, several different scales may be printed horizontally across the top and bottom of the spectrum. The scales are used to define the location of peaks in the spectrum relative to those of a reference compound, normally **tetramethylsilane,** $(CH_3)_4Si$, which is abbreviated as TMS. This compound, which is a volatile liquid that does *not* react with most organic compounds, may be added directly to solutions of the sample and thus serves as an **internal standard.** The ^1H NMR spectra in this textbook show only the *delta-* or δ-scale, and it is found at the bottom of each spectrum. This scale is in units of **parts per million,** ppm, for reasons discussed under the next heading. Other aspects of ^1H NMR spectra are presented under the headings **"Spin-Spin Splitting"** and **"Integration,"** respectively.

1. Chemical Shift. Three groups of peaks, centered at about 1.0, 2.0, and 4.4 on the δ-scale, are seen in the spectrum of Figure 8.26. These values are the **chemical shifts,** in ppm, of the three chemically distinct types of hydrogen nuclei in 1-nitropropane. Predicting whether the various protons in a molecule are chemically distinct and thus will *probably* appear at different chemical shifts, is an important part of understanding ^1H NMR spectra. Fortunately, determining how many chemically different protons are present in a molecule is straightforward.

The first step in the analysis is to group the protons according to their **connective equivalency,** placing in the same set all protons having identical molecular connectivities. The next step is to assess whether *connectively* equivalent nuclei are **chemically equivalent** and thus likely to have identical chemical shifts. This procedure is illustrated by further consideration of 1-nitropropane (**1**). Examining **1** reveals that it contains three connectively unique sets of protons, labeled H_a, H_b, and H_c, respectively. *Hydrogen nuclei that are connectively distinct are chemically nonequivalent* and are thus defined as being **heterotopic.** As a general rule, such nuclei have *different* chemical shifts from one another, although some of the peaks may overlap.

The second step in the analysis involves evaluating connectively *identical* nuclei for their possible chemical nonequivalency. This is done by a **substitution test** in which it is supposed that a hydrogen nucleus is replaced with a **probe nucleus** that can be distinguished from other nuclei in the same set. The stereoisomeric relationship between such substituted species is then assessed. The substitution procedure is presented in Figure 8.27 for protons of types H_a and H_c in **1**. As shown, the hypothetical replacement of H_{a1}, H_{a2}, and H_{a3} with deuterium (**D**) as the probe nucleus provides three new representations that are *identical* to one another; the identity is established by rotating about the bond joining the methyl group adjacent to the methylene group and seeing that the resulting molecules are *superimposable* upon one another. Such hydrogens are defined as being **homotopic** and are *chemically equivalent;* they will all have the *same* chemical shift in the ^1H NMR spectrum of **1**. Applying the same substitution process to H_{c1} and H_{c2} gives *enantiomeric* representations, and the two hydrogen atoms are thus described as being **enantiotopic.** Enantiotopic nuclei are also chemically equivalent and will have the same chemical shift, provided they are in an *achiral* environment, such as that provided by the solvents commonly used in NMR studies. It is left as an exercise to demonstrate that the hydrogen atoms of type H_b are also enantiotopic (see Exercise 2 at the end of this section).

FIGURE 8.27 Topicity analysis of 1-nitropropane by substitution.

(a)

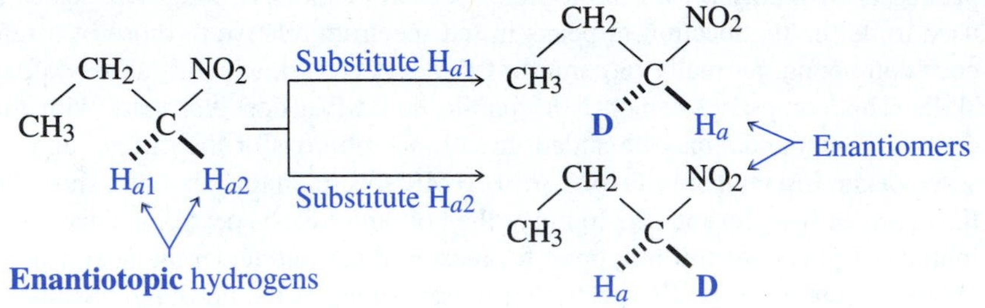

(b)

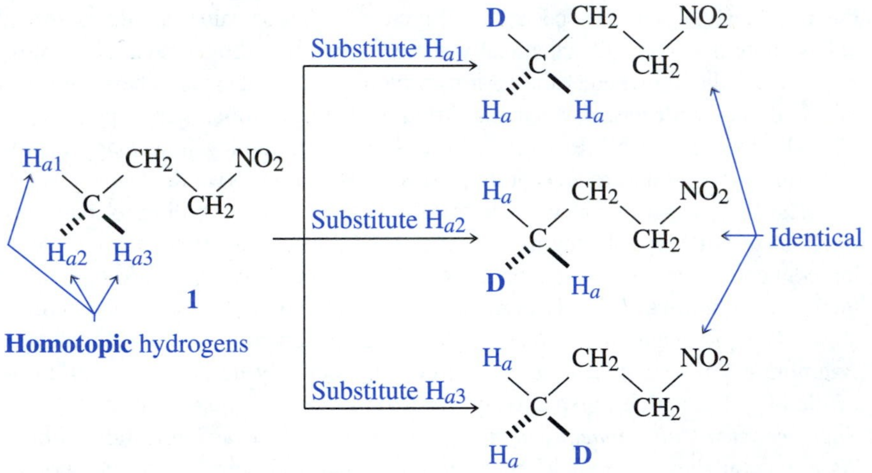

Another type of topicity is possible when a molecule contains a **chiral center** (see Sec. 7.1), or is subject to **restricted rotation,** a phenomenon usually associated with the presence of a π-bond or a ring. The first case is illustrated by considering 2-chlorobutane (**2,** Fig. 8.28); although only a single enantiomer is shown, the ensuing analysis gives the same result for the other one as well. The molecule has a chiral center at C-2, the effect of which makes the connectively equivalent hydrogen atoms at C-3 chemically *nonequivalent*. The nonequiva-

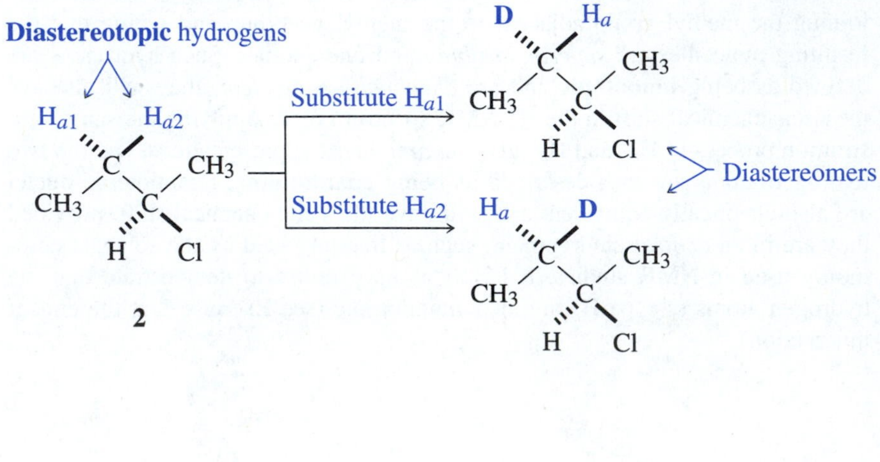

FIGURE 8.28 Topicity analysis of 2-chlorobutane by substitution.

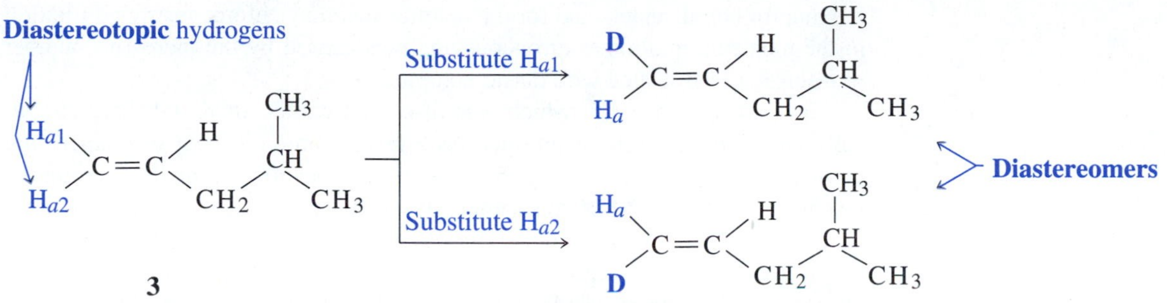

Diastereotopic hydrogens

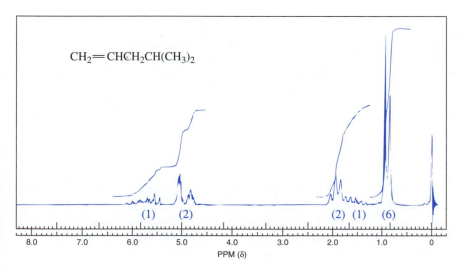

3

FIGURE 8.29   Topicity analysis of 4-methyl-1-pentene by substitution.

lency is revealed by performing the substitution test on these protons, as illustrated in Figure 8.28. This generates **diastereomers,** thereby defining the two protons at C-3 as **diastereotopic.** Because diastereotopic nuclei *may* have chemical shifts that are different from one another, there are *five* rather than four sets of chemically nonequivalent protons in **2**: the two different methyl groups, each of which contains three *homotopic* protons, the methine hydrogen at C-2, and the two *diastereotopic* nuclei at C-3.

Chemical nonequivalency arising from restricted rotation is seen in 4-methyl-1-pentene (**3**, Fig. 8.29). By applying the substitution test (Fig. 8.29), the two vinylic hydrogen atoms at C-1 are found to be *diastereotopic* and thus chemically nonequivalent. This means that **3** contains *three* distinct sets of vinylic protons, each of which appears at a different chemical shift, as seen in Figure 8.30. Applying the substitution test in this instance is simply a formal way of proving that the geminal vinylic protons of an *unsymmetrically* substituted alkene, $H_2C=CR^1R^2$, are not equivalent to one another. For example, one of these hydrogen atoms will be *cis* to R^1, and the other one *trans* to this substituent; their nonequivalency can be determined rapidly and simply by this stereochemical relationship.

The occurrence of diastereotopic hydrogens in a molecule may complicate the appearance of a ^1H NMR spectrum. However, the simple procedure of

$CH_2=CHCH_2CH(CH_3)_2$

(1)   (2)   (2) (1)   (6)

8.0   7.0   6.0   5.0   4.0   3.0   2.0   1.0   0

PPM (δ)

FIGURE 8.30 ^1H NMR spectrum of 4-methyl-1-pentene (60 MHz).

looking for chiral centers and for the features that may enforce restricted rotation in the molecule minimizes erroneous analyses caused by the increased number of resonances associated with diastereotopicity.

In 1-nitropropane (**1**), which was discussed earlier, there are three chemically nonequivalent sets of protons. As a result, three different resonances are predicted, as is observed (Fig. 8.26). The relative locations of the three groups of resonances in the ^{1}H NMR spectrum of **1** varies according to the magnetic environment of each hydrogen nucleus in the molecule. This primarily depends on two factors: (1) the externally applied magnetic field, and (2) circulation of the electrons within the molecule, which provides an **electronic shield** about the atoms. The field H_0 is of *uniform* strength throughout the molecule, so it *cannot* be responsible for the magnetic nonequivalence of the different nuclei. On the other hand, the shielding effect produced by circulation of electrons is *not* uniform because the electron density varies within the molecule. This is a critical point because the **internal electric field** of the molecule induces an **internal magnetic field,** H_i, about its nuclei. The induced magnetic field acts to *oppose* the applied magnetic field, so the **effective magnetic field,** H_e, at the nuclei is less than H_0 (Eq. 8.7); this phenomenon is the source of **diamagnetic shielding** the molecule. The strength of H_i varies in the same way as that of the electric field and thus makes H_e different for the various types of protons in the sample. For example, the greater is H_i the less is H_e, so a larger external field must be applied to achieve resonance. In other words, the more magnetic shielding experienced by a nucleus, the higher is the external field H_0 required to produce the resonance condition at a particular oscillator frequency.

$$H_e = H_0 - H_i \qquad \textbf{(8.7)}$$

The consequences of the magnitude of H_i on the location of resonances in ^{1}H NMR spectra are seen in Figure 8.26. The nitro group NO_2 serves as an electron sink, and this causes the electric field to be lower for those nuclei near it. Correspondingly, the value of H_i is less for these nuclei than for others more distant from the nitro function. The protons labeled H_c in **1** thus experience the *least* diamagnetic shielding and resonate at the *lowest* value of H_0. The methyl protons H_a are the *most* shielded because they are furthest from the nitro group and appear at the highest field (Fig. 8.26). Put another way, the nearer a proton is to the electronegative nitro group, the more **deshielded** it is relative to other hydrogen atoms in the molecule, and the further **downfield** it appears. On this basis, the various groups of peaks in the spectrum are assigned as shown below.

$$
\begin{array}{ccc}
\textbf{A} & \textbf{B} & \textbf{C} \\
CH_3 - CH_2 - CH_2 - NO_2 \\
\uparrow & \uparrow & \uparrow \\
\delta\ 1.0 & 2.05 & 4.38
\end{array}
$$

The locations of the three groups of peaks in 1-nitropropane are expressed in units of ppm on the **δ-scale**, as was noted earlier. This scale, calculated according to Equation (8.8), is defined such that measurement of chemical shifts is *independent* of the frequency of the rf oscillator of the NMR instrument. The numerator is the shift, in Hz, of a resonance of the sample relative to that of the reference compound, which is TMS for ^{1}H NMR spectra. The sign of the shift by convention is taken to be positive if the resonance appears *downfield* of TMS and negative if it is upfield. The shift is divided by the corresponding oscillator frequency, and for convenience in reporting chemical shifts, the result is multiplied by 10^6 to provide units of ppm (see Exercise 4 at the end of this section.) Alternatively, the relative positions of peaks in the spectrum could simply be expressed in Hz relative to TMS. This is unsatisfactory because the chemical shift would then depend on the frequency of oscillator being used in the spectrometer (see Eq. 8.6). Consequently, reporting shifts in Hz would require specifying the frequency at which the data were obtained.

$$\delta = \frac{\text{chemical shift (in Hz)} \times 10^6}{\text{oscillator frequency (in Hz)}} \qquad \text{(8.8)}$$

To illustrate the difference between measuring chemical shifts in terms of δ versus Hz, consider the case in which a resonance is observed *60 Hz* downfield of TMS when a spectrometer operating at 60 MHz is used. On the δ-scale, this corresponds to a shift of *1.0 ppm*. Had the spectrum been measured on a 300-MHz instrument, however, the resonance would appear *300 Hz* downfield of TMS. The shift would remain at *1.0 ppm* on the δ-scale.

Almost all types of protons appear *downfield* of TMS and thus have chemical shifts that are *positive* values because of the sign convention noted above. However, some organic compounds contain hydrogen nuclei that are *more* shielded than those in TMS and therefore appear *upfield* of it. The chemical shifts are *negative* in such cases.

The consistent effect that particular functional groups have on chemical shifts has led to tabulations correlating these shifts with the structural features responsible for them. Table 8.4 contains one such compilation, and Table 8.5 is an expanded listing of the chemical shifts typical of a variety of specific types of hydrogen atoms. Because the shifts are given in units of ppm, they apply to all NMR spectrometers regardless of the frequency at which the instrument is operating.

The *magnitude* of chemical shifts is primarily related to two factors, namely, the electronegativity of any functional groups that are near the proton being observed and the nature of induced magnetic fields, H_e, in molecules having π-bonds. Both effects are associated with circulation of electrons in the molecule and the interaction of the resulting induced magnetic field with the applied field H_0 (Eq. 8.7).

The influence of electronegativity is evident from comparison of the chemical shifts of the methyl halides, $CH_3 - X$, in which the δ-values are 4.3, 3.0, 2.7, and 2.2 for X = F, Cl, Br, and I, respectively (Table 8.5). That of π-electrons is seen in the relative chemical shifts of the aliphatic protons of alkanes versus the vinylic protons of alkenes, as in ethane (δ 0.9) and ethylene (δ 5.25). The hydrogen nuclei of ethylene are *deshielded* partly because an sp^2-hybridized carbon atom is more electronegative than one that is sp^3-hybridized. Moreover, the vinylic protons are further deshielded because they lie in a region of the induced field H_e where the lines of force *add* to H_0, as shown in Figure 8.31(a). A similar analysis of the induced field accounts for the fact that an alkynic proton, as in acetylene (δ 1.80), resonates at *higher* field than does a vinylic hydrogen, even

TABLE 8.4 Chemical Shifts of Hydrogen Atoms Attached to Various Functional Groups

Functional Group; Hydrogen Type Shown as H	Chemical Shift, ppm, δ	Functional Group; Hydrogen Type Shown as H	Chemical Shift, ppm, δ
TMS, $(CH_3)_4Si$	0	Alcohols, ethers	
Cyclopropane	0–1.0	$HO-C-H$	3.4–4
Alkanes		$RO-C-H$	3.3–4
$\quad RCH_3$	0.9		
$\quad R_2CH_2$	1.3	Acetals $\genfrac{}{}{0pt}{}{-O}{-O}\!\!\diagdown C \diagup H$	5.3
$\quad R_3CH$	1.5		
Alkenes		Esters	
$\quad -C=C-H$ (vinyl)	4.6–5.9	$R-\overset{O}{\overset{\|}{C}}-O-C-H$	3.7–4.1
$\quad -C=C-CH_3$ (allyl)	1.7	$RO-\overset{O}{\overset{\|}{C}}-C-H$	2–2.6
Alkynes		Carboxylic acids	
$\quad -C\equiv C-H$	2–3	$HO-\overset{O}{\overset{\|}{C}}-C-H$	2–2.6
$\quad -C\equiv C-CH_3$	1.8		
Aromatic		$R-\overset{O}{\overset{\|}{C}}-O-H$	10.5–12
$\quad Ar-H$	6–8.5		
$\quad Ar-C-H$ (benzyl)	2.2–3	Ketones	
Fluorides, $F-C-H$	4–4.5	$R-\overset{O}{\overset{\|}{C}}-C-H$	2–2.7
Chlorides		Aldehydes	
$\quad Cl-C-H$	3–4	$R-\overset{O}{\overset{\|}{C}}-H$	9–10
$\quad \overset{Cl}{\underset{}{Cl-C-H}}$	5.8	Amides	
Bromides, $Br-C-H$	2.5–4	$R-\overset{O}{\overset{\|}{C}}-N-H$	5–8
Iodides, $I-C-H$	2–4	Alcohols, $R-O-H$	4.5–9
		Phenols, $Ar-O-H$	4–12
Nitroalkanes, O_2N-C-H	4.2–4.6	Amines, $R-NH_2$	1–5

TABLE 8.5 Compilation of ^{1}H NMR Absorptions for Various Molecules

Listed below are the ^{1}H NMR chemical shifts observed for the protons of a number of organic compounds. The shifts are classified according to whether they are methyl, methylene, or methine types of hydrogen atoms. The atom shown in **bold** is responsible for the absorptions listed below.

METHYL ABSORPTIONS

Compound	Chemical Shift (ppm), δ	Compound	Chemical Shift (ppm), δ
CH_3NO_2	4.3	CH_3CHO	2.2
CH_3F	4.3	CH_3I	2.2
$(CH_3)_2SO_4$	3.9	$(CH_3)_3N$	2.1
$C_6H_5COOCH_3$	3.9	$CH_3CON(CH_3)_2$	2.1
$C_6H_5-O-CH_3$	3.7	$(CH_3)_2S$	2.1
CH_3COOCH_3	3.6	$CH_2=C(CN)CH_3$	2.0
CH_3OH	3.4	CH_3COOCH_3	2.0
$(CH_3)_2O$	3.2	CH_3CN	2.0
CH_3Cl	3.0	CH_3CH_2I	1.9
$C_6H_5N(CH_3)_2$	2.9	$CH_2=CH-C(CH_3)=CH_2$	1.8
$(CH_3)_2NCHO$	2.8	$(CH_3)_2C=CH_2$	1.7
CH_3Br	2.7	CH_3CH_2Br	1.6
CH_3COCl	2.7	$C_6H_5C(CH_3)_3$	1.3
CH_3SCN	2.6	$C_6H_5CH(CH_3)_2$	1.2
$C_6H_5COCH_3$	2.6	$(CH_3)_3COH$	1.2
$(CH_3)_2SO$	2.5	$C_6H_5CH_2CH_3$	1.2
$C_6H_5CH=CHCOCH_3$	2.3	CH_3CH_2OH	1.2
$C_6H_5CH_3$	2.3	$(CH_3CH_2)_2O$	1.2
$(CH_3CO)_2O$	2.2	$CH_3(CH_2)_3Cl, Br, I$	1.0
$C_6H_5OCOCH_3$	2.2	$CH_3(CH_2)_4CH_3$	0.9
$C_6H_5CH_2N(CH_3)_2$	2.2	$(CH_3)_3CH$	0.9

METHYLENE ABSORPTIONS

Compound	Chemical Shift (ppm), δ	Compound	Chemical Shift (ppm), δ
$EtOCOC(CH_3)=CH_2$	5.5	$EtCH_2Cl$	3.4
CH_2Cl_2	5.3	$(CH_3CH_2)_4N^+I^-$	3.4
CH_2Br_2	4.9	CH_3CH_2Br	3.4
$(CH_3)_2C=CH_2$	4.6	$C_6H_5CH_2N(CH_3)_2$	3.3
$CH_3COO(CH_3)C=CH_2$	4.6	$CH_3CH_2SO_2F$	3.3
$C_6H_5CH_2Cl$	4.5	CH_3CH_2I	3.1
$(CH_3O)_2CH_2$	4.5	$C_6H_5CH_2CH_3$	2.6

(continued)

TABLE 8.5 *(continued)*

Compound	Chemical Shift (ppm), δ	Compound	Chemical Shift (ppm), δ
$C_6H_5CH_2OH$	4.4	CH_3CH_2SH	2.4
$CF_3COCH_2C_3H_7$	4.3	$(CH_3CH_2)_3N$	2.4
$Et_2C(COOCH_2CH_3)_2$	4.1	$(CH_3CH_2)_2CO$	2.4
$HC{\equiv}C{-}CH_2CL$	4.1	$BrCH_2CH_2CH_2Br$	2.4
$CH_3COOCH_2CH_3$	4.0	Cyclopentanone (α-CH_2)	2.0
$CH_2{=}CHCH_2Br$	3.8	Cyclohexene (α-CH_2)	2.0
$HC{\equiv}CCH_2Br$	3.8	Cycloheptane	1.5
$BrCH_2COOCH_3$	3.7	Cyclopentane	1.5
CH_3CH_2NCS	3.6	Cyclohexane	1.4
CH_3CH_2OH	3.6	$CH_3(CH_2)_4CH_3$	1.4
		Cyclopropane	0.2

METHINE ABSORPTIONS

Compound	Chemical Shift (ppm), δ	Compound	Chemical Shift (ppm), δ
C_6H_5CHO	10.0	C_6H_5Cl	7.2
p-ClC_6H_4CHO	9.9	$CHCl_3$	7.2
p-$CH_3OC_6H_4CHO$	9.8	$CHBr_3$	6.8
CH_3CHO	9.7	p-Benzoquinone	6.8
Pyridine (α-**H**)	8.5	$C_6H_5NH_2$	6.6
p-$C_6H_4(NO_2)_2$	8.4	Furan (β-**H**)	6.3
$C_6H_5CH{=}CHCOCH_3$	7.9	$CH_3CH{=}CHCOCH_3$	5.8
C_6H_5CHO	7.6	Cyclohexene (vinylic **H**)	5.6
Furan (α-**H**)	7.4	$(CH_3)_2C{=}CHCH_3$	5.2
Naphthalene (β-**H**)	7.4	$(CH_3)_2CHNO_2$	4.4
p-$C_6H_4I_2$	7.4	Cyclopentyl bromide (**H** at C-1)	4.4
p-$C_6H_4Br_2$	7.3	$(CH_3)_2CHBr$	4.2
p-$C_6H_4Cl_2$	7.2	$(CH_3)_2CHCl$	4.1
C_6H_6	7.3	$C_6H_5C{\equiv}C{-}\mathbf{H}$	2.9
C_6H_5Br	7.3	$(CH_3)_3C{-}\mathbf{H}$	1.6

though an *sp*-hybridized carbon atom is the *most* electronegative of all carbon atoms found in organic compounds. In alkynes, the acetylenic proton is in a portion of the magnetic field induced by the circulating π-electrons where the lines of force *oppose* the applied field (Fig. 8.31b), thereby increasing the shielding of this type of nucleus. The protons of aromatic compounds, as in benzene (δ 7.27), resonate at lower fields than the vinylic protons of alkenes because the induced mag-

Deshielded   Shielded   Deshielded
region   region   region

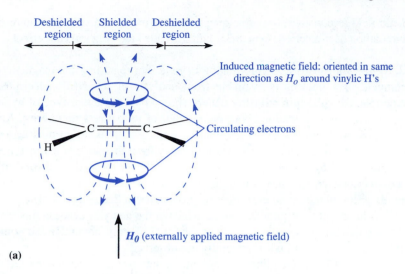

Induced magnetic field: oriented in same
direction as H_0 around vinylic H's

Circulating electrons

H_0 (externally applied magnetic field)

(a)

FIGURE 8.31   Effect on chemical shifts of induced field from circulation of π-electrons. (a) Deshielding of vinylic protons. (b) Shielding of acetylenic protons. (c) Deshielding of aromatic protons.

Deshielded   Shielded   Deshielded
region   region   region

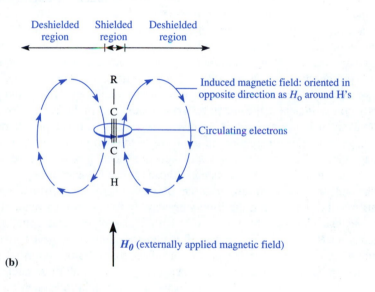

Induced magnetic field: oriented in
opposite direction as H_0 around H's

Circulating electrons

H_0 (externally applied magnetic field)

(b)

Deshielded   Shielded   Deshielded
region   region   region

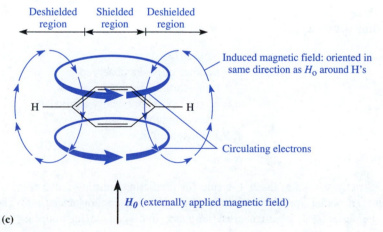

Induced magnetic field: oriented in
same direction as H_0 around H's

Circulating electrons

H_0 (externally applied magnetic field)

(c)

netic field responsible for deshielding is greater owing to the *cyclic* nature of the circulation of π-electrons (Fig. 8.31c); this is called the **ring-current effect.**

2. Spin-Spin Splitting.   Information beyond that regarding the magnetic environment of a nucleus is available from analysis of the NMR spectrum. For example, the **spin-spin splitting** pattern shown by a particular type of hydrogen atom provides information about the number of its **nearest neighbors.** Nearest neighbors are defined as elements that have nuclear spin and, in most cases, are *no more* than *three* bonds, or *two* atoms, away from the atom of interest. For our purposes, hydrogen is the primary element of interest for analyzing splitting patterns because compounds containing other magnetically active nuclei are only rarely encountered in the experimental procedures in this textbook. The presence of ^{13}C in organic compounds is *not* a factor in the analysis because this isotope, despite its nuclear spin, is present at the level of only about 1%, so spin-spin splitting associated with it is generally *not* observed.

Observation of coupling between atoms having nuclear spins requires that the nuclei must *not* be *magnetically* equivalent. By definition, **magnetic equivalency** means that *every* nucleus in a particular set of chemical shift-equivalent nuclei is coupled equally to every other nucleus of the spin system. Chemical shift-nonequivalent nuclei are always magnetically nonequivalent and couple with one another. As discussed earlier, this type of nonequivalence is generally seen for protons that are chemically distinct by virtue of being *heterotopic* or *diastereotopic*. However, that protons may be chemically equivalent but magnetically nonequivalent is illustrated by a discussion later in this subsection.

Analysis of 1-nitropropane (**1**) in this sense reveals that its type *a* hydrogen atoms have *two* nearest neighbors in the form of the two protons of type *b* that are magnetically active and chemically nonequivalent. Nuclei of type *c* also have two such nearest neighbors, whereas those of type *b* nuclei have *five*.

The **splitting** or **coupling pattern** of each chemically distinct type of nucleus may be predicted in the following way. In the general case, an atom *A* having nuclear spin I_z, where I_z = 1/2, 1, 3/2, and so on, that couples with another atom *B* splits the resonance of *B* into the number of peaks given by Equation (8.9a). When *A* is ^1H, ^{13}C, ^{19}F, or ^{31}P, this expression simplifies to the $n + 1$ rule expressed in Equation (8.9b), because $I_z = 1/2$ for all these nuclei. The expected splitting patterns for 1-nitropropane (**1**), according to Equation (8.1b), are *three* peaks each for H_a and H_c and *six* peaks for H_b. This is precisely the splitting pattern appearing in Figure 8.26.

$$N = 2nI_z + 1 \tag{8.9a}$$

where N = number of peaks observed
for absorbing atom *B*
n = number of magnetically equivalent
nearest neighbor atoms *A*
I_z = nuclear spin of *A*

$$N = n + 1 \tag{8.9b}$$

Strictly speaking, the $n + 1$ rule for predicting splitting patterns of coupled hydrogen nuclei applies only under the following circumstances: (1) The coupling constants, *J,* of all nearest neighbor hydrogen nuclei coupling with the nucleus of interest must have the same value and (2) the ratio of the difference in chemical shift Δv, in Hz, of the coupled nuclei to their coupling constant J must be greater than about 9–10 (Eq. 8.10). When these criteria are met, a **first-order analysis** of the multiplicity of each resonance is possible. For example, reconsidering the spectrum of Figure 8.26, we find that $J_{ab} = J_{bc} = 7$ Hz. The difference in chemical shifts for the sets of coupled nuclei are $\Delta v_{AB} = 140$ Hz, and $\Delta v_{BC} = 63$ Hz, respectively. Application of Equation (8.10) gives values of $140/7 = 20$ and $63/7 = 9$, respectively, for the ratios of interest. The fact that both ratios are close to or exceed the criterion of Equation (8.10) means that the splitting patterns may be predicted rather well from the $n + 1$ rule, as evidenced in Figure 8.26.

$$n + 1 \text{ rule valid if } \Delta v(\text{Hz})/J(\text{Hz}) \geq \sim 10 \qquad (8.10)$$

The relative intensities of the peaks in a multiplet that obeys Equation (8.10) may be predicted by applying Pascal's triangle (Fig. 8.32). For example, if a hydrogen atom has two nearest neighbors, then the triplet that results is predicted to have intensities in the ratio of about $1:2:1$. If there are five nearest neighbors, the corresponding ratio for the sextet is $1:6:15:20:15:6:1$. Examining the relevant multiplets in Figure 8.26 reveals the approximate nature of the prediction, which becomes more accurate as the ratio $\Delta v/J$ increases.

Systems not meeting the criteria of Equations (8.9) and (8.10) produce splitting patterns that are not predictable by the $n + 1$ rule. An example of this is seen in the ^{1}H NMR spectrum of 1-butanol (**4**), as shown in Figure 8.33. In this compound, nuclei of type H_e resonate at δ 0.95 ppm and nuclei of type H_d appear at about δ 1.5 ppm. The $\Delta\delta$ of 0.55 ppm translates to a Δv of 33 Hz with a 60-MHz spectrometer. Since J_{de} is about 7 Hz, the ratio $\Delta v/J_{de}$ is about 3.5, so the crite-

$$
\begin{array}{cccc}
H_e & H_d & H_c & H_b \\
| & | & | & | \\
H_e - C - C - C - C - OH_a \\
| & | & | & | \\
H_e & H_d & H_c & H_b
\end{array}
$$

4

1-Butanol

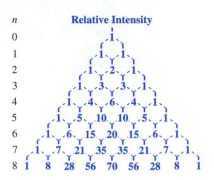

n	Relative Intensity
0	1
1	1 1
2	1 2 1
3	1 3 3 1
4	1 4 6 4 1
5	1 5 10 10 5 1
6	1 6 15 20 15 6 1
7	1 7 21 35 35 21 7 1
8	1 8 28 56 70 56 28 8 1

FIGURE 8.32 Pascal's triangle showing relative intensities of first-order multiplets; coupling constants of all nearest neighbors, n, assumed to be equal.

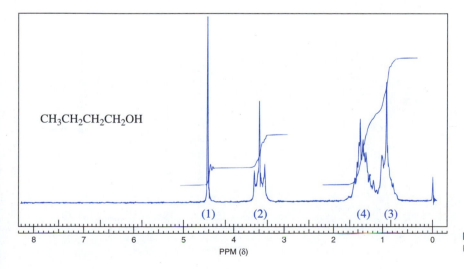

CH$_3$CH$_2$CH$_2$CH$_2$OH

(1)   (2)   (4) (3)

PPM (δ)

FIGURE 8.33 ^{1}H NMR spectrum of 1-butanol (60 MHz).

rion of Equation (8.9) is not satisfied. Consequently, the splitting pattern of the nuclei labeled H_e only remotely resembles a triplet. The situation with respect to the splitting pattern for protons of type H_c and H_d is even more complex: Δv between these types of nuclei is even less because the chemical shifts δ of both are about 1.5 ppm. The result is a multiplet of overlapping resonances that cannot be interpreted on the basis of the $n + 1$ rule.

Application of the $n + 1$ rule to nuclei of type H_b leads to a predicted splitting pattern of a quartet, on the assumption that $J_{bc} \equiv J_{ab}$. A somewhat broadened triplet is actually observed, however, as seen in Figure 8.33. In addition, proton H_a appears as a singlet rather than the triplet expected from the $n + 1$ rule. These two apparent deviations from the rule are *not* a result of a failure of $\Delta\delta/J_{de}$ to satisfy the criterion of Equation (8.10). The explanation in this case is that the spectrum was obtained under conditions in which the hydroxylic proton H_a is undergoing rapid exchange with other molecules of the alcohol or possibly with adventitious water present in the sample. As a result of this rapid exchange, the spin information carried by the hydroxylic proton is lost, so coupling between H_a and H_b is *not* seen. Spectra obtained under conditions that suppress this exchange, such as using DMSO-d_6 rather than CDCl$_3$ as the solvent, allow the coupling to be observed, and the splitting patterns for nuclei of types H_a and H_b become predictable from the $n + 1$ rule. This is illustrated in Figure 8.34.

Thus far, we have predicted splitting patterns by grouping hydrogens according to their chemical shift equivalence and then determining the number of nearest neighbors n. There are cases, however, in which nuclei may have equivalent chemical shifts but are *not* magnetically equivalent; this may complicate the appearance and interpretation of the spectrum. This phenomenon is often seen in the ^{1}H NMR spectra of aromatic compounds such as *p*-bromonitrobenzene (**5**). Analyzing the molecule by the substitution test leads to the con-

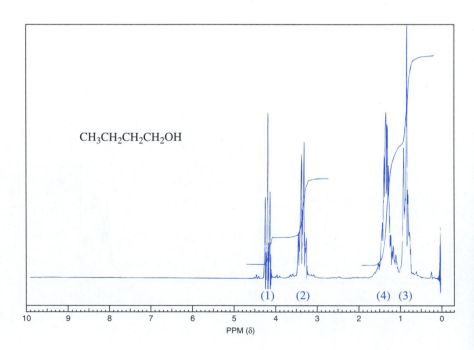

CH$_3$CH$_2$CH$_2$CH$_2$OH

FIGURE 8.34 ^{1}H NMR spectrum of 1-butanol in dimethyl sulfoxide-d_6 (90 MHz).

clusion that there are two sets of homotopic hydrogen atoms, $H_aH_{a'}$ and $H_bH_{b'}$. The two protons in each set have equivalent chemical shifts. However, the coupling constants between all members of the spin set are not equivalent. Namely, $J_{H_aH_b} \equiv J_{H_{a'}H_{b'}}$ and $J_{H_{ab'}} \equiv J_{H_{a'b}}$, but $J_{H_{a'}H_{b'}} \neq J_{H_{a'}H_b}$ and $J_{H_aH_b} \neq J_{H_aH_{b'}}$. This makes H_a and $H_{a'}$ **magnetically nonequivalent,** as are H_b and $H_{b'}$. It should be mentioned that the pairs of nuclei H_a-$H_{b'}$ and $H_{a'}$, -H_b are not nearest neighbors, yet coupling may occur between them. Nuclear spin interactions that occur over more than three bonds are termed **long-range coupling,** a phenomenon that is most frequently observed in systems having conjugated π-bonds.

$$
\begin{array}{c}
\text{Br} \\
H_a \!\!-\!\!\!\!\bigcirc\!\!\!\!-\!\! H_{a'} \\
H_b \!\!-\!\!\!\!\!\!\!\!-\!\! H_{b'} \\
\text{NO}_2
\end{array}
$$

5

p-Bromonitrobenzene

The effect of magnetic nonequivalency is evident in the ^{1}H NMR spectrum of **5** (Fig. 8.35). Careful examination of the resonances for the aromatic protons reveals additional splitting for each of the doublets that is the result of magnetic nonequivalency within the set. Nevertheless, the aromatic multiplets still have the overall appearance of the doublets expected from the simple first-order analysis using the $n + 1$ rule.

A first-order analysis may be used to determine the number, n, of nearest neighbors to a hydrogen nucleus in an unknown compound. For example, if a *sextet* is observed in a ^{1}H NMR spectrum, the proton(s) responsible for it have *five* nearest neighbors, since N of Equation (8.9b) is 6. This type of splitting pattern appears in Figure 8.47 of Exercise 8 at the end of this section.

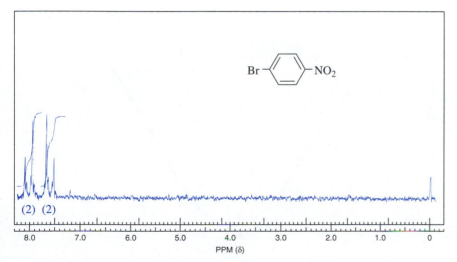

(2) (2)

8.0   7.0   6.0   5.0   4.0   3.0   2.0   1.0   0

PPM (δ)

FIGURE 8.35 ^{1}H NMR spectrum of *p*-bromonitrobenzene (60 MHz).

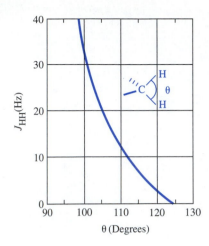

FIGURE 8.36 The *geminal* Karplus correlation. J_{HH} for CH_2 groups as a function of H–C–H angle, θ.

The *magnitude* of coupling constants, J_{HH}, is *independent* of the oscillator frequency, so units of Hz are used. The measured values depend on two main factors: the number of bonds intervening between the coupled nuclei, and the bond angle or dihedral angle between the nuclei. The angular dependence of **geminal coupling,** which is coupling between magnetically different protons on the *same* atom, is shown graphically in Figure 8.36, in which θ is the bond angle between the coupled nuclei. The situation for **vicinal coupling,** in which the coupled nuclei are on *adjacent* atoms and have a **dihedral angle** φ is presented in Figure 8.37. Some typical ranges for coupling constants are provided in Figure 8.38.

Coupling constants provide valuable information about the structure of organic compounds. If two different protons are on adjacent carbon atoms, the coupling constant between the two has only one value, that is, $J_{ab} \equiv J_{ba}$, because the protons are mutually coupled with each other. Thus, if the coupling constants of two multiplets suspected to be the result of two particular nuclei coupling with one another are *not* the same, the two nuclei are *not* coupled. This point is illustrated by the spectrum of 1-nitropropane (Fig. 8.26). Someone analyzing this spectrum who did not know what substance was responsible for it might think that multiplets *A* and *C* result from these two types of nuclei coupling with each another. However, careful measurement of the coupling constants of each of the two triplets shows that the *J*-value for group *A* is about 7 Hz whereas that for group *B* is approximately 6 Hz. These two types of protons are therefore *not* coupled and thus must be separated from one another by *more* than three bonds. This fact helps in making a structural assignment to the molecule.

It is difficult to measure coupling constants using the spectra reproduced in this textbook because the spectra have been photographically reduced, making measurement of the distance between peaks subject to large error. However, determining the values of coupling constants for simple splitting patterns is straightforward when full-sized spectra or printouts of chemical shifts of the var-

FIGURE 8.37 The *vicinal* Karplus correlation. Relationship between dihedral angle, φ, and coupling constant for vicinal protons.

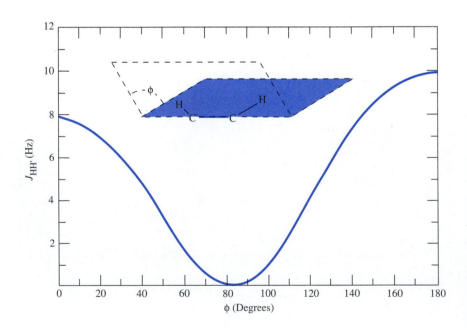

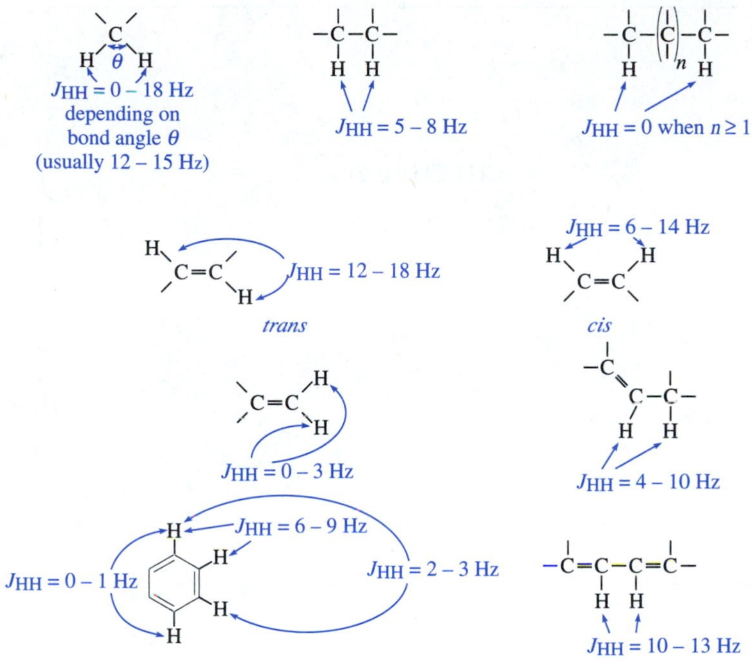

FIGURE 8.38 Examples of the values of coupling constants J_{HH} for selected functional groups.

ious peaks are available. It must be remembered that data provided in the form of $\Delta\delta$ must be converted to Hz before computation of the coupling constant, J. For example, if the peaks of a quartet are found to differ by 0.05 ppm in a spectrum obtained on a 200-MHz instrument, the coupling constant would be found to be 10 Hz by application of Equation (8.11), where $\Delta\delta$ is 0.05 and *instrument frequency* is 200×10^6 Hz.

$$J \text{ (Hz)} = (\Delta\delta \times 10^{-6}) \times \text{instrument frequency} \qquad \textbf{(8.11)}$$

3. Integration. The areas of the peaks in a ^{1}H NMR spectrum are important measures of the *relative* numbers of the different types of hydrogen nuclei in a molecule because these peak areas are a linear function of the number of nuclei producing the peaks. The areas are determined by electronic integration and are reported either in the form of a printout or as a stepped trace that appears on the spectrum itself, as seen in Figure 8.26.

In practice, the ^{1}H NMR spectrum is first recorded and the integration is then plotted on the same sheet of paper. The vertical distance that the integration rises over a peak or group of peaks is a direct measure of the area under the peak(s) and thus of the relative number of protons producing the resonances. The integrator may be reset to zero after each set of resonances has been measured, as in Figure 8.26, although this need not be done as seen in Figure 8.39. Resetting the integrator allows the height of each step in the integration curve to be higher, so the experimental error in measuring the step is less; setting the integrator at a sensitivity that would produce comparable heights for the sets of resonances without resetting it would run the pen off the vertical scale, making the integration useless.

FIGURE 8.39 ¹H NMR spectrum of pentane (90 MHz).

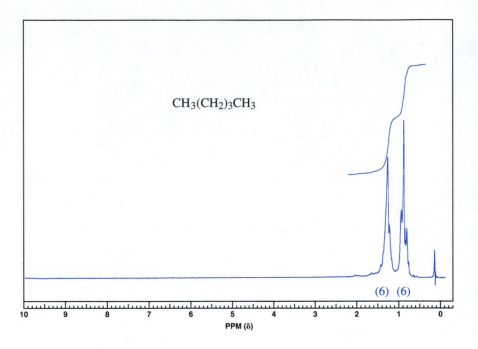

CH₃(CH₂)₃CH₃

(6) (6)

10   9   8   7   6   5   4   3   2   1   0
PPM (δ)

The analysis of the stepped plot is performed as follows. The height of each step is first measured, usually in millimeters, and the *relative* number of hydrogen atoms producing the steps is then computed by dividing the integration heights by the smallest height measured; this gives a ratio of the various types of protons. In the spectrum of 1-nitropropane (**1**) the heights of the steps in Figure 8.26 are represented by the distances *ab*, *ac*, and *ad*; in mm, these are 28, 28, and 44, respectively. The relative heights are computed by dividing all of these measurements by 28 to obtain the ratio 1.55 : 1.0 : 1.0 for groups *A*, *B*, and *C* in the spectrum. Because the number of hydrogen atoms present in a molecule must be an integer, this ratio is converted to whole numbers by multiplying *each* value by an *integer* that gives the **relative ratio** of the types of protons in whole numbers. *If* the molecular formula of the molecule is known, the integer is chosen to give a *sum* of the values in the ratio corresponding to the *absolute* number of hydrogen nuclei present. There are seven hydrogen atoms in 1-nitropropane, so *two* is used as the multiplier to provide the **absolute ratio** 2.0 : 2.0 : 3.10, the sum of which is 7.1. Because the accuracy of electronic integration is generally no better than 5–10%, this is within experimental error of the exact value for the total number of protons in the molecule.

Note that the integration of peak areas provides only the empirical formula with respect to the numbers of hydrogen atoms present in a molecule. Thus, a relative ratio of 2 : 2 : 3 : 1 for four different groups of peaks could be obtained for molecules that contain four magnetically distinct types of protons in any integral multiple of this ratio. For example, molecules having 8 or 24 hydrogen atoms could theoretically produce this same *relative* ratio; their *absolute* ratios would be different, however (see Exercise 4 at the end of this section).

8.3.3   ANALYSIS OF ^{1}H NMR SPECTRA

General Procedure.   The proper analysis of the ^{1}H NMR spectrum of a compound may provide a wealth of information regarding its structure. Determining the *chemical shift* of each peak or group of peaks allows speculation about the *type* of functional group to which the hydrogen nucleus producing the resonance is attached. The *spin-spin splitting pattern* and the *n* + 1 rule give information concerning the *number of nearest neighbor hydrogen atoms* for the proton(s) producing a particular absorption. *Integration* allows evaluation of the peak areas of each set of peaks and defines the *relative numbers of each type* of proton present in the molecule; if the molecular formula of the compound is known, the *absolute* number of each type may also be computed. However, without additional information about a substance, such as its molecular formula or the chemical reaction(s) by which the compound was formed, determining its molecule structure *solely* on the basis of a ^{1}H NMR spectrum is usually not possible.

Appropriate steps for complete analysis of a ^{1}H NMR spectrum are as follows:

1. Determine the relative numbers of the different types of hydrogen nuclei by measuring the integrated step height for each peak or group of peaks. Convert the resulting *relative* ratio to an *absolute* ratio if the molecular formula is known. The *absolute* values of the integrations are provided for most of the spectra in this textbook by the numbers in parentheses located above or below the integration curve. If such values are not given, peak heights must be measured with a ruler.

2. Measure the *chemical shifts* of the different sets of resonances and use the data to try to deduce the functional groups that might be present in the molecule. This process is greatly aided if the molecular formula and/or other spectral data, such as IR (Sec. 8.2) and ^{13}C NMR (Sec. 8.3.4) spectra, are available.

3. Analyze the *spin-spin splitting* patterns of each group of resonances to determine the number of *nearest neighbors* associated with each type of hydrogen nucleus.

Applying these steps *may* allow assignment of the structure of an unknown compound. However, it may only be possible to make a partial structural determination, and more information, such as other chemical or spectral data may be required to make a complete assignment. In any case, structural assignments based *exclusively* on interpretation of spectra *must* be confirmed by comparing the spectra of the unknown with those of an authentic specimen, if the compound is known, or by chemically converting the unknown to a known compound.

Analysis of an Unknown.   An example of the use of this procedure for making a structural assignment follows. Other examples may be found in the exercises at the end of this section.

Problem.   Provide the structure of the compound having the molecular formula C_4H_9Cl and whose ^{1}H NMR spectrum is given in Figure 8.40.

FIGURE 8.40 ¹H NMR spectrum for
unknown compound (90 MHz).

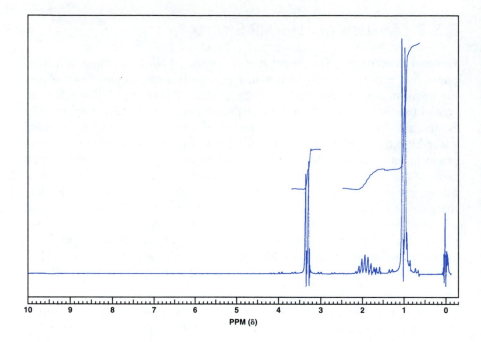

FIGURE 8.40 ¹H NMR spectrum for unknown compound (90 MHz).

Solution. For purposes of this analysis, the groups of peaks are referred to as representing nuclei of types *A–C* in going from left to right, or upfield, in the spectrum.

(a) Integration. The step heights are 11 mm, 5.2 mm, and 29.5 mm for H_a, H_b, and H_c, respectively. The *relative* ratio of the steps is $2.1:1.0:5.7$. Given the molecular formula, the *absolute* ratio must be $2:1:6$, for a total of nine hydrogen atoms. Thus, there are two hydrogen atoms of type *A*, *one* of type *B*, and *six* of type *C*.

(b) Chemical Shifts. The δ-values at the *center* of the individual multiplets are approximately 3.3 (H_a), 1.9 (H_b), and 1.0 (H_c). The low-field doublet for H_a presumably reflects the presence of the electronegative chlorine atom nearby. Since there are *two* such nuclei, the partial structure, $-CH_2Cl$, may be written. The appearance of H_c at about δ 1.0 is consistent with these nuclei being associated with a methyl group. Because there are six such hydrogen atoms, *two* methyl groups must be present. Another partial structure, $(CH_3)_2C-$, may then be written. Because these two partial structures account for *all* of the carbon atoms required by the molecular formula and all but *one* of the hydrogen atoms, the missing atom must be a *methine* hydrogen atom located on the carbon atom bearing the two methyl groups and the chloromethyl substituent. Although the chemical shift of this methine hydrogen atom is outside the range for such nuclei, as given in Table 8.4, reference to Table 8.5 shows that the methyl resonances are shifted *downfield* when there is a halogen β to the carbon atom bearing the nuclei of interest; this is evidenced in the δ-value of 1.6 for the methyl

group of ethyl bromide, CH_3CH_2Br. The same effect applies to methine hydrogen atoms.

Enough information is now available to enable assignment of the structure of the unknown as 2-methyl-1-chloropropane (isobutyl chloride), $(CH_3)_2CHCH_2Cl$. In this case, the determination of the structure may be made simply on the basis of molecular formula, chemical shifts, and integration; analysis of the spin-spin splitting patterns, as in Part (c), is required only to confirm the conclusion regarding the structure of the unknown and not for initially generating the structure. More typically, the information contained in the splitting patterns is crucial to a first formulation of the structure of an unknown.

(c) Spin-Spin Splitting Patterns. The fact that the nuclei of types *A* and *B* appear as a doublet means that they have *one* nearest neighbor, which is consistent with the fact that both H_a and H_c in 2-methyl-1-chloropropane should be coupled with the methine hydrogen atom H_b but *no* others. Conversely, H_b must be coupled with both the other types of protons, which total eight. Application of the $n + 1$ rule dictates nine peaks in the splitting pattern of the methine hydrogen atom. In fact only seven are seen in the spectrum, as the two peaks on the extreme right of the multiplet are actually **spinning side bands** arising from the methyl doublet; the cause of this phenomenon is beyond the scope of this discussion. The two "missing" members of this multiplet are not apparent because they are the weak, outermost peaks, and the amplification at which the spectrum was measured was insufficient for them to be seen. If the multiplet for H_b had been measured at a higher amplification, all nine peaks would appear. There is an important lesson to be learned from the fact that fewer peaks are observed than would be expected from the $n + 1$ rule: When multiplets of six or more are predicted, one or more of the outermost absorptions of the multiplet is frequently too weak (see Pascal's triangle, Figure 8.32) to appear under the normal operating conditions for the spectrometer. This fact must be taken into account when using spin-spin splitting patterns to interpret the spectrum of an unknown compound.

(d) Determination of Structure. Although the present problem was solved by using only two of the three basic types of information available from an integrated ^1H NMR spectrum, the usual case would require careful interpretation of all three. The strategy is to generate partial structures that are consistent with the available data and then to attach these pieces in various ways that are consistent with the observed splitting patterns and the valences of the bonded atoms. In a majority of cases, particularly with molecules having a total of only twenty atoms or so and one or more functional groups, the actual structure of the compound may be correctly assigned. Once proposed, a possible structure must be checked carefully to ascertain that it is consistent with all of the available data. If not, the structure is incorrect, and alternate possibilities must be explored and tested against the data until one is found that *is* consistent with *all* of the spectral information.

Most of the ^1H NMR spectra presented in this textbook will be of known rather than unknown compounds, so it will not be necessary to solve the structure. However, it will be valuable to interpret the spectra in terms of the known structures. Among other things, this will enhance your familiarity with the effect of various types of functional groups on chemical shifts and with the patterns of multiplets resulting from different sorts of spin-spin splitting. Your instructor may require that such interpretations be made part of the write-up associated with each experimental procedure.

The preparation of samples suitable for ^{1}H NMR analysis is described in Section 8.3.5.

EXERCISES

1. Define, explain, or give an example of the following:
   a. tetramethylsilane (TMS) as an internal standard
   b. delta (δ) scale
   c. upfield shift
   d. $n + 1$ rule
   e. chemical shift
   f. integration curve
   g. *relative* ratio of integrated peaks
   h. *absolute* ratio of integrated peaks
   i. homotopic hydrogen atoms
   j. enantiotopic hydrogen atoms
   k. diastereotopic hydrogen atoms
   l. coupling constant
   m. chemically equivalent protons
   n. diamagnetic shielding
   o. heterotopic hydrogen atoms

2. Demonstrate that the H_b hydrogen atoms in 1-nitropropane (**1**) are enantiotopic.

3. Label the sets of connectively equivalent sets of protons in the molecules (**a**)–(**j**) according to whether they are homotopic, enantiotopic, or diastereotopic. Using Tables 8.4 and 8.5, predict the approximate chemical shifts, in δ, and the splitting patterns expected for each chemically distinct type of hydrogen atom that you identify. Assume in this analysis that the coupling constants of all nearest neighbors are identical.

   a. 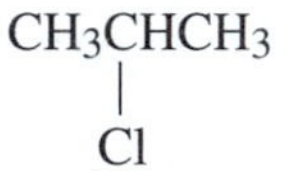 CH$_3$CHCH$_3$
        |
        Cl

   b. 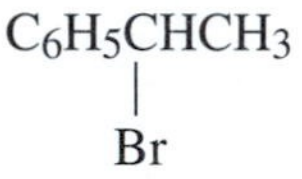 C$_6$H$_5$CHCH$_3$
           |
           Br

   c. (CH$_3$)$_2$CHCH(CH$_3$)$_2$

   d. 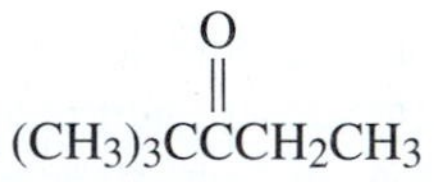 (CH$_3$)$_3$CCCH$_2$CH$_3$ (with O double-bonded to the second C)

   e. 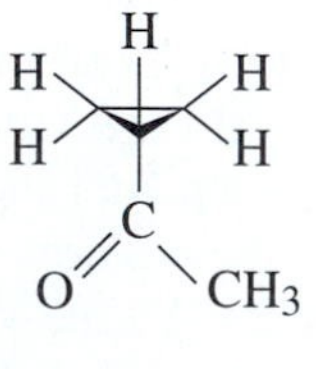 [structure showing a cyclopropane ring with H substituents and a C(=O)CH$_3$ group]

   f. ClCH$_2$CH$_2$CH$_2$Cl

   g. 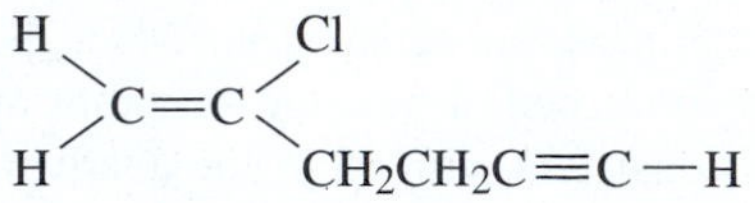 [structure: H and H on one alkene carbon; Cl and CH$_2$CH$_2$C≡C—H on the other]

   h. (CH$_3$)$_2$CHC—OH (with O double-bonded to C)

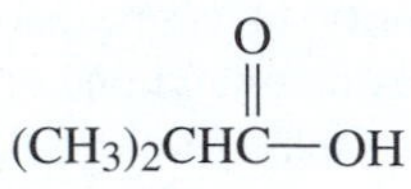

   i. 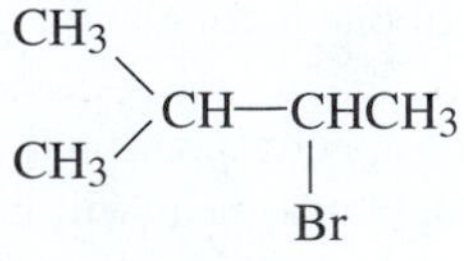 [structure: (CH$_3$)$_2$CH—CHCH$_3$ with Br below]

4. In principle, ^{1}H NMR chemical shifts independent of oscillator frequencies could be computed with the equation that follows. Using this equation, calculate the chemical shift for a resonance that appears 700 Hz downfield of TMS when a spectrometer having a 250 MHz rf oscillator is used and then discuss why the resulting value is less convenient to report than that obtained by application of Equation (8.8).

$$\delta' = \frac{\text{chemical shift (in Hz)}}{\text{oscillator frequency (in Hz)}}$$

5. By application of the criterion of Equation (8.10), predict whether the multiplicities of the following sets of coupled nuclei, H_a and H_b, could be predicted by a first-order analysis. Note the units used to report the chemical shifts of the nuclei and the frequency of the rf oscillator for the instrument used to obtain the shifts.

   a. δ_{H_a} 1.3 ppm, δ_{H_b} 3.9 ppm, J_{ab} 8.5 Hz, 200-MHz instrument.
   b. δ_{H_a} 0.8 ppm, δ_{H_b} 1.0 ppm, J_{ab} 2.5 Hz, 200-MHz instrument.
   c. δ_{H_a} 0.8 ppm, δ_{H_b} 1.0 ppm, J_{ab} 8.5 Hz, 200-MHz instrument.
   d. δ_{H_a} 0.8 ppm, δ_{H_b} 1.0 ppm, J_{ab} 8.5 Hz, 500-MHz instrument.
   e. δ_{H_a} 0.8 ppm, δ_{H_b} 1.0 ppm, J_{ab} 2.5 Hz, 200-MHz instrument.

6. The ^{1}H NMR chemical shifts, splitting patterns, and *relative* numbers of hydrogen atoms for three compounds are provided in **a–c.** Deduce one or more structures consistent with these data. You may find it helpful to sketch these spectra on a sheet of paper.

   a. C_4H_9Br: δ 1.04 (6 H) doublet, δ 1.95 (1 H) multiplet, δ 3.33 (2 H) doublet.
   b. $C_3H_6Cl_2$: δ 2.2 (2 H) quintet, δ 3.75 (1 H) triplet.
   c. $C_5H_{11}Br$: δ 0.9 (3 H) doublet, δ 1.8 (1.5 H) complex multiplet, δ 3.4 (1 H) triplet.

7. Compute the *relative* and *absolute* ratios for the different types of hydrogens in each pair of molecules shown below.

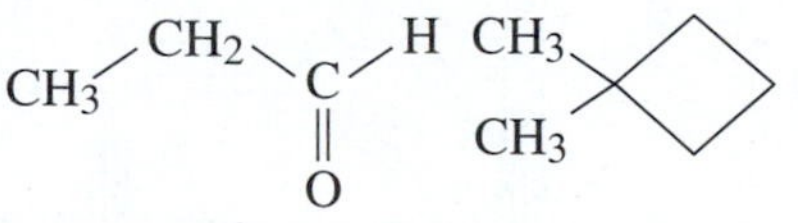

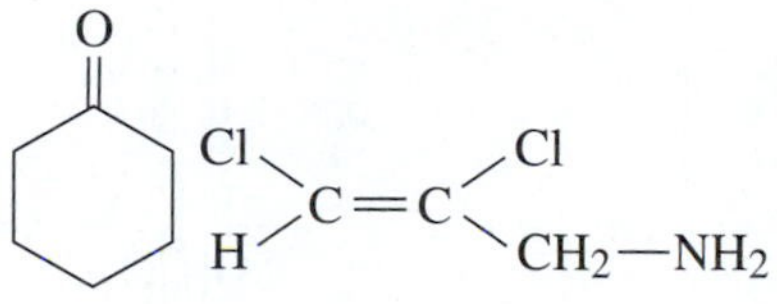

8. Figures 8.41–8.47 are of the compounds shown on the spectra. Interpret these spectra as completely as possible in terms of the observed chemical shifts, integrations, and splitting patterns.

FIGURE 8.41   ¹H NMR spectrum of methyl-cyclopentane (90 MHz).

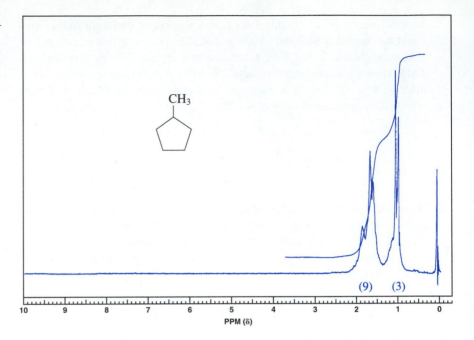

FIGURE 8.42   ¹H NMR spectrum of ethyl phenyl ether (90 MHz).

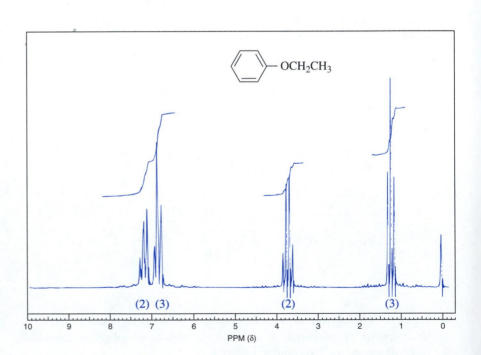

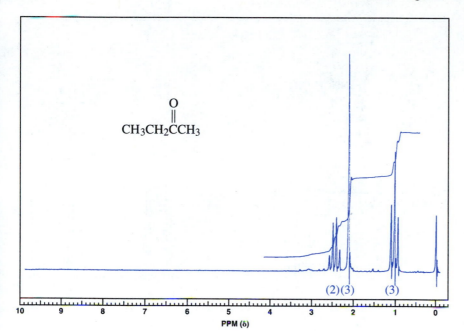

FIGURE 8.43 ^{1}H NMR spectrum of 2-butanone (90 MHz).

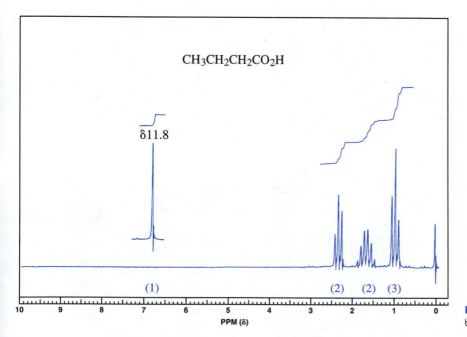

FIGURE 8.44 ^{1}H NMR spectrum of butanoic acid (butyric acid) (90 MHz).

FIGURE 8.45 ¹H NMR spectrum of *N*-ethylaniline (90 MHz).

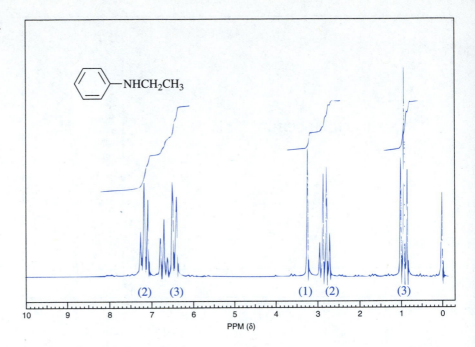

FIGURE 8.46 ¹H NMR spectrum of ethyl acetate (90 MHz).

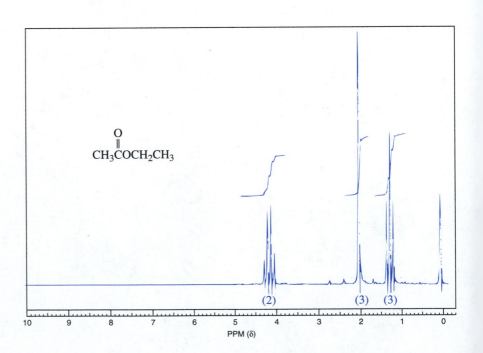

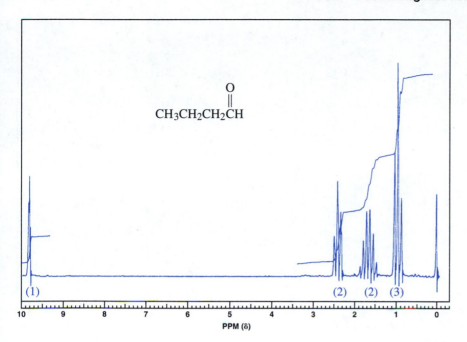

FIGURE 8.47 ^{1}H NMR spectrum of butanal (90 MHz).

9. Provide structures of the compounds **6–8**, whose ^{1}H NMR spectra are given in Figures 8.48–8.50, respectively.

   a. Compound **6**, $C_{10}H_{14}$
   b. Compound **7**, C_3H_8O
   c. Compound **8**, C_8H_9Cl

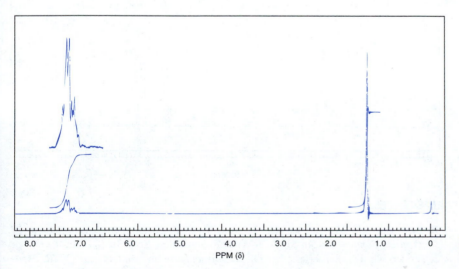

FIGURE 8.48 ^{1}H NMR spectrum (60 MHz) for compound **6**.

FIGURE 8.49 ^{1}H NMR spectrum (90 MHz) for compound **7.**

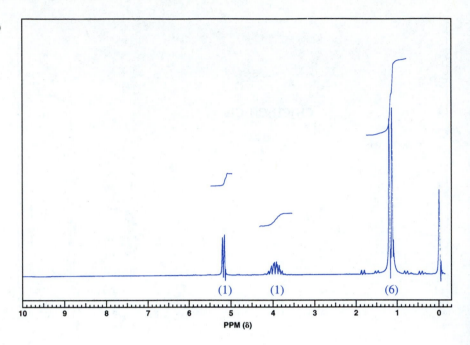

FIGURE 8.50 ^{1}H NMR spectrum (90 MHz) for compound **8.**

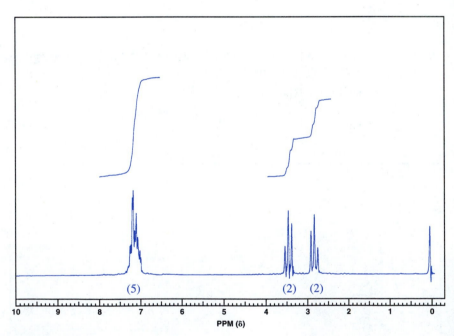

REFERENCES

1. Silverstein, R. M.; Bassler, G. C.; Morrill, T. C. *Spectrometric Identification of Organic Compounds*, 5th ed., John Wiley & Sons, New York, 1992, Chap. 4.
2. Yoder, C. H.; Schaeffer, Jr., C. D. *Introduction to Multinuclear NMR*, Benjamin/Cummings Publishing Co., Menlo Park, CA, 1987.
3. Abraham, R. J.; Fisher, J.; Loftus, P. *Introduction to NMR Spectroscopy*, 2nd ed., John Wiley & Sons, New York, 1988.

4. King, R. W.; Williams, K. R. "The Fourier Transform in Chemistry-NMR," *J. Chem. Ed.* **1989,** *66,* A213–A219, A243–A248; King, R. W.; Williams, K. R. "The Fourier Transform in Chemistry-NMR," *J. Chem. Ed.* **1990,** *67,* A100–A105, A125–A137; Williams, K. R.; King, R. W. "The Fourier Transform in Chemistry-NMR," *J. Chem. Ed.* **1990,** *67,* A93–A99.

5. *Sadtler Nuclear Magnetic Resonance Spectra*, vol. 1–101, Sadtler Research Laboratories, Philadelphia, PA. Compilation of 55,500 ^{1}H NMR spectra, as of 1992.

6. Poucher, C. J., ed. *Aldrich Library of NMR Spectra*, vol. I–II, Aldrich Chemical Co., Milwaukee, WI, 1983.

8.3.4   CARBON-13 NUCLEAR MAGNETIC RESONANCE (^{13}C NMR) SPECTROSCOPY

Carbon-12, ^{12}C, has no **nuclear spin,** but its carbon-13 isotope, ^{13}C, has a nuclear spin of 1/2, as does hydrogen, ^{1}H. The **natural abundance** of ^{13}C is about 1.1%, however, so unless substances are prepared from precursors having artificially high levels of this isotope, only 1.1% of the carbon atoms in a compound will undergo the *spin flip* or *resonance* that is characteristic of the NMR experiment (Sec. 8.3.1). In other words, only about one in a hundred carbon atoms of a sample having ^{13}C present at natural abundance will be the proper isotope for producing an absorption in the NMR spectrum. This low level of the NMR-active isotope of carbon makes it more difficult to obtain a suitable **carbon nuclear magnetic resonance** (^{13}C NMR) spectrum. For example, whereas it is usually possible to measure a ^{1}H NMR spectrum in a few minutes, it may take tens of minutes or even hours to accumulate enough data to produce a ^{13}C NMR spectrum in which the signal to noise ratio is high enough for the resonances due to the carbon atoms to be seen. Nonetheless, modern spectrometers and the sophisticated computers associated with them allow acquisition of the data necessary for a ^{13}C NMR spectrum on samples of 1–5 mg, which is about an order of magnitude greater than the amount on which a ^{1}H NMR spectrum may be obtained.

The principles of ^{13}C NMR spectroscopy are the same as those for ^{1}H NMR spectroscopy, which are presented in Section 8.3.1. When placed in an external magnetic field, ^{13}C nuclei adopt one of two spin states, whose energy difference is determined by the strength of the field (Fig. 8.25). Exposing such nuclei to electromagnetic radiation that has the appropriate energy produces the resonance condition (Eq. 8.5b). This means that the spin flip associated with transition from one energy level to another (Fig. 8.25) may occur. The ^{13}C nuclei exist in different electronic environments in the molecule, as do hydrogen nuclei, and this gives rise to *chemical shifts* that are characteristic of the magnetic environments of the various types of carbon atoms. As in ^{1}H NMR spectroscopy, tetramethylsilane, $(CH_3)_4Si$, is used as the reference compound for measuring chemical shifts in ^{13}C NMR spectra. Most ^{13}C resonances are *downfield* from that of TMS and are given *positive* values by convention. These chemical shifts are computed according to Equation (8.8) and are reported as parts per million (ppm) on a δ scale (Fig. 8.51), in analogy to the procedure used with ^{1}H NMR shifts.

Modern instruments are capable of measuring both ^{1}H and ^{13}C NMR spectra. When used for the latter purpose, the spectrometer is usually operated in a mode that permits decoupling of all of the proton(s) that are *nearest neighbors*

FIGURE 8.51 ^{13}C NMR spectrum of 2-butanone.

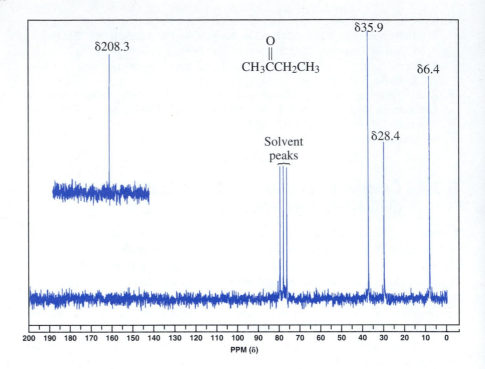

to the ^{13}C atom, a process termed **broadband proton decoupling.** Applying this technique provides ^{13}C NMR spectra that do *not* show the effects of the spin-spin splitting observed in ^{1}H NMR spectra. Moreover, possible coupling between adjacent ^{13}C atoms is of no concern: Because of the low natural abundance of this isotope in organic compounds, the probability of having two such atoms bound to one another is only *one* in *ten thousand*. Proton-decoupled ^{13}C NMR spectra are thus extremely simple, as they consist of a single sharp resonance line for each magnetically distinct carbon atom. This is seen in the spectrum of 2-butanone (Fig. 8.51), in which the four different types of carbon atoms produce four separate resonances. Note that three peaks in the spectrum are associated with the solvent, deuterochloroform, $CDCl_3$. The single carbon atom of this solvent produces more than one peak because of *deuterium*-carbon splitting, which is *not* eliminated by the broadband decoupling technique. Deuterium has a nuclear spin, I_z, of 1, so applying Equation (8.9a) predicts a *triplet* for the carbon atom, as is observed experimentally.

In contrast to decoupled spectra, *coupled* ^{13}C NMR spectra may be extremely complex and difficult to interpret because the magnitudes of ^{1}H–^{13}C coupling constants are large, on the order of 120–230 Hz, so overlapping of peaks may become a problem. However, special instrumental techniques provide the type of information that would normally be derived by analysis of splitting patterns, namely the number of hydrogen atoms bound to the carbon atom of interest. For example, **off-resonance decoupling** simplifies the spectrum as is illustrated in Figure 8.52, where the multiplets for each carbon atom correspond to the number of peaks predicted by the same $n + 1$ rule (Eq. 8.9b) that was used to predict spin-spin splitting patterns in ^{1}H NMR spectra. The difference is that n corresponds to the number of hydrogen nuclei that are *directly bound* to the

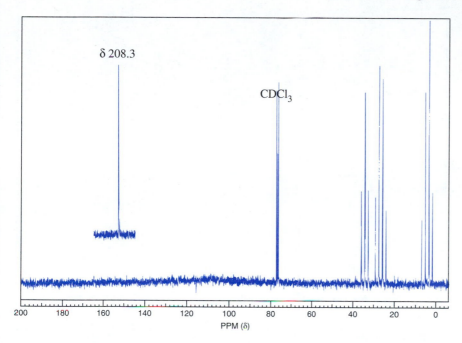

FIGURE 8.52 Proton-coupled ^{13}C NMR spectrum of 2-butanone.

δ 208.3

CDCl$_3$

PPM (δ)

carbon atom rather than the number of *nearest neighbors* as defined to predict the splitting patterns for proton spectra. The techniques available to furnish information regarding the degree of substitution of a carbon atom, as determined by the number of hydrogens bound to it, are not described here, but discussions of them are found in textbooks on organic chemistry and in the references given at the end of this subsection.

The chemical shifts of carbons atoms are much more sensitive to their molecular environments than are those of protons. Consequently, the range of chemical shifts for ^{13}C resonances is much greater than for ^{1}H. Specifically, whereas most types of hydrogens resonate within 10 ppm downfield of TMS, the chemical shifts of carbon atoms occur over a range of some 220 ppm downfield of TMS. This means that it is unlikely that two different types of carbon atoms will resonate at exactly the same chemical shift. Consequently, the number of peaks in a proton-decoupled ^{13}C NMR spectrum may *tentatively* be interpreted as being equal to the number of connectively different types of carbon atoms present in the molecule. This principle is illustrated both by the observation of four peaks in the ^{13}C NMR spectrum of 2-butanone (Fig. 8.52) and of six peaks in that of methyl benzoate (Fig. 8.53). In the latter case, the carbon atoms *ortho* and *meta* to the ester function are magnetically equivalent, so there are only four resonances for the carbon atoms of the aromatic ring.

As with ^{1}H NMR spectroscopy, tables of chemical shifts for carbon atoms in different environments have been developed; one example is provided by Table 8.6 (p. 259). Examination of the data in the table shows that some of the structural features in a molecule that produce downfield chemical shifts in ^{1}H NMR spectra do the same in ^{13}C NMR spectra. For example, an electronegative substituent such as a carbonyl group or heteroatom on a carbon atom causes a downfield shift in the resonance, relative to a saturated hydrocarbon, just as it would with hydrogen nuclei. This effect results from the *deshielding* effect of the electronegative

FIGURE 8.53 ^{13}C NMR spectrum of methyl benzoate.

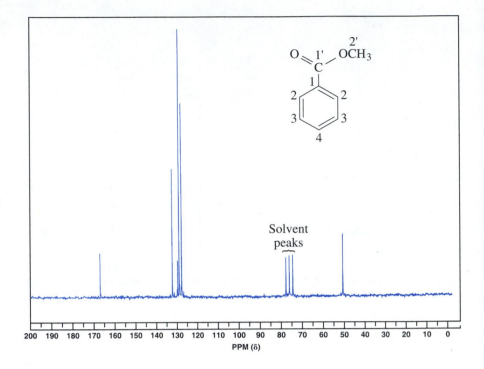

moiety. A second factor in defining chemical shifts in ^{13}C NMR spectra is the hybridization of the carbon atom, as is seen by comparing the range of chemical shifts for the sp^3 - (10–65 ppm), sp^2 - (115–210 ppm), and sp-hybridized (65–85 ppm) carbon atoms of the general structures shown in the table.

Finally, the nature of a substituent G that is zero (α-effect), one (β-effect), or two (γ-effect) atoms away from the carbon atom of interest, as shown in **9,** may also affect the chemical shift. Such substituent effects are remarkably additive in nature, so it becomes possible to predict with reasonable accuracy the expected position of the chemical shifts for carbon atoms in a molecule whose spectrum has not been reported. Table 8.7 (p. 260) is a compilation of some of these additivity effects for acyclic alkanes. These may be used in conjunction with the data in Table 8.8 to predict chemical shifts for a compound. Similar tables of additivity effects are available for other classes of organic compounds, such as alkenes and aromatic compounds, and are published in the references provided at the end of this subsection.

9

The following example illustrates the method for making these predictions. Consider 1-chlorobutane (**10,** $CH_3CH_2CH_2CH_2Cl$). The computed chemical shifts, δ, are determined from the observed chemical shifts (Table 8.8, p. 261) of 13.4 and 25.2 ppm for C-1 (C-4) and C-2 (C-3), respectively, for butane as follows:

TABLE 8.6 ^{13}C Chemical Shifts in Carbon-13 Magnetic Resonance (^{13}C NMR) Spectroscopy

Absorbing Carbon Atom (shown as C or Ar)	Approximate Chemical Shift (ppm), δ
RCH_2CH_3	13–16
RCH_2CH_3	16–15
R_3CH	25–38
$CH_3\overset{\displaystyle O}{\overset{\|}{C}}–R$	30–32
$CH_3\overset{\displaystyle O}{\overset{\|}{C}}–OR$	20–22
$RCH_2–Cl$	40–45
$RCH_2–Br$	28–35
$RCH_2–NH_2$	37–45
$RCH_2–OH$	50–65
$RC\equiv CH$	67–70
$RC\equiv CH$	74–85
$RCH=CH_2$	115–120
$RCH=CH_2$	125–140
$RC\equiv N$	118–125
ArH	125–150
$R\overset{\displaystyle O}{\overset{\|}{C}}–OR'$	170–175
$R\overset{\displaystyle O}{\overset{\|}{C}}–OH$	175–185
$R\overset{\displaystyle O}{\overset{\|}{C}}–H$	190–200
$R\overset{\displaystyle O}{\overset{\|}{C}}–CH_3$	205–210

C-1: $13.4 + 31 = 44.4$ ppm   C-2: $25.2 + 11 = 36.2$ ppm

C-3: $25.2 – 4 = 21.2$ ppm   C-4: $13.4 + 0 = 13.4$ ppm

The observed chemical shifts for **10** are provided below the structure. The agreement is remarkable and shows the impressive predictive power of this approach.

$$CH_3–CH_2–CH_2–CH_2–Cl$$
$$13.4 \quad 20.4 \quad 35.0 \quad 44.6$$

10

1-Chlorobutane

In the spectrum of 2-butanone presented earlier (Fig. 8.51), the four different resonances of the sample appear as singlets, owing to broadband decoupling

TABLE 8.7 Incremental Substituent Effects (ppm) on Replacement of H by Y in Alkanes. Y is Terminal or Internal[a] (+ downfield, – upfield)

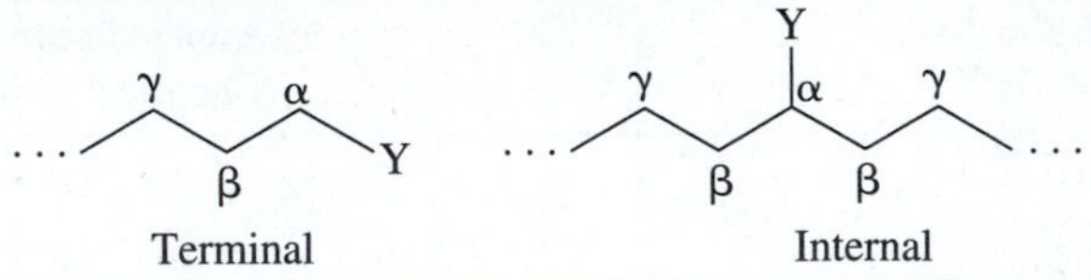

Terminal   Internal

Y	α Terminal	α Internal	β Terminal	β Internal	γ
CH_3	+ 9	+ 6	+10	+ 8	–2
$CH=CH_2$	+20		+ 6		–0.5
$C \equiv CH$	+4.5		+ 5.5		–3.5
COOH	+21	+16	+ 3	+ 2	–2
COOR	+20	+17	+ 3	+ 2	–2
COR	+30	+24	+ 1	+ 1	–2
CHO	+31		0		–2
Phenyl	+23	+17	+ 9	+ 7	–2
OH	+48	+41	+10	+ 8	–5
OR	+58	+51	+ 8	+ 5	–4
OCOR	+51	+45	+ 6	+ 5	–3
NH_2	+29	+24	+11	+10	–5
NHR	+37	+31	+ 8	+ 6	–4
NR_2	+42		+ 6		–3
CN	+ 4	+ 1	+ 3	+ 3	–3
F	+68	+63	+ 9	+ 6	–4
Cl	+31	+32	+11	+10	–4
Br	+20	+25	+11	+10	–3
I	– 6	+ 4	+11	+12	–1

[a] Add these increments to the shift values of the appropriate carbon atom in Table 8.8.

Source: Reference 1.

of the protons. With the exception of the carbonyl carbon atom itself, which appears far downfield at δ 208.3, the carbon atom responsible for each resonance may be assigned either (a) by recognizing that the carbonyl group causes the carbon atoms to which it is directly attached (α-effect) to be shifted downfield farther than those atoms farther removed from it or (b) by using the data of Tables 8.7 and 8.8 to compute the expected chemical shifts. These computations are provided below. Again, comparing with the experimentally determined shifts listed in Figure 8.51 shows the value of the method of substituent additivity shown on the spectrum.

TABLE 8.8 ^{13}C Shifts for Some Linear and Branched-Chain Alkanes (ppm from TMS)

Compound	C-1	C-2	C-3	C-4	C-5
Methane	−2.3				
Ethane	5.7				
Propane	15.8	16.3	15.8		
Butane	13.4	25.2	25.2		
Pentane	13.9	22.8	34.7	22.8	13.9
Hexane	14.1	23.1	32.2	32.2	23.1
Heptane	14.1	23.2	32.6	29.7	32.6
Octane	14.2	23.2	32.6	29.9	29.9
Nonane	14.2	23.3	32.6	30.0	30.3
Decane	14.2	23.2	32.6	31.1	30.5
2-Methylpropane	24.5	25.4			
2-Methylbutane	22.2	31.1	32.0	11.7	
3-Methylpentane	11.5	29.5	36.9	18.8 (3-CH$_3$)	

Source: Reference 1.

C-1: −2.3 + 30 = 27.7 ppm   C-2: cannot be computed from data in Tables 8.7 and 8.8

C-3: 5.7 + 31 = 36.7 ppm   C-4: 5.7 + 1 = 6.7 ppm

It is apparent in the ^{13}C NMR spectrum of 2-butanone that the relative intensities of the various peaks are *not* identical, even though a single carbon atom is responsible for each resonance. The physical basis for this is beyond the scope of this discussion, but the consequence is that there is *not* a 1:1 correlation between the number of carbon atoms producing a particular resonance and the intensity of that resonance. However there is a *rough* correlation between intensities as reflected in peak heights, and whether or not any hydrogen atoms are on the carbon atom resonating at a particular chemical shift: As a general rule, if there are *no* hydrogen atoms attached to the carbon atom, the intensity of the spectral peak associated with that atom will be relatively low. This relationship is illustrated in Figure 8.51, although it should be noted that the signal at δ 208.3, which is the resonance for the carbonyl carbon atom, has been electronically amplified by a factor of two relative to all other resonances in the spectrum. In contrast, note that the intensity of the methylene carbon atom is greater than that of either of the carbon atoms of the methyl groups in the molecule, a fact showing that there is *no general* relationship between the number of hydrogen nuclei on a carbon atom and the intensity of the peak for that atom.

Considering the ^{13}C NMR spectrum (Fig. 8.53) of methyl benzoate (**11**) shows how assigning peaks may be done on the basis of information in Table 8.7 and use of peak intensities as a measure of whether or not a carbon atom bears any hydrogens. First of all, there are a total of *six* magnetically distinct carbon atoms in the molecule because of its symmetry, and six peaks due to the sample are indeed observed. According to the data in Table 8.7, the resonance at δ 166.8 must be due to the carbonyl carbon atom, C-1', and, as expected, is of relatively low intensity. Similarly, the table supports assigning the absorption at δ 51.8 to the methoxy carbon atom, C-2'. The remaining four resonances are in the aromatic region of the ^{13}C NMR spectrum and may be assigned as follows. Based on its low intensity, the peak of δ 130.5 is probably due to C-1 of the aromatic ring. The carbon atoms *ortho* and *para* to an ester group should be *downfield* of those for the *meta* carbon atoms because the ester moiety deshields the *ortho* and *para* positions by delocalizing π-electrons, as illustrated by resonance structures **11a**–**11c** and by inductively withdrawing electrons from the ring via σ-bonds.

| **11** | **11a** | **11b** | **11c** |

Methyl benzoate

This means that the resonance at δ 128.5 is for the two meta carbon atoms. A *tentative* assignment of the remaining two resonances at δ 129.7 and 132.9 in the aromatic region is made on the basis that there are twice as many *ortho* as *para* carbon atoms. Thus the more intense resonance at δ 129.7 is for C-2 and that at δ 132.9 for C-4. These assignments are consistent with those calculated using the substituent additivity effects provided in Table 8.9. Confirmation of this is left as an exercise (see Exercise 1 at the end of this subsection).

With the exception of the illustrative spectra of this chapter, plots of ^{13}C NMR spectra are not reproduced in this textbook. Rather, only the chemical shifts derived from broadband decoupled analyses are enumerated for starting materials and products. Valuable practice in interpreting such spectra will be gained by attempting to make assignments of the peaks to the carbon atoms responsible for them. Reference to Tables 8.6 through 8.9 will be useful in this task. As noted earlier, more extensive compilations of substituent effects of ^{13}C NMR chemical shifts may be found in Reference 1 at the end of this subsection.

EXERCISES

1. Based on the substituent effects shown in Table 8.9, calculate the chemical shifts expected for the various carbon atoms of the aromatic ring of methyl benzoate (**11**).

2. The ^{13}C NMR spectrum of 3-methyl-2-butanone is provided in Figure 8.54. Provide assignments of the various resonances to the carbon atoms responsible for them, using Tables 8.6 through 8.8.

TABLE 8.9   Incremental Shifts of the Aromatic Carbon Atoms of Monosubstituted Benzenes (ppm from Benzene at 128.5 ppm, + downfield, – upfield). Carbon Atom of Substituents in parts per million from TMS

Substituent	C-1 (Attachment)	C-2	C-3	C-4	C of Substituent (ppm from TMS)
H	0.0	0.0	0.0	0.0	
CH_3	9.3	+0.7	–0.1	–2.9	21.3
CH_2CH_3	+15.6	–0.5	0.0	–2.6	29.2 (CH_2), 15.8 (CH_3)
$CH(CH_3)_2$	+20.1	–2.0	0.0	–2.5	34.4 (CH), 24.1 (CH_3)
$C(CH_3)_3$	+22.2	–3.4	–0.4	–3.1	34.5 (C), 31.4 (CH_3)
$CH=CH_2$	+9.1	–2.4	+0.2	–0.5	137.1 (CH), 113.3 (CH_2)
$CH{\equiv}CH$	–5.8	+6.9	+0.1	+0.4	84.0 (C), 77.8 (CH)
C_6H_5	+12.1	–1.8	–0.1	–1.6	
CH_2OH	+13.3	–0.8	–0.6	–0.4	64.5
OH	+26.6	–12.7	+1.6	–7.3	
OCH_3	+31.4	–14.4	+1.0	–7.7	54.1
OC_6H_5	+29.0	–9.4	+1.6	–5.3	
$O\overset{\text{O}}{\overset{\|}{C}}CH_3$	+22.4	–7.1	–0.4	–3.2	23.9 (CH_3), 169.7 (C=O)
$\overset{\text{O}}{\overset{\|}{C}}H$	+8.2	+1.2	+0.6	+5.8	192.0
$\overset{\text{O}}{\overset{\|}{C}}CH_3$	+7.8	–0.4	–0.4	+2.8	24.6 (CH_3), 195.7 (C=O)
$\overset{\text{O}}{\overset{\|}{C}}C_6H_5$	+9.1	+1.5	–0.2	+3.8	196.4 (C=O)
$\overset{\text{O}}{\overset{\|}{C}}OH$	+2.9	+1.3	+0.4	+4.3	168.0
$\overset{\text{O}}{\overset{\|}{C}}OCH_3$	+2.0	+1.2	–0.1	+4.8	51.0 (CH_3), 166.8 (C=O) 168.5
$\overset{\text{O}}{\overset{\|}{C}}Cl$	+4.6	+2.9	+0.6	+7.0	
$C{\equiv}N$	–16.0	+3.6	+0.6	+4.3	119.5
NH_2	+19.2	–12.4	+1.3	–9.5	
$N(CH_3)_2$	+22.4	–15.7	+0.8	–11.8	40.3
$NH\overset{\text{O}}{\overset{\|}{C}}CH_3$	+11.1	–9.9	+0.2	–5.6	
NO_2	+19.6	–5.3	+0.9	+6.0	
F	+35.1	–14.3	+0.9	–4.5	
Cl	+6.4	+0.2	+1.0	–2.0	
Br	–5.4	+3.4	+2.2	–1.0	
I	–32.2	+9.9	+2.6	–7.3	
SO_2NH_2	+15.3	–2.9	+0.4	+3.3	

Source: Reference 1.

FIGURE 8.54 ^{13}C NMR spectrum of 3-methyl-2-butanone for Exercise 2.

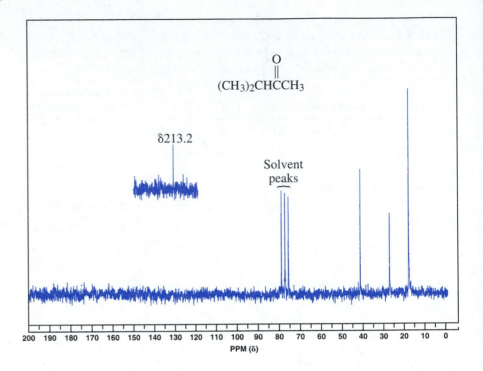

$\delta213.2$

Solvent peaks

REFERENCES

1. Silverstein, R. M.; Bassler, G. C.; Morrill, T. C. *Spectrometric Identification of Organic Compounds*, 5th ed., John Wiley & Sons, New York, 1992, Chap. 5.
2. Yoder, C. H.; Schaeffer, Jr., C. D. *Introduction to Multinuclear NMR*, The Benjamin/Cummings Publishing Co., Menlo Park, 1987.
3. Abraham, R. J.; Fisher, J.; Loftus, P. *Introduction to NMR Spectroscopy*, 2nd ed., John Wiley & Sons, New York, 1988.
4. Whitesell, J. K.; Minter, M. A. *Stereochemical Analysis of Alicyclic Compounds by C-13 NMR Spectroscopy*, Chapman & Hall, London, 1987.
5. Breitmaier, E.; Voelter, W. *Carbon-13 Nuclear Magnetic Resonance*, 3rd ed., VCH Publishers, New York, 1987.
6. Wehrli, F. W.; Marchand, A. P.; Wehrli, S. *Interpretation of Carbon-13 NMR Spectra*, 2nd ed., John Wiley & Sons, New York, 1988.
7. *^{13}C NMR Spectra*, vol. 1–170, Sadtler Research Laboratories, Philadelphia, PA. Compilation of 34,000 ^{13}C NMR spectra, as of 1992.

8.3.5   PREPARATION OF SAMPLES FOR NMR SPECTROSCOPY

Organic chemists normally obtain NMR spectra of liquid samples contained in special glass tubes of high-precision bore. Although spectra may be obtained of pure liquids having low viscosities, the substrate of interest, regardless of its normal physical state, is more generally dissolved in an appropriate solvent. The spectra of viscous samples are unsatisfactory because broad rather than sharp absorptions may be observed, so there is a loss of resolution between the peaks.

Most modern NMR spectrometers operate in the pulsed Fourier transform (FT) mode (Sec. 8.3); for technical reasons this means that deuterated solvents must be used. Deuterochloroform, $CDCl_3$, is the most common such solvent although a number of other, albeit generally more expensive, solvents such as acetone-d_6, $(CD_3)_2CO$, and benzene-d_6, C_6D_6, are also available. Deuterium oxide, D_2O, is often used if the sample is water-soluble. In older instruments that operate in the continuous wave (CW) mode, carbon tetrachloride, CCl_4, and other *aprotic* solvents may also be used.

Proton-containing solvents are generally not appropriate because the intense resonance due to the hydrogen atoms of the solvent may obscure absorptions of the sample itself. Although deuterium, ^2H, also possesses the property of nuclear spin, its resonance appears in a region of the NMR spectrum different from that of ^1H. Replacing all of the hydrogen atoms of solvent with deuterium removes such resonances from the ^1H NMR spectrum, but for practical reasons weak absorptions due to residual protium are still usually observed. This is because common organic solvents containing 100 atom % D are too expensive for routine spectral work. Thus, $CDCl_3$ of this purity costs about \$2.00/g, whereas material having 99 atom % D is only 19¢/g. Minor resonances due to solvent must therefore be taken into account when interpreting ^1H NMR spectra. Moreover, in solvents in which there is more than one deuterium atom on a carbon atom, the residual *protium* on that atom will have a splitting pattern that reflects the presence of the deuterium atoms. The resonance of the residual hydrogen present in deuteroacetone thus appears as a *quintet* centered at δ 2.17, owing to the coupling of a single residual hydrogen with the two deuteriums that are present on average on the same carbon atom in acetone-d_6 having 99 atom % or greater deuterium content.

Of course, the *carbon* atom(s) of an NMR solvent appear in the ^{13}C NMR spectrum, whether or not the solvent is deuterated. As with residual protium, splitting of the carbon resonance(s) of the solvent by attached deuterium atoms is observed. The appearance of *three* peaks for the carbon atom of $CDCl_3$ illustrates this (see Figs. 8.51–8.53). Care should therefore be exercised when interpreting ^{13}C NMR spectra to avoid mistaking peaks due to solvent for those of the sample itself. If overlap of sample and solvent peaks is suspected, it may be necessary to select a different solvent for a ^{13}C NMR spectrum.

The concentration of the solutions appropriate for obtaining NMR spectra is in the range of 5–15% by weight; about 1 mL of solution is needed to fill an NMR tube to the proper level, although as little as 0.6 mL can be used. The tube must be scrupulously clean and dry, and the solution must be free of undissolved solids arising from the sample itself or even dust. Furthermore, as a consequence of contact of the solute or the solvent with metals such as the iron in "tin" cans, trace amounts of ferromagnetic impurities may contaminate the solution; the result is a spectrum that has only broad, weak absorptions. A solution that contains any solid material must be filtered *prior* to measuring its NMR spectrum. This is done most easily by inserting a small plug of glasswool into a disposable pipet and then filtering the solution through the plug into the NMR tube. Following their use, NMR tubes should be cleaned and dried thoroughly. They are best stored in a closed container or in an inverted position so as to minimize the possibility of particles of dust entering them.

8.3.6 INTERPRETATION OF IR AND NMR SPECTRA

Section 8.2 and the preceding subsections of 8.3 contain descriptions of the principles and applications of IR and both ^{1}H NMR and ^{13}C NMR spectroscopy for the analysis of organic compounds. Individually, each of these types of spectroscopies is valuable for determining selected structural aspects in molecules, but their greatest power arises from their combined application to structural elucidation. The present discussion is intended to illustrate how proper interpretation of data from these three techniques can support the assigned structures of known substances and provide evidence for possible structures of unknown compounds. As the spectral interpretations are developed, it is important to bear in mind the type of information available from each form of spectroscopy. Specifically, the IR spectrum yields information concerning the *functional groups* that may be present, and ^{1}H NMR spectroscopy gives data regarding the *number*, *chemical environments*, and *magnetically active nearest neighbors* of the hydrogen nuclei in a molecule. Finally, ^{13}C NMR spectroscopy may be used to determine the *number of different types* and the *chemical environments* of the carbon atoms that are present.

The spectral analyses presented here are of two kinds, the first being for a *known* compound, the other for an *unknown* substance. Learning how to interpret spectral data for substances for which the structure is given provides valuable practice for the more challenging task of using such data to assign structures to unknowns. Consequently, we urge you to examine the spectra of reactants and products reproduced in this textbook carefully and to analyze them as completely as possible.

Analysis of IR, ^{1}H NMR and ^{13}C NMR Spectra of a Known Compound

Consider the spectra of 2-methyl-1-propanol (**12,** isobutyl alcohol) that are presented in Figure 8.55. The IR spectrum (Fig. 8.55a) contains three prominent bands in the region 5000–1250 cm^{-1}. Reference to Tables 8.2 and 8.3 allows assignment of the absorptions near 2900 and 1465 cm^{-1} to the C–H bonds in the molecule. The broad, intense band at 3250 cm^{-1} is due to the O–H stretching vibration, and the location and breadth of this absorption indicates that this functionality is involved in hydrogen-bonding. The strong absorption at 1050 cm^{-1} may be assigned to stretching of the C–O bond. The *absence* of bands characteristic of multiple bonds is entirely consistent with the fact that **12** is a saturated molecule.

As described in Section 8.3.3, the interpretation of the ^{1}H NMR (Fig. 8.55b) starts with analysis of the integration. Going *upfield* on the spectrum, the relative areas, in millimeters, are 2.6 : 5.6 : 2.8 : 15, which corresponds to the ratio of 1.0 : 2.2 : 1.1 : 5.8 when each value is divided by 2.6. The total number of hydrogens in **12** is 10, so the observed ratio is within experimental error of the absolute ratio of 1 : 2 : 1 : 6.

The integration data and reference to Tables 8.4 and 8.5 allow assignment of the observed chemical shifts to the four chemically nonequivalent sets of hydrogen atoms in 2-methyl-1-propanol (**12**). The doublet at δ 0.85 represents the six homotopic hydrogens of the methyl groups in **12;** the observed chemical

(a) IR spectrum.

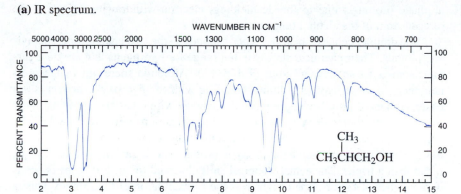

WAVENUMBER IN CM⁻¹

FIGURE 8.55   IR and NMR spectra of 2-methyl-1-propanol.

(b) ¹H NMR spectrum (60 MHz).

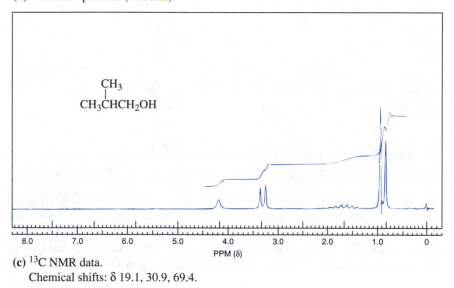

PPM (δ)

(c) ¹³C NMR data.
Chemical shifts: δ 19.1, 30.9, 69.4.

shift of these protons is consistent with that for *aliphatic* methyl groups. The splitting is due to the presence of *one* nearest neighbor, which is the tertiary hydrogen atom, $(CH_3)_2\mathbf{CH}$, as predicted by the *n + 1* rule (Eq. 8.9b). The multiplet centered at about δ 1.7 arises from this particular proton, with its *eight* nearest neighbors—the six methyl and the two methylene hydrogens. The intensity of the multiplet is too low to allow all nine peaks expected on the basis of the *n + 1* rule to be seen. This is not surprising inasmuch as reference to Pascal's triangle (Fig. 8.32) shows that the ratio of the outer peaks to the central peaks of a nine-peak multiplet should be only 1 : 126. Careful examination of the individual peaks reveals that they are broad rather than sharp, an indication that the methylene protons have a slightly different coupling constant with the methine hydrogen atom than do the methyl groups. The fact that the chemical shift of the methine proton is somewhat greater than that given in Table 8.4 for a tertiary aliphatic proton is due to the deshielding, electron-withdrawing effect of the oxygen atom of the alcohol function.

The doublet centered near δ 3.3 corresponds to the two enantiotopic methylene protons. Their chemical shift is in the range for hydrogens that are on a carbon atom α to an oxygen atom (Table 8.5). Although there are *two* nearest neighbors to these hydrogen atoms, coupling with the hydroxylic proton is *not* observed because it is undergoing rapid exchange with other such protons in the sample, a phenomenon that is discussed in connection with Figure 8.33. The exchange process also explains why the remaining resonance in the spectrum at δ 4.2, which is assigned to the hydroxylic proton, is a singlet, albeit a broad one, rather than a triplet. The actual chemical shift observed for such protons depends on a number of factors, one of which is the concentration of the sample. This is because the extent of hydrogen bonding in the sample is concentration-dependent, thus affecting the magnetic environment of the protons involved. Consequently, the resonance for such nuclei may occur over a wide range.

There are three chemically nonequivalent carbon atoms in the alcohol **12,** and three resonances are observed in the ^{13}C NMR spectrum. By reviewing the data in Table 8.6, it is possible to assign the resonance at δ 19.1 to the methyl carbon atoms, and those at δ 30.9 and 69.4 to the methine and methylene carbon atoms, respectively. It is left as an exercise to confirm these assignments by using the data in Tables 8.7 and 8.8 to compute the expected chemical shifts for the various carbon atoms of this molecule (see Exercise 1 at the end of this section).

Structural Assignment of an Unknown Compound on the Basis of Molecular Formula, IR, ^{1}H NMR, and ^{13}C NMR Spectra

One approach to the use of spectroscopic data for determining the structure of an unknown compound can be illustrated by the following example. Suppose that you are required to provide a structure for a compound **X** whose molecular formula is $C_9H_{10}O_2$, and which gives the spectra shown in Figure 8.56. Analysis of the molecular formula results in an **Index of Hydrogen Deficiency** (IHD) or degree of unsaturation for the compound of *five,* and thus mandates the presence of rings and/or multiple bonds. Examination of the IR spectrum reveals two intense absorptions in the functional group region. Reference to Tables 8.2 and 8.3 shows that the band at about 1700 cm^{-1} is consistent with the presence of a carbonyl group, whereas the broad absorption ranging from 3400–2300 cm^{-1} is characteristic of the hydroxyl moiety of a carboxylic acid function: the O–H stretch of an alcohol is not as broad as that of a carboxylic acid. Further evidence for the acid functionality is found in the absorption at 1320 cm^{-1}, which may be assigned to C–O bond stretching. Stretching vibrations associated with the C–H that must be present in the unknown appear to be buried in the O–H, so their absence should be of no concern. Finally, the presence at 1500 cm^{-1} of a sharp band of intermediate intensity may signify an aromatic ring, a possibility that seems reasonable in light of the value of IHD.

The ^{13}C NMR spectrum shows that there are seven magnetically different carbon atoms in **X.** The resonance at δ 179.5 has a chemical shift appropriate for the carbonyl carbon atom of a carboxylic function (Table 8.6), whereas those at δ 30.5 and 35.5 are clearly associated with *sp*^{3}-rather than *sp*^{2}- or *sp*-hybridized

(a) IR spectrum.

FIGURE 8.56   IR and NMR spectra for unknown compound **X**.

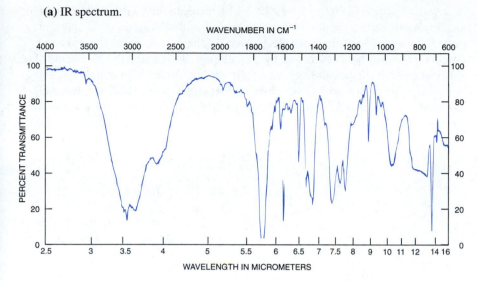

(b) ^{1}H NMR spectrum (90 MHz).

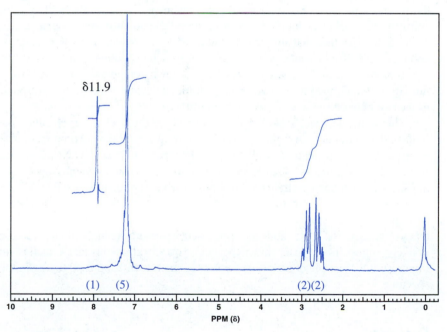

(c) ^{13}C NMR data.
    Chemical shifts: δ 30.5, 35.5, 126.3, 128.2, 128.5, 140.1, 179.5.

carbon atoms. The resonances appearing in the range of δ 126.3–140.1 have chemical shifts expected for sp^2-hybridized carbon atoms. Because there are only four such resonances and these must account for six carbon atoms, two of the peaks must each represent two magnetically identical carbon atoms. This requirement is met by proposing that **X** contains either a monosubstituted or a *p*-disubstituted benzene ring, **13** or **14**, respectively. Given the conclusions reached earlier about the nature of the other three carbon atoms present, only structures **15–17** are possible. Which of these is actually the unknown compound could be determined by calculating the expected ^{13}C chemical shifts of each of the possibilities, using the data in Tables 8.7 through 8.9, and comparing the results with those observed experimentally. However, we shall use the data available from the ^1H NMR to differentiate among the three possibilities.

The ^1H NMR spectrum reveals that there are *four* sets of chemically non-equivalent protons in the unknown. These are in the relative ratio of 1:5:2:2 when going upfield. This is also the absolute ratio because there are only 10 hydrogen atoms in **X.** The appearance of a resonance at very low field, δ 11.9, confirms the earlier conclusion that a carboxylic acid function is present (see Table 8.4). The fact that the broad singlet at δ 7.2, which is in the range of chemical shifts for aromatic protons, integrates for *five* hydrogens means that the ring is *monosubstituted*. Consequently, the unknown compound must have structure **15**, in which the pair of two-proton multiplets centered at δ 2.8 correspond to the two heterotopic sets of methylene protons.

This example illustrates a logical approach to the use of a molecular formula and spectral data in deducing the structure of an unknown compound. The interplay of the information available from IR, ^1H NMR, and ^{13}C NMR spectra is critical to the process of structural elucidation, and it is commonly necessary to use more than one spectral technique before an assignment is possible. It cannot be overemphasized that *all* of the spectral data available must be consistent with the proposed structure of the sample. The existence of even one inconsistency between the data and the proposed structure renders that structure unlikely, unless the inconsistency is produced by an extraneous factor such as an impurity in the sample or a malfunction in the instrumentation used to obtain the spectra.

EXERCISES

1. Compute the ^{13}C NMR chemical shifts expected for 2-methyl-1-propanol (**12**).

   The following exercises provide various combinations of spectral data for unknown compounds. Interpret the data as fully as possible and assign a structure or structures to the unknowns on the basis of your interpretation.

2. Compound **18**, $C_4H_8O_2$, whose IR and ^{13}C NMR spectra are provided in Figure 8.57.

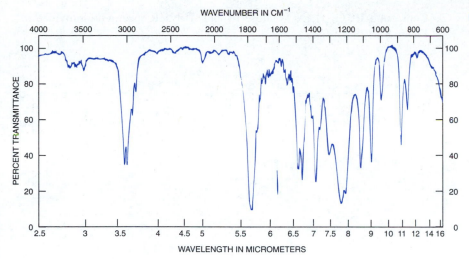

FIGURE 8.57 IR and ^{13}C NMR spectra of compound **18**.

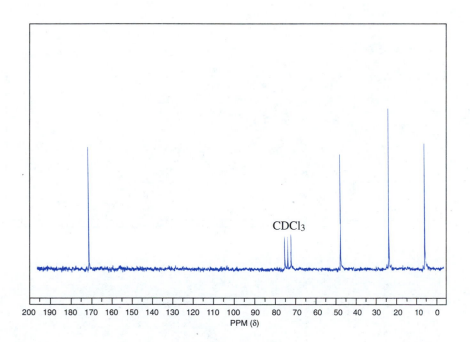

3. Compound **19**, $C_{13}H_{11}NO$, whose IR and ^1H NMR spectra are provided in Figure 8.58.

FIGURE 8.58 IR and ^1H NMR spectra of compound **19**. Note that all peaks in the ^1H NMR spectrum are offset by 1.1 ppm *upfield*.

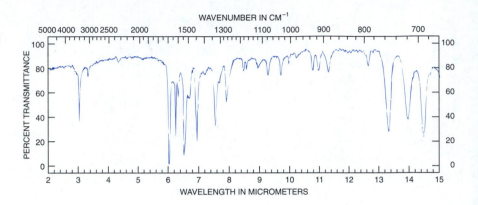

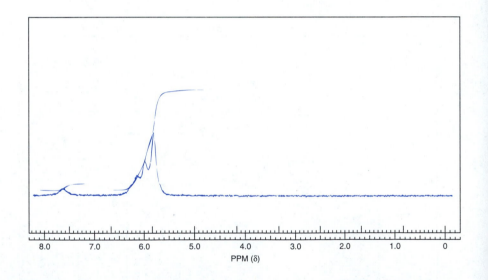

4. Compound **20**, $C_4H_{10}O$, whose IR and ^{13}C NMR spectra are provided in Figure 8.59.

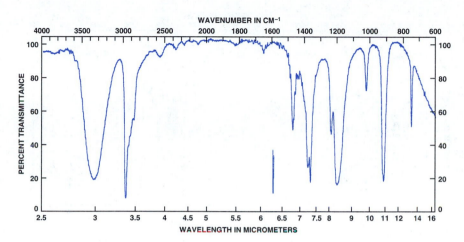

FIGURE 8.59   IR and ^{13}C NMR spectra of compound **20**.

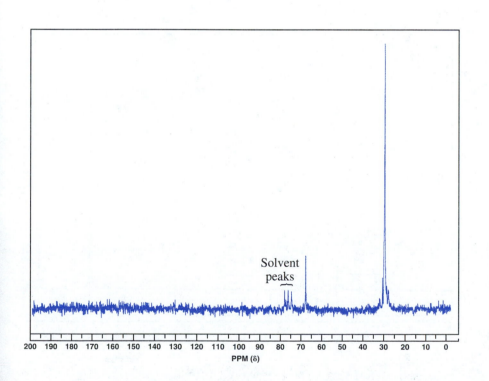

5. Compound **21**, $C_{10}H_{14}O$, whose IR and NMR spectra are provided in Figure 8.60.

FIGURE 8.60   IR and NMR spectra of compound **21**.

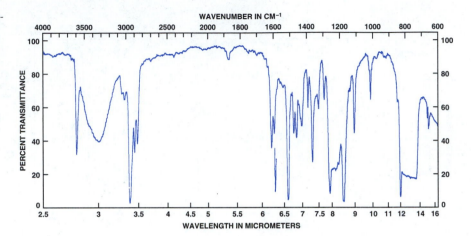

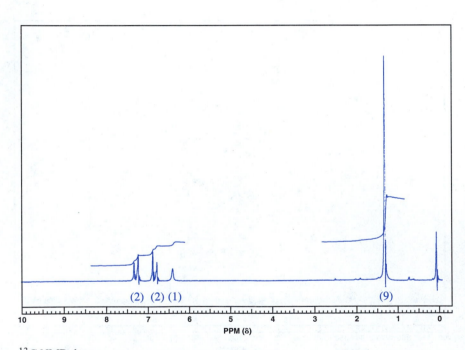

^{13}C NMR data.
Chemical shifts: δ 31.5, 33.7, 114.9, 125.9, 141.6, 154.7.

6. Compound **22**, $C_5H_{10}O_2$, whose IR and NMR spectra are provided in Figure 8.61.

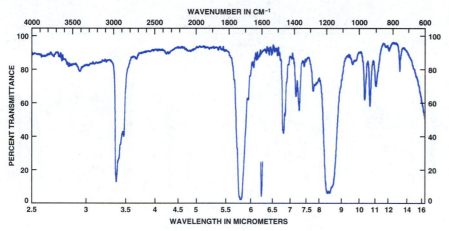

FIGURE 8.61   IR and NMR spectra of compound **22**.

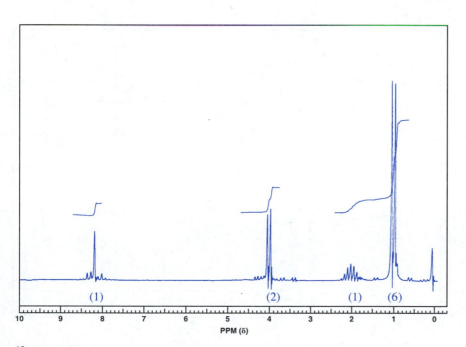

^{13}C NMR data.

Chemical shifts: δ 19.0, 27.9, 70.0, 161.2.

7. Compound **23**, C_7H_9N, whose IR and ^1H NMR spectra are provided in Figure 8.62.

FIGURE 8.62 IR and ^1H NMR spectra of compound **23**.

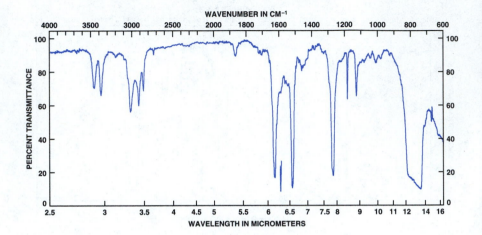

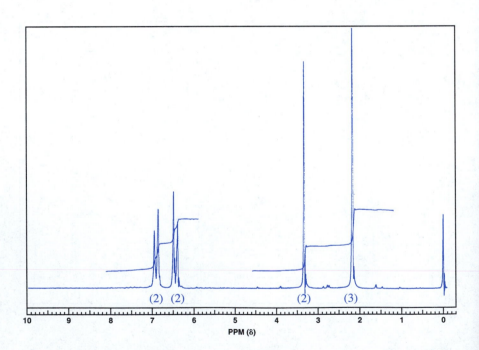

8. Compound **24**, $C_9H_{10}O$, whose IR and ^{13}C NMR spectra are provided in Figure 8.63.

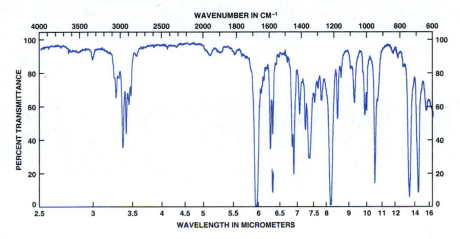

FIGURE 8.63 IR and ^1H NMR spectra of compound **24**.

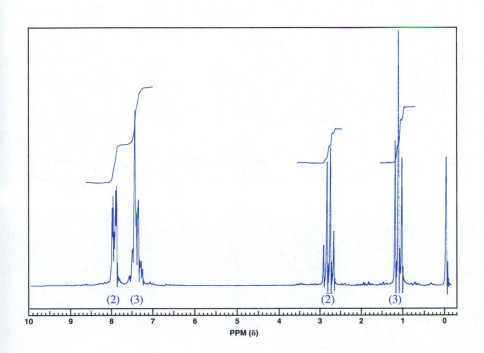

9. Compound **25**, $C_{12}H_{18}$, whose IR and ^1H NMR spectra are provided in Figure 8.64.

FIGURE 8.64 IR and ^1H NMR spectra of compound **25**.

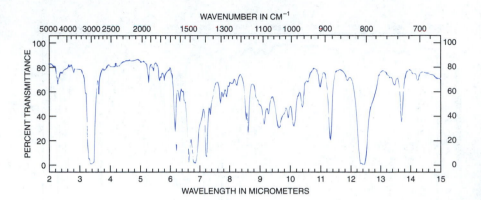

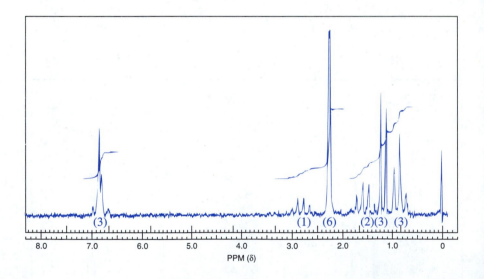

10. Compound **26,** C_3H_5Br, whose IR and ^{13}C NMR spectra are provided in Figure 8.65.

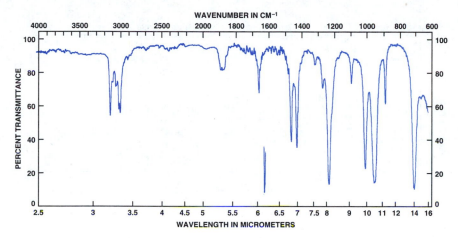

FIGURE 8.65 IR and ^{13}C NMR spectra of compound **26.**

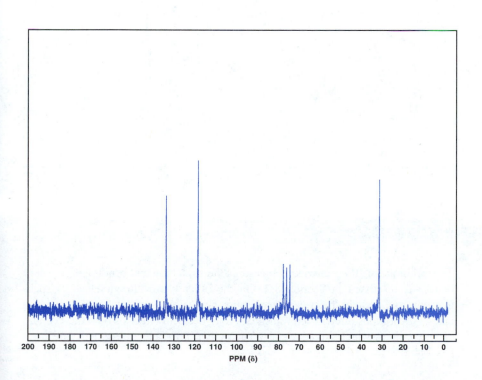

11. The IR and ¹H NMR spectra of citral, as a mixture of diastereomers (geometric isomers), are provided in Figure 8.66. Interpret these spectra as completely as possible.

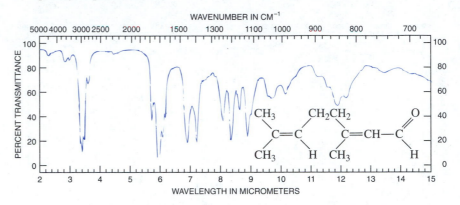

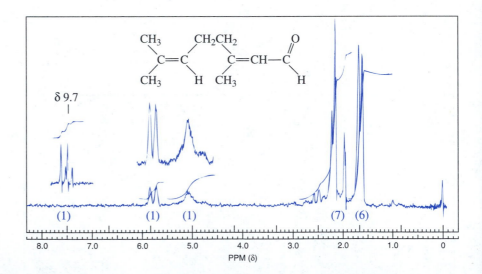

FIGURE 8.66 IR and ¹H NMR spectra of citral.

8.4 ULTRAVIOLET AND VISIBLE SPECTROSCOPY

The technique of IR spectroscopy measures the stretching and bending of covalent bonds, as discussed in Section 8.2. On the other hand, ultraviolet (UV) spectroscopy and visible spectroscopy measure **electronic transitions** within molecules. Absorptions due to such transitions may occur in either or both the UV and visible regions, which are adjacent to each other in the electromagnetic spectrum (see Table 8.1).

8.4.1 PRINCIPLES

Ultraviolet and visible spectroscopy both depend on the same fundamental molecular phenomenon, namely the excitation of an electron from a lower-energy to a higher-energy electronic state (see Fig. 8.1). As shown in Table 8.1, the energy

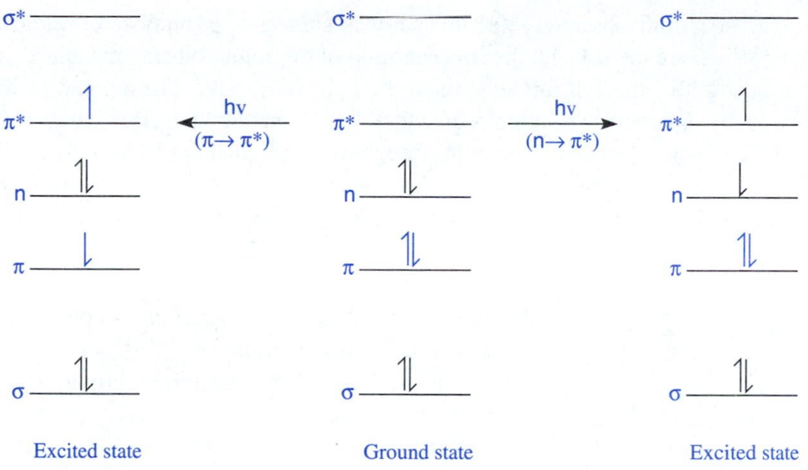

FIGURE 8.67   Energy diagram showing electronic transitions.

required for such an excitation ranges from about 38 to more than 100 kcal/mol and involves light in the wavelength range of 750 to 200 nm. These electronic changes are brought about by UV and visible light, and they generally involve either nonbonded electrons or electrons in π-bonds.

The two types of electronic excitations in this range that are of greatest interest in organic molecules are those involving promotion of an electron that originally occupied either a **nonbonding molecular orbital,** an *n*-electron, or a **bonding molecular orbital,** a σ-electron or a π-electron into an **antibonding** molecular orbital. However, the energy needed to promote a σ-type electrons to an antibonding molecular orbital is normally higher than 150 kcal/mol, which is too high to be observed at the short wavelength limit of the UV spectrum, namely, 200 nm. This kind of electronic excitation thus is not a feature of UV/visible spectra; rather, UV/visible spectroscopy is limited to excitations of *n*- and π-electrons. Furthermore, because π-type antibonding molecular orbitals, designated π^*, are considerably lower in energy than the corresponding σ^* orbitals, the transitions normally stimulated by absorption of light having a wavelength of the UV/visible region involve population of the π^* state. The electronic transitions may then be classified as $n \rightarrow \pi^*$ and $\pi \rightarrow \pi^*$, and are shown schematically in Figure 8.67.

8.4.2 PRACTICAL CONSIDERATIONS

Ultraviolet spectra are determined with a UV spectrometer, an instrument that is conceptually similar to an IR spectrometer, except that the former measures absorption of light in the UV region, whereas the latter measures IR radiation. The substance under study is dissolved in a suitable solvent, and the solution is placed in a cell. The same pure solvent is contained in another cell to serve as a reference. Methanol, ethanol, hexane, and water are common UV solvents because they do not absorb UV radiation; for the same reason, the optics and the cells that are used are made of quartz. The light source is usually a hydrogen lamp. The light passes through both cells, and the net amount of energy absorbed is measured and recorded. The spectrum that results is due entirely to the absorption of energy by the solute that is present, because dispersion of light by the sample cell and the solvent are compensated for by the reference cell.

The spectrum observed experimentally is subject to a number of variables, among which are the solvent, the concentration of the solution being examined, and the pathlength of the cell through which the light must pass. The amount of light absorbed by a particular solution is quantitatively defined by the Beer-Lambert law (Eq. 8.5), which is restated in terms of **absorbance** in Equation (8.12).

$$A = \log \frac{I_0}{I} = \epsilon c l \qquad \textbf{(8.12)}$$

where A = absorbance or optical density
I_0 = intensity of radiation incident on sample at particular wavelength
I = intensity of radiation transmitted through sample at same wavelength
ϵ = molar extinction coefficient
c = concentration of sample in mol/liter
l = length of cell in cm

The UV spectrometer measures **absorbance** A, as a function of the wavelength λ of the incident radiation; the resulting plot is the **UV spectrum** of a compound. The **concentration** c, and **cell length,** l, are known quantities; and the absorbance A may be experimentally determined from the spectrum at each wavelength. Thus, the **molar extinction coefficients,** ϵ, may be calculated according to Equation (8.12); these typically range from 10 to 100,000.

If the molecular weight of the compound is unknown, ϵ cannot be calculated; that is, it would not be possible to express the concentration c in *mol/liter* of solution. In this instance, the intensity of an absorption must be evaluated in a manner different from that described by Equation (8.12). One way of doing so is given in Equation (8.13), in which A and l have their usual meaning, but c is measured in g/100 mL of solution. Like ϵ, E is a characteristic measure of the absorptivity of the unknown substance.

$$E_{1\ cm}^{1\%} = \frac{A}{cl} \qquad \textbf{(8.13)}$$

The intensities of the peaks in a UV spectrum are affected by the experimental variables of concentration and pathlength, so it is common practice to eliminate these factors from the spectrum by plotting ϵ or log ϵ, rather than A, *versus* wavelength. The concentration of the solution and the pathlength of the cell will be shown on the spectra; in some instances a single spectrum will show traces made at more than one concentration so both strong and weak absorbances can be discerned, as in the UV spectrum of 4-methyl-3-penten-2-one (**27**), shown in Figure 8.68.

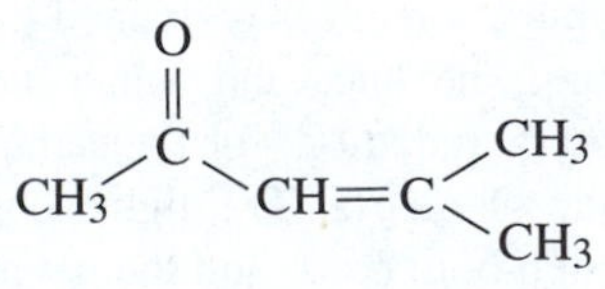

27
4-Methyl-3-penten-2-one

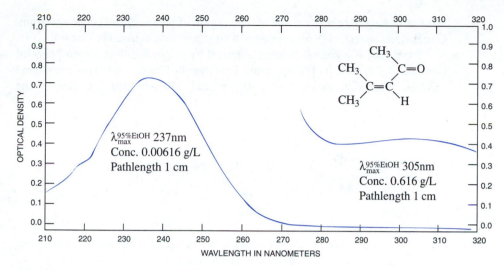

FIGURE 8.68 UV spectrum of 4-methyl-3-penten-2-one.

The entire UV visible spectrum of a compound is frequently not reported; rather only the wavelength, λ_{max}, and molar extinction coefficient, ϵ or $\log \epsilon$, of any maxima are given, along with the solvent in which the measurement was made. Thus, the crucial information contained in Figure 8.68 is expressed in the following way:

$$\lambda_{max}^{95\%EtOH} \quad 237 \text{ nm, } \log \epsilon \text{ 4.1; 305 nm, } \log \epsilon \text{ 1.8}$$

The solvent should be stated, since the values of both λ_{max} and ϵ are solvent-dependent; a discussion of these solvent effects and an interpretation of their origin is presented in References 1 and 2, respectively at the end of this subsection.

4-Methyl-3-penten-2-one (**27**) contains both *n*- and π-electrons, the former being the nonbonding electron pairs on oxygen. Note that the absorption bands in its UV spectrum (Fig. 8.68) are rather broad, and the maxima are relatively poorly defined. The diffuseness of the spectrum arises because electronic transitions occur from a variety of vibrational and rotational levels of the ground electronic state into a number of different such levels of the excited electronic state (Fig. 8.1). The transitions themselves are quantized, and only certain ones are allowed from a theoretical standpoint. However, sharp lines are not observed in the spectrum because closely spaced vibrational-rotational levels give rise to closely spaced lines that coalesce into the broad absorbance or **band envelope** that is experimentally observed.

The most important characteristics of a UV spectrum are the locations of any maxima and their corresponding intensities because these parameters are characteristic for a particular type of electronic excitation. For our purposes, they are used to identify a specific type of **chromophore,** a term that refers to the particular arrangement of atoms responsible for absorption of the incident light. In the case of **27** and its UV spectrum (Fig. 8.68), the chromophore is the α,β-unsaturated carbonyl moiety, with the longer-wavelength, less-intense maximum at 314 nm being assigned to the $n \rightarrow \pi^*$ excitation and the other maximum at 238 nm to the $\pi \rightarrow \pi^*$ process. Thus it is possible to detect various functional groups if they serve as chromophores in the UV visible region of the electromagnetic spectrum. Correlations between UV absorption and structure are discussed in Part (c).

Citral (**28**) is a natural product obtained by steam distilling lemon grass oil (see Sec. 4.2.7), and its UV spectrum is shown in Figure 8.69. This compound exhibits absorption maxima at 230 nm and 320 nm that are due to the

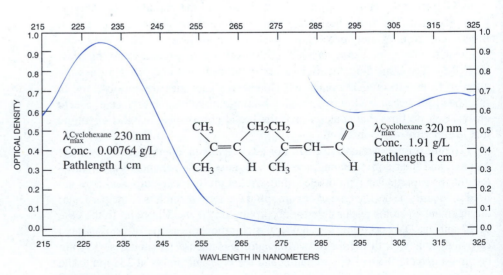

28
Citral

α,β-unsaturated carbonyl chromophore and the nature of the electronic transitions, $n \rightarrow \pi^*$ and $\pi \rightarrow \pi^*$, is also the same. This particular chromophore is the same as that contained in **27**. The observed $\lambda_{\max}$ for **28** are of comparable magnitudes and intensities as those for **27**, the differences being due to the type of carbonyl group, aldehyde versus ketone, and the nature of the substituents attached to the α,β-unsaturated carbonyl moieties. The second carbon-carbon double bond in **28** is *not* part of the conjugated α,β-unsaturated carbonyl chromophore, because it is isolated from the conjugated system by two sp^3-hybridized carbon atoms. Isolated π-bonds usually do not produce $\pi \rightarrow \pi^*$ maxima at wavelengths greater than 200 nm, and no maximum for this chromophore appears in Figure 8.69.

8.4.3 CORRELATION BETWEEN STRUCTURE AND UV SPECTRA

A compilation of a few of the chromophores commonly encountered by organic chemists, along with the wavelengths and intensities of their absorption maxima, is provided in Table 8.10. The intensities of maxima, as measured by ϵ or log ϵ,

FIGURE 8.69 UV spectrum of citral.

TABLE 8.10   Some Common Ultraviolet-Active Chromophores and Their Properties

Chromophore	Wavelength of λ_{max} (nm)	Type of Excitation	Log ϵ_{max}
	217–230	$\pi \rightarrow \pi^*$	4.0–4.3
	240–280	$\pi \rightarrow \pi^*$	3.5–4.0
	270–300	$n \rightarrow \pi^*$	1.1–1.3
(R' = H, alkyl)	200–235	$n \rightarrow \pi^*$	1.0–1.7
	215–250	$\pi \rightarrow \pi^*$	4.0–4.3
	310–330	$n \rightarrow \pi^*$	1.3–1.5
(R' = H, alkyl)	200–240	$\pi \rightarrow \pi^*$	4.0–4.1
(G = one or more alkyl groups)	256–272	$\pi \rightarrow \pi^*$	2.3–2.5
	200–210	$\pi \rightarrow \pi^*$	3.9–4.1
(Het = heteroatom)	265–290	$\pi \rightarrow \pi^*$	2.3–3.4
	210–230	$\pi \rightarrow \pi^*$	3.8–4.0

reflect the probability that a particular electronic transition will occur. Because these transitions are quantized, as noted earlier, some are theoretically "allowed" and others are "forbidden." Nonetheless, the latter type of excitations may be observed in some cases, for reasons that are not discussed in this text. Because of the lower probability for forbidden transitions, they generally provide less intense absorptions than the allowed transitions. Reference to Figure 8.69 and Table 8.10 shows that log ϵ for $\pi \rightarrow \pi^*$ transitions is relatively large compared to that for $n \rightarrow \pi^*$ excitations. Such relative intensities correspond to the prediction from theory that the latter transitions are "forbidden," whereas the former are "allowed."

The $\pi \rightarrow \pi^*$ transition of the carbonyl group is normally observed only in *conjugated* carbonyl compounds. The carbonyl group contains both *n*-electrons and π-electrons, but when it is not conjugated, such as in simple ketones, the $\pi \rightarrow \pi^*$ excitation generally occurs at a wavelength of less than 200 nm, which is out of the normal UV range. On the other hand, simple ketones do exhibit an absorption maximum at 270–300 nm due to the $n \rightarrow \pi^*$ excitation, but it appears as a *weak* band with log $\epsilon = 1.0$–1.7.

Many of the UV-active chromophores given in Table 8.10 contain isolated carbon-carbon or carbon-oxygen double bonds, or conjugated systems such as dienes, α,β-unsaturated carbonyl compounds, and aromatic ring systems. Electronic delocalization associated with conjugation of π-electrons *decreases* the energy gap between the occupied and unoccupied energy levels of the molecule (Figs. 8.1 and 8.67), and this results in λ_{max} being shifted to *longer* wavelengths. If this shift is great enough, light is absorbed in the visible region, and a substance appears colored. For example, lycopene (**29**), which contains a highly conjugated chromophore, is found in watermelons and tomatoes and is responsible for their red color. Many pH indicators, such as Congo Red (**30**), textile fabric dyes, such as Fast Red A (**31**), and food colorings owe their color to the conjugated systems that they contain.

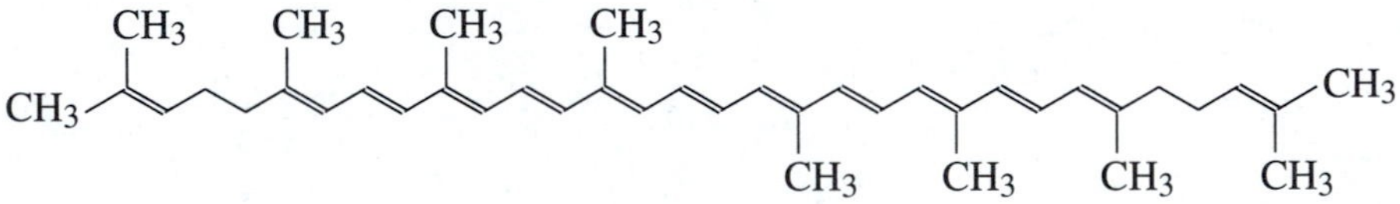

29
Lycopene

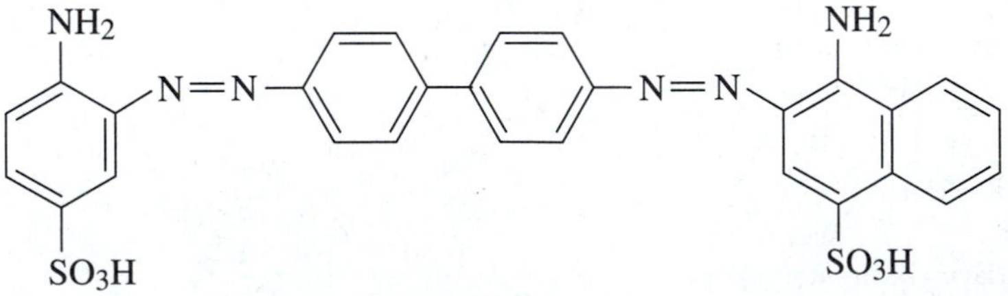

30
Congo Red

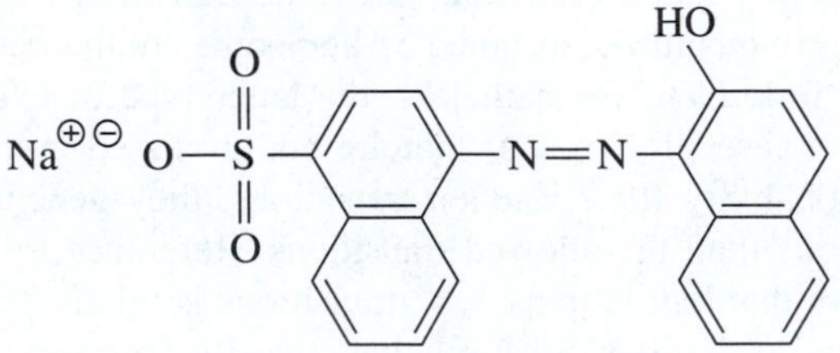

31
Fast Red A

Empirical generalizations have been developed that permit λ_{max} to be predicted as a function of the nature of the substituents and, in some instances, their location on the chromophore of interest. Although these rules are not provided here, they permit prediction of the approximate positions of λ_{max} in compounds as closely related as **32** and **33.** These types of correlations are discussed in Reference 1 at the end of this section.

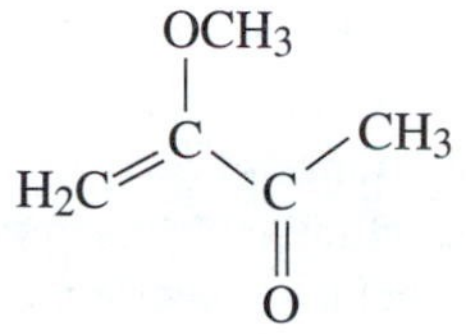

32	**33**
3-Methoxy-3-buten-2-one	*E*-4-Methoxy-3-buten-2-one

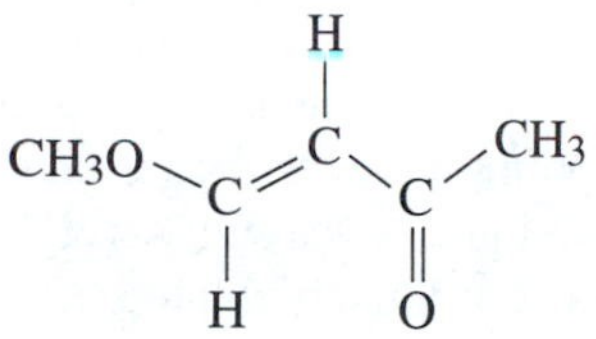

8.4.4   SAMPLE PREPARATION FOR UV-VISIBLE SPECTROSCOPY

Preparing samples for analysis by UV-visible spectroscopy is straightforward as most spectra are obtained in solutions. The cells for UV work are constructed of quartz and commonly have a pathlength of 1 cm, whereas those for visible spectroscopy are made of borosilicate glass, which is less expensive; this latter type of glass is opaque to light in the UV region and is therefore not suitable for UV cells. Both UV and visible cells typically require about 3 mL of solution.

A variety of organic solvents, as well as water, can be used for UV-visible spectroscopy. They all share the property of having no significant absorption at wavelengths greater than about 220 nm, as indicated in Table 8.11. The wavelengths given in the table constitute the **cut-off point** for the respective solvents. Below the specified wavelength the solvent begins to absorb appreciably, so solvents cannot be used if they absorb at the same wavelengths as the solute. "Technical" and "reagent-grade" solvents often contain light-absorbing impurities and must be purified before use, but more expensive "spectral-grade" solvents, which can be used without being purified, are available for UV-visible

TABLE 8.11   Solvents for UV-Visible Spectroscopy

Solvent	Useful Spectral Range in nm (Lower Limit)
Acetonitrile	<200
Chloroform	245
Cyclohexane	205
95% Ethanol	205
Hexane	200
Methanol	205
Water	200

spectroscopy. The choice of solvent will depend on the solubility of the solute of interest and the absence of chemical reactions between solvent and solute.

The concentration of the solution should provide an observed value of A in the range of about 0.3–1.5 because this allows the greatest accuracy in making the measurement. The approximate concentration should be estimated from the value of ϵ for any chromophores present (Table 8.10), from which c can be determined using Equation (8.12). As a rough rule of thumb, 0.01–0.001 M solutions produce absorbances having log ϵ of about 1.0, an appropriate magnitude for excitations of low intensity; diluting this solution will then permit more intense absorptions to be observed in the desired range of A.

The solute must be carefully weighed, quantitatively transferred to a volumetric flask, and accurately diluted in order to achieve precise measurements of A and therefore of ϵ. Furthermore, the accidental introduction of even a small amount of an intensively absorbing impurity into the solution may have a dramatic effect on the observed UV spectrum, so great care should be exercised in the handling and cleaning of all apparatus associated with the preparation of the solution. The cells should be rinsed thoroughly with the solvent both before and after use in order to minimize contamination in the sample and the cells. Quartz cells for UV work should *not* be rinsed with acetone for cleaning and drying. Trace residues of acid or base catalysts on the surface of the quartz may lead to the formation of trace quantities of 4-methyl-3-penten-2-one (**27**) through acid- or base-catalyzed aldol condensation; the UV absorptions of this contaminant could then invalidate precise measurements of ϵ for the solute of interest. The outer optical surfaces of the cells must be clean and free of fingerprints. The cells are best cleaned using the same pure solvent that is being used for the solute, and they must also be rinsed thoroughly when changing from one concentration to another.

EXERCISES

1. For the UV spectra of Figures 8.68 and 8.69, assign the chromophores responsible for each maximum and determine the value(s) of λ_{max} and of log ϵ.

2. Replot Figures 8.68 and 8.69, using values of log ϵ rather than optical density (absorbance, A) along the ordinate.

3. Repeat Exercises 1 and 2 for the following UV spectra, which are reproduced in this text.
   a. Figure 13.7
   b. Figure 18.14
   c. Figure 18.21
   d. Figure 18.36
   e. Figure 18.7

REFERENCES

1. Silverstein, R. M.; Bassler, G. C.; Morrill, T. C. *Spectrometric Identification of Organic Compounds*, 5th ed., John Wiley & Sons, New York, 1992, Chap. 7.
2. Haberfield, P., et al. *Journal of the American Chemical Society*, **1977,** *99,* 6828.
3. *Sadtler Standard Ultraviolet Spectra,* vol. 1–102, Sadtler Research Laboratories, Philadelphia, PA. Compilation of over 28,000 spectra, as of 1992.

ALKANES

9.1 INTRODUCTION

Alkanes are saturated hydrocarbons having the general molecular formula C_nH_{2n+2} and are typically unreactive. This characteristic lack of reactivity, even toward strong acids and bases, limits the use of alkanes as practical starting materials for most laboratory syntheses. However, some chemical reactions that occur by **free radical chain** mechanistic pathways do allow the introduction of functional groups such as NO_2, OOH, Cl, or Br onto an alkane. These functional groups can then be transformed into a variety of other functional groups. For example, nitroalkanes, RNO_2, and hydroperoxides, ROOH, are obtained by allowing an alkane, RH, to react with nitrogen tetroxide or molecular oxygen, respectively. Subsequent reduction of the nitroalkane produces an amine, whereas reduction of the hydroperoxide yields an alcohol (Eqs. 9.1 and 9.2). Although the *direct* introduction of an amino or hydroxyl group into the alkane is *not* possible, one *can* use a two-step sequence of the type described to circumvent this problem and prepare these useful classes of compounds.

$$R\text{—}H + N_2O_4 \xrightarrow{\text{heat}} R\text{—}NO_2 \xrightarrow[\text{catalyst}]{H_2} R\text{—}NH_2 \qquad \textbf{(9.1)}$$

| An alkane | A nitroalkane (mixture) | An amine (mixture) |

$$R\text{—}H + O_2 \xrightarrow{\text{initiator}} R\text{—}OOH \xrightarrow[\text{catalyst}]{H_2} R\text{—}OH$$

$$\textbf{(9.2)}$$

| An alkane | A hydroperoxide (mixture) | An alcohol (mixture) |

Alkanes may also be converted into alkyl chlorides or bromides, R–X (X = Cl or Br, respectively) by a free-radical process in which a mixture of the alkane and halogen is heated at 200–400 °C or is irradiated with ultraviolet light (Eq. 9.3). Under these conditions, the molecular chlorine or bromine is homolytically cleaved to generate chlorine and bromine atoms, which are free radicals. Generating chlorine and bromine atoms is essential to initiating the reaction between an alkane and molecular chlorine or bromine to form alkyl halides.

$$R\text{—}H + X_2 \xrightarrow{\text{heat or } h\nu} R\text{—}X + H\text{—}X \qquad (9.3)$$

An alkyl halide
(mixture)

$$X = Cl \text{ or } Br$$

The major drawback of the radical reactions depicted in Equations (9.1)–(9.3) is that *mixtures* of monosubstituted and polysubstituted products that are not easily separated are generally obtained. The experiments described in Sections 9.2 and 9.3 illustrate several methods for transforming alkanes to alkyl chlorides and bromides by free-radical substitution reactions. Some of these experiments give you an opportunity to analyze the mixtures of products that are obtained and to gain some insights about the relative reactivities of different types of hydrogen atoms toward chlorine and bromine radicals.

9.2 CHLORINATION USING SULFURYL CHLORIDE

Chlorine atoms may be generated from molecular chlorine under mild conditions using an **initiator,** In-In. Thus, homolysis of a molecule of initiator occurs upon heating or irradiation to give a pair of free radicals, In·, according to Equation (9.4). These free radicals, In·, may then react with molecular chlorine to produce In–Cl and a chlorine atom (Eq. 9.6), to initiate the **free-radical chain reaction.**

$$In\text{—}In \xrightarrow{\text{heat or } h\nu} In\cdot + In\cdot \qquad (9.4)$$

$$In\cdot + Cl\text{—}Cl \longrightarrow In\text{-}Cl + Cl\cdot \qquad (9.5)$$

Although thermolytic or photolytic generation of chlorine radicals may be readily implemented in industrial processes to produce alkyl chlorides according to Equation (9.3), this technique is not well suited for laboratory applications. Consequently, for safety and convenience, sulfuryl chloride, SO_2Cl_2, rather than molecular chlorine is used in this experiment as the source of chlorine radicals. The first step in initiating the free-radical chain reaction involves the homolysis of azobisisobutyronitrile (2,2′-azobis[2-methylpropionitrile]) (**1**), commonly abbreviated as AIBN, to form nitrogen and the free radical $(CH_3)_2(CN)C\cdot$ (Eq. 9.6); this reaction proceeds at a kinetically acceptable rate at 80–100 °C. This radical then attacks sulfuryl chloride to generate chlorine atoms and SO_2 according to Equations (9.7) and (9.8). The series of reactions depicted in Equations (9.6)–(9.8) comprise the **initiation** steps of the free-radical chain reaction.

Initiation

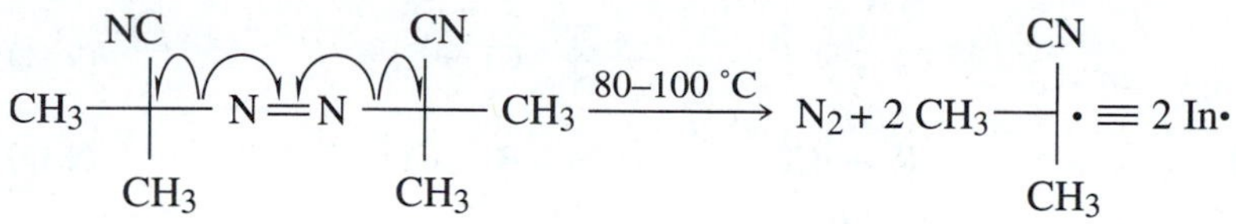

$$CH_3 \overset{NC}{\underset{CH_3}{\rule{0pt}{0pt}}}\!\!\!\!\!- N\!=\!N\!\!\!\!\!-\overset{CN}{\underset{CH_3}{\rule{0pt}{0pt}}}\!\!\!\!\!- CH_3 \xrightarrow{80\text{–}100\,^\circ C} N_2 + 2\,CH_3\!\!\!\!-\overset{CN}{\underset{CH_3}{\rule{0pt}{0pt}}}\!\!\!\!\!\cdot \equiv 2\,In\cdot \qquad (9.6)$$

1

AIBN

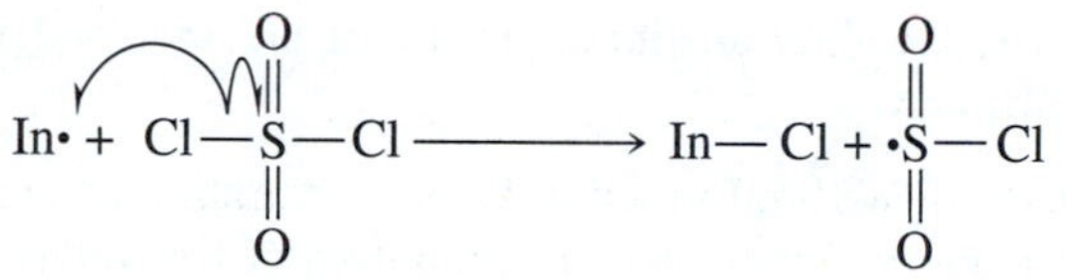

$$In\cdot + \ Cl\!-\!\overset{O}{\underset{O}{\overset{\|}{\underset{\|}{S}}}}\!-\!Cl \longrightarrow In\!-\!Cl + \cdot\overset{O}{\underset{O}{\overset{\|}{\underset{\|}{S}}}}\!-\!Cl \qquad (9.7)$$

$$\cdot\overset{O}{\underset{O}{\overset{\|}{\underset{\|}{S}}}}\!-\!Cl \longrightarrow SO_2 + Cl\cdot \qquad (9.8)$$

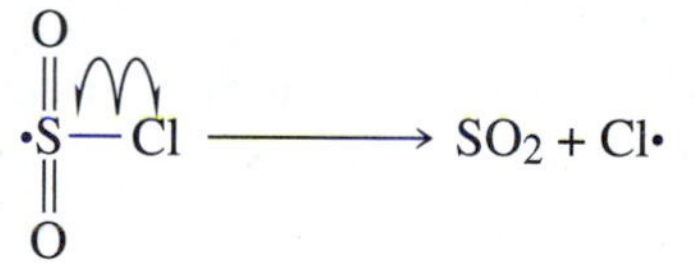

The next stage of the reaction involves the **propagation** steps. These include abstraction of a hydrogen atom from the hydrocarbon by a chlorine atom to produce a new free radical, R· (Eq. 9.9), followed by attack of this radical upon sulfuryl chloride to yield the alkyl chloride and the radical ·SO$_2$Cl (Eq. 9.10). The latter, as noted in Equation (9.8) and repeated in Equation (9.11), fragments into SO$_2$ and another chlorine atom as the final propagation step, thereby producing a radical chain reaction.

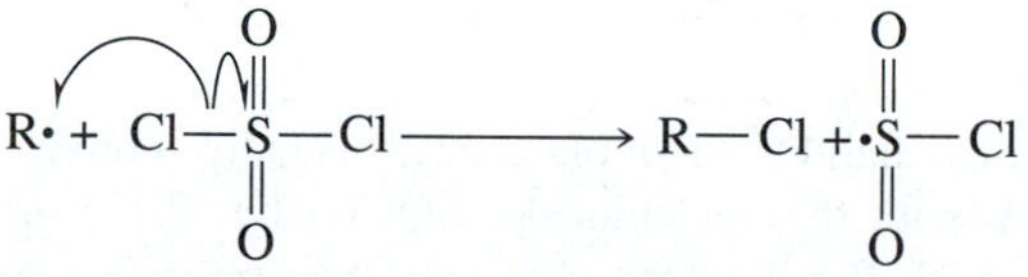

Propagation

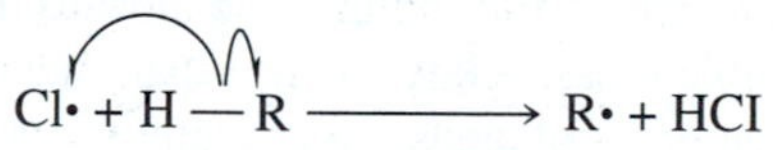

$$Cl\cdot + H\!-\!R \longrightarrow R\cdot + HCl \qquad (9.9)$$

$$R\cdot + \ Cl\!-\!\overset{O}{\underset{O}{\overset{\|}{\underset{\|}{S}}}}\!-\!Cl \longrightarrow R\!-\!Cl + \cdot\overset{O}{\underset{O}{\overset{\|}{\underset{\|}{S}}}}\!-\!Cl \qquad (9.10)$$

$$\cdot\overset{O}{\underset{O}{\overset{\|}{\underset{\|}{S}}}}\!-\!Cl \longrightarrow SO_2 + Cl\cdot \qquad (9.11)$$

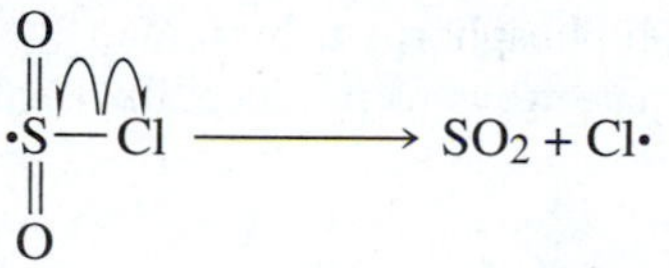

One might assume that, once initiated, this reaction continues until either sulfuryl chloride or the alkane, whichever is the **limiting reagent** (see Sec. 1.7), is completely consumed. In practice, the chain reaction is interrupted by a series of side reactions known as **termination** reactions (Eqs. 9.12–9.14), and the initiation process must be continued throughout the course of the reaction.

Termination

$$Cl\cdot + \cdot Cl \longrightarrow Cl-Cl \qquad (9.12)$$

$$R\cdot + \cdot Cl \longrightarrow R-Cl \qquad (9.13)$$

$$R\cdot + \cdot R \longrightarrow R-R \qquad (9.14)$$

In summary, the three distinct stages that are involved in a **free-radical chain mechanism** are

1. *initiation:* Radicals are formed in *low* concentration from neutral molecules resulting in an *increase* of the concentration of free radicals within the system.

2. *propagation:* Radicals produced in the first step react with molecules to yield new molecules and new radicals; there is *no net change* in radical concentration.

3. *termination:* Various radicals combine to give molecules resulting in a net *decrease* in radical concentration and a decrease in the rate of the reaction.

Free-radical halogenation of saturated hydrocarbons generally produces mixtures of several monohalogenated isomers as well as polyhalogenated products. The ratio of the different isomers is easily predicted by considering the number of hydrogens that can be replaced to provide a specific positional isomer and then correcting this statistical term by taking into account the relative reactivities of **primary** (1°), **secondary** (2°), and **tertiary** (3°) hydrogens toward halogen radicals (see Sec. 9.3, Fig. 9.11, and Exercises 10 and 15 at the end of this subsection). Although the separation of mixtures containing monohalogenated isomers and polyhalogenated products into pure components is usually difficult, such separations are performed in some industrial processes.

Because of the problems associated with separating the mixtures obtained by free-radical halogenation of alkanes, other methods have been developed for preparing pure alkyl halides in the laboratory. For example, the reaction of a hydrogen halide with an alkene (Eq. 9.15) or an alcohol (Eq. 9.16) provides the corresponding alkyl halide. Alcohols are also readily converted to alkyl chlorides by reaction with thionyl chloride, $SOCl_2$ (Eq. 9.17) or with phosphorus trichloride (Eq. 9.18, X = Cl); the corresponding bromides are produced when alcohols are treated with phosphorus tribromide (Eq. 9.19, X = Br). Procedures for some of these alternate ways to produce alkyl halides are given in Sections 10.4, 14.4, and 14.5.

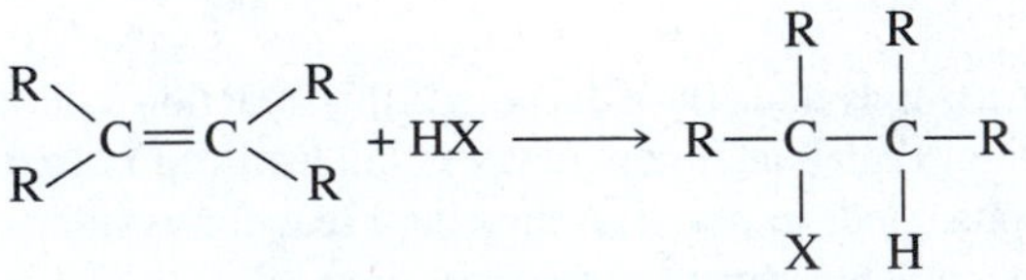

$$R_2C=CR_2 + HX \longrightarrow R-\underset{X}{\underset{|}{C}}-\underset{H}{\underset{|}{C}}-R \qquad (9.15)$$

$$R-OH + HX \longrightarrow R-X + H_2O \qquad (9.16)$$

$$RO-H + Cl-\overset{\overset{\textstyle O}{\|}}{S}-Cl \xrightarrow{-HCl} R-O-\overset{\overset{\textstyle O}{\|}}{S}-Cl \xrightarrow{heat} R-Cl + SO_2 \quad \textbf{(9.17)}$$

$$3 \;\; RO-H + PX_3 \longrightarrow 3 \;\; R-X + H_3PO_3 \quad \textbf{(9.18)}$$

Pre-Lab exercises for Section 9.2, "Chlorination Using Sulfuryl Chloride," are found on page PL. 29.

EXPERIMENTAL PROCEDURES

SAFETY ALERT

1. The sulfuryl chloride used in this experiment reacts violently with water. Be sure that your glassware is dry. Take special care to avoid getting sulfuryl chloride on your skin and do *not* breathe its vapors. Weigh out this chemical in a hood, and wear rubber gloves when transferring it.
2. The hydrocarbons used in this experiment are flammable, so avoid using burners if possible.
3. Be certain that all connections in your apparatus are tight prior to heating the reaction mixture.
4. When the reaction mixture is washed with aqueous sodium carbonate, carbon dioxide is generated in the separatory funnel. To relieve any gas pressure that develops, be sure to vent the funnel *frequently* when shaking it (see Sec. 2.18).

Procedures for the chlorination of three alkanes cyclohexane, heptane, and 1-chlorobutane are outlined below. It is not practical for a single student to perform all the parts of the experiment. However, important principles with respect to the number, type, and ratio of products that are obtained from each of the three starting materials are illustrated by each reaction, so each part should be performed by one group of students in the class and the results may be compared.

$$CH_3(CH_2)_5CH_3 \qquad\qquad CH_3CH_2CH_2CH_2Cl$$

Heptane   1-Chlorobutane

Cyclohexane

A. CYCLOHEXANE

Equip a 50-mL round-bottom flask with one or two boiling stones or a stirring bar, if magnetic stirring is available. Fit the flask with a water-cooled reflux condenser that is equipped at its top with a vacuum adapter connected as shown in Figure 9.1 to serve as a *trap* for the SO_2 and HCl produced in the reaction. Either fill the upper section of the adapter with calcium chloride, as in Figure 9.1, or attach a drying tube filled with calcium chloride to the female joint of the adapter. Place 7.0 g of cyclohexane drawn from a *glass* container, 6.0 g of sulfuryl chloride, and 0.04 g of azobisisobutyronitrile (AIBN) in the flask. If the cyclohexane has been stored in a metal container, it must *first* be purified by simple distillation to remove any traces of metal ions that may inhibit the reaction. Weigh the flask and its contents. Connect the condenser and trap, turn on the vacuum to produce a modest flow of air through the trap, and heat the mixture to a *gentle* reflux for 20 min using either a heating mantle or an oil bath. Cool the reaction mixture, disconnect the flask from the condenser, and weigh the flask and its contents.✳ If the loss of weight is less than expected, add another small portion of AIBN and heat the mixture under reflux for 10 min. The theoretical loss of weight is calculated on the basis that 1 mole each of HCl and of SO_2 is evolved for each mole of sulfuryl chloride consumed.

After the expected weight has been lost, cool the reaction mixture and, working at the hood, *cautiously* pour it into 20 mL of ice water with stirring. Transfer the resulting two-phase solution to a separatory funnel and separate the layers.✳ If separation into two layers does not occur readily, add some sodium chloride to the separatory funnel and shake it. Wash the organic layer in the funnel with 10 mL of 0.5 *M* sodium carbonate solution: Check the aqueous wash with litmus or pHydrion paper; if it is not basic, repeat washing the organic layer until the aqueous layer is basic. Wash the organic layer with 10 mL of saturated sodium chloride, then dry it over about 1–2 g of anhydrous sodium sulfate.

Decant most of the dried solution into a 25-mL round-bottom flask; complete the transfer of the remaining product mixture using a disposable glass pipet. Fit the flask with a column packed with glass or stainless steel for fractional distillation (see Sec. 2.8); carefully perform a fractional distillation to separate the chlorinated products from unreacted starting material. After the starting material has been distilled (Fraction 1), remove the fractionating column from the apparatus, replace it with a simple distillation head, and distil the chlorinated

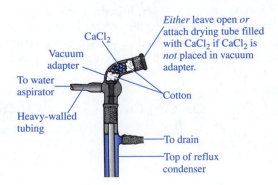

FIGURE 9.1   Details of gas trap arrangement.

product. To minimize heat loss and facilitate the distillation, the stillhead may be wrapped with glasswool and then aluminum foil. Suggested boiling ranges (760 torr) for collecting two fractions are as follows: Fraction 2, ambient–85 °C; Fraction 3, 85–145 °C.

Weigh each of the distillation cuts and either analyze them by gas chromatography (see Sec. 6.4) or submit samples to your instructor for such analysis. Estimate the approximate yields of the chlorinated products, taking into account the amount of unchanged starting material that was recovered.

Finishing Touches   Combine all of the *aqueous layers and washes;* if this solution is acidic, neutralize it carefully with sodium carbonate and flush it down the drain with a large excess of water. Discard the *calcium chloride from the drying tube* in a container for nonhazardous solids. Since the *calcium chloride that was used as a drying agent* will be contaminated with the chlorinated product, dispose of it in the container for halogenated solvent-contaminated drying agents. Place any unused or unreacted *starting material* in a container for nonhalogenated organic liquids.

B. HEPTANE

Follow the directions given in Part (A), using 8.8 g of heptane in place of cyclohexane. If the heptane has been stored in a metal container, it must first be purified by simple distillation to remove any traces of metal ions that may inhibit the reaction. Suggested boiling ranges (760 torr) are as follows: Fraction 2, ambient–105 °C; Fraction 3, 105–160 °C.

Finishing Touches   Combine all of the *aqueous layers and washes;* if this solution is acidic, neutralize it carefully with sodium carbonate and flush it down the drain with a large excess of water. Discard the *calcium chloride from the drying tube* in a container for nonhazardous solids. Since the *calcium chloride that was used as a drying agent* will be contaminated with the chlorinated product, dispose of it in the container for halogenated solvent-contaminated drying agents. Place any unused or unreacted *starting material* in a container for nonhalogenated organic liquids.

C. 1-CHLOROBUTANE

Follow the directions given in Part (A), using 8.2 g of 1-chlorobutane in place of cyclohexane. If the 1-chlorobutane has been stored in a metal container, it must be first purified by simple distillation to remove any traces of metal ions that may inhibit the reaction. Suggested boiling ranges (760 torr) are as follows: Fraction 2, ambient–82 °C; Fraction 3, 82–165 °C.

Finishing Touches   Combine all of the *aqueous layers and washes;* if this solution is acidic, neutralize it carefully with sodium carbonate and flush it down the drain with a large excess of water. Discard the *calcium chloride from the drying tube* in a container for nonhazardous solids. Since the *calcium chloride that was used as a drying agent* will be contaminated with the chlorinated product, dispose of it in the container for halogenated solvent-contaminated drying agents. Place any unused or unreacted *starting material* in a container for halogenated organic liquids.

EXERCISES

1. Suggest at least one termination process for halogenation with sulfuryl chloride that has no counterpart in halogenation with molecular chlorine.

2. Why is the amount of sulfuryl chloride used less than the amount theoretically required to convert all the starting materials to monochlorinated products?

3. From which of the three starting materials should it be easiest to prepare a single pure monochlorinated product? Why?

4. What factors determine the proportion of monochlorinated isomers of heptane? of 1-chlorobutane?

5. Why is only a catalytic amount of initiator needed in these halogenation reactions? What problems might attend the use of larger quantities of initiator?

6. Why is the potential contamination of the chlorinated hydrocarbons with $(CH_3)_2C(Cl)CN$, the product derived from azobisisobutyronitrile (**1**), of little concern in the halogenation experiments of this section?

7. How might a greater than theoretical loss of weight occur in a chlorination reaction performed according to the procedures of this section? Explain your reasoning.

8. Why does adding sodium chloride facilitate separating an emulsion consisting of water, chloroalkane(s), and starting alkane into layers?

9. Why is it prudent to use caution when pouring the crude reaction mixture into ice water? Why is ice water rather than tap water specified?

10. Calculate the percent of each monochlorination product expected from heptane based on a relative reactivity of primary (1°):secondary (2°):tertiary (3°) hydrogens of 1.0:3.3:4.4. Referring to Figure 9.2, calculate the observed ratio of 1°:2° chloroheptanes and compare the result with the theoretical ratio.

11. Referring to Figure 9.3, calculate the percentage of each dichlorobutane present in Fraction 2 (do not include the area of the peak due to 1-chlorobutane when making the calculation). Why does the observed ratio of products not agree with that predicted by using the relative reactivities given in Exercise 10?

12. Figure 9.10 is the ^{1}H NMR spectrum of a mixture of dichlorobutanes obtained by free-radical chlorination of 1-chlorobutane.

    a. Assign the multiplet at δ 5.80, the doublet at δ 1.55, and the triplet at δ 1.07 to the hydrogen nuclei of the three different isomers from which they arise. (*Hint:* Write the structures of the isomeric 1,*x*-dichlorobutanes and predict the multiplicity and approximate chemical shift of each group of hydrogens (see Sec. 8.3 and Tables 8.4 and 8.5 for data needed to estimate chemical shifts).

    b. Calculate the approximate percentage of each isomer present in the mixture.

13. Calculate the heat of reaction for the reaction between cyclohexane and chlorine to yield chlorocyclohexane and hydrogen chloride, using the following bond dissociation energies and bond energies (in kcal/mol): C–H, 98.7; C–Cl, 81.0; Cl–Cl, 58.0; H–Cl, 103.2.

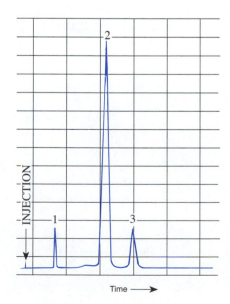

14. Chlorination of propene leads to high yields of 3-chloro-1-propene to the exclusion of products of substitution of the vinyl hydrogens. Why is this so?

15. Write the expected monochlorinated products of free-radical chlorination of methylcyclohexane with SO_2Cl_2. Predict which chlorinated isomer would be formed in highest yield, using the data on relative reactivities given in Exercise 10.

INJECTION

Time ⟶

FIGURE 9.2   GLC trace of fraction 2 resulting from chlorination of heptane. Peak 1: heptane; peak 2: 2-, 3-, 4-chloroheptanes; (bp at 760 torr 150 °C, 151 °C, and 152 °C, respectively); peak 3: 1-chloroheptane (bp 159 °C, 760 torr); column and conditions: 1.5 m, 5% silicone elastomer on Chromosorb® W; 80 °C, 40 mL/min.

FIGURE 9.3   GLC trace of fractions 1 and 2 resulting from chlorination of 1-chlorobutane. Peak 1: 1-chlorobutane; peak 2: 1,1-dichlorobutane (bp 114–115 °C, 760 torr); peak 3: 1,2-dichlorobutane (bp 121–123 °C, 760 torr); peak 4: 1,3-dichlorobutane (bp 131–133 °C, 760 torr); peak 5: 1,4-dichlorobutane (bp 161–163 °C, 760 torr); column and conditions: 1.5 m, 5% silicone elastomer on Chromosorb W; 60 °C, 40 mL/min.

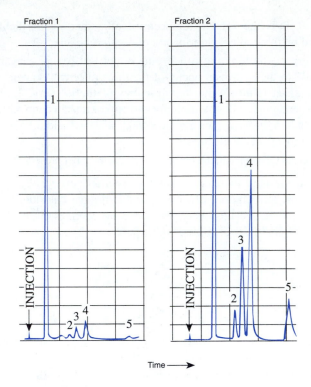

SPECTRA OF STARTING MATERIALS AND PRODUCTS

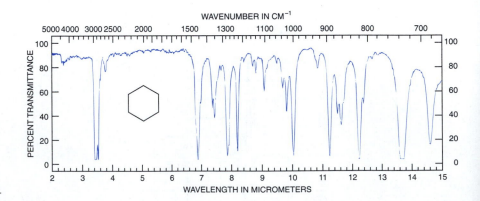

FIGURE 9.4   IR spectrum of cyclohexane.

(a) ^{1}H NMR spectrum (90 MHz).

FIGURE 9.5   NMR data for cyclohexane.

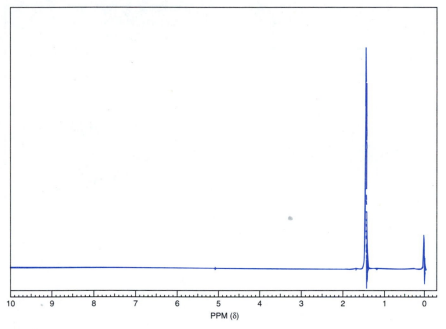

(b) ^{13}C NMR datum.
 Chemical shift: δ 27.3.

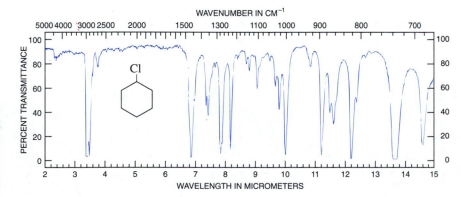

FIGURE 9.6   IR spectrum of chlorocyclo-hexane.

FIGURE 9.7   NMR data for chlorocyclohexane.

(a) ^{1}H NMR spectrum (90 MHz).

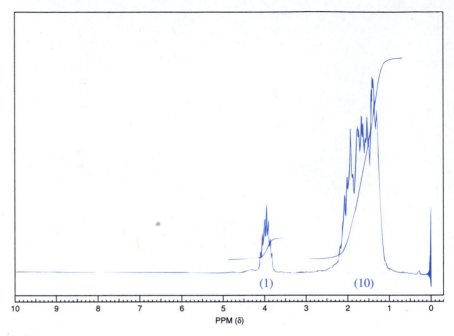

(b) ^{13}C NMR data.
Chemical shifts: δ 25.0, 25.4, 36.9, 59.9.

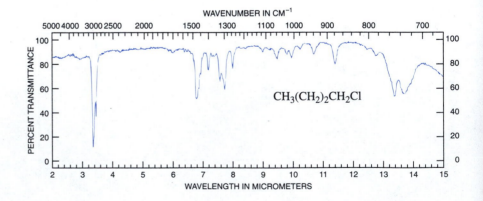

$CH_3(CH_2)_2CH_2Cl$

FIGURE 9.8   IR spectrum of 1-chlorobutane.

(a) ^{1}H NMR spectrum (60 MHz).

FIGURE 9.9 NMR data for 1-chlorobutane.

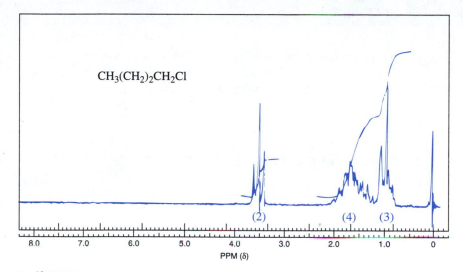

$CH_3(CH_2)_2CH_2Cl$

(2) (4) (3)

PPM (δ)

(b) ^{13}C NMR data.
Chemical shifts: δ 13.4, 20.3, 35.0, 44.6.

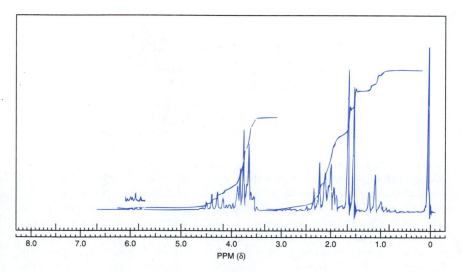

PPM (δ)

FIGURE 9.10 ^{1}H NMR spectrum of mixture of 1,x-dichlorobutanes (60 MHz).

9.3 BROMINATION: SELECTIVITY OF HYDROGEN ATOM ABSTRACTION AS A FUNCTION OF STRUCTURE

In one of the experiments in the preceding section, it is possible to compare the relative reactivity of different hydrogen atoms bonded to primary (1°) and secondary (2°) carbon atoms toward substitution by chlorine atoms. This is done simply by determining the relative amounts of primary and secondary chloroheptanes that are produced from heptane according to Equation (9.19) (see also Sec. 9.2, Exercise 10).

$$CH_3CH_2CH_2CH_2CH_2CH_2CH_3 \xrightarrow{Cl\cdot} \begin{cases} n\text{-}C_6H_{13}CH_2\text{-Cl} \\ + n\text{-}C_5H_{11}CHClCH_3 \\ + n\text{-}C_4H_9CHClCH_2CH_3 \\ + (n\text{-}C_3H_7)_2CHCl \end{cases} \tag{9.19}$$

In the present experiment, you will determine the relative rates of the radical chain reaction of bromine with several hydrocarbons containing different types of hydrogen atoms (Eq. 9.20).

The measurement of relative reactivities is easier for free-radical bromination than for free-radical chlorination. This is because bromine is a less reactive, more selective reagent than chlorine, so the rates of bromination of different hydrocarbons vary sufficiently that the qualitative order of reactivity is readily determined. Moreover, bromine is more highly colored than chlorine, so its disappearance may be more easily followed. Another advantage of bromine is that it is a liquid rather than a gas at room temperature and thus is easier to handle than chlorine.

$$R\text{—H} + Br_2 \xrightarrow{h\nu} R\text{—Br} + H\text{—Br} \tag{9.20}$$

Colorless   Reddish orange   Colorless   Colorless

As summarized by the structures in Figure 9.11, hydrogen atoms in hydrocarbons are classified in several ways according to the nature of the carbon atom to which they are attached. Specifically, hydrogens bound to sp^3-hybridized carbons are categorized as **aliphatic;** all hydrogen atoms in the hydrocarbon family of alkanes belong to this group. Aliphatic hydrogens may be further classified as being **primary (1°), secondary (2°),** or **tertiary (3°)** depending upon the number of other carbon atoms that are attached to the reference carbon atom; primary carbons are bonded to one other carbon, secondary carbons to two, and tertiary carbons to three. An aliphatic hydrogen atom attached to a carbon atom that in turn is bound to a vinylic or aromatic carbon is called **allylic** or **benzylic,** respectively. These various classifications can be combined, so that one refers to primary, secondary, or tertiary aliphatic, allylic, or benzylic hydrogens. Hydrogens attached to sp^2-hybridized carbons are classified as either **vinylic** or **aromatic,** depending upon whether the carbon atom is part of a normal double bond or an aromatic ring. Hydrogens are called **acetylenic** when bonded to a sp-hybridized carbon.

The experimental determination of relative reactivities is conceptually simple. An excess of a hydrocarbon reacts with molecular bromine to give the alkyl bromide (Eq. 9.20), and the time is measured for disappearance of the color due

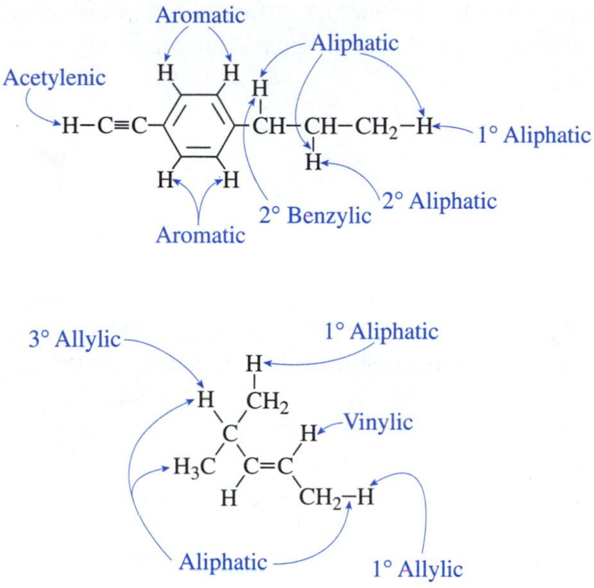

FIGURE 9.11 Classification of hydrogen atoms in hydrocarbons.

to bromine as it is consumed in the reaction. The period of time required for the decoloration of bromine depends upon the reaction conditions and the relative reactivities of the various hydrocarbons. The ratios of the different reaction times represent the relative rates of the bromination reaction. By carefully considering the results from this experiment, you should be able to deduce an order of reactivity toward bromine atom for seven different types of hydrogens. Determining the absolute rates of bromination of these same hydrocarbons could be done similarly, but the concentration of bromine as a function of time would have to be precisely measured, and the temperature of the reaction mixture would have to be more exactly controlled. This is a more difficult experiment and is not described.

The reactions are carried out in carbon tetrachloride solution using a sequence of three different conditions: (1) at room temperature without special illumination; (2) at 50 °C without special illumination; and (3) at 50 °C with strong illumination. Under all these conditions, replacement of hydrogen by bromine occurs by a free-radical chain mechanism analogous to that discussed for the chlorination reaction of Section 9.2. In this case, the initiation step is the thermally or photochemically promoted dissociation of molecular bromine into bromine atoms, which are radicals, as shown in Equation (9.21). The propagation steps are depicted in Equations (9.22) and (9.23), and the termination steps are analogous to those of Equations (9.12)–(9.14) in Section 9.2.

$$\text{Br}-\text{Br} \xrightarrow{\text{heat or } h\nu} \text{Br}\cdot + \text{Br}\cdot \qquad \textbf{(9.21)}$$

$$\text{Br}\cdot + \text{H}-\text{R} \longrightarrow \text{R}\cdot + \text{HBr} \qquad \textbf{(9.22)}$$

$$\text{Br}\cdot + \text{Br}-\text{Br} \longrightarrow \text{R}-\text{Br} + \cdot\text{Br} \qquad \textbf{(9.23)}$$

Pre-Lab exercises for Section 9.3, "Bromination: Selectivity of Hydrogen Atom Abstraction as a Function of Structure," are found on page PL. 31.

EXPERIMENTAL PROCEDURE

SAFETY ALERT

1. *Bromine is a hazardous chemical.* Do not breathe its vapors or allow it to come into contact with the skin because it may cause serious chemical burns. Perform all operations involving the transfer of the pure liquid or its solutions at a hood, and wear rubber gloves. If you get bromine on your skin, wash the area immediately with soap and warm water and soak the affected area in 0.6 M sodium thiosulfate solution for up to 3 h if the burn is particularly serious.

2. Dispense the 1 M bromine in carbon tetrachloride solution from burets fitted with Teflon stopcocks and located in hoods. This simplifies the precise measurement of 0.5-mL portions of the solution and provides for greater safety in transferring the bromine.

3. Bromine reacts with acetone to produce the powerful **lachrymator,** α-bromoacetone, $BrCH_2COCH_3$. Do *not* rinse glassware containing residual bromine with acetone!

4. Avoid excessive inhalation of the vapors of any of the materials being used in this experiment. To lower the concentration of vapors in your area, use an inverted funnel attached to a vacuum source and placed over the test tubes.

Construct a table in your notebook with the following four main headings: (1) "Hydrocarbon"; (2) "Types of Hydrogen," entries under which will include the terms, 1° aliphatic, 2° benzylic, and so on; (3) "Conditions," with the subheadings, 25 °C, 50 °C, 50 °C/$h\nu$; and (4) "Elapsed Time," entries under which will be the time required for reaction as measured by decoloration. In each of six 10-mm × 100-mm test tubes place a solution of 0.5 mL of hydrocarbon in 2.5 mL of carbon tetrachloride, using each of the following hydrocarbons: toluene, ethylbenzene, isopropylbenzene, *tert*-butylbenzene, cyclohexane, and methylcyclohexane. Carefully label these test tubes to avoid confusing them and jeopardizing the interpretation of the experimental results. Place 0.5 mL of a 1 M solution of bromine in carbon tetrachloride in each of seven other test tubes, and to one of them add 3 mL of carbon tetrachloride. Label this tube "Control," and label the others according to which hydrocarbon is to be placed in each tube. Now add the solutions of hydrocarbons to those containing bromine in rapid succession. This addition should be done with agitation to ensure good mixing. Note and record the mixing time and the elapsed times required for the reddish color of bromine to be discharged in each reaction mixture. Because the rates of bromination in some cases are quite fast, it is advisable to repeat the determination to be confident that the relative orders of decoloration are known.

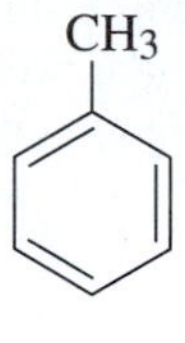

CH3

Toluene

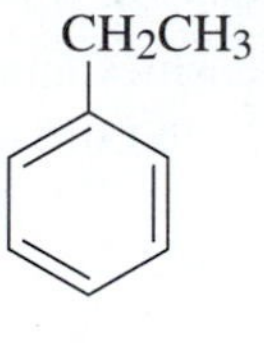

CH2CH3

Ethylbenzene

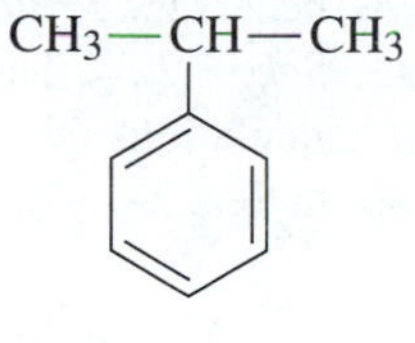

CH3—CH—CH3

Isopropylbenzene

CH3
|
CH3—C—CH3
|

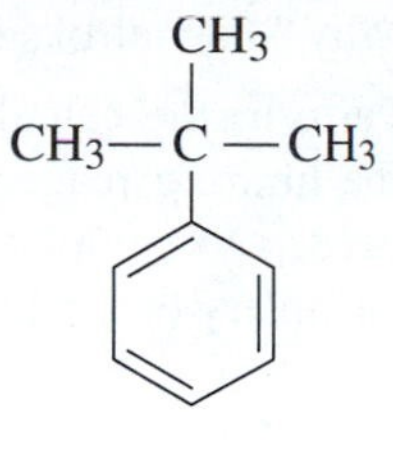

tert-Butylbenzene

Cyclohexane

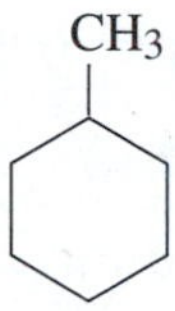

CH3

Methylcyclohexane

After watching for decoloration at room temperature for 20 min, place the Control and those reaction mixtures in which color remains in a beaker of water kept at 50 °C on a steam bath or hot plate. Do not heat the water much above this temperature because carbon tetrachloride has a boiling point of 77 °C. After 20–30 min at this temperature, suspend an unfrosted 100- or 150-watt light bulb over the test tubes at a distance of 10–13 cm. Continue to observe and record the times at which the color is discharged in any tubes still containing bromine. The experiment can be discontinued when only one colored solution, other than the Control sample, remains because it is obvious that this hydrocarbon reacts the slowest. On the basis of your results, complete the exercises that follow.

> **Finishing Touches**   Decolorize any *solutions* in which the color of bromine is visible by the dropwise addition of cyclohexene; then discard the resulting solutions together with all other solutions in a container for halogenated organic liquids.

EXERCISES

1. Arrange the six hydrocarbons in increasing order of reactivity toward bromination.

2. On the basis of the order of reactivity of the hydrocarbons, deduce the order of reactivity of the seven different types of hydrogens found in these compounds, that is, (1) primary aliphatic, (2) secondary aliphatic, (3) tertiary aliphatic, (4) primary benzylic, (5) secondary benzylic, (6) tertiary benzylic, and (7) aromatic.

3. Clearly explain how you arrived at your sequence in Exercise 2.

4. Draw the structure of the major monobromination product expected from each of the hydrocarbons used in this experiment.

5. Comment on the need for heat and/or light in order for bromination to occur with some of the hydrocarbons.

6. Why is a control sample needed in this experiment?

7. Perform the calculations necessary to demonstrate that bromine is indeed the limiting reagent (see Sec. 1.2.3). The densities and molecular weights needed to complete the calculations can be found in various handbooks of chemistry (see references in Sec. 24.1.3).

SPECTRA OF STARTING MATERIALS

The IR and NMR spectra of cyclohexane are provided in Figures 9.4 and 9.5, respectively.

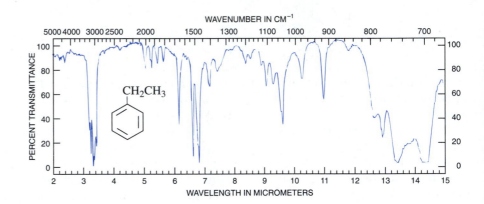

FIGURE 9.12 IR spectrum of ethylbenzene.

(a) ^{1}H NMR spectrum (60 MHz).

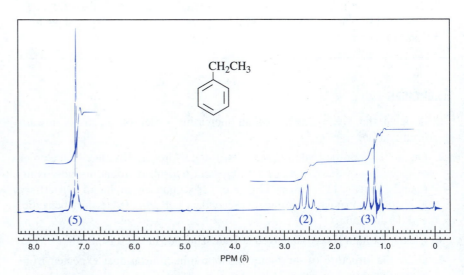

(b) ^{13}C NMR data.

Chemical shifts: δ 15.6, 29.1, 125.7, 127.9, 128.4, 144.2.

FIGURE 9.13 NMR data for ethylbenzene.

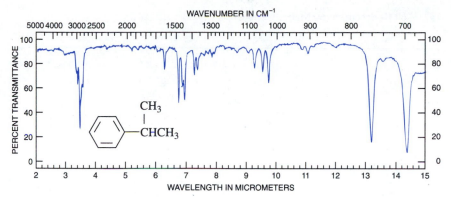

FIGURE 9.14 IR spectrum of isopropyl-benzene (cumene).

(a) ^{1}H NMR spectrum (60 MHz).

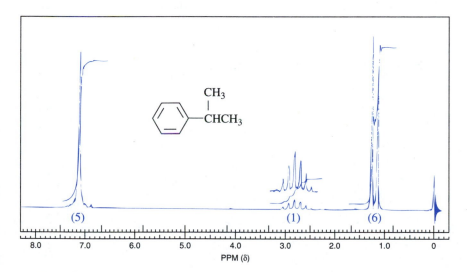

(b) ^{13}C NMR data.
   Chemical shifts: δ 24.1, 34.2, 125.8, 126.4, 128.4, 148.8.

FIGURE 9.15 NMR data for isopropylbenzene (cumene).

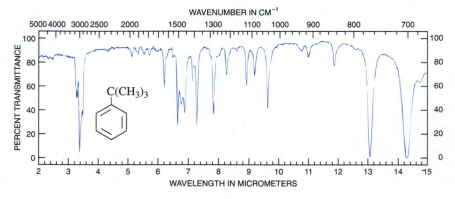

FIGURE 9.16 IR spectrum of *tert*-butylbenzene.

FIGURE 9.17   NMR data for *tert*-butylbenzene.

(a) ^{1}H NMR spectrum (60 MHz).

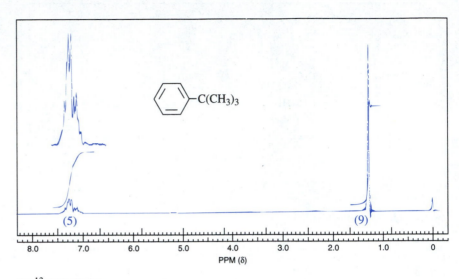

(b) ^{13}C NMR data.

Chemical shifts: δ 31.3, 34.5, 125.1, 125.3, 128.0, 150.8.

ALKENES

A functional group is an atom or group of atoms that governs the chemical and physical properties of a family of compounds. The introduction and manipulation of these functional groups are major objectives in modern organic chemistry. In this chapter, the chemistry of **alkenes,** which are organic compounds that possess a polarizable carbon-carbon π-double bond as the functional group, is explored. Methods for introducing a carbon-carbon double bond into a molecule from alkyl halides and alcohols are developed first, and some of the reactions characteristic of this functional group are then examined.

The carbon-carbon π-bond in an alkene **1** may be introduced in a variety of ways, but **elimination reactions** are among the most commonly used. For example, the elements of hydrogen halide, **H–X,** may be eliminated from an alkyl halide **2** in a reaction known as **dehydrohalogenation.** The functional group of an alkyl halide is a carbon-halogen, **C–X,** single bond, and the process by which the carbon-halogen bond is converted into a carbon-carbon π-bond is an example of a **functional group transformation.** A carbon-carbon π-bond may also be formed upon removing the elements of water from an alcohol **3,** in which a **C–OH** single bond is the functional group; this reaction is called **dehydration.** In both dehydrohalogenation and dehydration, the polarized carbon-heteroatom bond is cleaved heterolytically to give an alkene **1.** Although other aspects of the chemistry of alkyl halides and alcohols will be presented in Sections 14.1–14.5 and 16.2, a brief introduction to these families is essential to understand how they may be used as starting materials for the synthesis of alkenes.

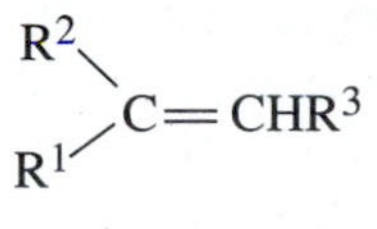

$$\underset{R^1}{\overset{R^2}{>}}C=CHR^3 \qquad \qquad R^1-\underset{\underset{R^3}{|}}{\overset{\overset{R^2}{|}}{CH}}-\overset{\delta\oplus}{CH}-\overset{\delta\ominus}{X} \qquad \qquad R^1-\underset{\underset{R^3}{|}}{\overset{\overset{R^2}{|}}{CH}}-\overset{\delta\oplus}{CH}-\overset{\delta\ominus}{OH}$$

1	**2**	**3**
An alkene	X = Cl, Br, I	An alcohol
	An alkyl halide	

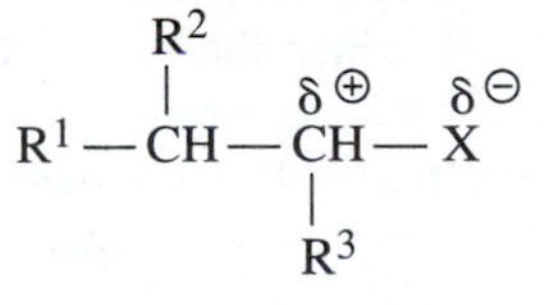

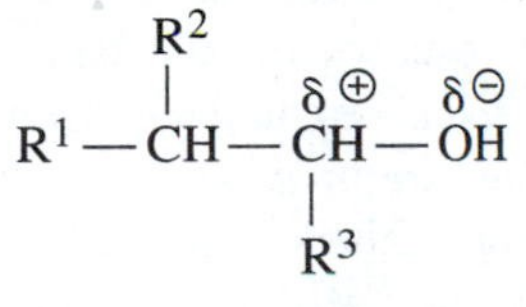

10.2 DEHYDROHALOGENATION OF ALKYL HALIDES

The electronegative halogen atom of an alkyl halide **2** polarizes the carbon-halogen bond so the carbon atom bears a partial positive charge and the halogen atom a partial negative charge. This bond may be further polarized by interaction of the lone pairs of electrons on the halogen atom, either by electrostatic attractions with solvent or by direct complexation with Lewis acids. This polarization may be transmitted through the σ-bond network over short distances, a phenomenon referred to as an **inductive effect,** to enhance the acidity of hydrogen atoms on the β-carbon atom. When a proton from this β-carbon atom is transferred to a strong base, B:⁻, the simultaneous departure of halide ion, which is a **leaving group,** from the α-carbon atom leads smoothly to the formation of the carbon-carbon π-bond of the alkene **4** in a **concerted reaction** as shown in Equation (10.1). An important characteristic of a leaving group in organic reactions is that it should be a weak base. The **transition state 4** for the reaction is shown in Equation (10.2), in which the curved arrows depict the flow of electrons. This transformation of alkyl halides to alkenes by the base-induced loss of a molecule hydrogen halide is a specific example of an elimination reaction called **dehydrohalogenation.**

$$\overset{\beta\ \ \alpha}{RCHCH_2} - X + B:^{\ominus} \longrightarrow RCH = CH_2 + B-H + X:^{\ominus} \quad \textbf{(10.1)}$$

(with an H below the β-carbon)

An alkyl halide   An alkene

4
Transition state

The reaction depicted in Equation (10.2) is classified as an **E2** process, where E stands for elimination and 2 refers to the molecularity of the rate-determining step of the reaction. The molecularity of the reaction is determined experimentally by measuring the dependence of the rate of the reaction on the concentration of the reactants. In the present case, the rate of the reaction is found to be equal to the rate constant k_2 times the product of the concentrations of each of the reactants (Eq. 10.3). Since the rate of the dehydrohalogenation depends upon the concentrations of the alkyl halide *and* the base, *both reactants must be involved in the transition state of the rate-determining step of the reaction.* This is illustrated in the transition state **4** of Equation (10.2).

$$\text{Rate} = k_2[\text{alkyl halide}][\text{B:}^-] \quad \textbf{(10.3)}$$

The electrophilic carbon atom of alkyl halides is also subject to attack by **nucleophiles,** which, as Lewis bases, are electron-rich and frequently anionic species. This process leads to displacement of halide ion as the **leaving group,** with the formation of **substitution products** (Eq. 10.4).

$$
\begin{array}{c}
\text{RCH}_2 \\
\overset{\delta\oplus}{\text{B}:^{\ominus}} \text{H} \cdots \text{C} - \text{X} \\
\quad\text{H}
\end{array}
\longrightarrow
\left[
\begin{array}{c}
\text{CH}_2\text{R} \\
\delta^{\ominus}\, \text{B} \cdots \text{C} \cdots \text{X}\, \delta^{\ominus} \\
\text{H}\ \text{H}
\end{array}
\right]^{\ddagger}
\longrightarrow
\begin{array}{c}
\text{CH}_2\text{R} \\
\text{B} - \text{C} \cdots \text{H} \\
\text{H}
\end{array}
+ \text{X}:^{\ominus}
\qquad (10.4)
$$

Transition state

Although the competition between *elimination* and *substitution* is discussed in detail in Section 14.3, it is useful to identify the factors that favor elimination and the formation of the desired alkene. Examining the transition states for the two reactions depicted in Equations (10.2) and (10.4) suggests that steric factors play a critical role in the competition. In support of this hypothesis, it is observed experimentally that elimination is favored as the degree of substitution on the carbon bearing the halogen atom increases. Steric hindrance at the carbon atom in more highly substituted halides inhibits direct attack of the base on the carbon atom and instead fosters removal of the β-proton. Thus, when the carbon atom bearing the leaving group is tertiary there is little competition between elimination and substitution; elimination to form alkenes is generally the only observable reaction. However, substitution competes favorably with elimination when the carbon atom bearing the leaving group is secondary, and substitution becomes the *dominant* process when the carbon atom is primary. Since a proton on the β-carbon will be less sterically hindered than the carbon atom bearing the leaving group, using a sterically hindered base also favors elimination relative to substitution. In the experiments in this section, a tertiary alkyl chloride is used as the alkyl halide, so there are no problems of substitution competing with the desired elimination. Dehydrohalogenation of tertiary alkyl halides proceeds cleanly using base/solvent combinations such as alkoxides, RO^-, in alcohols or potassium hydroxide in ethanol.

A second type of mechanism for the elimination of alkyl halides to give alkenes, shown in Equation (10.5), is an example of an **E1** reaction. This reaction involves generating an intermediate carbocation by rate-determining *heterolysis* of the carbon-halogen bond; the base then abstracts a β-proton to give the alkene **5.** Alkenes derived from carbocations formed by *rearrangement* of the initially formed carbocation may also be obtained. (See Sec. 10.3 for a more extensive discussion of the rearrangements of carbocations.) The rate of the reaction depends upon the ease of formation and stability of the intermediate carbocation, which follows the order $3° > 2° \gg 1°$. Although departure of the halide ion is assisted by solvation, the rate of reaction does *not* depend on the concentration of solvent because this quantity is essentially unchanged throughout the course of the reaction. This concentration is therefore incorporated as part of the rate constant k_1, so the rate of the reaction depends *only* on the concentration of the alkyl halide (Eq. 10.6), hence the designation of this unimolecular reaction as an **E1** process. As a general rule, elimination via an E1 mechanism is favored in polar nonnucleophilic solvents and in the absence of strong base. Solvent, SOH, may also serve as a nucleophile, resulting in substitution via a S_N1 reaction as illustrated in Equation (10.7). (See also Sec. 14.2.)

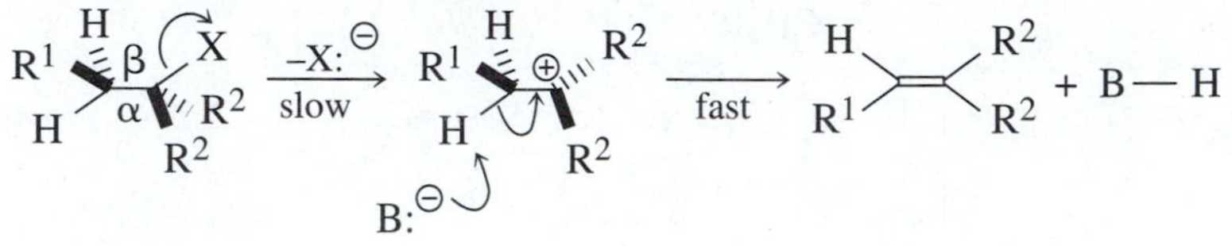

1		**5**
An alkyl halide	A carbocation	An alkene

$$\text{Rate} = k_1[\text{alkyl halide}] \qquad (10.6)$$

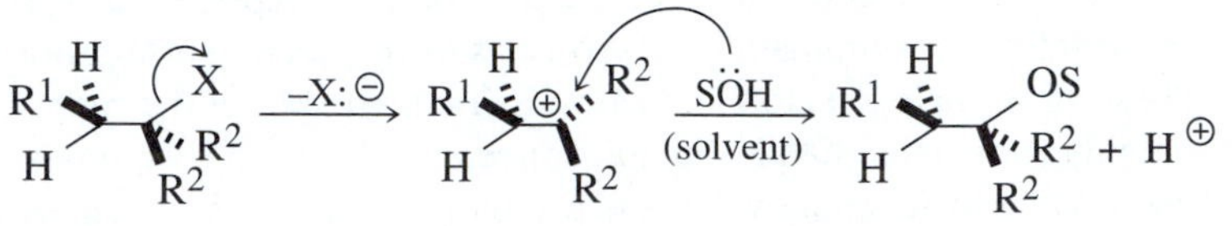

1

Comparing Equations (10.3) and (10.6) provides an important chemical insight. If E1 and E2 processes are proceeding simultaneously, as they do for secondary alkyl halides, the latter pathway will be favored by using an *excess* of strong base because the rate of an E1 reaction is *independent* of the concentration of base. Conversely, the *absence* of strong bases and the use of polar solvents that facilitate ionization of alkyl halides tend to favor elimination by an E1 mechanism.

Dehydrohalogenation may give a mixture of products if the halide leaving group is unsymmetrically located on the carbon skeleton. For example, 2-chloro-2-methylbutane (**6**, Eq. 10.8) yields both 2-methyl-2-butene (**7**) and 2-methyl-1-butene (**8**) on reaction with base. Elimination reactions are normally irreversible under the experimental conditions used, so the alkenes **7** and **8** are the products of two *competing* elimination reactions rather than of a single elimination followed by an equilibration that interconverts **7** and **8.** Consequently, the ratio of **7** and **8** is subject to the relative rates of two dehydrohalogenations and is therefore determined by the relative free energies of the two corresponding transition states, **kinetic control,** rather than by the relative free energies of the alkenes themselves, **equilibrium control** (see Sec. 13.1).

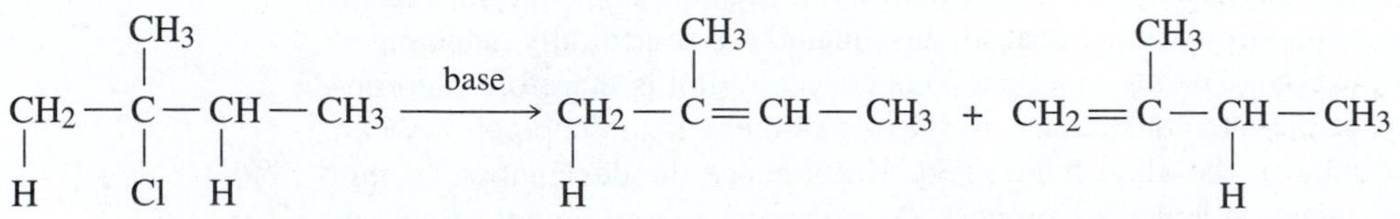

6	**7**	**8**
2-Chloro-2-methylbutane	2-Methyl-2-butene	2-Methyl-1-butene

The transition states for elimination of 2-chloro-2-methylbutane to give **7** and **8** are **9** and **10,** respectively. The dihedral angle formed by the H–C–C–Cl array is 180°, a relationship that is variously called *trans-coplanar* and *anti-periplanar*. It is from this geometry that E2 reactions occur most readily. If the dihedral angle between the β-hydrogen atom and the leaving group is much different from 180°, the free energy of activation for elimination, $\Delta G^{\ddagger}$, increases substantially.

$$\qquad\qquad \textbf{9} \qquad\qquad\qquad\qquad\qquad \textbf{10}$$

In the absence of complicating factors such as the steric factors discussed shortly, the predominant product in an E2 elimination is the *more highly substituted alkene.* An increase in the number of alkyl substituents on the double bond almost always increases the stability—lowers the free energy—of alkenes, so elimination occurs to give preferential formation of the more stable alkene. The relative free energies of the products **7** and **8** reflect the relative energies of the corresponding transition states **9** and **10** because those factors that stabilize the product alkenes also play a role in stabilizing the respective transition states in which partial double–bond character is developing between the two carbon atoms. Thus, the free energy of activation for forming the more stable alkene **7** is less than that for the less stable alkene **8,** making **7** the preferred product.

Relative free energies of transition states of competing elimination reactions may be influenced by steric factors, which increase the energies of some transition states relative to others. For instance, if steric factors along the reaction pathway to the more highly substituted alkene are more important than those for forming the less highly substituted one, the product distribution need *not* depend upon the relative stability of products; the less stable alkene may be the major product. Bulky substituents bound to the carbon atom from which the proton is removed cause greater hindrance to the approaching base than do substituents on the carbon atom bearing the leaving group. As more sterically demanding, bulky bases are used, abstraction of the more accessible proton will be further favored, thereby leading to the formation of increased amounts of the less substituted alkene. An illustrative example of these principles is depicted in Equation (10.9). In the transition state leading to the formation of **11,** there are unfavorable steric interactions between the bulky alkyl residue and the base as it approaches the internal methylene group; there is no such interaction when the base abstracts a proton from the methyl group to give **12.** Use of a highly hindered base may lead to the formation of **12** as the exclusive product.

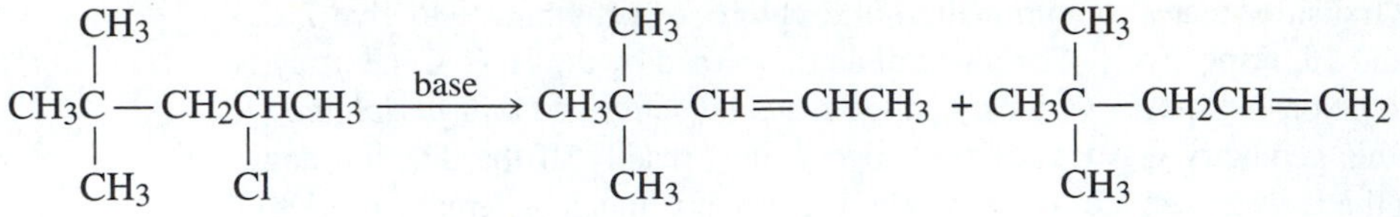

$$CH_3C(CH_3)(Cl)—CH_2CHCH_3 \xrightarrow{base} CH_3C(CH_3)—CH=CHCH_3 + CH_3C(CH_3)—CH_2CH=CH_2 \quad (10.9)$$

4-Chloro-2,2-dimethylpentane

11
4,4-Dimethyl-2-pentene
minor product

12
4,4-Dimethyl-1-pentene
major product

In the series of experiments that follows, the general techniques for performing base-promoted elimination reactions of 2-chloro-2-methylbutane, which may be prepared according to the procedure in Section 14.5, are demonstrated. The use of different bases to effect the elimination illustrates the steric effect as a parameter of the E2 reaction. To gain maximum benefit from these experiments, groups of students should be assigned different bases. The results may then be collected, averaged for each base, and presented to the class in order to define the trend in product ratio as a function of the size of the base used.

Pre-Lab exercises for Section 10.2, "Dehydrohalogenation of 2-Chloro-2-methylbutane," are found on page PL. 33.

EXPERIMENTAL PROCEDURES

SAFETY ALERT

1. Most of the chemicals used in this experiment are highly flammable, so do not handle these liquids near open flames.
2. Refer to Section 2.7 and the "Safety Alert" portion of Section 4.2.3 for precautions regarding simple distillations and the condition and handling of your glassware. Carefully follow the instructions for assembling the apparatus.
3. The solutions used in this experiment are *highly* caustic. *Do not allow them to come in contact with your skin.* If this should happen, flood the affected area with water and then thoroughly rinse the area with a solution of dilute acetic acid. Wear rubber gloves when preparing and transferring all solutions.
4. Sodium metal reacts violently with water with the formation, and possible explosive combustion, of hydrogen gas. Use only *dry* containers, forceps, and so on. Do not handle pieces of sodium metal with your bare fingers.
5. Avoid spilling solid sodium methoxide or potassium *tert*-butoxide while weighing these materials. They react readily with the moisture in air to form a strongly alkaline and corrosive residue. Clean up spillage with a water-soaked paper towel; then wash your hands.

A.   ELIMINATION WITH ALCOHOLIC POTASSIUM HYDROXIDE

In this procedure, a Hempel column serves two purposes. It first functions as a reflux condenser, and then acts as a fractional distillation column, its more usual function. This dual use minimizes glassware assembly.

Place 2.3 g of potassium hydroxide (commercial potassium hydroxide contains approximately 15% by weight of water) and 25 mL of absolute ethanol in a dry 50-mL round-bottom flask. Attach a calcium chloride drying tube (see Sec. 2.23) to the flask and warm the mixture in a hot-water or oil bath until the potassium hydroxide dissolves; magnetic stirring reduces the amount of time required for dissolution of the potassium hydroxide. Using an ice-water bath, cool the flask to room temperature and add 2.7 g of 2-chloro-2-methylbutane and a few boiling chips to the flask; if magnetic stirring is available do not add the boiling chips.

Continuation.   Equip the flask for fractional distillation by setting up an apparatus as shown in Figure 2.16. If you are using glassware equipped with ground-glass joints, be sure to lubricate the joint connecting the Hempel column to the flask with a hydrocarbon or silicone grease. Lubrication of these joints is *particularly* important in this experiment because the strong bases being used may cause the joints to freeze. In order to increase the efficiency of the Hempel column for fractionation, fill it with Raschig rings, coarsely broken glass tubing, coarse steel wool, or another packing material as directed by your instructor. Using a short piece of rubber tubing, fit the vacuum adapter holding a 10-mL receiving flask with a calcium chloride drying tube and immerse the receiving flask in an ice-water bath. Attach water hoses to the Hempel column and circulate water through the jacket of the column during the period of reflux for this reaction. Heat the reaction mixture under gentle reflux (see Sec. 2.19) with a heating mantle or an oil or hot-water bath (see Sec. 2.5) for a period of 2 h. This should be enough time to allow the reaction to go to about 95% completion. By the end of this period some solid should have precipitated, which may cause some bumping.✳

At the end of the reflux period, cool the flask containing the reaction mixture with an ice-water bath, turn off the cooling water, remove the water hoses from the Hempel column, and allow the water to drain from the jacket. Reconnect the water hoses to the condenser so that the apparatus is now set for fractional distillation. If any low-boiling distillate has condensed in the receiving flask during the reflux period, allow it to remain and continue to cool this flask in an ice-water bath. Distil the product mixture, collecting all distillate boiling below 45 °C; the reported boiling points (760 torr) of the alkenes are 2-methyl-1-butene, bp 31 °C, and 2-methyl-2-butene, bp 38 °C. Transfer the product to a preweighed sample bottle with a tight-fitting stopper or cap and determine the yield. Perform qualitative tests that will confirm the presence of alkenes in the distillate (see Sec. 23.5.2). Analyze your product by GLC or submit a sample of it for such analysis. After obtaining the results, calculate the relative percentages of the two isomeric alkenes formed (see Sec. 6.4). Typical GLC traces of the products from this elimination and from one in which potassium *tert*-butoxide was used as the base are shown in Figure 10.1(a). See Figure 10.4 for the ^{1}H NMR spectrum of the product mixture obtained from a representative experiment.

FIGURE 10.1 Typical GLC traces of the products of elimination of 2-chloro-2-methylbutane. Assignments: peak 1: 2-methyl-1-butene; peak 2: 2-methyl-2-butene. (a) Elimination with KOH, showing approximately 44% 2-methyl-1-butene. (b) Elimination with KOC(CH3)3, showing approximately 76% 2-methyl-1-butene. Analyses were performed at 45 °C on a 3-m column of 30% silicone gum rubber supported on Chromosorb P.

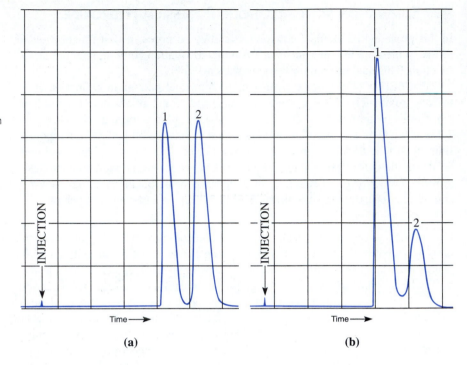

(a)   (b)

Finishing Touches   Neutralize the *residue remaining in the stillpot* with 10% aqueous hydrochloric acid; then flush it down the drain. Place the *carbon tetrachloride solution* from the bromide test for unsaturation in a container for halogenated organic liquids; put the *manganese dioxide* from the Baeyer test for unsaturation in a container for heavy metals.

B. ELIMINATION WITH SODIUM METHOXIDE

Prepare a solution of sodium methoxide in anhydrous methanol by either of the following methods. Owing to the hazards associated with handling sodium metal, we recommend Method 1 for larger classes.

Method 1 (Commercial Sodium Methoxide).   Dissolve 1.9 g of sodium methoxide powder in 25 mL of anhydrous methanol contained in a dry 50-mL round-bottom flask. Use a clean, dry spatula to transfer the powder, taking care to perform the weighing operation *as rapidly as possible* to minimize the reaction between the methoxide and atmospheric water vapor (Eq. 10.10). *The bottle from which the sodium methoxide is taken must be kept tightly closed* to ensure reproducible results.

$$NaOCH_3 + H_2O \longrightarrow CH_3OH + NaOH \qquad \textbf{(10.10)}$$

Method 2 (Sodium Metal).   Prepare a solution of 0.8 g of sodium metal in 25 mL of anhydrous methanol in the following way. In each of two 25-mL beakers place 10 mL of dry toluene and weigh one of the two. The amount of sodium metal required for the experiment must now be added to the weighed beaker. This metal is normally stored under mineral oil for protection from water and oxygen. Using a small dry knife or dry spatula, cut a piece of sodium metal about the size of a pea, stick the tip of the knife or spatula into the metal, and rinse off the mineral oil by swirling the metal in the unweighed beaker of toluene. Briefly blot the piece of sodium with a dry paper towel, and transfer the piece of metal to the weighed beaker. Repeat the sequence until the correct amount of sodium is contained in the weighed beaker.

Place 25 mL of anhydrous methanol in a dry 50-mL round-bottom flask fitted with a calcium chloride drying tube (see Sec. 2.23). Remove a piece of sodium from the weighed beaker with the aid of the knife or spatula used previously, briefly blotting it as before, and add it to the flask containing the methanol. *Caution: No flames in the vicinity; hydrogen gas is evolved!* Replace the drying tube on the flask and wait until the initially vigorous gas evolution subsides. Then continue adding the sodium *one piece at a time* until all of it has been added. Allow the metal to react completely before continuing.

Complete the experiment as follows: Cool the methanolic sodium methoxide solution to room temperature using an ice-water bath and add to this solution 2.7 g of 2-chloro-2-methylbutane and a few boiling stones or a magnetic stirbar. Follow the directions in the paragraph headed "Continuation" in Part A.

Finishing Touches   Neutralize the *residue remaining in the stillpot* with 10% aqueous hydrochloric acid; then flush it down the drain. Place the *carbon tetrachloride solution* from the bromine test for unsaturation in a container for halogenated organic liquids; put the *manganese dioxide* from the Baeyer test for unsaturation in a container for heavy metals.

C.   ELIMINATION WITH POTASSIUM *TERT*-BUTOXIDE

Place 3.9 g of solid potassium *tert*-butoxide and 25 mL of anhydrous *tert*-butyl alcohol in a dry 50-mL round-bottom flask. The precautions for handling potassium *tert*-butoxide are the same as those described for sodium methoxide in Part B. Attach a calcium chloride drying tube (see Sec. 2.23) and warm the flask with a water bath to assist dissolution of the base. If all of the solid does not dissolve in 5 min, continue the experiment with the heterogeneous mixture. Cool the flask to room temperature with an ice-water bath, and add 2.7 g of 2-chloro-2-methylbutane and a few boiling chips or a magnetic stirbar to the flask. Complete the experiment from this point by following the directions in the paragraph headed "Continuation" in Part A. See Figure 10.1(b) for a typical GLC tracing and Figure 10.5(a) for a typical ^{1}H NMR spectrum of the mixture of products of this elimination.

> **Finishing Touches**   Neutralize the *residue remaining in the stillpot* with 10% aqueous hydrochloric acid; then flush it down the drain. Place the *carbon tetrachloride solution* from the bromide test for unsaturation in a container for halogenated organic liquids; put the *manganese dioxide* from the Baeyer test for unsaturation in a container for heavy metals.

EXERCISES

1. What are the expected results of the alkoxide-promoted eliminations if the alcohols used as solvents contain water?

2. What is the solid material that precipitates as the eliminations proceed?

3. Why does the excess of base used in these eliminations favor the E2 over the E1 mechanism for elimination?

4. If all the elimination reactions in the experimental section had proceeded by the E1 mechanism, would the results have been different from those actually obtained? Why?

5. From the results of the experiments that were performed and/or from the data in Figure 10.1, what conclusions can be drawn concerning the effect of relative base size upon product distribution?

6. What differences in product distributions would be expected for the eliminations of 2-chloro-2-methylbutane and 2-chloro-2,3-dimethylbutane with excess methanolic sodium methoxide?

7. If the leaving group in the 2-methyl-2-butyl system were larger than a methyl group, why would 2-methyl-1-butene be expected to be formed in greater amounts than if the leaving group were smaller than methyl, regardless of which base is used? Use "sawhorse" structural formulas to support your explanation.

8. What is the reason for packing the Hempel column with Raschig rings or other packing materials in these experiments?

9. Referring to Figures 10.4, 10.5, 10.7 and 10.9, calculate the percentage compositions of the mixtures of isomeric methylbutenes obtained from the reaction of 2-chloro-2-methylbutane with potassium hydroxide and potassium *tert*-butoxide. Note that Figures 10.7(a) and 10.9(a) are ^{1}H NMR spectra of the pure alkenes and Figures 10.4 and 10.5 are ^{1}H NMR spectra of the mixtures. In the latter two spectra, the integration of the resonances in the region of 5.0 ppm (δ) has been electronically amplified in the upper stepped line so that the relative areas of the two multiplets in that region can be more accurately measured.

10. Commercial potassium hydroxide contains approximately 15% by weight of water. Verify that to obtain 35 mmol of potassium hydroxide, 2.3 g of the commercial material must be used.

SPECTRA OF STARTING MATERIALS AND PRODUCTS

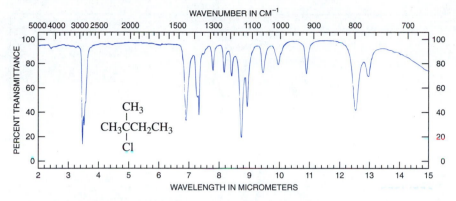

FIGURE 10.2 IR spectrum of 2-chloro-2-methylbutane.

(a) ^{1}H NMR spectrum (60 MHz).

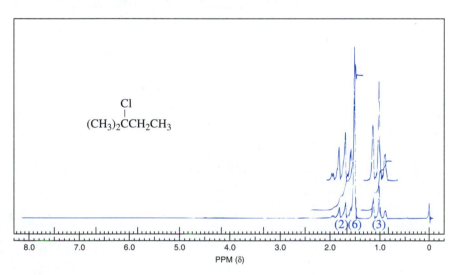

(b) ^{13}C NMR data.
   Chemical shifts: δ 9.5, 32.0, 39.0, 70.9.

FIGURE 10.3 NMR data for 2-chloro-2-methylbutane.

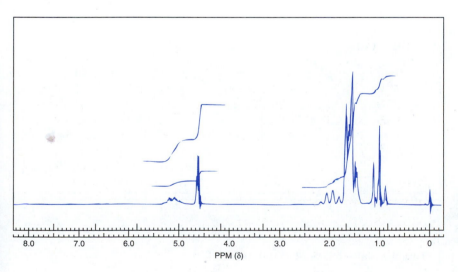

FIGURE 10.4 ^{1}H NMR spectrum (60 MHz) of the product mixture from the elimination of 2-chloro-2-methylbutane with potassium hydroxide.

FIGURE 10.5 ¹H NMR spectrum (60 MHz) of the product mixture from the elimination of 2-chloro-2-methylbutane with potassium *tert*-butoxide.

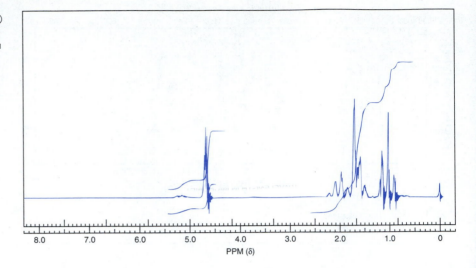

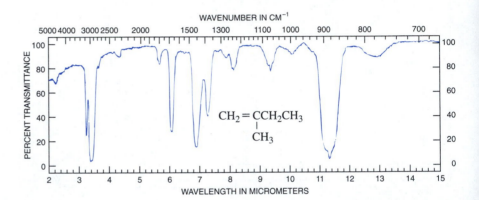

FIGURE 10.6 IR spectrum of 2-methyl-1-butene.

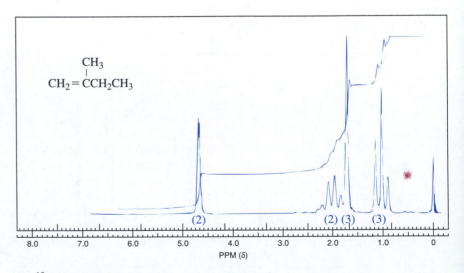

FIGURE 10.7 NMR data for 2-methyl-1-butene.

(b) ¹³C NMR data.
Chemical shifts: δ 12.5, 22.3, 31.0, 108.8, 147.5.

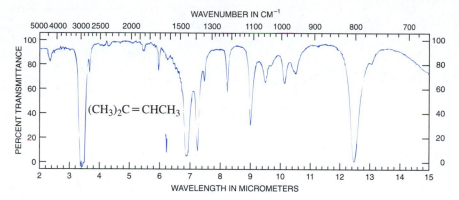

FIGURE 10.8 IR spectrum of 2-methyl-2-butene.

(CH₃)₂C=CHCH₃

(a) ¹H NMR spectrum (60 MHz).

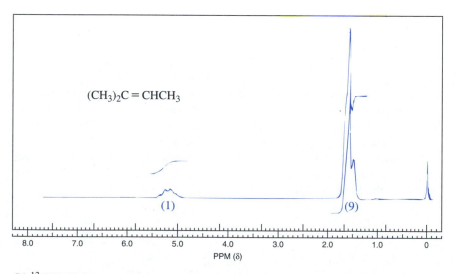

(CH₃)₂C=CHCH₃

(1)   (9)

(b) ¹³C NMR data.
   Chemical shifts: δ 13.4, 17.3, 25.6, 118.8, 132.0.

FIGURE 10.9 NMR data for 2-methyl-2-butene.

10.3 DEHYDRATION OF ALCOHOLS

The nonbonding lone pairs of electrons on the oxygen atom of the hydroxyl group, which is the functional group of alcohols, serve as a Lewis base and may complex with Brønsted and Lewis acids, which are defined as proton-donating and electron pair-accepting acids, respectively. For example, the protonation of an alcohol **13** is rapid and reversible and produces an oxonium ion **14** in which the positively charged oxygen atom further polarizes the carbon-oxygen single bond. In contrast to hydroxide ion, which is a strong base and a poor leaving group, water is a weak base and an excellent leaving group, and **14** may then undergo facile elimination and substitution reactions upon reaction with a base or nucleophile **Nu:⁻** as shown in Equation 10.11.

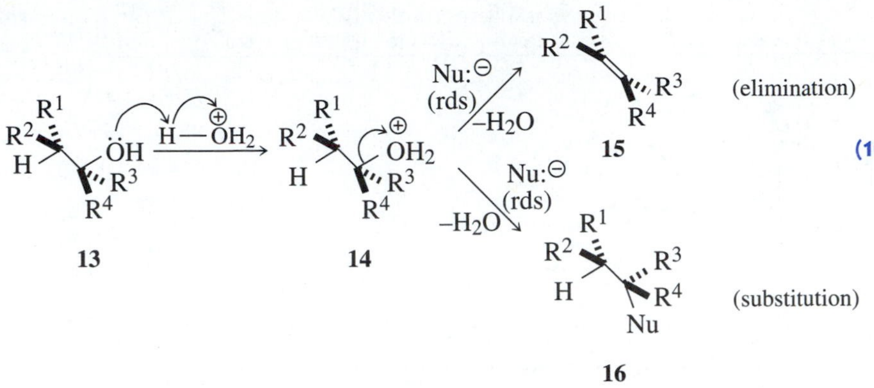

(10.11)

The **dehydration** of **14** to produce an alkene **15** by loss of water may occur by the two mechanistic pathways shown in Equations (10.12) and (10.13). With secondary and tertiary alcohols, the oxonium ion **14** undergoes **endothermic** heterolysis via the transition state **14‡** to produce a carbocation **17** and water. Rapid transfer of a proton from an adjacent carbon atom to a weak base such as the alcohol or water then gives the alkene **15** (Eq. 10.12). The fragmentation of **14** to give **17** is kinetically a **first-order** process and is the **rate-determining step** (rds) of the overall reaction which is classified as **E1.** Since the formation of the carbocation **17** is an endothermic process, the transition state **14‡** is productlike, extensive heterolysis of the carbon-oxygen bond with simultaneous development of positive charge on the carbon atom has already occurred. Thus, the ease of dehydration of alcohols parallels the relative stabilities of the intermediate carbocations following the familiar order $3° > 2° \gg 1°$.

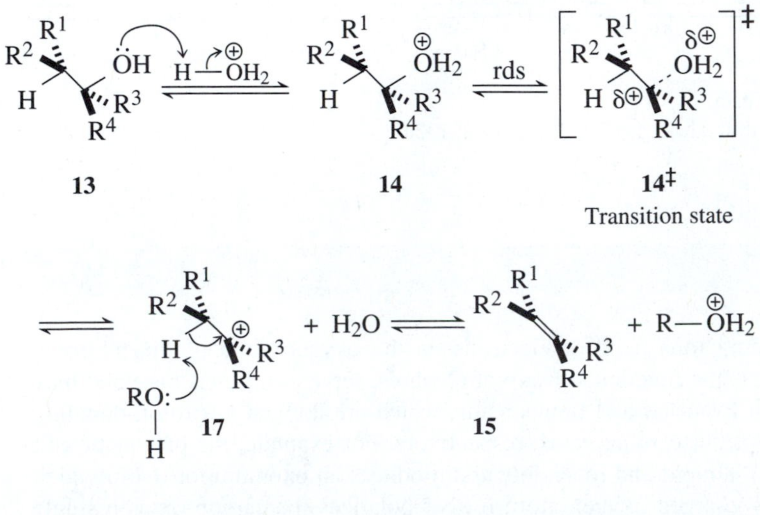

(10.12)

The energy of a primary carbocation is so high that an alternate, E2-type mechanism for elimination applies to the acid-catalyzed dehydration of primary alcohols, as shown in Equation (10.13). Here the formation of the primary oxonium ion **18** is followed by removal of a proton from the β-carbon atom by a molecule of water or alcohol, which serves as the Lewis base B: (Eq. 10.13). The reaction is kinetically a **second-order** process.

$$R^2 \overset{R^1}{\underset{H}{\overset{\cdots}{\bigvee}}} \overset{\ddot{O}H}{\underset{H}{\overset{H}{\bigvee}}} H \overset{\oplus}{\longrightarrow} \overset{R^1}{\underset{H}{\overset{\cdots}{\bigvee}}} \overset{\oplus}{\underset{H}{\overset{\cdots}{OH_2}}} \overset{rds}{\longrightarrow} R^2 \overset{R^1}{\underset{CH_2}{\overset{\cdots}{\bigvee}}} + H_2O + R \overset{\oplus}{-} \overset{\oplus}{OH_2} \quad \textbf{(10.13)}$$

RO:
|
H

18

As illustrated in Equations (10.12) and (10.13), each of the steps along the reaction pathway is reversible, so the desired alkene may also undergo acid-catalyzed hydration to form an alcohol. In practice, reversal of the dehydration may be avoided by removing the alkene, whose boiling point is always lower than the parent alcohol, from the reaction mixture by distillation. This technique shifts the equilibrium to the right and maximizes the yield of alkene. Shifting an equilibrium of a reversible reaction to the right by removing one of the products as it is formed follows from the **LeChatelier principle.**

As shown previously in Equation (10.11), elimination and substitution reactions are frequently competing pathways, but the relative importance of each pathway may often be controlled through proper choice of reaction conditions and reagents. In the case of the acid-catalyzed dehydration of secondary and tertiary alcohols, these competing reactions involve combining the intermediate carbocation with the **conjugate base** of the acid catalyst or with some other nucleophile such as solvent. For example, if hydrochloric acid were chosen as the catalyst for dehydration, the chloride ion could trap the carbocation to give an alkyl chloride (Eq. 10.14), a reaction that is used experimentally (see Sec. 14.5). This side reaction thus not only produces an undesired by-product, it also consumes the acid catalyst. Consequently, the desired dehydration process is terminated because formation of the alkyl chloride is irreversible under the conditions normally used for dehydration.

$$R \overset{R}{\underset{R}{\overset{|}{-}C-}} OH \xrightarrow[-H_2O]{HCl} R \overset{R}{\underset{R}{\overset{|}{-}C}} \oplus \longrightarrow R \overset{R}{\underset{R}{\overset{|}{-}C-}} Cl \quad \textbf{(10.14)}$$

Cl:$^{\ominus}$

The side reaction depicted in Equation (10.14) is avoided when the formation of the substitution product is reversible. For this reason, the catalysts recommended to effect the dehydration of alcohols in the laboratory are sulfuric or phosphoric acid, since the intermediate alkyl bisulfate (Nu:$^-$ = $^-OSO_3H$) or dihydrogen phosphate (Nu:$^-$ = $^-OPO_3H_2$) ions readily reionize to the intermediate carbocation under the reaction conditions (Eq. 10.15). Subsequent loss of a proton then gives the desired alkene, the removal of which by distillation drives the equilibrium.

$$R^1-\underset{\underset{\textstyle H}{|}}{\overset{\overset{\textstyle R^2}{|}}{C}}-\underset{}{\overset{\overset{\textstyle R^3}{|}}{CH}}-OH \underset{-H_2O}{\overset{H^{\oplus}}{\rightleftharpoons}} R^1-\underset{\underset{\textstyle H}{|}}{\overset{\overset{\textstyle R^2}{|}}{C}}-\underset{\oplus}{\overset{\overset{\textstyle R^3}{|}}{CH}} \underset{}{\overset{-H^{\oplus}}{\rightleftharpoons}} \underset{R^1}{\overset{R^2}{>}}C=CHR^3$$

Elimination
product

(10.15)

$$\Big\uparrow Nu:^{\ominus}$$

$$R^1-\underset{\underset{\textstyle H}{|}}{\overset{\overset{\textstyle R^2}{|}}{C}}-\underset{\underset{\textstyle Nu}{|}}{\overset{\overset{\textstyle R^3}{|}}{CH}}$$

Substitution
product

Two or more isomeric alkenes may be formed in E1 dehydration reactions. For example, loss of chemically nonequivalent protons from an unsymmetrical carbocation will lead to isomeric alkenes in which the carbon-carbon double bond is located between different carbon atoms; mixtures of geometric isomers (i.e., *E*- and *Z*-) may also be obtained (Eq. 10.16). The relative amounts of **21** and **22** that will be produced upon dehydration of **19** depend upon the relative free energies of the transition states for the loss of the respective protons H_A and H_B from the intermediate carbocation **20**. In the two transition states for the loss of H_A and H_B, the C–H bond is partially broken and the C–C double bond is

$$R^1-\underset{\underset{\textstyle H_A}{|}}{\overset{\overset{\textstyle H}{|}}{C}}-\underset{}{\overset{\overset{\textstyle OH}{|}}{CH}}-\underset{\underset{\textstyle H_B}{|}}{\overset{\overset{\textstyle R^2}{|}}{C}}-R^3 \underset{-H_2O}{\overset{H^{\oplus}}{\longrightarrow}} R^1-\underset{\underset{\textstyle H_A}{|}}{\overset{\overset{\textstyle H}{|}}{C}}-\overset{\oplus}{CH}-\underset{\underset{\textstyle H_B}{|}}{\overset{\overset{\textstyle R^2}{|}}{C}}-R^3$$

19   **20**

$-H_A^{\oplus}$ ↗

$-H_B^{\oplus}$ ↘

$$R^1-\overset{\overset{\textstyle H}{|}}{C}=CH-\underset{\underset{\textstyle H_B}{|}}{\overset{\overset{\textstyle R^2}{|}}{C}}-R^3$$

21

(10.16)

$$R^1-\underset{\underset{\textstyle H_A}{|}}{\overset{\overset{\textstyle H}{|}}{C}}-CH=\overset{\overset{\textstyle R^2}{|}}{C}-R^3$$

22

partially formed, as depicted in **23** and **24**. Since the major difference between these two transition states is the substitution on the incipient carbon-carbon double bond, their relative energies parallel the relative stabilities of the product alkenes. Consequently the lower energy transition state is the one leading to the more highly substituted, and hence more stable alkene, which in this case is **22**. Furthermore, for steric reasons the *E*-isomer of the more stable alkene should be the major product.

$$\left[\begin{array}{c} H \\ | \\ R^1 - \overset{\delta\oplus}{C} = CH - \overset{R^2}{\underset{|}{C}} - R^3 \\ \delta\oplus \overset{|}{H_A} \quad H_B \end{array} \right]^{\ddagger} \left[\begin{array}{c} H \\ | \\ R^1 - C - CH = \overset{R^2}{\underset{|}{C}} - R^3 \\ \overset{|}{H_A} \quad \overset{||}{\underset{\delta\oplus}{H_B}} \end{array} \right]^{\ddagger}$$

23   24

The distribution of olefinic products from the acid-catalyzed dehydration of alcohols cannot always be predicted based upon the orientational factors just described. Rather, the intermediate carbocation is prone to rearrangement by the migration of either a hydride, $H:^-$, or an alkyl group, $R:^-$, from a carbon atom adjacent to the cationic center if a carbocation of similar or greater stability is thereby produced. The subsequent loss of a proton from this rearranged carbocation according to the orientation effects discussed previously then leads to the major product.

The acid-catalyzed dehydration of 4-methyl-2-pentanol (25), which is one of the experiments in this section, is shown in Equation (10.17) and nicely illustrates the preceding principles. Dehydration of 25 produces a complex mixture of isomeric alkenes, including 4-methyl-1-pentene (26), *trans*-4-methyl-2-pentene (27),

$$\underset{25}{(CH_3)_2CHCH_2\overset{\overset{\displaystyle OH}{|}}{C}H-CH_3} \xrightarrow[\Delta]{H^{\oplus}} \underset{26}{(CH_3)_2CHCH_2CH=CH_2} + \underset{27}{\underset{H}{\overset{(CH_3)_2CH}{\diagdown}}C=C\underset{CH_3}{\overset{H}{\diagup}}} \qquad (10.17)$$

$$+ \quad \underset{28}{\underset{H}{\overset{(CH_3)_2CH}{\diagdown}}C=C\underset{H}{\overset{CH_3}{\diagup}}} \quad + \quad \underset{29}{\underset{H_3C}{\overset{H_3C}{\diagdown}}C=C\underset{H}{\overset{CH_2CH_3}{\diagup}}} \quad + \quad \underset{30}{\underset{H}{\overset{H}{\diagdown}}C=C\underset{CH_2CH_2CH_3}{\overset{CH_3}{\diagup}}}$$

cis-4-methyl-2-pentene (28), 2-methyl-2-pentene (29), and 2-methyl-1-pentene (30). A pathway to form 26–28 is by way of deprotonation of the intermediate carbocation 31. However, producing 29 and 30 requires intervention of carbocations 32 and 33. Possible mechanisms for forming these ions are the subject of Exercise 10.

$$\underset{31}{(CH_3)_2CH - CH_2 - \overset{\oplus}{C}H - CH_3} \qquad \underset{32}{(CH_3)_2CH - \overset{\oplus}{C}H - CH_2 - CH_3} \qquad \underset{33}{(CH_3)_2\overset{\oplus}{C} - CH_2 - CH_2 - CH_3}$$

Cyclohexanol undergoes acid-catalyzed dehydration without rearrangement to yield a single product, cyclohexene (Eq. 10.18). This dehydration reaction may be accompanied by some acid-catalyzed polymerization of the cyclohexene.

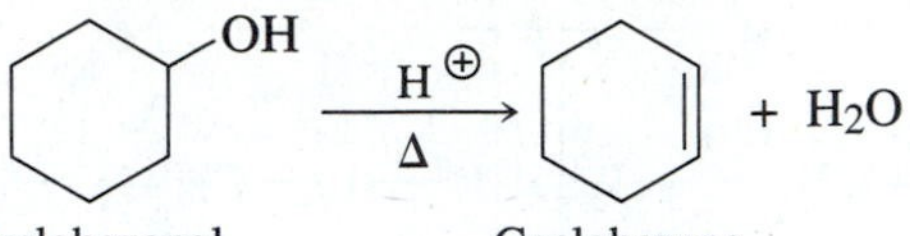

$$ \text{Cyclohexanol} \xrightarrow[\Delta]{H^{\oplus}} \text{Cyclohexene} + H_2O \qquad \textbf{(10.18)} $$

Cyclohexanol   Cyclohexene

Pre-Lab exercises for Section 10.3, "Dehydration of 4-Methyl-2-pentanol and Cyclohexanol," are found on page PL. 35.

EXPERIMENTAL PROCEDURES

SAFETY ALERT

1. The majority of materials, particularly the product alkenes, that will be handled during this experiment are highly flammable.
2. Refer to Sections 2.7 and 4.2.3 for precautions regarding simple distillations. Pay particular attention to those concerning the assembly and condition of your glassware.
3. Several experimental operations require pouring, transferring, and weighing chemicals and reagents that cause burns on contact with your skin. Wear rubber gloves when handling the strongly acidic catalysts and during the work-up and washing steps in which a separatory funnel is used. Should acidic solutions accidentally come in contact with your skin, immediately flood the affected area with water, then wash it with 5% sodium bicarbonate solution.

A. DEHYDRATION OF 4-METHYL-2-PENTANOL

Place 5.1 g of 4-methyl-2-pentanol in a 25-mL round-bottom flask, and then add 2.5 mL of 9 *M* sulfuric acid. Thoroughly mix the contents by swirling the flask gently. Add two or three boiling chips, and assemble the flask for fractional distillation according to Figure 2.16. The receiving flask should be immersed in an ice-water bath. Heat the reaction flask with a heating mantle or an oil bath (*no flames*) and collect all distillates while maintaining the head temperature below 90 °C. If the reaction mixture is not heated too strongly, the head temperature will stay below 80 °C for most of the reaction. When about 2.5 mL of liquid remains in the reaction flask, discontinue heating. Transfer the distillate to a dry 25-mL Erlenmeyer flask and add 1–2 g of anhydrous potassium carbonate to neutralize any acid and to dry the distillate (see Sec. 2.21).✱ Occasionally swirl the mixture during a period of 10–15 min.

Transfer all of the dried organic mixture into a dry 10-mL distilling flask by decanting or by using a Pasteur pipet. Add boiling stones and distil (*no flames*) the mixture through a predried simple distillation apparatus (see Secs. 2.7 and 2.24 and Fig. 2.15b). The expected products of the reaction and their boiling points (760 torr) are 4-methyl-1-pentene (53.9 °C), *cis*-4-methyl-2-pentene

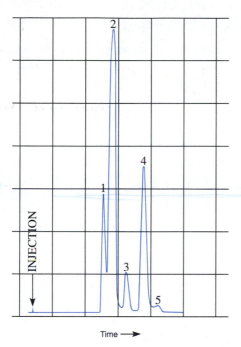

Time ⟶

FIGURE 10.10 GLC trace of the product mixture from dehydration of 4-methyl-2-pentanol. The peaks have been assigned as follows: (1) 4-methyl-1-pentene (14.4%), (2) *cis-* and *trans-*4-methyl-2-pentene (combined 52.1%), (3) 2-methyl-1-pentene (6.5%), (4) 2-methyl-2-pentene (26.9%), and (5) unidentified. This analysis was made with a 2.4-m column packed with 5% SF-96 on 60/80 Chromosorb W at 35 °C.

(56.4 °C), *trans-*4-methyl-2-pentene (58.6 °C), 2-methyl-1-pentene (61 °C), and 2-methyl-2-pentene (67.3 °C). Collect the fraction boiling between 53 and 69 °C in a preweighed, dry 10-mL receiving flask or dry sample bottle.✳ If a bottle is used, its top should be flush with the bottom of the ground-glass joint of the vacuum adapter during the distillation. Some product may be lost because an ice-water bath is not being used to cool the receiver, but such cooling tends to cause condensation of atmospheric water vapor with the alkenes. This problem can be avoided if a 10-mL round-bottom receiving flask is used and a calcium chloride drying tube is attached to the nipple of the vacuum adapter. Determine the weight of the distillate, and calculate the yield of alkenes obtained.

Test the distillate for unsaturation, using the bromine in carbon tetrachloride and Baeyer tests (see Sec. 23.5.2). Either analyze your product by GLC (see Sec. 6.4) or submit a sample of it to your instructor for such analysis. Figure 10.10 shows a typical GLC tracing of the products of this reaction. The retention times of authentic samples of the isomeric methylpentenes may be correlated with peaks in the gas chromatogram of the mixture. Measure the areas under the peaks, and calculate the percentage composition of your own product mixture.

Finishing Touches   Dilute the *residue remaining in the stillpot* with water, carefully neutralize it with sodium carbonate, flush it down the drain with large quantities of water. Dry the *potassium carbonate* on a tray in the hood and flush it down the drain or place it in a container for nonhazardous solids after drying it on a tray in the hood. Pour the *carbon tetrachloride solution* from the bromine test for unsaturation in a container for halogenated organic liquids, and put the *manganese dioxide* from the Baeyer test for unsaturation in a container for heavy metals.

B. DEHYDRATION OF CYCLOHEXANOL

Place 5.0 g of cyclohexanol in a 25-mL round-bottom flask, and add 2.5 mL of 9 M sulfuric acid. Thoroughly mix the contents by swirling the flask gently. Add two or three boiling chips, and assemble the flask for fractional distillation according to Figure 2.16. The receiving flask should be immersed in an ice-water bath. Heat the reaction flask with a heating mantle or an oil bath (*no flames*), and collect *all* distillates while maintaining the head temperature below 90 °C (760 torr). If the reaction mixture is not heated too strongly, the head temperature will remain below about 60–70 °C for most of the period of reaction. When about 2.5 mL of liquid remains in the reaction flask, discontinue heating. Transfer the distillate to a dry 25-mL Erlenmeyer flask, and add 1–2 g of anhydrous potassium carbonate to neutralize any acid and to dry the distillate (see Sec. 2.21).✱ Occasionally swirl the mixture for 15–20 min.

Transfer all of the dried organic mixture into a dry 10-mL distilling flask by decanting or using a Pasteur pipet. Because water and cyclohexene form a minimum-boiling azeotrope (see Sec. 4.2.4), the product must be dry at this stage for pure cyclohexene to be obtained. Add boiling stones and distil (*no flames*) the mixture through a predried simple distillation apparatus (see Secs. 2.7 and 2.24 and Fig. 2.15b). Collect the fraction boiling between 80 and 85 °C (760 torr) in a preweighed dry 10-mL receiving flask or dry sample bottle.✱ If the bottle is used, its top should be flush with the bottom of the ground-glass joint of the vacuum adapter during the distillation. Some product may be lost because an ice-water bath is not being used to cool the receiver, but such cooling tends to cause condensation of atmospheric water vapor along with the alkenes. This problem can be avoided if a 10-mL round-bottom receiving flask is used and a calcium chloride drying tube is attached to the nipple of the vacuum adapter.

Weigh the distillate, and assuming that it is pure cyclohexene, calculate the isolated yield of cyclohexene. Test the distillate for unsaturation, using both bromine in carbon tetrachloride and the Baeyer tests (see Sec. 23.5.2). Obtain IR and ^{1}H NMR spectra of your product and compare your spectra with those of an authentic sample (Figs. 10.25 and 10.26). Analyze your distillate by GLC and adjust your calculation of the yield of cyclohexene produced using the results of this analysis.

> **Finishing Touches** Dilute the *residue remaining in the stillpot* with water, carefully neutralize it with sodium carbonate, and flush it down the drain with large quantities of water. Dry the *potassium carbonate* on a tray in the hood and flush it down the drain or place it in a container for non-hazardous solids. Pour the *carbon tetrachloride solution* from the bromine test for unsaturation in a container for halogenated organic liquids, and put the *manganese dioxide* from the Baeyer test for unsaturation in a container for heavy metals.

EXERCISES

General Questions for Parts A and B

1. Why is the boiling point of the parent alcohol higher than that of the product alkene?

2. In principle the equilibrium in the dehydration of an alcohol could be shifted to the right by removal of water. Why is this tactic not a good option for the dehydration of 4-methyl-2-pentanol and cyclohexanol?

3. Why are the distillates obtained in the initial step of the dehydration reaction dried over anhydrous potassium carbonate?

4. Why is the head temperature kept below 90 °C in the initial step of the dehydration reaction?

5. What would be the consequence of distilling the slurry of alkenes and the drying agent? In other words, why is the organic solution separated from the drying agent prior to the final distillation?

6. Which of the following primary alcohols would be most likely to dehydrate by the E1 mechanism? By the E2 mechanism? Explain your reasoning.

$$CH_3CH_2OH \qquad\qquad (CH_3)_3CCH_2OH$$

7. Give structures for the products of dehydration of each of the following alcohols. For each, order the products with respect to preference of formation.

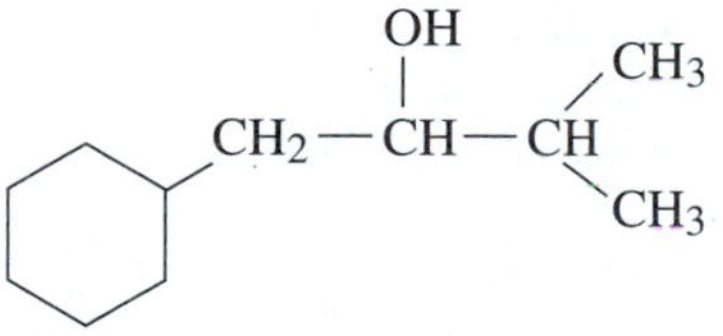

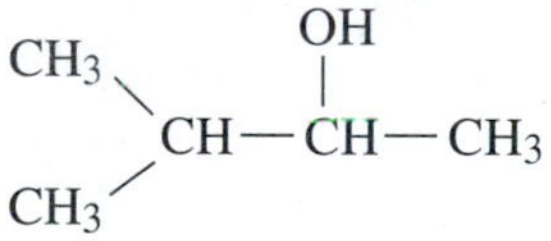

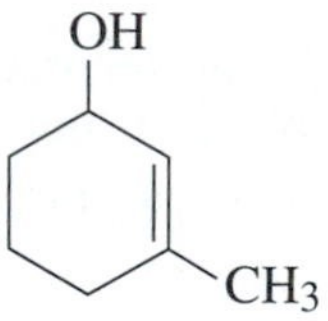

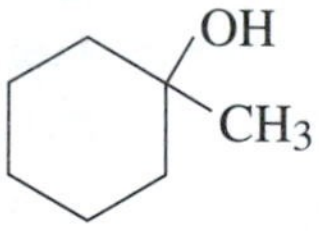

8. The loss of a proton from suitably substituted carbocations can provide both the *E*- and *Z*-isomers of the resulting alkene, but the *E*-isomer normally predominates. For example, deprotonation at C-3 of the 2-pentyl carbocation produces mainly *E*-2-pentene. By analyzing the relative energies of the conformational isomers of the carbocation that lead to the two isomeric 2-pentenes, explain why the *E*-isomer is formed preferentially. This analysis is aided by the use of Newman projections based on the partial structure shown.

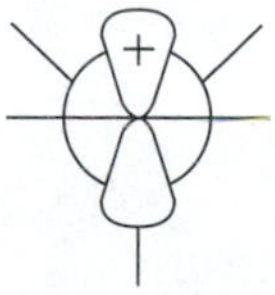

Newman
projection

Questions for Part A

9. Give a detailed mechanism explaining formation of **26** and **27** from the dehydration of 4-methyl-2-pentanol. Use curved arrows to symbolize the flow of electrons.

10. A mechanism for formation of **28–30** can be developed from rearrangement of the carbocation **31** by way of hydride shifts.

    a. Provide a mechanism by which **28** and **29** may be formed from **31** through a hydride shift followed by deprotonation.

    b. Show how your mechanism in part (a) can be elaborated to provide for formation of **30.**

    c. An alternative pathway for formation of the alkenes **28** and **29** involved addition of a proton to **26** or **27** and deprotonation of the carbocation that results. Write out this mechanism, using curved arrows to symbolize the flow of electrons.

    d. How might you experimentally distinguish the two types of mechanisms for the formation of **28** and **29?**

11. Near the end of the dehydration of 4-methyl-2-pentanol, a white solid may precipitate from the reaction mixture. What is the solid likely to be?

12. Give the structure, including stereochemistry, of the product of adding bromine to *cis*-4-methyl-2-pentene and that of adding *trans*-4-methyl-2-pentene (see Sec. 10.4). Relative to one another, are these products identical, enantiomeric, or diastereomeric (see Sec. 7.1)? Explain your reasoning.

Questions for Part B

13. Define a "minimum-boiling azeotrope."

14. Give the structure, including stereochemistry, of the product of addition of bromine to cyclohexene (see Sec. 10.4). Should it be possible, at least in principle, to resolve this dibromide into separate enantiomers? Explain your answer.

SPECTRA OF STARTING MATERIALS AND PRODUCTS

The ^{1}H NMR spectrum of 4-methyl-1-pentene is given in Figure 8.30.

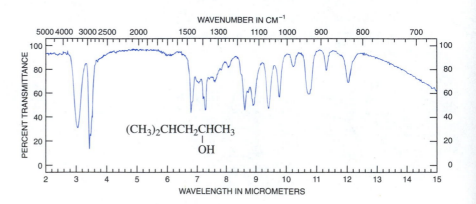

FIGURE 10.11 IR spectrum of 4-methyl-2-pentanol.

(a) ^{1}H NMR spectrum (60 MHz).

FIGURE 10.12 NMR data for 4-methyl-2-pentanol.

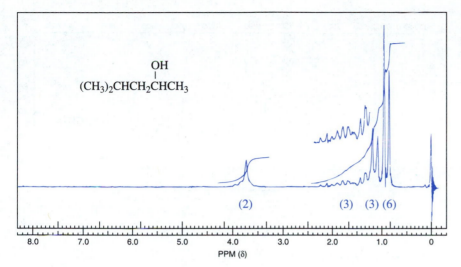

OH
|
(CH$_3$)$_2$CHCH$_2$CHCH$_3$

(2)   (3) (3) (6)

PPM (δ)

(b) ^{13}C NMR data.
Chemical shifts: δ 22.5, 23.2, 23.9, 24.9, 48.8, 65.8.

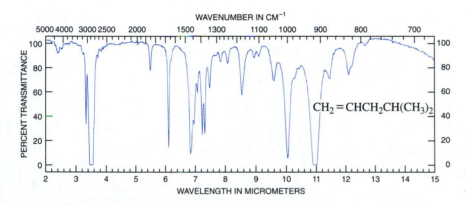

WAVENUMBER IN CM^{-1}

PERCENT TRANSMITTANCE

CH$_2$=CHCH$_2$CH(CH$_3$)$_2$

WAVELENGTH IN MICROMETERS

FIGURE 10.13 IR spectrum of 4-methyl-1-pentene.

(a) ^{1}H NMR spectrum (60 MHz).

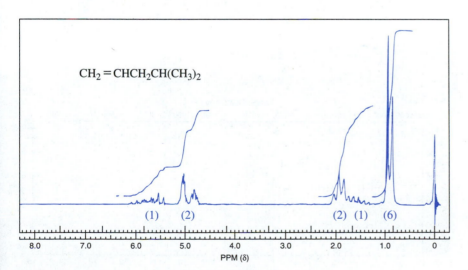

CH$_2$=CHCH$_2$CH(CH$_3$)$_2$

(1)   (2)   (2) (1)   (6)

PPM (δ)

(b) ^{13}C NMR data.
Chemical shifts: δ 22.3, 28.1, 43.7, 115.5, 137.8.

FIGURE 10.14 NMR data for 4-methyl-1-pentene.

FIGURE 10.15 IR spectrum of *trans*-4-methyl-2-pentene.

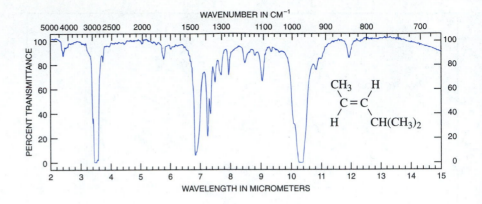

(a) ¹H NMR spectrum (60 MHz).

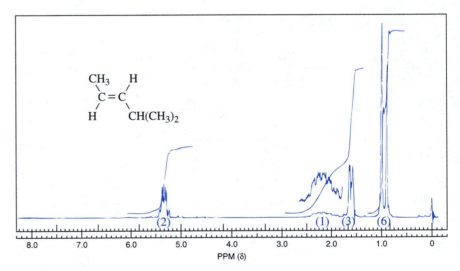

FIGURE 10.16 NMR data for *trans*-4-methyl-2-pentene.

(b) ¹³C NMR data.
   Chemical shifts: δ 17.6, 22.7, 31.5, 121.6, 139.4.

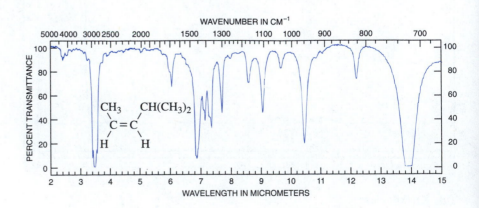

FIGURE 10.17 IR spectrum of *cis*-4-methyl-2-pentene.

(a) ^{1}H NMR spectrum (60 MHz).

FIGURE 10.18 NMR data for *cis*-4-methyl-2-pentene.

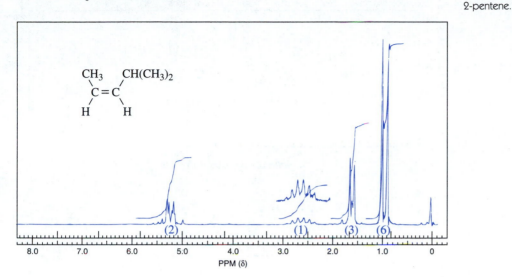

(b) ^{13}C NMR data.
Chemical shifts: δ 12.7, 23.1, 26.4, 121.4, 138.6.

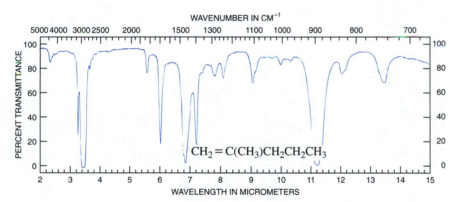

FIGURE 10.19 IR spectrum of 2-methyl-1-pentene.

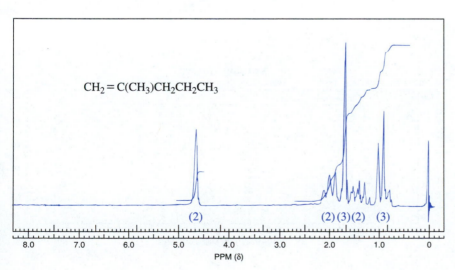

FIGURE 10.20 NMR data for 2-methyl-1-pentene.

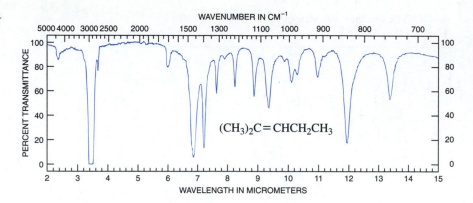

$(CH_3)_2C=CHCH_2CH_3$

(a) ^1H NMR spectrum (60 MHz).

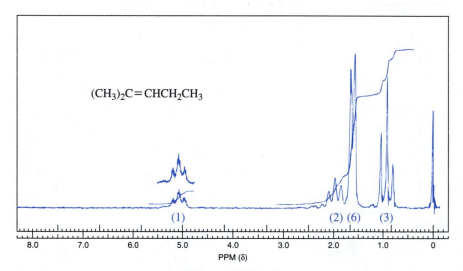

$(CH_3)_2C=CHCH_2CH_3$

(1)   (2) (6)   (3)

PPM (δ)

(b) ^{13}C NMR data.
Chemical shifts: δ 14.5, 17.5, 21.6, 25.7, 126.9, 130.6.

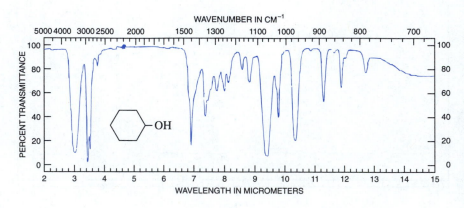

—OH

(a) ^{1}H NMR spectrum (60 MHz).

FIGURE 10.24 NMR data for cyclohexanol.

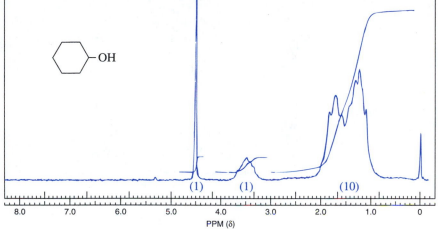

(b) ^{13}C NMR data.
Chemical shifts: δ 24.5, 25.9, 35.5, 70.1.

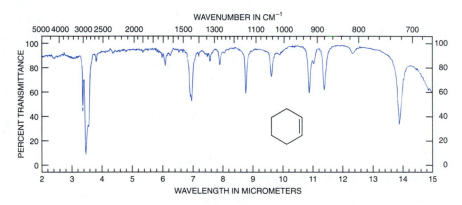

FIGURE 10.25 IR spectrum of cyclohexene.

FIGURE 10.26   NMR data for cyclohexene.

(a) ^{1}H NMR spectrum (60 MHz).

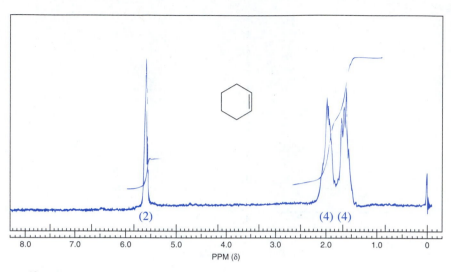

(b) ^{13}C NMR data.
   Chemical shifts: δ 22.9, 25.4, 127.3.

10.4   ADDITION REACTIONS OF ALKENES

Alkenes are useful starting materials for organic syntheses because they undergo a large variety of reactions involving their functional group, the carbon-carbon π-bond. One of the typical reactions of alkenes is the **addition** of a reagent X–Y across the π-bond of the alkene according to Equation 10.19. Such additions to the π-bond occur because of two principal factors. First, the strength of the carbon-carbon π-bond is 60-65 kcal/mol, a value substantially lower than the 80–100 kcal/mol that is typical for the σ-bond strengths of other atoms bound to carbon. Adding a reagent across the double bond is therefore usually *exothermic* because one π-bond is being replaced by two σ-bonds. Second, the π-electrons of the double bond are more loosely held than σ-electrons, so they are more polarizable. Consequently, the π-electron "cloud" is readily distorted through electrostatic interaction with an approaching, electron-deficient reagent, and this enhances the reactivity of the alkene toward attack.

$$\underset{R^2}{\overset{R^1}{}}C=C\underset{R^4}{\overset{R^3}{}} + X-Y \longrightarrow R^2-\overset{\overset{R^1}{|}}{\underset{\underset{X}{|}}{C}}-\overset{\overset{Y}{|}}{\underset{\underset{R^4}{|}}{C}}-R^3 \qquad (10.19)$$

Alkenes may undergo addition by either ionic or radical mechanisms depending upon the nature of X–Y and the conditions under which the reaction is performed. When unsymmetrical reagents X–Y add to the carbon-carbon bond of an alkene by an ionic mechanism, the reaction is termed an **electrophilic addition.** In such cases, the reagent X–Y may be designated as **E–Nu,** wherein Nu:⁻ is the more electronegative of the two atoms. The reaction is stepwise and involves sequential combinations of Lewis acids and Lewis bases. In the first step, the electron-rich double bond, a Lewis base, attacks the **electrophile** E^+, an electron-deficient species and a Lewis acid, to give either an acyclic planar carbocation **34** or a cyclic cation **35** (Eq. 10.20). If an acyclic carbocation **34** is formed as an intermediate in the electrophilic addition to an *unsymmetrically substituted* alkene, the electrophile E^+ generally adds to the terminus of the double bond that produces the *more stable carbocation* **34** according to the **Markovnikov rule.** Subsequent reaction of this cation, a Lewis acid, with the **nucleophile, Nu:⁻,** which is electron-rich and a Lewis base, then gives the addition product. When the positively charged carbon of **34** is substituted with three different groups, the nucleophile may attack from *both* faces of the planar ion by paths **a** and **b** to give a mixture of **36** and **37,** which are **stereoisomers** (see Sec. 7.1). Acyclic carbocations like **34** may also rearrange via 1,2-hydride or 1,2-alkyl migration to form a more stable carbocation (see Sec. 10.3); subsequent reaction of this isomeric carbocation with the nucleophilic partner produces adducts isomeric with **36** and **37.** Some electrophilic reagents add to a carbon-carbon double bond to form a cyclic carbocation **35;** such cations tend not to rearrange. The nucleophile Nu:⁻ may then attack **35** by paths **c** and **d** to give mixtures of isomers **38** and **39.**

(10.20)

The general principles set forth in Equation (10.20) may be illustrated by considering several examples. The addition of the unsymmetrical reagent hydrogen bromide, H–Br, to an alkene is shown in Equation (10.21), and the acid-catalyzed addition of water, or **hydration,** of an alkene is depicted in Equation (10.22). In both reactions, the electron-rich carbon-carbon double bond of the alkene, which is a Lewis base, is protonated to form the more stable intermediate carbocation according to the Markovnikov rule. This carbocation may then rearrange to a more stable carbocation (see Sec. 10.3), or, as shown in the present case, may undergo direct nucleophilic attack from both faces of the planar carbocation to give the observed alkyl bromide or alcohol. These two examples of electrophilic additions to alkenes are **regioselective** processes because the two unsymmetrical reactants combine predominantly in one orientational sense to give one **regioisomer** preferentially.

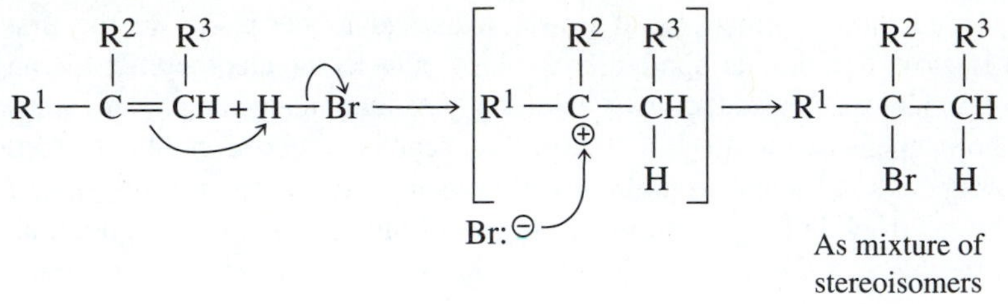

$$R^1 - \underset{\underset{R^2}{|}}{C} = CH + H - Br \longrightarrow \left[R^1 - \underset{\underset{\oplus}{|}}{\overset{\overset{R^2}{|}}{C}} - \underset{\underset{H}{|}}{\overset{\overset{R^3}{|}}{CH}} \right] \longrightarrow R^1 - \underset{\underset{Br}{|}}{\overset{\overset{R^2}{|}}{C}} - \underset{\underset{H}{|}}{\overset{\overset{R^3}{|}}{CH}} \quad \textbf{(10.21)}$$

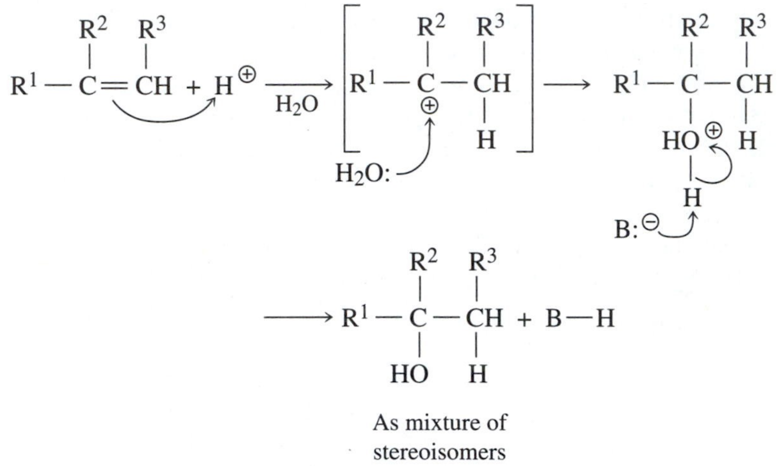

Br:$^\ominus$

As mixture of
stereoisomers

$$R^1 - \underset{\underset{R^2}{|}}{C} = CH + H^\oplus \underset{H_2O}{\longrightarrow} \left[R^1 - \underset{\underset{\oplus}{|}}{\overset{\overset{R^2}{|}}{C}} - \underset{\underset{H}{|}}{\overset{\overset{R^3}{|}}{CH}} \right] \longrightarrow R^1 - \underset{\underset{HO^\oplus}{|}}{\overset{\overset{R^2}{|}}{C}} - \underset{\underset{H}{|}}{\overset{\overset{R^3}{|}}{CH}}$$

H$_2$O:

H

B:$^\ominus$ **(10.22)**

$$\longrightarrow R^1 - \underset{\underset{HO}{|}}{\overset{\overset{R^2}{|}}{C}} - \underset{\underset{H}{|}}{\overset{\overset{R^3}{|}}{CH}} + B - H$$

As mixture of
stereoisomers

Since the ionic addition of hydrogen bromide and the acid-catalyzed addition of water to an alkene proceed via open, planar carbocations, the formation of products derived from rearranged carbocations is relatively common. In contrast, the bromination of alkenes, which is illustrated by the addition of bromine to cyclopentene in Equation (10.23), occurs via a **cyclic bromonium ion** related to **35;** skeletal rearrangements of such cations are *not* generally observed. Moreover, the intervention of this cyclic ion dictates that the stereochemistry of the addition is completely in the *anti* sense as shown. The red-brown color of bromine is discharged upon addition to an alkene, making this reaction a useful qualitative test for unsaturation (see Sec. 23.5.2).

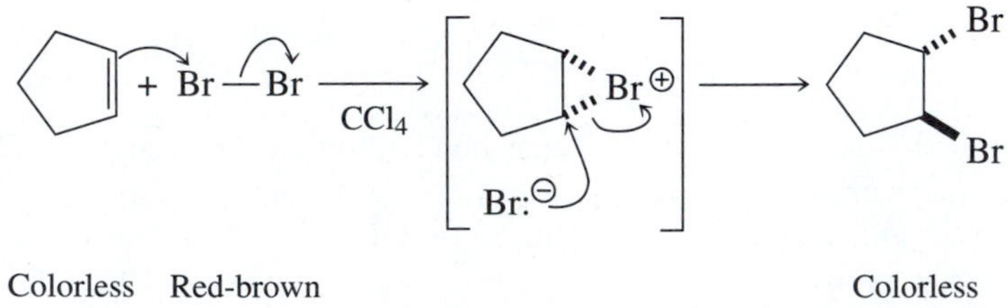

Colorless   Red-brown   Colorless

Some reagents X–Y may react with alkenes by the alternate, mechanistic pathway known as **free-radical addition** if free-radical initiators are present or if the reaction mixture is exposed to heat or light of the proper wavelength (see Sec. 7.3). In such reactions the radical X· adds to the double bond to yield the intermediate carbon-centered radical **40** (Eq. 10.24); rearrangements of such radicals are generally not observed. The ensuing reaction between **40** and X–Y provides the product and regenerates the radical X· to propagate the **radical chain reaction** (see Sec. 9.2). When unsymmetrical alkenes are the reactants, X· becomes bound to the terminus of the carbon-carbon double bond that yields the more substituted and thus more stable of the two possible radicals, as illustrated by the radical addition of hydrogen bromide to an alkene in Equation (10.25). In this reaction the initiator abstracts a hydrogen atom from a molecule of H–Br. The resulting bromine atom then adds to the carbon-carbon double bond to form a carbon-centered radical that leads to formation of the observed product.

$$\underset{R^1}{\overset{R^2}{>}}C=CHR^3 \longrightarrow \left[R^1 - \underset{|}{\overset{R^2}{\underset{|}{C}}} - \underset{\underset{X}{|}}{\overset{R^3}{\underset{|}{CH}}} \right] \longrightarrow R^1 - \underset{\underset{Y}{|}}{\overset{R^2}{\underset{|}{C}}} - \underset{\underset{X}{|}}{\overset{R^3}{\underset{|}{CH}}} + X· \quad \textbf{(10.24)}$$

40

$$\underset{R^1}{\overset{R^2}{>}}C=CHR^3 + H-Br \xrightarrow[\text{(initiator)}]{\text{Peroxides}} R^2 - \underset{\underset{H}{|}}{\overset{R^1}{\underset{|}{C}}} - \underset{\underset{Br}{|}}{\overset{R^3}{\underset{|}{CH}}} \quad \textbf{(10.25)}$$

Comparing the reactions in Equations (10.21) and (10.25) reveals that the regiochemistries of polar and of free-radical addition of hydrobromic acid to an alkene are opposite, and different products are formed as a result. One must therefore carefully control reaction conditions in order to obtain pure products. Reaction conditions that favor **electrophilic addition** are low temperatures, the presence of ionic salts, and the absence of light, peroxides, and other radical initiators. Conversely, **free-radical addition** is fostered by performing the reaction in the gas phase or in nonpolar solvents and in the presence of strong light, peroxides, or other radical initiators.

The addition of reagents X–Y to carbon-carbon double bonds may also proceed via a concerted mechanism in which each new σ-bond is formed simultaneously on the *same* face of the π-bond. The stereochemistry of such reactions is *syn*. For example, the reaction of potassium permanganate, which is purple, with an alkene such as cyclohexene proceeds via initial *syn*-addition of permanganate ion across the π-bond to give **41,** which is colorless. Subsequent decomposition of **41** gives a *cis*-1,2-diol and manganese dioxide, the brown precipitate that is observed as the other product of the reaction (Eq. 10.26). This decoloration of potassium permanganate by alkenes forms the basis of the Baeyer qualitative test for the presence of carbon-carbon double bonds (see Sec. 23.5.2). However, this reaction is not of general preparative use because the 1,2-diol thus formed usually undergoes further oxidation by permanganate ion to give diketones and carboxylic acids. Other examples of *syn*-additions to carbon-carbon double bonds are hydroboration (see Sec. 10.4), the Diels-Alder reaction (see Chap. 12) and catalytic hydrogenation (see Sec. 17.2).

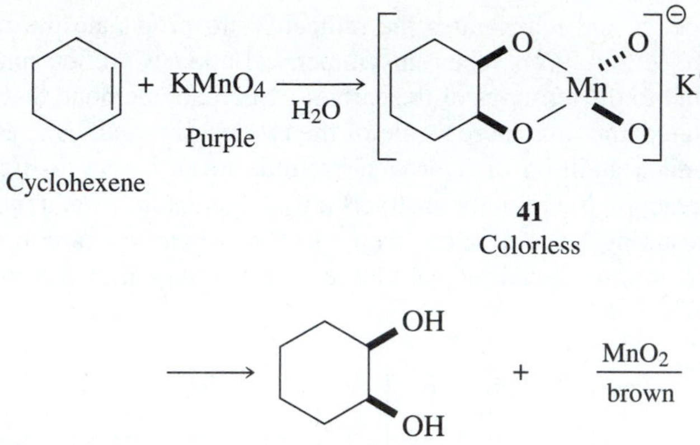

Cyclohexene + KMnO₄ →(H₂O) [structure] K⊕

Purple

41

Colorless (10.26)

+ MnO₂
brown

cis-Cyclohexane-1,2-diol

10.4.1 ADDING H–BR TO ALKENES

In this experiment, the **ionic addition** of H-Br to 1-hexene to yield 2-bromo-hexane (Eq. 10.27) according to Markovnikov's rule is examined. This type of reaction is normally rather difficult to perform in the undergraduate laboratory for several reasons. First, common alkenes are immiscible with the concentrated aqueous hydrobromic acid, and the reaction is sluggish if the layers are not mixed efficiently. Second, hydrobromic acid, H–Br, is a strong acid that protonates water extensively to give the hydronium ion, H_3O^+, which is a weaker acid than undissociated hydrobromic acid and is unable to protonate the alkene rapidly under mild reaction conditions. Moreover, the presence of water in the reaction mixture introduces the possibility of competing acid-catalyzed addition of water to the alkene (see Eq. 10.22, Sec 10.2). These problems are reduced by using anhydrous hydrogen bromide, but the highly corrosive nature of this gas makes it difficult to handle and use.

$$CH_3(CH_2)_3CH{=}CH_2 + H{-}Br \longrightarrow CH_3(CH_2)_3CH{-}CH_2 \quad (10.27)$$
$$Br \quad H$$

A convenient solution to these experimental difficulties entails the addition of a catalytic amount of a quaternary ammonium salt such as methyltrioctylam-monium chloride, $CH_3(n\text{-}C_8H_{17})_3N^+Cl^-$, also known as Aliquat 336®, to the heterogeneous mixture of the aqueous acid and the alkene. The tetraalkyl ammonium salt partitions between the aqueous and organic phases because of the amphoteric nature of the catalyst: It is **lipophilic** or **non-polar-loving** due to the alkyl groups and **hydrophilic** or **polar-loving** because of the ionic ammonium function, respectively. By forming a complex with the H–Br, for example **42,** the quaternary ammonium salt drags H–Br out of the aqueous phase and transports it, in only a partially hydrated form, into the organic phase and the presence of the alkene, as depicted in Scheme 10.1. The quaternary ammonium salt repartitions into the aqueous phase to complete the catalytic cycle. The transfer of the H–Br into the organic phase essentially dehydrates the acid, making it more reactive toward the alkene so that the addition becomes possible.

SCHEME 10.1

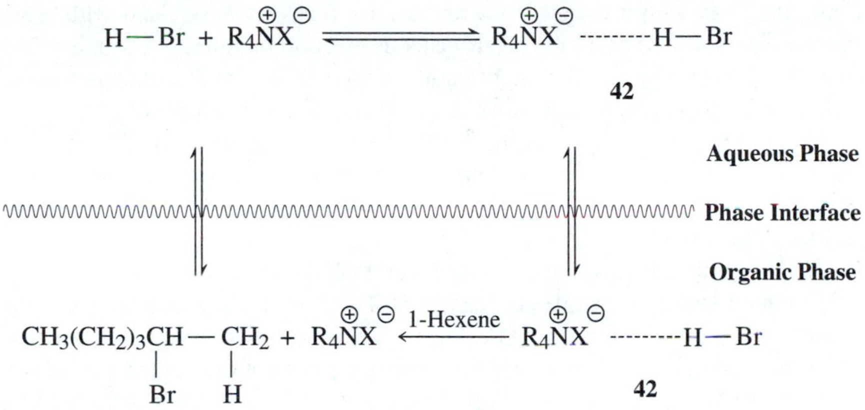

$$H\!-\!Br \; + \; R_4\overset{\oplus}{N}\overset{\ominus}{X} \; \rightleftharpoons \; R_4\overset{\oplus}{N}\overset{\ominus}{X} \text{ --------} H\!-\!Br$$

42

Aqueous Phase

Phase Interface

Organic Phase

$$CH_3(CH_2)_3\underset{\underset{Br}{|}}{CH}\!-\!\underset{\underset{H}{|}}{CH_2} \; + \; R_4\overset{\oplus}{N}\overset{\ominus}{X} \;\overset{\text{1-Hexene}}{\longleftarrow}\; R_4\overset{\oplus}{N}\overset{\ominus}{X} \text{ --------} H\!-\!Br$$

42

Compounds that promote the transport of reagents between immiscible layers by means of ion pairs like **42** are called **phase-transfer catalysts.** Their presence can have dramatic effects on the rates of bimolecular reactions between reagents contained in immiscible phases: Rate accelerations of 10^4 to 10^9 are common. One factor that determines the overall rate of a reaction involving phase-transfer catalysis is the efficiency of the partitioning of the reagents and reactants between the two phases. This is a function of the total surface area at the interface of the phases. To increase this area, the reaction mixture is agitated vigorously by stirring or shaking to promote emulsification and the formation of tiny droplets of the immiscible layers.

Pre-Lab exercises for Section 10.4.1, "Addition of H–Br to 1-Hexene," are found on page PL. 37.

EXPERIMENTAL PROCEDURE

SAFETY ALERT

1. Concentrated hydrobromic acid (47–49%) is a *highly corrosive* and *toxic* material. Measure out the amount required inside a hood and avoid inhaling its vapors. Use rubber gloves when transferring the acid between containers. If the acid comes in contact with your skin, flood the affected area immediately and thoroughly with water and rinse it with 5% sodium bicarbonate solution.

2. Quaternary ammonium salts are *toxic* substances and can be absorbed through the skin. Should they accidentally come in contact with your skin, wash the affected area immediately with large amounts of water.

3. If a flame *must* be used in the distillation step, be certain that all joints in the apparatus are well lubricated and tightly mated because 1-hexene and the solvents used, particularly petroleum ether, are highly flammable. An oil or sand bath is a safer heating technique (see Sec. 2.5).

In a 50-mL round-bottom flask combine 2.1 g of 1-hexene, 13.9 mL of 47–49% aqueous hydrobromic acid, and 1.0 g of methyltrioctylammonium chloride. Equip the flask with a magnetic stirrer and a reflux condenser, and with rapid stirring (see Sec. 2.6) bring the heterogeneous reaction mixture to a gentle reflux (see Sec. 2.19) using an oil or sand bath (see Sec. 2.5). Continue stirring under reflux for 2–3 h and then allow the mixture to cool to room temperature.✱

Carefully transfer the two-phase mixture to a separatory funnel, rinse the reaction flask with 15 mL of petroleum ether, and add the rinse to the separatory funnel. Shake the funnel thoroughly in order to effect extraction (see Sec. 2.18) and then allow the layers to separate. Three layers should form! Verify that the lowest one is an aqueous phase and remove it. Wash the two phases remaining in the funnel with two 15-mL portions of 10% sodium bicarbonate solution; the mixture will separate into only two layers. Vent the funnel *frequently* because gas is evolved in this step and excessive pressure must not develop in the funnel. Transfer the organic layer to a 50-mL Erlenmeyer flask and dry it with swirling over 1–2 g of anhydrous sodium sulfate (see Sec. 2.21) for at least 0.5 h.✱

Decant (see Sec. 2.15) or filter (see Sec. 2.12) the dried solution into a 25-mL round-bottom flask, equip the flask for simple distillation (see Sec. 2.7, Fig. 2.15b), and distil the product using an oil or sand bath or, if necessary, a flame (see Sec. 2.5). Carefully control the rate of heating throughout the course of this distillation because severe foaming can occur. Do *not* attempt to remove the solvent too rapidly, as this may result in excessive loss of product due to foaming. After the first fraction containing solvent and unreacted 1-hexene is removed over a range of ambient–80 °C (760 torr),✱ change the receiving flask and collect the bromohexane as a single fraction. Because the volume of product is likely to be only 1–2 mL, an accurate boiling point may be difficult to obtain, and all material that distils above 110–115 °C (760 torr) should be collected in order to obtain a reasonable yield. The pure product is a colorless, mobile liquid that boils at 146–147 °C (760 torr).

Verify that the product is 2-bromohexane rather than 1-bromohexane or a mixture of the two by spectroscopic methods (IR and NMR) and/or by GLC, assuming authentic samples of the two isomers are available (see Fig. 10.27 for reference). Alternatively, demonstrate whether or not Markovnikov addition has occurred by subjecting the product to the silver nitrate or sodium iodide/acetone tests for classifying alkyl halides (see Sec. 23.5.3) and by determining the boiling point of the product more precisely with the aid of a micro boiling-point apparatus (see Sec. 2.4).

Finishing Touches   Slowly combine the *aqueous layers and washes*, neutralize if necessary with sodium carbonate, and flush them down the drain with excess water. Since the *first distillation fraction* may contain some halogenated product, pour it in a container for halogenated organic solvents. Evaporate the petroleum ether from the *sodium sulfate used as drying agent* in the hood, and place the solid in a container for nonhazardous solids. Dilute the *silver nitrate test solution* with water and flush it down the drain with excess water.

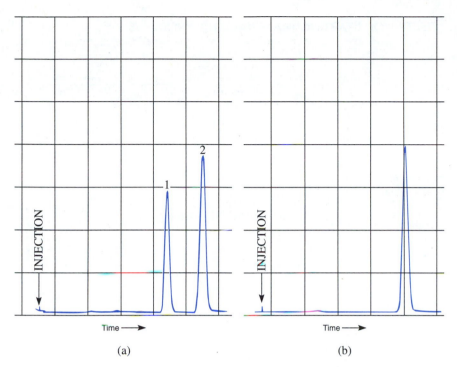

FIGURE 10.27 GLC traces of the bromo-hexanes. (a) Mixture of 1-bromohexane (peak 2) and 2-bromohexane (peak 1). (b) Distillate from addition of HBr to 1-hexene. Analysis on 3-m 15% FFAP on 60/80 Chromosorb P/A at 130 °C, flowrate of 60 mL/min.

EXERCISES

1. In general, the yield of addition product is higher when the reaction mixture is stirred than when it is not. Why is this?

2. Devise an experiment that would demonstrate that a phase-transfer catalyst accelerates the rate of reaction between 1-hexene and aqueous hydrobromic acid.

3. When methyltrioctylammonium chloride is used as the phase transfer catalyst, some 1- or 2-chlorohexane could be obtained as a by-product. Why would contamination of the desired addition product by such a by-product be a minor concern in this experimental procedure?

4. Why is methyltrioctylammonium chloride only partially soluble in 1-hexene? In aqueous hydrobromic acid?

5. Outline a procedure that would allow monitoring of the course of this addition reaction as a function of time. In other words, how might the reaction mixture be analyzed periodically so that you could determine when the reaction was complete?

6. Why would it be difficult to perform the ionic addition of hydrogen bromide to 1-pentene using the procedure outlined in this experiment?

7. Refer to Figure 10.27(a) and determine whether the fatty acid polymer (FFAP) stationary phase used for the GLC analysis shown separates the bromohexanes on the basis of their relative boiling points or on the basis of some other property. Explain your answer.

8. The addition of H–Br to 1-hexene in the presence of peroxides and light proceeds by a radical chain reaction to give 1-bromohexane. Outline a mechanism for this transformation, showing the initiating, propagating, and two possible terminating steps. Symbolize the flow of electrons with curved arrows.

SPECTRA OF STARTING MATERIAL AND PRODUCTS

FIGURE 10.28   IR spectrum of 1-hexene.

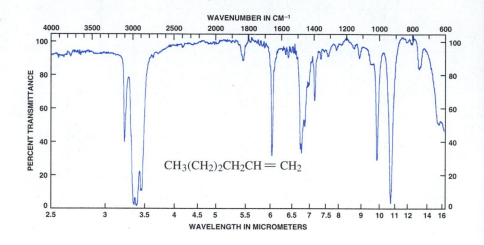

(a) ^{1}H NMR spectrum (90 MHz).

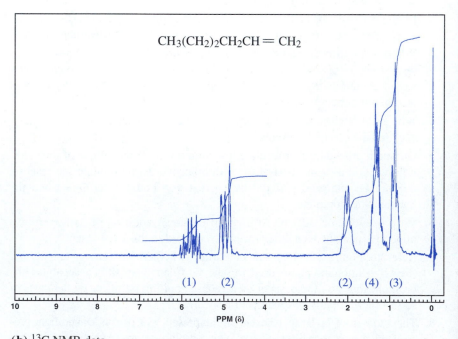

(b) ^{13}C NMR data.
Chemical shifts: δ 14.0, 22.5, 31.6, 33.8, 114.3, 139.2.

FIGURE 10.29   NMR data for 1-hexene.

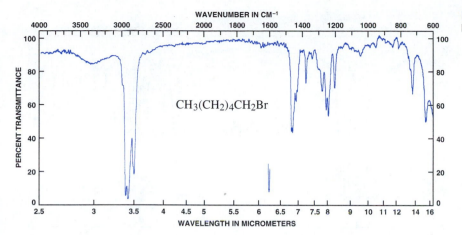

FIGURE 10.30   IR spectrum of 1-bromo-hexane.

$CH_3(CH_2)_4CH_2Br$

(a) ^1H NMR spectrum (90 MHz).

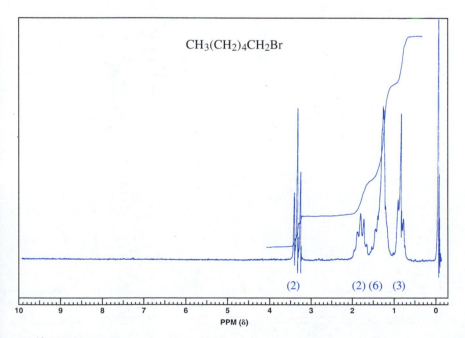

$CH_3(CH_2)_4CH_2Br$

(2)   (2) (6)   (3)

(b) ^{13}C NMR data.
Chemical shifts: δ 14.0, 22.7, 28.1, 31.2, 33.1, 33.4.

FIGURE 10.31   NMR data for 1-bromo-hexane.

FIGURE 10.32 IR spectrum of 2-bromo-hexane.

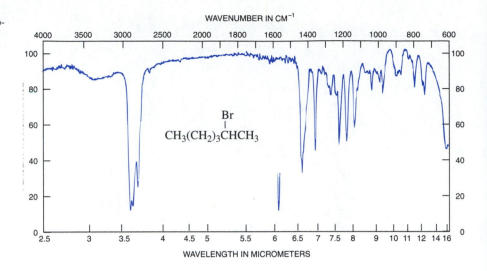

(a) ^{1}H NMR spectrum (90 MHz).

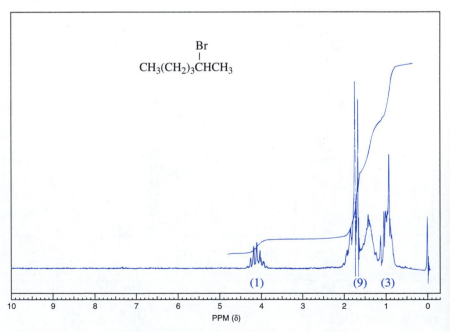

FIGURE 10.33 NMR data for 2-bromo-hexane.

(b) ^{13}C NMR data.
  Chemical shifts: δ 13.9, 22.1, 26.5, 29.9, 41.0, 51.0.

10.4.2   ACID-CATALYZED HYDRATION OF ALKENES

The acid-catalyzed addition of the elements of water across a carbon-carbon double bond to give an alcohol is referred to as **hydration** of an alkene (see Eq. 10.22). Mechanistically, this process is simply the reverse of the acid-catalyzed **dehydration** of alcohols, which is discussed in Section 10.3. The position of the equilibrium for these two competing processes depends upon the reaction conditions. Hydration of a double bond requires the presence of excess water to drive the reaction to completion, whereas the dehydration of an alcohol requires the removal of water to complete the reaction. In the experiment that follows, the acid-catalyzed hydration of norbornene (**43**) to give exclusively *exo*-norborneol (**44**) (Eq. 10.28) is examined.

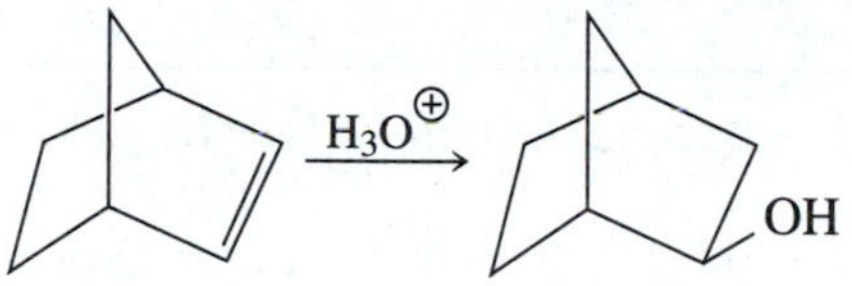

(10.28)

The mechanism of this reaction and the structure of the intermediate carbocation have been the subject of intense controversy for over 30 years. When the double bond of norbornene (**43**) is protonated, a carbocation is produced. The question of whether this cation is **45a** and is in rapid equilibrium with the isomeric cation **45b**, or whether **45a** and **45b** are simply two contributing resonance structures to the resonance hybrid **46**, has been the focus of the debate. Note that the contributing structures **45a** and **45b** *differ in the location of a* σ-bond, not in the location of π-bonds as is more commonly encountered in resonance structures. The delocalized structure **46** is referred to as a **nonclassical carbocation,** and most chemists now accept this formulation as the more likely representation of this intermediate. It is evident from empirical results that the sterically more accessible side of the carbocation is the one away from, or *exo* to the larger bridge as shown. Nucleophilic attack of water solely from this side leads to *exo*-norborneol (**44**) rather than *endo*-norborneol (**47**).

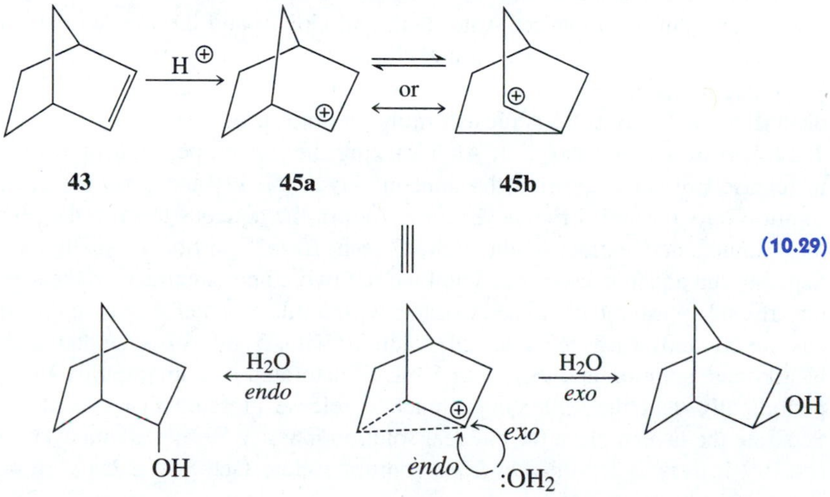

(10.29)

Pre-Lab exercises for Section 10.4.2, "Hydration of Norbornene," are found on page PL. 39.

EXPERIMENTAL PROCEDURES

SAFETY ALERT

1. Norbornene is a volatile solid. Perform all weighings in a hood if possible.
2. If the *acidic solution* used in the first part of the experiment comes in contact with your skin, *immediately* flood the affected area with water and thoroughly rinse it with 5% sodium bicarbonate solution.
3. If the *basic solution* used in this experiment comes in contact with your skin, immediately flood the affected area with water and thoroughly rinse the area with a *dilute* solution (1%) of acetic acid.

Slowly add 2 mL of concentrated sulfuric acid *to* 1 mL of water in a 25-mL Erlenmeyer flask and cool the solution to room temperature. To this solution *in the hood* add 1.0 g of norbornene in small pieces. Swirl the mixture until all of the norbornene dissolves, cooling the flask briefly in an ice-water bath if the mixture becomes noticeably warm. However, *do not cool the contents of the flask below room temperature.* Allow this solution to stand (ca 15–30 min) while you prepare a solution of 1.5 g of potassium hydroxide in 7.5 mL of water. Cool both solutions in an ice-water bath and slowly add the base to the acidic reaction mixture to *partially* neutralize the acid.

Place the combined mixture in a small separatory funnel, and add 15 mL of diethyl ether. If any solid is present in the bottom of the separatory funnel, add 1–2 mL of water to dissolve it. After shaking the funnel, periodically venting it to release pressure, separate the aqueous layer (*Save!*) and pour the ethereal solution into a 50-mL Erlenmeyer flask. Return the aqueous layer to the separatory funnel, and extract again with a fresh 10-mL portion of diethyl ether. Separate the aqueous layer, and combine the two ethereal extracts in the separatory funnel. Wash the combined extracts with 5 mL of water. Separate the aqueous layer, and wash the ethereal solution with 5 mL of saturated sodium bicarbonate solution and then with 5 mL of saturated sodium chloride solution; periodically vent the separatory funnel to release pressure that may develop. Separate the layers, place the ethereal solution in a dry 50-mL Erlenmeyer flask, and dry it over 1–2 g of *anhydrous* sodium sulfate (see Sec. 2.21) with occasional swirling for 10–15 min.∗

Decant the solution into a 125-mL filter flask and concentrate the solution to an oil (see Sec. 2.25). Do this by placing a rubber stopper on the flask and connecting the sidearm to a water aspirator; place a trap *between* your flask and the aspirator (see Fig. 2.37). Gently swirl the flask and warm it with your hands while applying vacuum; all of the diethyl ether should be removed after 5–10 min.

Purify the crude *exo*-norborneol by sublimation (see Sec. 2.17) as follows. Break the vacuum to the flask and replace the solid rubber stopper with one containing a 13-mm × 120-mm test tube that extends to within about 2 cm of the bottom of the flask (see Fig. 2.30b). Half-fill the test tube with chipped ice and evacuate the flask using a water aspirator equipped with a trap. *Gently* heat the flask using a steam bath or the sweeping motion of a heat gun or *small* flame. When the alcohol has completely sublimed, cool the flask, break the vacuum to the aspirator, and disconnect the flask. Carefully decant the water out of the cold finger before carefully removing it from the flask. Maintain the cold finger in a horizontal or slightly inverted position. Scrape the soft crystalline product from the test tube and weigh it. Determine the melting point by placing the sample in a capillary tube (see Sec. 2.3.1) that has been sealed about 2 cm from the open end using a small flame. The reported melting point of norborneol is 124–126 °C.

Finishing Touches   Slowly combine the *aqueous layers and washes*, if necessary neutralize them with sodium carbonate, and flush them down the drain with excess water. After the diethyl ether evaporates from the *sodium sulfate* in the hood, place the solid in a container for nonhazardous solids.

EXERCISES

1. What purpose is served by washing the ethereal solution of product with bicarbonate solution? Saturated sodium chloride solution?

2. Why is it necessary to seal the capillary tube before taking the melting point of norborneol?

3. Norborneol undergoes reaction with concentrated H–Br to form a single diastereomer. Write the structure of this bromide and propose a stepwise mechanism for its formation. Why is only one diastereoisomer produced?

4. Norborneol contains three chiral centers, but the product obtained from the hydration of norbornene in this experiment is optically inactive. Why?

5. For each of the alkenes (a)-(e), write the structure of the alcohol that should be the major product obtained upon acid-catalyzed hydration.

(a) Me Me Me

(b) Me

(c) Me

(d) Me

(e) NO₂ MeO

SPECTRA OF STARTING MATERIAL AND PRODUCT

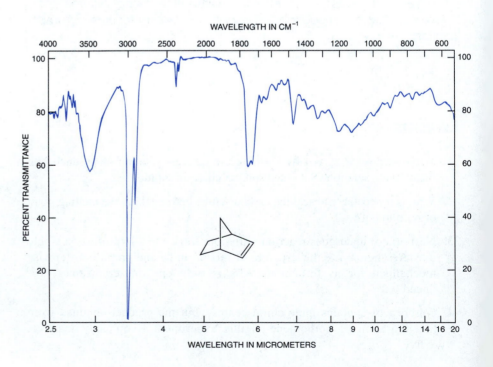

FIGURE 10.34   IR spectrum of norbornene.

(a) ^{1}H NMR spectrum (90 MHz).

FIGURE 10.35 NMR data for norbornene.

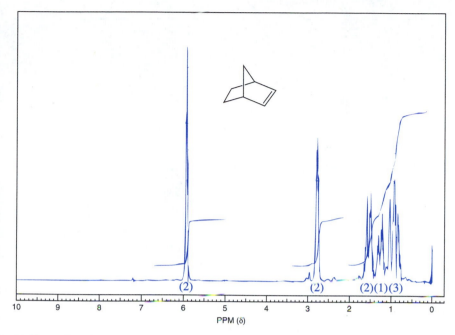

(b) ^{13}C NMR data.
Chemical shifts: δ 24.8, 42.0, 48.8, 135.4.

WAVELENGTH IN CM^{-1}

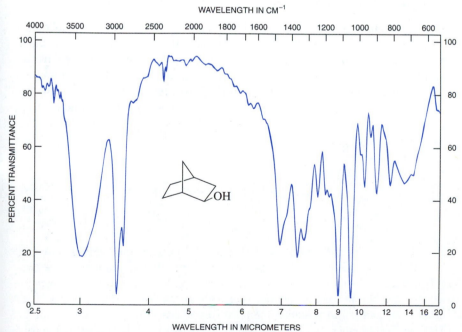

WAVELENGTH IN MICROMETERS

FIGURE 10.36 IR spectrum of *exo*-norborneol.

FIGURE 10.37 NMR data for *exo*-norbornanol.

(a) ^{1}H NMR spectrum (90 MHz).

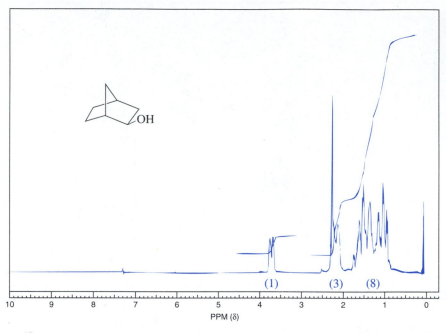

(b) ^{13}C NMR data.
Chemical shifts: δ 24.7, 28.4, 34.5, 35.5, 42.2, 44.2, 74.6.

10.4.3 HYDROBORATION-OXIDATION OF ALKENES

The **regiochemistry** of acid-catalyzed hydration of alkenes to give alcohols is that predicted by Markovnikov's rule (see Sec. 10.4) because the more stable intermediate carbocation is preferentially formed. Sometimes it is desirable to add the elements of water across a carbon-carbon double bond in the opposite regiochemical sense to provide the **anti-Markovnikov** product (Eq. 10.30). In order to accomplish this goal, a process termed **hydroboration-oxidation** is available that involves the reaction of an alkene sequentially with diborane, B_2H_6, and basic hydrogen peroxide.

$$\underset{R^1}{\overset{R^2}{>}}C=CHR^3 \longrightarrow R^1-\underset{\underset{H}{|}}{\overset{\overset{R^2}{|}}{C}}-\underset{\underset{OH}{|}}{\overset{\overset{R^3}{|}}{CH}} \tag{10.30}$$

Diborane is a dimer of borane, BH_3 (Eq. 10.31). The bonding in diborane is unusual because the hydrogen atoms bridge the two boron atoms with the two monomeric BH_3 subunits being bound with *two-electron, three-center bonds*. Because the boron atom in borane possesses an empty *p*-orbital, borane is a Lewis acid, and it forms stable complexes upon reaction with tetrahydrofuran (THF) and other ethers, which function as Lewis bases as illustrated by the formation of a borane-THF complex (Eq. 10.32).

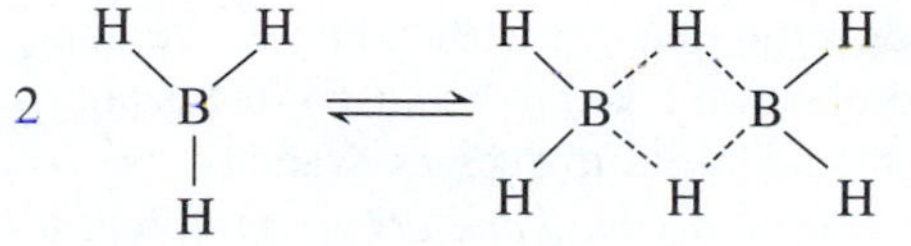

(10.31)

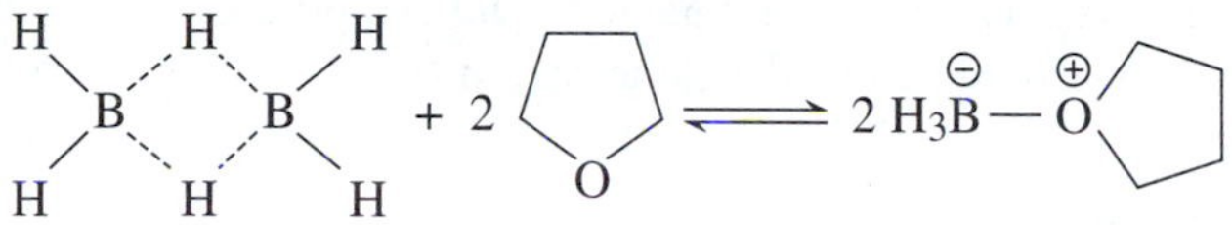

(10.32)

A borane-THF complex

Alkenes may serve as Lewis bases and react with borane, a Lewis acid, by the net addition of the boron-hydrogen bond across the carbon-carbon double bond to give a monoalkylborane as the initial product. This process, known as **hydroboration,** is illustrated by the hydroboration of 1-butene in Equation (10.33). Depending upon steric factors and the degree of substitution on the double bond, the intermediate monoalkylborane, **48,** may add successively to additional molecules of the same alkene to form a dialkylborane **49** or a trialkylborane **50.**

As observed in the electrophilic additions to alkenes depicted in Equations (10.21) and (10.22), the addition of borane to alkenes is *regioselective* (see Sec. 10.4) and occurs so the *electrophilic boron atom* becomes preferentially attached to the *less substituted carbon atom* of the carbon-carbon double bond.

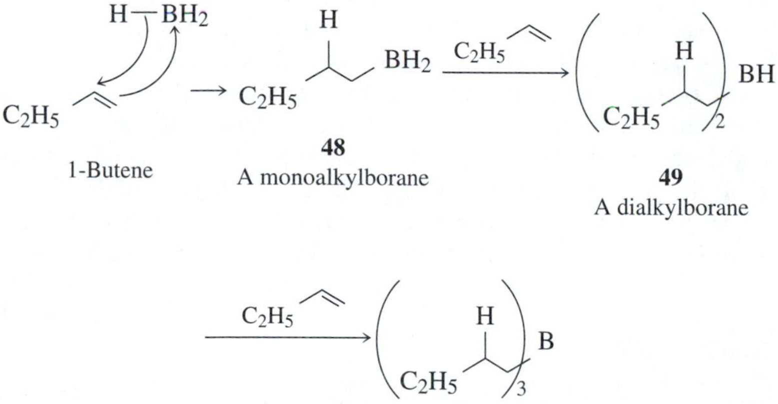

1-Butene

48
A monoalkylborane

49
A dialkylborane

(10.33)

50
A trialkylborane

The reason for this selectivity is understood by considering the mechanism of hydroboration. As borane approaches the double bond, a Lewis acid–Lewis base interaction occurs in which the π-electrons of the double bond begin to interact with the empty *p*-orbital on the Lewis acidic boron atom, resulting in the development of a partial positive charge on the more highly substituted carbon atom. The boron atom then assumes a partial negative charge that facilitates the transfer of the hydride ion, H:⁻, from the boron to the carbon atom bearing the partial positive charge; this occurs through the four-centered transition state **51,** in which all four atoms undergo changes in bonding at the same time. Thus, the regiochemistry of hydroboration is predicted by the same general rule that applies to all electrophilic additions to alkenes. Namely, *the reaction of an electrophile with a carbon-carbon double bond occurs preferentially via the transition state in which a partial positive charge develops on that carbon atom better able to accommodate it.* Geometric constraints inherent in the cyclic transition state **51** require that the addition of borane to the alkene proceed so that both the boron and the hydrogen add from the same face of the double bond, a process called *syn* **addition.**

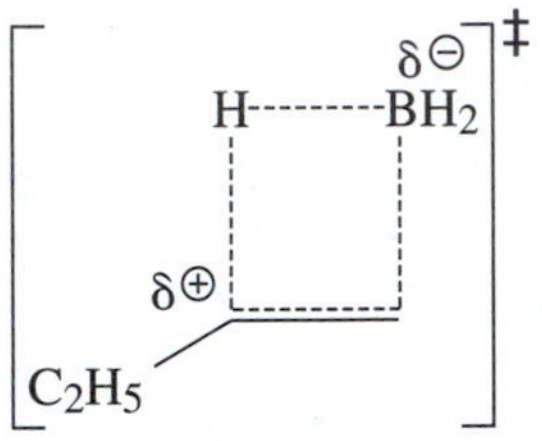

51

Treating a trialkylborane with alkaline hydrogen peroxide promotes oxidative cleavage to form 3 mole of an alcohol and boric acid, as exemplified in the conversion of **50** into 1-butanol (**55**), shown in Scheme 10.2. The first step of the reaction involves nucleophilic attack of the hydroperoxide anion on the boron atom of **50** to give **52.** In the next step, migration of the alkyl group from boron to oxygen proceeds with displacement of hydroxide ion and cleavage of the weak oxygen-oxygen bond to give a borate monoester **53.** An important consequence of the *intramolecularity* of this rearrangement is that it proceeds with *retention* of configuration at the carbon atom. The sequence is repeated until all of the boron-carbon bonds have been converted into oxygen-carbon bonds, whereupon the intermediate trialkoxyborane **54** undergoes alkaline hydrolysis to give the alcohol **55** and boric acid. Thus, hydroboration-oxidation of 1-butene gives 1-butanol, the product of overall *anti-Markovnikov* addition of the elements of H–OH to the carbon-carbon double bond. Note, however, that the key step determining the regiochemistry of the alcohol produced is the *Markovnikov* addition of the hydrogen-boron bond across the double bond.

SCHEME 10.2

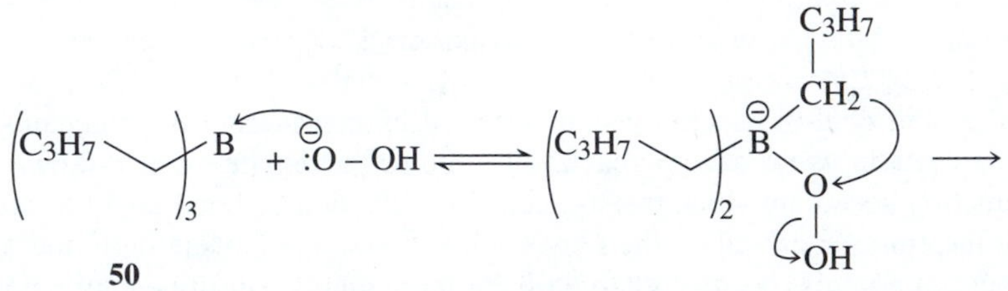

50   **52**

Scheme 10.2 (Continued)

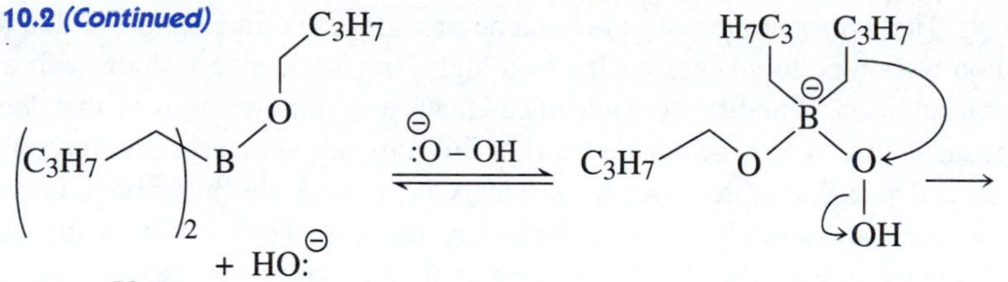

53

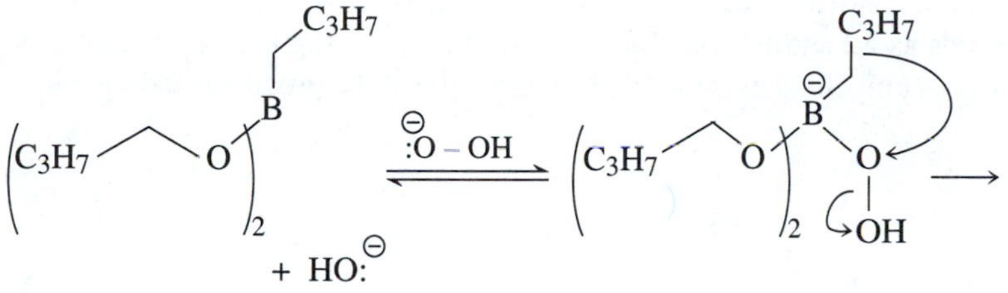

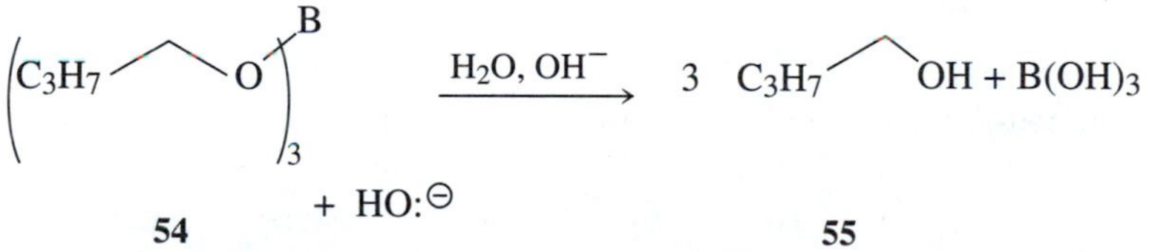

54
1-Butanol

55
Boric Acid

In this experiment, the regiochemistry and stereochemistry of the sequence of hydroboration-oxidation are examined using (+)-α-pinene (**56**) as the alkene and a solution of borane-tetrahydrofuran complex as the hydroborating reagent.

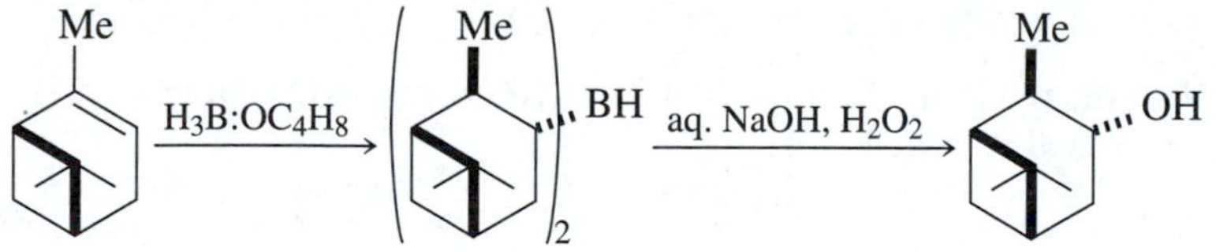

$$(10.34)$$

56
(+)-α-Pinene

57

58
(−)-Isopinocampheol

Using this complex avoids the need of handling diborane, B_2H_6, which is potentially hazardous. Since the carbon-carbon double bond in **56** is trisubstituted and sterically hindered, only two molecules of this alkene react with each molecule of borane to form the intermediate dialkylborane **57**. Subsequent oxidation of **57** with alkaline hydrogen peroxide then gives (−)-isopinocampheol (**58**) as the sole product. The complex of borane with tetrahydrofuran is generated *in situ* by the reaction of sodium borohydride with iodine in tetrahydrofuran according to Equation (10.35).

$$NaBH_4 + I_2 + C_4H_8O \longrightarrow H_3B{:}OC_4H_8 + H_2 + NaI \qquad (10.35)$$

A borane-THF complex

The hydroboration of (+)-α-pinene provides experimental proof that the addition of borane to alkenes occurs in a highly regioselective fashion with exclusive *syn* addition. Another stereochemical feature of this reaction is that the borane reagent approaches the double bond of **56** from only one face. Because only one of several possible stereoisomeric products is formed, the hydroboration of (+)-α-pinene is an example of a **stereoselective** reaction. The identity of the product of this reaction is verified by determining its melting point, the melting point of a suitable derivative, and the sign of optical rotation. Comparing these data with those for the alternative Markovnikov products **59–61** and the possible anti-Markovnikov products **62** and **63** (see Table 10.1 in the following experimental procedure) provides compelling evidence that the structure of the product is indeed **58.**

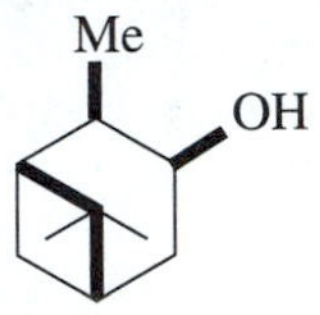

59
(+)-Neoisopinocampheol

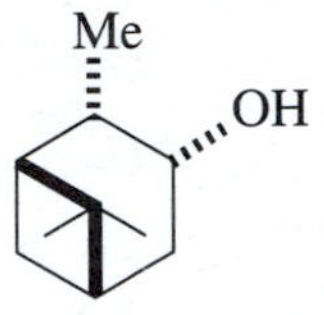

60
(–)-Neopinocampheol

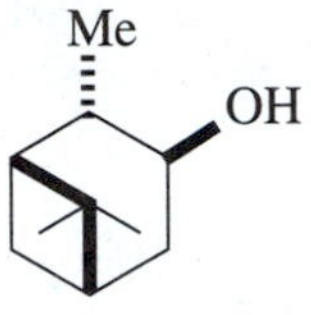

61
(+)-Pinocampheol

62
(+)-*cis*-2-Pinanol

63
(+)-*trans*-2-Pinanol

Pre-Lab exercises for Section 10.4.3, "Hydroboration-Oxidation of (+)-α-Pinene," are found on page PL. 41.

EXPERIMENTAL PROCEDURE

SAFETY ALERT

1. Tetrahydrofuran is a potentially toxic and flammable solvent. Do not use this solvent near open flames.
2. The 30% hydrogen peroxide used is a strong oxidant and may blister the skin on contact. If you accidentally spill some on your skin, wash the affected area with copious amounts of water.
3. The complex of borane and tetrahydrofuran reacts vigorously with water. Use care when disposing of any excess reagent and when quenching the reaction.

Equip a 125-mL round-bottom flask with a Claisen connecting tube, and prepare a drying tube filled with calcium chloride for attachment to the sidearm of the connecting tube; fit the other standard taper joint, which is directly in line with the lower male joint, with a rubber septum cap. Using a small flame, gently heat the flask to remove traces of water from the apparatus (see Sec. 2.24); do not overheat the joint with the septum cap. After the flask has been heated for a few minutes, discontinue the heating, add a magnetic stirbar, attach the drying tube, and allow the flask to cool to room temperature. Weigh out 0.26 g of NaBH$_4$ and measure 15 mL of dry tetrahydrofuran (THF) into a dry graduated cylinder. Remove the septum cap from the Claisen connecting tube, and quickly transfer first the NaBH$_4$ and then the THF to the round-bottom flask. Reattach the septum, cool the flask and its contents in an ice-water bath while stirring the mixture using a magnetic stirrer. Draw a solution of 0.85 g of iodine dissolved in 10 mL of dry THF into a glass syringe. Pierce the septum cap with the needle of the syringe, being careful not to press down on the plunger, and *slowly* add the iodine solution directly to the flask. Perform this addition at a rate such that all of this solution is added in 0.5 h; maintain the temperature of the reaction mixture at about 0 °C throughout the course of the addition.

Once this addition is complete, use a clean glass syringe to transfer a solution of 1.4 g of (+)-α-pinene in 2 mL of dry THF to the solution of borane-tetrahydrofuran complex; this addition should require about 5 min. Remove the ice-water bath immediately and allow the reaction mixture to warm to room temperature with continued stirring. After the reaction mixture has been at room temperature for about 1.5 h, remove the septum from the flask and *slowly* add 2 mL of water dropwise to the mixture to destroy the unreacted borane-THF complex; then add an additional 10 mL of THF to the mixture.✻

Oxidize the intermediate dialkylborane by dropwise addition of 20 mL of a solution of alkaline hydrogen peroxide that has been prepared by mixing 10 mL of 30% aqueous hydrogen peroxide with 10 mL of 3 *M* aqueous NaOH. Do this addition cautiously, as vigorous gas evolution may occur. Once the addition of alkaline peroxide is completed, stir the resulting mixture for about 5 min and then transfer it to a separatory funnel. Separate the layers, and extract the aqueous layer with three 10-mL portions of diethyl ether. Wash the combined organic layers sequentially with two 10-mL portions each of water and saturated brine and dry them over anhydrous sodium sulfate.✻ Filter or decant the dried solution into a 50-mL round-bottom flask and remove the volatile solvents by simple distillation. The final traces of solvents may be removed by attaching the flask to a vacuum source, and gently swirling the contents as the vacuum is applied (see Sec. 2.25, Fig. 2.37).✻

If you are instructed to do so, recrystallize the crude (–)-isopinocampheol (**50**) thus obtained by dissolving it in a *minimum* volume of warm ethanol and adding water until the mixture just becomes cloudy; complete the crystallization by cooling the mixture in an ice-water bath. Collect the crystals and determine their melting point. Verify the identity of the product by comparing its properties and the melting point of its α-naphthylcarbamate derivative, prepared according to the procedure in Section 23.5.5D, with the data summarized in Table 10.1.

TABLE 10.1. Physical Data for Possible Hydration Products of (+)-α-Pinene

Structure	Melting Point,°C	Melting Point,°C of α-Naphthylcarbamate	Sign of Rotation of Polarized Light
58	51–53 (55–56)	87.5–88	–
59	45–47 (48)	88	+
60	27	148	–
61	67	91	+
62	78–79	not formed	+
63	58–59	not formed	+

Finishing Touches Dilute the *aqueous layer* with water, neutralize it with acetic acid, and then flush it down the drain with excess water. Place the *distillate* containing THF and ether in the container for nonhalogenated organic solvents.

EXERCISES

1. Why is it necessary to maintain the pH of the reaction mixture above 7 during the oxidation of an alkyl borane with hydrogen peroxide?

2. Why was it necessary to use a dry apparatus and solvents for the preparation and reaction of the complex of borane with THF?

3. Predict the structure of the major product that would be formed upon hydroboration-oxidation of each of the alkenes (a)–(e). Include the correct stereochemistry wherever appropriate.

(a) (b) 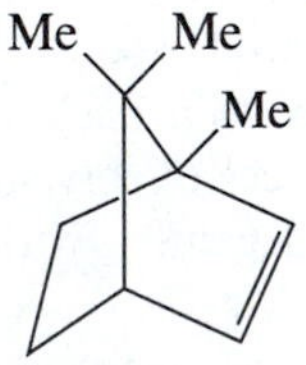 (c) 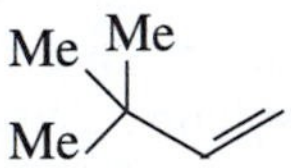 (d) 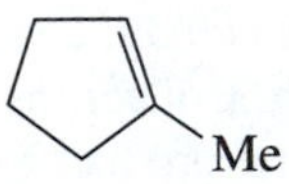 (e)

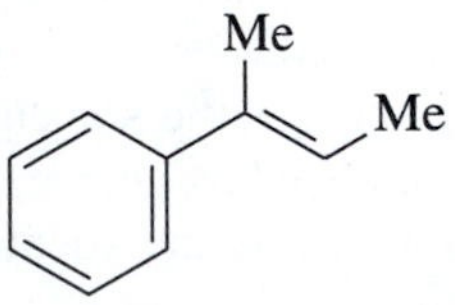

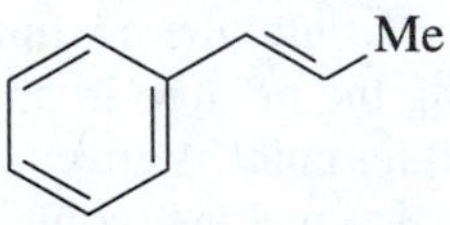

4. For each of the alkenes in Exercise 3, write the structure of the alcohol that should be the major product upon acid-catalyzed hydration.

5. One of the advantages of the sequence of hydroboration-oxidation is that in contrast to the acid-catalyzed hydration of alkenes, rearrangements of the carbon skeleton do not occur. For example, acid-catalyzed hydration of (+)-α-pinene (**56**) proceeds with skeletal rearrangements to produce a number of isomeric alcohols.

   a. Draw the structure(s) of the product(s) you would expect to obtain upon hydration of (+)-α-pinene if *no* skeletal rearrangements occurred.
   b. Draw the structures of the isomeric alcohols having the formula $C_{10}H_{18}O$ that you would expect to obtain upon treatment of (+)-α-pinene with aqueous acid.

6. The hydroboration of (+)-α-pinene (**56**) is a stereoselective reaction in which borane approaches from only one face of the carbon-carbon bond to give the dialkylborane **57**. Provide a rationale for this observed stereoselectivity.

7. Neither of the alcohols **62** or **63** forms an α-naphthylcarbamate derivative, whereas the isomers **58–61** each do. Explain the difference in reactivities.

SPECTRA OF STARTING MATERIAL AND PRODUCT

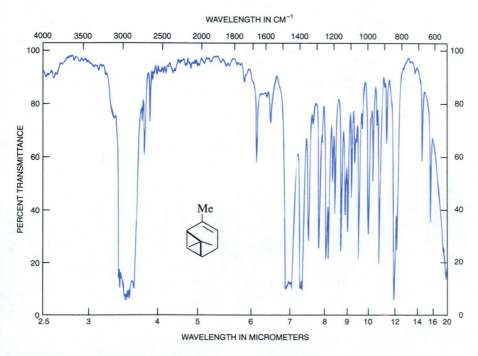

FIGURE 10.38   IR spectrum of (+)-α-pinene.

FIGURE 10.39 NMR data for (+)-α-pinene.

(a) ^{1}H NMR spectrum (90 MHz).

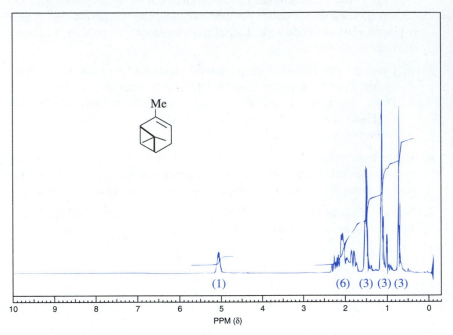

(b) ^{13}C NMR data.
Chemical shifts: δ 20.8, 22.8, 26.8, 31.4, 31.5, 38.0, 41.5, 42.2, 116.2, 144.1.

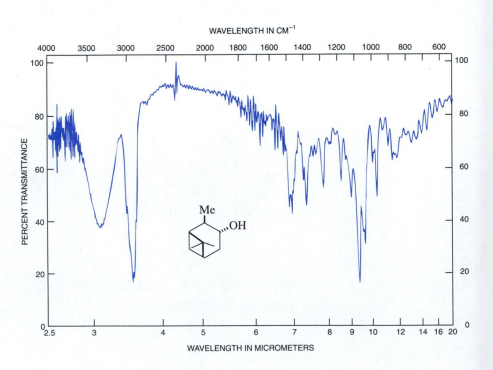

FIGURE 10.40 IR spectrum of (−)-isopinocampheol.

(a) ^{1}H NMR spectrum (90 MHz).

FIGURE 10.41   NMR data for (−) - isopinocampheol.

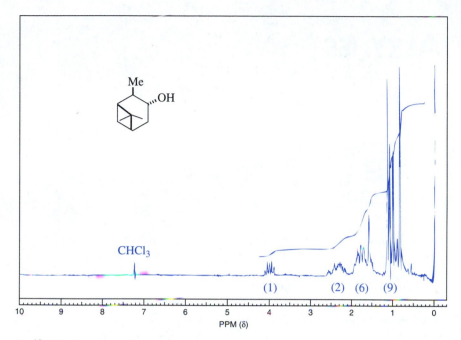

(b) ^{13}C NMR data.
  Chemical shifts: δ 20.7, 23.7, 27.7, 34.4, 38.1, 39.1, 41.8, 47.8, 47.9, 71.7.

ALKYNES

Organic compounds that contain a carbon-carbon triple bond as the functional group are called **alkynes.** Acetylene (ethyne), H–C≡C–H, is the simplest alkyne and is widely used in industry as a fuel and as a chemical feedstock for the preparation of other organic compounds such as vinyl chloride (CH_2=CHCl), a monomer used in the manufacture of polyvinyl chloride, and chloroprene (CH_2=CCl–CH=CH_2), which polymerizes to give neoprene.

The carbon-carbon triple bond is formed by the overlap of two orthogonal pairs of p-orbitals on adjacent carbon atoms. Like the π-electrons found in the double bond in alkenes, the π-electrons of the carbon-carbon triple bond in alkynes are loosely held and provide a Lewis base site for interaction with electrophiles, which are Lewis acids. One of the general classes of reactions that characterizes the chemistry of alkynes as well as alkenes (see Chap. 10) is the **electrophilic addition** of reagents generalized as **E–Nu** (Eq. 11.1). Under carefully controlled reaction conditions using only one equivalent of electrophilic reagent **E–Nu,** it is sometimes possible to stop the addition reaction at the alkene stage.

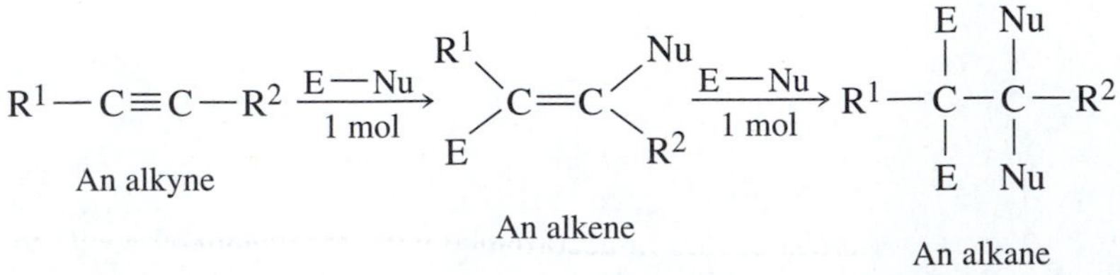

$$R^1-C\equiv C-R^2 \xrightarrow[\text{1 mol}]{E-Nu} \underset{\underset{\text{An alkene}}{E}}{\overset{R^1}{>}}C=C\overset{Nu}{\underset{R^2}{<}} \xrightarrow[\text{1 mol}]{E-Nu} R^1-\overset{\overset{E}{|}}{\underset{\underset{E}{|}}{C}}-\overset{\overset{Nu}{|}}{\underset{\underset{Nu}{|}}{C}}-R^2 \quad \textbf{(11.1)}$$

An alkyne   An alkane

The electrophilic addition of unsymmetrical reagents to terminal alkynes follows Markovnikov's rule. Thus, the electrophilic component of the reagent, which is a Lewis acid, adds to the electron-rich triple bond to form the more stable of the two possible intermediate vinyl carbocations, which then undergoes nucleophilic attack by the Lewis base component of the reagent (Eq. 11.2). Note that the addition of the second mole of E–Nu to the alkene has the same *regiochemistry* (see Sec. 10.4) as the initial addition of E–Nu to the alkyne (Eqs. 11.1 and 11.2).

$$R-C\equiv C-H \xrightarrow[\text{1 mol}]{E-Nu} \underset{Nu}{\overset{R}{\diagdown}}C=C\underset{H}{\overset{E}{\diagup}} \xrightarrow[\text{1 mol}]{E-Nu} R-\underset{\underset{Nu}{|}}{\overset{\overset{Nu}{|}}{C}}\equiv\underset{\underset{E}{|}}{\overset{\overset{E}{|}}{C}}-H \quad (11.2)$$

A terminal
alkyne

Because alkynes undergo electrophilic addition reactions, the same characterization tests that were employed to detect unsaturation in alkenes may be used for alkynes (see Sec. 23.5.2). For example, alkynes decolorize a solution of bromine in carbon tetrachloride, but two moles of bromine are required for complete reaction. Alkynes also decolorize potassium permanganate solutions in the Baeyer test. The acidity of the acetylenic hydrogen of **terminal alkynes** ($-C\equiv C-H$), pKa $\approx$ 25, provides the basis of a simple test for such compounds. Treating a terminal alkyne with a solution containing silver ammonia complex, prepared by dissolving silver nitrate in ammonia, provides a solid silver salt, as illustrated by the general reaction shown in Equation (11.3). This test differentiates terminal alkynes from alkenes, nonterminal alkynes, dienes, and allenes.

$$R-C\equiv C-H + [Ag(NH_3)_2]^{\oplus} \longrightarrow R-C\equiv C-Ag + NH_3 + NH_4 \quad (11.3)$$

A terminal
alkyne   A silver acetylide

This unique reaction of terminal alkynes may be used to separate them from nonterminal alkynes, which do not form insoluble silver salts and remain in solution. The silver acetylide salts can be reconverted to the terminal alkynes on treatment with hydrochloric acid as shown in Equation (11.4). *Dry silver salts of this type are quite sensitive to shock and tend to decompose explosively.* Therefore they should never be allowed to dry out before being decomposed by hydrochloric acid.

$$R-C\equiv C-Ag + HCl \longrightarrow R-C\equiv C-H + AgCl \quad (11.4)$$

An important reaction involving the carbon-carbon triple bond is **hydration** to give an **aldehyde** or **ketone,** a process of considerable value in synthetic organic chemistry. The reaction is catalyzed by mercuric sulfate in the presence of sulfuric acid and proceeds by the initial formation of a substituted vinyl alcohol, which is unstable and tautomerizes to a **carbonyl group, C=O.** The process is illustrated by the hydration of the terminal alkyne 2-methyl-3-butyn-2-ol (**1**) to give 3-hydroxy-3-methyl-2-butanone (**2,** Eq. 11.5). The hydration of terminal alkynes occurs in accordance with *Markovnikov's rule* to give alkyl methyl ketones, rather than in the *anti-Markovnikov* (see Sec. 10.4) sense that would give the corresponding aldehyde, 3-methyl-3-hydroxybutanal (**3**). Although the role that mercuric ion plays in this reaction is not completely understood, one hypothesis is that the mercuric ion forms a π-complex with the triple bond, thus rendering the alkyne more soluble in water. It is also possible that the mercuric ion, which is a polarizable Lewis acid, initially coordinates with the triple bond and thereby facilitates the attack of water. Subsequent loss of mercuric ion could then occur by a number of mechanistically different pathways.

$$
\underset{\substack{\textbf{1}\\ \text{2-Methyl-3-butyn-2-ol}}}{CH_3-\underset{\underset{CH_3}{|}}{\overset{\overset{OH}{|}}{C}}-C\equiv C-H} + H_2O \xrightarrow[\text{HgSO}_4]{\text{H}_2\text{SO}_4} \left[\underset{\substack{\text{A substituted vinyl alcohol (enol)}\\ (\textit{unstable; not isolated})}}{CH_3-\underset{\underset{CH_3}{|}}{\overset{\overset{HO}{|}}{C}}-\overset{\overset{OH}{|}}{C}=CH_2} \right] \rightleftharpoons \underset{\substack{\textbf{2}\\ \text{3-Hydroxy-3-methyl-}\\ \text{2-butanone}}}{CH_3-\underset{\underset{CH_3}{|}}{\overset{\overset{HO}{|}}{C}}-\overset{\overset{O}{\|}}{C}-CH_3}
$$

$$\tag{11.5}$$

$$
\underset{\substack{\textbf{3}\\ \text{(not formed)}}}{CH_3-\underset{\underset{CH_3}{|}}{\overset{\overset{HO}{|}}{C}}-CH_2-\overset{\nearrow O}{\underset{\searrow H}{C}}}
$$

The alkyne that we use as a substrate to study the chemistry of alkynes is 2-methyl-3-butyn-2-ol (**1**), which has a terminal triple bond. The presence of a hydroxyl group in **1** has little effect on the chemical properties of the carbon-carbon triple bond. Rather, the main effect of the polar hydroxyl group is on the physical properties of the molecule, with the melting and boiling points of **1** being considerably higher than those for acetylenic hydrocarbons having the same molecular weight.

That the hydration of **1** does give **2** and not **3** is verified by confirming the identity of the product using one of several general techniques that are commonly used in organic chemistry to characterize organic compounds. The first of these involves determining the spectroscopic properties of the molecule by the various analytical methods outlined in Chapter 8 and comparing the spectra of the product of the reaction with known standard samples. When all of the spectral properties match, the identity of the product is established. A second common method of assigning a structure to the product of a reaction depends on preparing a solid derivative and comparing its melting point with the melting point of the same derivative of a series of known compounds. If the melting points of the derivatives of the unknown and known samples are identical, then a mixed melting point is taken to confirm the mutual identity of the two samples (see Secs. 3.2.2 and 3.3.3).

Both of these general techniques may be illustrated by applying them to confirm the identity of the product formed by the hydration of **1.** For example, the IR, ^{1}H and ^{13}C NMR spectra of **2** and **3** are clearly distinct. A solid derivative of the product of the reaction could also be made for comparison with the same derivatives of authentic samples of **2** and **3** that had been synthesized by independent methods; see Section 23.5.1 for further information concerning identification of carbonyl compounds by preparing derivatives. Forming the semicarbazone **4,** which melts at 162.5 °C, from **1** proceeds smoothly, but converting the aldehyde **3** to its semicarbazone occurs with concomitant dehydration to give **5,** which melts at 222–223 °C; **5** is the semicarbazone of the unsaturated aldehyde $(CH_3)_2C{=}CHCHO$.

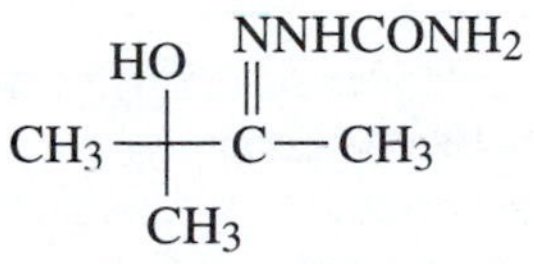

HO NNHCONH$_2$

CH$_3$ — C — CH$_3$

CH$_3$

4

CH$_3$ NNHCONH$_2$

C = CH — C

CH$_3$ H

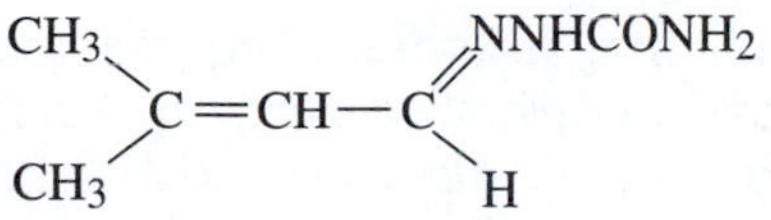

5

Pre-Lab exercises for Chapter 11, "Hydration of 2-Methyl-3-butyn-2-ol," are found on page PL. 43.

EXPERIMENTAL PROCEDURES

SAFETY ALERT

1. Handle the concentrated sulfuric acid carefully. Be sure that you add the acid slowly *to* water and *not* the reverse.
2. Take special care to destroy all of the silver salt with hydrochloric acid *before* discarding it.

A. QUALITATIVE TESTS FOR THE TRIPLE BOND

Reaction with Bromine in Carbon Tetrachloride.   Follow the experimental procedure given in Section 23.5.2A to test for the presence of the carbon-carbon triple bond in 2-methyl-3-butyn-2-ol. Note the **SAFETY ALERT** given there.

Baeyer Test for Unsaturation.   Follow the experimental procedure given in Section 23.5.2B to test for the presence of unsaturation in 2-methyl-3-butyn-2-ol. Note the **SAFETY ALERT** given there.

Reaction with Silver Ammonia Complex; Formation of a Silver Acetylide and its Decomposition.   Prepare a solution of silver ammonia complex from 2.5 mL of 0.1 *M* silver nitrate solution by adding ammonium hydroxide solution dropwise. Brown silver oxide forms first; add *just enough* ammonium hydroxide to dissolve the silver oxide. Dilute the solution by adding 1.5 mL of water. Add 3 mL of the diluted silver ammonia complex solution to about 0.1 mL of 2-methyl-3-butyn-2-ol, and note the formation of the silver acetylide salt. Collect the silver salt by vacuum filtration of the aqueous solution; *be careful not to let the silver salt dry because the dry salt is explosive.* Treat the silver salt with a small amount of dilute hydrochloric acid, and observe what changes occur, especially in the color and form of the precipitate. Destroy *all* of the salt according to the procedure given in **FINISHING TOUCHES.**

> **Finishing Touches** Pour the *carbon tetrachloride solution* from the bromine test for unsaturation in a container for halogenated organic liquids, and place the *manganese dioxide* from the Baeyer test in a container for heavy metals. Dilute the *silver ammonia complex test solution* that remains with water and flush it down the drain with water. Destroy all *solid silver acetylide salt* by treatment with hydrochloric acid. Collect the *silver chloride* and put it in a container for heavy metals, and then flush the *filtrate* down the drain.

B. Preparation of 3-Methyl-3-hydroxy-2-butanone

Carefully add 3 mL of *concentrated* sulfuric acid *to* 20 mL of water contained in a 100-mL round-bottom flask. Dissolve 0.2 g of reagent grade mercuric oxide in the resulting warm solution and then cool the flask to about 50 °C. Attach a reflux condenser to the flask, and add 4.2 g of 2-methyl-3-butyn-2-ol in one portion through the top of the condenser. The cloudy white precipitate that forms is presumably the mercury complex of the alkyne. Shake or stir the reaction flask to mix the contents thoroughly, whereupon a mildly exothermic reaction may ensue. Allow the reaction to proceed by itself for about 2 min; then heat the mixture to reflux. The precipitate will dissolve, and the solution will turn light brown. Continue heating the mixture at reflux for 30 min; then allow it to cool to room temperature.✱ While the reaction mixture is cooling, prepare an apparatus for steam distillation *with internal steam generation* as in Figure 2.22. Add about 35–40 mL of water to the reaction flask. Heat the flask and steam distil the product until about 30–35 mL of distillate is collected.✱

Transfer the distillate to a small separatory funnel, add 2–3 g of potassium carbonate and then carefully saturate the solution with sodium chloride; avoid adding more sodium chloride than will dissolve readily. If a second layer forms at this point, do not separate it but continue with the extraction as described. Extract the mixture with two 10-mL portions of dichloromethane and combine the extracts. Dry the organic extracts over *anhydrous* sodium sulfate until they are clear.

Decant the solution into a 50-mL round-bottom flask and remove the dichloromethane by simple distillation using the short path apparatus shown in Figure 2.15(b); cool the receiving flask in an ice-water bath. Using a micropipet to minimize material loss, transfer the crude product into a 10-mL round-bottom flask. Remove the last traces of dichloromethane from the crude product by connecting the flask to a water aspirator; then continue the distillation using the same apparatus fitted with a clean, tared receiving flask. Since the volume of product is likely to be only 1–2 mL, an accurate boiling point may be difficult to obtain, and all material that distils above 110–115 °C should be collected in order to obtain a reasonable yield; pure product, which should be a colorless liquid, boils between 138–141 °C (760 torr). Characterize the crude or purified product by spectroscopic or chemical methods as described in Sections C and D that follow.

Finishing Touches   Since the *residue in the stillpot* from the steam distillation contains soluble mercury salts, place it in a special container for solutions of mercuric salts; do *not* pour it down the drain. If directed to do so by your instructor, you may convert the *soluble mercuric salts* into insoluble mercuric sulfide by reaction with sodium sulfide or hydrosulfide; by vacuum filtration collect the *mercuric sulfide* for recycling. Flush the *other aqueous solutions* down the drain with water. Pour the *dichloromethane* recovered as the forerun in the distillation in a container for halogenated organic solvents.

C.   IDENTIFICATION OF THE PRODUCT: SPECTROSCOPIC METHODS

The IR and NMR spectra of both the starting material **1** and the product **2** are provided at the end of this chapter. Although it is possible to identify the product by converting it to a known solid derivative, it is also quite easy to prove the structure using spectral methods. An examination of the infrared spectra provided indicates the absence of the C=O group in **1,** whereas an intense absorption for this group is clearly present in **2.** With the aid of Tables 8.2 and 8.3, identify the absorption attributed to this functional group. The ^{1}H NMR spectrum also shows a terminal acetylenic hydrogen for **1,** whereas the hydrogens of the newly formed CH_3 adjacent to the C=O are observed for **2.** Using Tables 8.4 and 8.5, identify the peaks in the ^{1}H NMR spectra.

D.   IDENTIFICATION OF THE PRODUCT: SEMICARBAZONE FORMATION

Add 0.5 mL of **2** to a solution prepared from 0.5 g of semicarbazide hydrochloride and 0.8 g of sodium acetate dissolved in 2.5 mL of water. Shake or stir the mixture vigorously. Collect the crude solid semicarbazone that forms; use a glass rod to scratch at the air-liquid interface if necessary to initiate precipitation. Dissolve the solid product in a minimum volume of hot 2-propanol, cool the mixture to room temperature and then in an ice-water bath, and collect the crystals by vacuum filtration. Dry the crystals and determine the melting point. Use this information to confirm the identity of the product of hydration of 2-methyl-3-butyn-2-ol.

Finishing Touches   Dilute the *filtrate* from the reaction and the *mother liquor* from the recrystallization with water, and flush the solutions down the drain.

EXERCISES

1. Suggest a method for preparing 2-methyl-3-butyn-2-ol from two simple organic compounds.

2. How might you separate, and obtain in pure form, 1-octyne and 2-octyne from a mixture containing both of them?

3. Give the structures for the products that you would expect to obtain on hydration of 1-octyne and of 2-octyne.

4. It is well known that alcohols can be oxidized to ketones by oxidizing agents such as potassium permanganate. An examination of alkyne **1** shows that it contains a hydroxyl group as well as a triple bond. Therefore, a student might conclude that the positive permanganate test is due to the oxidation of the hydroxyl group and not to reaction of the triple bond. Could this be the case? If so, what additional control experiments might you suggest to eliminate this possibility from further consideration?

5. Consider the compounds hexane, 1-hexene, and 1-hexyne. What similarities and what differences would you expect in the reactions of these compounds with (a) bromine in carbon tetrachloride; (b) an aqueous solution of potassium permanganate; (c) an aqueous solution of sulfuric acid; and (d) a solution of silver ammonia complex? Give the structures of the products, if any, that would be obtained from each of these reactions.

6. ^{1}H NMR spectroscopy could also be used to determine whether **2** or **3** is produced by the hydration of **1.** Sketch the ^{1}H NMR spectrum you would expect from **3,** and discuss the differences that you would expect in the ^{1}H NMR spectra of **2** and **3.**

7. IR spectroscopy may be used to determine whether **2** or **3** is produced by the hydration of **1.** Describe the important difference(s) you would expect in the IR spectra of the two compounds.

8. The reaction of 1–butyne with hydrogen bromide gives 2,2-dibromobutane, as shown in Equation (11.6). Write a stepwise mechanism for this process, using curved arrows to symbolize flow of electrons. Also provide a rationale for the observed regioselectivity for addition of each mole of hydrogen bromide.

$$CH_3CH_2C \equiv C - H + 2\ H - Br \longrightarrow CH_3CH_2CBr_2CH_3 \qquad (11.6)$$

SPECTRA OF STARTING MATERIAL AND PRODUCT

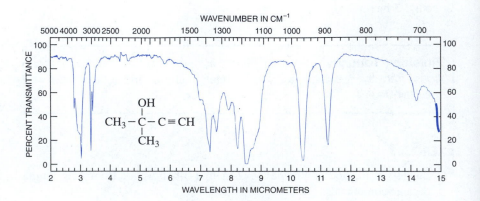

FIGURE 11.1 IR spectrum of 2-methyl-3-butyn-2-ol.

(a) ^{1}H NMR spectrum (60 MHz).

FIGURE 11.2 NMR data for 2-methyl-3-butyn-2-ol.

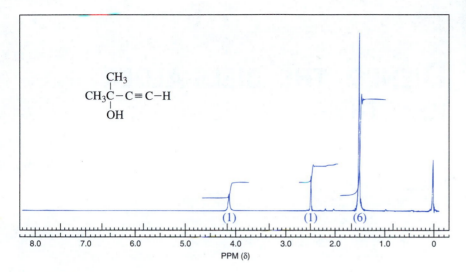

(b) ^{13}C NMR data.
Chemical shifts: δ 31.3, 64.8, 70.4, 89.1.

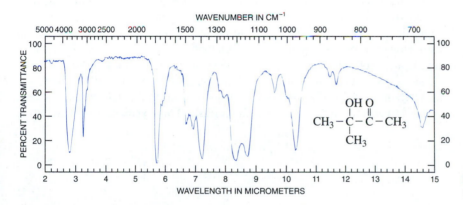

FIGURE 11.3 IR spectrum of 3-hydroxy-3-methyl-2-butanone.

(a) ^{1}H NMR spectrum (60 MHz).

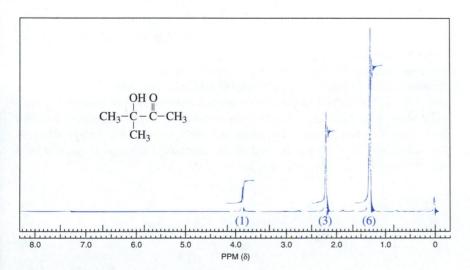

(b) ^{13}C NMR data.
Chemical shifts: δ 23.8, 26.4, 76.7, 213.2.

FIGURE 11.4 NMR data for 3-hydroxy-3-methyl-2-butanone.

DIENES: THE DIELS-ALDER REACTION

12.1 INTRODUCTION

In synthetic organic chemistry, reactions that produce carbocyclic rings from acyclic precursors are of great importance because they lead to the easy formation of complex molecular structures. Rings may be produced from the cyclization of an acyclic starting material by forming a single carbon-carbon bond. However, one of the most useful methods for constructing six-membered rings from acyclic starting materials involves the reaction of a 1,3–diene with an alkene to give a derivative of cyclohexene with the formation of *two* new carbon–carbon bonds, as illustrated by Equation (12.1). In this reaction, the alkene partner is generally referred to as a **dienophile.** This **cycloaddition** reaction is

$$\text{(12.1)}$$

1,3-Diene Dienophile

called the Diels-Alder reaction in honor of Otto Diels and Kurt Alder, the two German chemists who recognized its importance and shared the 1950 Nobel Prize in chemistry for their extensive development of this reaction. A more detailed account of their elegant work is included in the historical highlight in this chapter. The Diels-Alder reaction also belongs to a class of reactions termed 1,4-additions because the two new carbon-carbon σ-bonds are formed between the 1- and 4-carbon atoms of the diene and the two π-bonded carbon atoms of the dienophile. If an alkyne is used as the dienophile rather than an alkene, a derivative of 1,4–cyclohexadiene is produced (Eq. 12.2).

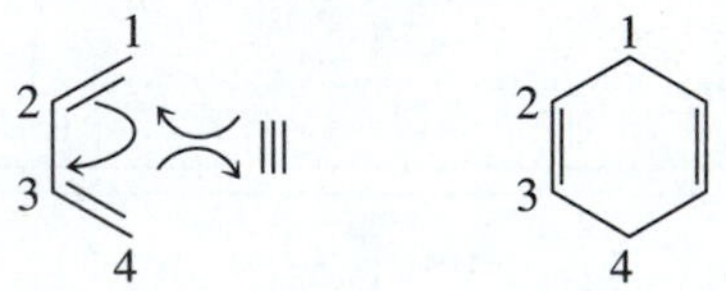

$$\text{(12.2)}$$

The scope of the Diels-Alder reaction is broad, and many combinations of dienes and dienophiles furnish cycloadducts in good yields. The presence of electron-releasing substituents such as alkyl, R, and alkoxy, OR, groups on the diene and electron-withdrawing substituents such as cyano, C≡N, and carbonyl, C=O, groups on the dienophile increases the rate and the yield of the reaction. When the dienophile is electron-deficient, Lewis acids may be used to catalyze the Diels-Alder reaction. The reaction can also be made to occur when the diene is electron-poor, and the dienophile is electron-rich. The Diels-Alder reaction is remarkably free of complicating side reactions, and the yields of the desired product are often high. Probably the single most important side reaction that may be encountered is dimerization or polymerization of the diene (see Ex. 2g).

12.2 MECHANISTIC AND STEREOCHEMICAL ASPECTS

The mechanism of the Diels-Alder reaction is now fairly well understood. The accumulated evidence favors a picture in which reorganization of the π-electrons is occurring in the transition state for the reaction so the two new σ-bonds are formed in a more or less **synchronous** or **concerted** fashion; little charge or free-radical character is developed at any of the terminal carbon atoms (Eq. 12.3). The reaction is one of a number of **pericyclic reactions** that are controlled by **orbital symmetry.** Detailed discussions of orbital symmetry and its ramifications may be found in all modern organic textbooks, as well as the references at the end of this chapter.

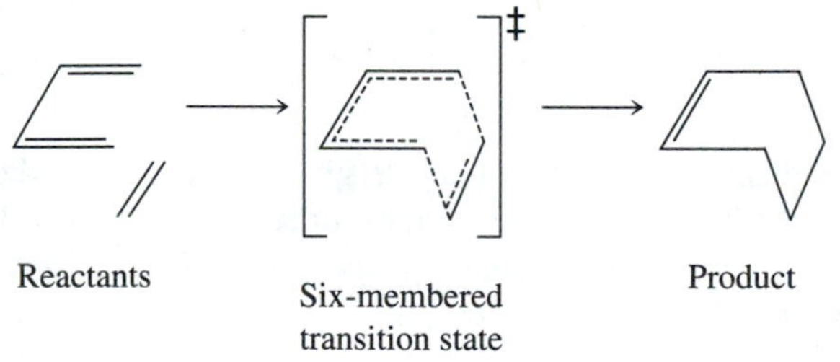

| Reactants | Six-membered transition state | Product | (12.3) |

A consequence of the concerted nature of the Diels-Alder reaction is that the diene must be able to adopt a planar conformation in which the dihedral angle between the two double bonds is 0°. Such a conformation is designated as *s-cis*, meaning **cisoid** about the single bond. Dienes that have difficulty achieving this conformation are expected to undergo Diels-Alder reactions slowly or not at all. A concerted reaction of a dienophile with a diene in its *s-trans* conformation would lead to a six-membered ring containing a highly strained *trans* double bond, an improbable structure. An attempt to construct a molecular model of *trans*-cyclohexene should convince you on this point.

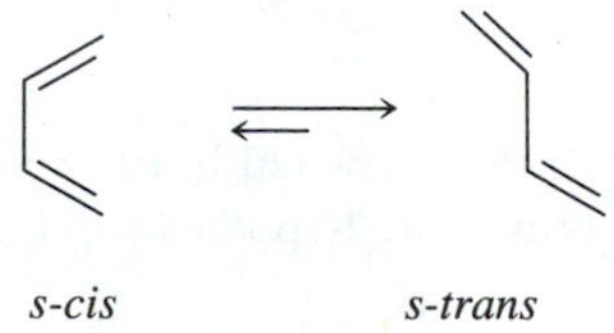

s-cis   *s-trans*

A consequence of the highly ordered relationship between the diene and dienophile in the transition state depicted in Equation (12.3) is that Diels-Alder reactions are highly **stereoselective** with the formation of the new carbon-carbon σ-bonds occurring on the *same* face of *both* the diene and the dienophile. Such addition reactions are termed **syn-additions** (see also Sec. 10.4). This selectivity is readily verified. For example, the cycloadduct resulting from reaction of *trans,trans*-2,4-hexadiene and a dienophile is *exclusively cis*-3,6-dimethylcyclohexene (Eq. 12.4), whereas the product obtained from *cis, trans*-2,4-hexadiene and a dienophile is *trans*-3,6-dimethylcyclohexene. Both these reactions proceed by syn-1,4-addition to the diene, with the stereochemical relationship between the two methyl groups on the diene being maintained. Similarly, the reaction

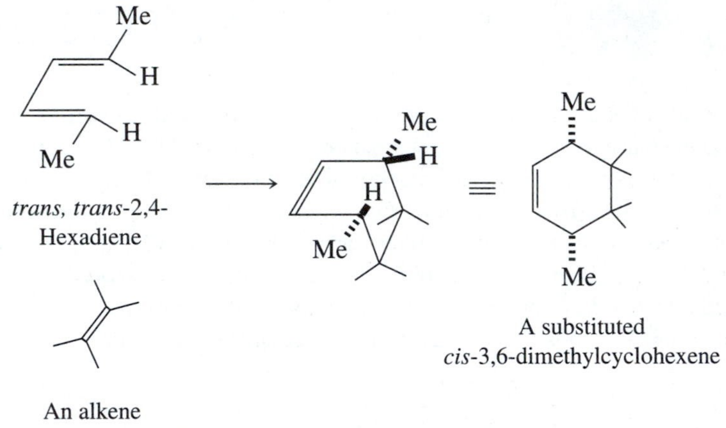

trans, trans-2,4-Hexadiene

An alkene

A substituted *cis*-3,6-dimethylcyclohexene

(12.4)

between 1,3-butadiene and dimethyl maleate gives solely the *cis*-diester **1** (Eq. 12.5). This example illustrates that the Diels-Alder reaction proceeds by exclusive *syn*-1,2-addition to the double bond of the dienophile; no isomerization to the more stable *trans*-diester is observed.

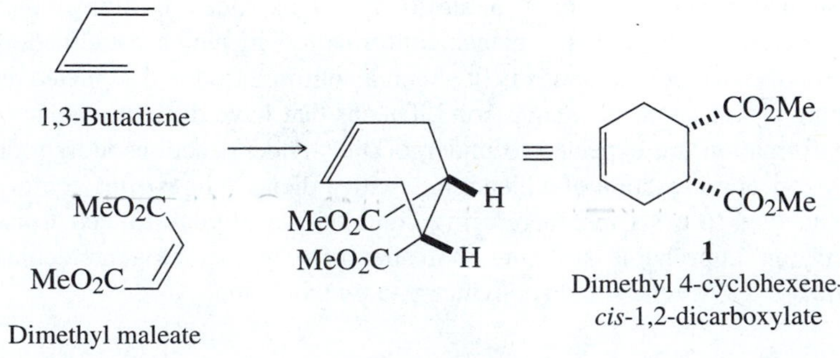

1,3-Butadiene

Dimethyl maleate

1
Dimethyl 4-cyclohexene-*cis*-1,2-dicarboxylate

(12.5)

The results of the reactions depicted in Equations (12.4) and (12.5) are in accord with, but do not prove, the hypothesis that the Diels-Alder reaction is concerted, with both new carbon-carbon bonds forming simultaneously. For example, they are also consistent with a stepwise mechanism in which the intermediate produced after the first carbon-carbon single bond is formed undergoes cyclization to the product *faster* than rotation about any of the bonds.

There is a second type of stereoselectivity that is characteristic of the Diels-Alder reaction. The addition of a dienophile such as maleic anhydride to a *cyclic* diene like 1,3–cyclopentadiene could in principle provide two products, the *endo*-adduct **2** and the *exo*-adduct **3** (Eq. 12.6). However, only the *endo*-cycloadduct **2** is observed experimentally, and its preferential formation follows what is now commonly termed the **Alder rule.** The basis for this result is believed to be stabilization of the transition state **4** by **through-space, or secondary, orbital interactions** between the *p*-orbitals on the internal carbons of the diene and the carbonyl carbon atoms of the dienophile. Analogous stabilization is not possible in transition state **5.** Transition state **4** is characterized as the one having the *maximum accumulation of double bonds,* an alternate way of stating that stabilization results from development of the through-space overlap of the *p*-orbitals. It should be noted that not all Diels-Alder reactions are as stereoselective as the one between 1,3-cyclopentadiene and maleic anhydride; mixtures of *endo-* and *exo*-products are sometimes obtained.

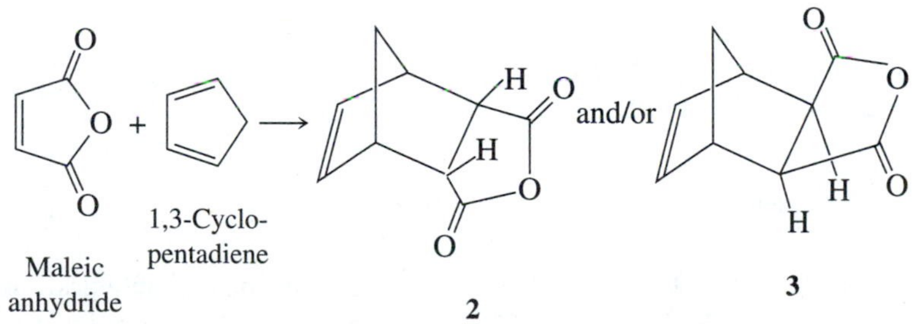

Maleic anhydride   1,3-Cyclo-pentadiene   2   and/or   3   (12.6)

2 ⟵   or   ⟶ 3

4   5

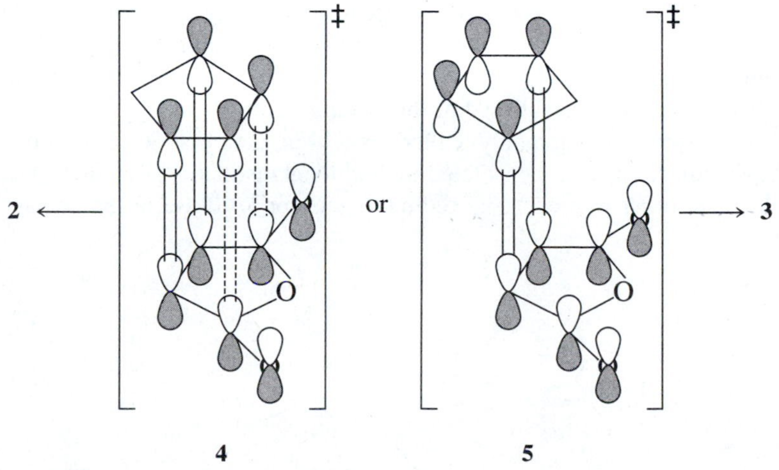

12.3 APPLICATIONS OF DIELS-ALDER REACTIONS

Three experimental procedures that illustrate various features of the Diels-Alder reaction are presented in this section. In all cases the highly reactive dienophile maleic anhydride is used. In Part A, the Diels-Alder reaction of 1,3-butadiene with maleic anhydride to give 4-cyclohexene-*cis*-1,2-dicarboxylic anhydride is performed (Eq. 12.7). Since 1,3-butadiene is a gas at room temperature (bp −4.4 °C, 760 torr), Diels-Alder reactions involving this diene are normally performed in sealed steel pressure vessels into which the diene is introduced under pressure. However, 1,3-butadiene is generated conveniently by the thermal decomposition of 3-sulfolene (Eq. 12.8), and it then reacts immediately with a dienophile that is present. This technique for the *in situ* preparation of 1,3-butadiene is used in this experiment to prepare 4-cyclohexene-*cis*-1,2-dicarboxylic anhydride.

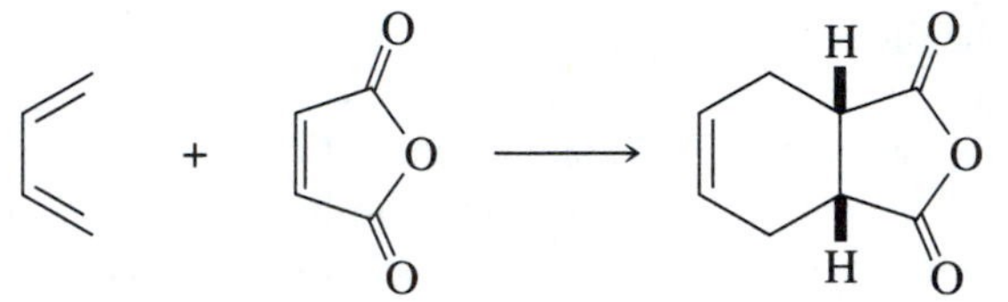

1,3-Butadiene   Maleic anhydride   4-Cyclohexene-*cis*-1,2-
dicarboxylic anhydride

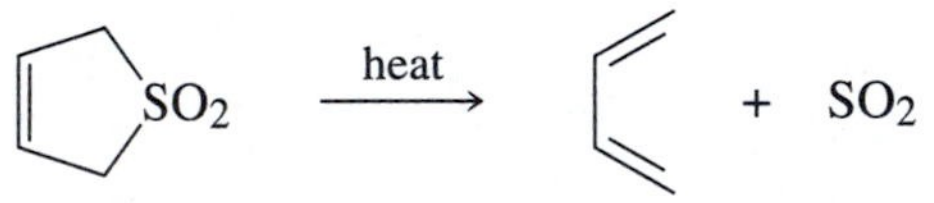

(12.7)

3-Sulfolene   1,3-Butadiene

(12.8)

In Part B, the cycloaddition of cyclopentadiene with maleic anhydride is carried out according to Equation 12.6. This experiment provides an opportunity to examine the stereoselectivity of the Diels-Alder reaction. Monomeric 1,3-cyclopentadiene cannot be purchased because it readily dimerizes by a Diels-Alder reaction at room temperature to give dicyclopentadiene (Eq. 12.9), which is commercially available. Fortunately, the equilibrium between the monomer and the dimer can be established at the boiling point of the dimer (170 °C, 760 torr) by a process commonly called **cracking,** and the lower-boiling 1,3-cyclopentadiene can then be isolated by fractional distillation. The diene *must* be kept cold in order to prevent its redimerization prior to use in the Diels-Alder reaction.

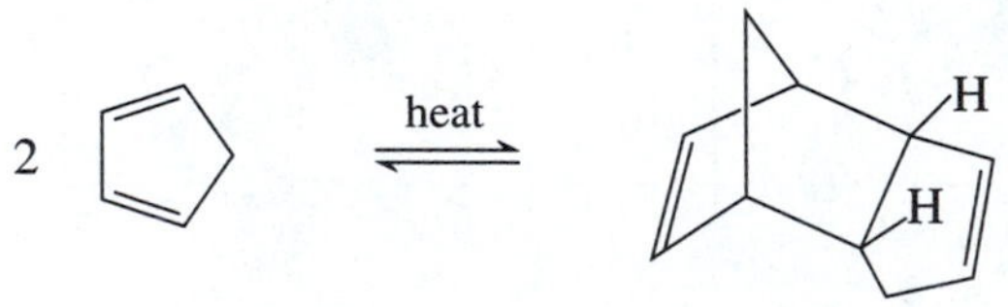

1,3-Cyclopentadiene   Dicyclopentadiene

(12.9)

In Part C the reaction of maleic anhydride with a commercially available mixture of *cis-* and *trans-*1,3-pentadiene is examined (Eq. 12.10). The goal of this experiment is to determine the relative reactivity of these two dienes in this Diels-Alder reaction. Assuming that the rule of maximum accumulation of double bonds, the Alder rule, applies to this reaction, the cycloadducts **6** and **7** are the expected products. Analysis of the mixture of *cis-* and *trans-*1,3-pentadiene by GLC prior to and after reaction with maleic anhydride gives a preliminary indication of which isomer of 1,3-pentadiene reacts preferentially (Fig. 12.1).

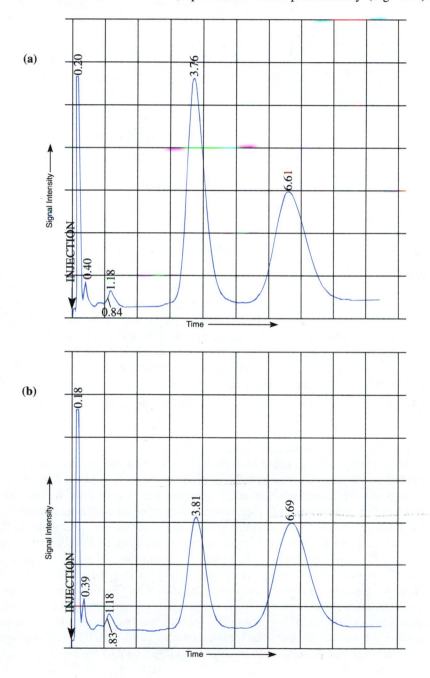

FIGURE 12.1   GLC traces of *cis-* and *trans-*1,3-pentadienes. (a) Commercial sample. (b) Distillate following reaction with maleic anhydride. Analysis on 1 *M* 30% AgNO₃-ethylene glycol on 60/80 Chromosorb P, at room temperature. Numbers above peaks are retention times, measured in minutes.

Since the melting points of **6** and **7** are different (40–41 °C and 62–63 °C, respectively), it is possible to identify the preferred product simply by isolating the solid and determining its melting point. As successfully completing Exercises 13–15 shows, it is then possible to assess the relative reactivities of these two dienes toward maleic anhydride.

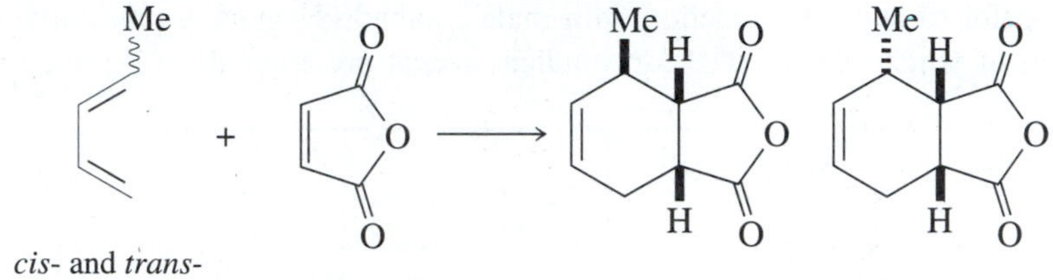

cis- and *trans-*
1,3-Pentadiene   Maleic anhydride   **6**   **7**   (12.10)

EXPERIMENTAL PROCEDURES

Pre-Lab exercises for Section 12.3, "A. Reaction of 1,3-Butadiene and Maleic Anhydride," are found on page PL. 45.

A. REACTION OF 1,3-BUTADIENE AND MALEIC ANHYDRIDE
SAFETY ALERT

> **1.** Be certain that all joints in the apparatus are tight and well lubricated before heating the reaction mixture. The organic solvents used are highly flammable, and the sulfur dioxide that is evolved is toxic and foul-smelling. Be sure that the gas trap is functioning properly before starting to heat the reaction mixture to prevent emission of gaseous sulfur dioxide into the laboratory.

Place 2.5 g of 3-sulfolene, 1.5 g of finely pulverized maleic anhydride, and 1 mL of dry xylene in a 25-mL round-bottom flask equipped with a water-cooled reflux condenser. Fit the condenser with the gas trap as shown in Figure 9.1. Warm the flask gently while swirling it to effect solution; and then heat the mixture to a gentle reflux, using a small burner, heating mantle, or oil bath; continue heating for about 30 min.✻ Cool the solution and add about 10 mL of xylene to the residue. Heat the mixture on a steam bath with swirling to dissolve all of the solids, and transfer the hot solution to an Erlenmeyer flask. If any of the product crystallizes in the flask during this operation, heat the mixture gently until all of the solid redissolves; carefully add petroleum ether until cloudiness develops, and set the solution aside to cool to room temperature.✻ Collect the crystals by vacuum filtration, and dry them thoroughly. Record their weight and determine the melting point of the product. The reported melting point is 103–104 °C. Perform the qualitative tests for unsaturation described in Section 23.5.2.

Convert the anhydride to the corresponding diacid in the following manner. Place 1.0 g of anhydride and 5 mL of distilled water in a 25-mL Erlenmeyer flask, and heat the mixture over a Bunsen burner until it boils and all the oil that forms initially dissolves. Allow the solution to cool to room temperature; then induce crystallization by scratching the flask at the air-liquid interface.* After crystallization begins, cool the flask to 0 °C to complete the process and collect the solid by vacuum filtration. Determine the yield and melting point of the diacid.

Recrystallize the diacid from water. The reported melting point of 4-cyclo-hexene-cis-1,2-dicarboxylic acid is 164–166 °C. Perform qualitative tests to show whether hydrolysis has destroyed the carbon-carbon double bond (see Sec. 23.5.2). Also test a saturated aqueous solution of the diacid with litmus or pHydrion paper and record the result.

The anhydride may also be *directly* hydrolyzed to the diacid in the following way. After the initial 30-min period of reflux, allow the reaction to cool for several minutes and add about 5 mL of water to the flask through the top of the condenser. Heat the resulting mixture under gentle reflux with swirling or stirring for at least 30 min. Cool the mixture to room temperature. If no crystals form, acidify the mixture with about 0.3 mL (6 drops) of concentrated sulfuric acid; swirl to mix and cool the mixture to 0 °C. Collect the resulting crude diacid by vacuum filtration; the crude diacid has mp 160–164 °C. Purify it by recrystallization from water, if necessary.

Finishing Touches   Pour the *mixture of xylenes and petroleum ether* recovered as the initial mother liquor from the reaction into a container for nonhalogenated organic solvents. Neutralize the *aqueous mother liquors* from hydrolysis of the anhydride with sodium carbonate, dilute the solution, and flush it down the drain. Place the *carbon tetrachloride solution* from the bromine test for unsaturation in a container for halogenated organic liquids, and put the *manganese dioxide* from the Baeyer test for unsaturation in a container for heavy metals.

Pre-Lab exercises for Section 12.3, "B. Reaction of 1,3-Cyclopentadiene and Maleic Anhydride," are found on page PL. 47.

B. REACTION OF 1,3-CYCLOPENTADIENE AND MALEIC ANHYDRIDE

SAFETY ALERT

1. 1,3-Cyclopentadiene is a mildly toxic and rather volatile substance. Prepare and use it in a hood if possible; avoid inhaling its vapors. Keep the diene cold at all times to minimize vaporization and dimerization.
2. Use open flames only in those steps in which you are directed to do so. The organic solvents and the 1,3-cyclopentadiene used in this experiment are highly flammable. Be certain that all joints in the apparatus are tight before heating the dicyclopentadiene to produce the monomer.

Place 7 mL of dicyclopentadiene in a 25-mL round-bottom flask, and attach the flask to a fractional distillation apparatus that is equipped with an *ice-cooled* receiver and fitted with a calcium chloride drying tube and an *unpacked* fractionating column. Using either a small burner or a heating mantle, *gently* heat the dimer until the solution is briskly refluxing (*Caution*: occasional foaming) and the monomer begins to distil in the range 40–42 °C. Distil the monomer as rapidly as possible, but do not permit the head temperature to exceed 43–45 °C. About 1.5–2.0 g of monomer should be obtained after the distillation, which will require about 30 min. Terminate the distillation when approximately 1 mL of residue remains in the flask. If the distilled 1,3-cyclopentadiene is cloudy because of condensation of moisture in the cold receiver, add about 0.5 g of *anhydrous* calcium chloride to dry it. To simplify this experiment, the cracking of dicyclopentadiene may be done by your instructor.

While the distillation is in progress, place 1.5 g of maleic anhydride in a 25-mL Erlenmeyer flask and dissolve it in 5 mL of ethyl acetate by heating on a steam or hot-water bath. Add 5 mL of petroleum ether and cool the solution thoroughly in an ice-water bath. Be sure the solution is still *homogeneous* prior to executing the next step. To this cooled solution add 1.2 g of dry 1,3-cyclopentadiene, and swirl the resulting solution gently until the exothermic reaction subsides and the cycloadduct separates as a white solid.✱ Heat the mixture on a steam or hot water bath until the solid has redissolved; then let the solution cool slowly to room temperature.✱ Filter the solution, and determine the yield and melting point of the solid anhydride obtained. The reported melting point is 164–165 °C. Test the product for unsaturation (see Sec. 23.5.2).

Hydrolyze 1.0 g of the anhydride to the diacid according to the procedure given in Part A. The reported melting point of the dicarboxylic acid is 180–182 °C. Perform qualitative tests to show whether hydrolysis has destroyed the carbon-carbon double bond (see Sec. 23.5.2). Also test a saturated aqueous solution of the diacid with litmus or pHydrion paper and record the result.

Finishing Touches   Pour the *pot residue* of dicyclopentadiene and any unused 1,3-cyclopentadiene into a container for recovered dicyclopentadiene or a container for nonhalogenated organic solvents. If calcium chloride was used to dry the cyclopentadiene, allow the *calcium chloride* to dry on a tray in the hood, and then place it in a container for nonhazardous solids. Since the *recrystallization solvent* contains only a small amount of solid product, put it in the container for nonhalogenated organic solvents. Neutralize the *aqueous mother liquors* with sodium carbonate, dilute the solution, and then flush it down the drain. Pour the *carbon tetrachloride solution* from the test for unsaturation in a container for halogenated organic liquids, and place the *manganese dioxide* from the test for unsaturation in a container for heavy metals.

Pre-Lab exercises for Section 12.3, "C. Reaction of *cis*- and *trans*-1,3-Pentadiene and Maleic Anhydride," are found on page PL. 49.

C. REACTION OF *CIS-* AND *TRANS-***1,3-PENTADIENE AND MALEIC ANHYDRIDE**

SAFETY ALERT

1. Use open flames only in those steps in which you are directed to do so. The organic solvents and the 1,3-pentadiene used in this experiment are highly flammable. Be certain that all joints in the apparatus are well lubricated and tight before heating the reaction mixture to reflux and before distilling the solvent.
2. 1,3-Pentadiene is a low-boiling, unpleasant-smelling liquid. Avoid release of its vapor into the laboratory by keeping vessels containing it closed as much as possible. Handle the diene in a hood, if possible.

Fit a 25-mL round-bottom flask with a reflux condenser and a drying tube containing calcium chloride, and dry the apparatus by heating it with a Bunsen burner (see Sec. 2.24). Place 1.0 g of powdered maleic anhydride in the flask and add to it a solution containing 2.5 g of a commercial mixture of *cis-* and *trans-*1,3-pentadiene (about 65–70% purity) and 5 mL of toluene. Warm this mixture *gently* with a Bunsen burner, oil bath, or heating mantle, and swirl the flask to dissolve most of the maleic anhydride. Continue to heat the resulting mixture under *gentle* reflux for 30 min, during which time everything should dissolve. Allow the mixture to cool to room temperature, equip the reaction flask for simple distillation and distil about 4–5 mL of volatiles from the reaction mixture. Allow the turbid residue in the stillpot to cool to room temperature, and add about 3–5 mL of diethyl ether to effect coagulation of any gummy polymer that forms. Remove the solids by vacuum filtration through a pad of a filter-aid (see Sec. 2.16); concentrate the filtrate to a volume of about 3 mL, either on a steam cone in a hood or by simple distillation (*no flames*). Cool the residue in an ice-water bath to effect crystallization; use a glass rod to scratch at the air-liquid interface or seed the solution to induce crystallization if necessary. Collect the slightly off-white product by vacuum filtration.

Reserve a small portion of the product to determine a melting point of the crude material, and recrystallize the remainder by dissolving it in a minimum amount of warm cyclohexane and then letting the solution cool to room temperature. Cool the mixture in an ice-bath to complete recrystallization and collect the crystals. Determine the weight and melting point of the purified material, and use the latter as a basis for deciding which of the two possible Diels-Alder adducts has been obtained. The reported melting points of **6** and **7** are 40–41 °C and 62–63 °C, respectively.

Hydrolyze the anhydride to the diacid according to the procedure given in Part A. The reported melting points of the diacids derived from **6** and **7** are 161–162 °C and 156–158 °C, respectively. Perform qualitative tests to show whether hydrolysis has destroyed the carbon-carbon double bond (see Sec. 23.5.2). Also test a saturated aqueous solution of the diacid with litmus or pHydrion paper and record the result.

> **Finishing Touches**   Pour the *distillate from the reaction mixture and the mother liquors from crystallization* in a container for nonhalogenated organic solvents. Place the *solid polymer side product* from the reaction in a container for nonhazardous organic solids. Neutralize the *aqueous mother liquors* with sodium carbonate, dilute the solution, and then flush it down the drain. Pour the *carbon tetrachloride solution* from the bromine test for unsaturation in a container for halogenated organic liquids, and place the *manganese dioxide* from the Baeyer test for unsaturation in a container for heavy metals.

EXERCISES

General Questions

1. Why does a 1,3-diene like 1,3-cyclopentadiene dimerize much more readily than does an acyclic one like *trans*-1,3-pentadiene?

2. Write structures for the products expected in the following possible Diels-Alder reactions. If no reaction is anticipated, write "N.R."

(a)

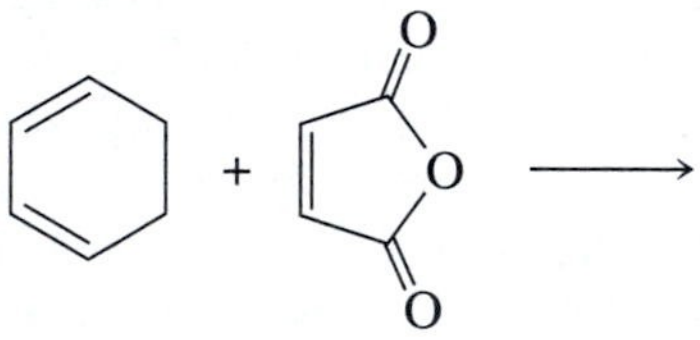

(b)

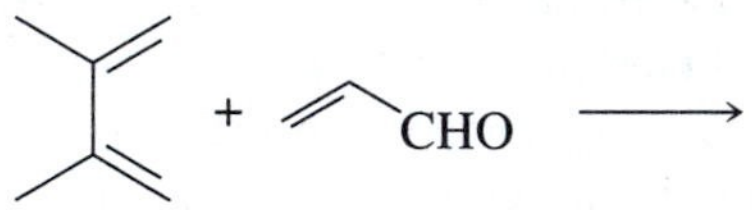

(c)

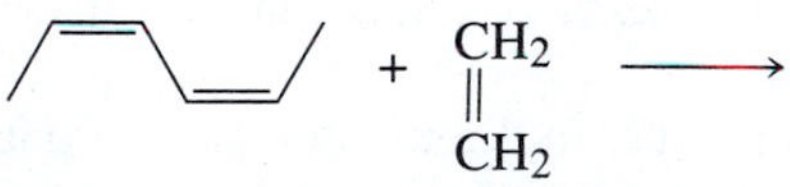

(d)

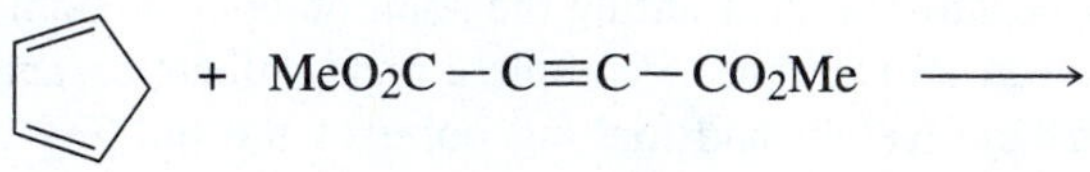

(e)

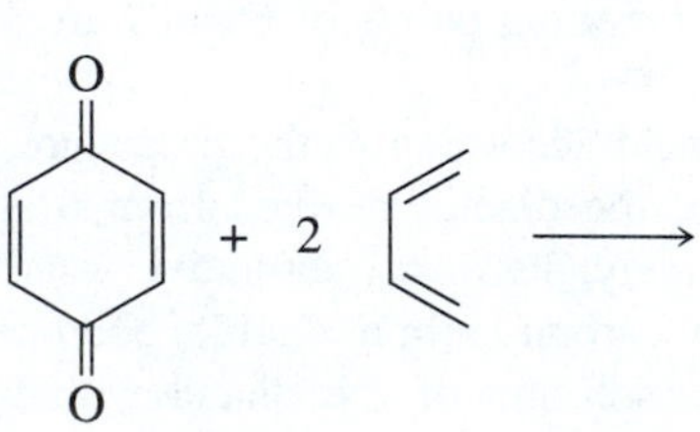

(f) 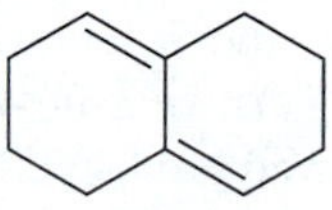 + $MeO_2C-C\equiv C-CO_2Me$ $\longrightarrow$

(g) 2 $\longrightarrow$

3. Write a detailed mechanism for the acid-catalyzed hydrolysis of maleic anhydride to give maleic acid. Use curved arrows to symbolize the flow of electrons.

Questions for Part A

4. Why should 3-sulfolene and maleic anhydride be completely dissolved in the xylene before heating the mixture to decompose the sulfolene?

5. Write the structure, including stereochemistry, of the expected addition product of bromine to the Diels-Alder adduct obtained by this procedure.

Questions for Part B

6. Why should the head temperature be maintained below about 45 °C when cracking dicyclopentadiene?

7. What problems might arise if the 1,3-cyclopentadiene is not dry prior to its reaction with maleic anhydride?

8. Why is it good technique to induce crystallization of the diacid derived from the Diels-Alder adduct *before* cooling the solution to 0 °C?

9. Why does 1,3-cyclopentadiene react more rapidly with maleic anhydride than with another molecule of itself?

10. The "cracking" of dicyclopentadiene to two moles of 1,3-cyclopentadiene (Eq. 12.9) is an example of a reverse, or retro, Diels-Alder reaction. Predict the products to be anticipated from an analogous reaction with the compounds (a)–(d).

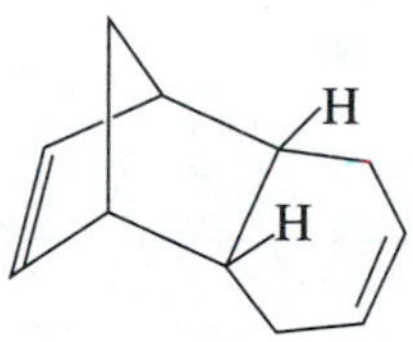

(a) 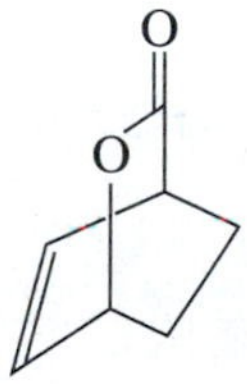   **(b)** 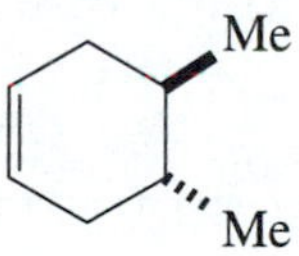   **(c)**    **(d)**

Questions for Part C

11. Which cycloadduct, **6** or **7**, is formed by the reaction of the mixture of *cis*- and *trans*-1,3-pentadiene with maleic anhydride?

12. Based on the Alder rule, which was illustrated by the reaction of 1,3-cyclopentadiene and maleic anhydride (Eq. 12.6), did *cis-* or *trans*-1,3-pentadiene react with maleic anhydride to form the observed product? Draw the transition states for the reaction of each diene with maleic anhydride to illustrate your reasoning.

13. Why does one geometric isomer of 1,3-pentadiene react faster than the other in a Diels-Alder reaction? You should consider carefully the factors that are responsible for increasing the energy of activation for one process relative to the other.

14. In Figure 12.1 are given GLC traces of (a) a commercial sample of "65-70%" *cis-* and *trans*-1,3-pentadiene and (b) the distillate boiling below 60 °C (760 torr) obtained after reaction of the mixture of dienes with maleic anhydride.

    a. Calculate the ratio of the peaks at 3.8 min and 6.6 min relative to one another in the two traces. These two components constitute 83 and 86% of the total volatiles in the sample before and after reaction, respectively.

    b. Assign the peaks to *cis-* and to *trans*-1,3-pentadiene and defend your assignment. Look up the boiling points of *cis-* and *trans*-1,3-pentadiene. Do the two isomers emerge from the GLC column in the order of their increasing boiling points? For your answer, assume that any nonvolatile polymeric by-product that was produced is constituted of equal proportions of both of the isomeric dienes.

    c. Why is it necessary in part (b) that the composition of the polymeric by-product be considered?

    d. Do the results of the analysis by GLC correspond to what you would have expected based on your answer in Exercises 11 and 12?

REFERENCES

1. Sauer, J. *Angewandte Chemie, International Edition English,* **1966,** *5,* 211. A general review of the Diels-Alder reaction.
2. Carruthers, W. *Some Modern Methods of Organic Synthesis,* 3rd ed., Cambridge University Press, Cambridge, UK, 1986. A review emphasizing synthetic applications of the Diels-Alder reaction.
3. Woodward, R.B.; Hoffmann, R. *The Conservation of Orbital Symmetry,* Academic Press, New York, 1970.

SPECTRA OF STARTING MATERIALS AND PRODUCTS

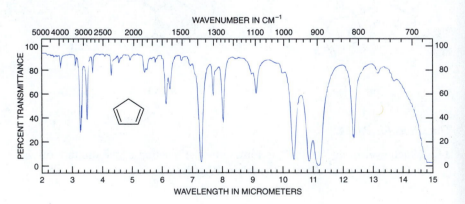

FIGURE 12.2   IR spectrum of 1,3-cyclopentadiene.

(a) ^{1}H NMR spectrum (60 MHz).

FIGURE 12.3 NMR data for 1,3-cyclopentadiene.

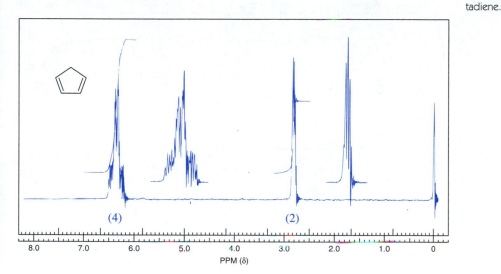

(b) ^{13}C NMR data.
Chemical shifts: δ 42.2, 133.0, 133.4.

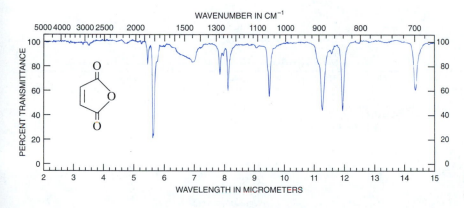

FIGURE 12.4 IR spectrum of maleic anhydride (in CCl$_4$ solution).

^{1}H NMR data. Chemical shift: δ 7.10.

^{13}C NMR data. Chemical shifts: δ 137.0, 165.0.

FIGURE 12.5 NMR data for maleic anhydride.

FIGURE 12.6 IR spectrum of 3-sulfolene (in CCl₄ solution).

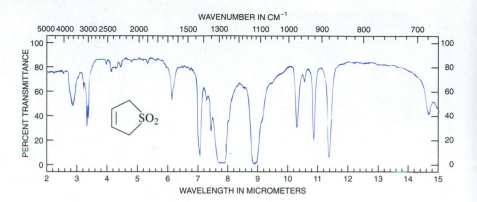

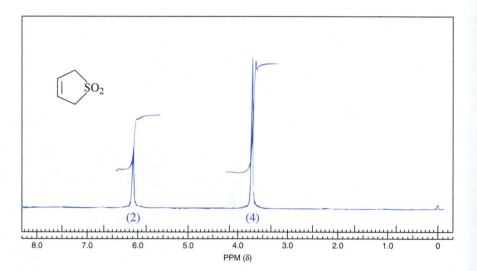

FIGURE 12.7 NMR data for 3-sulfolene.

(b) ^{13}C NMR data.
Chemical shifts: δ 55.7, 124.7.

FIGURE 12.8 IR spectrum of 4-cyclo-hexene-*cis*-1,2-dicarboxylic anhydride (KBr pellet).

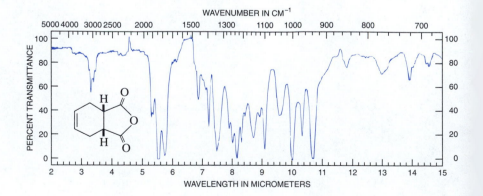

(a) ^{1}H NMR spectrum (60 MHz).

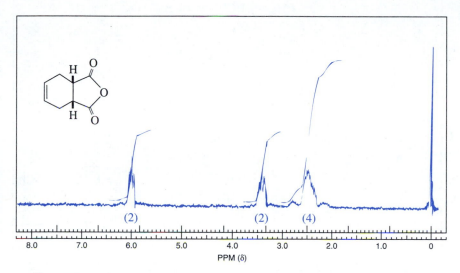

(b) ^{13}C NMR data.
   Chemical shifts: δ 26.1, 39.3, 125.5, 174.7.

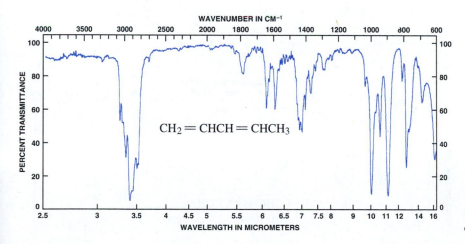

$$CH_2 = CHCH = CHCH_3$$

FIGURE 12.10   IR spectrum of a commercial mixture of 1,3-pentadienes.

FIGURE 12.11 ^{1}H NMR spectrum (90 MHz) of a commercial mixture of 1,3-pentadienes.

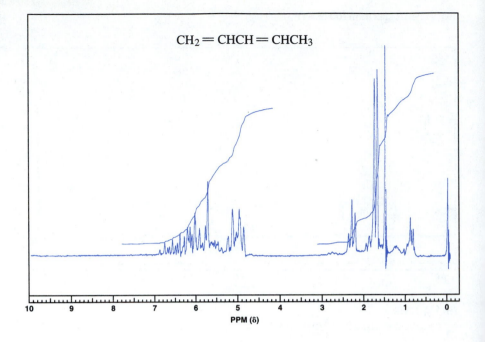

$$CH_2 = CHCH = CHCH_3$$

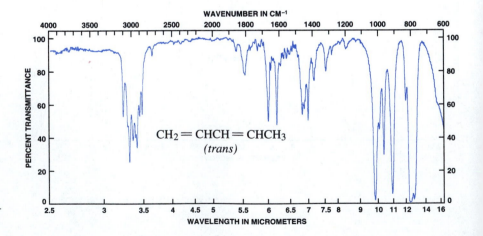

$$CH_2 = CHCH = CHCH_3$$
$$(trans)$$

FIGURE 12.12 IR spectrum of *trans*-1,3-pentadiene.

(a) ^{1}H NMR spectrum (90 MHz).

FIGURE 12.13 NMR data for *trans*-1,3-pentadiene.

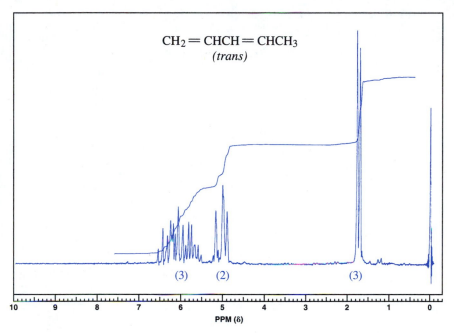

$$CH_2 = CHCH = CHCH_3$$
$$(trans)$$

(3)   (2)   (3)

PPM (δ)

(b) ^{13}C NMR data.
  Chemical shifts: δ 17.6, 113.9, 129.0, 132.9, 137.5.

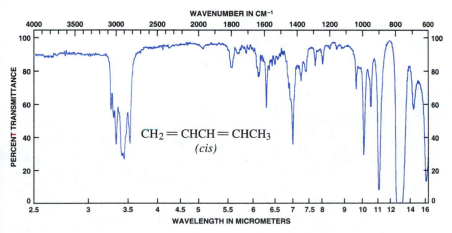

WAVENUMBER IN CM^{-1}

$$CH_2 = CHCH = CHCH_3$$
$$(cis)$$

PERCENT TRANSMITTANCE

WAVELENGTH IN MICROMETERS

FIGURE 12.14 IR spectrum of *cis*-1,3-pentadiene.

FIGURE 12.15 NMR data for *cis*-1,3-pentadiene.

(a) ^{1}H NMR spectrum (90 MHz).

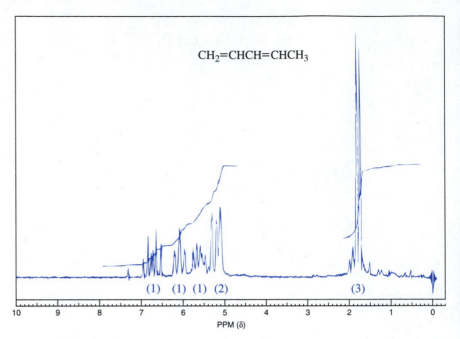

(b) ^{13}C NMR data.
Chemical shifts: δ 12.9, 116.1, 126.1, 130.6, 132.2.

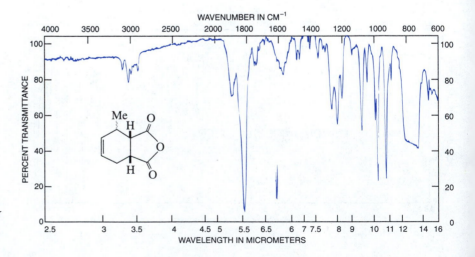

FIGURE 12.16 IR spectrum of Diels-Alder adduct of maleic anhydride and a 1,3-pentadiene (in CCl$_4$ solution).

(a) ^{1}H NMR spectrum (90 MHz).

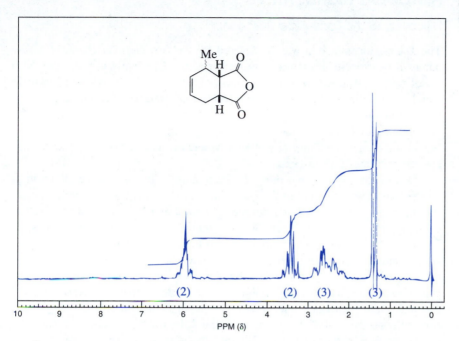

(b) ^{13}C NMR data.

Chemical shifts: δ 17.1, 24.1, 30.3, 41.3, 45.5, 127.7, 135.3, 172.2, 175.0.

HISTORICAL HIGHLIGHT:

Discovery of the Diels-Alder Reaction

The Diels-Alder reaction was discovered by the two German chemists whose names it bears. Otto Diels was a professor at the University of Kiel, and Kurt Alder was a postdoctoral student working with him. Diels, who was born in Hamburg in 1866, studied at the University of Berlin, where he received a Ph.D.; he later worked as an assistant to the world-famous chemist Emil Fischer. Alder, born in Upper Silesia in 1902, studied chemistry at the University of Berlin and then at the University of Kiel, where he became associated with Diels and received his Ph.D. in 1926.

One of the subjects that intrigued Diels and Alder was a novel addition of one or two molecules of 1,3-cyclopentadiene to one of *p*-benzoquinone, a reaction that had been investigated in other laboratories for over twenty years. For example, H. Albrecht, another German chemist, was apparently the first to observe this reaction, and he proposed in 1906 that the structures of the products were **1** and **2**. Between 1906 and 1928, this reaction was studied in several laboratories by a number of renowned chemists, and several other structures were proposed for the products. Diels and Alder conducted extensive investigations of this and similar reactions, and in a classic paper published in 1928, they questioned the assignment of the structures **1** and **2**. In a series of reports in succeeding years, they showed conclusively that the structures proposed by Albrecht and others were incorrect and that the products from the reaction of 1,3-cyclopentadiene with *p*-benzoquinone were in fact **3** and **4**.

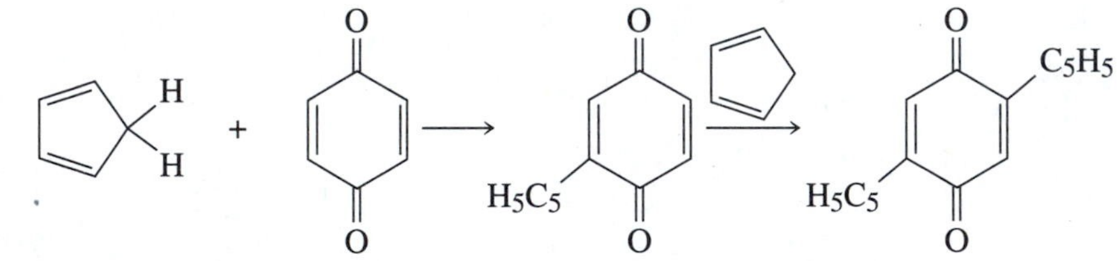

1,3-Cyclopentadiene   *p*-Benzoquinone   **1**   **2**

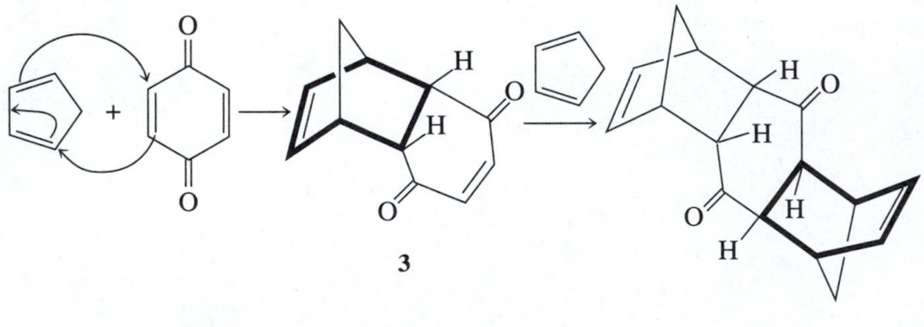

3   **4**

As a result of their experiments, Diels and Alder suggested that this reaction and numerous other related processes belonged to common reaction type in which a 1,3-diene undergoes cycloaddition to a π-bond of a dienophile to form a product containing a new six-membered ring. This key feature is illustrated by the highlighted bonds in the structures of **3** and **4.** At the time of its discovery, this was an entirely new and important type of organic reaction for the construction of such rings.

The Diels-Alder reaction is considered to be one of the few truly new organic reactions of the 20th century, and it ranks in synthetic importance with the Wittig and Grignard reactions. In recognition of their research and development of this process, Diels and Alder were awarded the Nobel Prize in 1951. In his Nobel Prize lecture, Diels commented that he thought it plausible that nature might use this type of reaction to fabricate some of the complicated polycyclic compounds that are so important to living organisms. The Diels-Alder reaction often does not require temperatures that lie outside of the range at which life exists, and it frequently proceeds in almost quantitative yield without using any special catalysts, reagents or solvents. Despite Diels' suggestion that the Diels-Alder reaction would be a common transformation in nature, the biosyntheses of only a few natural products appear to involve such a process. On the other hand, this reaction has been extensively used as a key step in the laboratory syntheses of a number of important naturally occurring substances—including alkaloids, terpenes, dyes, vitamins, and hormones.

An idea of the growth in the importance of the Diels-Alder reaction is gained by looking at the Cumulative Indexes of *Chemical Abstracts*. Only two entries were listed under the name "Diels-Alder reaction" in the ten-year period, 1927–1936, whereas such entries had increased to 180 by the decade spanning 1947–1956, during which time Diels and Alder shared the Nobel Prize in Chemistry. By the five-year period of 1982–1986, a total of about 1300 entries appeared under this heading, and this figure nearly doubled to some 2400 in the most recent cumulative index, which covers 1987–1991.

A recent and more extensive description of the history of the Diels-Alder reaction may be found in: Berson, J. *Tetrahedron* **1992,** *48*, 3–17.

Relationship of Historical Highlight to Experiments

The procedures in this chapter illustrate the basic principles of the Diels-Alder reaction, many of which were first unveiled by the extensive research efforts of Diels and Alder. In all cases, the reactions form products containing new, unsaturated six-membered rings, which is the fundamental structural outcome of the Diels-Alder process.

KINETIC AND THERMODYNAMIC CONTROL OF A REACTION

13.1 INTRODUCTION

Organic chemists frequently need to predict which of several possible competing reactions will predominate when a specified set of reactants is used. The problem is illustrated by considering the general situation where reagents **W** and **X** can combine to produce either **Y** (Eq. 13.1a) or **Z** (Eq. 13.1b). Predicting which product will be formed preferentially under a particular set of conditions requires developing an **energy diagram** for the two reactions. Strictly speaking, such a diagram should be based on the **free energy changes, ΔG,** that occur during the conversion of reactants to products. To generate the needed diagram would require knowledge of both the **enthalpic, ΔH,** and **entropic, ΔS,** changes involved. If the reactions being considered are similar, as we assume is the case for the processes of Equation (13.1), entropic changes are not too important, so using values for ΔH gives a close approximation to the changes occurring in ΔG. This is the basis on which our potential energy diagram will be constructed.

$$\text{W} + \text{X} \quad \rightleftharpoons \quad \text{Y} \tag{13.1a}$$

$$\text{W} + \text{X} \quad \rightleftharpoons \quad \text{Z} \tag{13.1b}$$

First assume for the purposes of this discussion that both of the reactions of Equation (13.1) are *exothermic and* that **Y** is *thermodynamically less stable* than **Z.** These assumptions mean that (1) the energy levels of both products are below that of the starting materials and (2) the heat of reaction for formation of Z is greater than that for **Y,** as symbolized by ΔH_Z and ΔH_Y in Figure 13.1. Now let the activation barriers for conversion of the reagents to the two transition states, $\text{Y}^{\ddagger}$ and $\text{Z}^{\ddagger}$, for forming **Y** and **Z,** be measured by $\Delta H_Y^{\ddagger}$ and $\Delta H_Z^{\ddagger}$, respectively. The latter barrier has been set *lower* on the diagram than that for **Y.** This ordering of the relative energies of the two transition states is somewhat arbitrary but is consistent with the *empirical* observation that when comparing two *similar* organic reactions, the *more exothermic process often has the lower enthalpy of activation, $\Delta H^{\ddagger}$.* In other words, the relative energies of the transition states reflect the relative energies of the products that are produced from the transition states. This generalization is useful because it allows us to complete the potential energy diagram and to predict that because less energy is required

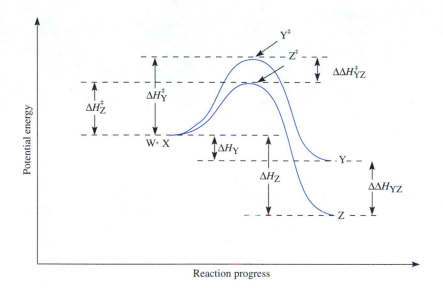

FIGURE 13.1 Typical reaction profile for competing reactions.

for reactants **W** and **X** to reach the transition state leading to **Z,** this product will be formed faster than **Y.**

For all practical purposes, most organic reactions are either irreversible or are performed under conditions such that equilibrium between products and starting materials is *not* attained. The ratio of products obtained from reactions run under these circumstances is said to be subject to **kinetic control.** This means that the relative amounts of the various products are determined by the relative rates of their formation rather than by their relative thermodynamic stabilities. The major product of such a reaction is often called the **kinetically controlled product** even though other compounds may be formed. In the example of Figure 13.1, **Z** would represent the kinetically controlled product; note that $\Delta H_Z^{\ddagger} < \Delta H_Y^{\ddagger}$.

Organic reactions that are performed under conditions in which *equilibrium* is established between reactants and products produce product ratios that are subject to **thermodynamic control.** In such cases, the predominant product is called the **thermodynamically controlled product** and would be the more stable **Z** under the circumstances portrayed in Figure 13.1. Note that the ratio **Y:Z** would probably *not* be the same under both sets of reaction conditions because it is likely that the difference in enthalpies of activation, $\Delta\Delta H_{YZ}^{\ddagger}$, between the two transition states would not be equal to the enthalpy difference, $\Delta\Delta H_{YZ}$, between the two ground states of the products.

The situation shown in Figure 13.1 is the one most frequently encountered in organic chemistry; namely, that the major product of kinetic control and thermodynamic control is the same. There are exceptions to this, however, and the reaction profiles of Figure 13.2 depict just such a case. To generate these profiles, it is again assumed that the reactions of Equation (13.1) are exothermic and that **Z** is more stable than **Y.** What is changed is that the activation barrier, $\Delta H_Y^{\ddagger}$, for forming **Y** is now *less* than that for forming **Z,** $\Delta H_Z^{\ddagger}$. The lower barrier means that **Y** would now be the product of kinetic control, although **Z** would remain the product of thermodynamic control.

Note that in the development of Figures 13.1 and 13.2, conclusions about the preferred products of kinetic control and thermodynamic control do not

FIGURE 13.2 Reaction profile that predicts different products from kinetic and equilibrium control of competing reactions.

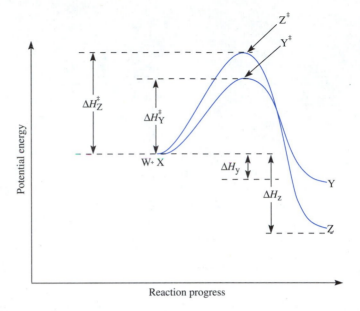

depend on the exothermicity or endothermicity of the overall reactions. *All that matters are the relative energies of the transition states leading to products, if the reaction is under kinetic control, or the relative energies of the products themselves, if the conditions are those of thermodynamic control.* Whether different major products are produced under reversible as compared to irreversible reaction conditions is determined by measuring the ratios of products formed under the two different circumstances. Each product can also be isolated and then subjected to the reaction conditions under which they are produced. If equilibration is occurring, analysis of the product mixture will demonstrate this. Both types of procedures are described in this experiment.

13.2 FORMATION OF SEMICARBAZONES UNDER KINETIC AND THERMODYNAMIC CONTROL

The example chosen to illustrate the principle of kinetic and thermodynamic control of products involves the competing reactions of semicarbazide (**1**) with two carbonyl compounds, cyclohexanone (**2**, Eq. 13.2) and 2-furaldehyde (**3**, Eq. 13.3), to give two different products. In this case, one compound, semicarbazide,

$$H_2NCNHNH_2 \ + \ O=\!\!\!\bigcirc \ \rightleftharpoons \ H_2NCNHN=\!\!\!\bigcirc \ + \ H_2O \qquad \textbf{(13.2)}$$

1	**2**	**3**
Semicarbazide	Cyclohexanone	Cyclohexanone semicarbazone

$$\text{H}_2\text{NCNHNH}_2 \; + \; \underset{\underset{\text{H}}{|}}{\overset{\overset{\text{O}}{\|}}{\text{C}}}\!\!\diagup\!\!\diagdown\!\!\text{O} \;\rightleftharpoons\; \text{H}_2\text{NCNHN}\!=\!\underset{\underset{\text{H}}{|}}{\text{C}}\!\!\diagup\!\!\diagdown\!\!\text{O} \;+\; \text{H}_2\text{O} \quad \textbf{(13.3)}$$

1   **4**   **5**

2-Furaldehyde   2-Furaldehyde semicarbazone

reacts with two different compounds, cyclohexanone and 2-furaldehyde to give two different **semicarbazones, 4** and **5,** respectively. Both are crystalline solids and have distinctive melting points by which they may be easily identified. This makes it simple to determine experimentally which compound is the product of kinetic control and which is the product of thermodynamic control.

The rates and equilibrium constants of reactions of carbonyl compounds with nucleophiles such as semicarbazide are affected significantly by variations in the pH of the reaction medium. This may be understood by examining the results of protonating the reactants. First, let us consider the effect that protonation of the carbonyl function has on the rate of reaction of a carbonyl compound with nucleophiles. The addition of a proton to the oxygen atom of a carbonyl group produces **6** (Eq. 13.4). This makes the carbonyl carbon atom much more *electrophilic* than the unprotonated compound because the partial positive charge on the carbon atom is increased, as shown by consideration of the resonance structures in Equation (13.4). Since the rate-determining step for the formation of semicarbazones is the nucleophilic attack of semicarbazide (**1**) on the carbonyl carbon atom of the substrate, the enhanced electrophilicity of **6** *increases* the rate of formation of the semicarbazone. Now let us assess what happens if the nucleophile is protonated. Addition of a proton to **1** produces its conjugate acid **7** (Eq. 13.5), which is *not* nucleophilic. Consequently the rate of the desired reaction (Eq. 13.6) is *decreased,* an outcome just the *reverse* of that described for protonation of the carbonyl compound. If the medium in which the reaction was performed is sufficiently acidic, essentially all of the **1** will be converted to **7,** thereby lowering the concentration of the nucleophilic **1.** The formation of the semicarbazone would then become extremely slow or might not occur at all. Both the rate and equilibrium constant for the reaction of Equation (13.6) are thus affected significantly by the pH at which the process is conducted.

$$\text{R}_2\text{C}=\ddot{\text{O}} + \text{H}^{\oplus} \;\rightleftharpoons\; \left[\text{R}_2\text{C}\!=\!\overset{\oplus}{\ddot{\text{O}}}\text{H} \;\longleftrightarrow\; \text{R}_2\overset{\oplus}{\text{C}}\!-\!\ddot{\text{O}}\text{H} \right] \quad \textbf{(13.4)}$$

Carbonyl   **6**
compound

$$\underset{\textbf{1}}{\text{H}_2\text{NCNHNH}_2} \; + \; \text{H}^{\oplus} \;\rightleftharpoons\; \underset{\textbf{7}}{\text{H}_2\text{NCNH}\overset{\oplus}{\text{NH}}_3} \quad \textbf{(13.5)}$$

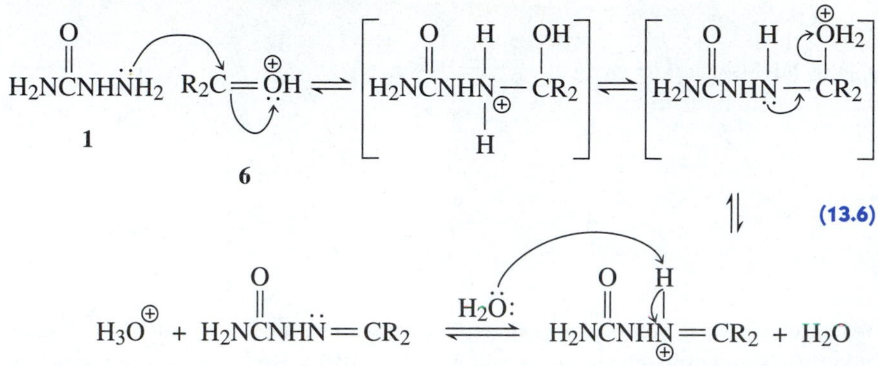

$$H_2NCNHNH_2 \quad R_2C=\overset{\oplus}{O}H \;\rightleftharpoons\; \left[H_2NCNHN\underset{H}{\overset{\oplus}{-}}CR_2 \right] \;\rightleftharpoons\; \left[H_2NCNHN\overset{\oplus}{-}CR_2 \right]$$

1

6

$$\text{(13.6)}$$

$$H_3\overset{\oplus}{O} \;+\; H_2NCNHN=CR_2 \;\rightleftharpoons\; H_2O: \quad H_2NCNH\overset{\oplus}{N}=CR_2 \;+\; H_2O$$

The preceding discussion makes clear that with respect to Equation (13.6), the concentration of the nucleophile **1** is reduced at low pH whereas that of the activated electrophile **6** is reduced at high pH. It follows that there is an optimum pH at which the mathematical product of the concentrations of the conjugate acid of the carbonyl compound and the nucleophilic reagent, [**1**] × [**6**], is maximized. For reactions of aldehydes and ketones with reagents such as semicarbazide (**1**), phenylhydrazine (**8**), and hydroxylamine (**9**), the desired pH is produced and maintained by *buffer solutions*.

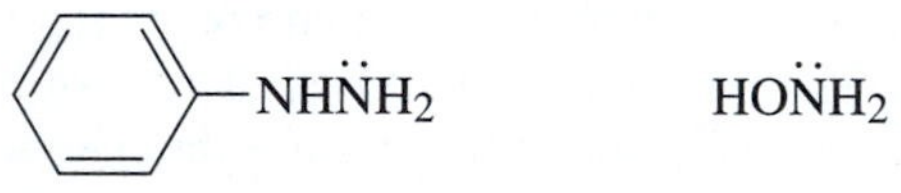

$$\langle\rangle-NHNH_2 \qquad\qquad HONH_2$$

8

9

A buffer solution resists changes in its pH. In general, this requires the presence of a pair of substances in the solution, one that neutralizes hydroxide ions and another that neutralizes protons. For example, the $H_2PO_4^-/HPO_4^{2-}$ *buffer system* is produced by the addition of dibasic potassium phosphate, K_2HPO_4, to semicarbazide hydrochloride. The $H_2PO_4^-$ component of the buffer system is produced as shown in Equation (13.7). The function of the two ions in neutralizing hydroxide and hydrogen ions is illustrated in Equations (13.8) and (13.9). Each different buffer system functions to maintain the pH within a rather narrow range characteristic of the weak acid and weak base of that particular system. In the case of the phosphate system, the range is pH 6.1–6.2. By comparison, a carbonate/bicarbonate buffer system (CO_3^{2-}/HCO_3^-) maintains a pH of 7.1–7.2.

$$H_2NCNHN\overset{\oplus}{H}_3 \;\overset{\ominus}{Cl} \;+\; HPO_4^{2-} \;\rightleftharpoons\; H_2NCNHNH_2 \;+\; H_2PO_4^{\ominus} \;+\; Cl^{\ominus} \quad \text{(13.7)}$$

1

$$H_2PO_4^{\ominus} \;+\; HO^{\ominus} \;\rightleftharpoons\; HPO_4^{2-} \;+\; H_2O \qquad\qquad \text{(13.8)}$$

$$HPO_4^{2-} \;+\; H_3O^{\oplus} \;\rightleftharpoons\; H_2PO_4^{\ominus} \qquad\qquad \text{(13.9)}$$

The maximum rates of the reactions of most aldehydes and ketones with semicarbazide occur in the pH range of 4.5–5.0, and the rates decrease at pHs either above or below this value. For the purpose of making derivatives of carbonyl compounds (Sec. 27.5.1), semicarbazide is best used in an acetate buffer ($CH_3CO_2H/CH_3CO_2^-$) solution, which maintains a pH in the maximum rate range, 4.5–5.0. However, to demonstrate the principle of kinetic and thermodynamic control of reactions, buffers that maintain higher pHs, and thus produce lower rates, are more desirable. Parts A–C of the experimental procedure involve a phosphate buffer system, whereas the bicarbonate system is used in Part D, a strategy that allows a comparison of how the difference in rates in the two buffer systems affects the product ratio. Analysis of the products from the various parts of these experiments provides strong clues as to which of the semicarbazones is the product of kinetic control and which is the product of thermodynamic control.

In Section 13.1, it was noted that implicit in the theory of kinetic and thermodynamic control is the fact that the *kinetic product* (**Z** in Fig. 13.1 and **Y** in Fig. 13.2), which is produced more rapidly, is also reconverted to starting material more rapidly. The *thermodynamic product*, on the other hand, being more stable, is not so readily reconverted to starting material. Experimental procedure E provides tests of the relative stabilities of the two semicarbazone products toward the reverse reaction. The results of these experiments should provide additional evidence of the identity of one of the semicarbazones as the product of kinetic control and the other as the product of thermodynamic control.

Pre-Lab exercises for Section 13.2, "Formation of Semicarbazones under Kinetic and Thermodynamic Control," are found on page PL. 51.

EXPERIMENTAL PROCEDURES

SAFETY ALERT

Because the melting point of one of the compounds is over 200 °C, do not use a liquid-filled apparatus such as a Thiele tube *unless* the heating medium is silicone oil. Heating mineral oil to this temperature may cause it to smoke or even catch fire. A metal block melting-point apparatus is satisfactory for determining the melting points of the semicarbazones in this experiment.

Note: Your instructor may divide up the class and assign you to perform only certain parts of these procedures rather than requiring that you do all parts. If this is done, you will be provided with the results obtained by other students for the parts that you did not do. This should enable you to draw your own conclusions and write them up accordingly.

In the following experiments it is important that the isolated crystals be completely dry prior to determining their melting points!

A.   PREPARATION OF CYCLOHEXANONE SEMICARBAZONE

Dissolve 1.0 g of semicarbazone hydrochloride and 2.0 g of dibasic potassium phosphate (K_2HPO_4) in 12 mL of water contained in a 50-mL Erlenmeyer flask. Using a 1-mL graduated pipet, deliver 1.0 mL of cyclohexanone into a test tube containing 5 mL of 95% ethanol. Pour the ethanolic solution into the aqueous semicarbazide solution, and swirl or stir the mixture immediately. Allow 5 or 10 min for crystallization of the semicarbazone to reach completion; then collect the crystals by vacuum filtration and wash them on the filter with a little cold water. Dry the crystals in air✷ and determine their weight and melting point. The reported melting point of cyclohexanone semicarbazone is 166 °C. Recrystallization of the product is unnecessary.

B.   PREPARATION OF 2-FURALDEHYDE SEMICARBAZONE

For best results the 2-furaldehyde should be distilled just before use. Prepare the semicarbazone of 2-furaldehyde by following the procedure of Part A exactly, except use 0.8 mL of 2-furaldehyde instead of 1.0 mL of cyclohexanone.✷ The reported melting point of 2-furaldehyde semicarbazone is 202 °C. Recrystallization of the product is unnecessary.

C.   REACTIONS OF SEMICARBAZIDE WITH CYCLOHEXANONE AND 2-FURALDEHYDE IN PHOSPHATE BUFFER SOLUTION

Dissolve 3.0 g of semicarbazide hydrochloride and 6.0 g of dibasic potassium phosphate in 75 mL of water. Because this is an aqueous solution, it is referred to as *solution W*. Prepare a solution of 3.0 mL of cyclohexanone and 2.5 mL of 2-furaldehyde in 15 mL of 95% ethanol. Because this is an ethanolic solution, it is referred to as *solution E.*✷

1. Cool a 25-mL portion of solution W and a 5-mL portion of solution E separately in an ice-water bath to 0–2°C. Add solution E to solution W and swirl the mixture; crystals should form almost immediately. Place the mixture in an ice-water bath for 3–5 min; then collect the crystals by vacuum filtration and wash them on the filter with 2–3 mL of cold water. Dry the crystals✷ and determine their weight and melting point.

2. Add a 5-mL portion of solution E to a 25-mL portion of solution W at room temperature; crystals should be observed in 1–2 min. Allow the mixture to stand at room temperature for 5 min, cool it in an ice-water bath for about 5 min, and then collect the crystals by vacuum filtration and wash them on the filter with 2–3 mL of cold water. Dry the crystals✷ and determine their weight and melting point.

3. Warm a 25-mL portion of solution W and a 5-mL portion of solution E *separately* on a steam bath or in a water bath to 80–85 °C; add solution E to solution W and swirl the mixture. Continue to heat the solution for 10–15 min, cool it to room temperature, and then place it in an ice-water bath for about 5–10 min. Collect the crystals by vacuum filtration, and wash them on the filter with 2–3 mL of cold water. Dry the crystals✷ and determine their weight and melting point.

D. REACTIONS OF SEMICARBAZIDE WITH CYCLOHEXANONE AND 2-FURALDEHYDE IN BICARBONATE BUFFER SOLUTION

Dissolve 2.0 g of semicarbazide hydrochloride and 4.0 g of sodium bicarbonate in 50 mL of water. Prepare a solution of 2.0 mL of cyclohexanone and 1.6 mL of 2-furaldehyde in 10 mL of 95% ethanol. Divide each of these solutions into two equal portions.

1. Mix half of the aqueous solution and half of the ethanolic solution and allow the mixture to stand at room temperature for 5 min. Collect the crystals by vacuum filtration and wash them on the filter with 2–3 mL of cold water. Dry the crystals✳ and determine their weight and melting point.

2. Warm the other portions of the aqueous and ethanolic solutions *separately* on a steam bath or in a water bath to 80–85 °C; then combine them and continue heating the mixture for 10–15 min. Cool the solution to room temperature, and place it in an ice-water bath for 5–10 min. Collect the crystals by vacuum filtration and wash them on the filter with a 2–3 mL of cold water. Dry the crystals✳ and determine their weight and melting point.

E. TESTS OF REVERSIBILITY OF SEMICARBAZONE FORMATION

1. Place 0.3 g of cyclohexanone semicarbazone, prepared in Part A, 0.3 mL of 2-furaldehyde, 2 mL of 95% ethanol, and 10 mL of water in a 25-mL Erlenmeyer flask. Warm the mixture until a homogeneous solution is obtained (about 1 or 2 min should suffice), and continue warming an additional 3 min. Cool the mixture to room temperature and then place in an ice-water bath. Collect the crystals on a filter, and wash them with 2–3 mL of cold water. Dry the crystals✳ and determine their weight and melting point.

2. Repeat the preceding experiment, but use 0.3 g of 2-furaldehyde semicarbazone (prepared in Part B) and 0.3 mL of cyclohexanone in place of the cyclohexanone semicarbazone and 2-furaldehyde.✳

On the basis of your results from experiments C, D, and E, deduce which semicarbazone is the product of kinetic control and which is the product of thermodynamic control. To do this, first use the observed melting points of the crystals produced in parts C1, C2, and C3 to deduce whether the product in each part is the semicarbazone of cyclohexanone, 2-furaldehyde, or a mixture of the two. Note that in C1 the crystals of product separate almost immediately, in C2 after 1 or 2 min, and in C3 only after 10–15 min at a higher temperature; thus the reaction time is shortest in C1, intermediate in C2, and longest in C3. Then compare the product of D1 with that of C2 and the product of D2 with that of C3.

When analyzing the results of the experiments in part E, remember that the thermodynamic product, being the more stable, is not easily transformed into the less stable kinetic product. In contrast, the kinetic product is more readily converted into the more stable thermodynamic product. These circumstances arise because the reverse reaction from the less stable kinetic product has a lower activation energy than the reverse reaction of the thermodynamic product.

Your completed laboratory report should include the diagram called for in Exercise 1 and answers to Exercises 2 and 3 as well, unless you were instructed to omit some parts of the experiment.

> **Finishing Touches** Neutralize the various *filtrates* and then flush them down the drain.

EXERCISES

1. Why is it particularly important in this experiment that the semicarbazones be dry before their melting points are determined?

2. On the basis of the results from experiments C, D, and E, draw a diagram similar to Figure 13.2 and clearly label the products corresponding to those obtained in experiments A and B.

3. On the basis of the results from the experiments of Part D, explain the effect of the higher pH on the reactions between semicarbazide and the two carbonyl compounds.

4. What results are expected if sodium acetate buffer, which provides a pH of approximately 5, is used in experiments analogous to those of Part C? Explain.

5. Show how you could calculate the exact pH of a buffer solution prepared by adding equimolar amounts of glacial acetic acid and sodium acetate to distilled water; the pK_a of acetic acid is 4.75.

6. What different result, if any, is expected if the heating period at 80–85 °C is extended to one hour in experiment D2?

7. There are two NH_2 groups in semicarbazide (**4**), yet only one of them reacts with the carbonyl group to form a semicarbazone. Explain.

8. Figure 13.7 is the UV spectrum of 2-furaldehyde.

   a. Calculate ϵ_{max} for the absorption bands present in the spectrum.
   b. Assuming that the UV spectrum of 2-furaldehyde is the same in 95% ethanol as it is in methanol and that cyclohexanone does not absorb significantly in the UV spectrum, show how the concentration of 2-furaldehyde in solution E could accurately be determined.

SPECTRA OF STARTING MATERIALS AND PRODUCTS

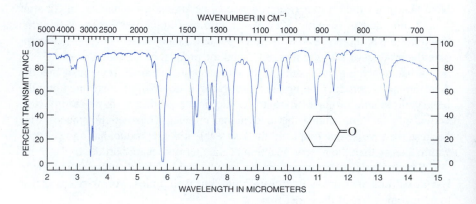

FIGURE 13.3 IR spectrum of cyclohexanone.

(a) ^{1}H NMR Spectrum (60 MHz).

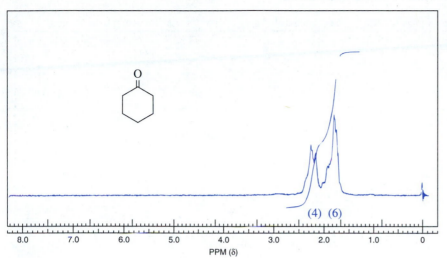

FIGURE 13.4 NMR data for cyclohexanone.

(4) (6)

(b) CMR data.

Chemical shifts: δ 25.1, 27.2, 41.9, 211.2.

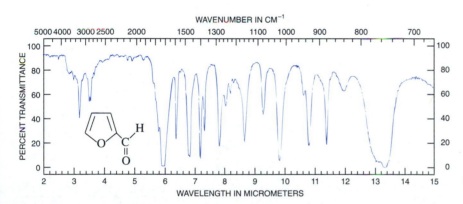

WAVENUMBER IN CM^{-1}

FIGURE 13.5 IR spectrum of 2-furaldehyde.

(a) ^{1}H NMR spectrum (60 MHz).

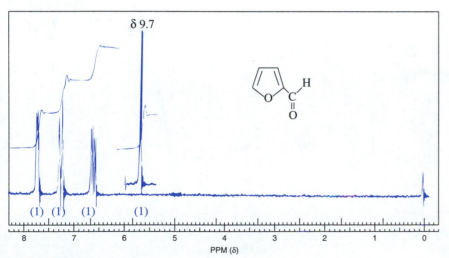

δ 9.7

(1) (1) (1) (1)

(b) ^{13}C NMR data.

Chemical shifts: δ 112.9, 121.9, 148.6, 153.3, 178.1.

FIGURE 13.6 NMR data for 2-furaldehyde.

FIGURE 13.7 UV spectrum of 2-furaldehyde.

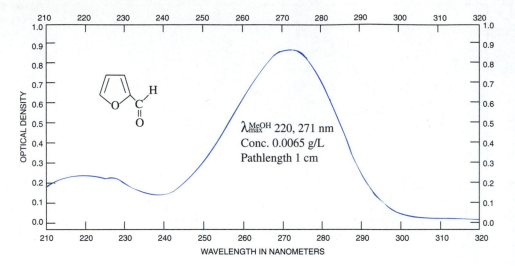

λ_{max}^{MeOH} 220, 271 nm
Conc. 0.0065 g/L
Pathlength 1 cm

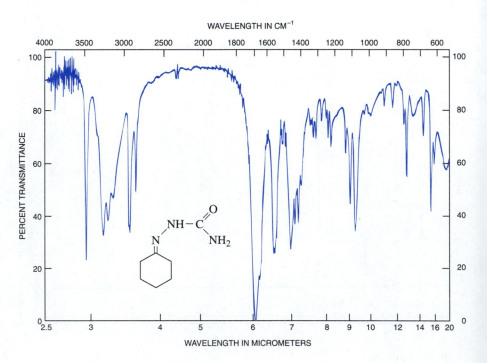

FIGURE 13.8 IR spectrum of cyclohexanone semicarbazone (KBr pellet).

(a) ^{1}H NMR Spectrum (90 MHz).

FIGURE 13.9   NMR data for cyclohexa-none semicarbazone.

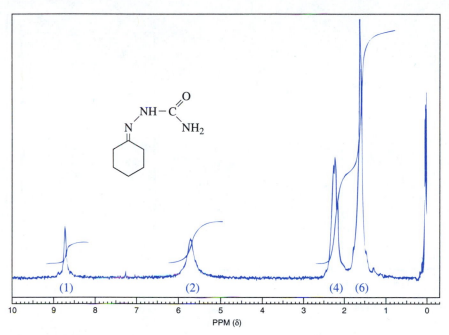

(b) ^{13}C NMR data.

Chemical shifts: δ 25.6, 25.8, 26.5, 27.0, 35.4, 153.4, 158.8.

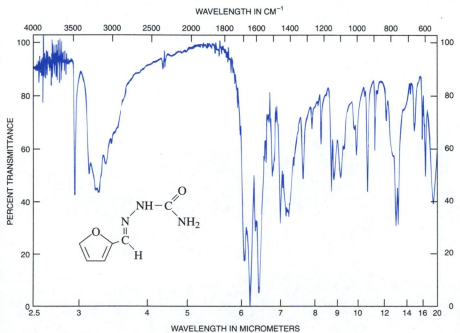

FIGURE 13.10   IR spectrum of 2-furaldehyde semicarbazone (KBr pellet).

FIGURE 13.11   NMR data for 2-
furaldehyde semicarbazone.

(a) ^{1}H NMR spectrum (90 MHz).

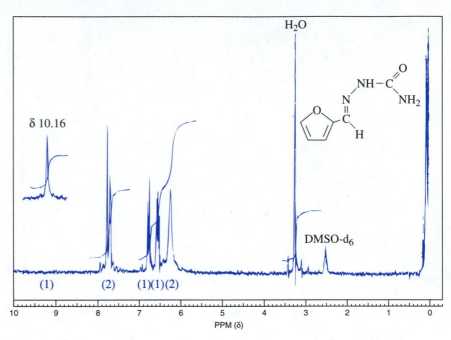

(b) ^{13}C NMR data.
Chemical shifts: δ 110.6, 111.8, 129.8, 143.8, 149.9, 156.4.

NUCLEOPHILIC ALIPHATIC SUBSTITUTION

PREPARATION OF ALKYL HALIDES

Nucleophilic aliphatic substitution reactions constitute an important class of reactions in organic chemistry. Compounds containing a wide diversity of different functional groups can frequently be prepared from a single starting material simply by varying the nature of the nucleophilic partner. In this chapter we review the general features of this reaction and conduct experiments to study the conversion of alcohols into alkyl halides, one representative process of this type.

14.1 GENERAL CONCEPTS

The substitution of one group for another at a saturated, sp^3-hybridized, carbon atom is a reaction commonly used as a means to interconvert different functional groups. This process is named **nucleophilic aliphatic substitution** and is exemplified in Equation (14.1), where **Nu:** is a symbol for a generalized nucleophile and represents a molecule or ion that has Lewis basic character, and **L:** represents a leaving group. The reaction may be considered to be a Lewis acid-base reaction in which the carbon atom bound to the electronegative leaving group possesses Lewis acid character since it bears partial positive charge because of the polarization of the C–L bond.

$$\overset{\ominus}{Nu:} \; + \; R - \overset{|}{\underset{|}{C}} - \overset{\delta\oplus \; \delta\ominus}{L} \;\longrightarrow\; Nu - \overset{|}{\underset{|}{C}} - R \; + \; L:^{\ominus} \qquad (14.1)$$

Nucleophiles have in common the property of bearing *at least one nonbonding pair of electrons* and being either *neutral or negatively charged*. During the course of the substitution reaction, the nonbonding pair of electrons of the nucleophile is donated to an electrophilic, Lewis acidic, carbon atom with concomitant formation of a new covalent bond. Examples of some typical nucleophiles include $Cl:^-$, $Br:^-$, $I:^-$, $HO:^-$, $N\equiv C:^-$, $H_2O:$, $:N_3^-$, $:NR_3$, and $RS:^-$.

The leaving group, L:, which like Nu: may be neutral or negatively charged, must be able to accept the pair of bonding electrons from the carbon atom as the C–L bond breaks. The ease with which various groups leave in nucleophilic substitution reactions has been determined experimentally by studying the rates of reaction of a specific nucleophile with substrates having different leaving groups. Based upon these investigations, it is now clear that the leaving ability of a particular group L: correlates with the strength of its conjugate acid: The better leaving groups are those that are **conjugate bases** of the stronger acids. For example, a leaving group such as Cl^- is the conjugate base of the strong acid, HCl, and so Cl^- is a good leaving group. On the other hand, HO^- is a poor leaving group because it is the conjugate base of the weak acid water; however, H_2O is a good leaving group because it is the conjugate base of the strong acid H_3O^+.

14.2 CLASSIFICATION OF NUCLEOPHILIC SUBSTITUTION REACTIONS

Nucleophilic substitution is a general reaction for aliphatic compounds in which the leaving group is attached to an sp^3-hybridized carbon (see Eq. 14.1); however, the mechanism for a given transformation depends upon the structure of the alkyl group bearing the leaving group. The two different mechanistic pathways that apply to such substitutions are designated by the symbols S_N1 (S for substitution, N for nucleophilic, and 1 for unimolecular) and S_N2 (2 for bimolecular). These two mechanisms are depicted in general terms in Equations (14.2) and (14.3), respectively.

S_N1 mechanism

$$R_3C-L \xrightleftharpoons[\text{slow}]{\text{rds}} L:^{\ominus} + R_3C^{\oplus} \xrightarrow[\text{fast}]{Nu:^{\ominus}} R_3C-Nu \qquad \textbf{(14.2)}$$

S_N2 mechanism

$$Nu:^{\ominus} \begin{array}{c} R^1 \\ | \\ H^{\text{''''}}C-L \\ | \\ R^2 \end{array} \xrightarrow{\text{rds}} \left[\begin{array}{c} R^1 \\ | \\ \delta^{\ominus}Nu\cdots C\cdots L\delta^{\ominus} \\ | \quad | \\ H \quad R^2 \end{array} \right]^{\ddagger} \longrightarrow Nu-C\begin{array}{c} R^1 \\ {}^{\text{''''}}H \\ R^2 \end{array} + L:^{\ominus} \qquad \textbf{(14.3)}$$

Transition state

When nucleophilic substitution occurs by an S_N1 mechanism, the reaction proceeds in two successive steps, as illustrated in Equation (14.2). The first step involves the **heterolytic cleavage,** or **ionization,** of the bond between the carbon atom and the leaving group. This step is assisted by the polar interactions between solvent molecules and the incipient cationic and anionic centers. Since the leaving group acquires the pair of bonding electrons, the organic fragment becomes a **carbocation,** with the carbon atom formerly bound to L now bearing a positive charge. The intermediate carbocation may then undergo the usual reactions of carbocations: (1) rearrangement to a more stable carbocation (see

Sec. 10.3); (2) loss of a proton to give an alkene by net **elimination** of the elements of **H–L** (see Sec. 14.3); or (3) combination with a nucleophile to form the substitution product, as indicated in Equation (14.2). Normally, the concentration of the nucleophile, Nu:⁻, is high compared to that of the L:⁻ that has been produced, so the reaction of the carbocation with L:⁻ to give the starting material is relatively unimportant. If the nucleophile is the solvent, the reaction is known as a **solvolysis.**

The first step of an S_N1 reaction is slower than the second because it involves breaking of the C–L bond, an endothermic process; the second step is a fast, exothermic process involving bond formation. Thus, the first step of an S_N1 reaction is the **rate-determining step** (rds) of the reaction, and the rate of the reaction depends only on the concentration of the substrate, R–L, and is independent of the concentration of the nucleophile. This is stated mathematically in Equation (14.4), where k_1 is the first-order rate constant.

$$\text{Rate} = k_1[\text{R–L}] \qquad \textbf{(14.4)}$$

When substitution occurs by an S_N2 mechanism, the nucleophile attacks the substrate directly, with the angle of approach being 180° to the C–L bond; this is called "backside attack." The C–L bond is being broken concurrently with the formation of the C–Nu bond, so both the substrate, R–L, *and* the nucleophile are involved in the transition state of the rate-determining step. Reactions in which two reactants are involved in the transition state of the rate-determining step are bimolecular reactions, and the rate of such processes depends on the concentration of the substrate and the nucleophile, as shown in Equation (14.5), where k_2 is the second-order rate constant.

$$\text{Rate} = k_2[\text{R–L}][\text{Nu:}] \qquad \textbf{(14.5)}$$

It is useful to summarize some of the important factors that have been found to play a role in dictating whether a particular substrate undergoes substitution preferentially by a S_N1 or S_N2 mechanism.

1. As more alkyl groups are attached to the carbon atom, **C–L**, undergoing substitution, it becomes more difficult for the nucleophile to attack from the backside, thereby decreasing the ease with which the S_N2 process can occur.

2. With increasing substitution of alkyl groups on the carbon atom **C–L**, the incipient carbocation in the S_N1 reaction becomes more stable, thereby increasing its ease of formation along the S_N1 pathway.

These two effects reinforce one another and yield the following general trends:

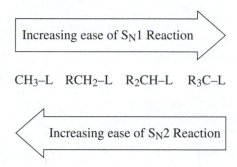

Increasing ease of S_N1 Reaction

CH_3–L $\quad$ RCH_2–L $\quad$ R_2CH–L $\quad$ R_3C–L

Increasing ease of S_N2 Reaction

Primary (1°, RCH$_2$–L) substrates undergo nucleophilic substitution predominantly by an S$_N$2 mechanism, whereas tertiary (3°, R$_3$C–L) substrates tend to react via an S$_N$1 mechanism. Secondary substrates (2°, R$_2$CH–L) may react by both mechanisms, and the specific pathway is dictated by factors such as solvent, reaction conditions, and nature of the nucleophile. Such generalizations must always be applied cautiously, and the mechanism by which a particular reaction occurs must be confirmed *experimentally*. For example, if the structures of the starting substrate and the product(s) are determined, it may be possible to deduce the mechanism by which the process occurred. Since carbocations are involved in the S$_N$1 mechanism, the formation of a product arising from a rearranged carbocation points to this mechanism; however, the lack of rearrangement does not necessarily exclude an S$_N$1 process.

14.3 COMPETITION BETWEEN SUBSTITUTION AND ELIMINATION

In all reactions in which nucleophilic aliphatic substitution occurs, elimination reactions may compete to produce alkenes. Unimolecular elimination reactions, **E1,** compete with S$_N$1 substitutions, and bimolecular elimination processes, **E2** (E stands for elimination and 2 for bimolecular), compete with S$_N$2 transformations. These competitions are shown in Equations (14.6) and (14.7). The nature of E1 reactions is discussed in detail in Section 10.2 and that of E2 processes in Section 10.1.

S$_N$2 versus E2

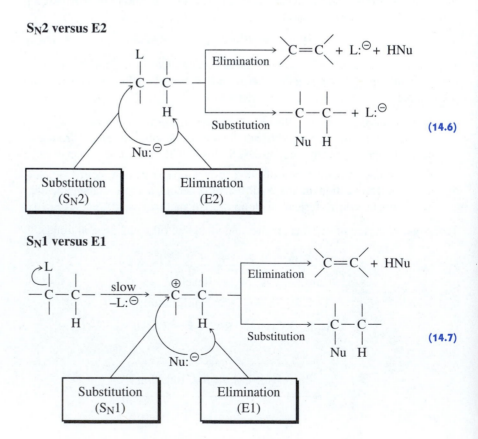

(14.6)

S$_N$1 versus E1

(14.7)

The course of the reaction is influenced to some extent by the nature of the nucleophile or Lewis base that is present. *Substitution is favored* with weakly basic and highly polarizable nucleophiles such as I$^-$, Br$^-$, Cl$^-$, H$_2$O, and CH$_3$CO$_2^-$, whereas *elimination is favored* when strongly basic and only slightly polarizable nucleophiles such as RO:$^-$, H$_2$N:$^-$, H:$^-$, and HO:$^-$ are used. **Polarizability** is a measure of the ease of distortion of the electron cloud of the species by a nearby charged center, which is a partial or full positive charge in substitution reactions. Furthermore, bulky nucleophiles are more prone toward inducing elimination because the hydrogen atom is more sterically accessible than the carbon atom bearing the leaving group.

14.4   PREPARATION OF 1-BROMOBUTANE: AN S$_N$2 REACTION

A common technique for converting a primary alcohol to an alkyl halide involves treating the alcohol with a hydrogen halide H–X (H–X = H–Cl, H–Br, or H–I) as shown in Equation (14.8). This reaction is *reversible,* and displacing the equilibrium to the right normally involves using a large excess of the acid, a strategy in accord with the **LeChatlier principle.**

$$H-X + R-OH \underset{\Delta}{\rightleftharpoons} R-X + H_2O \qquad \text{(14.8)}$$

where X = Cl, Br, I

1-Bromobutane may be prepared by heating 1-butanol with hydrobromic acid, H–Br, in the presence of sulfuric acid according to Equation (14.9). The mechanism for this reaction has been shown to occur in two steps, as shown in

$$CH_3CH_2CH_2CH_2-OH + HBr \underset{\Delta}{\overset{H_2SO_4}{\rightleftharpoons}} CH_3CH_2CH_2CH_2-Br + H_2O \quad \text{(14.9)}$$

1-Butanol	1-Bromobutane
(*n*-butyl alcohol)	(*n*-butyl bromide)

Equation (14.10). The alcohol is protonated in the first step via a Lewis acid-base reaction to give the oxonium ion **1.** This oxonium ion then undergoes displacement by the bromide ion to form 1-bromobutane and water. This process is an S$_N$2 reaction in which *water is the leaving group and bromide ion is the nucleophile.* The sulfuric acid serves two important purposes: (1) It is a dehydrating agent that reduces the activity of water and shifts the position of equilibrium to the right, and (2) it provides an added source of hydrogen ions to increase the concentration of oxonium ion **1.** The use of *concentrated* hydrobromic acid also helps to establish a favorable equilibrium.

$$n\text{-}C_3H_7CH_2 - \overset{..}{\underset{..}{O}}H + H^{\oplus} \rightleftharpoons n\text{-}C_3H_7CH_2 - \overset{\overset{\displaystyle H}{|}}{\underset{\displaystyle H}{O^{\oplus}}} \quad \overset{Br:^{\ominus}}{\rightleftharpoons} \quad n\text{-}C_3H_7CH_2 - Br + H_2O \qquad \text{(14.10)}$$

No reaction occurs between 1-butanol and NaBr in the absence of strong acid, because leaving groups in nucleophilic substitution reactions must be *weakly* basic, as is water in Equation (14.10) (see Sec. 14.1). If the reaction of 1-butanol and NaBr were to occur (Eq. 14.11), the leaving group would necessarily be the *strongly basic* hydroxide ion; thus the forward reaction depicted in Equation (14.11) does not occur. On the other hand, the reverse reaction between 1-bromobutane and hydroxide proceeds readily, since the leaving group in this reaction is the weakly basic bromide ion.

$$CH_3CH_2CH_2CH_2-OH + NaBr \rightleftharpoons CH_3CH_2CH_2CH_2Br + NaOH \quad \textbf{(14.11)}$$

Preliminary insights regarding the mechanism of a reaction may be obtained by analysis of the products. In the present case, GLC (see Sec. 6.4) may be used to determine whether 1-bromobutane, 2-bromobutane, or a mixture of the two is formed in the reaction. If only 1-bromobutane is formed, the reaction probably follows the S_N2 mechanism. Substitution via an S_N1 process necessarily involves formation of the highly unstable 1-butyl cation. If this primary carbocation were formed in solution, it would undergo molecular rearrangement to the more stable secondary butyl cation, which would then react with bromide ion to form 2-bromobutane, as shown in Equation (14.12). Thus, reaction by the S_N1 mechanism would lead to a mixture of products.

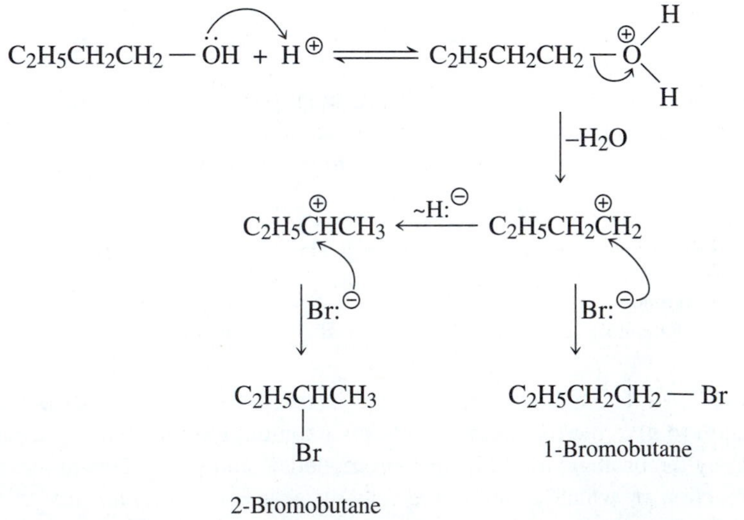

Care must be exercised when using the results of simple product analysis to gain mechanistic information, as pitfalls await the unwary. For example, if 2-bromobutane were detected in the reaction mixture, it would be tempting to conclude that it was produced from 1-butanol via an S_N1 mechanism. This conclusion would not be unequivocal, however, because alternative reaction pathways can also lead to the formation of 2-bromobutane. One such pathway begins with the acid-catalyzed dehydration of 1-butanol (see Sec. 10.3) to furnish a mixture of 1-butene and 2-butene. Subsequent addition of H–Br across the carbon-carbon double bonds of these alkenes gives 2-bromobutane as the major product (see Sec. 10.4). Thus, before making reliable conclusions regarding the mechanism of a reaction based solely upon product analysis, it is necessary to ensure that other reaction pathways that could also lead to the observed products are *not* operative.

The mixture of hydrobromic acid and sulfuric acid may be prepared in two ways. One method is to add concentrated sulfuric acid to concentrated hydrobromic acid; the second is to generate the hydrobromic acid *in situ* by adding concentrated sulfuric acid to aqueous sodium bromide (Eq. 14.13). Both of these methods work well and give good yields of the alkyl bromide when low-molecular-weight alcohols are used. With higher-molecular-weight alcohols, the method of generating H–Br *in situ* is not effective because of the low solubility of these alcohols in concentrated salt solutions. Concentrated hydrobromic acid is used in these instances. With very high-molecular-weight alcohols, the method of choice is to add hydrogen bromide gas to the alcohol at elevated temperatures or to select a different brominating reagent, such as PBr$_3$.

$$NaBr + H_2SO_4 \rightleftharpoons HBr + NaHSO_4 \qquad \text{(14.13)}$$

Although the presence of concentrated sulfuric acid facilitates the formation of the alkyl bromide, several side reactions involving sulfuric acid and the alcohol can occur. For example, esterification of the alcohol by sulfuric acid can form an alkyl hydrogen sulfate **2** (Eq. 14.14). This reaction is reversible, and the

$$RO-H + H_2SO_4 \rightleftharpoons RO-SO_3H + H_2O \qquad \text{(14.14)}$$

$$\mathbf{2}$$

An alkyl hydrogen
sulfate

position of equilibrium is shifted to the left, regenerating the alcohol from **2** as the alkyl bromide is produced. The formation of the alkyl hydrogen sulfate **2** itself does not directly decrease the yield of alkyl bromide, but **2** undergoes other reactions to give undesired by-products. For example, **2** suffers elimination on heating to give a mixture of alkenes (Eq. 14.15), or it reacts with another molecule of alcohol to give a dialkyl ether by an S$_N$2 reaction in which the nucleophile is ROH (Eq. 14.16). Both of these side reactions consume alcohol and decrease the yield of alkyl bromide. These side reactions are minimized for *primary* alcohols by controlling the temperature of the reaction.

$$RO-SO_3H \xrightarrow{\text{heat}} \text{alkenes} + H_2SO_4 \qquad \text{(14.15)}$$

$$RO-SO_3H + ROH \xrightarrow{\text{heat}} R-O-R + H_2SO_4 \qquad \text{(14.16)}$$

It is necessary to use different procedures to prepare secondary alkyl bromides from *secondary* alcohols because such alcohols are easily dehydrated by concentrated sulfuric acid to give alkenes (Eq. 14.15). In fact, the acid-catalyzed dehydration of secondary and tertiary alcohols is a common method for synthesizing alkenes (see Sec. 10.3). This problem is frequently circumvented by using concentrated (48%) hydrobromic acid; however, secondary alkyl halides are generally better prepared by the reactions of secondary alcohols with phosphorus trihalides, PX_3 (X = Br or I), as shown in Equation (14.17) or with thionyl chloride, $SOCl_2$, in the presence of pyridine (Eq. 14.18).

$$3\ R\!-\!OH + PX_3 \longrightarrow 3\ R\!-\!X + H_3PO_3 \qquad (14.17)$$

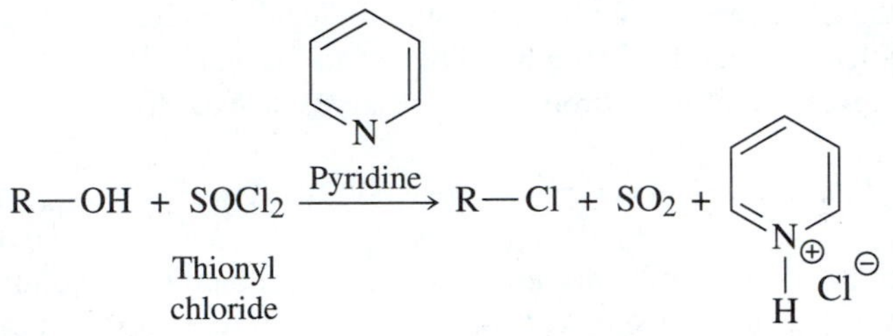

Phosphorus
trihalide

$$R\!-\!OH + SOCl_2 \xrightarrow{\text{Pyridine}} R\!-\!Cl + SO_2 + \qquad (14.18)$$

Thionyl
chloride

Whenever the secondary alcohol is susceptible to *carbocationic rearrangements* (see Secs. 10.3 and 15.2), thionyl chloride must be used with pyridine, which serves both as a solvent and as a base to neutralize the hydrochloric acid as it is formed (Eq. 14.18).

Pre-Lab exercises for Section 14.4, "Preparation of 1-Bromobutane," are found on page PL. 53.

EXPERIMENTAL PROCEDURE

SAFETY ALERT

1. Examine your glassware for cracks and chips. This experiment involves heating concentrated acids, and defective glassware could break under these conditions and spill hot corrosive chemicals on you and those working around you.

2. Concentrated sulfuric acid and water mix with the evolution of substantial quantities of heat. *Always add the acid to the water* in order to disperse the heat through warming of the water. Add the acid slowly and with swirling to ensure continuous and thorough mixing.

3. If possible, wear rubber gloves in this experiment. Be very careful when handling concentrated sulfuric acid. When it is poured from the reagent bottle, some may run down the outside of the bottle. If any concentrated sulfuric acid comes in contact with your skin, immediately wash it off with copious amounts of cold water and then with dilute sodium bicarbonate solution.

Place 10.3 g of sodium bromide in a 100-mL round-bottom flask, and add about 10 mL of water and 7.5 g of 1-butanol. Mix the contents of the flask thoroughly by swirling it; then cool the flask in an ice-water bath. *Slowly* add 10 mL of *concentrated* sulfuric acid to the cold mixture with swirling and continuous cooling. Remove the flask from the ice-water bath, add several boiling chips or a stirbar, if magnetic stirring is available, attach a reflux condenser, warm the flask gently until most of the salts have dissolved, and then heat the mixture under gentle reflux. A noticeable reaction occurs, and two layers form. Continue heating under reflux for 45 min.✻

Equip the flask for simple distillation. Distil the mixture rapidly, and collect the distillate, which contains water and 1-bromobutane, in an ice-cooled receiver. Continue the distillation until the distillate is clear; the head temperature should be around 115 °C at this point. The increased boiling point is caused by codistillation of sulfuric acid and hydrobromic acid with water.✻

Transfer the distillate to a separatory funnel and add about 10 mL of water. Shake the funnel gently with venting. Separate the layers, and decide which of these is the organic layer (see Sec. 2.18). Wash the organic layer with 4 mL of 2 *M* aqueous sodium hydroxide solution and then with about 10 mL of saturated ~~sodium chloride solution.~~ H₂O 10mL Transfer the cloudy 1-bromobutane layer to a 25-mL Erlenmeyer flask, and dry it over about 2 g of *anhydrous* sodium sulfate.✻ Swirl the flask occasionally for a period of 10–15 min until the crude 1-bromobutane is clear; add further small portions of anhydrous sodium sulfate if the solution does not become clear.

Using a micropipet to minimize material loss, carefully transfer all of the crude 1-bromobutane, without the drying agent, to a clean, dry 25-mL round-bottom flask. Alternatively, the drying agent may be removed by gravity filtration, but the material loss is greater. Add two or three boiling chips or a magnetic stirbar, equip the flask for simple distillation (see Fig. 2.15b), and carefully distil the product; pure 1-bromobutane boils at 100–104 °C (760 torr). Because of the relatively small quantity of material, an accurate boiling point may be difficult to measure, and you should collect the fraction having a boiling point of 90–104 °C (760 torr).

Compute the yield of the product isolated. Obtain the retention times of standard samples of 1-bromobutane and of 2-bromobutane on a nonpolar column (SF-96 or SE-30 is satisfactory), and analyze your product by GLC on the same column. In your report discuss the relative percentages of 1- and 2-bromobutanes in your product in terms of the mechanism(s) of the reaction. If possible, obtain IR and NMR spectra of your product and of 1-butanol. Characterize your alkyl bromide by the procedures of Section 23.5.3.

Finishing Touches   Carefully dilute the *stillpot residue from the reaction* with water and then slowly combine this with the *water and the sodium hydroxide washes*. Neutralize the *combined aqueous mixture* with sodium carbonate, and flush it down the drain with *excess water*. The *sodium sulfate* used as the drying agent is contaminated with product, so place it in the container for halogenated solvent-contaminated drying agents. Place the *residue in the stillpot from the distillation* into the container for halogenated organic solvents.

EXERCISES

1. In the procedure some water was added to the initial reaction mixture. How might the yield of bromobutane be affected if the water were not added, and what product(s) would be favored? How might the yield of product be affected by adding twice as much water as is called for while keeping the quantities of the other reagents the same?

2. In the purification process, the organic layer is washed with 2 M NaOH and then with saturated sodium chloride solution. What is the purpose of these washes?

3. After the washes described in Exercise 2, the 1-bromobutane is treated with anhydrous sodium sulfate. Why is this done?

4. The final step of the purification process involves adding some dichloromethane, followed by a simple distillation.
   a. Why is simple rather than fractional distillation used?
   b. Why is this final distillation performed? Suggest what is removed by this distillation.

5. The following reaction was carried out and the indicated products were isolated.

$$(CH_3)_2CHCHCH_3 \xrightarrow{\text{HBr}} (CH_3)_2CHCHCH_3 + (CH_3)_2CCH_2CH_3$$

$$\overset{|}{\text{OH}} \qquad\qquad \overset{|}{\text{Br}} \qquad\qquad \overset{|}{\text{Br}}$$

$$\qquad\qquad\qquad\qquad\qquad \textit{10\%} \qquad\qquad \textit{90\%}$$

   Suggest reasonable mechanisms for this reaction, and indicate whether each is S_N1 or S_N2.

6. Discuss the differences observed in the IR and NMR spectra of 1-butanol and 1-bromobutane.

7. Discuss the differences observed in the ^1H and ^{13}C NMR spectra of 1-bromobutane, which appear in Figure 14.4(a) and (b), with those of 1-chlorobutane, which appear in Figures 9.8 and 9.9. To what factor do you attribute these differences?

SPECTRA OF STARTING MATERIAL AND PRODUCT

The ^1H NMR spectrum of 1-butanol appears in Figure 8.33.

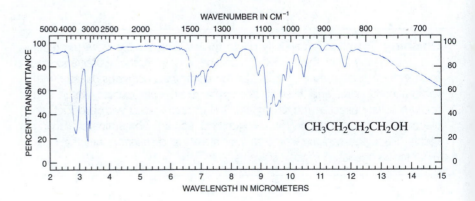

$CH_3CH_2CH_2CH_2OH$

FIGURE 14.1   IR spectrum of 1-butanol.

CH$_3$CH$_2$CH$_2$CH$_2$OH ^{13}C NMR data.   **FIGURE 14.2** ^{13}C NMR data for 1-butanol.
                                    Chemical shifts: δ 13.9, 19.2, 35.0, 62.2.

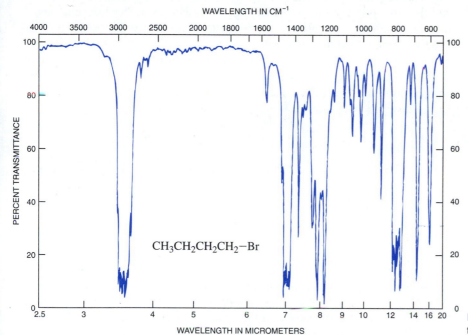

CH$_3$CH$_2$CH$_2$CH$_2$—Br

FIGURE 14.3   IR spectrum of 1-bromobutane.

(a) ^{1}H NMR spectrum (90 MHz).

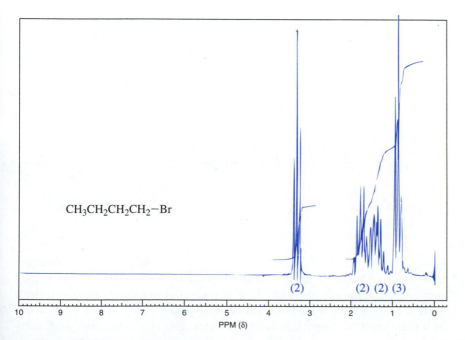

CH$_3$CH$_2$CH$_2$CH$_2$—Br

(2)   (2) (2) (3)

(b) ^{13}C NMR data.
Chemical shifts: δ 13.2, 21.5, 33.1, 35.0.

FIGURE 14.4   NMR data for 1-bromobutane.

14.5 PREPARATION OF TERTIARY ALKYL CHLORIDES: AN S_N1 REACTION

Different reagents such as HX, PX_3, and SOX_2 can be used to prepare alkyl halides from primary and secondary alcohols. However, because elimination reactions predominate when tertiary alcohols are treated with phosphorous trihalides or thionyl halides, preparing tertiary alkyl halides from tertiary alcohols proceeds with good yields only if concentrated hydrogen halides, HX, are used. The reaction of 2-methyl-2-butanol (*tert*-amyl alcohol) with hydrochloric acid to produce 2-chloro-2-methylbutane (*tert*-amyl chloride) as shown in Equation 14.19 illustrates this process:

$$
\underset{\substack{\text{2-Methyl-2-butanol}\\(tert\text{-amyl alcohol})}}{CH_3CH_2-\overset{\overset{\displaystyle CH_3}{|}}{\underset{\underset{\displaystyle CH_3}{|}}{C}}-OH} + HCl \rightleftharpoons \underset{\substack{\text{2-Chloro-2-methylbutane}\\(tert\text{-amyl chloride})}}{CH_3CH_2-\overset{\overset{\displaystyle CH_3}{|}}{\underset{\underset{\displaystyle CH_3}{|}}{C}}-Cl} + H_2O \qquad \text{(14.19)}
$$

Experiments have shown that the conversion of tertiary alcohols to the corresponding tertiary alkyl halide using concentrated hydrohalic acids proceeds via an S_N1 mechanism as shown in Equations (14.20–14.22). The first step of the reaction involves protonation of the hydroxyl group of the alcohol by combination of a Lewis acid with a Lewis base (Eq. 14.20). In the second step of the reaction, ionization occurs, and a molecule of water is lost (Eq. 14.21). Owing partially to steric hindrance at the tertiary carbon atom bearing the protonated

$$
CH_3CH_2-\overset{\overset{\displaystyle CH_3}{|}}{\underset{\underset{\displaystyle CH_3}{|}}{C}}-\ddot{\underset{}{O}}\overset{\displaystyle H}{\diagup} + H^{\oplus} \overset{fast}{\rightleftharpoons} CH_3CH_2-\overset{\overset{\displaystyle CH_3}{|}}{\underset{\underset{\displaystyle CH_3}{|}}{C}}-\overset{\oplus}{\underset{\underset{\displaystyle H}{}}{O}}\overset{\displaystyle H}{\diagup} \qquad \text{(14.20)}
$$

3

$$
CH_3CH_2-\overset{\overset{\displaystyle CH_3}{|}}{\underset{\underset{\displaystyle CH_3}{|}}{C}}-\overset{\oplus}{\underset{\underset{\displaystyle H}{}}{O}}\overset{\displaystyle H}{\diagup} \overset{rds}{\rightleftharpoons} \underset{\substack{\text{Relatively stable}\\\text{tertiary carbocation}}}{CH_3CH_2-\overset{\overset{\displaystyle CH_3}{|}}{\underset{\underset{\displaystyle CH_3}{|}}{C}}\oplus} + H_2O \qquad \text{(14.21)}
$$

hydroxyl group, attack of chloride on **3** in an S_N2 process as discussed in Section 14.4 does not occur. This difference in reactivity between the oxonium ions derived from tertiary alcohols and those of secondary or primary alcohols reflects the relative stabilities of the three types of carbocations (3° > 2° > 1°) that would be formed upon the loss of a molecule of water. This step is the slow, *rate-determining,* step in this sequence. In the final step of the reaction, chloride ion attacks the intermediate carbocation to give 2-chloro-2-methylbutane (Eq. 14.22).

$$\text{CH}_3\text{CH}_2-\underset{\underset{\text{CH}_3}{|}}{\overset{\overset{\text{CH}_3}{|}}{\text{C}}}\oplus \ + \ :\ddot{\text{Cl}}\!:^{\ominus} \ \underset{\text{fast}}{\rightleftharpoons} \ \text{CH}_3\text{CH}_2-\underset{\underset{\text{CH}_3}{|}}{\overset{\overset{\text{CH}_3}{|}}{\text{C}}}-\text{Cl} \qquad \text{(14.22)}$$

The principal side reaction in S_N1 reactions of this type is E1 elimination, which results from the loss of a proton from the carbocation to give 2-methyl-1-butene, together with 2-methyl-2-butene according to Equations (14.23) and

$$\text{CH}_3\text{CH}_2-\underset{\underset{\text{CH}_3}{|}}{\overset{\overset{\text{CH}_2-\text{H}}{|}}{\text{C}}}\oplus \ + \ :\ddot{\text{Cl}}\!:^{\ominus} \ \longrightarrow \ \text{CH}_3\text{CH}_2-\underset{\underset{\text{CH}_3}{|}}{\overset{\overset{\text{CH}_2}{\diagup\!\diagup}}{\text{C}}} \ + \ \text{H}-\text{Cl} \qquad \text{(14.23)}$$

$$\text{CH}_3\text{CH}-\underset{\underset{\text{CH}_3}{|}}{\overset{\overset{\text{CH}_3}{|}}{\text{C}}}\oplus \ \longrightarrow \ \text{CH}_3\text{CH}\!=\!\text{C}\!\!\begin{array}{c}\diagup \text{CH}_3 \\ \diagdown \text{CH}_3\end{array} \ + \ \text{H}-\text{Cl} \qquad \text{(14.24)}$$

$$:\ddot{\text{Cl}}\!:^{\ominus}$$

(14.24). However, under the reaction conditions of the present experiment, elimination is reversible through Markovnikov addition of HCl to the 2-methyl-1-butene or the 2-methyl-2-butene thus produced to give the desired product, 2-chloro-2-methylbutane (Eq. 14.25). A more extensive discussion of the ionic addition of hydrogen halides to alkenes is presented in Section 10.4.

$$\text{CH}_3\text{CH}_2-\underset{\underset{\text{CH}_3}{|}}{\overset{\overset{\text{CH}_2}{\|}}{\text{C}}}\text{H}\!-\!\text{Cl} \rightarrow \text{CH}_3\text{CH}_2-\underset{\underset{\text{CH}_3}{|}}{\overset{\overset{\text{CH}_3}{|}}{\text{C}}}\oplus \ + \ :\ddot{\text{Cl}}\!:^{\ominus} \rightarrow \text{CH}_3\text{CH}_2-\underset{\underset{\text{CH}_3}{|}}{\overset{\overset{\text{CH}_3}{|}}{\text{C}}}-\text{Cl}$$

$$\text{(14.25)}$$

or

$$\text{CH}_3\text{CH}\!=\!\text{C}\!\!\begin{array}{c}\text{H}\!-\!\text{Cl} \\ \diagup \text{CH}_3 \\ \diagdown \text{CH}_3\end{array}$$

Pre-Lab exercises for Section 14.5, "Preparation of 2-Chloro-2-methylbutane," are found on page PL. 55.

EXPERIMENTAL PROCEDURE

SAFETY ALERT

If possible, wear rubber gloves throughout the experiment because concentrated hydrochloric acid is being used. If any acid spills on your skin, wash it off with large volumes of water and then with dilute sodium bicarbonate solution.

Place 8.8 g of 2-methyl-2-butanol and 25 mL of reagent-grade *concentrated* (12 *M*) hydrochloric acid in a 60-mL separatory funnel. Swirl the contents of the separatory funnel gently *without* the stopper on the funnel to mix the reactants. After swirling the funnel for about 1 min, stopper the funnel and invert it; with the funnel inverted, release the excess pressure by *carefully* opening the stopcock. *Do not shake the funnel until the pressure has been equalized*. Now close the stopcock and shake the funnel for several minutes, with intermittent venting. Allow the mixture to separate into two distinct layers.

Separate the layers, determine which is the organic layer, and wash the organic layer with 10 mL of saturated sodium chloride solution and then with 10 mL of saturated sodium bicarbonate solution. On initial addition of the bicarbonate solution, vigorous gas evolution will normally occur; gently swirl the *unstoppered* separatory funnel until the vigorous gas effervescence ceases. Stopper the funnel and invert it carefully; vent the funnel immediately to release gas pressure. Shake the separatory funnel gently, with *frequent* venting; then shake it vigorously, again with frequent venting. Separate the organic layer, wash it with about 10 mL of water, and then with 10 mL of saturated sodium chloride solution. Carefully remove the aqueous layer, transfer the 2-chloro-2-methylbutane to a 25-mL Erlenmeyer flask, and dry the product over *anhydrous* sodium sulfate.✷ Swirl the flask occasionally for 10–15 min until the product is dry; add several small portions of anhydrous sodium sulfate until the product is clear.

Using a micropipet to minimize material losses, transfer the crude product, without the drying agent, to a 25-mL round-bottom flask, add several boiling chips or a magnetic stirring bar, and equip the flask for simple distillation (see Fig. 2.15a). Collect the fraction boiling from 83–85 °C (760 torr) in a receiver cooled in an ice-water bath. Determine the percent yield of the product.

Finishing Touches   Carefully dilute the *aqueous layer* from the first separation with water; then neutralize it with sodium carbonate. Combine this solution with the other *aqueous washes* (water, saturated sodium bicarbonate, and saturated sodium chloride) and *layers,* and flush them down the drain with excess water. The *sodium sulfate* used as the drying agent is contaminated with product, so place it in the container for halogenated solvent-contaminated drying agents. Pour the *residue in the stillpot* into the container for halogenated organic solvents.

EXERCISES

1. Draw the structures of all the other alcohols that are isomeric with 2-methyl-2-butanol. Arrange these alcohols in order of increasing reactivity toward concentrated hydrochloric acid. Which, if any, of these other alcohols would you expect to give a reasonable yield of the corresponding alkyl chloride under such reaction conditions?

2. The work-up procedures for the reactions in this section call for washing the crude alkyl chlorides with sodium bicarbonate solution.

   a. What purpose does this wash serve?

b. This washing procedure is accompanied by vigorous gas evolution, which increases the difficulty of handling and requires considerable caution. Alternatively, one might consider using a dilute solution of sodium hydroxide instead of the sodium bicarbonate. Discuss the relative advantages and disadvantages of using these two basic solutions in the workup. On the basis of these considerations, why were you instructed to use sodium bicarbonate, even though it is more difficult to handle?

3. The *rate* of reaction of alcohols with concentrated hydrochloric acid is increased when anhydrous zinc chloride is added to the acid. For all practical purposes, primary alcohols are unreactive toward pure concentrated hydrochloric acid, but they react at a reasonable rate when zinc chloride is added. Explain this fact. (*Hint:* zinc chloride is a *Lewis acid*.)

4. On prolonged heating with concentrated hydrochloric acid, 2,2-dimethyl-1-propanol (neopentyl alcohol), a primary alcohol, reacts as shown in Equation (14.26). Provide a stepwise mechanism for this reaction, using curved arrows to symbolize flow of electrons. Also account for the fact that the observed product is a tertiary alkyl chloride even though the starting alcohol is primary.

$$CH_3-\underset{\underset{CH_3}{|}}{\overset{\overset{CH_3}{|}}{C}}-CH_2-OH \xrightarrow{\;HCl\;} CH_3-\underset{\underset{Cl}{|}}{\overset{\overset{CH_3}{|}}{C}}-CH_2CH_3 + H_2O \qquad \textbf{(14.26)}$$

2,2-Dimethyl-1-propanol   2-Chloro-2-methylbutane
(neopentyl alcohol)   (*tert*-pentyl chloride)

5. Why is *saturated* sodium chloride solution, rather than dilute solution, specified for the final washing of the product?

6. The alkyl chloride is dried over with anhydrous sodium sulfate in this procedure. Could solid sodium hydroxide or potassium hydroxide be used for this purpose? Explain.

SPECTRA OF STARTING MATERIAL AND PRODUCT

The IR, ^{1}H and ^{13}C NMR spectra of 2-chloro-2-methylbutane are provided in Figures 10.2 and 10.3, respectively.

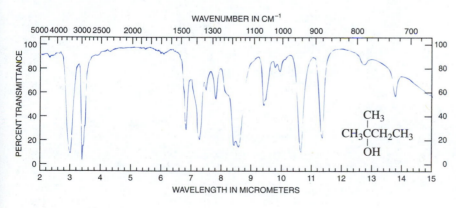

FIGURE 14.5 IR spectrum of 2-methyl-2-butanol.

FIGURE 14.6 NMR data for 2-methyl-2-butanol.

(a) ^{1}H NMR spectrum (60 MHz).

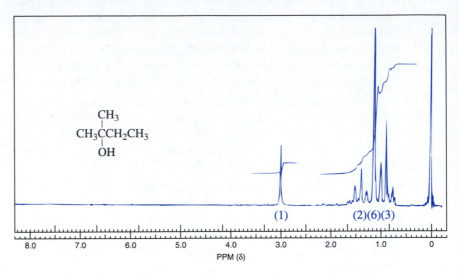

(b) ^{13}C NMR data.

Chemical shifts: δ 8.7, 28.6, 36.5, 71.1.

CHAPTER

15

ELECTROPHILIC AROMATIC SUBSTITUTION

15.1 INTRODUCTION

Electrophilic aromatic substitution is an important reaction in organic chemistry because it allows the introduction of many different functional groups onto an aromatic ring. A general form of the reaction is given by Equation (15.1), where Ar–H is an **aromatic compound,** an **arene,** and E^+ represents an **electrophile** that replaces H on the ring. This equation is oversimplified because the electrophile is usually generated during the reaction, and a Lewis base assists in the removal of H^+.

$$\underset{\text{An arene}}{\text{Ar—H}} + \underset{\text{An electrophile}}{E^{\oplus}} \longrightarrow \underset{\text{A substituted arene}}{\text{Ar—E}} + H^{\oplus} \qquad \text{(15.1)}$$

The rates of electrophilic aromatic substitutions generally, but not always, are *second-order* overall, being first *order* in both the aromatic component *and* the electrophile, as shown in Equation (15.2). A general mechanism that is consistent with this kinetic expression is depicted in Equations (15.3–15.5).

$$\text{rate} = k_2[\text{Ar—H}][E^{\oplus}] \qquad \text{(15.2)}$$

Step 1: Formation of electrophile:

$$\text{E—Nu} \xrightleftharpoons{\text{catalyst}} E^{\oplus} + \text{Nu}^{\ominus} \qquad \text{(15.3)}$$

Step 2: Reaction of electrophile:

$$\text{Ar—H} + E^{\oplus} \xrightleftharpoons{\text{rds}} \left[\overset{\oplus}{\text{Ar}}\begin{smallmatrix} \diagup \text{H} \\ \diagdown \text{E} \end{smallmatrix} \right] \qquad \text{(15.4)}$$

Step 3: Loss of proton to give product:

$$\left[\begin{array}{c} \overset{\oplus}{Ar} \overset{\overline{H}}{\diagdown}\diagup \\ \diagdown \\ E \end{array} \right] \overset{fast}{\rightleftharpoons} Ar-E \ + \ H^{\oplus} \qquad (15.5)$$

The electrophile is usually produced by the reaction between a catalyst and a compound containing a potential electrophile (Eq. 15.3). The second-order nature of the reaction arises from the step shown in Equation (15.4) in which one molecule each of arene and electrophile react to give the positively charged intermediate; the formation of this cation is the **rate-determining step** (rds) in the overall reaction. The subsequent deprotonation of the cation (Eq. 15.5) is fast. The bimolecular nature of the transition state for the rate-limiting step and the fact that an electrophile is involved in attacking the aromatic substrate classifies the reaction as S_E2 (**S**ubstitution **E**lectrophilic **Bi**molecular).

Experiments involving four different such reactions are given in this chapter: Friedel-Crafts alkylation (Sec. 15.2) and acylation (Sec. 15.5), nitration (Sec. 15.3), and bromination (Sec. 15.4).

15.2 FRIEDEL-CRAFTS ALKYLATION OF *P*-XYLENE WITH 1-BROMOPROPANE

The **Friedel-Crafts alkylation reaction** using alkyl halides (Eq. 15.6) is a classic example of electrophilic aromatic substitution (see the Historical Highlight concerning this reaction at the end of the chapter). As a versatile method for directly attaching alkyl groups to aromatic rings, it is a process of great industrial importance. The three main limitations to this reaction as a synthetic tool are (1) the substitution fails if the aromatic ring carries strongly **ring-deactivating groups,** such as NO_2, R_3N^+, C(O)R, and CN, (2) the introduction of more than one alkyl group onto the aromatic ring, a process termed polyalkylation, and (3) the formation of mixtures of isomeric products that result from rearrangement of the alkyl group. The first restriction is associated with the fact that aromatic rings bearing electron-withdrawing substituents are not sufficiently nucleophilic to react with the electrophiles generated under the reaction conditions. The second problem arises because addition of the first alkyl group activates the ring toward further substitution, but this difficulty is minimized by using a large excess of the arene, as is done in our experiment. The final limitation results from rearrangements of the alkyl group, R (Eq. 15.6), and is discussed further in the following paragraphs.

$$Ar-H \ + \ R-X \ \overset{catalyst}{\rightleftharpoons} \ Ar-R \ + \ HX \qquad (15.6)$$

An arene   An alkyl halide   An alkylarene
(X = halogen)

A generally accepted mechanism for Friedel-Crafts alkylation using alkyl halides is illustrated in equations (15.7–15.9). In the first step, the active electrophile, written as R^+ for the present purpose, is produced from an alkyl halide by reaction with a Lewis acid such as $AlCl_3$ (Eq. 15.7). The rate-determining step (rds) involves attack of the arene, which functions as a Lewis base, on the electrophile to yield a positively charged, resonance-stabilized **sigma complex** (Eq. 15.8). Finally, a base such as a halide ion from $XAlCl_3^-$ deprotonates this complex to give HX, thereby re-forming the aromatic system, and regenerating the catalyst (Eq. 15.9).

$$R-X \ + \ AlCl_3 \ \rightleftharpoons \ R^{\oplus} \ + \ \overset{\ominus}{X}AlCl_3 \qquad \textbf{(15.7)}$$

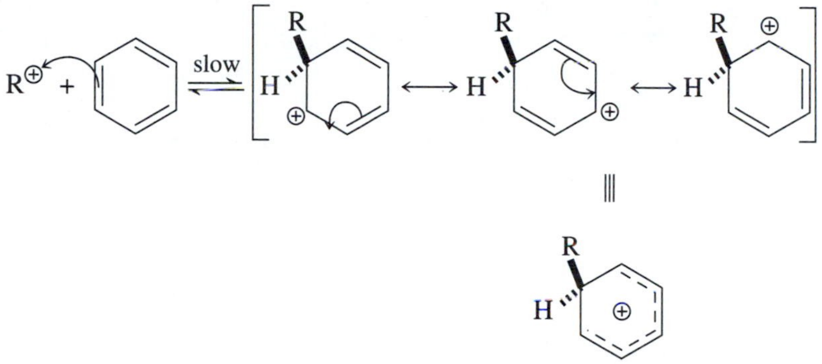

$$\textbf{(15.8)}$$

σ–Complex

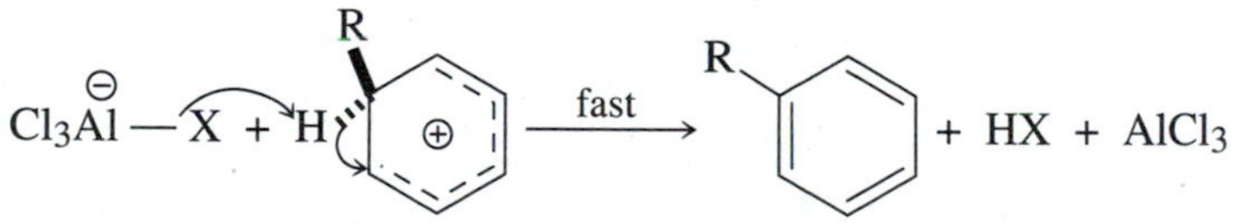

$$\textbf{(15.9)}$$

Representing the electrophile as a carbocation, as in Equation (15.7), is reasonable when *secondary* and *tertiary* alkyl halides are used. However, this is probably *not* the case with the electrophile derived from a *primary* alkyl halide, where the reactive species is better written as a polarized complex such as $R^{\delta+}$----X----$Al^{\delta-}Cl_3$. Nevertheless, the electrophiles from all three types of alkyl halides are represented as carbocations, R^+, to simplify the ensuing discussion.

Molecular rearrangements involving hydride, H^-, and alkyd, R^-, shifts become possible when a carbocation or a polarized cation-like complex is an intermediate in a reaction. As discussed in Section 10.2, such processes accompany the formation of carbocations during elimination reactions. Isomerization of a less stable carbocation to a more stable one is *thermodynamically favorable* of course, but a rearrangement may be *kinetically unfavorable* if it involves little gain in stability for the cation or requires generating a higher-energy intermediate along the pathway to the more stable cation.

These concepts are illustrated by the reactions of the isomeric butyl halides in Friedel-Crafts alkylation of benzene (Scheme 15.1). An *n*-butyl halide, **1**, reacts with benzene to give a mixture of *n*-butylbenzene (**3**), which is the unrearranged product, and *sec*-butylbenzene (**6**), a rearranged product, in a ratio of about 1:2. In this case, the primary *n*-butyl carbocation (**2**), or more likely its

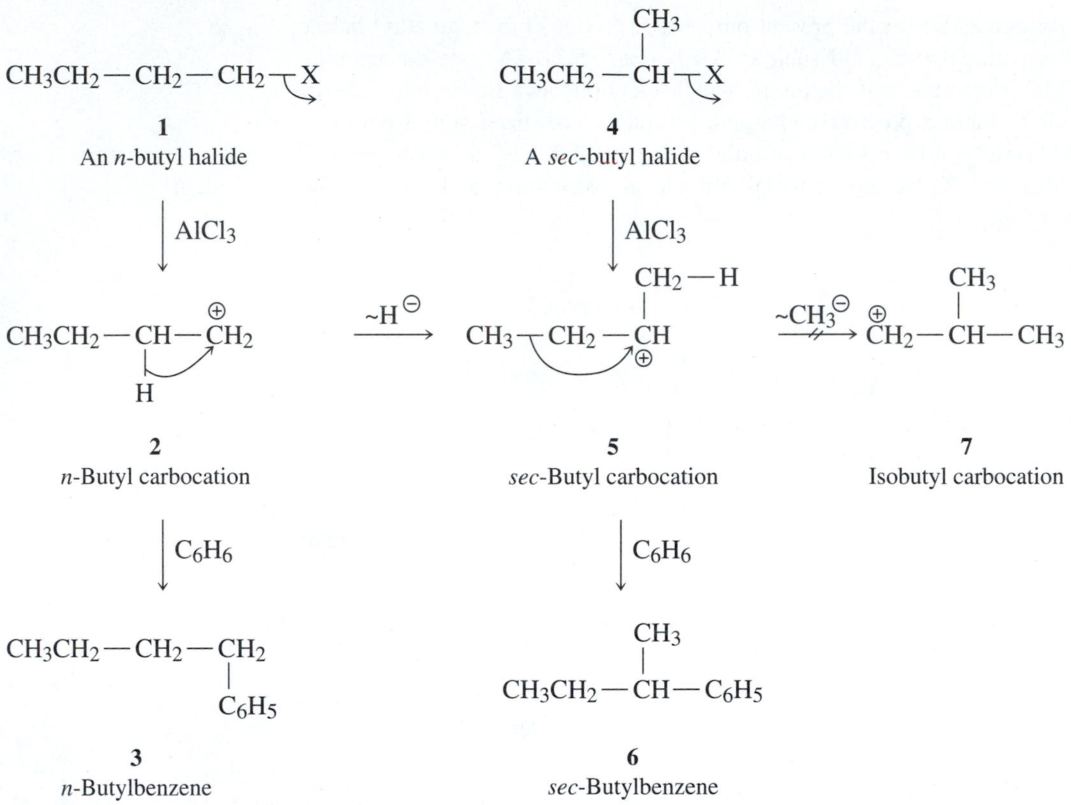

$$CH_3CH_2-CH_2-CH_2\!\!\!\diagup X$$

1

An *n*-butyl halide

$$\underset{CH_3CH_2\,-\,CH\,\diagup X}{\overset{\overset{\textstyle CH_3}{|}}{}}$$

4

A *sec*-butyl halide

$$\Big\downarrow AlCl_3$$

$$\Big\downarrow AlCl_3$$

$$CH_3CH_2-\underset{\underset{\textstyle H}{|}}{CH}-\overset{\oplus}{CH_2}$$

2

n-Butyl carbocation

$$\xrightarrow{\;\sim H^{\ominus}\;}$$

$$CH_3\diagup CH_2-\underset{\overset{\oplus}{}}{\overset{\overset{\textstyle CH_2-H}{|}}{CH}}$$

5

sec-Butyl carbocation

$$\xrightarrow{\;\sim CH_3^{\ominus}\;}\!\!\!\!/\!\!\!/$$

$$\overset{\oplus}{CH_2}-\underset{\overset{\textstyle CH_3}{|}}{CH}-CH_3$$

7

Isobutyl carbocation

$$\Big\downarrow C_6H_6$$

$$\Big\downarrow C_6H_6$$

$$CH_3CH_2-CH_2-\underset{\underset{\textstyle C_6H_5}{|}}{CH_2}$$

3

n-Butylbenzene

$$CH_3CH_2-\underset{\overset{\textstyle CH_3}{|}}{CH}-C_6H_5$$

6

sec-Butylbenzene

SCHEME 15.1

equivalent polarized complex, undergoes a 1,2-hydride shift to give the more stable secondary ion (**5**). The proportion of **3:6** is determined by the relative rate of reaction of **2** and the rate of rearrangement of **2** to **5** *before* reaction with benzene.

Alkylation of benzene with the secondary butyl halide **4,** affords only unrearranged product, *sec*-butylbenzene (**6**). The possible rearrangement of the intermediate carbocation **5** to the isobutyl carbocation (**7,** Scheme 15.1), which then would isomerize to the tertiary species **11** and lead to *tert*-butylbenzene (**12**) does *not* occur (Scheme 15.2). The rearrangement of **5** to **7** is too endothermic, because secondary carbocations are much more stable than primary ones, so this process is not kinetically competitive with reaction between **5** and benzene.

In contrast, alkylation with an isobutyl halide, **8,** gives only the rearranged product *tert*-butylbenzene (**12,** Scheme 15.2). The electrophilic intermediate initially formed from **8** and AlCl$_3$ is probably the polarized complex mentioned earlier rather than the primary cation **7**—it rearranges so rapidly by a 1,2-hydride shift to give the more stable tertiary carbocation **11** that alkylation *prior* to rearrangement is not a competitive process. Consequently, none of product **9** is obtained.

Finally, use of a *tert*-butyl halide **10** in the Friedel-Crafts alkylation of benzene yields only the unrearranged product **12.** This is the expected result

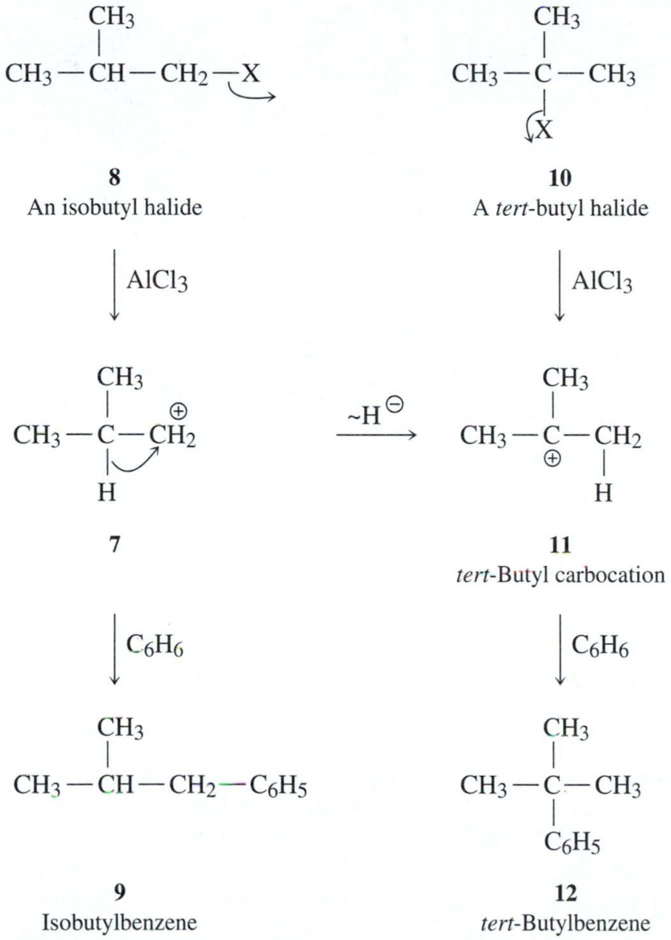

$$
\begin{array}{ccc}
& \text{CH}_3 & \\
& | & \\
\text{CH}_3-\text{CH}-\text{CH}_2-\text{X} & \longrightarrow & \\
\end{array}
\qquad
\begin{array}{c}
\text{CH}_3 \\
| \\
\text{CH}_3-\text{C}-\text{CH}_3 \\
| \\
\text{X}
\end{array}
$$

8
An isobutyl halide

10
A *tert*-butyl halide

↓ AlCl₃ ↓ AlCl₃

$$
\begin{array}{c}
\text{CH}_3 \\
| \\
\text{CH}_3-\overset{\oplus}{\text{C}}-\text{CH}_2 \\
| \\
\text{H}
\end{array}
\qquad
\xrightarrow{\;\sim\text{H}^{\ominus}\;}
\qquad
\begin{array}{c}
\text{CH}_3 \\
| \\
\text{CH}_3-\underset{\oplus}{\text{C}}-\text{CH}_2 \\
\qquad\quad | \\
\qquad\quad \text{H}
\end{array}
$$

7

11
tert-Butyl carbocation

↓ C₆H₆ ↓ C₆H₆

$$
\begin{array}{c}
\text{CH}_3 \\
| \\
\text{CH}_3-\text{CH}-\text{CH}_2-\text{C}_6\text{H}_5
\end{array}
\qquad
\begin{array}{c}
\text{CH}_3 \\
| \\
\text{CH}_3-\text{C}-\text{CH}_3 \\
| \\
\text{C}_6\text{H}_5
\end{array}
$$

9
Isobutylbenzene

12
tert-Butylbenzene

SCHEME 15.2

because the tertiary carbocation **11** produced from **10** is not subject to rearrangement to **7**.

The amount of rearrangement accompanying alkylation also depends on factors other than the structure of the alkyl halide. These include the nature of the aromatic substrate, the temperature at which the reaction is run, the solvent used, if any, and the nature and concentration of the catalyst. The effect that the first of these factors has in determining the outcome of a Friedel-Crafts alkylation with the same alkyl halide is the subject of the present experiment.

It is known that the reaction of 1-bromopropane with benzene in the presence of aluminum chloride gives a mixture of *n*-propylbenzene (**15**) and isopropylbenzene (**16**, cumene), as shown in Scheme 15.3. Moreover, there is little variation in the proportion of the two isomers when the reaction is performed at different temperatures. Given that the ratio of **15:16** is 1:2, it follows that the rate of direct reaction of **13**, or the equivalent complex, with benzene to give **15** *must* be competitive with that for isomerization of the primary cation **13** to the secondary ion **14**, which then reacts with benzene to afford **16**. In mathematical terms, $k_2[\mathbf{13}][\text{C}_6\text{H}_6]$ must be comparable to $k_1[\mathbf{13}]$.

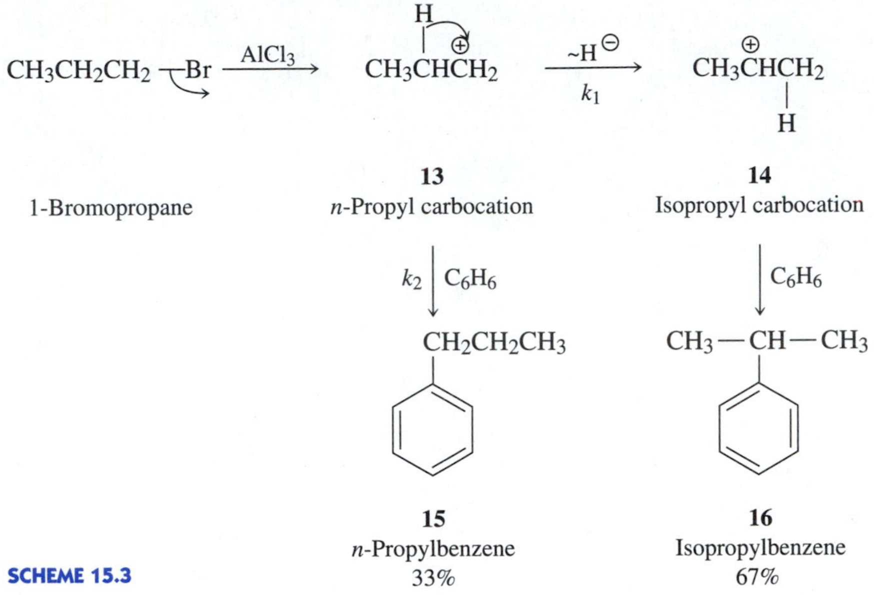

CH3CH2CH2⏜Br $\xrightarrow{\text{AlCl}_3}$ CH3CHCH2⊕ $\xrightarrow[k_1]{\sim\text{H}^{\ominus}}$ CH3CHCH2⊕
                                                                              |
                                                                              H

13   **14**

1-Bromopropane   *n*-Propyl carbocation   Isopropyl carbocation

k_2 | C6H6 | C6H6

CH2CH2CH3   CH3—CH—CH3

15   **16**

SCHEME 15.3   *n*-Propylbenzene   Isopropylbenzene

33%   67%

It is reasonable that the value of the rate constant k_1 would be *independent* of the nature of the arene that is present because the rearrangement with which k_1 is associated does *not* involve the arene. In contrast, the value of k_2 would depend significantly on the arene being used because this rate constant is for a reaction directly involving the aromatic substrate. The magnitude of k_2 increases as the arene becomes more nucleophilic, so that the overall rate of the electrophilic substitution involving this rate constant also increases.

The importance of k_2 in determining the ratio of unrearranged to rearranged products can be studied by using *p*-xylene rather than benzene in the Friedel-Crafts reaction with 1-bromopropane (Eq. 15.10). Because methyl groups activate the ring toward the S_E2 reaction, thereby making $k_{2(p\text{-xylene})} > k_{2(\text{benzene})}$, we expect the ratio of unrearranged product **17** to its rearranged isomer **18** to be greater than that of **15:16**. The experiment you are to perform tests this prediction.

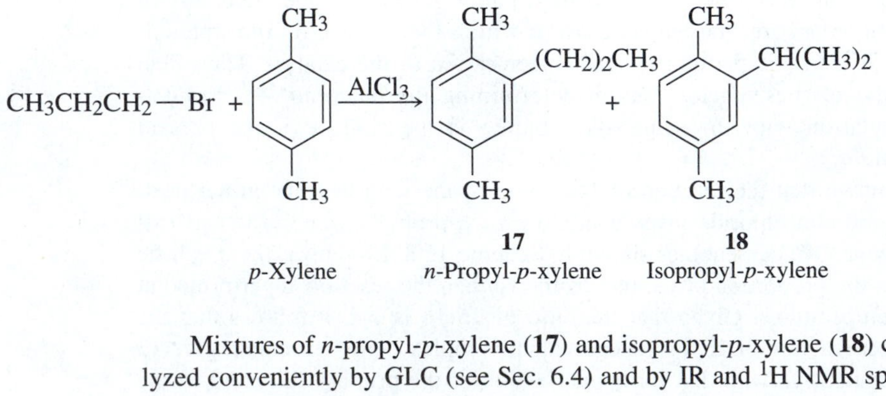

CH3CH2CH2—Br + [*p*-Xylene ring with CH3 top and CH3 bottom] $\xrightarrow{\text{AlCl}_3}$ [ring with CH3 top, (CH2)2CH3, CH3 bottom] + [ring with CH3 top, CH(CH3)2, CH3 bottom]   **(15.10)**

p-Xylene   **17**   **18**

n-Propyl-*p*-xylene   Isopropyl-*p*-xylene

Mixtures of *n*-propyl-*p*-xylene (**17**) and isopropyl-*p*-xylene (**18**) can be analyzed conveniently by GLC (see Sec. 6.4) and by IR and ^{1}H NMR spectroscopy (see Secs. 8.2 and 8.3.2), respectively.

Pre-Lab exercises for Section 15.2, "Friedel-Crafts Alkylation of *p*-Xylene with 1-Bromopropane," are found on page PL. 57.

Pre-Lab exercises for Section 15.2, "Friedel-Crafts Alkylation of *p*-Xylene with 1-Bromopropane," are found on page PL. 57.

EXPERIMENTAL PROCEDURE

SAFETY ALERT

1. Anhydrous aluminum chloride is very hygroscopic and reacts rapidly with water, even the moisture on your hands, producing fumes of hydrogen chloride, which are highly corrosive. Do not allow aluminum chloride to come in contact with your skin. If it does, flush the affected area with copious amounts of water.

2. Because aluminum chloride is a powdery solid that easily becomes airborne, weigh and transfer it into the reaction flask *in the hood*.

3. *p*-Xylene is flammable. Assemble the apparatus carefully and be sure that all joints are tightly mated. Have your instructor inspect your set-up before you begin the distillation. Because the boiling points of *p*-xylene and the propylxylenes are higher than 100 °C (760 torr), steam heating is inadequate. Electrical rather than gas heating is preferred, but if a burner is used, keep the flame away from the distillate.

If instructed to do so, dry the *p*-xylene for this experiment by azeotropic distillation. Do this by a simple distillation (see Sec. 2.7); discard the first portion of the distillate, which may be cloudy if the *p*-xylene contains some water and continue distilling until the required amount of *p*-xylene is obtained

Equip a 50-mL round-bottom flask with a Claisen connecting tube, a water-cooled condenser, a gas trap, and an addition funnel, as shown in Figure 15.1. If magnetic stirring is available, put a stirring bar in the flask. Place 0.7 g of anhydrous powdered aluminum chloride in the flask, and immediately cover it with 12 mL of dry *p*-xylene. Because the success of this experiment is highly dependent on the quality of the aluminum chloride that is used, obtain it from a *freshly* opened bottle and *weigh it quickly*. Alternatively, you may be provided with a weighed amount of aluminum chloride. In either case, *minimize* exposure of this chemical to the atmosphere.

Measure 6.2 g of 1-bromopropane in a 10-mL graduated cylinder, and pour it into the separatory funnel (*stopcock closed*). Prepare an ice-water bath in case it is needed to cool the reaction mixture. Turn on the water to the aspirator, or connect the trap to the house vacuum source; then begin adding the 1-bromopropane *dropwise* to the mixture of *p*-xylene and aluminum chloride. Loosen the clamp holder of the lower clamp so that the reaction mixture can be swirled gently from time to time as the 1-bromopropane is being added. If magnetic stirring is available, it will not be necessary to swirl the reaction mixture manually. If the evolution of hydrogen bromide becomes too vigorous, reduce the rate at which the 1-bromopropane is being added and raise the ice-water bath to cool the reaction mixture. The addition should take about 10 min. After all the 1-bromopropane has been added, allow another 30 min for completing the reaction, swirling the mixture every few minutes. Pour the mixture into a 50-mL beaker containing 10 g of crushed ice; work at a hood, because HBr and HCl are evolved. After stirring the mixture until all the ice has melted, pour it into a 125-mL separatory funnel and separate the aqueous layer. Pour the organic layer into a 125-mL Erlenmeyer flask containing *anhydrous* sodium sulfate and swirl it for about 1 min.✴

Decant the dried solution into a 25-mL round-bottom flask and equip the flask for fractional distillation with either a Vigreux or an unpacked Hempel column (see Fig. 2.17); use an air-cooled condenser. Make sure that all connections are tight and then distil, using either electrical or flame heating. Collect the excess *p*-xylene and forerun in one receiver (labeled A) until the distillation temperature reaches 180 °C (760 torr). Discontinue heating and allow any liquid in the column to drain into the flask.✴ Refit the distillation flask for short-path distillation (see Fig. 2.15b) and resume the distillation, collecting the fraction boiling between 180 and 207 °C (760 torr) in a second receiver (labeled B). Receiver A should be used to collect any liquid boiling below 180 °C before receiver B is attached to the apparatus. Isomers **17** and **18** have bp 204 °C and 196 °C (760 torr), respectively.

Record the weights of fractions A and B. Submit or save samples of each fraction for GLC, IR, and/or ^{1}H NMR analysis (see Figs. 15.2 to 15.6). Using the weights of fractions A and B and the analytical results, calculate the yield of propylxylenes, and estimate the proportion of the isomers.

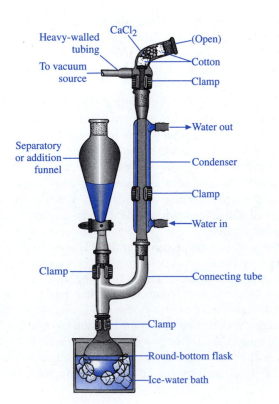

FIGURE 15.1   Apparatus for Friedel-Crafts alkylation.

> **Finishing Touches**   Neutralize the *aqueous layer* from hydrolysis of the reaction mixture with sodium carbonate and flush it down the drain. Spread the *calcium chloride* from the gas trap and the *sodium sulfate* used for drying on a tray in the hood to allow evaporation of the volatiles; then put these agents in the container for nontoxic solids. Place the *distillate* collected in flask A and the *pot residue* from the distillation in the container for nonhalogenated organic liquids.

EXERCISES

1. What compound(s) might be present in the higher-boiling residue remaining in the distillation flask after collection of the propylxylene isomers?

2. Calculate the molar ratio of 1-bromopropane to *p*-xylene used in the experiment. What would be the effect on the experimental results of using a ratio twice that actually used? Consider this in connection with your answer to Exercise 1.

3. Analyze your product by GLC and calculate the proportion of isopropyl-*p*-xylene and *n*-propyl-*p*-xylene in the mixture. Alternatively, use the GLC trace of Figure 15.2 to make this calculation. Discuss whether these calculations support the prediction regarding the relative amounts of unrearranged versus rearranged product to be expected from *p*-xylene as compared to benzene.

4. Account for the higher boiling point of **17** as compared to that of **18**. Are the boiling points consistent with the relative retention times of the two isomers as seen in the GLC trace of Figure 15.2? Explain your answer.

5. Why do you think it is recommended that the distillation fraction containing the mixture of **17** and **18** be taken over such a wide temperature range?

6. On the assumption that relative cost was not an overriding consideration, give a reason why *p*-xylene is used in preference to *m*-xylene as the aromatic substrate for this experiment.

7. Alkylation of toluene with 1-bromopropane gives a mixture of four isomeric propyltoluenes. Write formulas for these compounds, and based on the results of the alkylation of *p*-xylene predict the relative amounts in which they would be formed.

8. Propose a way to prepare pure *n*-propyl-*p*-xylene that does not require separation from isopropyl-*p*-xylene. (*Hint:* Consider Friedel-Crafts acylation, as discussed in Section 15.4.)

9. The choice of 1-bromopropane rather than 1-chloropropane as the alkylating agent in this experiment is based solely on the relative cost of these starting materials. Recent prices for these two chemicals are $28.90 and $90.75, respectively, for 500 mL. Show that the proper choice was made by proving that the bromo compound is less expensive on a *per mole* basis than its chloro relative.

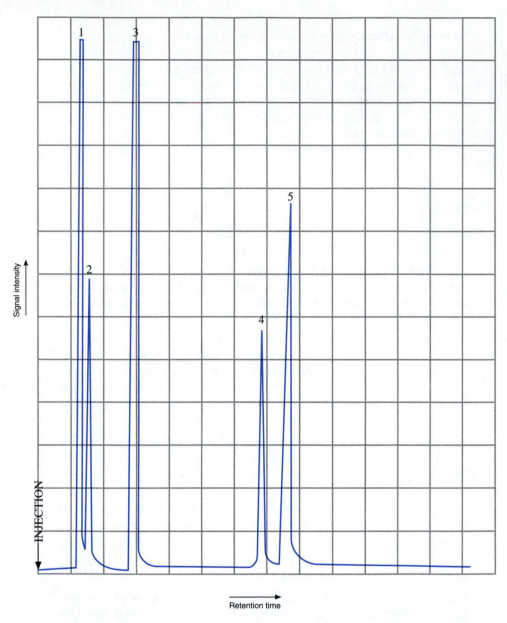

Signal intensity →

INJECTION

Retention time →

FIGURE 15.2 GLC trace of reactions mixture from alkylation of *p*-xylene with 1-bromopropane. The peaks are identified as (1) diethyl ether (solvent), (2) benzene and/or toluene (present as impurity in *p*-xylene), (3) *p*-xylene, (4) isopropyl-*p*-xylene, and (5) *n*-propyl-*p*-xylene. Column: 15-m, 2% silicone oil on DC-550 HiPak®.

10. **a.** Recent prices of 1-chlorobutane and 1-bromobutane are $17.90/L and $29.45/kg, respectively. Assuming that these both gave the same yield of products **3** and **6** in the Friedel-Crafts alkylation of benzene, determine which is better to use on the basis of economics.

   b. The corresponding costs of 2-chloro- and 2-bromobutane are $16.30/250 mL and $35.85/500 g. Make the same determination as you did in Part (a).

SPECTRA OF STARTING MATERIAL AND PRODUCTS

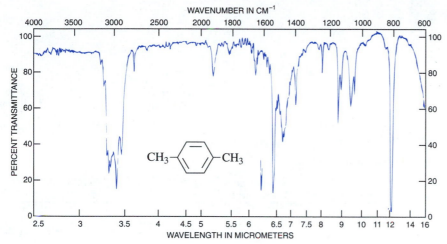

FIGURE 15.3 IR spectrum of *p*-xylene.

(a) ^{1}H NMR spectrum (90 MHz).

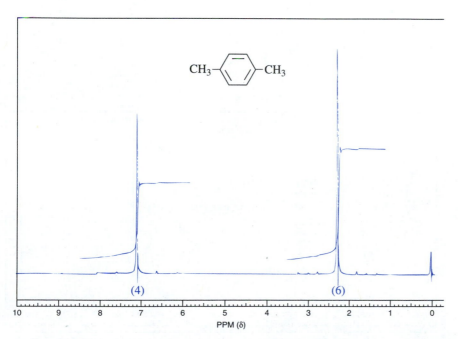

(b) ^{13}C NMR data.
Chemical shifts: δ 20.9, 129.0, 134.6.

FIGURE 15.4 NMR data for *p*-xylene.

FIGURE 15.5 IR spectra of the propyl-*p*-xylenes.

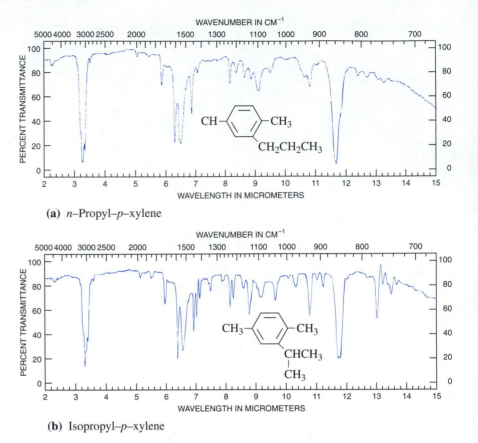

(a) *n*–Propyl–*p*–xylene

(b) Isopropyl–*p*–xylene

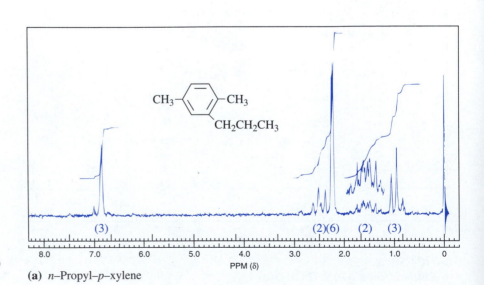

FIGURE 15.6 ^{1}H NMR spectra (60 MHz) of the propyl-*p*-xylenes.

(a) *n*–Propyl–*p*–xylene

FIGURE 15.6 (Continued)

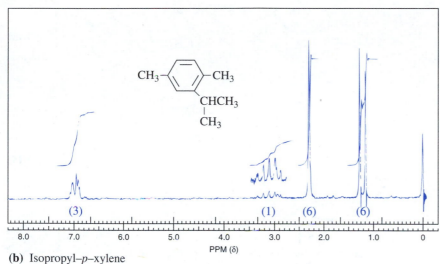

(b) Isopropyl–*p*–xylene

15.3 NITRATION OF BROMOBENZENE

Reaction of an aromatic compound such as benzene with a mixture of concentrated sulfuric and nitric acids introduces a nitro group on the ring by way of electrophilic aromatic substitution (see Sec. 15.1.) as depicted in Equation (15.11). The electrophilic species is the **nitronium ion,** NO_2^+, which is produced by reaction of sulfuric acid with nitric acid (Eq. 15.12); note that nitric acid, the *weaker* of these two strong mineral acids, is serving as a *base*, rather than an acid in the equilibrium!

$$\text{Benzene} \xrightarrow[\text{H}_2\text{SO}_4]{\text{HNO}_3} \text{Nitrobenzene} + \text{H}_2\text{O} \qquad \textbf{(15.11)}$$

$$\text{HO}-\text{NO}_2 \rightleftharpoons \overset{\oplus}{\text{O}}-\text{NO}_2 \rightleftharpoons \text{H}_2\text{O} + {}^{\oplus}\text{NO}_2 \qquad \textbf{(15.12)}$$

Nitric acid · · · · · · · · · · · · · · · · · · Nitronium ion

The rate-determining step (rds) in the nitration reaction involves nucleophilic attack of the aromatic ring, a Lewis base, on the nitronium ion, a Lewis acid, to form the delocalized **sigma σ-complex 19** (Eq. 15.13). In the final step of the reaction, a base such as water or bisulfate deprotonates this complex to regenerate the aromatic ring (Eq. 15.14).

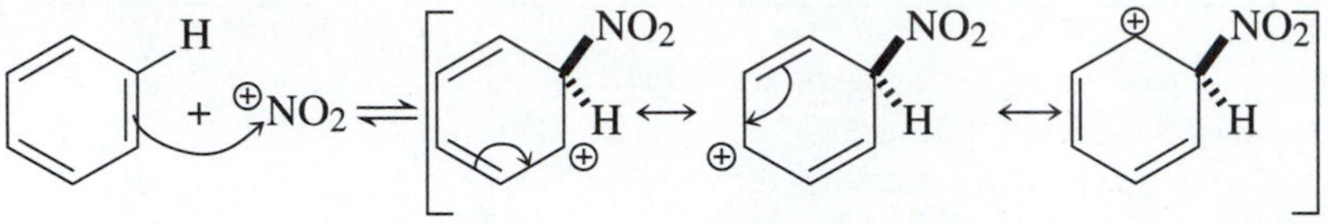

(15.13)

19
σ-Complex

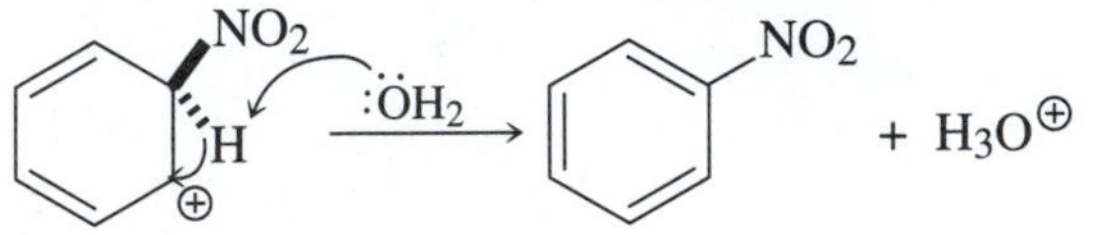

(15.14)

19
(One resonance form)

The mechanism of the reaction of a nitronium ion with a *substituted* benzene such as bromobenzene (**20**), the arene used in this experiment, is similar to that with benzene. Now, however, *three* different products can be formed because of the presence of the bromine atom on the ring (Eq. 15.15). The activation energies for attack at the positions *ortho* and *para* to the substituent are

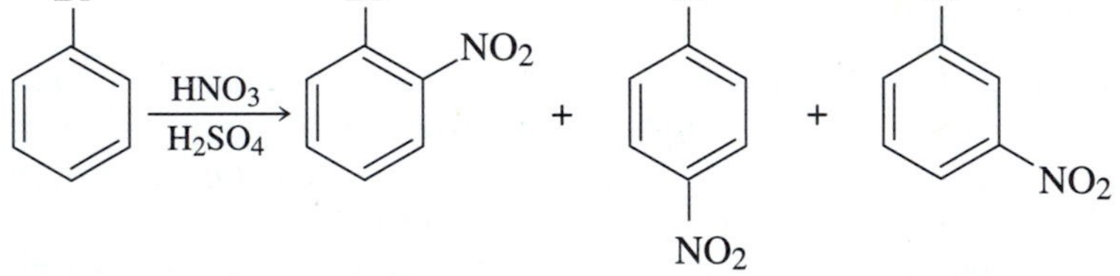

(15.15)

20	**21**	**22**	**23**
Bromobenzene	*o*-Bromo-nitrobenzene	*p*-Bromo-nitrobenzene	*m*-Bromo-nitrobenzene

lower than that at the *meta* position. This is because the σ complexes **24** and **25,** which are the precursors to the *ortho* and *para* products **21** and **22,** respectively, are more stable than complex **26,** which gives the *meta* compound **23.** The increased stability (lower energy) of **24** and **25** relative to **26** is associated with the participation of nonbonding electrons on the bromine atom in delocalizing the positive charge in the complexes **24** and **25** (see Exercise 2). Thus, like an alkyl group, a bromine atom is an example of an ***ortho-para* director,** since it favors formation of these isomers over the *meta* isomer; the preference is about 100:1 experimentally.

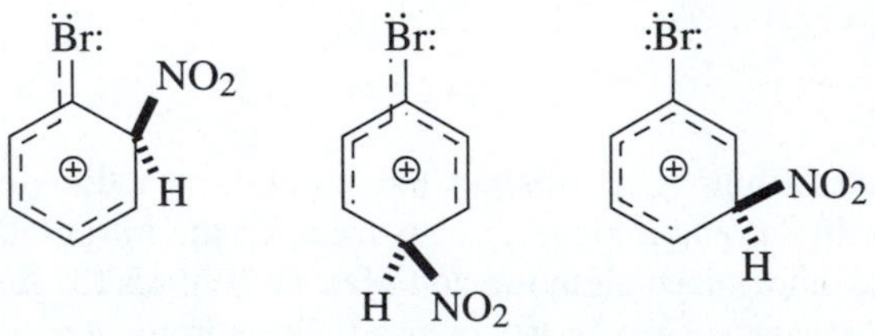

24   **25**   **26**

The *o:p* ratio itself is of interest because isolating pure disubstituted products requires separation of these isomers, which might be difficult with some pairs of substituents. Assuming that attack of the electrophile at the two positions occurs on a purely statistical basis, you would predict this ratio to be 2:1, since there are two *ortho* carbon atoms and only one *para* carbon atom in bromobenzene. This prediction neglects the possibility that the steric bulk and electron-withdrawing inductive effect of the bromine atom might inhibit approach of the electrophile to the *ortho* relative to the *para* position, however. These two factors appear to be important in determining the *o:p* ratio, which is approximately 1:2. Consequently, *o*-bromonitrobenzene (**21**) is an important by-product in the preparation of *p*-bromonitrobenzene (**22**), and this must be taken into consideration in the procedure used to isolate pure product.

Another type of by-product results if further nitration of **21** and **22** occurs. Considering the directing effects of the bromo and nitro substituents leads to the prediction that the major product from both isomers would be 1-bromo-2,4-dinitrobenzene (**27**), as shown in Equation (15.16). The possible formation of 1-bromo-2,6-dinitrobenzene (**28**) from **22** is less likely because of the steric and inductive effects of the bromo substituent. Fortunately, it is relatively easy to suppress dinitration by performing the nitration at temperatures below 60 °C. Controlling the temperature suppresses dinitration because the strongly deactivating nitro substituent in **21** and **22** raises the activation energy for their nitration above that of **20.** Another factor decreasing the dinitration of the desired major product **22** is precipitation of this isomer during the course of the reaction, thereby removing it from the nitrating medium.

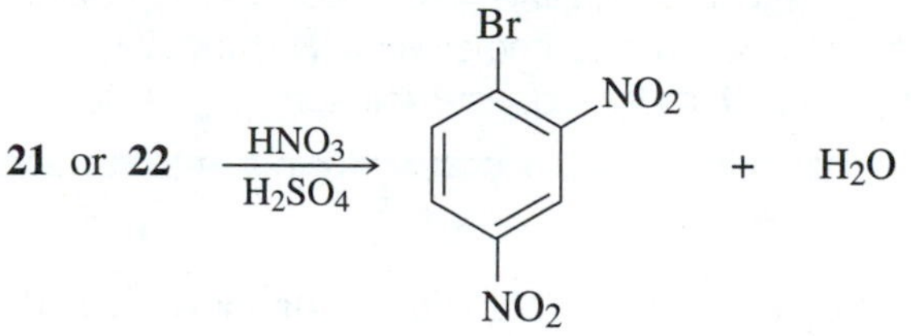

27
1-Bromo-2,4-dinitrobenzene

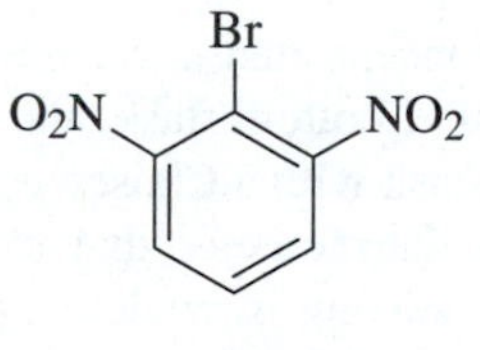

28
1-Bromo-2,6-dinitrobenzene

Based on the preceding discussion, the major obstacle to isolating pure *p*-bromonitrobenzene (**22**) is removal of *o*-bromonitrobenzene (**21**). The strategy for doing this relies on the greater polarity of **21,** which makes this isomer more soluble than **22** in polar solvents (see Exercise 1). This is dramatically demonstrated by their relative solubilities in ethanol. At room temperature, the *ortho* isomer **21** is very soluble, whereas the *para* isomer is slightly soluble, dissolving to the extent of only 1.2 g/100 mL. This large difference in solubilities allows separation of the isomers by the technique of **fractional crystallization.** The mixture of products obtained from nitration of bromobenzene is dissolved in hot 95% ethanol, and the solution is allowed to cool. The less soluble *para* isomer selectively crystallizes from solution and is isolated by filtration. Concentration of the filtrate allows isolation of a second crop of **22.**

In a fractional crystallization it is commonly possible to induce crystallization of the more soluble component, in this case the *ortho* isomer **21,** once the less soluble material has been mostly removed. However, the low melting point of **21** means that this isomer is difficult to crystallize in the presence of impurities. Consequently, isolating this isomer involves column chromatography (see Sec. 6.2).

EXPERIMENTAL PROCEDURES

SAFETY ALERT

Because concentrated sulfuric and nitric acids each may cause severe chemical burns, do not allow them to contact your skin. Wear rubber gloves when handling these reagents. Wipe off any drips and runs on the outside surface of reagent bottles and graduated cylinders before picking them up. Wash any affected area immediately and thoroughly with cold water, and apply 5% sodium bicarbonate solution.

Pre-Lab exercises for Section 15.3, "Nitration of Bromobenzene (Part A)," are found on page PL. 59.

A. NITRATION OF BROMOBENZENE

Prepare a mixture of 4.0 mL of concentrated nitric acid and 4.0 mL of concentrated sulfuric acid in a 25-mL round-bottom flask, and cool it to room temperature with a water bath. Equip the flask with a Claisen connecting tube fitted with a water-cooled condenser and a thermometer that extends into the flask, as shown in Figure 15.7. If magnetic stirring is available, place a stirring bar in the flask before attaching the connecting tube. Over a period of about 10 min, add to the stirred mixture 3.0 g of bromobenzene through the top of the condenser in portions of approximately 0.5 mL. If magnetic stirring is unavailable, loosen the clamp to the flask and carefully swirl the contents of the flask vigorously and frequently during the addition. Do *not* allow the temperature of the reaction mixture to exceed 50–55 °C during the addition. Control the temperature by allowing more time between the addition of successive portions of bromobenzene and by cooling the reaction flask with an ice-water bath.

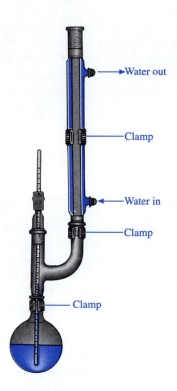

FIGURE 15.7 Apparatus for nitration of bromobenzene.

→ Water out

— Clamp

← Water in

— Clamp

— Clamp

After the addition is complete and the exothermic reaction has subsided, heat the flask with a steam or oil bath for 15 min, keeping the temperature of the reaction mixture *below* 60 °C. Cool the flask to room temperature; then pour the reaction mixture *carefully* and with stirring into 40 mL of cold water in a 100-mL beaker. Isolate the mixture of crude bromonitrobenzenes by vacuum filtration. Wash the filter cake thoroughly with cold water until the washes are neutral to litmus paper; allow the crystals to drain under vacuum until nearly dry.✴

Transfer the filter cake to a 50-mL Erlenmeyer flask containing approximately 1 mL of 95% ethanol per gram of crystals. Heat this mixture to boiling in order to dissolve the crude product. Allow the solution to cool slowly to room temperature; then cool it to 0 °C in an ice-water bath. Isolate the crystalline product by vacuum filtration. Wash the crystals with a little *ice-cold* alcohol, allowing the washes to drain into the filter flask with the mother liquors. When the crystals are dry, determine their weight and melting point.✴ The reported melting point of *p*-bromonitrobenzene is 125–126 °C.

Evaporate the mother liquors to a volume of about 10 mL on a steam bath, preferably in the hood, and allow the solution to cool to room temperature. A second crop of *p*-bromonitrobenzene is obtained. Either add this crop to the first one and recrystallize the combined product, or submit it separately to your instructor.

Further concentrate the mother liquors from the second crop to a volume of 3–4 mL. Cooling results in an oil containing crude *o*-bromonitrobenzene (**22**). Separate the oil from the two-phase mixture by means of a micropipet. Purify the *o*-bromonitrobenzene by column chromatography, as described in Part B.

> **Finishing Touches**   Neutralize the *aqueous filtrates* with sodium carbonate and flush them down the drain. Pour any *oil* obtained by concentration of the ethanolic filtrate into the container for halogenated liquids, unless the oil is to be subjected to chromatography.

Pre-Lab exercises for Section 15.3, "Nitration of Bromobenzene (Part B)," are found on page PL. 61.

B. THIN-LAYER AND COLUMN CHROMATOGRAPHY

Prepare solutions in two small vials of *p*-bromonitrobenzene and of the oil containing *o*-bromonitrobenzene in about 0.5 mL of dichloromethane. Obtain a 3-cm × 10-cm strip of a silica gel chromatogram sheet, and apply spots of each of these two solutions to the sheet according to the procedure of Section 6.3. Allow the spots to dry and develop the chromatogram in a TLC chamber with 9 : 1 (v:v) hexane:ethyl acetate as the eluting solvent.

When the solvent is within about 0.5 cm of the top of the plate, remove the developed chromatogram from the chamber, quickly mark the solvent front with a pencil, and allow the plate to dry. The spots may be visualized by placing the dry plate under a uv lamp or in a chamber whose atmosphere is saturated with iodine vapor. Calculate the R_f values of the spots observed, and identify them as either *o*- or *p*-bromonitrobenzene. A small orange spot may be observed very near the origin for the oil; this spot is 1-bromo-2,4-dinitrobenzene.

Following the procedure given in Section 6.2, prepare a column using a 25- or 50-mL buret and 5 g of silica gel. With a micropipet, apply a 0.2 g-sample of the oil containing the *o*-bromonitrobenzene to the head of the column and rinse the inside of the buret with 1 mL of ethyl acetate. Open the stopcock and allow the liquid to drain just to the top of the sand. Fill the buret with 9:1 (v:v) hexane:ethyl acetate and elute the column until a total of 40 mL of the solvent has passed through the column. Do not allow the level of liquid to drain below the sand at the top of the column. Collect the eluent in 5- to 8-mL fractions in a series of numbered test tubes.

Analyze each fraction by TLC using 9:1 (v:v) hexane:ethyl acetate as the eluting solvent. With careful spotting it should be possible to analyze four different fractions on each 3-cm × 8-cm strip of silica gel chromatogram sheet that is used. You may have to respot more than once to introduce sufficient sample onto the plate.

On a steam bath in the hood, evaporate to dryness those fractions containing only *o*-bromonitrobenzene. Characterize the residue obtained by either melting point determination or spectroscopic analysis. *o*-Bromonitrobenzene has a reported melting point of 40–41 °C and thus may be isolated as an oil.

> **Finishing Touches**   After first allowing any residual solvent to evaporate from the silica gel in the hood, put the *silica gel* in the container for non-toxic solids. Pour all *eluents* containing bromonitrobenzenes into the container for halogenated organic compounds; put all other *eluents* in the container for nonhalogenated solvents.

EXERCISES

1. Explain why *p*-bromonitrobenzene (**22**) is less polar than *o*-bromonitrobenzene (**21**).

2. Write out the resonance structures that contribute to the delocalized structure (σ-complex) **24.**

3. Find the pK_as of sulfuric and nitric acids in a reference book and confirm that the latter is the weaker acid. What is the approximate value of the ratio, K_a (H_2SO_4)/K_a (HNO_3)?

4. Explain why the melting point of *p*-bromonitrobenzene (**21**) is considerably higher than that of the *ortho*-isomer **22.**

5. **a.** Provide the resonance structures that contribute to the resonance hybrids **25** and **26.**
   b. Use these resonance structures to explain why the formation of *m*-bromonitrobenzene is disfavored relative to *p*-bromonitrobenzene.

6. Explain how TLC may be used to select the most appropriate solvent for use in a column chromatographic separation.

7. Why does *p*-bromonitrobenzene (**21**) have a larger R$_f$-value in the TLC analysis than does the *o*-isomer **22?**

8. The *o:p* ratio in the mononitration of bromobenzene has been reported to be 38:62. Use this ratio and the amount of *p*-bromonitrobenzene actually isolated to estimate the experimental yield of mononitration in the reaction. What errors are there in using this method to calculate the extent of mononitration?

SPECTRA OF STARTING MATERIAL AND PRODUCTS

The ^{1}H NMR spectrum for *p*-bromonitrobenzene is presented in Figure 8.35.

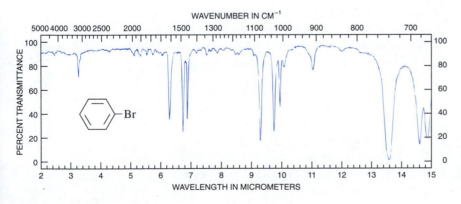

FIGURE 15.8   IR spectrum of bromobenzene.

FIGURE 15.9   NMR data for bromobenzene.

(a) ^{1}H NMR spectrum (90 MHz).

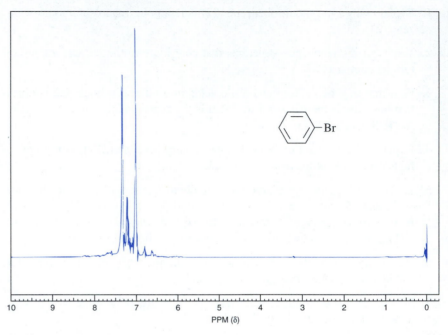

(b) ^{13}C NMR data.
   Chemical shifts: δ 122.5, 126.7, 129.8, 131.4.

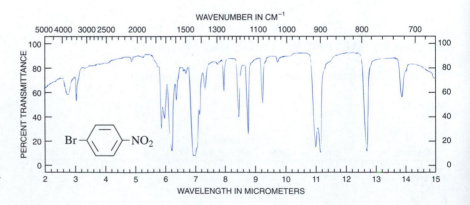

FIGURE 15.10   IR spectrum of *p*-bromonitrobenzene (KBr pellet).

FIGURE 15.11 ^{13}C NMR data for *p*-bromonitrobenzene

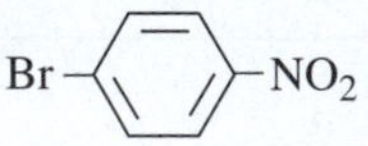

Br—◯—NO$_2$ Chemical shifts: δ 124.9, 129.8, 132.5, 147.0.

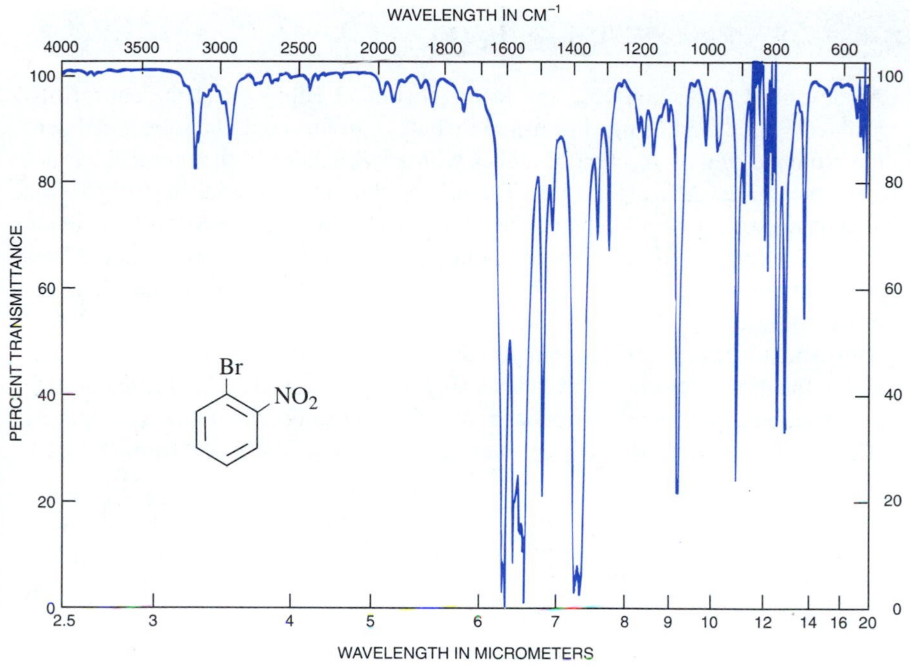

WAVELENGTH IN CM^{-1}

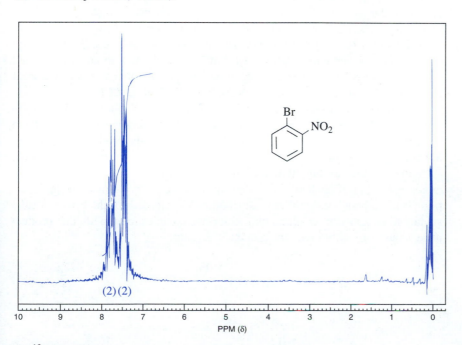

FIGURE 15.12 IR spectrum of *o*-bromonitrobenzene.

(a) ^{1}H NMR spectrum (90 MHz).

(b) ^{13}C NMR data.
  Chemical shifts: δ 114.1, 125.4, 128.2, 133.2, 134.8, 136.4.

FIGURE 15.13 NMR data for *o*-bromonitrobenzene.

15.4   FRIEDEL-CRAFTS ACYLATION OF *m*-XYLENE WITH PHTHALIC ANHYDRIDE

As described in Section 15.2 and in the Historical Highlight at the end of this chapter, Friedel and Crafts discovered that alkyl groups could be substituted onto the aromatic ring by reaction of arenes with alkyl halides in the presence of aluminum chloride, $AlCl_3$ (Eq. 15.6). The role of aluminum chloride, a strong Lewis acid, is to convert the alkyl halide into a reactive electrophilic intermediate, in the form of a carbocation or a highly polarized carbon-halogen bond, that effects electrophilic substitution on aromatic arenes (see Sec. 15.1), which can function as Lewis bases (Eq. 15.7). They also explored the reaction of aluminum chloride with acid chlorides, **29,** and anhydrides, **30,** to produce **acylonium ions, 31,** which function as electrophiles, just as alkyl carbocations do. Upon reaction with an arene, an acyl group is introduced onto the ring to provide an aryl ketone, **32** (Eq. 15.17). The overall reaction is named the **Friedel-Crafts acylation.**

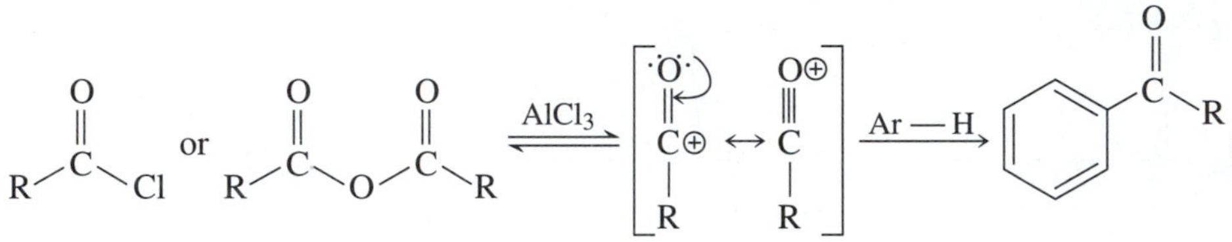

29	30	31	32
An acid chloride	An acid anhydride	An acylonium ion	An aryl ketone

The resonance structures in Equation (15.17) symbolize distribution of the positive charge in acylonium ions. This charge delocalization stabilizes the cation so that acylonium ions such as **31** do *not* undergo structural rearrangements as do alkyl carbocations (see Sec. 15.2). The increased stability of acylonium ions also makes them less electrophilic and thus less reactive than alkyl carbocations. Nevertheless, cation **31** is sufficiently reactive to undergo the S_E2 reaction (see Sec. 15.1) with arenes, provided the aromatic ring does not bear strongly deactivating groups such as NO_2, R_3N^+, C(O)R, and CN.

An important difference between the Friedel-Crafts alkylation and acylation reactions is that the latter process requires use of a *stoichiometric* rather than a *catalytic* amount of aluminum chloride. This is because the product aryl ketones, **32,** undergo a Lewis acid-base reaction with aluminum chloride to form a strong 1:1 complex, **33;** the aluminum chloride involved in the complex no longer promotes formation of acylonium ions and in essence is "consumed" in the reaction. Thus, although generation of acylonium ions needed for the Friedel-Crafts acylation is catalytic in aluminum chloride, complexation with the product makes aluminum chloride a stoichiometric reagent.

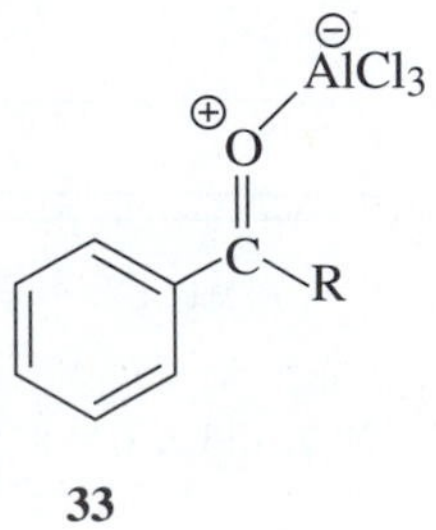

33

The acylation reaction performed in this section involves reaction of phthalic anhydride (**34**), a *cyclic* anhydride, with *m*-xylene (**35**) and aluminum chloride to afford ketoacid **37** (Scheme 15.4), which is named 2-(2′,4′-dimethyl-benzoyl)benzoic acid. As shown in Scheme 15.4, the initial product of the acylation process is **36,** in which one mole of aluminum chloride has reacted with the acid function to form the salt RCO_2AlCl_2 and a second mole is complexed to the carbonyl group. Adding ice and hydrochloric acid decomposes the complex to produce **37** and water-soluble aluminum salts.

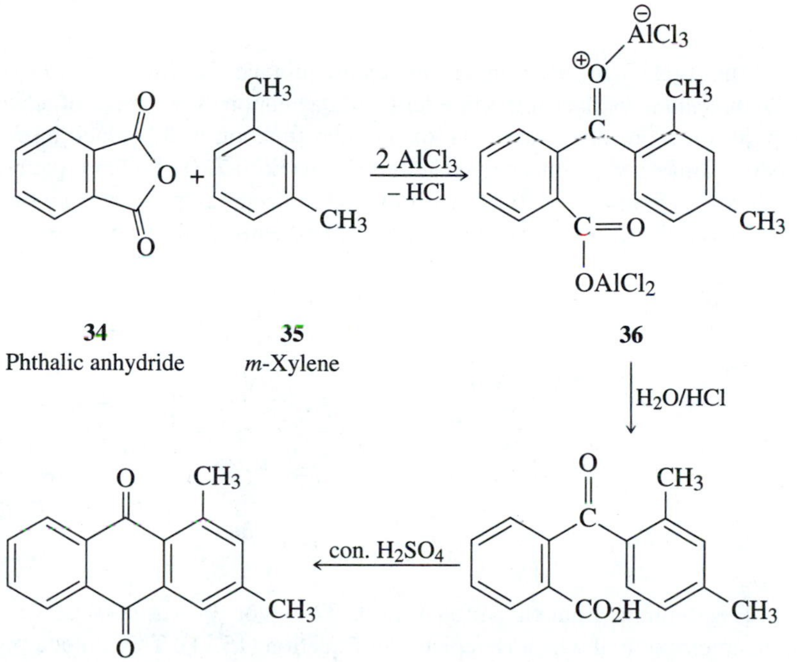

| **34** | **35** | **36** |
| Phthalic anhydride | *m*-Xylene | |

H_2O/HCl

| **38** | **37** |
| 2,4-Dimethylanthraquinone | 2,(2′4′-Dimethylbenzoyl)-benzoic acid |

SCHEME 15.4

The formation of **36** involves reaction of aluminum chloride with anhydride **34** to give the electrophile **39** (Eq. 15.18); this species then reacts with *m*-xylene (**35**). An issue in this step of the sequence is which of the three different positions of *m*-xylene undergoes attack by **39**. Because methyl groups are *o,p*-directors, reaction should occur preferentially at C-2 and C-4 of the ring. Steric hindrance impedes attack at C-2 so **35** reacts selectively at C-4 to give **36,** (Eq. 15.19), which provides the ketoacid **37** upon hydrolysis.

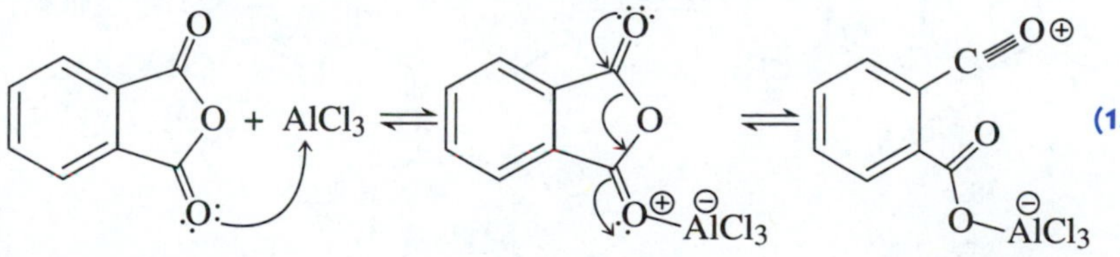

(15.18)

| **34** | **39** |
| | An acylonium ion (electrophile) |

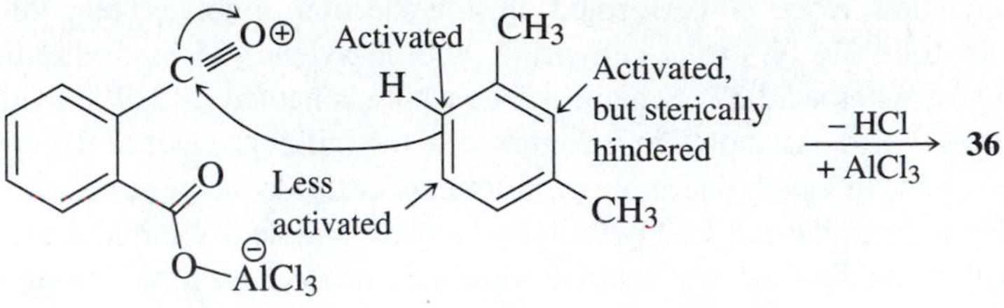

39   **35**   (15.19)

The final step in forming the anthraquinone **38** (Scheme 15.4) is an *intramolecular* variant of the Friedel-Crafts acylation. The source of acylonium ion **40** for the reaction is the carboxylic acid function of **37**, which loses water upon treatment with concentrated sulfuric acid (Eq. 15.20). For entropic reasons, intramolecular attack to effect cyclization then occurs in preference to the alternate possibility of *intermolecular* reaction of **38** with another molecule of **37**.

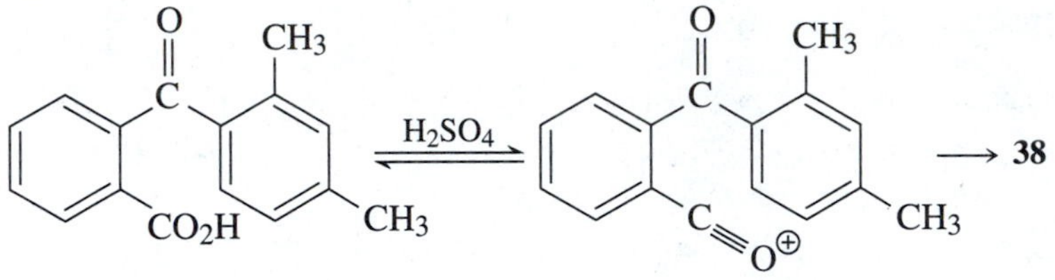

37   **40**   (15.20)

Electrophilic aromatic sulfonation of **37** and/or **38** is a possible side reaction, an example of which is depicted in Equation (15.21). This is not a problem with **37** because conversion to **40** is faster than is sulfonation of either ring, both of which are deactivated toward electrophilic attack by the presence of the carbonyl group. The quinone **38** can be sulfonated, but this process requires more forcing conditions than are used in this procedure as a consequence of the ring-deactivating effect of the two carbonyl groups. A relatively high temperature can thus be used to shorten the time for reaction between **37** and sulfuric acid without affecting the yield of **38**. Adding water to the hot reaction mixture precipitates the desired 2,4-dimethylanthraquinone (**38**) as golden yellow crystals.

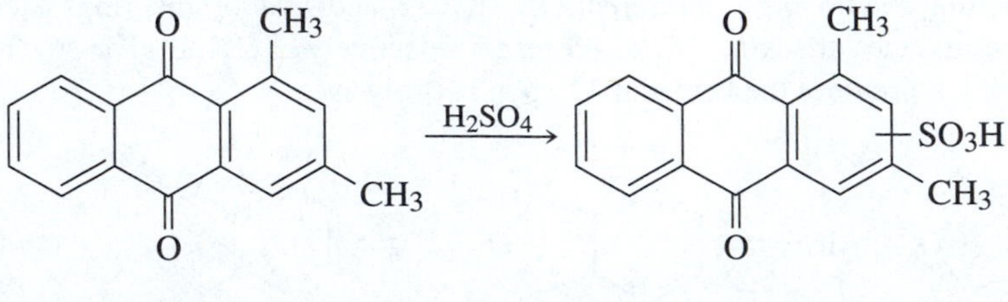

38   (15.21)

The carbonyl functions of quinones can be reduced, but the type of product formed depends on the reducing agent used. In the procedure of this section, anthraquinone **38** is reduced with tin(II) chloride in acetic acid to give a mixture of anthrones **41** and **42** (Eq. 15.22).

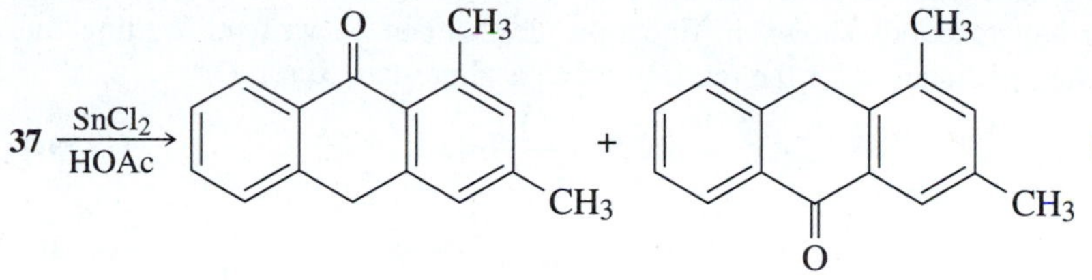

41
1,3-Dimethylanthrone

42
2,4-Dimethylanthrone

(15.22)

The reaction presumably occurs via a mixture of diastereomeric diols **43,** which undergo dehydration to the phenols **44** and **45** shown in Equation (15.23); subsequent tautomerization of these phenols produces the thermodynamically more stable anthrones. That the equilibria of Equations (15.23a) and (15.23b) favor the anthrones over the phenols is noteworthy. The isomeric pairs **41,44** and **42,45** represent keto–enol tautomers, and in aliphatic systems, it is the keto form that typically is favored thermodynamically (see Chap. 11). This situation is normally reversed with phenols because these enolic forms are aromatic. In the present instance, however, the stability afforded to **41** and **42** by the conjugated carbonyl function *exceeds* that associated with the aromaticity present in **44** and **45** (see Exercise 21).

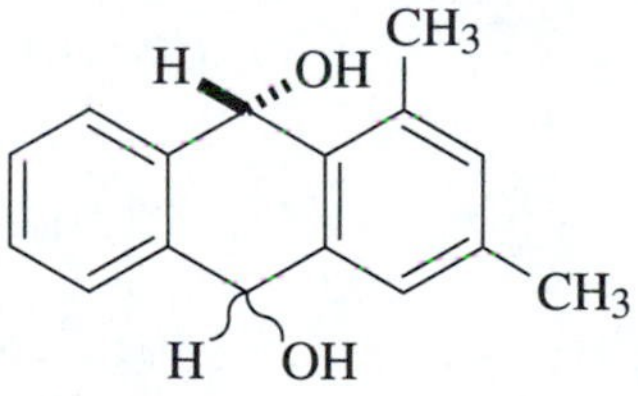

43

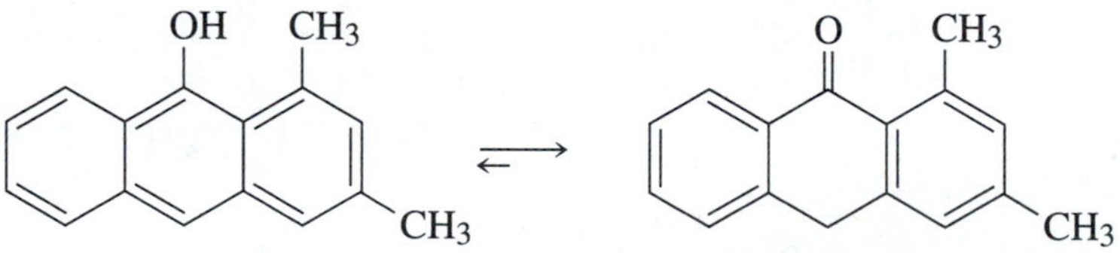

44
1,3-Dimethylanthrol

41

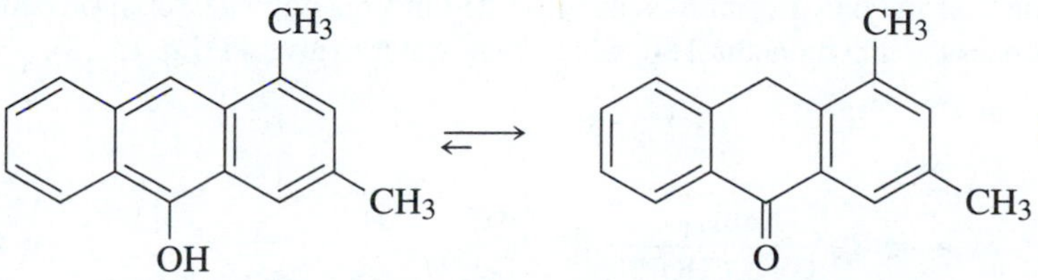

(15.23a)

45
2,4-Dimethylanthrol

42

(15.23b)

The stoichiometry of the reduction of **38** to **43** is determined in the usual way for a redox reaction (see Sec. 16.1). The oxidation state of the two carbon atoms changes from +2 to 0, whereas that of tin goes from +2 to +4. As a result, the two half-reactions shown in Equation (15.24) can be written. Writing the fully balanced equation for the reaction is left as a pre-lab exercise.

$$\textbf{38} + 4\,e^- + 2\,H^+ \rightleftharpoons \textbf{43} \tag{15.24a}$$

$$Sn^{+2} \rightleftharpoons Sn^{+4} + 2\,e^- \tag{15.24b}$$

The acid-catalyzed dehydration of **43** is presumably an S_N1 reaction that proceeds through the benzylic carbocations **46** and **47** (Eq. 15.25), the precursors of phenols **44** and **45**. The more stable of these two cations should be formed faster, and consequently the product from it should be preferred. Thus, by analysis of **46** and **47** you should be able to predict which of the two anthrones is formed as the major reduction product (see Exercise 21). This prediction can be tested with the aid of ^{1}H NMR spectral analysis of the product mixture.

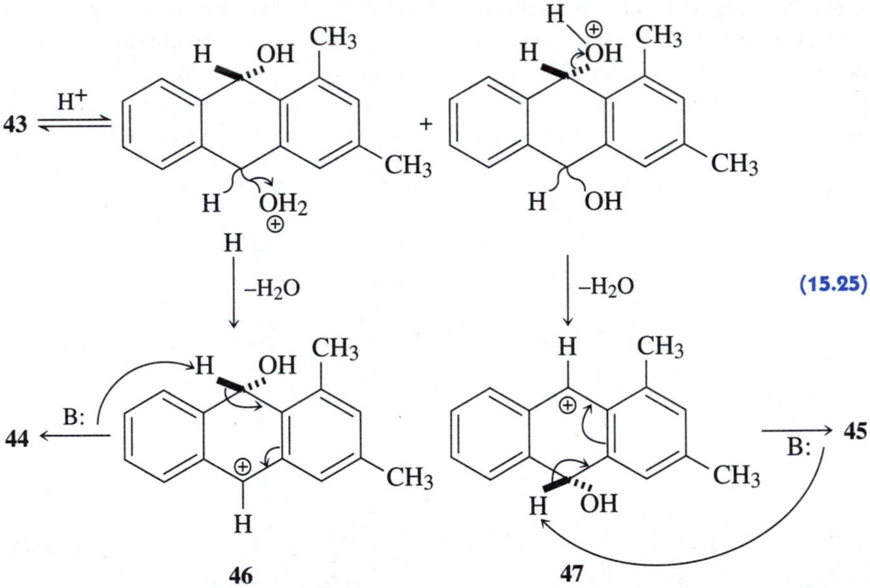

(15.25)

The further reduction of the mixture of anthrones **41** and **42** to 1,3-dimethylanthracene (**48**) is readily accomplished using sodium borohydride in isopropyl alcohol (Eq. 15.26). The reduction proceeds by hydride ion (H$^-$) transfer to **41** and **42**, forming the anion of the dihydroanthrols, as illustrated in Equation (15.27). Protonation of **49** produces the dihydroanthrol **50**, which loses water to give **48** (Eq. 15.28).

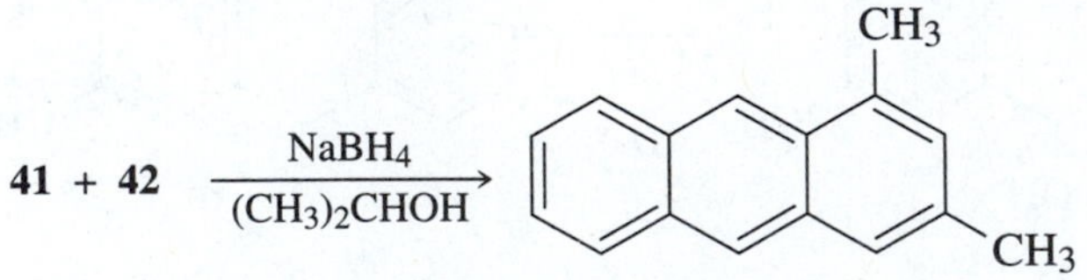

$$\textbf{41} + \textbf{42} \xrightarrow[\text{(CH}_3\text{)}_2\text{CHOH}]{\text{NaBH}_4} \tag{15.26}$$

48
1,3–Dimethylanthracene

The formation of a single isomer of dimethylanthracene by reducing the mixture of **41** and **42** is instructive because it shows that **41** and **42** have *identical* carbon skeletons, so structural isomerization has *not* accompanied the formation of these isomers from **38.** This conversion illustrates a concept of general use to organic chemists: If unknown compounds are thought to have the same carbon skeleton, this can be confirmed by converting them to a common product, in this case the dimethylanthracene **48.**

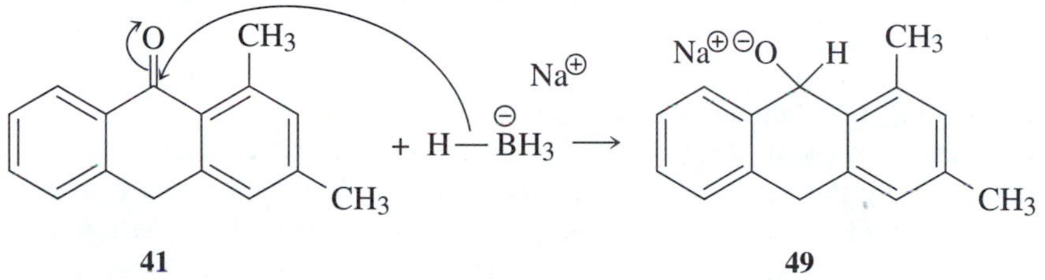

(15.27)

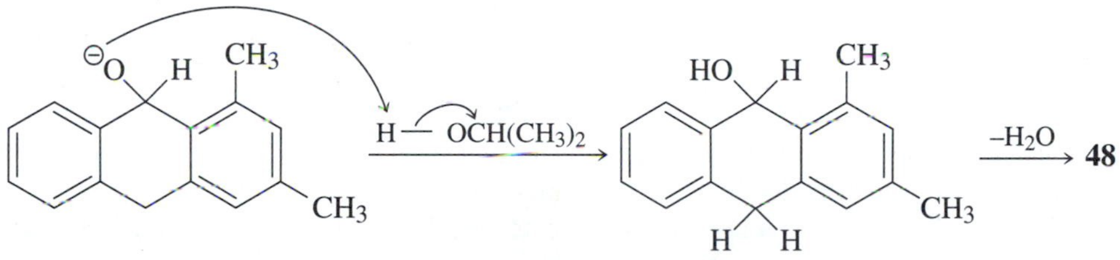

41 **49**

50
A dihydroanthrol

(15.28)

EXPERIMENTAL PROCEDURES

Pre-Lab exercises for Section 15.4, "Preparation of 2-(2′, 4′-Dimethylbenzoyl) benzoic Acid," are found on page PL. 63.

A. PREPARATION OF 2-(2′, 4′-DIMETHYLBENZOYL) BENZOIC ACID
SAFETY ALERT

1. Anhydrous aluminum chloride is very hygroscopic and reacts vigorously with water, even the moisture on your hands, producing fumes of hydrogen chloride that are highly corrosive if inhaled. Do not allow aluminum chloride to contact your skin. If it does, flush the affected area with copious amounts of water.
2. Because aluminum chloride is a powdery solid that easily becomes airborne, weigh and transfer it into the reaction flask *in the hood*.

Place 3.5 g of phthalic anhydride and 18 mL of *m*-xylene in a dry 250-mL round-bottom flask equipped with a magnetic stirring bar, if magnetic stirring is available, and a reflux condenser fitted with a trap for removal of HCl; a vacuum adapter filled with $CaCl_2$ and attached to the water aspirator is a suitable trap (see Fig. 2.20). Cool the mixture to 0 °C in an ice-water bath, remove the condenser from the flask, and, in one portion, quickly add 7 g of anhydrous aluminum chloride that has previously been weighed into a stoppered and dry 25-mL Erlenmeyer flask. Because the success of this experiment is highly dependent on the quality of the aluminum chloride that is used, obtain it from a *freshly* opened bottle and *weigh it quickly*. Alternatively, you may be provided with a weighed amount of aluminum chloride. In either case, *minimize* exposure of this chemical to the atmosphere.

Replace the condenser on the flask, and carefully stir or swirl the reaction mixture while maintaining its temperature at 0 °C; then allow it to warm to room temperature. The start of the reaction will be evidenced by evolution of hydrogen chloride. If this does not occur, gently warm the flask with a microburner or an oil bath until gas evolution begins. Be prepared to immerse the flask into an ice-water bath if the reaction becomes too vigorous. Continue gently heating the mixture until the reaction is proceeding smoothly enough to permit heating under reflux.

Continue heating and stirring or swirling the reaction mixture for a total of 1 h. Watch for the possible sudden separation of the aluminum salt of the product; the heat of crystallization of this salt may make it necessary to cool the flask in the ice-water bath to moderate the rate of reflux. At the end of the reflux period, cool the flask to 0 °C with stirring or swirling.

Remove the condenser, stopper the flask,✶ and transfer the flask and the ice-water bath to the hood. Obtain about 25 g of ice and add a few chunks of ice to the reaction mixture, swirling and cooling the flask in the ice-water bath as necessary. Once the ice has melted, add more of it, a few pieces at a time, with swirling and cooling, until all the ice has been added. Then carefully add 5 mL of *concentrated* HCl, followed by 25 mL of water, swirling the mixture vigorously and cooling it as necessary to keep it near room temperature during the additions. Add an additional 10 mL of water with vigorous swirling; cool the resulting mixture to room temperature or below,✶ and add 15 mL of diethyl ether. With the aid of a stirring rod, scrape any solid from the neck and walls of the flask and break up any lumps at the bottom. Stopper the flask and shake it vigorously for several minutes to complete decomposition of the aluminum salt of the product, extraction of the organic product into diethyl ether, and dissolution of the inorganic aluminum salts.✶

Transfer the mixture to a separatory funnel, rinse the reaction flask with 10 mL of diethyl ether, and add the rinse to the funnel. Shake the funnel, separate the layers, and extract the aqueous layer with two additional 20-mL portions of diethyl ether. If a grayish, fluffy precipitate appears below the organic layer, remove it by adding 10 mL of 6 *M* HCl, shaking the mixture for a few minutes and separating the layers. If any precipitate or grayish emulsion remains with the ethereal solution, remove it by vacuum filtration through a pad of a filter aid (see Sec. 2.16).

Place the combined ethereal extracts in an Erlenmeyer flask, cool the solution in an ice-water bath, add 15 mL of 3 *M* sodium hydroxide solution with stirring. Stir or swirl the mixture thoroughly and decant the ethereal layer from the thick slurry of salts that forms. Extract the ethereal decantate with 10 mL of 3 *M* sodium hydroxide solution, and combine this basic extract with the slurry. Cool the resulting mixture to 0 °C and, with vigorous stirring, carefully acidify it to pH 2 with concentrated HCl. Be sure to break up any lumps of solid that develop to ensure neutralization of all of the sodium salt that has been formed. The desired product should appear as a precipitate. If it oils out instead, scratch at the oil/glass interface with a stirring rod to induce crystallization.

Isolate the carboxylic acid **35** by vacuum filtration and reserve a sample of it for purification by recrystallization, using 50% aqueous ethanol. The reported melting point of **35** is 142.5–142.8 °C. The crude product may be used successfully in the next procedure for forming 1,3-dimethylanthraquinone.

Finishing Touches   Discard the *calcium chloride* from the drying tube and the *grayish solid* obtained by gravity filtration in the container for non-hazardous solids. Put the *ethereal solution* remaining after basic extraction in a container labeled for nonhalogenated organic liquids. Neutralize the *acidic aqueous filtrate* by adding solid sodium carbonate; then flush it down the drain. Flush the *filtrate from the recrystallization* and the *aqueous filtrate* containing aluminum salts down the drain.

Pre-Lab exercises for Section 15.4, "Preparation of 1,3-Dimethyl-anthraquinone," are found on page PL. 65.

B. PREPARATION OF 1,3-DIMETHYLANTHRAQUINONE
SAFETY ALERT

Concentrated sulfuric may cause severe chemical burns if it is allowed to come into contact with your skin. Take proper precautions to avoid this. Wipe off any drips and runs on the outside surface of reagent bottles and graduated cylinders before picking them up. Wash any affected areas immediately with copious amounts of cold water, and apply 5% sodium bicarbonate solution.

Note to Student: If you are *not* to convert the 1,3-dimethylanthraquinone obtained in this experiment to the mixture of **41** and **42**, the scale of the following procedure may be *decreased* by a factor of *three*.

Place 3.0 g of dry benzoylbenzoic acid **35** in a 125-mL round-bottom flask and add 15 mL of concentrated H_2SO_4. Gently heat the flask with a microburner, steam, or an oil bath until the solid dissolves; occasionally swirl the mixture to aid dissolution. Clamp the flask in place, insert a thermometer in the solution, and heat the contents at 110–120 °C for 45 min with a microburner or an oil bath. Remove the heat source, allow the mixture to cool to 100 °C, and remove the thermometer, first allowing any liquid to drain from it. Add 3 mL of water dropwise while swirling the reaction mixture to facilitate the formation of the crystalline product. Allow this mixture to cool until it is just warm to the touch and add an additional 50 mL of water.✳ Cool the resulting mixture in an ice-water bath and isolate the crude product by vacuum filtration, washing it well with cold water until the washings are neutral to pHydrion paper. Allow the solid to dry at room temperature or in the oven.✱ The desired quinone should be golden yellow and has a reported melting point of 161–162 °C.

If the solid is highly discolored, decolorize (see Sec. 2.14) it as follows. Dissolve the crude product in boiling acetone (*no flames*). About 30 mL of solvent/g of solid is required, but to ensure that crystallization does not occur during the hot filtration step add about 25% more than the minimum amount of solvent needed. Allow the solution to cool below the boiling point; then add about 0.5 g of decolorizing carbon and bring the mixture to boiling (*no flames*) for a period of about 5 min; swirl the flask frequently. Remove the heat source, add about 1 g of a filter aid to the mixture with swirling, reheat the mixture to boiling, and perform a hot filtration (see Sec. 2.13). Evaporate approximately one-fourth of the acetone from the filtrate, and allow the solution to cool to room temperature and then to 0 °C in an ice-water bath. Isolate the decolorized product by vacuum filtration. The procedure may have to be repeated if the solid is still highly colored and gives an unacceptably low melting point.

Assess the purity of the 1,3-dimethylanthraquinone by TLC or spectroscopic methods.

Finishing Touches   Neutralize the *acidic filtrate* by addition of solid sodium carbonate, and then flush it down the drain. Flush the *filtrate* from the recrystallization and the small quantity of *acidified extract* from the test for the presence of unchanged starting material down the drain.

Pre-Lab exercises for Section 15.4, "Reduction of 1,3-Dimethylanthraquinone," are found on page PL. 67.

C. REDUCTION OF 1,3-DIMETHYLANTHRAQUINONE

SAFETY ALERT

Concentrated hydrochloric acid is corrosive. Do not allow it to contact the skin. If this occurs, immediately wash the area with copious amounts of water and then with 5% sodium bicarbonate solution.

Note to Student: If you are to convert the mixture of **41** and **42** to **44,** *increase the scale of the following procedure by a factor of four.*

Prepare a solution by warming 2.0 g of $SnCl_2 \cdot 2H_2O$ and 2.5 mL of conc. HCl and add it to a mixture of 0.5 g of 2,4-dimethylanthraquinone (**38**) and 4–5 mL of acetic acid in a 25-mL round-bottom flask equipped for reflux. Heat the heterogeneous mixture under gentle reflux, using a burner or an oil bath, until the solid has dissolved; then continue heating at reflux for an additional 30 min. Record the total reflux time required. Allow the mixture to cool below the reflux temperature; then pour it into 50 mL of water.✱

Isolate the crystalline pale yellow product by vacuum filtration and, after drying it, determine its melting point; this will be broad (92–100 °C) owing to the presence of the isomeric anthrones. The reported melting point of the 2,4-dimethyl isomer **41** is 119–120 °C; that of the 1,3-dimethyl isomer **42** is 156–157 °C.

Analyze the product by TLC (see Sec. 6.3) to demonstrate whether two products are present. An appropriate solvent system for this analysis is 70:30 hexane:ethyl acetate; ethyl acetate is suitable for dissolving the sample to be spotted on the TLC plate.

Obtain a ^{1}H NMR spectrum of the product mixture and use the data to determine the ratio of the two anthrones produced.

Finishing Touches   Neutralize the *acidic filtrate* by addition of solid sodium carbonate, and then flush it down the drain. Put the *TLC plate* in the container for nonhazardous solid materials and pour the *solvents used for the TLC analysis* into the container for nonhalogenated organic liquids. Pour the *solution prepared for ^{1}H NMR analysis* into the container for halogenated organic liquids.

Pre-Lab exercises for Section 15.4, "Preparation of 1,3-Dimethyl-anthracene," are found on page PL. 69.

D. PREPARATION OF 1,3-DIMETHYLANTHRACENE
SAFETY ALERT

Toluene, a flammable solvent, is used in the recrystallization step. Extinguish *all* flames to avoid a fire.

Combine 0.5 g of the anthrones, 0.8 g of sodium borohydride, and 65 mL of iso-propyl alcohol in a 125-mL round-bottom flask fitted for reflux. With the aid of a microburner or oil bath, heat the mixture under reflux for 1 h; then allow it to cool to room temperature, and pour it into about 150 mL of ice water. Carefully acidify the reaction mixture with 6 *M* HCl (*caution:* hydrogen gas is evolved). Isolate the precipitate by vacuum filtration and thoroughly wash the filter cake with water.✱ To remove most of the residual water from the crude product, release the vacuum, rinse the walls of the funnel with methanol, using enough methanol to cover the cake, and then apply vacuum. Repeat this process once again applying vacuum until the filter cake is dry.✱

Transfer the filter cake to a 50-mL Erlenmeyer flask and add 12 mL of toluene; this amount of solvent should preclude crystallization in the funnel during the hot filtration step to follow. Heat the mixture (*no flames*) to dissolve the anthracene; if necessary, decolorize the solution (see Sec. 2.14) and perform a hot filtration (see Sec. 2.13). Otherwise, allow the solution to cool, first to room temperature and then to 0 °C, using an ice-water bath; isolate the 1,3-dimethyl-anthracene by vacuum filtration. Determine the melting point (reported mp 78–79 °C) and yield of product. Analyze the product by TLC (see Sec. 6.3) to determine whether only a single isomer has been formed. If instructed to do so, obtain IR and ^{1}H NMR spectra of the product.

When cleaning your equipment with acetone, you may observe fluorescence in the collected rinses; the source of this is the 1,3-dimethylanthracene. If you do not see this phenomenon, it may be because the fluorescence is readily quenched by the presence of traces of impurities.

> **Finishing Touches** Neutralize the *aqueous solution* with sodium carbonate, and flush the *filtrate* down the drain with copious amounts of water. Pour the *toluene* and *wash acetone* in the container for nonhalogenated organic liquids.

EXERCISES

Part A

1. Explain why a catalytic amount of $AlCl_3$ is insufficient to promote the Friedel-Crafts reaction of **34** and **35**.

2. Why is loss of carbon monoxide (decarbonylation) from the acylonium ion **39** an *unlikely* reaction?

3. Why is it important that the apparatus used for the Friedel-Crafts acylation be dry?

4. Write the products of the reaction of $AlCl_3$ and H_2O (excess).

5. What gas is evolved in the early stages of the reaction between *m*-xylene, phthalic anhydride, and aluminum chloride?

6. In the purification of **37**, what purpose is served by extraction of the ethereal solution of product with aqueous base?

7. In which of the two solvents, water or ethanol, should the product first be dissolved in the mixed solvent recrystallization? Explain.

8. Why does the acylonium ion **39** react more rapidly with **35** than with **34**?

9. What evidence do you have from this experiment that substitution of phthalic *acid* for phthalic anhydride would *not* provide acylonium ion **39** required for the acylation reaction. (*Hint:* Write the product of reaction of aluminum chloride with one of the carboxylic acid groups of phthalic acid and compare it to the carboxylate **36**.)

10. Explain how ^1H NMR and ^{13}C NMR spectra could be used to differentiate between **37** and its isomer **51**.

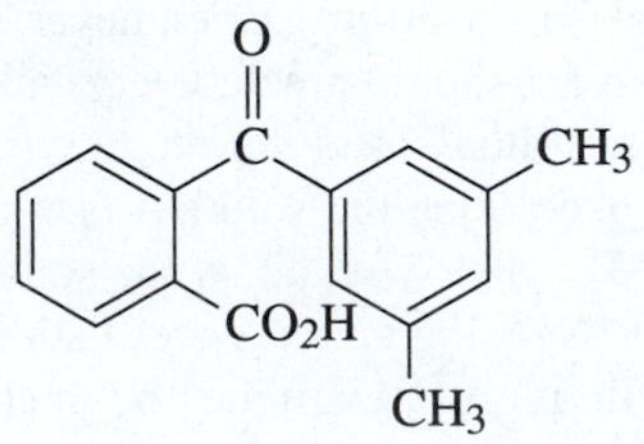

51

Part B

11. Why is decarbonylation of the acylonium ions **39** and **40** an *unlikely* reaction?

12. Why is the anthraquinone **38** *less* prone to electrophilic aromatic sulfonation than is *m*-xylene?

13. Write a possible intermolecular acylation product that could be obtained in the reaction involving **37** with concentrated H_2SO_4.

14. Explain why entropic considerations favor formation of **38** rather than the product you wrote in the preceding exercise.

15. Outline an experimental protocol for removing starting ketoacid **37,** were it present, from the anthraquinone **38.**

16. Note the chemical shifts of the methyl groups in the ^1H NMR spectrum of **38.** Predict which of the two methyl groups in **38** should have the larger value of δ and explain your prediction.

Part C

17. Predict which of the two carbocations, **46** or **47,** derived from **43** is the more stable and justify this prediction.

18. Given that steric hindrance *decreases* the ability of a polar functional group to interact with silica gel or alumina, the coatings on commercially available TLC plates, predict which of the anthrones **41** or **42** would have the *higher* R_f-value and defend your prediction.

19. Given your analysis in the preceding exercise, predict which of the two anthrones is the major product, based on TLC analysis.

20. Analyze the ^1H NMR spectrum of the mixture of **41** and **42** to determine which isomer predominates. Justify your answer. It will be helpful in the analysis to realize that protons subject to steric compression, which is caused by a nearby relatively bulky group, appear at lower fields than comparable protons not so positioned.

21. The aromatic stabilizations of a benzene ring and of an anthracene ring system are 36 kcal/mol 84 kcal/mol, respectively.

    a. Assuming that the substituents on the rings in **41** and **44** do not affect these values significantly, compute the difference in aromatic stabilization of these two molecules. Considering only the relative aromatic stabilization of the two isomers, which one is predicted to be more stable?

    b. Another way to assess the relative stabilities of the two molecules is to disregard aromaticity as a factor and to say that a C–H and a carbon-oxygen π-bond in **44** are replaced by an O–H and a carbon-carbon π-bond in **41.** Assuming that the approximate dissociation energies for these bonds are C–H (80 kcal/mol), O–H (94 kcal/mol), C=C π-bond (50 kcal/mol), and C=O π-bond (65 kcal/mol) compute the difference in energy between **41** and **44** based on the different types of bonds present in each. Considering only the relative dissociation energies of the specified bonds of the two isomers, which one is predicted to be more stable?

c. Considering both the factors calculated in parts (a) and (b), which of the two isomers is predicted to be more stable? By how much, in kcal/mol?

d. A factor that has not been taken into consideration in estimating the relative stability of **41** is the fact that the carbonyl function is conjugated with the aromatic rings of the molecule. How does this affect your conclusion in Part (c)?

Part D

22. What is the driving force for formation of 1,3-dimethylanthracene from the dihydroanthrol **50?**

23. The dehydration of **50** to 1,3-dimethylanthracene may occur either before or after acidification of the reaction mixture.
    a. Provide a reaction mechanism for dehydration assuming it occurs prior to acidification. Use curved arrows to symbolize the flow of electrons.
    b. Provide a reaction mechanism for dehydration assuming it occurs after acidification. Use curved arrows to symbolize the flow of electrons.

24. Why is hydrogen gas evolved upon acidification of the reaction mixture?

25. Anthracene has a reported melting point of 216–218 °C, whereas that of 1,3-dimethylanthracene is only 78–79 °C. What might account for the significant difference in melting points of these structurally similar molecules?

SPECTRA OF STARTING MATERIALS AND PRODUCTS

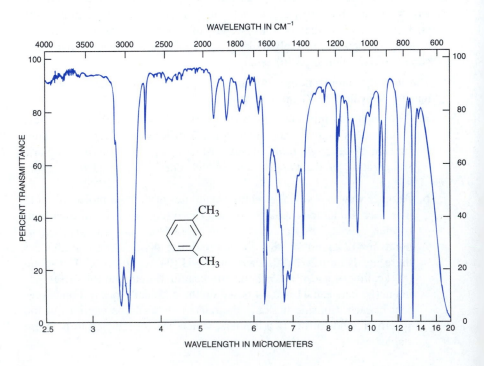

FIGURE 15.14   IR spectrum of *m*-xylene.

(a) ^{1}H NMR spectrum (90 MHz).

FIGURE 15.15   NMR data for *m*-xylene.

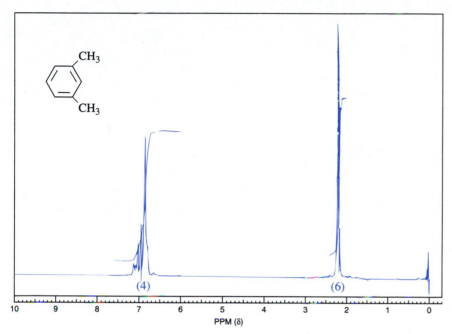

(b) ^{13}C NMR data.
   Chemical shifts: δ 21.3, 126.2, 128.2, 130.0, 137.6.

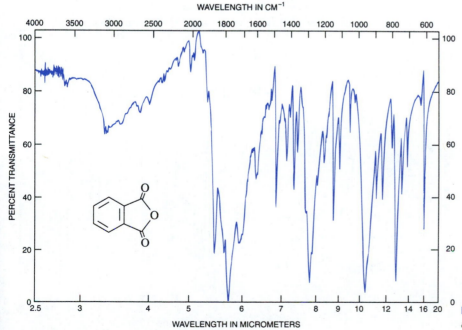

WAVELENGTH IN CM^{-1}

FIGURE 15.16   IR spectrum of phthalic anhydride (KBr pellet).

FIGURE 15.17 NMR data for phthalic anhydride.

(a) ^{1}H NMR spectrum (90 MHz).

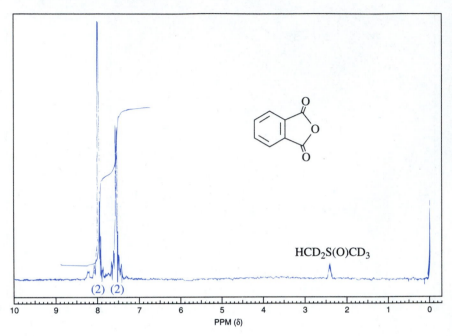

$HCD_2S(O)CD_3$

(2) (2)

PPM (δ)

(b) ^{13}C NMR data.
Chemical shifts: δ 125.3, 131.1, 136.1, 163.1.

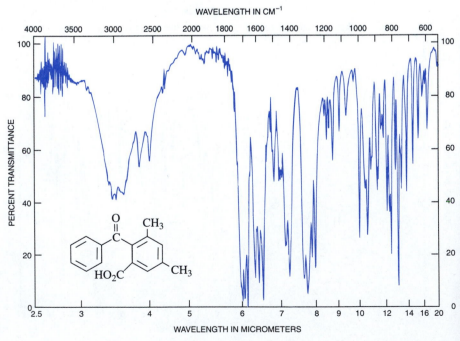

WAVELENGTH IN CM^{-1}

PERCENT TRANSMITTANCE

WAVELENGTH IN MICROMETERS

FIGURE 15.18 IR spectrum of 2-(2′,4′-dimethylbenzoyl) benzoic acid (KBr pellet).

(a) ^{1}H NMR spectrum (90 MHz).

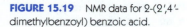
FIGURE 15.19   NMR data for 2-(2′,4′-dimethylbenzoyl) benzoic acid.

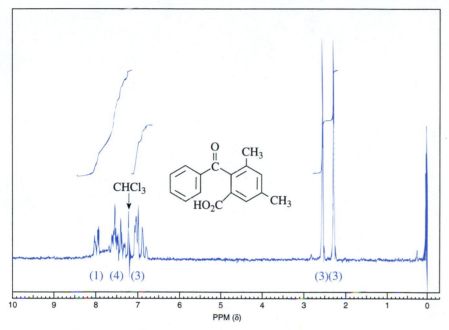

(b) ^{13}C NMR data.
Chemical shifts: δ 21.3, 21.4, 125.7, 128.0, 128.1, 129.4, 130.6, 131.8, 132.7, 132.8, 133.8, 140.3, 142.4, 143.9, 171.2, 198.4.

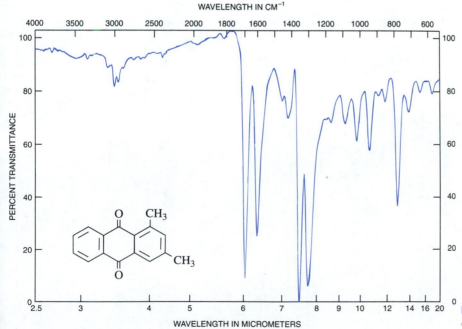

FIGURE 15.20   IR spectrum of 2,4-dimethylanthraquinone (KBr pellet).

FIGURE 15.21   NMR data for 2,4-dimethylanthraquinone.

(a) ^{1}H NMR spectrum (90 MHz).

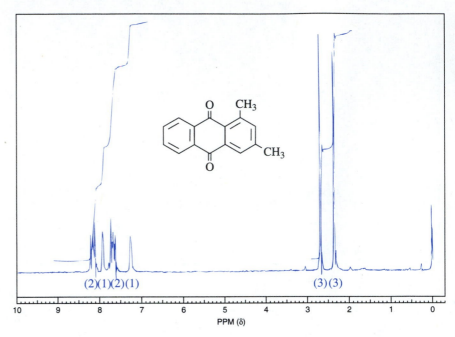

(2)(1)(2)(1)   (3)(3)

10   9   8   7   6   5   4   3   2   1   0
PPM (δ)

(b) ^{13}C NMR data.

Chemical shifts: δ 21.6, 23.3, 126.5, 126.6, 127.1, 128.9, 132.9, 133.3, 134.0, 134.9, 138.9, 142.2, 144.0, 183.9, 184.8.

WAVELENGTH IN CM^{-1}

4000   3500   3000   2500   2000   1800   1600   1400   1200   1000   800   600

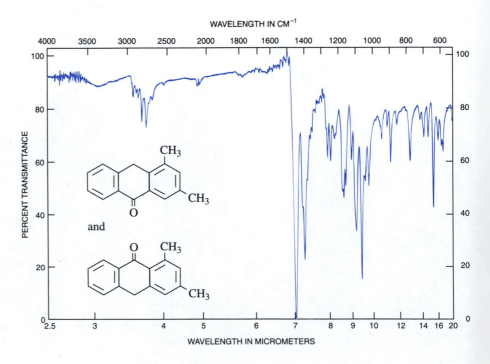

PERCENT TRANSMITTANCE

and

2.5   3   4   5   6   7   8   9   10   12   14 16 20
WAVELENGTH IN MICROMETERS

FIGURE 15.22   IR spectrum of dimethyl-anthrones (KBr pellet).

(a) ¹H NMR spectrum (90 MHz).

FIGURE 15.23   NMR data for dimethyl-anthrones.

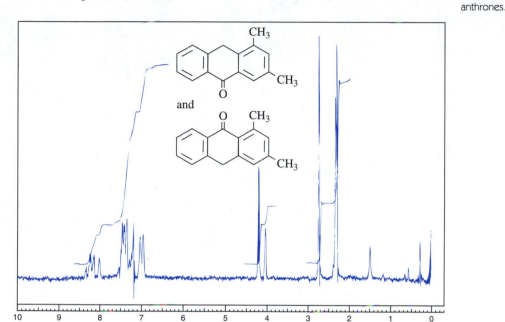

(b) ¹³C NMR data.

   Chemical shifts: δ 19.0, 21.0, 21.4, 23.6, 30.2, 33.3, 115.1, 115.6, 125.5, 126.6, 126.8, 126.9, 127.0, 127.1, 127.3, 127.4, 127.7, 128.7, 131.7, 131.8, 131.9, 132.5, 133.8, 135.3, 135.9, 136.3, 139.2, 141.8, 141.9, 142.3, 184.6, 186.1.

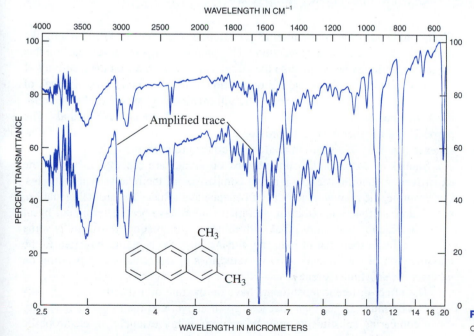

FIGURE 15.24   IR spectrum of 1,3-dimethylanthracene (KBr pellet).

FIGURE 15.25 NMR data for 1,3-dimethylanthracene.

(a) ^{1}H NMR spectrum (90 MHz).

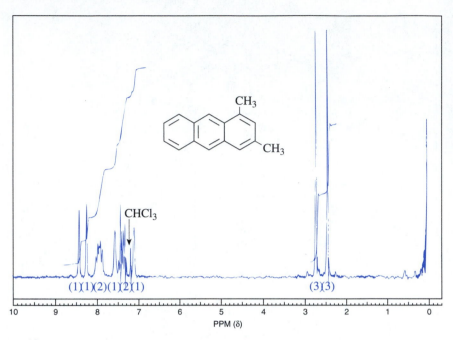

(b) ^{13}C NMR data.

Chemical shifts: δ 19.6, 21.9, 122.5, 124.8, 125.2, 125.7, 127.8, 128.5, 128.6, 130.0, 131.1, 131.6, 132.3, 134.0, 134.6.

15.5 RELATIVE RATES OF ELECTROPHILIC AROMATIC SUBSTITUTION

Electrophilic aromatic substitution (S_E2) reactions (see Sec. 15.1) are among the best understood of all organic reactions. The *qualitative* aspects of the reaction that are discussed in textbooks include the effect substituents have on the reactivity of arenes toward electrophiles and the orientation, *ortho, meta,* or *para,* of attack of such species on the ring. Relatively little information is given about the *quantitative* differences in rates and reactivities of substituted aromatic compounds. The purpose of this experiment is to provide a semiquantitative measure of the differences in reactivity of a series of arenes toward the bromonium ion, Br^+.

Studies of electrophilic substitutions on arenes are reported in which the experimental conditions allow a direct comparison of the relative reaction rates. For example, the relative reactivity of benzene and toluene toward halogenation, acetylation, sulfonation, nitration, and methylation have been determined. In *all* cases the reaction was more rapid with toluene; bromination of toluene is some 600 times faster than that of benzene, for instance. Such studies have led to the classification of substituents as **ring activators** or **deactivators,** depending on whether the substituted arene reacts *faster* or *slower* than benzene itself.

The effect different substituents have on the rate of a specific type of reaction can be much larger than that just given above, and the range of reactivity when comparing substituents may be enormous. For example, in electrophilic bromination of arenes, phenol reacts 10^{12} times *faster* than benzene, whereas nitrobenzene reacts 10^5 times *slower* than benzene. The hydroxy group is thus a powerful ring activator and the nitro group a comparably potent ring deactivator, the range of difference in reactivity due to these two substituents being a remarkable 10^{17}!

The relative rates of bromination of several monosubstituted benzenes are explored in this section (Eq. 15.29). This particular reaction is selected because qualitative evaluation of relative reaction rates can be done visually by measuring the time required for disappearance of the characteristic reddish color of molecular bromine.

$$\text{Ar}-\text{H} \quad + \quad \text{Br}_2 \quad \xrightarrow{\text{HOAc}} \quad \text{Ar}-\text{Br} \quad + \quad \text{HBr} \qquad \textbf{(15.29)}$$

$$\text{An arene} \qquad \text{(red–brown)} \qquad \qquad \text{An aryl bromide}$$
$$\text{(colorless)}$$

Pre-Lab exercises for Section 15.4, "Relative Rates of Electrophilic Aromatic Substitution," are found on page PL. 71.

EXPERIMENTAL PROCEDURE

SAFETY ALERT

1. *Bromine is a hazardous chemical.* Do not breathe its vapors or allow it to come into contact with your skin because it may produce severe chemical burns. Perform all operations involving the transfer of the pure liquid or its solutions in a hood, and wear rubber gloves for these manipulations. Even though the solutions used in the experiments are dilute, take proper precautions. If you get bromine on your skin, wash the area immediately with soap and warm water and soak any affected area in 0.6 *M* sodium thiosulfate solution, for up to 3 h if the burn is particularly serious.

2. Do not use your mouth to fill pipets in the transfer of *any* of the solutions in this experiment; this applies to both the bromine solutions and the acetic acid solutions of the substrates.

In the following experiments it is convenient for two students to work together.

Prepare stock solutions that are 0.2 *M* in aromatic substrate by dissolving the appropriate quantity of the arene in 15 *M* (90%) acetic acid; such solutions may have been prepared for you. Set up a water bath using a 1-L beaker and adjust and maintain its temperature at 35 ± 2 °C. Place 2 mL of the stock solutions of each of the following substrates in separate labeled test tubes: (a) phenol, (b) anisole, (c) diphenyl ether, (d) acetanilide, (e) *p*-bromophenol, and (f) α-naphthol. If stock solutions are provided, use special care to ensure that one solution is not contaminated by another. Suspend the test tubes partially in the water bath by looping a piece of copper wire around the neck of the test tube and over the rim of the beaker.

Prepare several disposable micropipets with a capacity of 2.0 mL, and equip them with rubber bulbs. Calibrate these by comparison with known volumes, and mark the droppers by scratching the glass with a file. These are used to introduce the bromine solution into the test tubes containing the substrates. It is essential that the addition be done as rapidly as possible and that the volumes be fairly accurate.

Transfer about 45 mL of a solution containing 0.05 M bromine in 15 M acetic acid to an Erlenmeyer flask, and allow thermal equilibration of the solution in the water bath for a few minutes. Add 2.0 mL of the bromine solution to one of the test tubes containing the substrates. Make the addition rapidly, mix the solution quickly, and note the exact time of addition. Observe the reaction mixture and note how long it takes for the bromine color to become faint yellow or to disappear. Repeat this procedure with each of the substrates, making sure that you use the same endpoint color for each one.

When the reaction is slow—that is, no decoloration occurs within 5 min—go on to another compound while waiting for the endpoint to be reached. Record the reaction times, and on the basis of these observations arrange the compounds in order of increasing reactivity toward bromine.

If it is not possible to determine the relative rates accurately at 35 °C, repeat the experiment with the compounds in doubt at 0 °C, using an ice-water bath for temperature control.

> **Finishing Touches**   Neutralize the combined *acetic acid solutions* by addition of saturated aqueous sodium bicarbonate and filter the resulting mixture under vacuum. Flush the *filtrate* down the drain; put the *filter cake*, which contains aryl bromides, in the container for halogen-containing solid waste.

EXERCISES

1. Write the structures for the major monobromination products that would be formed from each of the substrates used. Explain your predictions.

2. Based on the results obtained in this experiment, arrange the substrates in order of *decreasing* order of reactivity toward bromine. Explain the reactivity order in terms of inductive and/or resonance effects.

3. **a.** Compare the *initial* concentrations of bromine and of substrate in the reaction mixture, assuming that no reaction had occurred.
   b. Which reagent, bromine or substrate, is present in excess?
   c. Explain why the experimental procedure calls for one reagent to be present in excess over the other.
   d. Discuss whether you would be able to determine the rates of reaction if the concentrations of substrate and of bromine had been reversed over what is specified in the experimental procedure?

SPECTRA OF STARTING MATERIALS

IR and NMR spectra of acetanilide are contained in Figures 3.3 and 3.4, respectively.

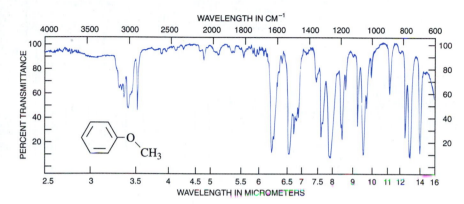

FIGURE 15.26   IR spectrum of anisole.

(a) ^{1}H NMR spectrum (90 MHz).

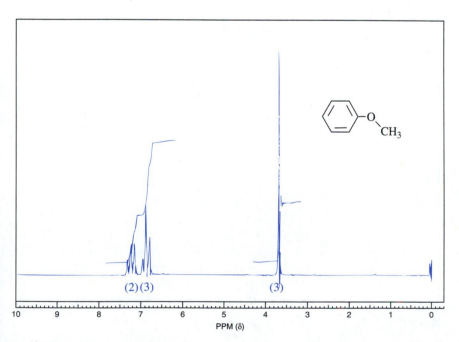

(b) ^{13}C NMR data.
   Chemical shifts: δ 54.8, 114.1, 120.7, 129.5, 159.9.

FIGURE 15.27   NMR data for anisole.

FIGURE 15.28 IR spectrum of *p*-bromophenol.

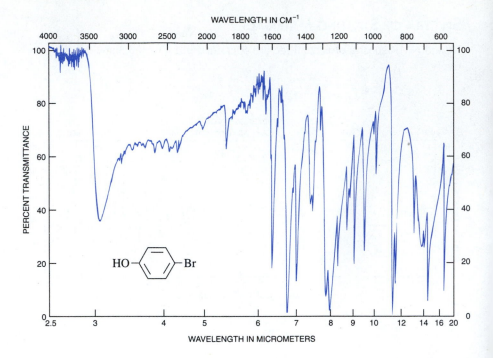

(a) ^{1}H NMR spectrum (90 MHz).

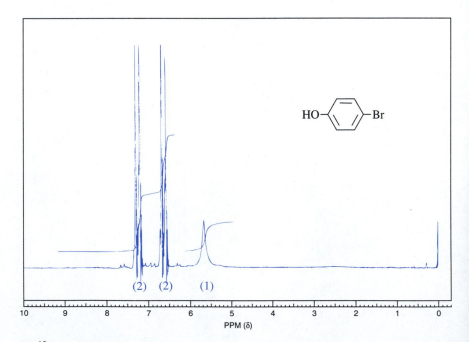

FIGURE 15.29 NMR data for *p*-bromophenol.

(b) ^{13}C NMR data.
Chemical shifts: δ 113.2, 117.2, 132.5, 153.9.

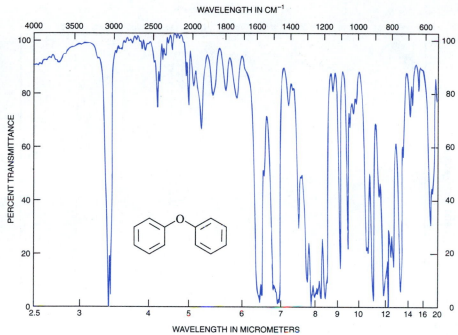

WAVELENGTH IN CM⁻¹

WAVELENGTH IN MICROMETERS

PERCENT TRANSMITTANCE

FIGURE 15.30 IR spectrum of diphenyl ether.

(a) ^{1}H NMR spectrum (90 MHz).

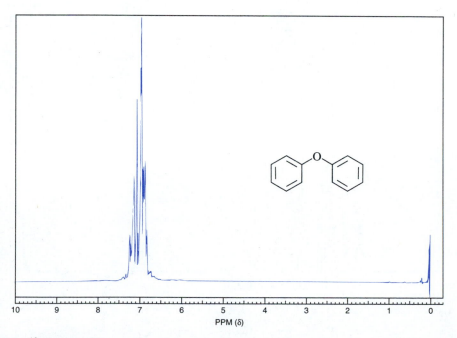

PPM (δ)

(b) ^{13}C NMR data.
   Chemical shifts: δ 119.0, 123.2, 129.8, 157.6.

FIGURE 15.31 NMR data for diphenyl ether.

FIGURE 15.32 IR spectrum for α-naphthol.

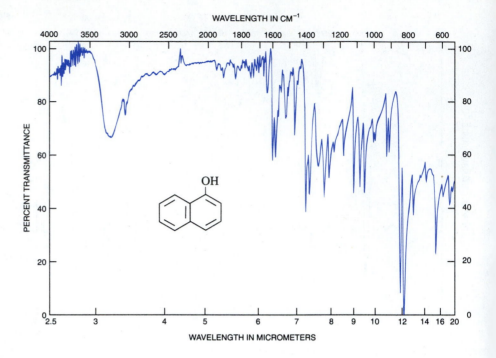

(a) ¹H NMR spectrum (90 MHz).

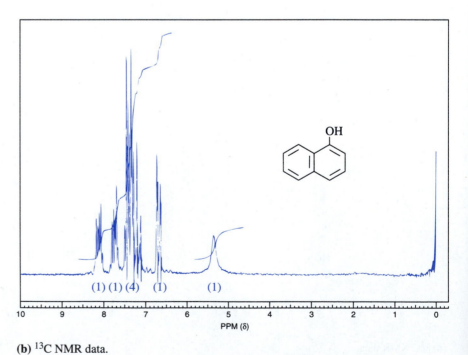

(b) ¹³C NMR data.
Chemical shifts: δ 108.8, 119.8, 122.1, 124.9, 126.2, 127.6, 134.9, 152.4.

FIGURE 15.33 NMR data for α-naphthol.

FIGURE 15.34   IR spectrum of phenol.

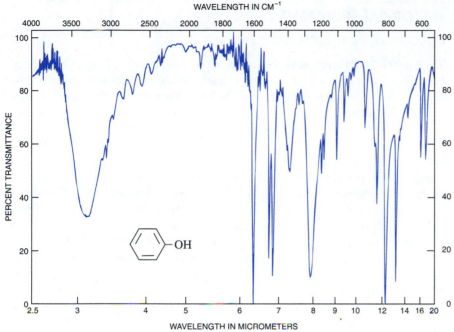

(a) ¹H NMR spectrum (90 MHz).

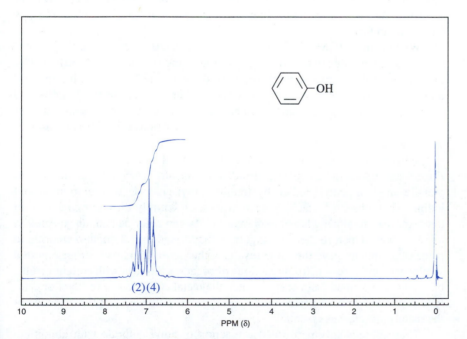

(b) ¹³C NMR data.
Chemical shifts: δ 115.4, 121.0, 129.7, 154.9.

FIGURE 15.35   NMR data for phenol.

HISTORICAL HIGHLIGHT:
Discovery of the Friedel-Crafts Reaction

The Friedel-Crafts reaction bears the names of two chemists, Charles Friedel, a Frenchman, and James Crafts, an American, who accidentally discovered this reaction in Friedel's laboratory in Paris in 1877. Friedel and Crafts quickly recognized the potential practical significance of their discovery, and immediately secured patents in France and England on procedures for preparing hydrocarbons and ketones using the reaction. Their judgment was accurate. Probably no other organic reaction has been of more practical value. Major industrial processes for producing high-octane gasoline, synthetic rubber, plastics, and synthetic detergents are representative applications of Friedel-Crafts chemistry. All of this came from an unexpected laboratory result and the brilliance of the two scientists who observed, interpreted, and extended it.

Crafts was born in 1839 in Boston. After graduating from Harvard University at the age of 19, he spent a year studying mining engineering, and then decided to go to Europe for further education. Crafts became fascinated with chemistry while studying metallurgy in Freiburg, Germany, and he subsequently secured positions in the laboratories of Bunsen in Heidelberg and Wurtz in Paris. It was in Wurtz's laboratory that he met Friedel and, owing to their similar scientific interests, the two began a collaboration in 1861.

Crafts returned to the United States in 1865 and, after a brief tenure as a mining inspector in Mexico and California, took a position as Professor of Chemistry at Cornell University shortly after this institution had been started. Three years later he moved to M.I.T., where he introduced numerous improvements in teaching and in the laboratories. Because of poor health he decided in 1874 to return to Europe and resume his collaboration with Friedel in Wurtz's laboratory in Paris.

When Crafts left M.I.T., he had expected to return in a short time, presumably after recovering his health. Owing to the change of climate or, perhaps, the excitement of the discovery he shared with Friedel in 1877, his health improved dramatically, yet he remained in Paris for another 17 years. Friedel and Crafts conducted an energetic research collaboration that by 1888 had produced over 50 publications, including patents, related to the reactions of aluminum chloride with organic compounds.

What was the accidental research discovery that engraved the names of Friedel and Crafts in scientific and industrial annals? They were attempting to prepare amyl (*n*-pentyl) iodide by treating amyl chloride with aluminum and iodine. They observed that the reaction took an entirely different course from that expected, producing large amounts of hydrogen chloride and, unexpectedly, *hydrocarbons*. Upon further investigation, they found that aluminum chloride in place of aluminum gave the same results. Other researchers had earlier reported somewhat similar results from reactions of organic chlorides with certain metals, but had not explained them or implicated the metal chloride as a reactant or catalyst. It was the work of Friedel and Crafts that first showed that the presence of the metal chloride was essential.

They subsequently performed a reaction of amyl chloride with aluminum chloride, using benzene as the solvent. Once again, they observed evolution of hydrogen chloride, but this time they found that *amylbenzene was a major product*! In describing their discovery to the Chemical Society of France in 1877, they reported, "With a mixture of [alkyl] chloride and hydrocarbon [an arene], the formation is established, in good yield, of hydrocarbons from the residues of the hydrocarbon less H and from the chloride less Cl. It is thus that ethylbenzene, amylbenzene, benzophenone, etc., are obtained." A general representation of the transformation is shown in the following equation.

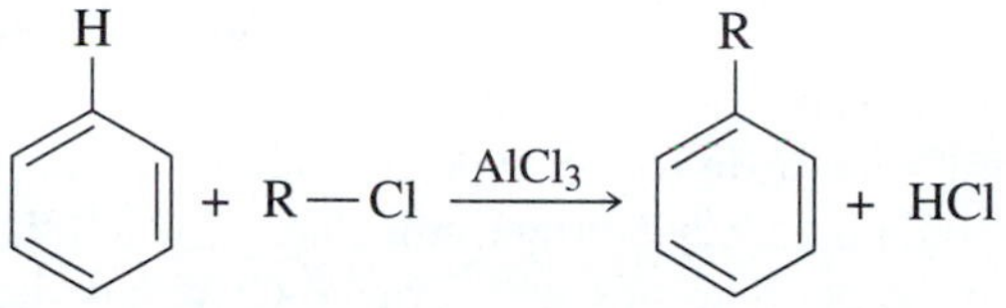

Friedel and Crafts recognized that their unexpected result promised the possibility of synthesizing a wide variety of hydrocarbons and ketones, and they immediately proved this by experiment. In the ensuing years the voluminous research papers and patents that came from Friedel and Crafts established a whole new area of research and practice in organic chemistry and laid the foundation for some important modern industrial chemical processes.

The Friedel-Crafts chemistry described here has touched our lives in many important but perhaps unrecognized ways. For example, Winston Churchill, in referring to the winning of the air war over Britain by the fighter pilots, once said "Never in the field of human conflict was so much owed by so many to so few." What was not recognized by many at the time, perhaps even Churchill, was that the victory in the air war was due not only to the skill and daring of the British pilots, but to the superiority of their aviation gasoline over that of the Germans. It has been generally acknowledged that the German fighter planes were mechanically superior to the British planes, but their fuel was not. The gasoline used in the British planes was of higher quality because it contained *toluene* and other alkylated aromatic hydrocarbons, which provided higher octane ratings. Production of these types of hydrocarbons was a direct outgrowth of Friedel-Crafts chemistry.

Similarly, "synthetic rubber" was vital to the ground-war effort in World War II, after the Japanese cut off the Allies from sources of natural rubber in Southeast Asia. A synthetic rubber was developed in an amazingly short time in a remarkable display of cooperation between government officials and industrial and academic scientists. This rubber was a copolymer (see Sec. 21.1) that contained styrene as a key component. Styrene is made from ethylbenzene, which is prepared industrially by the Friedel-Crafts reaction of benzene and ethylene, as shown here.

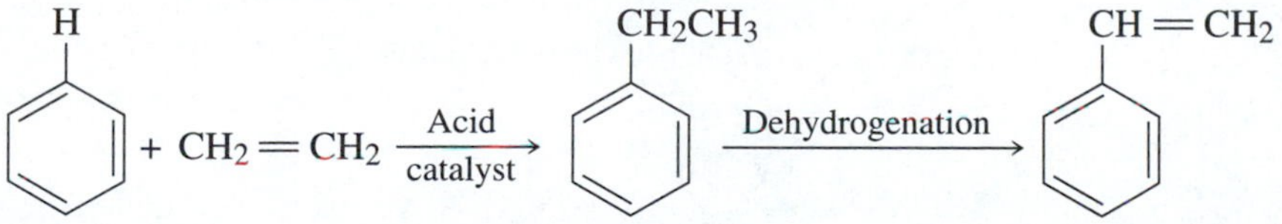

Styrene can also be homopolymerized to produce *polystyrene* (see Sec. 21.2). This material is one of the most versatile plastics ever invented. For example, it can be molded into rigid cases for radios and batteries and formulated in a more flexible form useful in the manufacture of toys and all kinds of containers. Polystyrene can also be produced in the form of a lightweight foam known as Styrofoam®. This foam is used for insulation in buildings and is molded into ice chests and disposable cups for hot and cold drinks.

The Friedel-Crafts reaction is a key part of the manufacture of many synthetic detergents, cleaning agents that have revolutionized the way we live today. We wash dishes and clothes with them, and, in contrast to soap, they work as well in hard water as in soft. A typical example of a *biodegradable* synthetic detergent is sodium dodecylbenzenesulfonate, in which the 12-carbon side chain is attached to the benzene ring by a Friedel-Crafts alkylation reaction, as illustrated here.

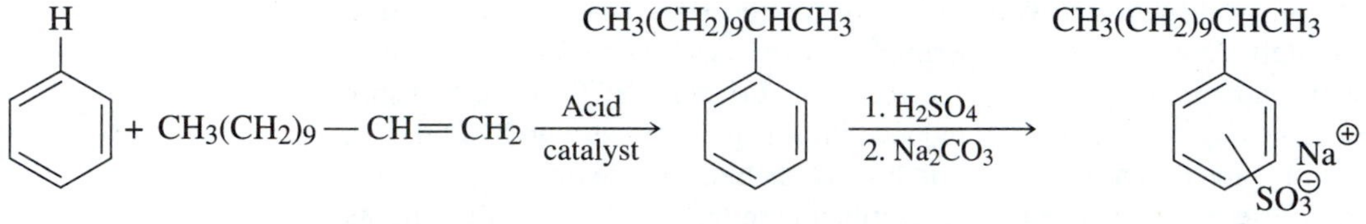

$$\text{H} - \bigcirc + CH_3(CH_2)_9 - CH = CH_2 \xrightarrow[\text{catalyst}]{\text{Acid}} \underset{\substack{\text{A dodecylbenzene} \\ \text{(other isomers also formed)}}}{CH_3(CH_2)_9CHCH_3 - \bigcirc} \xrightarrow[\text{2. Na}_2\text{CO}_3]{\text{1. H}_2\text{SO}_4} \underset{\substack{\text{A sodium dodecylbenzene-} \\ \text{sulfonate}}}{CH_3(CH_2)_9CHCH_3 - \bigcirc - SO_3^{\ominus} Na^{\oplus}}$$

These are just a few of the practical applications that have been developed in the years since Friedel and Crafts made their initial observation of an unexpected experimental result and by proper interpretation of the result have earned credit for a major discovery.

Relationship of Historical Highlight to Experiments

The experiments in Sections 15.2 and 15.4 illustrate the Friedel-Crafts alkylation and acylation of aromatic hydrocarbons, respectively. Although alkenes may serve in place of arenes in such reactions, aromatic compounds are far more commonly used as the reacting hydrocarbon. A complication of the Friedel-Crafts reactions is apparent in the procedure of Section 15.2, wherein rearrangements of the carbocations generated from the alkyl halide utilized provide mixtures of substitution products. The acylation reaction of Section 15.3 provides an example of how Friedel-Crafts chemistry is used in a multistep sequence to form polycyclic compounds.

OXIDATION OF ALCOHOLS, CARBONYL COMPOUNDS, AND ARENES

16.1 INTRODUCTION

T he **oxidation** of organic compounds is an important class of reactions that involves interaction of an oxidizing agent, often symbolized as [O], with the organic substrate. Oxidation may induce either simple functional group transformations—such as conversion of alkenes to glycols (Eq. 16.1) and of alkanes to alkyl halides (Eq. 16.2 and Chap. 9)—or cleavage of carbon-carbon bonds (Eq. 16.3), as in more complex processes such as those involving degradation of molecules. Indeed, the processes of life depend on oxidation of organic substrates,

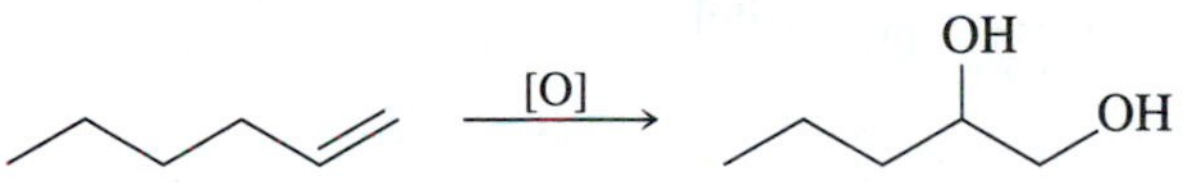

	[O]		
1-Hexene		1,2-Hexanediol	(16.1)

Cyclohexane $+$ SO_2Cl_2 $\xrightarrow{\text{Initiator}}$ Chlorocyclohexane $-Cl$ $+$ SO_2 $+$ HCl (16.2)

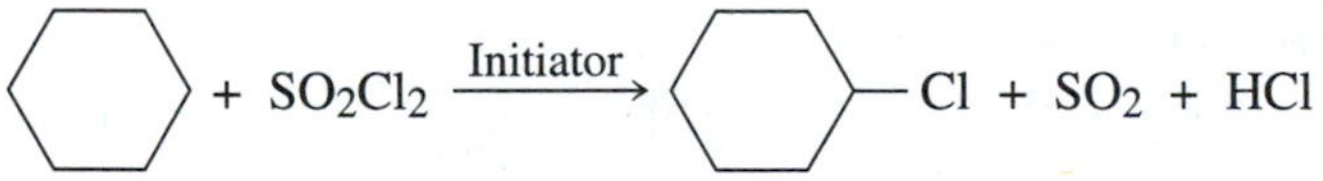

whereby metabolic energy is derived from the overall oxidation of carbohydrates, fats, and proteins to carbon dioxide and water, among other products, as illustrated in Equation 16.4. Potentially poisonous substances are commonly detoxified by biological oxidation to more benign substances. Thus, nicotine, which is toxic to humans if present in sufficiently high concentration, is oxidized in the liver to cotinine (Eq. 16.5), a substance having low toxicity.

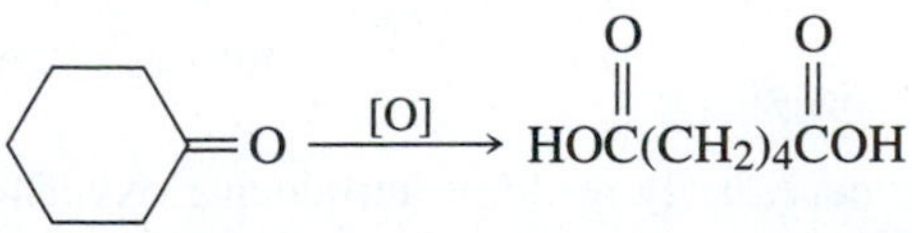

Cyclohexanone $=O$ $\xrightarrow{[O]}$ $HOC(CH_2)_4COH$ (16.3)

Hexanedioic acid

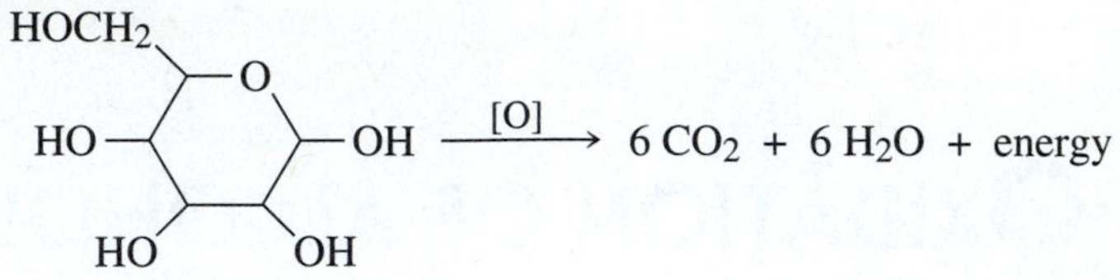

$$\text{A carbohydrate} \xrightarrow{[O]} 6\,CO_2 + 6\,H_2O + \text{energy} \qquad \textbf{(16.4)}$$

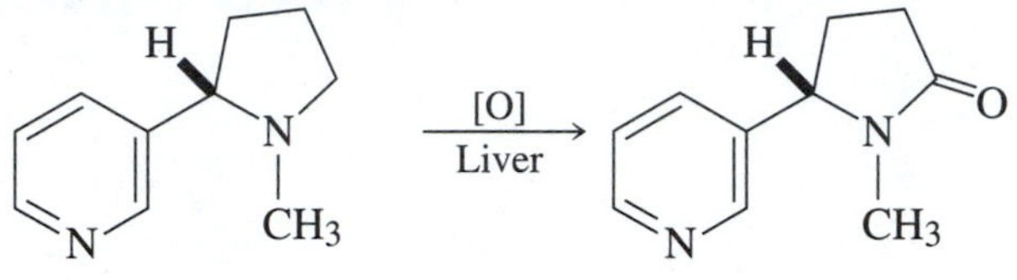

A carbohydrate

$$\text{Nicotine} \xrightarrow[\text{Liver}]{[O]} \text{Cotinine} \qquad \textbf{(16.5)}$$

Nicotine   Cotinine

Oxidations are also valuable processes in the organic laboratory, for they make it possible to introduce a variety of functional groups into organic substrates. For example, primary and secondary alcohols can be converted to aldehydes and ketones (Eq. 16.6 and Sec. 16.2), and these in turn may be transformed into a myriad of other products such as carboxylic acids, by further oxidation (Eqs. 16.3 and 16.7 and Sec. 16.2), and alkenes, by a Wittig reaction (Eq. 16.8 and Sec. 18.1.2).

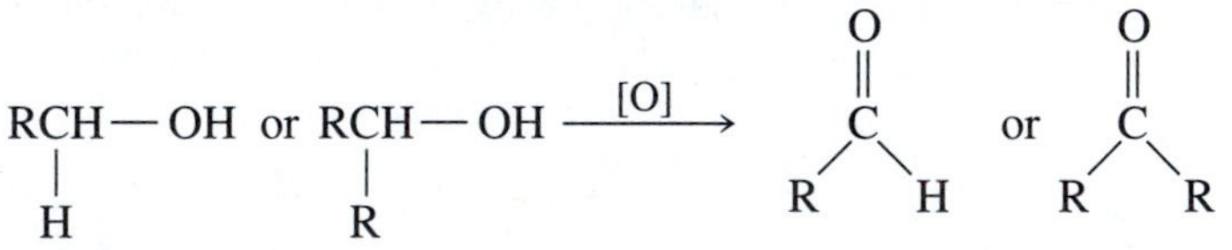

$$\underset{\text{An alcohol}}{RCH{-}OH \text{ or } RCH{-}OH} \xrightarrow{[O]} \underset{\text{An aldehyde}}{\overset{O}{\underset{R\quad H}{\parallel}{C}}} \text{ or } \underset{\text{A ketone}}{\overset{O}{\underset{R\quad R}{\parallel}{C}}} \qquad \textbf{(16.6)}$$

$$\underset{R\quad H}{\overset{O}{\parallel}{C}} \text{ or } \underset{R\quad R}{\overset{O}{\parallel}{C}} \xrightarrow{[O]} \underset{\text{A carboxylic acid}}{\overset{O}{\parallel}{RC}{-}OH} \qquad \textbf{(16.7)}$$

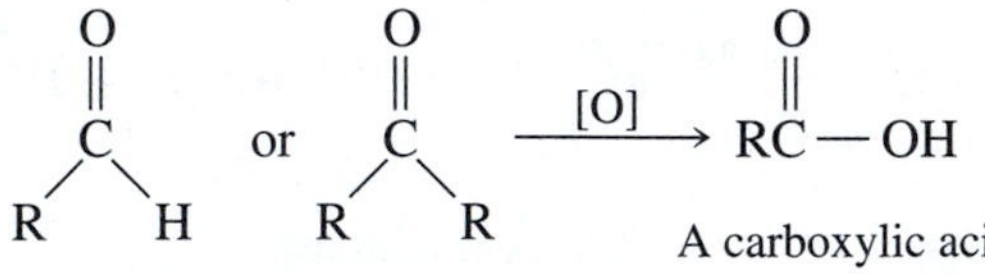

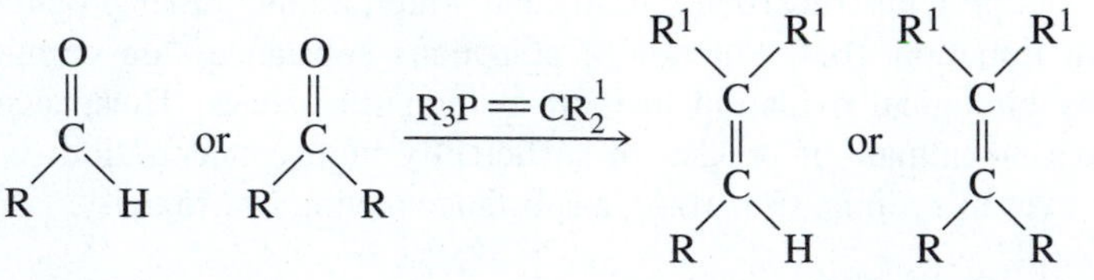

$$\underset{R\quad H}{\overset{O}{\parallel}{C}} \text{ or } \underset{R\quad R}{\overset{O}{\parallel}{C}} \xrightarrow{R_3P{=}CR_2^1} \underset{R\quad H}{\overset{R^1\quad R^1}{C{=}C}} \text{ or } \underset{R\quad R}{\overset{R^1\quad R^1}{C{=}C}} \qquad \textbf{(16.8)}$$

Alkenes

Oxidation does *not* necessarily result in introducing oxygen into an organic molecule (see Eq. 16.2, for example). In the most general sense, and in analogy to concepts of oxidation as applied to inorganic compounds, oxidation of an organic substance simply requires an increase in the **oxidation number,** or **oxidation state,** of carbon. There are several ways of defining the oxidation states of carbon atoms, but the one used here involves the following sequence.

1. *Select the carbon atom* whose oxidation number is to be defined.

2. *Assign oxidation numbers* to the atoms attached to this carbon atom using the following values:

   a. +1 for hydrogen
   b. −1 for halogen, nitrogen, oxygen and sulfur
   c. 0 for carbon

   In other words, with respect to the atoms bound to the carbon atom, those that are *more* electronegative are assigned a value of −1, whereas those that are less electronegative are given a value of +1.

3. *Sum the oxidation numbers* of these atoms. If a heteroatom (see 2b) is multiply bound to a carbon atom, multiply its oxidation number by the number of bonds linking it to the carbon atom involved.

4. The sum of the number from step 3 and the oxidation number of the carbon atom under consideration must equal the charge on the carbon atom, which is zero unless it bears a positive or negative charge.

Application of this method shows that the conversion of an alkene to the corresponding dibromide (Eq. 16.9) and that of an alkane to an alkyl halide (Eq. 16.10), reactions discussed in Section 10.4 and Chapter 9, respectively, are oxidations even though oxygen is not incorporated in these reactions. Determining the changes in the oxidation number of the various carbon atoms during the conversion of acetic acid to carbon dioxide and water (Eq. 16.11) illustrates how the method is used when heteroatoms are multiply bound to a carbon atom.

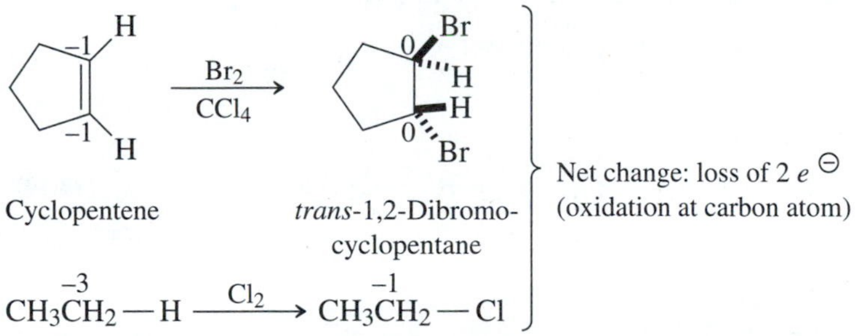

Cyclopentene   trans-1,2-Dibromo-   Net change: loss of 2 $e^{\ominus}$
                cyclopentane   (oxidation at carbon atom)

$$\overset{-3}{CH_3}CH_2 - H \xrightarrow{Cl_2} \overset{-1}{CH_3}CH_2 - Cl$$

(16.9)

(16.10)

Ethane   1-Chloroethane
                (ethyl chloride)

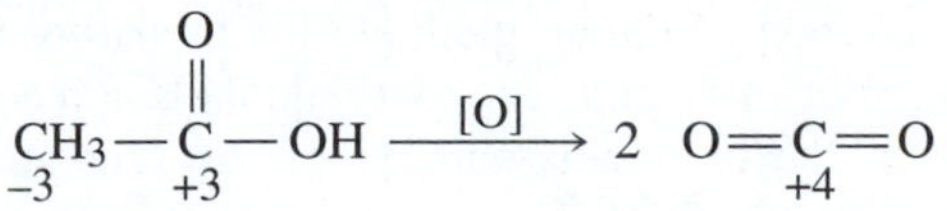

$$\underset{-3}{CH_3} - \overset{O}{\underset{+3}{\overset{\|}{C}}} - OH \xrightarrow{[O]} 2 \ \underset{+4}{O{=}C{=}O}$$

(16.11)

Acetic acid   Carbon dioxide

The oxidation reactions described in this chapter all involve an increase in the oxygen content of products relative to reactants. The source of oxygen in our experiments is not molecular oxygen but rather hypohalite, XO^-, or an oxidized form of a metal; the latter form of oxidant is most commonly used in the laboratory.

16.2 PREPARATION OF ALDEHYDES AND KETONES BY OXIDATION OF ALCOHOLS

Aldehydes and ketones have a central role in organic synthesis, and efficient methods for their preparation are of great importance. Such compounds are synthesized in a variety of ways, including hydration or hydroboration/oxidation of alkynes (Eqs. 16.12 and 16.13, respectively, and Chap. 11), and from carboxylic acids or their derivatives by reaction with organometallic reagents or reducing agents (Eqs. 16.14 and 16.15). The latter two reactions involve net **reduction** of the carbonyl carbon atom, whereas the transformations shown in Equations (16.12) and (16.13) are neither net oxidation nor reduction of the alkyne, as can be shown by determining the net change in oxidation state at the carbon atoms involved in the transformations (see Exercise 1 at the end of this section).

$$R^1\!-\!C\!\equiv\!C\!-\!R^2(H) \xrightarrow[\substack{H_2SO_4 \\ H_2O}]{HgSO_4} R^1\!-\!\overset{\overset{\displaystyle O}{\|}}{C}\!-\!CH_2\!-\!R^2(H) \qquad \textbf{(16.12)}$$

An alkyne A ketone

$$R^1\!-\!C\!\equiv\!C\!-\!R^2(H) \xrightarrow[\substack{2.\ H_2O_2/ \\ HO^-}]{1.\ B_2H_6} R^1\!-\!CH_2\!-\!\overset{\overset{\displaystyle O}{\|}}{C}\!-\!R^2(H) \qquad \textbf{(16.13)}$$

$$R^1\!-\!\overset{\overset{\displaystyle O}{\|}}{C}\!-\!OH \xrightarrow[\substack{2.\ H_3O^+}]{1.\ 2\ R^2Li} R^1\!-\!\overset{\overset{\displaystyle O}{\|}}{C}\!-\!R^2 \qquad \textbf{(16.14)}$$

A carboxylic acid

$$R^1\!-\!\overset{\overset{\displaystyle O}{\|}}{C}\!-\!OR^2 \xrightarrow{(i\text{-Bu})_2AlH} R^1\!-\!\overset{\overset{\displaystyle O}{\|}}{C}\!-\!H \qquad \textbf{(16.15)}$$

An ester

One of the most common synthetic methods for preparing aldehydes and ketones is the oxidation of primary and secondary alcohols, a reaction involving a two-electron change in the oxidation number of the functionalized carbon atom (Eq. 16.16). Chromic ion, Cr^{VI}, frequently in the form of chromic acid, H_2CrO_4, is a common oxidant, although other oxidants, such as halonium ion, X^+, as found in hypohalous acids, HOX, and permanganate ion, Mn^{VI}, as in potassium permanganate, $KMnO_4$, can be used. A description of the oxidation of alcohols to aldehydes or ketones with chromic or hypohalous acids follows, and procedures for their use and that of potassium permanganate are included in the Experimental Procedures.

$$R-\overset{\overset{\displaystyle OH}{|}}{C}H-R(H) \xrightarrow[\substack{\text{or HOX} \\ \text{or KMnO}_4}]{\text{H}_2\text{CrO}_4} R-\overset{\overset{\displaystyle O}{\|}}{C}-R(H) \qquad \textbf{(16.16)}$$

Oxidation number 0 (2° alcohol)   +2 (ketone)
                    −1 (1° alcohol)   +1 (aldehyde)

Chromic acid is not stable for extended periods and must be prepared as needed. This is easily done by reaction of sodium or potassium dichromate with aqueous sulfuric acid (Eq. 16.17) or by dissolution of chromic anhydride in water (Eq. 16.18). An excess of sulfuric acid is normally used in the former preparation because the rate of oxidation of alcohols by chromic acid is faster in acidic solution. For oxidation of substrates that are sensitive to acids, various other oxidizing agents may be used that are known to work under neutral or even basic conditions. These are generally more expensive, however, and are not discussed here.

$$\text{K}_2\text{CrO}_7 + 2\,\text{H}_2\text{SO}_4 \longrightarrow [\text{H}_2\text{Cr}_2\text{O}_7] \xrightarrow{\text{H}_2\text{O}} 2\,\text{H}_2\text{CrO}_4 + 2\,\text{NaHSO}_4 \quad \textbf{(16.17)}$$

Potassium   Chromic acid
dichromate

$$\text{CrO}_3 + \text{H}_2\text{O} \longrightarrow \text{H}_2\text{CrO}_4 \qquad\qquad \textbf{(16.18)}$$

Chromium
trioxide

The mechanism of oxidation with chromic acid is well understood and is depicted in a general way in Scheme 16.1. Alcohols rapidly form chromate esters **1** in the presence of chromic acid, a reaction that is analogous to the acid-catalyzed formation of esters from alcohols and carboxylic acids (see Sec. 19.2). Ester **1** is relatively unstable and undergoes elimination to produce the carbonyl compound; the elimination is normally the **rate-determining step** (rds) in the reaction.

The stoichiometry of the oxidation of alcohols by chromic acid is determined from the changes in oxidation numbers of the oxidizing agent and substrate alcohol. During the oxidation, chromic ion, which is Cr^{VI} in chromic acid, undergoes a two-electron reduction to give Cr^{IV} in the form of H_2CrO_3 (Scheme 16.1); the carbon atom of the alcohol similarly undergoes a two-electron oxidation (Eq. 16.14). If this corresponded to the overall course of the reaction, a one-to-one ratio of chromic acid to alcohol would be required, as is shown in Scheme 16.1. However, Cr^{IV} is an *unstable* oxidation state of chromium, so it rapidly disproportionates with Cr^{VI} to produce Cr^{V}, a new oxidizing agent shown as $HCr^{V}O_3$ (Eq. 16.19). The 2 mol of Cr^{V} thus

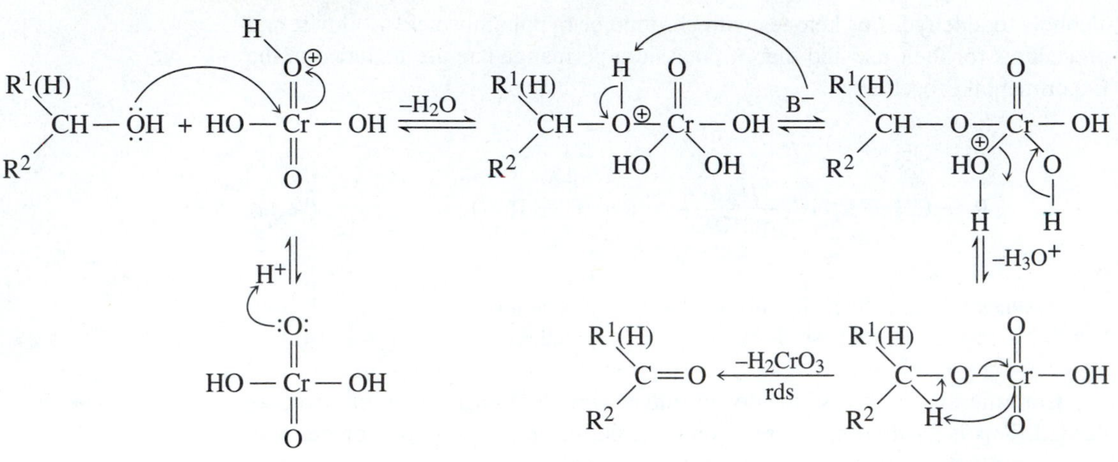

SCHEME 16.1

formed oxidize 2 mol more of alcohol to the carbonyl compound, and Cr^{III}, a stable oxidation state written here as H_3CrO_3, results (Eq. 16.20). In the presence of sulfuric acid, the latter species is converted to $Cr_2(SO_4)_3$ (Eq. 16.21). Summing Equations (16.19–16.21) results in Equation (16.22), which shows that the chromic acid oxidation of alcohols to aldehydes or ketones requires *two* equivalents of chromic acid to oxidize *three* equivalents of an alcohol. It is noteworthy that the oxidation also consumes *six* equivalents of acid.

$$H_2Cr^{IV}O_3 + H_2Cr^{VI}O_4 \longrightarrow 2\,HCr^{V}O_3 + H_2O \qquad (16.19)$$

$$2\,HCr^{V}O_3 + 2\,\underset{R^2}{\overset{R^1(H)}{\diagup}}CH-OH \longrightarrow 2\,\underset{R^2}{\overset{R^1(H)}{\diagup}}C=O + 2\,H_3CrO_3 \qquad (16.20)$$

$$2\,H_2CrO_3 + 3\,H_2SO_4 \longrightarrow Cr_2(SO_4)_3 + 6\,H_2O \qquad (16.21)$$

$$3\,R^2-\overset{OH}{\underset{|}{CH}}-R^1(H) + 2\,H_2CrO_4 + 3\,H_2SO_4 \longrightarrow 3\,R^2-\overset{O}{\overset{\|}{C}}-R^1(H) + Cr_2(SO_4)_3 + 8\,H_2O \quad (16.22)$$

A less detailed approach to determination of the stoichiometry of this oxidation may be developed as follows. The carbon atom of the alcohol undergoes a two-electron oxidation, whereas chromium has a net change involving a three-electron reduction from Cr^{VI} to Cr^{III}. Thus, the electronic change for each equivalent of chromic ion is sufficient to oxidize 1.5 equivalents of alcohol, or in whole numbers, the oxidation of 3 mol of alcohol consumes 2 mol of chromic ion (Eq. 16.23), the same result as shown in Equation (16.22).

$$3\ R^1 \overset{\displaystyle OH}{\underset{\displaystyle |}{-CH-}} R^2(H) + 2\ Cr^{VI} \longrightarrow 3\ R^1 \overset{\displaystyle O}{\underset{\displaystyle \|}{-C-}} R^2(H) + 2\ Cr^{III} \quad \textbf{(16.23)}$$

oxidation number 0 (2° alcohol)   +6   +2 (ketone)   +3
                  −1 (1° alcohol)   +1 (aldehyde)

Chromic acid may be prepared from other sources, and it is important to determine what quantities of these precursors are required to produce the necessary amount of oxidant. Reference to Equation (16.17) shows that *two* equivalents of chromic acid are produced from *one* equivalent of dichromate. Thus, if either sodium or potassium dichromate is used to generate chromic acid, only one equivalent of dichromate is required to oxidize three equivalents of alcohol. In the case of chromic anhydride, however, only one equivalent of chromic acid is provided per equivalent of the anhydride (Eq. 16.18), so two equivalents of chromic anhydride per three equivalents of alcohol are necessary to achieve the stoichiometry shown in Equations (16.22) and (16.23).

Some major side reactions complicate the chromic acid oxidation of a primary alcohol to an aldehyde, the most important of which is the further conversion of the aldehyde to a carboxylic acid (Eq. 16.24), a process that is presumably initiated by acid-catalyzed formation of the **hydrate** of the aldehyde

$$3\ R \overset{\displaystyle O}{\underset{\displaystyle \|}{-C-}} H + 2\ H_2CrO_4 + 3\ H_2SO_4 \longrightarrow 3\ R \overset{\displaystyle O}{\underset{\displaystyle \|}{-C-}} OH + Cr_2(SO_4)_3 + 5\ H_2O \qquad \textbf{(16.24)}$$

as shown in Equation (16.25). Subsequent steps in the oxidation are then analogous to those for converting an alcohol to an aldehyde or ketone (Scheme 16.1). This undesired overoxidation is minimized by adding the chromic acid *to* the primary alcohol so an excess of the oxidizing agent is not present in the reaction mixture *and* by distilling the aldehyde from the reaction mixture as it is formed. In order to distil an aldehyde from an oxidant such as chromic acid, the aldehyde must have a boiling point of no more than 125–150 °C if it is to be prepared in good yield. With less-volatile aldehydes, other more selective oxidizing agents must be used.

$$R \overset{\displaystyle O}{\underset{\displaystyle \|}{-C-}} H \overset{H^+}{\rightleftharpoons} R-\overset{\overset{\displaystyle H}{\displaystyle \diagdown}}{\underset{\displaystyle |}{C}}\overset{\oplus}{O}-H \quad :\ddot{O}H_2 \rightleftharpoons R-\overset{\displaystyle OH}{\underset{\underset{H}{\displaystyle \overset{\oplus}{\diagup}OH}}{\underset{\displaystyle |}{C}}}-H \overset{-H^+}{\rightleftharpoons} R-\overset{\displaystyle OH}{\underset{\displaystyle \underset{OH}{|}}{C}}-H \quad \textbf{(16.25)}$$

A hydrate

A poor yield of product may be obtained even when the aldehyde is volatile. This is due to the facile acid-catalyzed conversion of the aldehyde to the **hemiacetal 2,** which is subsequently oxidized to the ester of a carboxylic acid (Eq. 16.26). Occasionally this "side" reaction can be turned into a useful synthesis of the ester, as is the case in the preparation of *n*-butyl butyrate by the chromic acid oxidation of 1-butanol (Eq. 16.26, R = $CH_3(CH_2)_2-$).

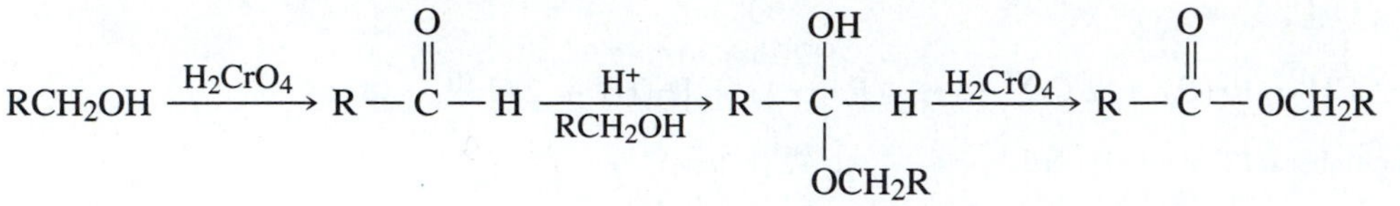

$$RCH_2OH \xrightarrow{H_2CrO_4} R-\overset{\overset{\displaystyle O}{\|}}{C}-H \xrightarrow[RCH_2OH]{H^+} R-\overset{\overset{\displaystyle OH}{|}}{\underset{\underset{\displaystyle OCH_2R}{|}}{C}}-H \xrightarrow{H_2CrO_4} R-\overset{\overset{\displaystyle O}{\|}}{C}-OCH_2R \quad (16.26)$$

2

A hemiacetal   An ester

Ketones are much more stable toward oxidizing agents in mildly acidic media than are aldehydes, and side reactions are usually not significant under the conditions typically used to oxidize secondary alcohols. Under alkaline or strongly acidic conditions, however, *enolizable* ketones will undergo oxidation with cleavage of a carbon-carbon bond to give two carbonyl fragments. For

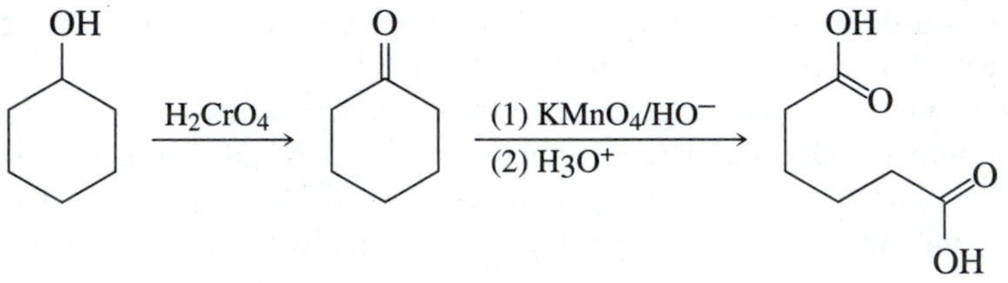

(16.27)

Cyclohexanol   Cyclohexanone   Hexanedioic acid
(adipic acid)

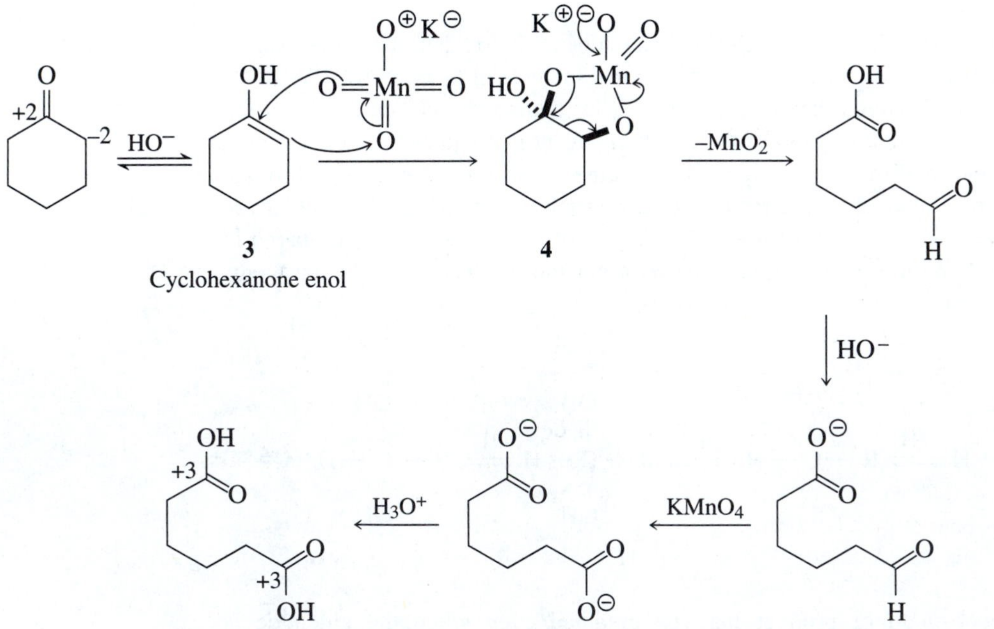

3
Cyclohexanone enol   **4**

A dicarboxylate   An aldehyde-carboxylate   **5**

SCHEME 16.2

example, chromic acid oxidation of cyclohexanol provides cyclohexanone in good yield, but this ketone can be oxidized further to hexanedioic acid, commonly known as adipic acid, on treatment with potassium permanganate under mild alkaline conditions (Eq. 16.27). The reaction involves conversion of the ketone to the **enol 3,** which reacts with permanganate ion to give the cyclic ester **4** (Scheme 16.2). Decomposition of **4** ensues with cleavage of a carbon-carbon bond to yield the aldehyde-carboxylate **5** and manganese dioxide. The aldehyde group of product **5** is readily oxidized by permanganate to give a dicarboxylate salt, which gives the diacid upon acidification of the reaction mixture.

The stoichiometry of the oxidation of cyclohexanone by permanganate ion is determined by considering the changes in oxidation states that occur. The two carbon atoms of interest undergo a net six-electron oxidation, as indicated on the structures of cyclohexanone and hexanedioic acid (Scheme 16.2). Reference to the half-reaction for the oxidant (Eq. 16.28) shows that permanganate ion undergoes a three-electron reduction to manganate ion, in the form of manganese dioxide. Two moles of the oxidant per mole of ketone are thus required to effect the six-electron change of Equation (16. 27).

$$Mn^{VII}O_4^{\ominus} + 0.5\ H_2O \longrightarrow \underline{Mn^{VI}O_2} + HO^{\ominus} + 1.5\ \ddot{O}\cdot \qquad \text{(16.28)}$$
$$\text{(brown precipitate)}$$

Cyclohexanone is a symmetrical ketone and gives *only* enol **3** (Scheme 16.2). If a ketone is *not* symmetrical, it can produce two isomeric enols (Eq. 16.29), each of which is oxidized by permanganate ion to different products, which may be difficult to separate. The formation of complex mixtures from oxidation of unsymmetrical ketones is a complication that detracts from the synthetic utility of this reaction.

$$\underset{H}{\overset{O}{\underset{R^1CH}{\parallel}}}\underset{H}{\overset{C}{\diagup}}\underset{H}{\overset{}{\diagdown}}CHR^2 \underset{}{\overset{HO^-}{\rightleftharpoons}} R^1CH{=}C\underset{CH_2R^2}{\overset{OH}{\diagup}} + \underset{R^1CH_2}{\overset{HO}{\diagdown}}C{=}CHR^2 \quad \text{(16.29)}$$

Chromic and permanganate ions are powerful, inexpensive oxidizing agents. However, both of these oxidants are derived from **heavy metals,** a class of elements which commonly are toxic and thus are environmentally hazardous. The safe disposal of these metals and their derivatives is of considerable concern, and avoiding their use is desirable. Fortunately, for oxidizing alcohols an alternative to chromic ion is available in the form of the hypohalites, XO^- (X = halogen), which are made by reaction of the corresponding halogen with aqueous base (Eq. 16.30). The most familiar sources of this type of oxidizing agent are household bleach, which is an aqueous solution of sodium hypochlorite, NaOCl, and the chlorinating agent for swimming pools, which usually is calcium hypochlorite, $Ca(OCl)_2$, in granular or tablet form.

$$X-X + HO^{\ominus} \rightleftharpoons X^{\ominus} + XO-H \xrightarrow{HO^-} XO^{\ominus} + H_2O \qquad \textbf{(16.30)}$$

(X = halogen)   Hypohalous   Hypohalite
                              acid

The mechanism by which hypohalites oxidize alcohols probably involves initial formation of an alkyl hypohalite, **6** (Eq. 16.31). This product arises from reaction of the alcohol with the hypohalous acid that is in equilibrium with hypohalite ion in the aqueous medium (Eq. 16.30). Base-promoted E2 elimination of the elements of H–X from **6** leads directly to either an aldehyde or ketone. The advantage of using hypohalite as an oxidant is immediately obvious upon review of Equation (16.31). The inorganic by-product derived from the oxidant is a halide salt (specifically sodium chloride when commercial bleach is employed) and this can safely be flushed down the drain.

$$R-\underset{\underset{H}{|}}{\overset{\overset{R}{|}}{C}}-\ddot{O}H + X-\ddot{O}H \underset{-H_2O}{\rightleftharpoons} R-\underset{\underset{H}{|}}{\overset{\overset{R}{|}}{C}}-O \xrightarrow{} \underset{R}{\overset{R}{\diagdown}}C=O + X^{\ominus} \qquad \textbf{(16.31)}$$

                 +1   +2   –1

6
Alkyl hypohalite

The stoichiometry of the reaction is determined in the manner described previously. The carbon atom of the substrate alcohol undergoes a two-electron change in oxidation states, from 0 to +2 in the case of a secondary alcohol, and so does the halogen (Eq. 16.31). Consequently a one-to-one ratio of the two reagents is required.

An aldehyde may undergo reaction with an additional equivalent of hypohalite to form a carboxylic acid (compare Eq. 16.24), and it may be difficult to suppress this further oxidation. Thus, unless carboxylic acids are the desired end products, the use of hypohalite as an oxidant is limited to the conversion of secondary alcohols to ketones. Another side reaction that may occur when hypohalite is used as an oxidizing agent is the α-halogenation of the desired carbonyl compound (Eq. 16.32). This process involves base-catalyzed formation of the enolate ion, followed by its reaction with hypohalous acid or some other source of halonium ion, such as X_2 (X = halogen). The replacement of the α-hydrogens of a ketone or aldehyde is the basis for the **haloform test,** described in Section 23.5.4.

$$R-\overset{O}{\overset{\|}{C}}-\overset{H}{\overset{|}{C}}HR \rightleftharpoons R-\overset{O^{\ominus}}{\overset{|}{C}}=CHR \longleftrightarrow R-\overset{O}{\overset{\|}{C}}-\overset{\ominus}{C}HR \xrightarrow{X-\ddot{O}H} R-\overset{O}{\overset{\|}{C}}-\overset{X}{\overset{|}{C}}HR \qquad \textbf{(16.32)}$$

An α-halocarbonyl
compound

SAFETY ALERT

1. The preparation of chromic acid requires diluting concentrated sulfuric acid with water. Be certain to add the acid *slowly* to the water and swirl the mixture to ensure continuous mixing. Swirling keeps the denser sulfuric acid from layering at the bottom of the flask and avoids possible splattering because of the heat generated when the two layers are suddenly mixed.

2. To keep the acids from contacting your skin, wear rubber gloves when preparing and handling solutions of potassium permanganate or chromic acid. These solutions will cause unsightly stains on your hands that persist for several days, and the chromic acid–sulfuric acid solution may cause severe chemical burns. If these oxidants come in contact with your skin, wash the affected area thoroughly with warm water. In the case of chromic acid, also rinse the area with 5% sodium bicarbonate solution.

3. Wear rubber gloves when handling solutions of bleach to keep this oxidant from contacting your skin. Should you spill bleach on your skin, wash the affected area thoroughly with water.

4. Residues of manganese dioxide are difficult to remove from glassware. Rinse stained glassware first with 10% sodium bisulfite solution; if this fails, add 6 M HCl to the flask and heat it in a hood until the stain has disappeared.

5. In the oxidation of 2-methyl-1-propanol, be certain that all joints in the apparatus are lubricated and tightly mated before the solution is heated to boiling.

Pre-Lab exercises for Section 16.2, "A. Preparation of 2-Methylpropanal," are found on page PL. 73.

A. PREPARATION OF 2-METHYLPROPANAL

Prepare a solution of chromic acid by dissolving 9.7 g of potassium dichromate in 50 mL of water and then slowly adding, with swirling of the solution, 7 mL of concentrated sulfuric acid. Using an ice-water bath, cool this solution to room temperature. If a stock solution of chromic acid has been prepared for your use, obtain 60 mL of the solution. Fit a 125-mL round-bottom flask with a Claisen adapter, and equip the adapter with a dropping funnel *directly above* the flask and an unpacked Hempel column on the parallel sidearm. Add boiling chips or a magnetic stirring bar if magnetic stirrers are available. To the top of the Hempel column attach a stillhead bearing a thermometer and a water-cooled condenser. Position a 50-mL graduated cylinder to collect the distillate that will be produced. Place 7.4 g of 2-methyl-1-propanol and 10 mL of water in the flask, and heat the mixture until gentle boiling begins. Discontinue heating and immediately begin to add the red-orange solution of chromic acid from the dropping funnel to the *hot* mixture of alcohol and water so that all of the acid is added in 10–15 min. The addition rate should be adjusted to cause the reaction mixture to boil vigorously, so a mixture of alcohol, aldehyde, and water steam-distils (see Sec. 2.10), giving a head temperature of 80–85 °C. This temperature should be maintained by appropriate adjustment of the rate of addition of the solution of chromic acid. After all of this solution has been added and the mixture has stopped distilling, heat and distil the dark green chromium(III) reaction mixture for an additional 15 min.*

Note the amount of water contained in the two-phase distillate, transfer the mixture to a separatory funnel, and add 0.2 g of sodium carbonate. Shake the funnel thoroughly, with venting to relieve any pressure; then saturate the aqueous layer by adding about 0.2 g of sodium chloride for each milliliter of water to salt out any dissolved product. Shake the funnel again to effect solution of the salt, remove the organic layer, and dry it over anhydrous sodium sulfate.*

Filter or decant the crude organic product into a 25-mL distilling flask equipped for fractional distillation (see Sec. 2.8), and isolate the 2-methylpropanal (bp 63–66 °C, 760 torr). If the fractionation is continued after the aldehyde has distilled, unchanged alcohol (bp 107–108 °C, 760 torr) and some isobutyl isobutyrate (bp 148–149 °C, 760 torr) can be isolated.

Prepare the 2,4-dinitrophenylhydrazone of 2-methylpropanal (mp 186–188 °C) by the procedure given in Section 23.5.1A. Apply the chromic acid in acetone test described in Section 23.5.1C to 2-methylpropanal.

If instructed to do so, prepare the sodium bisulfite addition product of the aldehyde (Eq. 16.33) by adding 3 mL of saturated sodium bisulfite solution to 0.5 mL of the aldehyde contained in an Erlenmeyer flask. Swirl the flask; then allow the mixture to stand for 10 min. The solution should become warm as the reaction proceeds. Add 8 mL of 95% ethanol, swirl the mixture well, and cool the resulting solution in an ice-salt bath. Collect the sodium bisulfite addition product by vacuum filtration, washing the filter cake once with 95% ethanol and once with diethyl ether. Do *not* attempt to determine the melting point of this solid because it may decompose explosively on heating. To regenerate 2-methylpropanal, add 3–5 mL of either 1 *M* sodium carbonate solution or 3 *N* hydrochloric acid to the addition product and gently warm the mixture.

$$(CH_3)_2CHC\overset{\displaystyle O}{\overset{\|}{}}\!-H \;+\; Na^{\oplus \ominus}SO_3H \;\longrightarrow\; (CH_3)_2CH\underset{\underset{HO}{\overset{\|}{SO_2}}}{\overset{\overset{Na^{\oplus}\ominus O}{}}{C}}\!-H \;\longrightarrow\; (CH_3)_2CH\underset{\underset{Na^{\oplus}\ominus O}{\overset{\|}{SO_2}}}{\overset{\overset{H\diagdown O}{}}{C}}\!-H \qquad \textbf{(16.33)}$$

2-Methylpropanal   Sodium bisulfite
(isobutyraldehyde)   addition product

Sodium bisulfite addition products are generally *not* good derivatives for *characterizing* carbonyl compounds, but they are often used to *purify* aldehydes, methyl ketones, and some cyclic ketones.

> **Finishing Touches**   Add solid sodium sulfite to the *aqueous solution of chromium salts* in order to destroy excess Cr^{VI}. Make the solution slightly basic, to form *chromium hydroxide*, and isolate this salt by vacuum filtration. Place the *filter paper* and the *filter cake* in the container for heavy metals; flush the filtrate down the drain with a large amount of water. Spread the *sodium sulfate* on a tray in the hood to evaporate residual solvent from it, and put the used drying agent in the nonhazardous solid waste container. Flush the *aqueous layer* from the steam distillation down the drain.

Pre-Lab exercises for Section 16.2, "B. Preparation of Cyclohexanone," are found on page PL. 75.

B.   PREPARATION OF CYCLOHEXANONE

Prepare a solution of chromic acid by dissolving 7.9 g of potassium dichromate in 40 mL of water and then slowly adding, with swirling, 7 mL of concentrated sulfuric acid. Using an ice-water bath, cool this solution to room temperature. If a stock solution of chromic acid has been prepared for your use, obtain 50 mL of the solution. Add the solution of chromic acid in one portion to a mixture of 7.0 g of cyclohexanol and 25 mL of water in a 125-mL Erlenmeyer flask. Thoroughly mix the solutions and determine the temperature of the reaction mixture. Do *not* use the thermometer for purposes of stirring! The mixture should quickly become warm. When the temperature of the solution reaches 55 °C, periodically cool the flask in an ice-water bath or a pan of cold water if necessary to maintain a temperature of 55–60 °C. When the oxidation has proceeded to the point that the temperature of the solution does not rise above 60 °C, allow the flask to stand for 1 h and swirl it with occasionally; its temperature should drop during this time.∗

Transfer the reaction mixture to a 250-mL round-bottom flask, add 35 mL of water and boiling chips or a magnetic stirring bar if magnetic stirrers are available. Equip the flask for fractional distillation through an unpacked column (Fig. 2.16). Distil the mixture until approximately 35 mL of distillate, which should consist of an aqueous and an organic layer, has been collected.∗ Transfer the distillate to a separatory funnel, and saturate the aqueous layer by adding sodium chloride (about 0.2 g of salt per milliliter of water will be required) and swirling the mixture to effect solution. Separate the layers, and extract the aqueous layer with 5 mL of diethyl ether. Combine this extract with the organic layer, and dry the solution over anhydrous sodium sulfate.∗

Filter or decant the solution into a 50-mL round-bottom flask, and equip the flask for short-path distillation (Fig. 2.15b). Wrap the flask and stillhead with some insulating material such as a towel or glasswool to optimize the efficiency of the distillation; cool the collection flask in an ice-water bath. First distil (*no flames*) the low-boiling diethyl ether; then continue the distillation, collecting cyclohexanone as a colorless liquid boiling in the range of 152–155 °C (760 torr). Calculate the percentage yield of product obtained.

Following the procedure given in Section 23.5.1A, prepare the 2,4-dinitrophenylhydrazone of cyclohexanone (mp 162–163 °C). Also apply the chromic acid test described in Section 23.5.1C to your product.

Finishing Touches Add sodium sulfite to the *aqueous solution* of *chromium salts* in order to destroy excess Cr^{VI}. Make the solution slightly basic, to form *chromium hydroxide*, and isolate this salt by vacuum filtration. Place the *filter paper* and the *filter cake* in the container for heavy metals; flush the filtrate down the drain. Spread the *sodium sulfate* on a tray in the hood to evaporate residual solvent from it; then put the used drying agent in the nonhazardous solid waste container. Transfer the *diethyl ether* to the container for nonhalogenated organic liquids. Flush the *aqueous layer* from the steam distillation down the drain with a large amount of water.

Pre-Lab exercises for Section 16.2, "C. Oxidation of Cyclohexanone to Hexanedioic Acid," are found on page PL. 77.

C. OXIDATION OF CYCLOHEXANONE TO HEXANEDIOIC ACID

Combine 2.5 g of cyclohexanone and a solution of 7.9 g of potassium permanganate in 60 mL of water in a 125-mL Erlenmeyer flask. Add 0.5 mL of aqueous 3 *M* sodium hydroxide to this mixture, and note the temperature of the resulting solution. When the temperature of the solution reaches 45 °C, periodically cool the flask in an ice-water bath or a pan of cold water to maintain this temperature. When the oxidation has proceeded to the point that the temperature of the solution does not rise above 45 °C, allow the solution to stand for an additional 5–10 min; its temperature should drop during this time. Complete the oxidation by heating the mixture to boiling for a few minutes with a hot plate or a burner. Test the solution for the presence of permanganate ion by placing a drop of the reaction mixture on a piece of filter paper; any unchanged potassium permanganate will appear as a purple ring around the spot of brown manganese dioxide. If potassium permanganate remains, add small portions of solid sodium bisulfite to the reaction mixture until the spot test is negative.*

Filter the mixture with vacuum through a pad of a filter aid (see Sec. 2.16), thoroughly wash the brown filter cake with water,* and then concentrate the aqueous filtrate to a volume of about 15 mL by heating it with a hot plate or a burner.* If the concentrate is colored, add decolorizing carbon, reheat the solution to boiling for a few minutes, and then filter by gravity. Carefully add concentrated hydrochloric acid to the filtrate until the solution tests acidic to pHydrion paper; then add an additional 4 mL of acid.* Allow the solution to cool to room temperature and isolate the precipitated hexanedioic acid by vacuum filtration. The acid is a white solid (mp 152–153 °C). Recrystallize it from ethanol-water if necessary.

Finishing Touches Place any *filter papers* and the *filter cake* containing *manganese salts* in the container for toxic solids. Flush all *filtrates* down the drain. Dispose of the *filter paper* used for isolation of the product in the container for nontoxic solid waste.

Pre-Lab exercises for Section 16.2, "D. Oxidation of Cyclododecanol to Cyclododecanone," are found on page PL. 79.

D. OXIDATION OF CYCLODODECANOL TO CYCLODODECANONE

Place 0.5 g of cyclododecanol, 1.2 mL of acetone, and 0.35 mL of glacial acetic acid in a 25-mL round-bottom flask. Equip the flask with a magnetic stirring bar, if magnetic stirring is available, and a reflux condenser. Warm the mixture to approximately 45 °C with a steam or oil bath and maintain this temperature within ±5 °C throughout the course of the reaction. Stir or swirl the contents of the flask to hasten dissolution of the alcohol.

With continued stirring or swirling of the reaction mixture, add 4.6 mL of commercial bleach (ca. 5.3% sodium hypochlorite) dropwise through the top of the condenser over a period of about 0.5 h. Upon completing the addition, stop agitating the reaction mixture so the layers may separate, and then remove a small portion of the aqueous layer with the aid of a micropipet. Place a drop or two of this solution on a dampened piece of starch-iodide paper to determine whether sufficient hypochlorite has been added; this is necessary because the concentration of oxidant in bleach depends upon the age of the solution. The indicator paper will turn blue-black in color if sufficient bleach has been added; the color is due to formation of a complex between starch and the *iodine* produced by oxidation of *iodide* by hypochlorite. If this color does not develop, add an additional 0.4 mL of bleach to the reaction mixture. Stir or swirl the resulting mixture for 2–3 min and repeat the test for hypochlorite. Add additional 0.4-mL portions of bleach until a positive test for oxidant is observed. Then stir or swirl the reaction mixture for an additional 10 min and retest for hypochlorite. If the test is negative, add an additional 0.4 mL of bleach. Whether this last test is positive or negative, stir or swirl the mixture for 10 min more.

Allow the reaction mixture to cool to room temperature, add 3 mL of diethyl ether, and stir or swirl the two-phase mixture well to extract the desired product from the aqueous layer. Transfer the organic layer to a 25-mL round-bottom flask, extract the aqueous layer with an additional 3-mL portion of diethyl ether, and add this extract to the original one. Wash the combined organic extracts with 2.5 mL of saturated sodium bicarbonate, by first stirring the mixture with a glass rod until evolution of carbon dioxide ceases, then capping the flask and shaking it with frequent venting to relieve any pressure that might develop. Remove as much of the aqueous layer as possible with the aid of a micropipet,✶ and wash the organic solution sequentially with 2.5-mL portions of saturated sodium bisulfite and water. Transfer the organic solution to a 10-mL Erlenmeyer flask, and add 0.5–1 g of anhydrous sodium sulfate.✶ Occasionally swirl the mixture over a period of 10–15 min to facilitate drying, decant the dried solution into a tared 25-mL Erlenmeyer flask, and allow the diethyl ether to evaporate. Do this by warming the solution at 40–50 °C *(no flames)* under a stream of dry air in the hood or under an inverted funnel attached to the source of a vacuum (see Sec. 2.25). Alternatively, leave the flask in the hood until the solvent has evaporated.

Once the solvent has been removed, weigh the flask to determine the yield of solid cyclododecanone. The reported melting point of this ketone is 61–62 °C. Prepare the oxime (mp 131–132 °C) or semicarbazone (mp 218–219 °C) according to the procedures given in Section 23.5.1. If necessary, recrystallize the derivatives from methanol.

Finishing Touches   Flush all *aqueous solutions* down the drain. Spread the *sodium sulfate* on a tray in the hood to evaporate residual solvent from it and then put the used drying agent in the nonhazardous solid waste container.

EXERCISES

General Questions

1. Prove that no *oxidation* occurs at the carbon atoms in the transformations of Equations (16.10)–(16.13).

Questions for Part A

2. Show that *reduction* occurs at the carbonyl carbon atom in the reactions of Equations (16.14) and (16.15)

3. Demonstrate that oxidation occurs at the carbon atom of the organometallic reagent of Equation (16.14).

4. What is the purpose of washing the steam distillate with sodium carbonate in the preparation of 2-methylpropanal?

5. Write the structure of the 2,4-dinitrophenylhydrazone of 2-methylpropanal. Is it possible for geometric isomers to be produced in the formation of this product? Explain your answer.

6. When the preparation of 2-methylpropanal is being attempted, why is the chromic acid solution added to the alcohol rather than the reverse?

7. Why is the two-phase distillate obtained in the oxidation of 2-methyl-1-propanol saturated with sodium chloride before the organic layer is separated?

8. What modifications of this procedure might be made in the oxidation of 2-methyl-2-propanol with chromic acid to maximize the yield of isobutyl isobutyrate?

9. Write a mechanism for the formation of isobutyl isobutyrate in this experiment.

10. Write the balanced equation for the oxidation of 2-methyl-1-propanol to isobutyl isobutyrate by chromic acid in aqueous sulfuric acid. Show all work.

11. Recent prices of sodium dichromate dihydrate, potassium dichromate, and chromium anhydride (CrO_3) are \$15.00/lb, \$4.10/lb, and \$6.70/lb, respectively. Determine which of these reagents is the most economical source of chromic acid to use for oxidizing 1-methyl-1-propanol to 2-methylpropanal.

Questions for Part B

12. When the preparation of the ketone is being attempted, why is the chromic acid solution added to the alcohol rather than the reverse?

13. Why is the two-phase distillate obtained in the oxidation of cyclohexanol saturated with sodium chloride before the organic layer is separated?

14. Write the structure of the 2,4-dinitrophenylhydrazone of cyclohexanone. Is it possible for geometric isomers to be produced in the formation of this product? Explain your answer.

15. Write out the balanced equation for the oxidation of cyclohexanol to cyclohexanone by potassium dichromate in aqueous sulfuric acid. Show all work.

16. Recent prices of sodium dichromate dihydrate, potassium dichromate, and chromium anhydride (CrO_3) are \$15.00/lb, \$4.10/lb, and \$6.70/lb, respectively. Determine which of these reagents is the more economical source of chromic acid to use for oxidation of cyclohexanol to cyclohexanone.

Questions for Part C

17. In the oxidation of cyclohexanone to hexanedioic acid, why is the reaction mixture made alkaline?

18. Why does acidification of the concentrated reaction mixture cause precipitation of the hexanedioic acid?

19. Write the balanced equation for the oxidation of cyclohexanone to hexanedioic acid with basic potassium permanganate. Show all work.

20. Give the products to be expected on oxidation of 2-methylcyclohexanone with alkaline potassium permanganate.

Questions for Part D

21. What is the function of sodium bisulfite in the procedure for isolating cyclododecanone?

22. Would cyclododecanone be expected to give a positive iodoform test (see Sec. 23.5.1D)? Explain.

23. Explain why bottles of commercial bleach that have been open to the atmosphere for long periods of times are likely to have lower concentrations of aqueous sodium hypochlorite than are freshly opened bottles.

24. Determine whether the conversion of iodide ion to iodine is an oxidation or a reduction.

SPECTRA OF STARTING MATERIALS AND PRODUCTS

The IR and NMR spectra of 2-methyl-1-propanol are provided in Figure 8.55. Spectra of cyclohexanol and cyclohexanone are presented in Figures 10.23 and 10.24, and in Figures 13.3 and 13.4, respectively.

FIGURE 16.1 IR spectrum of 2-methyl-propanal (isobutyraldehyde).

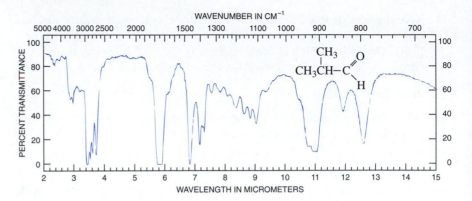

(a) ^{1}H NMR spectrum (60 MHz).

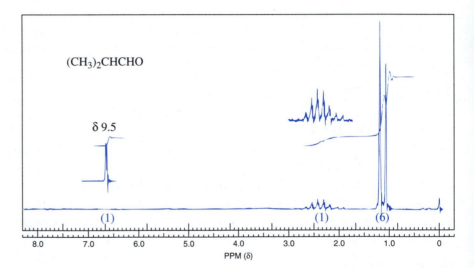

FIGURE 16.2 NMR data for 2-methyl-propanal (isobutyraldehyde).

(b) ^{13}C NMR data.
Chemical shifts: δ 15.5, 41.2, 204.7.

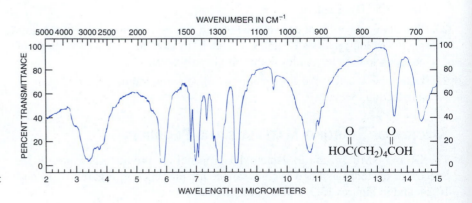

FIGURE 16.3 IR spectrum of hexanedioic acid (adipic acid) (KBr pellet).

(a) ¹H NMR spectrum (90 MHz).

FIGURE 16.4   NMR data for hexanedioic acid (adipic acid).

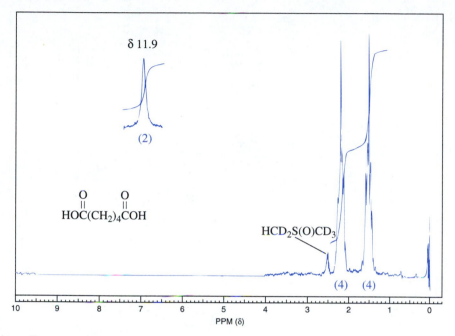

δ 11.9

(2)

$$\underset{\substack{\\ \text{O} \\ \| \\}}{\text{HOC}}(\text{CH}_2)_4\underset{\substack{\\ \text{O} \\ \| \\}}{\text{COH}}$$

HCD₂S(O)CD₃

(4)   (4)

10   9   8   7   6   5   4   3   2   1   0

PPM (δ)

(b) ¹³C NMR data.

Chemical shifts: δ 24.3, 33.6, 174.8.

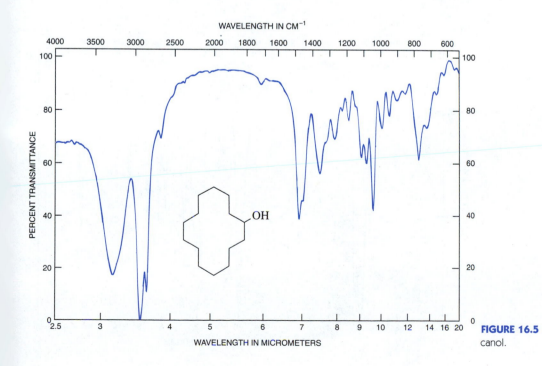

WAVELENGTH IN CM⁻¹

4000   3500   3000   2500   2000   1800   1600   1400   1200   1000   800   600

PERCENT TRANSMITTANCE

OH

2.5   3   4   5   6   7   8   9   10   12   14 16 20

WAVELENGTH IN MICROMETERS

FIGURE 16.5   IR spectrum of cyclododecanol.

FIGURE 16.6 NMR data for cyclodode-canol.

(a) ^{1}H NMR spectrum (90 MHz).

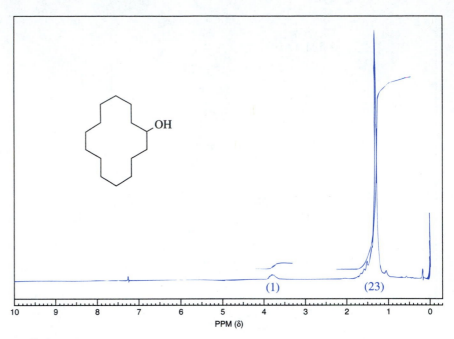

(b) ^{13}C NMR data.
Chemical shifts: δ 21.0, 23.5, 24.0, 24.4, 32.5, 69.1.

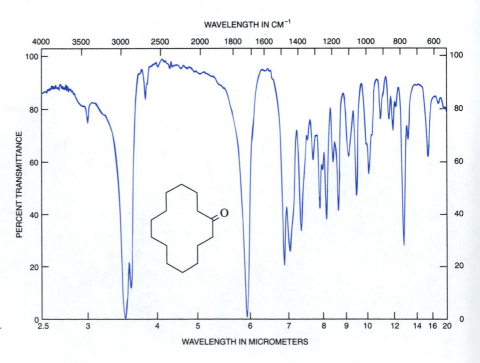

FIGURE 16.7 IR spectrum of cyclodode-canone.

(a) ^{1}H NMR spectrum (90 MHz).

FIGURE 16.8   NMR data for cyclodode-canone.

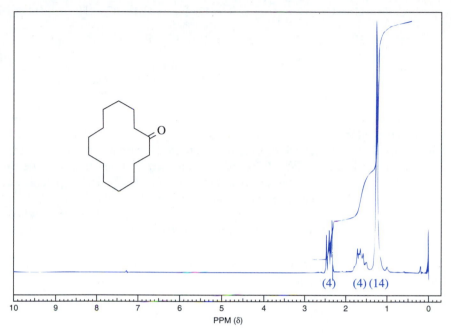

(4)   (4) (14)

(b) ^{13}C NMR data.

Chemical shifts: δ 22.7, 24.4, 24.9, 40.4, 211.4.

16.3   BASE-CATALYZED OXIDATION-REDUCTION OF ALDEHYDES: THE CANNIZZARO REACTION

Aldehydes having no hydrogens on the **α-carbon atom,** the carbon atom adjacent to the carbonyl group, undergo mutual oxidation and reduction in the presence of strong alkali (Eq. 16.34). In contrast, aldehydes with hydrogen atoms on the **α-carbon atom** preferentially undergo other types of base-promoted reactions such as the **aldol condensation** (Eq. 16.35), which is described in Section 18.2. The mutual oxidation-reduction, called the **Cannizzaro reaction,** is a consequence of the fact that an aldehyde is intermediate in oxidation state between an alcohol and a carboxylic acid and is converted into either one by a decrease or gain, respectively, of two in the oxidation number of its carbonyl carbon atom, as indicated in Equation (16.34).

$$
2\,R-\underset{\underset{R}{|}}{\overset{\overset{R}{|}}{C}}-\overset{O}{\overset{\|}{\underset{+1}{C}}}-H \xrightarrow[\text{2. H}_3\text{O}^+]{\text{1. HO}^-} R-\underset{\underset{R}{|}}{\overset{\overset{R}{|}}{C}}-\underset{\underset{H}{|}}{\overset{O-H}{\overset{|}{\underset{-1}{C}}}}-H \; + \; R-\underset{\underset{R}{|}}{\overset{\overset{R}{|}}{C}}-\overset{O}{\overset{\|}{\underset{+3}{C}}}-OH \qquad \textbf{(16.34)}
$$

An aldehyde   An alcohol   A carboxylic acid

$$2\ R-\overset{\overset{\displaystyle H}{|}}{\underset{\underset{\displaystyle H}{|}}{C}}-\overset{\overset{\displaystyle O}{\|}}{C}-H \xrightarrow[\text{2. }H_3O^+]{\text{1. }HO^-} R-\overset{\overset{\displaystyle H}{|}}{\underset{\underset{\displaystyle H}{|}}{C}}-\overset{\overset{\displaystyle OH}{|}}{\underset{\underset{\displaystyle H}{|}}{C}}-\overset{}{\underset{\underset{\displaystyle R}{|}}{CH}}-\overset{\overset{\displaystyle O}{\|}}{C}-H \qquad (16.35)$$

An aldol

(β-hydroxycarbonyl compound)

The accepted mechanism of the reaction follows logically from the ease with which nucleophiles add to the carbonyl group, particularly that of an aldehyde. The first step is attack of hydroxide ion on the carbonyl group to give a tetrahedral intermediate (Eq. 16.36); this is followed by transfer of hydride to the carbonyl group of another aldehyde function (Eq. 16.37). The step depicted in Equation (16.37) is the one in which both oxidation and reduction occur, as is shown by determining the oxidation numbers (see Sec. 16.1) of the carbon atoms involved (see Exercise 1 at the end of this section). The remaining steps (Eqs. 16.38 and 16.39) simply illustrate the types of acid-base chemistry expected to occur in the strongly basic medium in which the reaction is performed. Summation of Equations (16.36)–(16.39) gives Equation (16.34).

$$\underset{R}{\overset{O}{\underset{}{\diagup}}}\overset{\|}{C}\underset{H}{\diagdown} + {}^{\ominus}OH \longrightarrow R-\overset{\overset{\displaystyle O^{\ominus}}{|}}{\underset{\underset{\displaystyle OH}{|}}{C}}-H \qquad (16.36)$$

$$R-\overset{\overset{\displaystyle O^{\ominus}}{|}}{\underset{\underset{\displaystyle OH}{|}}{C}}-H + \underset{R}{\overset{O}{\diagup}}\overset{\|}{C}\underset{H}{\diagdown} \longrightarrow \underset{R}{\overset{O}{\diagup}}\overset{\|}{C}\underset{OH}{\diagdown} + R-\overset{\overset{\displaystyle O^{\ominus}}{|}}{\underset{\underset{\displaystyle H}{|}}{C}}-H \qquad (16.37)$$

$$\underset{R}{\overset{O}{\diagup}}\overset{\|}{C}\underset{O-H}{\diagdown} + {}^{\ominus}OH \longrightarrow \underset{R}{\overset{O}{\diagup}}\overset{\|}{C}\underset{O^{\ominus}}{\diagdown} + H-OH \qquad (16.38)$$

$$R-\overset{\overset{\displaystyle O^{\ominus}}{|}}{\underset{\underset{\displaystyle H}{|}}{C}}-H + H-OH \longrightarrow R-\overset{\overset{\displaystyle O-H}{|}}{\underset{\underset{\displaystyle H}{|}}{C}}-H + {}^{\ominus}OH \qquad (16.39)$$

Aromatic aldehydes, formaldehyde, and trisubstituted acetaldehydes all undergo the Cannizzaro reaction. In the experiment of this section, benzaldehyde, an aromatic aldehyde, is converted to benzyl alcohol and potassium benzoate, which gives benzoic acid upon acidification of the reaction mixture (Eq. 16.40).

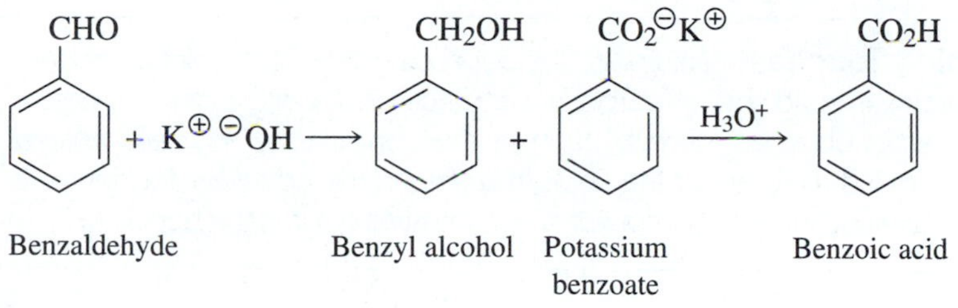

$$\text{Benzaldehyde} + K^{\oplus}{}^{\ominus}OH \longrightarrow \text{Benzyl alcohol} + \text{Potassium benzoate} \xrightarrow{H_3O^+} \text{Benzoic acid} \quad (16.40)$$

Pre-Lab exercises for Section 16.3, "Base-Catalyzed Oxidation-Reduction of Aldehydes: The Cannizzaro Reaction," are found on page PL. 81.

EXPERIMENTAL PROCEDURE

SAFETY ALERT

The solution of potassium hydroxide is corrosive. Should it come in contact with your skin, flood the affected area immediately with water and then rinse with 1% acetic acid.

Dissolve 1 g of solid potassium hydroxide in 1 mL of distilled water by swirling it in a 10-mL beaker or Erlenmeyer flask; then cool the mixture to room temperature. Put 1 mL of benzaldehyde in an 18-mm × 150-mm test tube and add the concentrated potassium hydroxide solution. Cork the tube securely and shake the tube vigorously until an emulsion is formed. Allow the stoppered tube to stand in your desk until the next laboratory period. Crystallization should occur in the interim.

Add about 0.5 mL of water to the mixture, stopper the tube, and shake it. If all the crystals originally present do not dissolve, add a little more water, break up the solid mass with a glass rod, stopper the tube and again shake it. Repeat this procedure until the solid is all in solution. Extract this solution three times with 5-mL portions of diethyl ether, removing the organic layer with the aid of a micropipet; shake the mixture gently to avoid formation of an emulsion.

If possible, dry the ethereal solution over sodium sulfate and examine it by GLC to determine the proportion of benzyl alcohol and unchanged benzaldehyde present. A 1.5-m column packed with 5% silicone gum rubber as the stationary phase is satisfactory for this analysis. In one experiment about 20% of the original benzaldehyde remained after 24 h at room temperature.

Pour the alkaline aqueous solution into a mixture of 2.5 mL of concentrated hydrochloric acid, 2 mL of water, and about 3 g of crushed ice. Stir the mixture vigorously during the addition. Cool the reaction mixture in an ice-water bath and collect the crystalline benzoic acid.✱ Dry this product and determine its melting point. Save a sample of the benzoic acid for comparison with the product of the experiment of Section 16.4.

> **Finishing Touches**   Neutralize the *aqueous filtrate* from isolation of benzoic acid with solid sodium carbonate and flush it down the drain. Place the *diethyl ether* obtained by distillation in the container for nonhalogenated organic liquids. Dispose of the *sodium sulfate* in the container for nontoxic solids after allowing residual solvent to evaporate from it in the hood.

EXERCISES

1. Confirm that oxidation and reduction of benzaldehyde occur in the Cannizzaro reaction by finding the changes in oxidation numbers associated with the first step of Equation (16.40).

2. When the Cannizzaro reaction is carried out on benzaldehyde in D_2O solution, no deuterium becomes attached to carbon in the benzyl alcohol formed. How does this support the mechanism given in Equations (16.36)–(16.39)?

3. By what means would aqueous sodium bisulfite remove unchanged benzaldehyde from the reaction mixture?

4. What is the solid that is formed after benzaldehyde has been allowed to react with aqueous potassium hydroxide?

5. Write an equation for the reaction of a mixture of benzaldehyde and formaldehyde with concentrated potassium hydroxide solution followed by acidification of the reaction mixture. Show all organic products.

6. The Cannizzaro reaction occurs much more slowly in dilute than in concentrated potassium hydroxide solution. Why is this?

7. How would propanal react with potassium hydroxide solution under the conditions of this experiment? Answer this question for the reaction of 2,2-dimethylpropanal.

8. Predict the products that would be formed by Cannizzaro reaction between equimolar amounts of *p*-chlorobenzaldehyde and *p*-bromobenzaldehyde.

SPECTRA OF STARTING MATERIAL AND PRODUCTS

The IR spectrum of benzyl alcohol is given in Figure 8.11. The IR and NMR spectra of benzoic acid are presented in Figures 3.1 and 3.2, respectively.

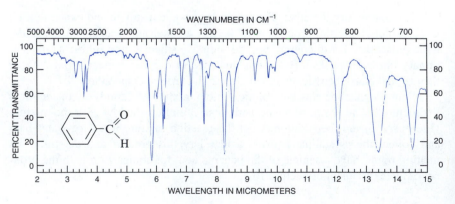

FIGURE 16.9   IR spectrum of benzaldehyde.

(a) ^{1}H NMR spectrum (60 MHz).

FIGURE 16.10   NMR data for benzalde-hyde.

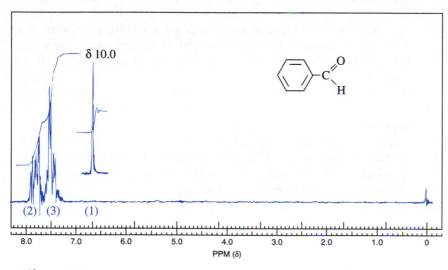

(b) ^{13}C NMR data.
    Chemical shifts: δ 129.0, 129.7, 134.4, 136.6, 192.0.

(a) ^{1}H NMR spectrum (90 MHz).

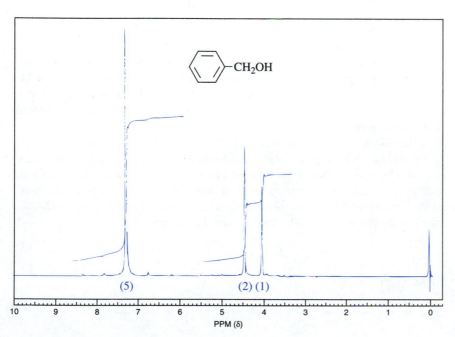

(b) ^{13}C NMR data.
    Chemical shifts: δ 64.5, 126.9, 127.2, 128.3, 141.0.

FIGURE 16.11   NMR data for benzyl alcohol.

16.4 CARBOXYLIC ACIDS

Carboxylic acids are synthesized in a variety of ways, including oxidation of an aldehyde or ketone (see Sec. 16.2) and hydrolysis of nitriles, esters, and amides. The carbonation of a Grignard reagent (see Sec. 19.3.2) is another useful synthetic route to such acids.

A special method available for the preparation of *aryl* carboxylic acids involves the oxidation of the alkyl group of an alkyl-substituted benzene with alkaline permanganate or with chromic acid solution (Eq. 16.41). Carbon-carbon as well as carbon-hydrogen bonds are cleaved in this process. Moreover, if more than one alkyl side chain is present, each will be oxidized to a carboxyl group, and a polycarboxylic acid results (Eq. 16.42). Identification of the mono- or polycarboxylic acid obtained from this side-chain oxidation thus allows the positions of attachment of the side chains on the aromatic ring to be determined.

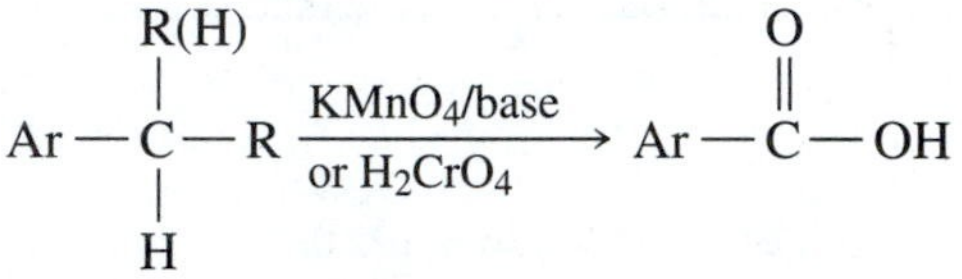

$$
\underset{\substack{\text{An aromatic}\\\text{hydrocarbon}}}{\text{Ar} - \overset{\displaystyle R(H)}{\underset{\displaystyle H}{\overset{|}{\underset{|}{C}}}} - R} \quad \xrightarrow[\text{or } H_2CrO_4]{KMnO_4/base} \quad \underset{\substack{\text{An aromatic}\\\text{carboxylic acid}}}{\text{Ar} - \overset{\displaystyle O}{\overset{\|}{C}} - OH} \qquad \textbf{(16.41)}
$$

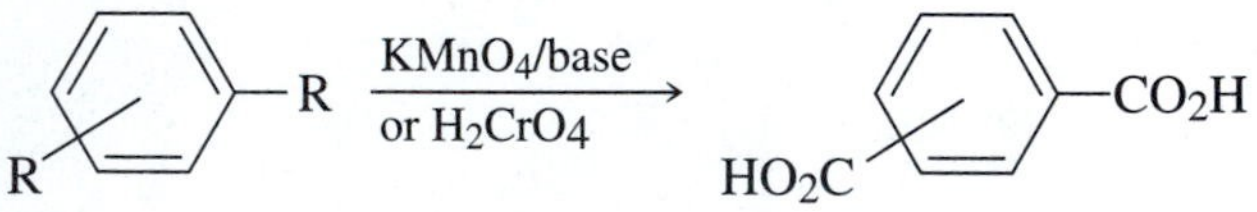

$$
\underset{(o,\ m,\ \text{or}\ p)}{R - \!\!\!\bigcirc\!\!\!- R} \quad \xrightarrow[\text{or } H_2CrO4]{KMnO_4/base} \quad \underset{(o,\ m,\ \text{or}\ p)}{HO_2C - \!\!\!\bigcirc\!\!\!- CO_2H} \qquad \textbf{(16.42)}
$$

The mechanism of this type of oxidation is not well understood, but it may involve removal of a benzylic hydrogen atom in a free-radical process initiated by permanganate ion. Supporting this idea is the fact that the benzylic carbon atom must bear at least one hydrogen atom for oxidation of the side-chain to be efficient. Subsequent reactions lead first to the alcohol and then to the ketone, which is further oxidized to the carboxylic acid (Eq. 16.43). The mechanism of oxidation of the alcohol is analogous to that developed when chromic acid is the oxidant (Eq. 16.17), whereas that of the final oxidation step parallels the one given for the oxidation of cyclohexanone to hexanedioic acid (see Sec. 16.2). The stoichiometry of the reaction is determined by using the methods developed in Section 16.1 and by recalling Equation (16.28). (See Exercise 1 at the end of this section.)

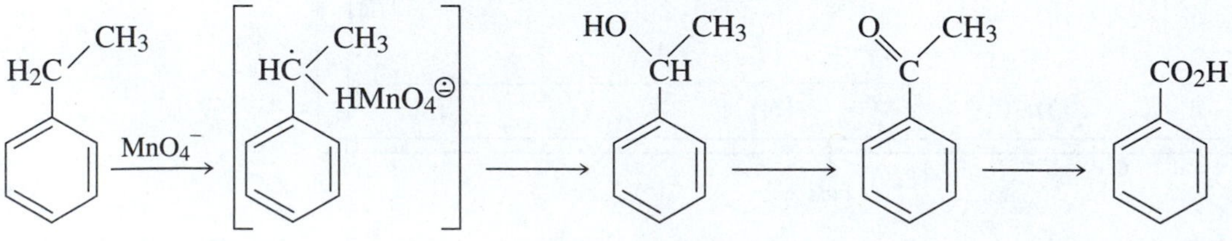

Ethylbenzene   1-Phenylethanol   Acetophenone   Benzoic acid

$$\textbf{(16.43)}$$

In the side-chain oxidation described in this experiment, ethylbenzene is converted to benzoic acid by reaction with alkaline potassium permanganate (Eq. 16.43). The sodium carbonate produced in the process arises from reaction of carbon dioxide with aqueous sodium hydroxide. The carbon dioxide itself is derived from the methyl group lost by the oxidation of the intermediate acetophenone.

Pre-Lab exercises for Section 16.4, "Carboxylic Acids," can be found on page PL. 83.

EXPERIMENTAL PROCEDURE

SAFETY ALERT

1. When preparing and handling solutions of potassium permanganate, wear rubber gloves and avoid contact of the solutions with your skin. These solutions will cause unsightly stains on your hands that persist for several days. If this oxidant comes in contact with your skin, wash the affected area thoroughly with warm water.
2. Residues of manganese dioxide can be difficult to remove from glassware. Rinse stained glassware with 10% sodium bisulfite solution; if this fails, add 6 M HCl to the flask and heat it in a hood until the stain has disappeared.

Place 5.6 g of potassium permanganate, 60 mL of water, 0.8 mL of 3 M aqueous sodium hydroxide, and 0.9 g of ethylbenzene in a 250-mL round-bottom flask fitted with a reflux condenser. Add boiling chips or a magnetic stirring bar if magnetic stirrers are available. *Gently* heat this mixture at reflux for at least 2 h (or up to 3.5 h if time permits). Severe bumping occurs if the mixture is heated too strongly.＊ Test the hot solution for unchanged permanganate by placing a drop of the reaction mixture on a piece of filter paper; if a purple ring appears around the brown spot of manganese dioxide, potassium permanganate remains. Destroy any excess permanganate ion by adding small amounts of solid sodium bisulfite to the mixture until the spot test is negative. Do not add a large excess of bisulfite. Add 1 g of a filter-aid to the hot mixture to promote its rapid filtration (Secs. 2.12 and 2.13)—the manganese dioxide formed in the reaction is finely dispersed and would otherwise clog the pores of the filter paper. Filter the mixture by vacuum, and rinse the reaction flask and filter cake with two 5-mL portions of hot water.＊

Concentrate the filtrate to about 15 mL by simple distillation. If the concentrate is turbid, filter it by gravity; the filtrate should be clear.＊ Acidify the aqueous residue with concentrated hydrochloric acid, adding acid slowly until no more benzoic acid separates; about 3–5 mL of acid may be required.＊ Cool this mixture in an ice-water bath, and isolate the benzoic acid by vacuum filtration. Purify the crude acid by recrystallization from hot water or by sublimation. If the latter procedure is used, the crude acid must be dry. Determine the yield and melting point of the benzoic acid obtained.

> **Finishing Touches**   Place any *filter papers* and the *filter cake* containing *manganese salts* in the container for toxic solids. Neutralize all *filtrates* as necessary and flush them down the drain. Dispose of the *filter paper* used for isolation of the product in the container for nontoxic solid waste.

EXERCISES

1. Write a balanced equation for conversion of ethylbenzene and permanganate to benzoic acid and manganese dioxide in the presence of hydroxide ion.

2. Would it be feasible to prepare benzaldehyde by oxidation of toluene with permanganate? Explain your answer.

3. Would you expect a carboxylic acid to be more or less soluble in water than in water containing mineral acids? Explain.

4. *tert*-Butylbenzene is *not* oxidized to benzoic acid by permanganate ion. Why not?

5. Write a balanced equation for the reaction of sodium bisulfite with potassium permanganate.

6. What effect on yield of isolated product is expected if the concentration step is omitted prior to acidification? Explain.

SPECTRA OF STARTING MATERIAL AND PRODUCT

The IR and NMR spectra for ethylbenzene are presented in Figures 9.10 and 9.11, and those for benzoic acid are given in Figures 3.1 and 3.2.

REDUCTION REACTIONS OF DOUBLE BONDS

ALKENES, CARBONYL COMPOUNDS, AND IMINES

17.1 INTRODUCTION

The general term **reduction** refers to those reactions in which hydrogen or electrons are added to a functional group present in a reactant; it is the *opposite* of **oxidation,** which involves the addition of oxygen or the net loss of electrons from a substrate. In organic chemistry, *reduction results in a decrease in the oxidation state of carbon and hence a net gain of hydrogen or electrons by the carbon atom.* For example, from the discussion in Section 16.1, we know that the **oxidation number** of each carbon atom in ethylene, $CH_2=CH_2$, is –2; however, in ethane, CH_3CH_3, the oxidation number is –3. Thus, conversion of ethylene to ethane involves a net gain of 1 e^- by *each* carbon atom and is therefore a reduction.

Organic functional groups containing double and triple bonds undergo reduction by the *net* addition of the elements of one or two molecules of hydrogen across the π-bond(s). The starting compounds in these processes are termed **unsaturated,** whereas the final products are **saturated** with respect to hydrogen, since they contain only single bonds and can absorb no more hydrogen. Equations (17.1)–(17.4) illustrate some types of reduction reactions that are commonly performed by organic chemists. The symbol [**H**] is generally used to represent the **catalytic** and **chemical** methods that are used to effect reduction. For example, a catalytic reduction involves adding hydrogen to a carbon-carbon or carbon-heteroatom double or triple bond in the presence of metal **catalysts** such as platinum, palladium, nickel, or rhodium. The selection of the best catalyst for a specific reduction is sometimes difficult, but this problem is typically solved by analogy with similar reactions that have been reported in the literature. Chemical methods of reduction include the use of hydride-donating reagents such as lithium aluminum hydride, $LiAlH_4$, and sodium borohydride, $NaBH_4$, and electron-transfer processes such as dissolving metal reductions. In addition to the functional groups shown in Equations (17.1)–(17.4), many other functional groups containing π-bonds can be reduced to form products containing only single σ-bonds. These include carboxylic acids and their derivatives (esters, amides, anhydrides, and nitriles), nitro groups, and aromatic rings.

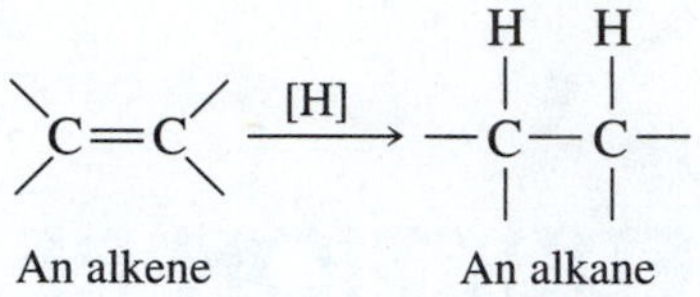

$$\underset{\text{An alkene}}{\overset{}{\text{C=C}}} \xrightarrow{\text{[H]}} \underset{\text{An alkane}}{\overset{\text{H}\ \text{H}}{-\text{C}-\text{C}-}} \tag{17.1}$$

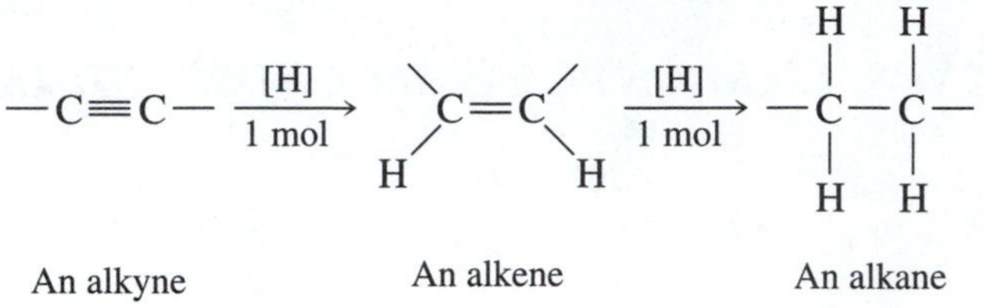

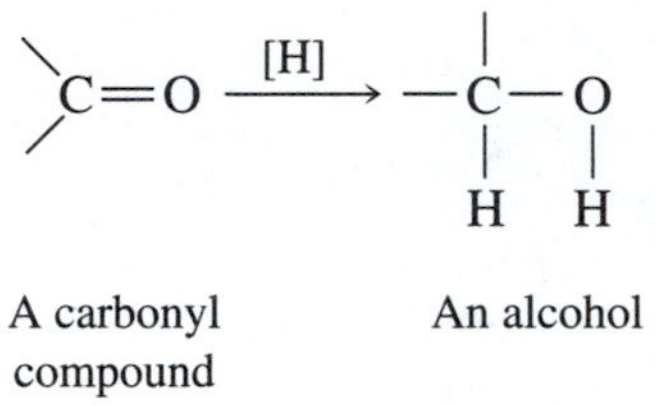

$$\underset{\text{An alkyne}}{-\text{C}\equiv\text{C}-} \xrightarrow[\text{1 mol}]{\text{[H]}} \underset{\text{An alkene}}{\overset{}{\text{C=C}}} \xrightarrow[\text{1 mol}]{\text{[H]}} \underset{\text{An alkane}}{\overset{\text{H}\ \text{H}}{-\text{C}-\text{C}-}} \tag{17.2}$$

$$\underset{\substack{\text{A carbonyl}\\\text{compound}}}{\overset{}{\text{C=O}}} \xrightarrow{\text{[H]}} \underset{\text{An alcohol}}{\overset{}{-\text{C}-\text{O}}} \tag{17.3}$$

$$\underset{\text{An imine}}{\overset{}{\text{C=N}}} \xrightarrow{\text{[H]}} \underset{\text{An amine}}{\overset{}{-\text{C}-\text{N}}} \tag{17.4}$$

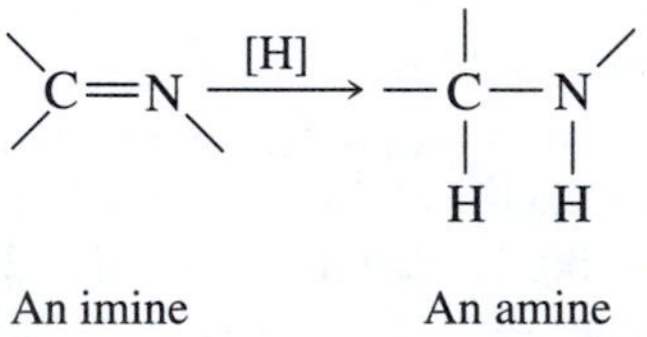

Catalytic hydrogenation is widely used in industry and in the research laboratory to reduce the π-bonds in a variety of functional groups. Analytical methods have been developed in which catalytic hydrogenations are performed on a small scale, and the amounts of hydrogen consumed and the heat evolved are accurately measured. Using such techniques, the degree of unsaturation in the form of the number of double and/or triple bonds in a compound is easily ascertained. The relative stabilities of different alkenes may also be determined from their heats of hydrogenation. Catalytic hydrogenation is frequently used in synthetic organic chemistry to prepare novel compounds that may be biologically active or exhibit interesting structural or physical properties. The catalytic hydrogenation of a compound containing a carbon-carbon double bond is performed in the experimental procedure in Section 17.2.

The **chemical reduction** of a number of organic functional groups is commonly effected using **metal hydride** reducing agents, which have different reactivities toward specific functional groups. This property is of considerable practical importance because one of the challenges in contemporary synthetic organic chemistry is developing new reducing reagents that react *selectively* with only one type of functional group in the presence of other reducible groups.

For example, lithium aluminum hydride, $LiAlH_4$, is *not* a selective reducing agent; as a very reactive hydride donor, it reduces aldehydes, ketones, and esters with equal ease. Sodium borohydride, $NaBH_4$, on the other hand, is less reactive and hence more selective; it reduces aldehydes and ketones but *not* esters. Not only is chemical selectivity a concern in choosing the appropriate reagent, but experimental factors such as ease of handling and safety often play important roles. Thus, if both lithium aluminum hydride and sodium borohydride could be used to effect a specific reduction, the latter is preferred because it is safer and easier to handle. Section 17.3 includes experiments for several chemical reductions that utilize sodium borohydride.

The nitro group, $-NO_2$, is reduced to the amino group, $-NH_2$, by catalytic hydrogenation or by using lithium aluminum hydride. However, this conversion is also accomplished by an alternative method of chemical reduction known as **dissolving metal reduction.** These reductions are typically performed either using finely divided iron or tin with hydrochloric acid or an alkaline earth metal such as lithium, sodium, and potassium in liquid ammonia. The reduction of 1,3-dimethylanthraquinone into a mixture of dimethylanthrols using $SnCl_2$ in hydrochloric acid (see Sec. 15.4), and the reduction of nitrobenzene, $C_6H_5NO_2$, to aniline, $C_6H_5NH_2$, using tin and hydrochloric acid (see Sec. 20.2.2) are examples of such reductions.

17.2 CATALYTIC HYDROGENATION OF THE CARBON-CARBON DOUBLE BOND

Catalytic hydrogenation of alkenes belongs to the general class of reactions known as **addition reactions,** which were introduced in Section 10.3. In these reactions, one mole of hydrogen is *stereospecifically* added to the *same* side of the carbon-carbon bond, as exemplified by the hydrogenation of 1,2-dimethylcyclopentene (**1**) to give *cis*-1,2-dimethylcyclopentane (**2**) (Eq. 17.5); such processes are termed *syn*-**additions.** Although the detailed mechanism by which this reaction occurs is not fully understood, it is generally believed that hydrogen gas is adsorbed on the surface of the finely divided metal catalyst to produce hydrogen atoms. The π-electrons of the double bond also complex with the surface of the catalyst, and the hydrogen atoms are transferred from the surface to the carbon atoms to form the saturated product. Catalytic hydrogenation is the method of choice for reducing alkenes to alkanes and alkynes to *cis*-alkenes.

(17.5)

| **1** | **2** |
| 1,2-Dimethylcyclopentene | *cis*-1,2-Dimethylcyclopentane |

Catalytic hydrogenation is accomplished in several ways in the laboratory. Generally, the appropriate metal catalyst is suspended in a solution containing the reactant in a suitable solvent and the resulting mixture is stirred or shaken under an atmosphere of hydrogen gas at pressures ranging from 1 to 60 atmospheres. Hydrogen, a highly flammable gas, is normally purchased in pressurized gas cylinders, and special apparatus and precautions must be used to deliver the hydrogen from the cylinder to the apparatus. It is more convenient in simple experiments to generate hydrogen gas *in situ*, and the reaction of sodium borohydride with concentrated hydrochloric acid provides a ready source of hydrogen via the union of H^+ from HCl with the hydride ions, $H:^-$, contained in $NaBH_4$ (Eq. 17.6). The catalyst, platinum on carbon, may be prepared in the same flask by reduction of chloroplatinic acid, H_2PtCl_6, with hydrogen in the presence of decolorizing carbon, which serves as a solid support for the finely divided metallic platinum produced (Eq. 17.7).

$$NaBH_4 + HCl + 3\,H_2O \longrightarrow 4\,H_2 + B(OH)_3 + NaCl \qquad (17.6)$$

$$H_2PtCl_2 + 2\,H_2 \longrightarrow Pt^\circ + 6\,HCl \qquad (17.7)$$

The experiment that follows involves the conversion of 4-cyclohexene-*cis*-1,2-dicarboxylic acid (**3**) to cyclohexane-*cis*-1,2-dicarboxylic acid (**4**), as shown in Equation (17.8). The catalytic hydrogenation of **3** is selective, because the carboxylic acid groups, –CO_2H, are *not* reduced under the reaction conditions. Moreover, the relative stereochemistry of the two carboxyl groups remains unchanged even though the *trans* diacid is more stable. The starting material for this reduction may be simply prepared by the hydrolysis of 4-cyclohexene-*cis*-1,2-dicarboxylic acid anhydride, which is the product of the Diels-Alder reaction between butadiene and maleic anhydride, according to the procedure outlined in Section 12.3, Part A.

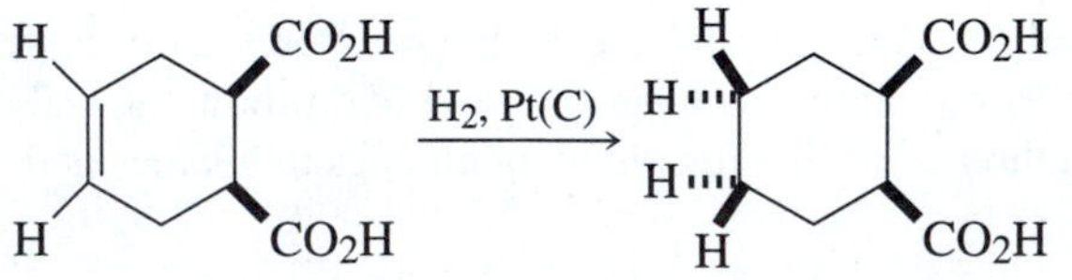

$$(17.8)$$

3
4-Cyclohexene-*cis*-
1,2-dicarboxylic acid

4
Cyclohexane-*cis*-
1,2-dicarboxylic acid

Pre-Lab exercises for Section 17.2, "Hydrogenation of 4-Cyclohexene-*cis*-1,2-dicarboxylic Acid," are found on PL. 85.

SAFETY ALERT

1. *Hydrogen gas is extremely flammable.* Use no flames in this experiment, and make certain that there are no open flames nearby.
2. The platinum catalyst prepared in this experiment may be *pyrophoric* and combust spontaneously in air. Do *not* allow it to become dry!
3. When removing the diethyl ether by simple distillation, use a steam bath or hot-water bath; *do not use a hot plate or a flame.*
4. Wear rubber gloves throughout the experiment to avoid skin contamination by the solutions you are using. If you do get these solutions on your hands, wash them thoroughly with water. In the case of contact with the sodium borohydride solution, rinse the affected areas with 1% acetic acid solution. In the case of acid burns, apply a paste of sodium bicarbonate to the area for a few minutes and then rinse with large amounts of water.

Prepare a reaction vessel for hydrogenation by using rubber bands or wire to attach a heavy-walled balloon to the sidearm of a 50-mL filter flask. If magnetic stirrers are available, place a magnetic stirring bar in the flask. Also prepare a 1 M aqueous solution of sodium borohydride by dissolving 0.2 g of sodium borohydride in 5 mL of 1% aqueous sodium hydroxide. Place 5 mL of water, 0.5 mL of a 5% solution of chloroplatinic acid ($H_2PtCl_6 \cdot 6H_2O$) and 0.2 g of decolorizing carbon in the reaction flask and slowly add 1.5 mL of the 1 M sodium borohydride solution with swirling or stirring. Let the resulting slurry stand for 5 min to allow the catalyst to form. During this time dissolve 0.5 g of 4-cyclohexene-*cis*-1,2-dicarboxylic acid (see Sec. 12.3, Part A) by heating it with 5 mL of water.

Pour 2 mL of *concentrated* hydrochloric acid into the reaction flask containing the catalyst; add the hot aqueous solution of the diacid to the flask. Seal the flask with a serum cap, and wire the cap securely in place. Draw 0.8 mL of the 1 M sodium borohydride solution into a plastic syringe, push the needle of the syringe through the serum cap, and inject the solution dropwise while swirling the flask. If the balloon on the sidearm of the flask becomes inflated, stop adding sodium borohydride solution until the balloon deflates. When the syringe is empty, remove it from the flask and refill it with an additional 0.8 mL of 1 M sodium borohydride solution. The addition should cause the balloon to inflate somewhat, indicating a positive pressure of hydrogen in the system. Remove the syringe from the serum cap, and allow the flask to stand, occasionally swirling or continuing to stir its contents for 5 min. Then heat and swirl the flask on a steam cone for about 15–20 min to complete the reaction.

Release the pressure from the reaction flask by pushing a syringe needle through the serum cap. Remove the catalyst by vacuum filtration of the hot reaction mixture using a Hirsch funnel;* treat the catalyst according to the protocol given in the FINISHING TOUCHES section. Cool the filtrate and saturate it with solid sodium chloride. Extract the filtrate with three 15-mL portions of diethyl ether; dry the combined ethereal extracts by swirling them over solid anhydrous sodium sulfate. Carefully decant the solution into a tared round-bottom flask, and remove the diethyl ether by simple distillation *(no flames)*; remove the final traces of solvent by connecting the flask to a water aspirator. The product remains as a solid residue in the flask.✷ After the crude product is dry, determine its weight and melting point.

Recrystallize the crude product from a *minimum* amount of water as follows. Carefully scrape the bulk of the crude diacid from the round-bottom flask into a 10-mL Erlenmeyer flask. Add no more than 1 mL of water to the round-bottom flask, heat to dissolve any residual diacid, and pour the hot solution into the Erlenmeyer flask. Heat the aqueous mixture to boiling; if necessary, bring all the diacid into solution by *slowly* adding water dropwise as needed. Do not add more than 1 mL of water, as the diacid is somewhat soluble in water. Determine the pH of the aqueous solution using pHydrion paper to ascertain whether hydrogenation has affected the acidic nature of the molecule. When all of the diacid has dissolved, add 2 drops of *concentrated* hydrochloric acid to decrease its solubility. Allow the mixture to cool to room temperature, and then place the flask in an ice-water bath to effect complete crystallization of the product.✷ Isolate the product and determine its melting point and yield. The reported melting point is 192–193 °C. Perform the qualitative tests for unsaturation described in Section 23.5.2 on your product and record your observations in your notebook.

Finishing Touches   Wet the *catalyst* and the *filter paper* immediately with water, and remove them from the Hirsch funnel; place them in the container reserved for recovered catalyst, which should be kept wet with water at all times. Neutralize the combined *aqueous filtrates and washes* with sodium carbonate, and then flush them down the drain with excess water. Place the *recovered ethereal distillate* in the container for nonhalogenated organic solvents. After allowing the diethyl ether to evaporate from the *sodium sulfate* on a tray in the hood, place the used drying agent in the container for nonhazardous solids.

EXERCISES

1. Why is the sodium borohydride solution prepared in basic rather than acidic aqueous media?

2. Aqueous solutions of sodium borohydride can be prepared and kept for short periods of time, as in this experiment. On the other hand, lithium aluminum hydride reacts explosively with water. Suggest a possible reason for these great differences in reactivity with water.

3. Why does adding hydrochloric acid to an aqueous solution of a carboxylic acid decrease its solubility in water?

4. What is the purpose of swirling the ethereal extracts over sodium sulfate?

5. Determine the oxidation numbers of the carbon atoms undergoing change in the following reactions, and indicate the net gain or loss of electrons in each.

  a. $CH_3Cl \rightarrow CH_4$
  b. $CH_3CHO \rightarrow CH_3CH_2OH$
  c. acetylene $\rightarrow$ ethylene
  d. $CH_3CO_2H \rightarrow CH_3CHO$
  e. $CH_3CH=NH \rightarrow CH_3CH_2NH_2$

6. Draw the structure of the product of catalytic hydrogenation of

  a. 1,2-dimethylcyclohexene
  b. *cis*-2,3-dideutero-2-butene
  c. *trans*-2,3-dideutero-2-butene
  d. *cis*-3,4-dimethyl-3-hexene

SPECTRA OF STARTING MATERIAL AND PRODUCT

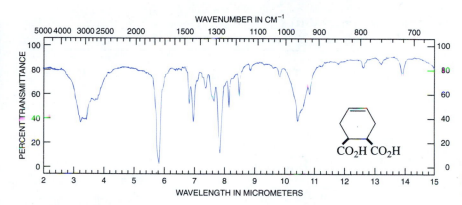

FIGURE 17.1   IR spectrum of 4-cyclohex-ene-*cis*-1,2-dicarboxylic acid.

(a) ^1H NMR spectrum (90 MHz).

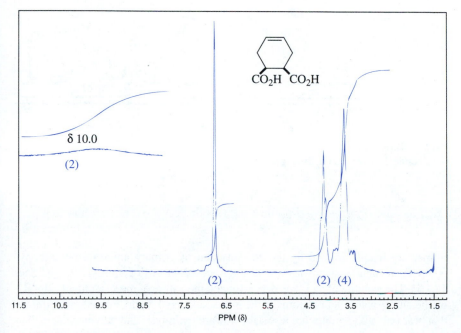

FIGURE 17.2   NMR data for 4-cyclohex-ene-*cis*-1,2-dicarboxylic acid.

(b) ^{13}C NMR data.
   Chemical shifts: δ 25.9, 39.2, 125.2, 174.9.

FIGURE 17.3   IR spectrum of cyclo-hexane-*cis*-1,2-carboxylic acid.

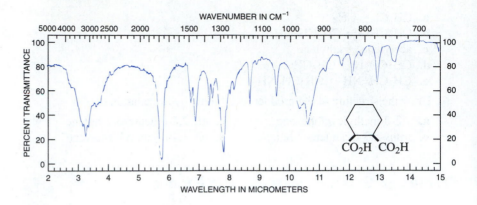

(a) ^{1}H NMR spectrum (90 MHz).

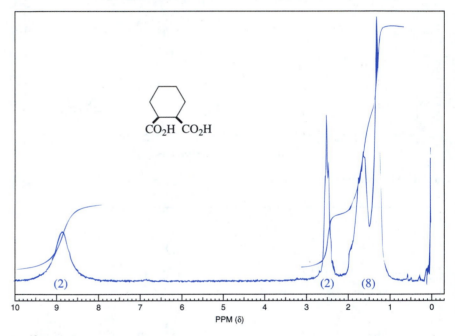

FIGURE 17.4   NMR data for cyclohexane-cis-1,2-carboxylic acid.

(b) ^{13}C NMR data.
Chemical shifts: δ 23.8, 26.3, 42.3, 175.6.

17.3   Reduction of Imines; Preparation of Amines

Imines are readily formed by the condensation of primary amines with carbonyl compounds such as aldehydes and ketones according to Equation (17.9) (see Sec. 18.1.1). The stability of an imine depends upon the nature of the substituents, R^1–R^3, on the nitrogen and carbon atoms of the imine functional group. For example, imines that bear only alkyl groups are less stable toward hydrolysis and polymerization than those that possess one or more aromatic substituents on the carbon-nitrogen double bond; the most stable imines are those having only aryl substituents.

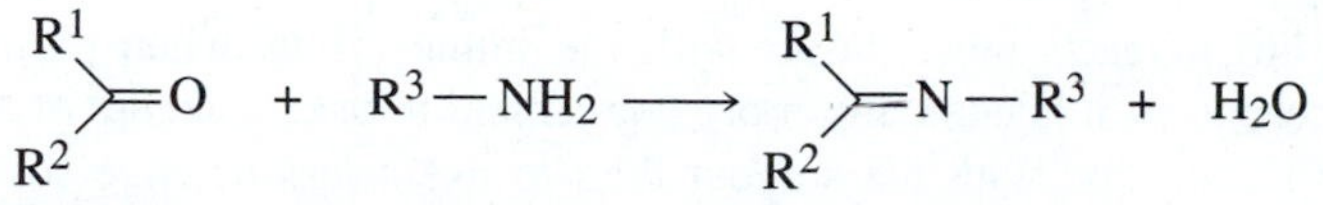

$$\underset{\substack{\text{An aldehyde or} \\ \text{a ketone}}}{\overset{R^1}{\underset{R^2}{\Large>}}{=}O} \ + \ \underset{\substack{\text{A primary amine}}}{R^3{-}NH_2} \longrightarrow \underset{\substack{\text{An imine}}}{\overset{R^1}{\underset{R^2}{\Large>}}{=}N{-}R^3} + \ H_2O \qquad \textbf{(17.9)}$$

Imines undergo addition of one mole of hydrogen in the presence of a catalyst such as palladium, platinum or nickel to produce secondary amines (Eq. 17.10). This process is analogous to the hydrogenation of alkenes to alkanes and of carbonyl compounds to alcohols.

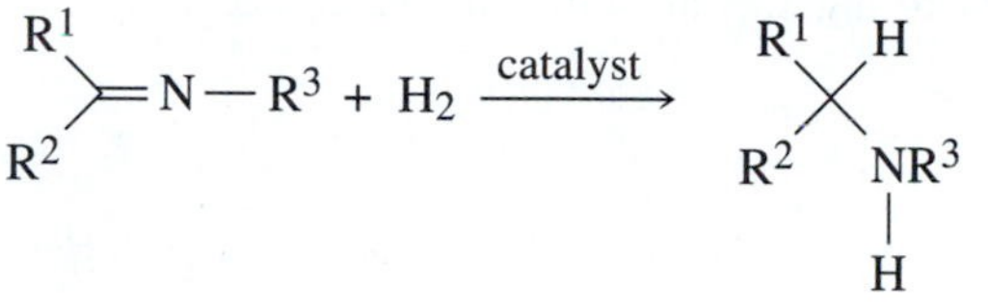

$$\underset{\substack{\text{An imine}}}{\overset{R^1}{\underset{R^2}{\Large>}}{=}N{-}R^3} + \ H_2 \ \xrightarrow{\text{catalyst}} \ \underset{\substack{\text{A secondary amine}}}{\overset{R^1\quad H}{\underset{R^2\quad NR^3}{\Large\times}}}{\underset{H}{\overset{|}{}}} \qquad \textbf{(17.10)}$$

In a useful laboratory process known as **reductive amination** or **reductive alkylation,** the reactions depicted in Equations (17.9) and (17.10) are executed in a single reaction vessel *without* isolating the intermediate imine. For example, the mixture obtained by combining an aliphatic or aromatic carbonyl compound with ammonia or an aliphatic or aromatic amine may be subjected directly to catalytic hydrogenation to produce primary, secondary, and tertiary amines in excellent yields (Eqs. 17.11 and 17.12). Polyalkylation of the starting amine, which is a common problem encountered in the alkylations of amines with alkyl halides, can usually be avoided in reductive aminations.

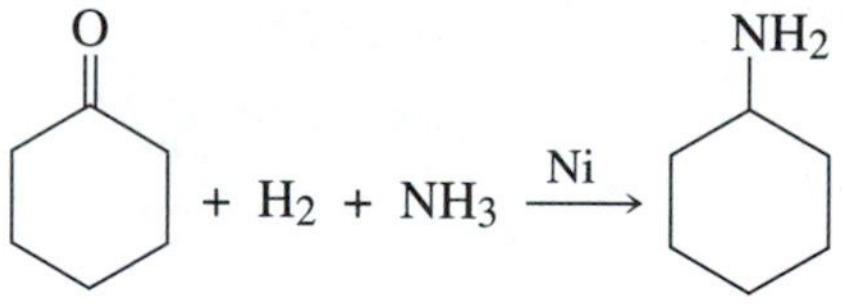

$$\underset{\text{Cyclohexanone}}{\bigcirc\!\!=\!\!O} \ + \ H_2 \ + \ NH_3 \ \xrightarrow{\text{Ni}} \ \underset{\text{Cyclohexylamine}}{\bigcirc\!\!-\!\!NH_2} \qquad \textbf{(17.11)}$$

$$\underset{\text{Benzaldehyde}}{C_6H_5{-}\overset{O}{\overset{\|}{C}}\!\diagdown_{H}} \ + H_2 + \underset{\text{Aniline}}{C_6H_5{-}NH_2} \ \xrightarrow{\text{Ni}} \ \underset{N\text{-Benzylphenylamine}}{C_6H_5{-}CH_2{-}NH{-}C_6H_5} \quad \textbf{(17.12)}$$

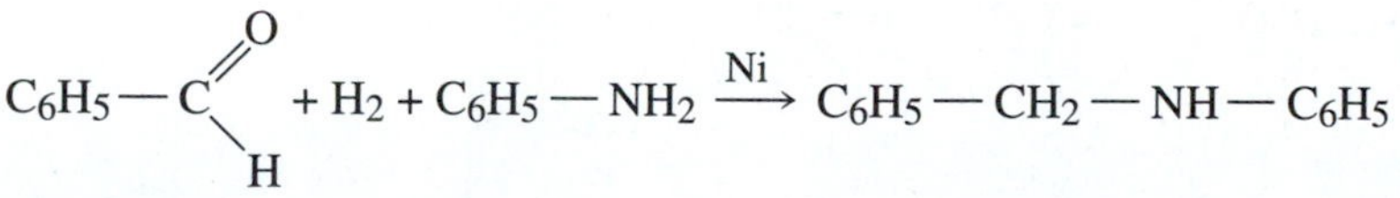

Other functional groups that are present in the starting materials and undergo catalytic hydrogenation, such as C=C, C≡C, N=O (as in NO$_2$), or C–X, may also be reduced under the conditions of the reaction. However, this limitation is readily overcome if the reduction is performed using a *selective* metal hydride, such as sodium cyanoborohydride, NaBH$_3$CN, or sodium borohydride, NaBH$_4$. Sodium borohydride reacts with methanol, which is frequently used as the solvent, but it reacts much faster with the imine. Thus, when performing small-scale reactions, it is generally more convenient to use an excess of $NaBH_4$ to allow for its reaction with the solvent than to use a less-reactive solvent in which it is less soluble.

In this section, we prepare *N*-cinnamyl-*m*-nitroaniline (**9**) by a sequence beginning with the condensation of cinnamaldehyde (**5**) with *m*-nitroaniline (**6**), followed by reduction of the intermediate imine **7** thus produced with sodium borohydride as shown in Equations (17.13)–(17.15). The formation of the imine is reversible, but the reaction is driven to completion by the use of **azeotropic distillation.** Because cyclohexane and water form a minimum-boiling azeotrope (see Sec. 4.2.4), the water generated by the condensation of **5** and **6** is continuously removed by distilling the cyclohexane-water azeotrope throughout the course of the reaction.

(17.13)

5
Cinnamaldehyde

6
m-Nitroaniline

7
N-Cinnamylidene-*m*-nitroaniline

(17.14)

7

8

(17.15)

8

9
N-Cinnamyl-*m*-nitroaniline

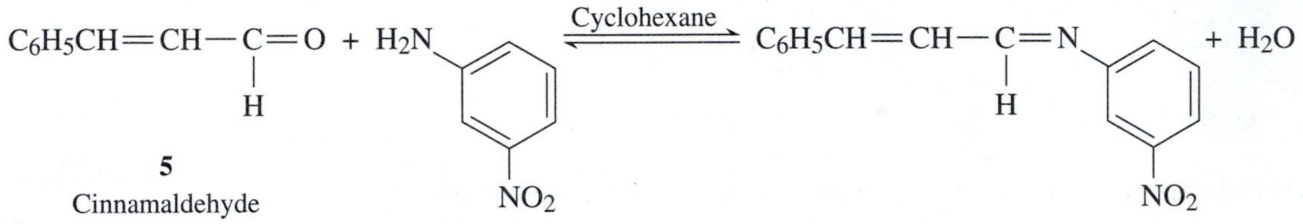

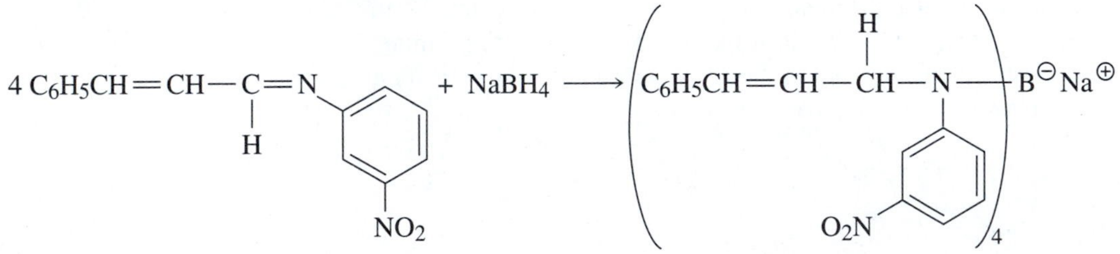

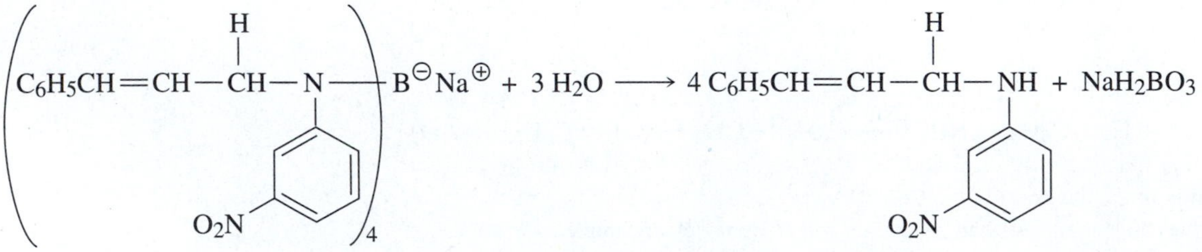

The reaction of sodium borohydride with the imine **7** (Eq. 17.14) is analogous to the addition of sodium borohydride (see Sec. 17.4) or a Grignard reagent (see Secs. 19.1 and 19.2) to a carbonyl compound. A nucleophilic hydride ion ($H:^-$) is transferred from the borohydride anion (BH_4^-) to the electrophilic carbon atom of the carbon-nitrogen double bond, and the electron-deficient boron atom becomes attached to nitrogen. All four hydrogens of the borohydride anion may be transferred to the imine carbon in this way to produce the organoborate anion **8,** which is subsequently decomposed with water to yield the secondary amine **9** as shown in Equation (17.15). This experiment nicely illustrates the selectivity of sodium borohydride, which reduces the intermediate imine **7** but not the carbon-carbon double bond, the nitro group, or the benzene rings that are also present.

Pre-Lab exercises for Section 17.3, "Formation and Reduction of *N*-Cinnamylidene-*m*-nitroaniline," are found on PL. 87.

EXPERIMENTAL PROCEDURE

SAFETY ALERT

1. Avoid contact of methanolic solutions of sodium borohydride with your skin; they are *highly caustic*. If possible, wear rubber gloves when handling these solutions. If these solutions accidentally get on your skin, wash the area with copious quantities of water.
2. Do *not* stopper flasks containing methanolic sodium borohydride; the solution slowly evolves hydrogen gas, and dangerous buildup of pressure could occur in a stoppered flask.
3. Because cyclohexane is flammable, exercise care if flames are used in this experiment.

Place 0.6 g of freshly distilled cinnamaldehyde, 0.6 g of *m*-nitroaniline, 5 mL of cyclohexane, and boiling stones in a 25-mL round-bottom flask. Assemble an apparatus for simple distillation (see Sec. 2.7) using a graduated cylinder as a receiver, and heat the reaction mixture on a steam, sand, or oil bath or with a heating mantle. If a small burner must be used, take proper precautions since cyclohexane is flammable. Distil until most of the cyclohexane is removed; about 4 mL of distillate should be obtained in approximately 5 min. *Discontinue* heating when the distillation rate decreases. Add another 5 mL of cyclohexane, and resume the distillation; stop heating when the rate of distillation decreases.✳

When the distillation is complete, remove the heat source, and take about 0.2 mL of the residual liquid from the stillpot with a micropipet. Transfer this aliquot to a small test tube, and add 1 mL of methanol; swirl the tube to effect solution and place the test tube in an ice-water bath. Collect any crystals that separate, dry them, and determine their melting point. The reported melting point of the *N*-cinnamylidene-*m*-nitroaniline is 92–93 °C. Purify this imine if necessary by recrystallization from a small volume of methanol.

Add 4 mL of methanol to the stillpot containing the remainder of the crude imine, and attach a water-cooled reflux condenser. Prepare a solution of 0.15 g of sodium borohydride in 3 mL of methanol. Because sodium borohydride reacts with methanol and slowly evolves hydrogen gas, this solution should be prepared in an *unstoppered* vessel *immediately* before it is used. Using a micropipet, transfer the methanolic solution of sodium borohydride dropwise through the top of the reflux condenser to the solution of the imine at a rate such that the addition is completed within 1 min. Try to add the solution *directly* into the flask without touching the walls of the condenser and swirl the reaction mixture periodically while making the addition. After all of the borohydride solution is added, heat the reaction mixture at reflux for 15 min.

Cool the reaction mixture to room temperature and add 10 mL of water.✻ Stir the mixture, and allow it to stand with occasional stirring for 10–15 min. Collect the orange crystals and wash them with water.✻ Dry the product, and determine its yield and melting point. *N*-Cinnamyl-*m*-nitroaniline may be recrystallized from 95% ethanol; the melting point of the pure secondary amine is 106–107 °C.

Finishing Touches   Pour the *cyclohexane distillates* in the container for nonhalogenated organic solvents. Dilute the *aqueous methanolic filtrate* with water, neutralize it with acetic acid to destroy any excess sodium borohydride, and flush the mixture down the drain with excess water.

EXERCISES

1. What causes the turbidity in the distillate collected while heating the carbonyl compound and primary amine in cyclohexane solution?

2. Why was the use of *dry* cyclohexane not specified?

3. Although 95% ethanol is a satisfactory solvent for recrystallizing *N*-cinnamyl-*m*-nitroaniline, it is not suitable for recrystallizing the corresponding imine from which it was produced. Why?

4. Determine the molar ratio of NaBH₄ to imine that you actually used in the experiment. Why is it necessary to use a greater molar ratio than theoretical?

5. Explain how adding sodium hydroxide stabilizes sodium borohydride toward reaction with methanol.

6. In Figures 17.9 and 17.11, assign as many of the IR absorption peaks to the components of the molecules as you can. Which peaks provide evidence of the conversion of the imine **7** into the secondary amine **9**? Which peaks show the retention of the carbon-carbon double bond and of the nitro group?

7. Compare the ^{1}H and ^{13}C NMR spectra of **7** and **9** in Figures 17.10 and 17.12 and identify the characteristic features in each that provide evidence that the imine **7** has been converted into the secondary amine **9**. Which peaks show that the carbon-carbon double bond has not been reduced?

8. After the reaction between sodium borohydride and the imine is complete, the reaction mixture is treated with water to produce the desired secondary amine. Explain this reaction by indicating the source of the hydrogen that ends up on the amino nitrogen.

9. Although sodium borohydride is fairly unreactive toward methanol, adding a mineral acid to this solution results in the rapid destruction of the NaBH₄. Explain.

10. Suppose that each of the following pairs of compounds were subjected to the reactions provided in the experimental procedures. Draw the structures

of the intermediate imines that would be formed, as well as those of the final amine product.

   a. cinnamaldehyde and aniline
   b. benzaldehyde and *m*-nitroaniline
   c. benzaldehyde and aniline

11. Propose a mechanism to show how imine **7** can be hydrolyzed by water at pH 6 to produce **5** and **6**. Symbolize the flow of electrons with curved arrows.

12. Explain why aliphatic imines are much less stable than those having at least one aromatic group attached to the carbon-nitrogen double bond.

SPECTRA OF STARTING MATERIALS AND PRODUCT

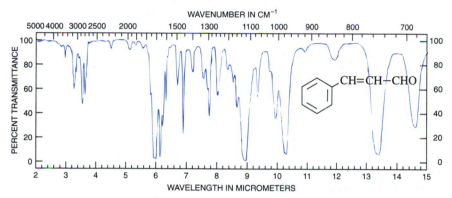

FIGURE 17.5   IR spectrum of cinnamaldehyde.

(a) ¹H NMR spectrum (60 MHz).

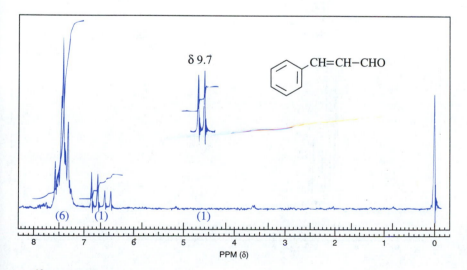

(b) ¹³C NMR data.
   Chemical shifts: δ 128.5, 129.0, 131.1, 134.1, 152.3, 193.2.

FIGURE 17.6   NMR data for cinnamaldehyde.

FIGURE 17.7 IR spectrum of *m*-nitroaniline.

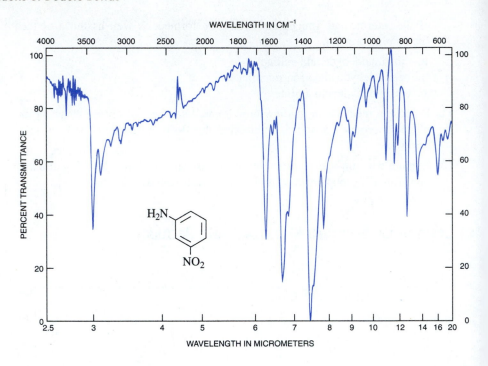

(a) ^{1}H NMR spectrum (90 MHz).

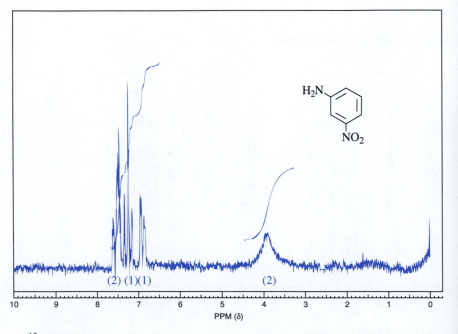

(b) ^{13}C NMR data.

Chemical shifts: δ 107.8, 110.5, 120.2, 129.6, 149.1, 149.8.

FIGURE 17.8 NMR data for *m*-nitroaniline.

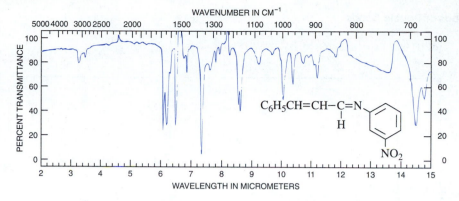

FIGURE 17.9 IR spectrum of *N*-cinnamyli-
dene-*m*-nitroaniline (in CCl$_4$ solution).

(a) ^{1}H NMR spectrum (90 MHz).

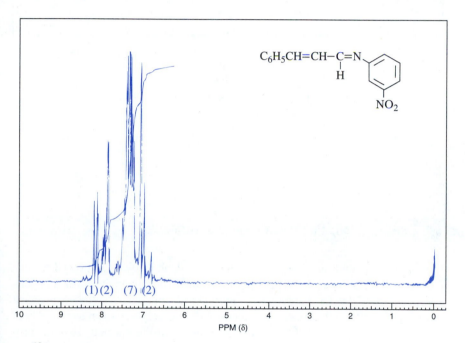

(b) ^{13}C NMR data.
Chemical shifts: δ 114.8, 120.1, 127.4, 128.6, 129.5, 129.8, 129.9, 130.2, 134.8,
145.7, 148.6, 152.6, 163.4.

FIGURE 17.10 NMR data for *N*-cinnamyli-
dene-*m*-nitroaniline.

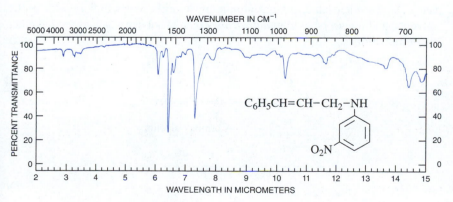

FIGURE 17.11 IR spectrum of *N*-cin-
namyl-*m*-nitroaniline (in CCl$_4$ solution).

FIGURE 17.12 NMR data for *N*-cinnamyl-*m*-nitroaniline.

(a) ^{1}H NMR spectrum (60 MHz).

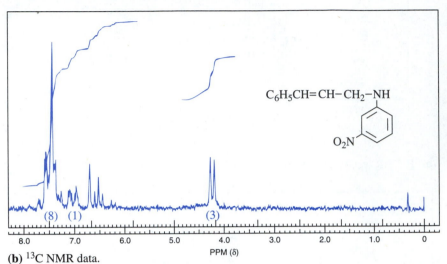

(b) ^{13}C NMR data.

Chemical shifts: δ 45.7, 106.5, 111.9, 118.8, 125.4, 126.3, 127.7, 128.6, 129.6, 132.1, 136.4, 148.7, 149.3.

17.4 REDUCTION OF CARBONYL COMPOUNDS; PREPARATION OF ALCOHOLS

Carbonyl compounds are commonly reduced to alcohols by catalytic hydrogenation or with metal hydrides. When applied to aldehydes, reduction, which is generally indicated by the symbol [H], provides a convenient route to primary alcohols (Eq. 17.16), whereas the reduction of ketones gives secondary alcohols (Eq. 17.17). Although catalytic hydrogenation of carbonyl groups is frequently the method of choice in industrial processes, lithium aluminum hydride, sodium borohydride, and their derivatives are generally used in the research laboratory. Sodium borohydride may be used in alcoholic and even aqueous solutions, because it reacts much more rapidly with the carbonyl group than with the solvent. On the other hand, lithium aluminum hydride reacts rapidly with protic solvents, and it *must* be used in *anhydrous* ethereal solvents such as diethyl ether and tetrahydrofuran.

$$R-C\overset{O}{\underset{H}{\Big|}} \xrightarrow{\text{[H]}} R-CH_2OH \qquad \textbf{(17.16)}$$

An aldehyde   A primary alcohol

R = alkyl or aryl

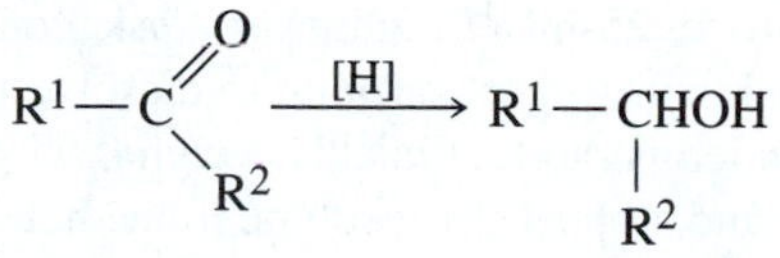

$$R^1 - \overset{\displaystyle O}{\underset{\displaystyle R^2}{C}} \xrightarrow{\text{[H]}} R^1 - \underset{\displaystyle R^2}{CHOH} \qquad \text{(17.17)}$$

A ketone

A secondary alcohol

R^1 and R^2 = alkyl or aryl

In this experiment, we examine the reduction of fluorenone (**10**) by sodium borohydride to give fluorenol (**12**), as shown in Equation (17.18). This reaction is mechanistically analogous to the reduction of imines with sodium borohydride (see Sec. 17.3) and involves the transfer of hydride ion (H:⁻) from BH_4^- to the electrophilic carbonyl carbon with concomitant transfer of the electron-deficient boron atom to the carbonyl oxygen. All four of the hydrogen atoms attached to boron may be transferred in this way to produce the intermediate borate salt **11**, which is decomposed upon addition of water and acid to yield fluorenol (**12**).

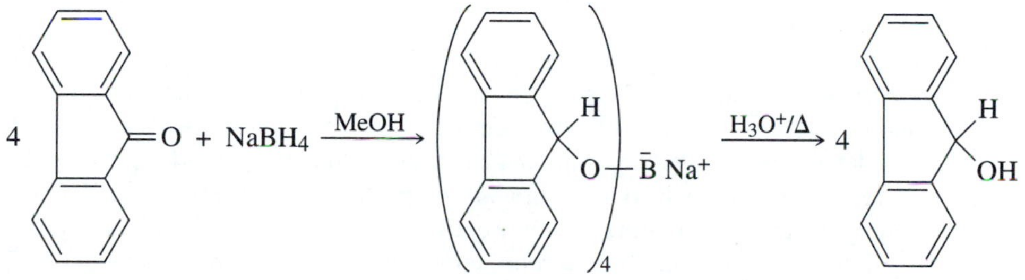

| **10** | **11** | **12** |
| Fluorenone | | Fluorenol |

Pre-Lab exercises for Section 17.4, "Reduction of Fluorenone," are found on page PL. 89.

SAFETY ALERT

1. Avoid contact of methanolic solutions of sodium borohydride with your skin; they are *highly caustic*. If possible, wear rubber gloves when handling these solutions. If these solutions accidentally get on your skin, wash the area with copious quantities of water.

2. Do *not* stopper flasks containing methanolic sodium borohydride; the solution slowly evolves hydrogen gas, and dangerous buildup of pressure could occur in a stoppered flask.

Add 0.6 g of fluorenone to a 25-mL Erlenmeyer flask containing 6 mL of methanol, and swirl the flask with slight warming to dissolve the ketone. Allow the solution to cool to room temperature. Quickly weigh 0.05 g of sodium borohydride into a dry test tube and stopper the test tube immediately to avoid undue exposure of the hygroscopic reagent to atmospheric moisture. Add the reducing agent in one portion to the solution of fluorenone in methanol, and swirl the mixture vigorously to dissolve the reagent. After all of the sodium borohydride dissolves, allow the solution to stand at room temperature for 20 min. During this time the yellow color of fluorenone should fade, and the solution should become colorless. If it does not become colorless, add an additional small portion of sodium borohydride to complete the reaction.✳

Add 2 mL of 3 *M* sulfuric acid to the reaction mixture. Hydrogen is evolved, and a white precipitate should appear. Place the flask containing the mixture on a steam bath, and heat the mixture gently and intermittently for 5–10 min. Stir the mixture occasionally with a glass rod to help dissolve the solid; maintain the internal temperature just below the reflux point to *minimize the loss of solvent*. If all of the solids do not dissolve, *gradually* add methanol in about 0.5 to 1 mL portions with continued heating until a solution is obtained. When all of the precipitated solids redissolve, remove the flask from the heating bath; allow the solution to cool to room temperature,✳ and then in an ice-water bath for 10–15 min. Collect the solid product by vacuum filtration, and *wash it thoroughly* with water until the filtrate is *neutral*.✳ Dry the solid, and determine the melting point and the yield of the crude product. Purify the crude fluorenol by dissolving it in a *minimum* volume of boiling methanol and adding water to the boiling solution until the solution turns cloudy or crystals begin to form. Cool the flask to room temperature and then in an ice-water bath for 10–15 min and collect and dry the crystals.✳ The melting point of pure fluorenol is 153–154 °C.

> **Finishing Touches**   Dilute the combined *aqueous methanol filtrates* with water, neutralize the resulting solution with sodium carbonate, and flush the mixture down the drain with excess water.

EXERCISES

1. Determine the molar ratio of $NaBH_4$ to fluorenone that you actually used in the experiment. Why is it necessary to use a greater molar ratio than theoretical?

2. After the reaction between sodium borohydride and the ketone is complete, the reaction mixture is treated with water and acid to produce the desired secondary alcohol. Explain this reaction by indicating the source of the hydrogen that ends up on oxygen.

3. Sodium borohydride is fairly unreactive toward methanol, but adding a mineral acid to this solution results in the rapid destruction of the $NaBH_4$. Explain.

4. Suggest a structure for the white precipitate formed in the reaction of fluorenone with sodium borohydride.

5. Besides NaBH$_4$ and LiAlH$_4$, what other reagents might be expected to reduce fluorenone to fluorenol?

6. Explain why fluorenone is colored whereas fluorenol is colorless.

7. Draw the structure of the product that results from complete reduction of the following compounds by NaBH$_4$:

   a. cyclohexanone   **b.** 3-cyclohexen-1-one   **c.** 1,4-butanedial
   d. 4-oxohexanal   **e.** acetophenone

8. Draw the structure of the product that would be formed from allowing each of the compounds in Exercise 7 to react with excess hydrogen gas in the presence of a nickel catalyst.

SPECTRA OF STARTING MATERIAL AND PRODUCT

The IR, ^{1}H and ^{13}C NMR spectra of fluorenone are given in Figures 6.6 and 6.7.

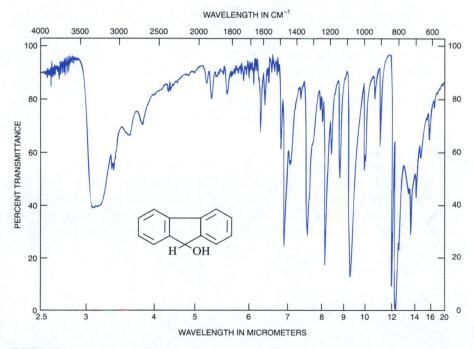

FIGURE 17.13   IR spectrum of fluorenol.

FIGURE 17.14 NMR data for fluorenol. **(a)** ^{1}H NMR spectrum (90 MHz).

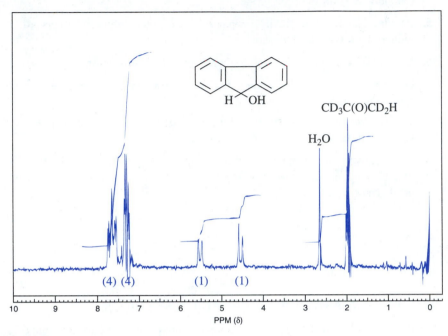

(b) ^{13}C NMR data.
Chemical shifts: δ 73.8, 119.6, 125.0, 127.2, 128.2, 139.5, 146.8.

17.5 ENZYMATIC REDUCTION: A CHIRAL ALCOHOL FROM A KETONE

In modern synthetic organic chemistry, it is often necessary to use optically pure starting materials to prepare a target molecule such as a drug in enantiomerically pure form. However, reduction, [H], of a **prochiral** ketone **13,** which bears two *different* alkyl or aryl residues on the carbonyl carbon atom, by catalytic hydrogenation or with hydride reducing agents such as sodium borohydride or lithium aluminum hydride gives equal amounts of the chiral alcohol **14** and its enantiomer ***ent*-14** (Eq. 17.19). This **racemate** is produced because these *achiral* reducing reagents attack **both** faces, *a* and *b,* of the planar prochiral carbonyl function with *equal* probability. However, if the reducing agent is chiral, it is possible to obtain the resulting chiral alcohol as an optically pure substance. For example, the reduction of prochiral ketones by catalytic hydrogenation in the presence of chiral catalysts or by the use of chiral hydride reducing agents

(17.19)

may produce secondary alcohols with high levels of enantioselectivity. There are also certain enzymes, which are polypeptides composed of L-amino acids and thus have chiral active sites, that can perform such **enantioselective** reductions.

When a chiral product is formed from an achiral starting material, the **optical yield** or **optical purity,** not just the chemical yield, of the product is important. Optical purity is commonly evaluated by calculating the **enantiomeric excess** (ee). For example, reduction of a prochiral ketone **13** with sodium borohydride produces 50% *R*- and 50% *S*-alcohol **14** and ***ent*-14;** the ee for this process is 0%. If 90% of the *S*- and 10% of the *R*-isomer were produced by the reaction, the ee would be 80%.

In this experiment, methyl acetoacetate (**15**) is reduced to methyl (*S*)-(+)-3-hydroxybutanoate (**16**), using one of the reducing enzymes found in baker's yeast (Eq. 17.20). The enzyme is one of many that are involved in the metabolism of D-glucose to ethanol. Enantiomeric excesses ranging from 70–97% have been reported for this reaction. The hydroxy ester **16** is a building block that has been used as a starting material in a number of syntheses of optically pure natural products.

15		**16**
Methyl acetoacetate		Methyl (*S*)-(+)-3-hydroxybutyrate

Pre-Lab exercises for Sec. 17.5, "Enzymatic Reduction of Methyl acetoacetate," are found on page PL. 91.

EXPERIMENTAL PROCEDURE

Dissolve 80 g of sucrose and 0.5 g of disodium hydrogen phosphate, which buffers the mixture to maintain an optimal pH, in 150 mL of warm (35 °C) tap water contained in a 500-mL Erlenmeyer flask; add two packets (16 g) of dry baker's yeast. Swirl the mixture vigorously for about 15 min to suspend all of the yeast and to initiate the fermentation. Add 5 mL of methyl acetoacetate and continue to swirl the flask for about 15 min to mix the contents thoroughly. Store the flask in a warm place, ideally at 30–35 °C, for at least 48 h or until the next laboratory period.✻ Although you may simply be instructed to cover the flask with a watchglass, the product will have higher optical purity if you exclude oxygen during the fermentation; such a fermentation is termed **anaerobic.** The apparatus shown in Figure 17.15 is well suited for this purpose.

According to the procedure of Section 2.16, prepare a bed of a filter aid on a 10-cm Büchner funnel, using 20 g of filter aid in 75 mL of water. Add another

FIGURE 17.15 Apparatus for the anaerobic fermentation of methyl acetoacetate.

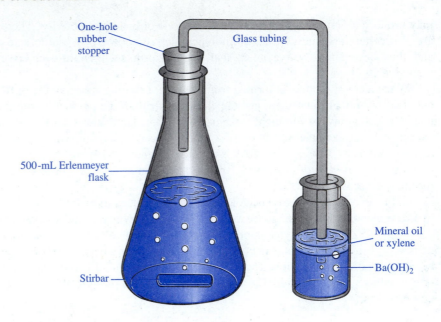

One-hole rubber stopper

Glass tubing

500-mL Erlenmeyer flask

Stirbar

Mineral oil or xylene

Ba(OH)$_2$

20 g of filter aid to the reaction mixture containing the yeast cells, and swirl the mixture thoroughly to mix the filter aid with the cells. Remove the yeast cells by vacuum filtration through the bed of filter aid. Wash the cells with 50 mL of water; then saturate the filtrate with solid sodium chloride. Extract the resulting solution five times with 50-mL portions of dichloromethane by *gently* shaking the separatory funnel to mix the layers. Do *not* shake the funnel too vigorously, because a bad emulsion forms. If such an emulsion does form, pass the contents of the separatory funnel through a *fresh* bed of filter aid by vacuum filtration; this will remove most of the fine particles that cause the emulsion. Wash the bed with 25 mL of dichloromethane. Again transfer the filtrate to the separatory funnel and continue the extractions; the fine emulsion that remains may be included with the aqueous phase. Dry the combined dichloromethane layers with anhydrous sodium sulfate.✽

Decant the dichloromethane solution into a *tared* distilling flask, and remove most of the dichloromethane by simple distillation.✽ Cool the stillpot to room temperature and remove the last traces of solvent by connecting the flask to a water aspirator pump; be sure to use a trap.✽ The residue, which is crude methyl (*S*)-(+)-3-hydroxybutanoate, should weigh about 3.5 g, and it should give a negative ferric ion test (see Sec. 23.5.6). Analyze it by TLC using dichloromethane as the eluting solvent or by GLC using a Carbowax® 20M column at 100 ºC to determine whether unreacted methyl acetoacetate is present. If starting material is present, the optical yield of methyl (*S*)-(+)-3-hydroxybutanoate as measured by specific rotation will be lower (see Sec. 17.6).

> **Finishing Touches**   Flush the filtrate from preparation of the bed of filter aid down the drain. Dilute the *aqueous layer* with water and flush it down the drain. After the dichloromethane evaporates from the *sodium sulfate* in the hood, place the solid in the container for nonhazardous solids. Place the recovered *dichloromethane* that was used as the extraction solvent and as the eluting solvent for the TLC in the container for halogenated organic solvents.

EXERCISES

1. Why is it necessary to saturate the aqueous mixture with sodium chloride before extracting the product with dichloromethane?

2. Predict the structure of the alcohols obtained by reducing the β-ketoesters a–c with baker's yeast.

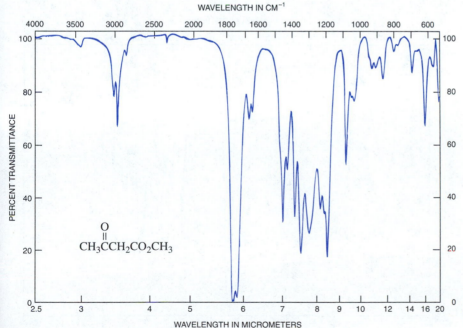

SPECTRA OF STARTING MATERIAL AND PRODUCT

WAVELENGTH IN CM^{-1}

$$CH_3CCH_2CO_2CH_3$$

PERCENT TRANSMITTANCE

WAVELENGTH IN MICROMETERS

FIGURE 17.16   IR spectrum of methyl acetoacetate.

FIGURE 17.17   NMR data for methyl ace-toacetate.

(a) ^{1}H NMR spectrum (90 MHz).

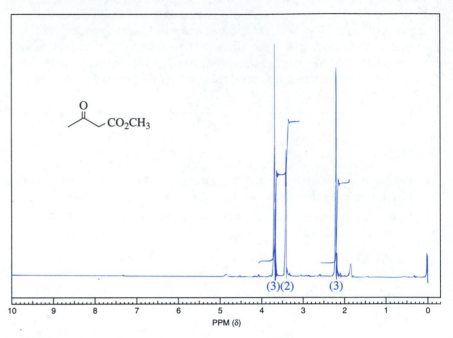

(b) ^{13}C NMR data.
   Chemical shifts: δ 29.9, 49.8, 52.1, 168.0, 200.8.

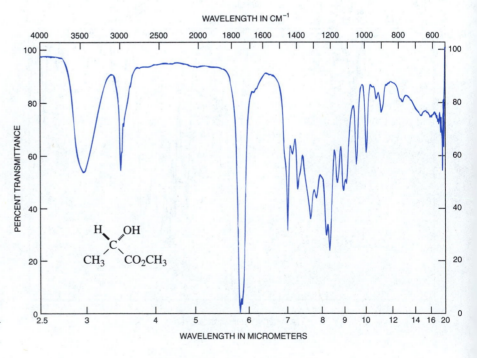

FIGURE 17.18   IR spectrum of methyl (*S*)-(+)-3-hydroxybutanoate.

(a) ¹H NMR spectrum (90 MHz).

FIGURE 17.19   NMR data for methyl (*S*)-(+)-3-hydroxybutanoate.

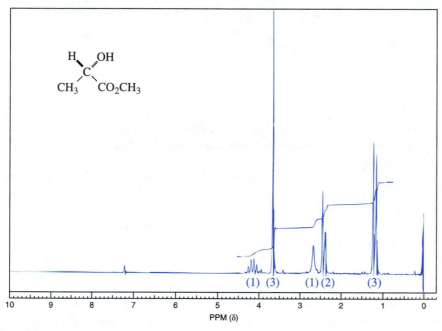

(1) (3)   (1) (2)   (3)

PPM (δ)

(b) ¹³C NMR data.
   Chemical shifts: δ 22.6, 42.8, 51.7, 64.3, 173.2.

17.6 DETERMINING OPTICAL PURITY

There are several ways to determine the optical purity of a compound obtained as the product of an asymmetric synthesis. The simplest involves measuring the optical rotation using a polarimeter (see Sec. 7.6). The specific rotation, $[\alpha]_D^{25}$, of pure methyl (*S*)-(+)-3-hydroxybutanoate is +38.5° (*c* = 1.80, chloroform), but the specific rotation of the product you obtain might range from +27° to +36°. For example, a measured specific rotation of 32.7° (*c* = 1.3, chloroform) for the product of this enzymatic reduction would correspond to an ee of 85%.

When only small quantities of a material are available, the use of ¹H NMR and a chiral shift reagent is a more accurate method for determining optical purity. In an *achiral* environment, the ¹H NMR spectra of methyl (*S*)-(+)-3-hydroxybutanoate and methyl (*R*)-(−)-3-hydroxybutanoate are *identical* and appear as that shown in Figure 17.19(a). However, a Lewis acid having chiral ligands can form *diastereomeric complexes* with the two enantiomers, and these complexes often give different ¹H NMR spectra. In the present case, the Lewis acid is the chiral shift reagent tris[3-(heptafluoropropylhydroxymethylene)-(+)-camphorato]europium (III) or Eu(hfc)₃ (**17**); the europium atom coordinates

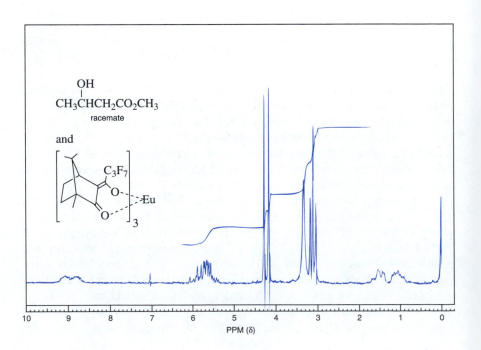

17

with the hydroxyl group in methyl 3-hydroxybutanoate and its camphor-derived ligand to form the diastereomeric complexes, each of which gives a different ^{1}H NMR spectrum.

An important property of europium shift reagents is that they spread the peaks in the ^{1}H NMR spectrum over a wider range, generally moving all of the peaks downfield. The magnitude of the shift of each peak depends on the concentration of the shift reagent and the proximity of the protons corresponding to that peak to the europium atom. Since the complexes that are formed between **17** and methyl (S)-(+)-3-hydroxybutanoate and methyl (R)-(+)-3-hydroxybutanoate are diastereomeric, peaks from one of the enantiomers move downfield more than those of the other, and the two diastereomeric complexes may be easily distinguished in the ^{1}H NMR spectrum, as is shown in Figure 17.20. It is then possible to integrate the peak areas corresponding to each of the enantiomers and determine the enantiomeric excess. Determining the ee using this NMR method gives results that are accurate to within approximately 2–3%.

In this experiment, you will determine the enantiomeric excess of the product that you obtained upon asymmetric reduction of methyl acetoacetate with baker's yeast using the chiral europium chiral shift reagent **17.**

FIGURE 17.20 ^{1}H NMR spectrum (90 MHz) of racemic methyl 3-hydroxybutanoate in the presence of chiral shift reagent **17.**

Pre-Lab exercises for Section 17.6, "Determining Optical Purity of Methyl 3-Hydroxybutanoate," are found on page PL. 93.

Dissolve 25–30 mg of racemic methyl 3-hydroxybutanoate obtained from your instructor in 0.75 mL of deuterochloroform and transfer the solution using a micropipet to an NMR tube; measure the ^{1}H NMR spectrum. Add 20 mg of tris[3-(heptafluoropropylhydroxymethylene)-(+)-camphorato]europium(III) shift reagent to this solution, shake the mixture to effect dissolution, and allow the solution to stand for about 20 min. Measure the ^{1}H NMR spectrum. The peaks of primary interest for determining the optical purity are the methoxy hydrogens of the ester. Continue the procedure of adding shift reagent and measuring the spectrum until these peaks are separated and accurate integrals may be obtained as shown in Figure 17.21; no more than about 80 mg of shift reagent should be required. Now perform the same series of measurements on the sample of enantiomerically enriched methyl 3-hydroxybutanoate prepared by the enzymatic reduction in the first part of this experiment. Determine the ratio of the two enantiomers in both of the samples by comparing the heights of the singlets that correspond to the methoxy protons of the ester, and calculate the percentages of each enantiomer in the two samples based upon this ratio. Calculate the ee of methyl (S)-(+)-3-hydroxybutanoate, which should be in the range 80–95%.

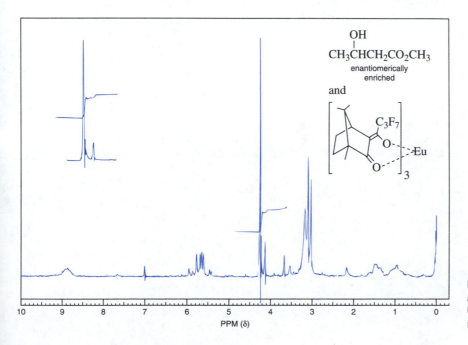

FIGURE 17.21 ^{1}H NMR spectrum (90 MHz) of enriched methyl (S)-(+)-3-hydroxy-butanoate in the presence of chiral shift reagent **17**.

Finishing Touches   Place the *solution* of deuterochloroform and shift reagent in the container for halogenated liquids.

EXERCISES

1. The ^{1}H NMR spectrum in Figure 17.21 is of mixture of the chiral shift reagent **17** and a sample of methyl (*S*)-(+)-3-hydroxybutanoate that was produced by the enzymatic reduction of methyl acetoacetate using baker's yeast. Determine the enantiomeric excess of the methyl (*S*)-(+)-3-hydroxybutanoate contained in this sample.

2. Why is it necessary to obtain a ^{1}H NMR spectrum of *racemic* methyl 3-hydroxybutanoate in the presence of **17**?

3. Which of the following compounds will form diastereomeric complexes with the chiral europium shift reagent **17**?

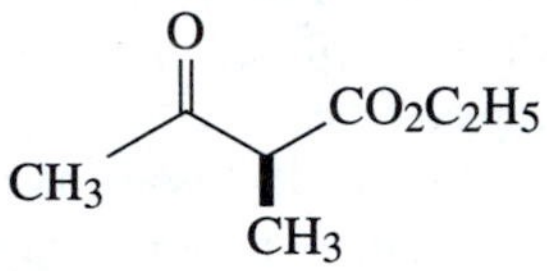

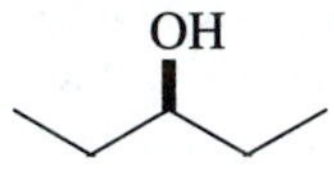

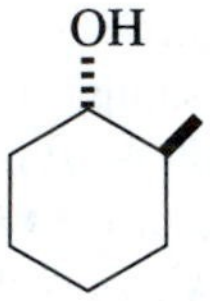

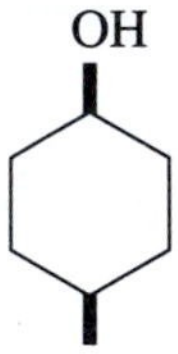

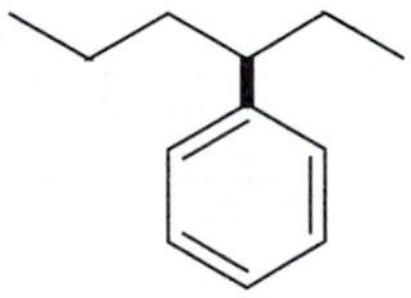

REFERENCES

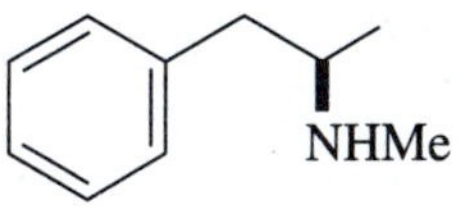

1. Amstutz, R.; Hungerbühler, E.; Seebach, D., *Helvetica Chimica Acta,* **1981,** *64,* 1796. Seebach, D., *Tetrahedron Letters,* **1982,** 159. Seebach, D.; Sutter, M. A.; Weber, R. H.; Züger, M. F., *Organic Synthesis, Collective Vol. VII,* **1990,** 215.
2. Lipkowitz, K. B.; Mooney, J. L., *Journal of Chemical Education,* **1987,** *64,* 985.

REACTIONS OF CARBONYL COMPOUNDS

T he **carbonyl group, 1,** is a rich source of many important reactions in
organic chemistry, with two fundamental properties of this functionality
being primarily responsible for its diverse chemistry. The first is the polarization
of the carbon-oxygen π-bond, owing to the relatively high electronegativity of
the oxygen atom. In terms of resonance theory, this polarization arises from the
contribution of the dipolar resonance structure **1b** to the resonance hybrid for
this functional group. Consequently, the carbonyl group undergoes a variety of

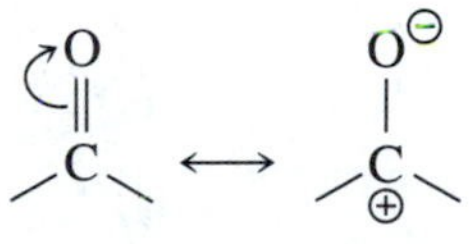

1a   **1b**

reactions in which the electrophilic carbonyl carbon atom is attacked by **nucleophiles,** which are Lewis bases; the oxygen atom, in turn, reacts with **electrophiles** or **Lewis acids.** The net effect is *addition* of a reagent *Nu–E* across the
π-bond of the carbonyl function, as illustrated in Equation (18.1). The sequence
by which the elements of Nu–E add to the carbonyl group varies according to
the particular reagents and reaction conditions used, but in a general sense follow the pathway of Equation (18.1a) in basic media and that of Equation (18.1b)
in acidic media.

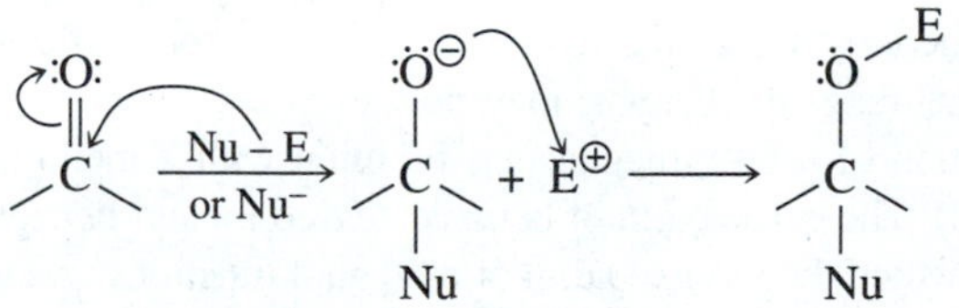

(18.1a)

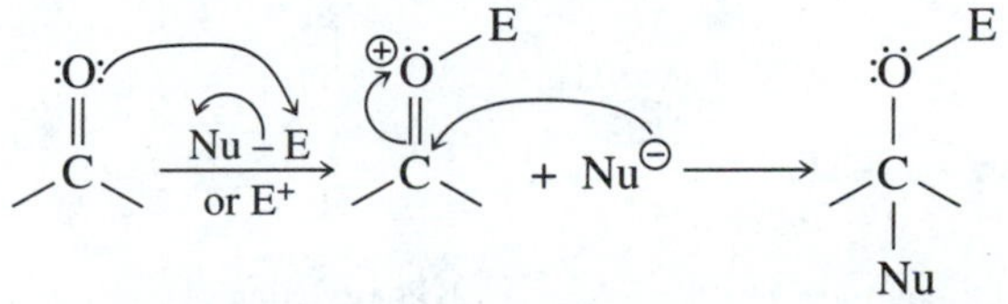

(18.1b)

The second property of a carbonyl moiety is to increase the acidity of the **α-hydrogen atoms,** which are the hydrogens on the carbon atom(s) attached to the carbonyl group. This enhanced acidity of the α-hydrogens means that the **α-carbon atoms** can become **nucleophilic.** This nucleophilicity arises in one of two general ways: through keto–enol equilibration, called **tautomerization** (Eq. 18.2), or by deprotonation to form an **enolate ion** (Eq. 18.3). As shown in Equation (18.2), an enol, **2,** can react with electrophiles, E⁺, at an α-carbon atom to give net substitution of the electrophile for an α-hydrogen atom. A similar result attends reaction of an enolate, as illustrated in Equation (18.3).

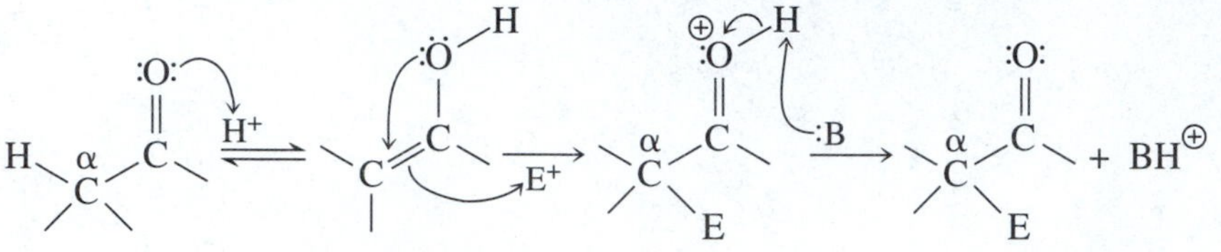

(18.2)

2
An enol

(18.3)

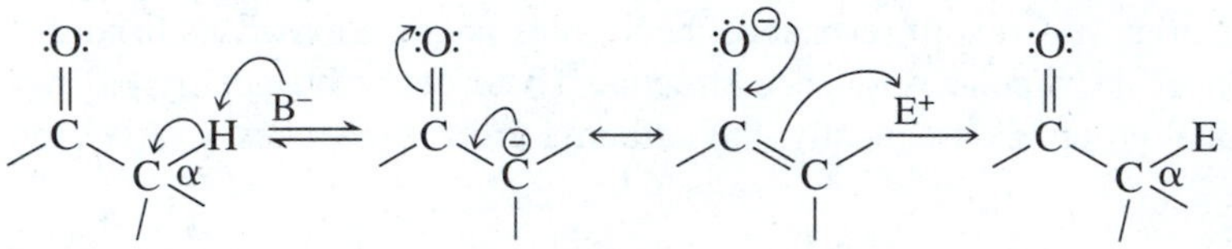

3a   **3b**
An enolate

The experiments described in the following sections involve reactions in which nucleophiles add to the electrophilic carbonyl carbon atom and in which the α-carbon atom functions as a nucleophile.

18.1 WADSWORTH-EMMONS MODIFICATION OF THE WITTIG REACTION

The attack of a nucleophilic atom on a carbonyl carbon atom produces a new bond between the two atoms (Eq. 18.1). The nature of this bond depends on the identity of the nucleophilic species, so it may be a carbon-hydrogen, carbon-carbon, or carbon-heteroatom linkage. For example, sodium borohydride, $NaBH_4$, may be considered as a source of hydride ion, H^-, for the reduction of acetophenone (Eq. 18.4 and Sec. 17.4), and a Grignard reagent, RMgBr, may be viewed as a source of a carbanion (R^-), as in the reaction of *n*-butylmagnesium bromide with 2-methylpropanal (Eq. 18.5 and Sec. 19.3.3). The conversion of benzoic acid to methyl benzoate (Eq. 18.6) represents a case in which the oxygen atom of methanol functions as a nucleophile.

(18.4)

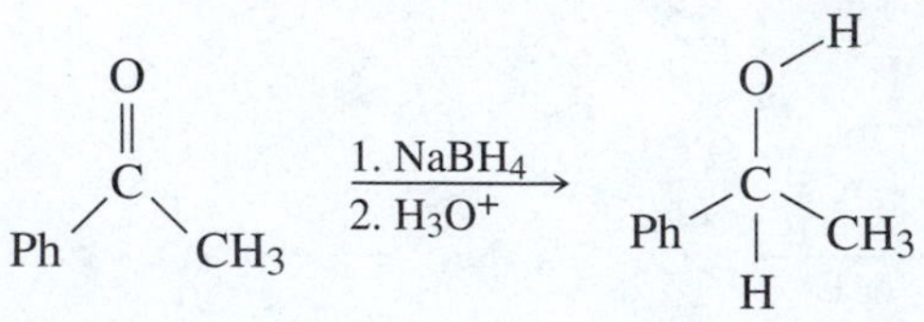

Acetophenone   1-Phenylethanol

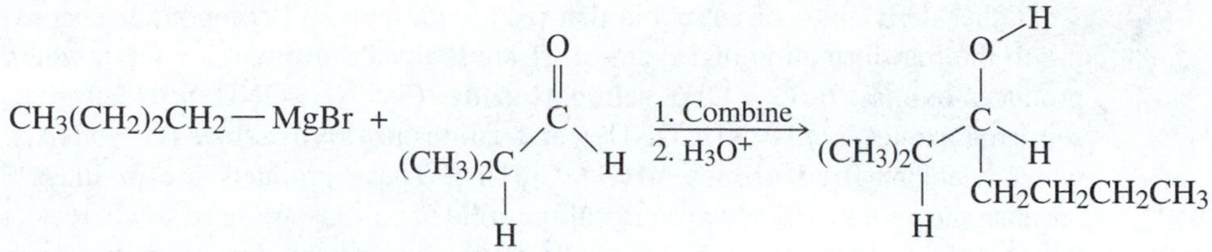

$$\text{(18.5)}$$

n-Butylmagnesium bromide   2-Methylpropanal   2-Methylheptan-3-ol

$$\text{(18.6)}$$

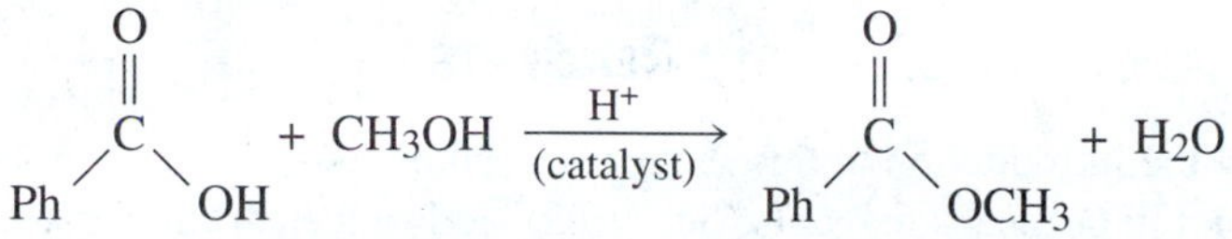

Benzoic acid   Methanol   Methyl benzoate

18.1.1 ADDITION OF DERIVATIVES OF AMMONIA

Nitrogen-containing compounds G–NH$_2$ (**4,** Scheme 18.1) may be classified as derivatives of ammonia and are another important type of nucleophilic reagent, Nu-E, that adds to carbonyl compounds. With **primary amines, 4** (G = alkyl or aryl), the addition product, **5,** readily loses a molecule of water to produce an **imine, 6.** The imines derived from **aliphatic amines, 4** (G = alkyl), are sometimes unstable and may polymerize, but those produced from **aromatic amines, 4** (G = aryl), yield stable imines, also called **Schiff bases,** that can be isolated. These types of imines may be easily hydrolyzed to regenerate the carbonyl compound and the amine, a reaction that is the reverse of that shown in Scheme 18.1.

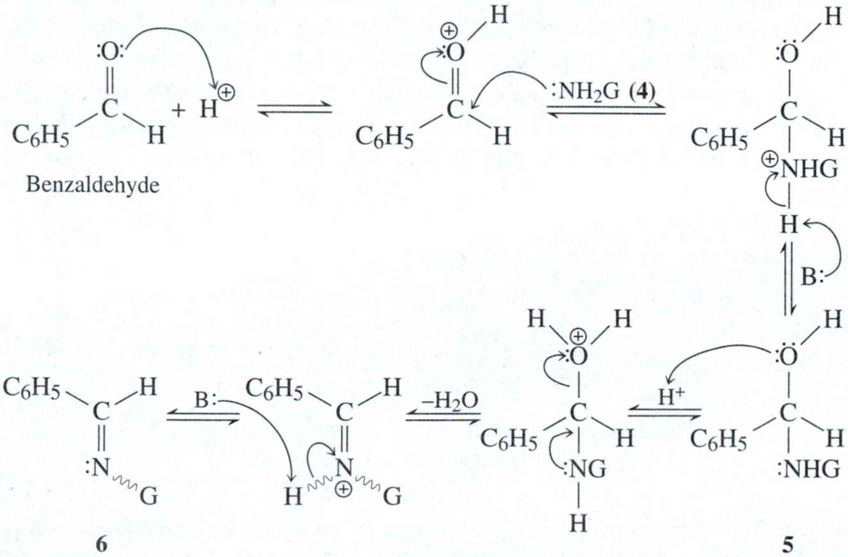

Benzaldehyde

6   **5**

SCHEME 18.1

Other derivatives of ammonia that react with carbonyl compounds according to the transformation of Scheme 18.1 are **hydroxylamine** (G = OH), which produces **oximes, 6** (G = OH), **semicarbazide** (G = NHCONH$_2$), which gives **semicarbazones, 6** (G = NHCONH$_2$), and various **arylhydrazines** (G = NHAr), which yield **arylhydrazones, 6** (G = NHAr). These products are of interest because they are nearly always crystalline solids and thus are used to characterize unknown aldehydes and ketones. General procedures for preparing such derivatives are provided in Section 23.5.1.

18.1.2 ADDITION OF WITTIG REAGENTS

It is noted in the introduction to this section that nucleophilic carbon atoms, such as those found in Grignard reagents (Sec. 19.2), add to the carbonyl carbon atom (Eq. 18.5). Another type of carbon nucleophile is prepared by deprotonating a variety of phosphorus-containing compounds. For example, treatment of the phosphonium salt **7** with *n*-butyllithium produces the nucleophilic species **8,** which is called an **ylide** because of the dipolar character reflected in resonance structure **8a** (Eq. 18.7). Phosphorus ylides like **8** are frequently referred to as

$$(C_6H_5)_3\overset{\oplus}{P}CH_2\!-\!H + Li\!-\!CH_2(CH_2)_2CH_3 \xrightarrow[-LiBr]{-C_4H_{10}} \left[(C_6H_5)_3\overset{\oplus}{P}\!-\!\overset{\ominus}{C}H_2 \longleftrightarrow (C_6H_5)_3P\!=\!CH_2\right] \quad \textbf{(18.7)}$$

| **7** | *n*-Butyllithium | **8a** | **8b** |
| Triphenylmethyl phosphonium bromide | | Triphenylmethylenephosphorane (**8**) | |

Wittig reagents, in recognition of their inventor, Georg Wittig, who showed that they react with aldehydes and ketones to produce alkenes, as depicted in Equation (18.8). The net effect of the Wittig reaction is conversion of a carbon-oxygen double bond to a carbon-carbon double bond. This transformation represents a general method for preparing alkenes that has two important advantages over other methods for preparing this functional group: (1) the carbonyl group is replaced *specifically* by a carbon-carbon double bond, without forming isomeric alkenes having the π-bond at other positions; (2) the reactions are carried out under mild conditions.

$$\textbf{8} + \underset{\text{Cyclohexanone}}{\overset{\text{O}}{\bigcirc}} \longrightarrow \underset{\substack{\text{Methylene-}\\\text{cyclohexane}}}{\overset{\text{CH}_2}{\bigcirc}} + \underset{\substack{\text{Triphenylphosphine}\\\text{oxide}}}{(C_6H_5)_3P\!=\!O} \quad \textbf{(18.8)}$$

The Wittig reaction involving highly reactive phosphoranes related to **8** is not suited for the introductory laboratory. For one thing, the triphenylphosphine, (C$_6$H$_5$)$_3$P, required to prepare the phosphonium salt **7** is expensive. Moreover, very strong bases, such as *n*-butyllithium, that are normally used to produce **8,** require special handling techniques, and it is generally necessary to perform experimental operations under a dry, inert atmosphere. Fortunately, various modifications of the Wittig reaction have been developed that circumvent these problems. The one used here involves formation of a **phosphonate ester, 9,** followed by its deprotonation with an alkoxide base to give the anion, **10** (Eq. 18.9). This anion serves in place of the ylide **8** in the **Wadsworth-Emmons modification** of the Wittig reaction. The ester **9** is prepared from relatively inexpensive reagents, and the base needed for the reaction is much easier to manipulate in the laboratory.

| **9** | | **10a** | **10b** |
| Phosphonate ester | Alkoxide | | |

(18.9)

The phosphonate ester **14** used in this experimental procedure is produced according to the sequence in Equation (18.10). In the first step, the phosphorus atom of triethyl phosphite (**11**) serves as the nucleophile in an S_N2 reaction with benzyl chloride (**12**) to produce the phosphonium salt **13**. This phosphonium salt is unstable toward halide ion and undergoes an S_N2 displacement to afford the desired ester **14** and ethyl chloride, which distils from the reaction mixture. The overall conversion illustrated in Equation (18.10) is called the **Arbuzov reaction.**

| **11** | **12** | **13** | **14** |
| Triethyl phosphite | Benzyl chloride | | Diethyl benzylphosphonate |

(18.10)

Treatment of **14** with sodium methoxide produces the nucleophilic species **15** (Eq. 18.11), setting the stage for the Wadsworth-Emmons reaction. Adding benzaldehyde (**16**) to a solution of **15** initiates the sequence of events shown in Scheme 18.2 to yield the stilbenes **17** and sodium diethyl phosphate (**18**). The latter is water-soluble, as is *N, N*-dimethylformamide, $(CH_3)_2NCHO$ (DMF), which is the solvent for the reaction. Consequently, both of these materials are removed by adding water to the reaction mixture and washing the product with aqueous methanol.

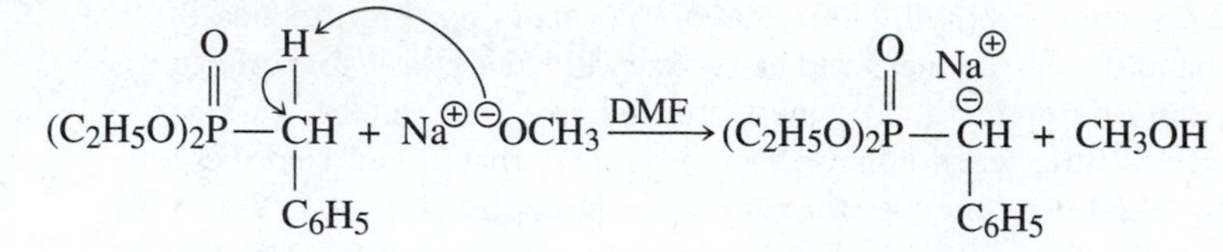

$$(C_2H_5O)_2P\overset{\displaystyle O}{\overset{\|}{\underset{\displaystyle C_6H_5}{-}}}\overset{\displaystyle H}{\underset{}{CH}} + Na^{\oplus}\,{}^{\ominus}OCH_3 \xrightarrow{\ DMF\ } (C_2H_5O)_2P\overset{\displaystyle O}{\overset{\|}{\underset{\displaystyle C_6H_5}{-}}}\overset{Na^{\oplus}}{\underset{}{\overset{\ominus}{CH}}} + CH_3OH \qquad (18.11)$$

14   **15**

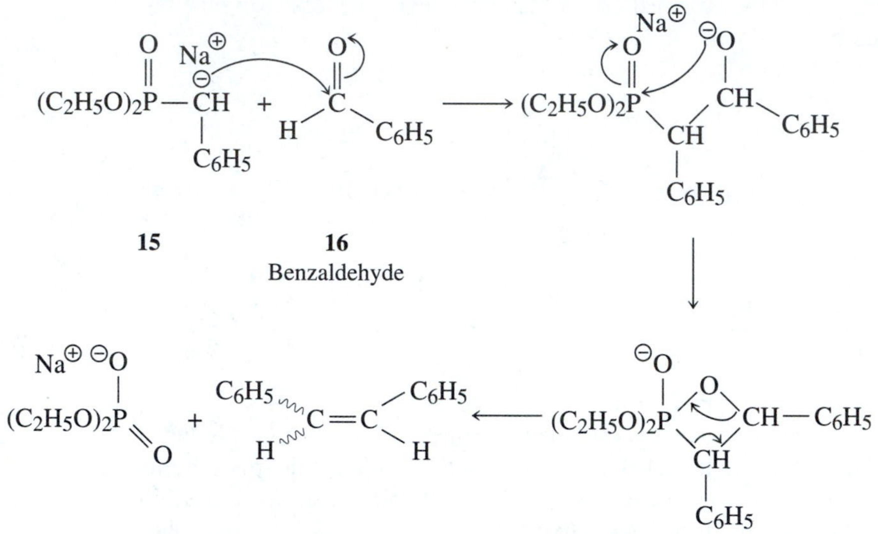

15   **16**
Benzaldehyde

18   **17**
Stilbenes

SCHEME 18.2

Examination of the general structure, **17,** written for the stilbenes reveals that two diastereomers or geometric isomers (see Sec. 7.1) are possible. In fact, both isomers are known, with the *trans* being a solid (mp 126–127 °C) and the *cis* being a liquid at room temperature (mp 6 °C). The phosphonate ester modification of the Wittig reaction is reported to provide only the *trans* isomer, even though the use of the Wittig reagent analogous to ylide **8** yields a mixture of *cis-* and *trans*-stilbene in a ratio of 30:70. It is possible to determine whether a mixture is formed in your synthesis not only by the melting-point behavior of the product but also because the IR, NMR, and UV spectra of the two isomers are distinctly different.

Optional Experiment

Aldehydes other than benzaldehyde (**16**) may be used in the phosphonate modification of the Wittig reaction. For example, the product from *trans*-cinnamaldehyde (**19**) is *trans, trans*-1,4-diphenyl-1,3-butadiene (**20**). Comparison of the UV spectrum of **20** with those of *cis-* and *trans*-stilbene (**17**), which are shown in Figure 18.7, is interesting because it shows the effect that an additional conjugated π-bond has on the wavelength of the absorption maxima that are present. If it is not practical to record the UV spectrum of **20** in your laboratory, a published spectrum is available in the reference given at the end of this section.

Analysis of the IR spectrum of the diene will assist in confirming that the newly formed carbon-carbon double bond of **20** is indeed *trans* rather than *cis* (see Ex. 12 at the end of this section).

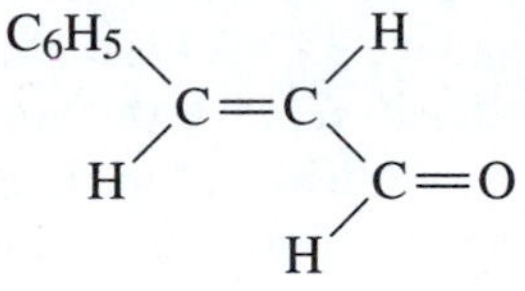

19	**20**
trans-Cinnamaldehyde	*trans, trans*-1,4-Diphenyl-1,3-butadiene

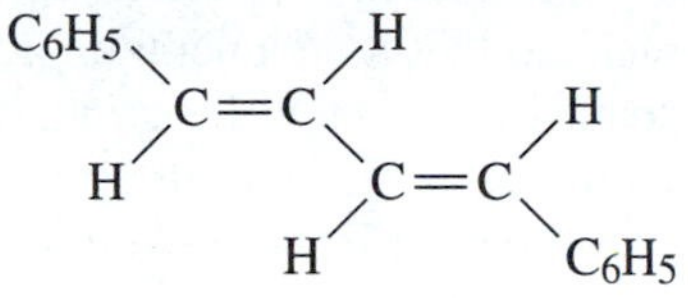

Pre-Lab exercises for Section 18.1, "Wadsworth-Emmons Modification of the Wittig Reaction," are found on page PL. 95.

EXPERIMENTAL PROCEDURES

SAFETY ALERT

1. Organophosphorus compounds are toxic, and benzyl chloride is irritating and lachrymatory. Avoid skin contact with these substances and inhaling their vapors. Perform experiments using these reagents in the hood if possible, and wear rubber gloves, particularly when measuring out the reagents. Should these substances come into contact with your skin, wash the area thoroughly with soap and warm water, and rinse it with large quantities of water.

2. Sodium methoxide is a powdery solid that easily becomes airborne. Take care to avoid inhaling this chemical as it irritates the mucous membranes.

A. *TRANS*-STILBENE

Note to Student. The sodium methoxide used in this experiment is very hygroscopic; so minimize its exposure to atmospheric moisture.

Working at the hood, measure 1.7 g of triethyl phosphite into a dry 10-mL round-bottom flask containing a stirring bar, if magnetic stirring is available, or several boiling stones. Add 1.3 g of benzyl chloride to the flask and then connect to it a water-cooled condenser protected with a drying tube containing calcium chloride. Heat the mixture under gentle reflux for 0.5 h. While this reaction mixture is cooling to room temperature, place 0.6 g of sodium methoxide in a dry 50-mL round-bottom flask and immediately add 5 mL of *N,N*-dimethylformamide (DMF). Protect this solution with a drying tube containing calcium chloride. Pour the *cooled* reaction mixture containing the phosphonate into the flask containing the sodium methoxide. Rinse the flask with 5 mL of DMF, and add this rinse to the reaction mixture. Return the drying tube to the flask and cool the reaction mixture to below 20 °C in an ice-water bath with swirling or stirring.

Remove the drying tube from the flask and slowly add 1.1 g of benzaldehyde while continuing to swirl or stir the cooled solution, cooling the mixture as necessary so that its temperature does not rise above 35 °C. Once the addition is complete, reattach the drying tube to the flask, remove the apparatus from the cooling bath, and allow the reaction mixture to stand at room temperature for about 15 min.＊ Add 10 mL of water with stirring, collect the precipitated product by vacuum filtration, and wash it with cold 1:1 methanol-water.

Determine the weight and melting point of the crystalline product. The reported melting point of *trans*-stilbene is 122–124 °C. If necessary, recrystallize the stilbene from acetone.

Finishing Touches   Flush all *filtrates* down the drain.

B. *TRANS, TRANS*-1,4-DIPHENYL-1,3-BUTADIENE

Note to Student. The sodium methoxide used in this experiment is very hygroscopic; minimize its exposure to atmospheric moisture.

Working at the hood, measure 1.7 g of triethyl phosphite into a dry 10-mL round-bottom flask containing a stirring bar, if magnetic stirring is available, or several boiling stones. Add 1.3 g of benzyl chloride to the flask and then connect to it a water-cooled condenser protected with a drying tube containing calcium chloride. Heat the mixture under gentle reflux for 0.5 h. While this reaction mixture is cooling to room temperature, place 0.6 g of sodium methoxide in a dry 50-mL round-bottom flask and immediately add 5 mL of *N,N*-dimethylformamide (DMF). Protect this solution with a drying tube containing calcium chloride. Pour the *cooled* reaction mixture containing the phosphonate into the flask containing the sodium methoxide. Rinse the flask with 5 mL of DMF, and add this rinse to the reaction mixture. Return the drying tube to the flask, and cool the reaction mixture to below 20 °C in an ice-water bath with swirling or stirring.

Remove the drying tube from the flask and, while continuing to swirl or stir the cooled solution, slowly add 1.3 g of cinnamaldehyde, cooling the mixture as necessary so that its temperature does not rise above 35 °C. Once the addition is complete, reattach the drying tube to the flask, remove the apparatus from the cooling bath and allow the reaction mixture to stand at room temperature for about 15 min.＊ Add 10 mL of water with stirring, collect the precipitated product by vacuum filtration, and wash it first with cold water and then with methanol until the washings are colorless.

Weigh the product and determine its melting point. The diene has mp 150–151 °C. If necessary, recrystallize impure product from ethyl acetate.

> **Finishing Touches**   Flush the *filtrates* down the drain. Place the *solvents* used for recrystallization in the container for nonhalogenated organic liquids.

EXERCISES

1. Compare the mechanism of aldol addition (see Sec. 18.2.1) to that of the Wittig synthesis, pointing out similarities and points of difference.

2. Why are ammonium salts of the type $(C_6H_5)_3\overset{\oplus}{N}CHR_2 \ \overset{\ominus}{Br}$ much less acidic than phosphonium salts such as **7**?

3. Write equations for the preparation of the following alkenes by the Wadsworth-Emmons modification of the Wittig reaction. Start with triethyl phosphite and any other organic or inorganic reagents that you require.
   a. $C_6H_5CH=C(CH_3)C_6H_5$
   b. $CH_2=CH-CH=CH-C_6H_5$
   c. $(CH_3)_2C=CH-CO_2C_6H_5$

4. Explain why you would expect an anion of type **8** to be *more* stable if one of the R-groups is cyano (CN) rather than alkyl.

5. Why should the sodium methoxide be exposed to the atmosphere for a minimal length of time?

6. Why should the aldehydes used as starting materials in the Wittig syntheses be free of contamination by carboxylic acids?

7. What peaks in the IR spectra of the stilbenes (Figs. 18.3 and 18.5) would be most useful for quantitative analysis of a mixture of the *cis-* and *trans-* isomers?

8. *N,N*-Dimethylformamide (DMF) is very water-soluble. Why?

9. DMF effectively solvates cations, which is one reason why sodium methoxide readily dissolves in it. Account for this property of DMF.

10. a. Refer to Figure 18.7 and calculate log ϵ for the three maxima in the UV spectrum of *trans*-stilbene and for the two maxima in the UV spectrum of *cis*-stilbene.
    b. Rationalize the differences in the UV spectra exhibited by these two diastereomers.

11. a. Determine whether the double bond in *trans*-stilbene is *E-* or *Z-*.
    b. Answer this question for *trans, trans*-1,4-diphenyl-1,3-butadiene.

12. Which peak(s) in the spectrum of *trans, trans*-1,4-diphenyl-1,3-butadiene (Fig. 18.8) support assignment of the geometry of the diene function as *trans?*

SPECTRA OF STARTING MATERIALS AND PRODUCTS

The IR spectrum and NMR data for benzaldehyde are given in Figures 16.9 and 16.10, respectively. The IR spectrum and NMR data for *trans*-cinnamaldehyde are presented in Figures 17.5 and 17.6, respectively.

FIGURE 18-1 IR spectrum of benzyl chloride.

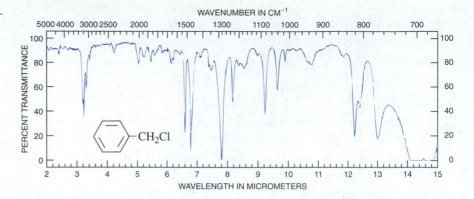

(a) ^{1}H NMR spectrum (90 MHz).

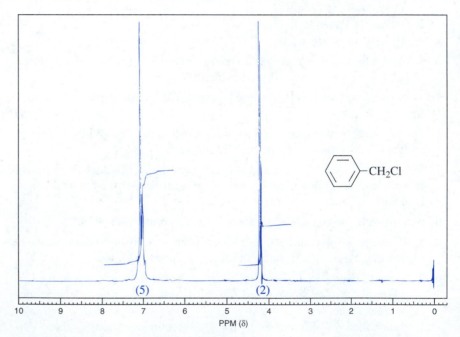

FIGURE 18-2 NMR data for benzyl chloride.

(b) ^{13}C NMR data.
Chemical shifts: δ 46.1, 128.2, 128.6, 137.5.

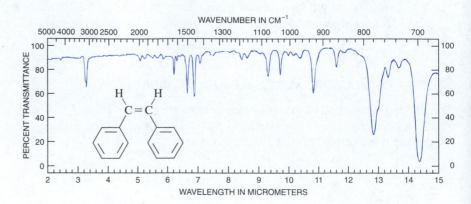

FIGURE 18-3 IR spectrum of *cis*-stilbene.

(a) ^{1}H NMR spectrum (60 MHz).

FIGURE 18-4   NMR data for *cis*-stilbene.

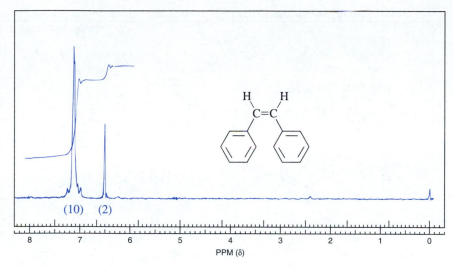

(b) ^{13}C NMR data.
Chemical shifts: δ 127.3, 128.4, 129.1, 130.5, 137.5.

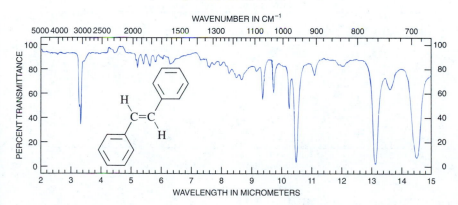

FIGURE 18-5   IR spectrum of *trans*-stilbene (KBr pellet).

(a) ^{1}H NMR spectrum (60 MHz).

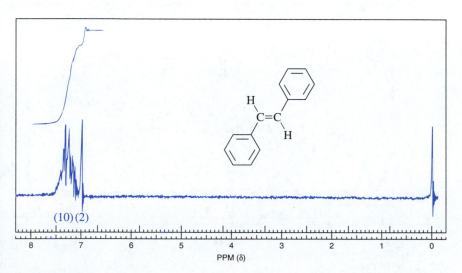

(b) ^{13}C NMR data.
Chemical shifts: δ 126.3, 127.8, 128.9, 129.0, 137.6.

FIGURE 18-6   NMR data for *trans*-stilbene.

FIGURE 18-7 UV spectrum of *cis*-stilbene (solid line) and *trans*-stilbene (dashed line). Concentration 0.0500 g/L, pathlength 1.0 cm.

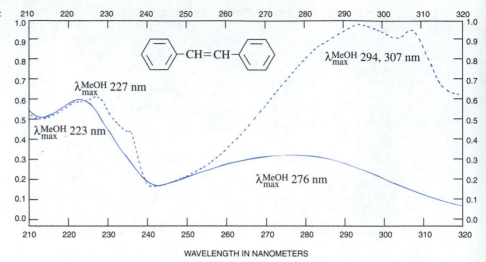

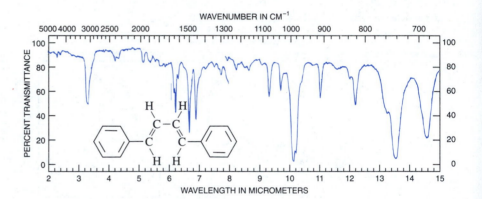

FIGURE 18-8 IR spectrum of *trans, trans*-1,4-diphenyl-1,3-butadiene (KBr pellet).

(a) ^{1}H NMR spectrum (60 MHz).

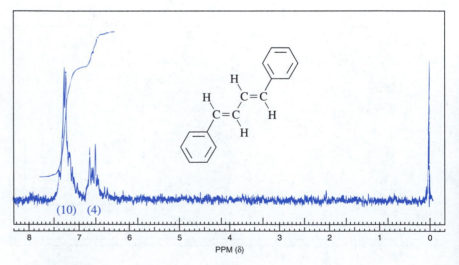

FIGURE 18-9 NMR data for *trans, trans*-1,4-diphenyl-1,3-butadiene.

(b) ^{13}C NMR data.
Chemical shifts: δ 127.6, 128.8, 129.9, 130.4, 134.0, 138.6.

REFERENCE

Pinckard, J. H.; Willie, B.; Zechmeister, L. *Journal of the American Chemical Society,* **1948,** *70,* 1939.

18.2 REACTIONS OF STABILIZED CARBANIONS DERIVED FROM CARBONYL COMPOUNDS

As described in the introduction to this chapter, the presence of the carbonyl function in a molecule allows the α-carbon atom to become nucleophilic by forming an **enol** (Eq. 18.2) or an **enolate ion** (Eq. 18.3). The nucleophilic property of an enol is the basis for the experiment in Section 18.3. Enolate ions, which are considerably more reactive nucleophiles than enols, are the subject of this section.

Formation of enolate ions (Eq. 18.3) involves the use of a base, and the strength of the base required will depend on the **pK_a** of the hydrogen atom α to the carbonyl carbon atom. The pK_a is defined as the negative log of K_a, the equilibrium constant for the ionization shown in Equation (18.12). Thus, $pK_a = -\log K_a$. Water is the normal reference solvent for reporting values of pK_a, which are known to be solvent dependent. Experimentally, the **pK_a-values** of aldehydes and ketones are in the range 18–20. This is a remarkably low value, when compared to those for saturated hydrocarbons, which are in excess of 50!

$$\underset{\alpha}{\overset{O}{\parallel}}\text{H}_{(aq)} \overset{K_a}{\rightleftharpoons} \text{H}^{\oplus}{}_{(aq)} + \underset{(aq)}{\overset{O}{\parallel}}{}^{\ominus} \qquad \textbf{(18.12)}$$

23
Enolate ion

The dramatic acidifying effect of the carbonyl group is due mainly to its ability to delocalize, and therefore to stabilize, the negative charge in the enolate ion, as reflected in **23** (Eq. 18.12), which is the **resonance hybrid** of the **resonance structures 3a** and **3b** (see Sec. 18.1). Because the pK_a-values of aldehydes and ketones fall in the range of water (pK_a 15.7) and alcohols (pK_a 15.5–18), it is possible to generate enolate ions using anions such as hydroxide or alkoxide (Eq. 18.3, $B^- = HO^-$ and RO^-, respectively); these are the bases utilized in the experimental procedures of this section.

18.2.1 ALDOL CONDENSATIONS

An important general reaction of enolate ions involves **nucleophilic addition** to the electrophilic carbonyl carbon atom of the aldehyde or ketone from which the enolate was derived. A dimeric anion **24** results, which may then be neutralized by abstraction of a proton to produce a β-hydroxycarbonyl compound, **25** (Eq. 18.13). As illustrated in Equation (18.13), if the reaction is performed in hydroxylic solvents such as water or an alcohol, the source of the proton may be the solvent, whose deprotonation will regenerate the base required for forming the enolate ion. The overall process is thus *catalytic* in the base that is used.

$$\left[\overset{\ominus}{\underset{}{O}}\!\!\diagup\!\!\diagup \longleftrightarrow \overset{O}{\underset{}{\diagup}}\!\!\ominus \right] + \overset{O}{\underset{}{\diagup}} \rightleftharpoons \overset{O\quad O^{\ominus}}{\diagup} \underset{H\text{-}OR}{\overset{}{\longrightarrow}} \overset{O\quad OH}{\underset{\alpha\ \beta}{\diagup}} + RO^{\ominus} \qquad \textbf{(18.13)}$$

3b   3a   **24**   **25**

Enolate ion   An aldol
(β-Hydroxycarbonyl compound)

The β-hydroxycarbonyl compound **25** is termed an **aldol,** and the reaction leading to its formation is referred to as an **aldol addition.** The origin of the word "aldol" is understood by realizing that use of an aldehyde in the reaction provides a dimerization product **25** containing an **aldehyde function** and an **alcohol group.** The isolation of such addition products depends on the reaction conditions used, because dehydration to an α,β-unsaturated carbonyl compound **26** may occur *if* **25** still has a hydrogen atom on the α-carbon atom that served as the nucleophile in the addition reaction (Eq. 18.14). The overall reaction described by Equations (18.13) and (18.14) is called an **aldol condensation.**

$$\overset{O\ \ OH}{\underset{\alpha\ \ \ \ \ }{\diagup}}\underset{\overset{|}{H}}{\diagup}\ \underset{^{-}OR}{\longrightarrow}\ \overset{O}{\underset{\alpha}{\diagup}}\overset{\beta}{\diagdown} + ROH + HO^{\ominus} \qquad \textbf{(18.14)}$$

25   **26**

α,β–Unsaturated
compound

The experiments described in this section are examples of **mixed** or **crossed aldol condensations.** This term refers to cases in which two *different* carbonyl compounds are the reactants. Such reactions are synthetically practical, selectively producing a single major condensation product in several circumstances. For example, two different *aldehydes* may be used, but only one of them has any α-hydrogen atoms and thus may serve as the nucleophile (Eq. 18.15). Alternatively, a ketone may condense selectively with an aldehyde because self-addition of two molecules of the ketone does not readily occur. This is because the carbonyl group of ketones is sterically and electronically not as susceptible to nucleophilic attack as is the carbonyl function of aldehydes.

$$\overset{O}{\underset{}{\overset{\|}{\text{—CH}}}} + C_6H_5\text{—}\overset{\alpha}{\underset{\overset{|}{H}}{\text{CHCHO}}} \xrightarrow{\text{Base}} \text{—CH}\text{=}\underset{\overset{|}{C_6H_5}}{\text{CCHO}} + H_2O \qquad \textbf{(18.15)}$$

Benzaldehyde   Phenylacetaldehyde   2-Phenylcinnamaldehyde
(No α-hydrogen atoms)

An example of the latter case is the reaction of acetophenone (**27**) with *p*-anisaldehyde (*p*-methoxybenzaldehyde, **29**) to give *p*-anisalacetophenone (**31**) according to the sequence outlined in Scheme 18.3. For steric reasons, the enolate ion **28** reacts preferentially with **29** rather than **27,** leading to the aldol **30.** This product then dehydrates in the presence of base, in the manner outlined in

Equation (18.14), to yield the condensation product **31.** Part of the thermodynamic driving force for this dehydration is associated with formation of a new carbon-bond π-bond that is **conjugated** with the aromatic ring as well as with the carbonyl function. As a general rule, formation of an extended conjugated system, as occurs here, increases the stability of a molecule.

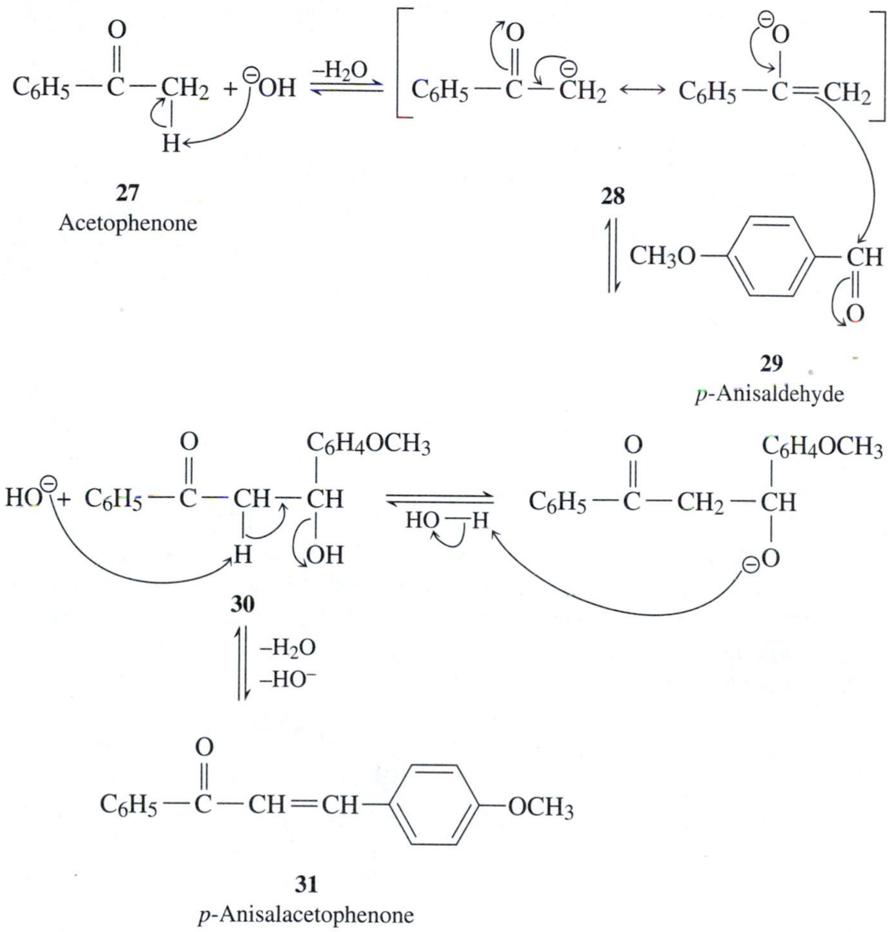

27
Acetophenone

28

29
p-Anisaldehyde

30

31
p-Anisalacetophenone

SCHEME 18.3

A second example of a mixed aldol condensation is that of 2-furaldehyde (**32**) with *m*-nitroacetophenone (**33**) to yield 3-(2-furyl)-1-(3-nitrophenyl)-propenone (**34,** Eq. 18.16). The mechanistic steps in this transformation are analogous to those shown in Scheme 18.3. Procedures for forming both **31** and **34** are provided in the experimental section.

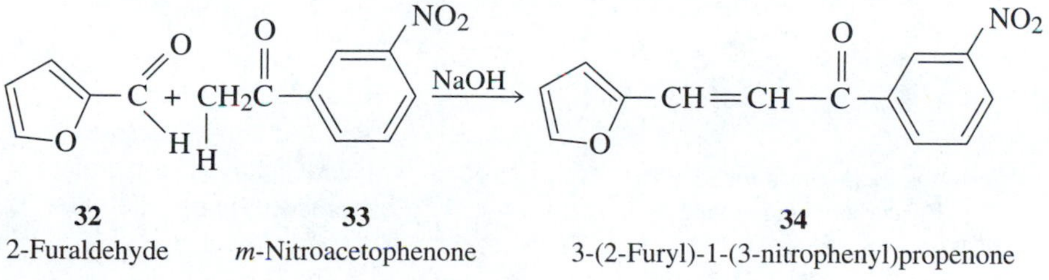

(18.16)

32
2-Furaldehyde

33
m-Nitroacetophenone

34
3-(2-Furyl)-1-(3-nitrophenyl)propenone

18.2.2 REACTIONS OF α,β-UNSATURATED KETONES

As expected, the aldol condensation products **31** and **34** undergo reactions that are typical of both the carbonyl function and the carbon-carbon double bond. For example, chemistry of the carbonyl group in **31** may be illustrated by preparation of the semicarbazone **35a** and the oxime **35b** by the procedures of Section 27.5.1.

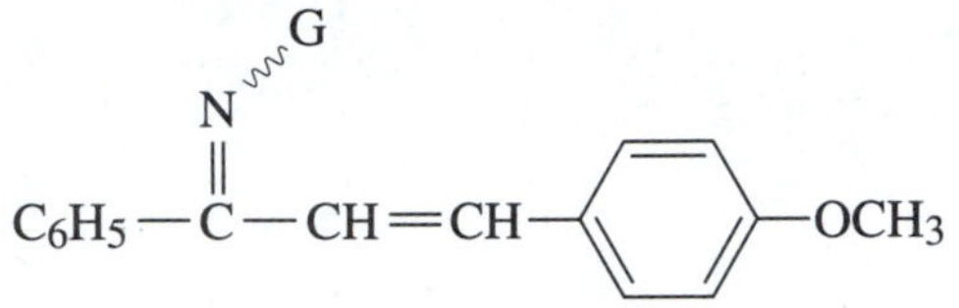

35

a. G = NHNHCNH$_2$
b. G = OH

A reaction that is characteristic of the carbon-carbon double bond of **31** is the addition of bromine (Eq. 18.17). Two chiral carbon atoms are present in the dibromide, so four stereoisomers, consisting of two pairs of enantiomers, may be formed by the addition. The product reported in the chemical literature (mp 140 °C) is presumed to have the *erythro* structure, **36** and would result from *anti* addition of bromine to *trans*-**31.** Because of its greater thermodynamic stability relative to *cis*-**31,** the *trans* diastereomer of **31** is the one expected to be formed preferentially by the mixed aldol condensation (see Ex. 7 at the end of this subsection). It is interesting to note that *cis*-**31** (mp 33–33.5 °C) may be obtained by irradiating *trans*-**31** with ultraviolet light.

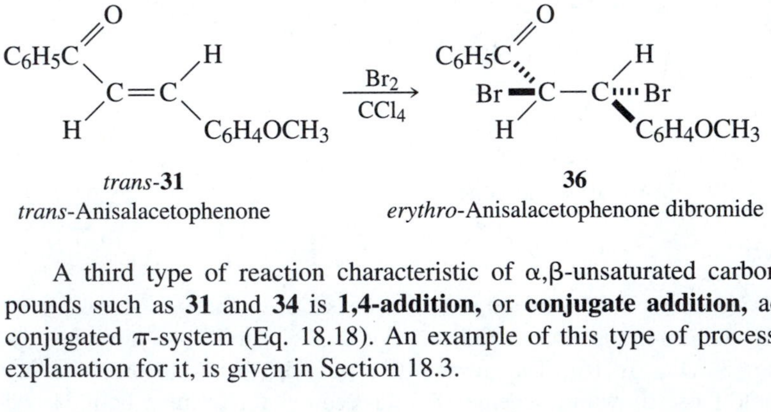

trans-**31**
trans-Anisalacetophenone

36
erythro-Anisalacetophenone dibromide

(18.17)

A third type of reaction characteristic of α,β-unsaturated carbonyl compounds such as **31** and **34** is **1,4-addition,** or **conjugate addition,** across the conjugated π-system (Eq. 18.18). An example of this type of process, and an explanation for it, is given in Section 18.3.

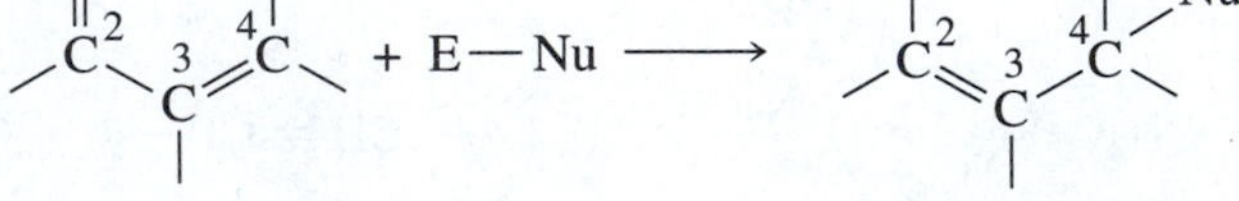

α,β–Unsaturated
carbonyl compound

1,4-Addition
product

(18.18)

Pre-Lab exercises for Section 18.2.1, "Aldol Condensations," are found on page PL. 97. Those for Section 18.2.2, "Reactions of α,β-Unsaturated Ketones," are found on page PL. 99.

EXPERIMENTAL PROCEDURES

SAFETY ALERT

1. The aqueous solution of concentrated sodium hydroxide is highly caustic. Take care not to get it on your skin; if you do, wash the area with copious amounts of water.
2. If you warm the solutions in 95% ethanol to hasten solution or recrystallize your products from this solvent, do *not* use a flame and take care that there are no flames in the vicinity.

The Aldol Condensation

1. Preparation of *p*-Anisalacetophenone. Place 1.4 g of *p*-anisaldehyde and 1.2 g of acetophenone in a 25-mL Erlenmeyer flask. Add 4 mL of 95% ethanol, and swirl the flask to dissolve the reactants. Put a pellet of NaOH in another Erlenmeyer flask containing 0.5 mL of water. Mash and stir the mixture with a glass rod to expedite dissolution; exercise care not to break the rod or the flask. Using a micropipet, transfer the NaOH solution into the ethanolic solution of the carbonyl compounds, swirl the mixture for a minute or two until a homogeneous solution results, and allow it to stand at room temperature for 10 min.✴ Cool the reaction mixture, in which crystals may have already begun to form, in an ice-water bath, and then collect the precipitate by vacuum filtration. Wash the crystals with a few milliliters of cold 95% ethanol, air-dry them,✴ and determine their melting point. The reported melting point of pure *trans*-anisalacetophenone is 77–78 °C. Recrystallize the product from 95% ethanol if necessary.

> **Finishing Touches** Neutralize the *ethanolic filtrates* with dilute hydrochloric acid before flushing them down the drain.

2. Preparation of 3-(2-Furyl)-1-(3-nitrophenyl)propenone. Follow the same procedure as that described in part 1 *except* dissolve 1.0 g of 2-furaldehyde and 1.7 g of *m*-nitroacetophenone in 10 mL of 95% ethanol. The 2-furaldehyde should be freshly distilled before use. Product **34** has mp 100–101 °C, and should be light yellow after being washed free of the dark by-products derived from 2-furaldehyde. Recrystallize the product if necessary from 95% ethanol.

> **Finishing Touches**   Neutralize the *ethanolic filtrates* with dilute hydrochloric acid before flushing them down the drain.

Reactions of α,β-Unsaturated Ketones

1. Preparations of a Semicarbazone and an Oxime of *p*-Anisalacetophenone. See Section 23.5.1 for directions for preparing these derivatives.

2. Preparation of *p*-Anisalacetophenone Dibromide.

SAFETY ALERT

> Handle the Br_2/CCl_4 solution with care because bromine is a hazardous chemical, even in solution. When using the reagent or solutions of it, perform the work at the hood, if possible, and wear rubber gloves. Do not breathe the vapors of the solution or spill it on your skin. If bromine or its solutions come in contact with your skin, wash the area *immediately* with warm water and soak the affected area in 0.6 *M* sodium thiosulfate solution, for up to 3 h if the burn is particularly serious.

Dissolve 0.4 g of *p*-anisalacetophenone in 1 mL of chloroform and add 2.5 mL of 1 *M* bromine in CCl_4 dropwise. Allow the mixture to stand at room temperature for 30–40 min✱ and collect the precipitate by vacuum filtration. Wash the crystals with a little cold CCl_4, allow them to air-dry,✱ and determine their melting point. The reported melting point of the dibromide, presumably the *erythro* diastereomer, is 139–140 °C.

> **Finishing Touches**   Pour the *filtrates* into the container for halogenated liquids.

EXERCISES

1. Compute the equilibrium constant K_{eq} for the reaction of equimolar amounts of acetophenone, $C_6H_5COCH_3$, and hydroxide ion to generate the enolate ion. The pK_a-values of the ketone and of water are 19.0 and 15.7, respectively.

2. Explain why the main reaction between acetophenone and *p*-anisaldehyde is the mixed aldol reaction rather than (a) self-condensation of acetophenone or (b) the Cannizzaro reaction (see Sec. 16.3) of *p*-anisaldehyde.

3. No more than about 0.5 mL of water should be used to dissolve the pellet of NaOH. Why should a larger volume of water *not* be used?

4. *p*-Anisalacetophenone dibromide is much more soluble in $CHCl_3$ than in CCl_4. Give a reason for this difference.

5. Crystals of pure *erythro-p*-anisalacetophenone dibromide are *white*, whereas crystals of pure *p*-anisalacetophenone are light *yellow*. Explain this difference in color.

6. Write a mechanism that rationalizes the stereochemistry of the formation of *erythro-p*-anisalacetophenone dibromide from the reaction of bromine and *trans-p*-anisalacetophenone.

7. Explain why *trans-p*-anisalacetophenone would be expected to more stable than the corresponding *cis* isomer.

8. Refer to Figures 18.16 and 18.23 and calculate log ϵ for the two maxima in the UV spectrum of *trans-p*-anisalacetophenone and for the single maximum in its dibromide.

9. Why does *trans-p*-anisalacetophenone exhibit a maximum in its UV spectrum (Fig. 18.18) at a substantially longer wavelength than does its dibromide? What electronic excitation is responsible for the absorption at longer wavelength?

SPECTRA OF STARTING MATERIALS AND PRODUCTS

The IR, NMR, and UV spectra of 2-furaldehyde are provided in Figures 13.5–13.7.

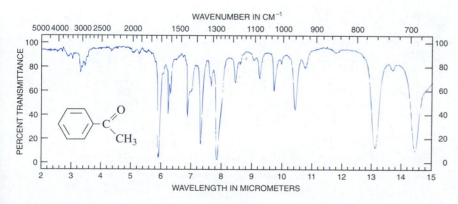

FIGURE 18-10   IR spectrum of acetophenone.

FIGURE 18-11 NMR data for acetophenone.

(a) ^{1}H NMR spectrum (60 MHz).

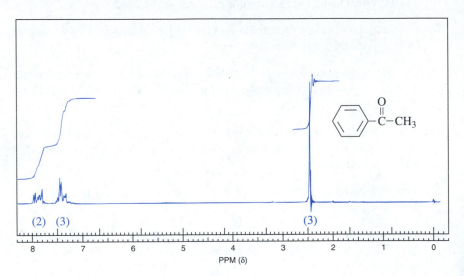

(b) ^{13}C NMR data.
Chemical shifts: δ 26.3, 128.6, 128.3, 133.0, 137.3, 197.4.

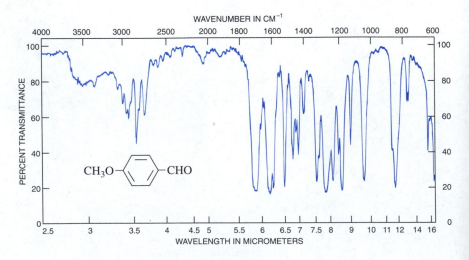

FIGURE 18-12 IR spectrum of *p*-anisaldehyde.

(a) ^{1}H NMR spectrum (90 MHz).

FIGURE 18-13   NMR data for *p*-anisalde-hyde.

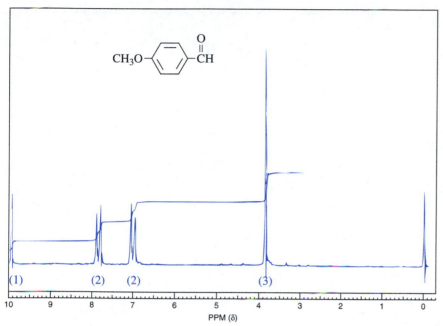

(b) ^{13}C NMR data.
   Chemical shifts: δ 55.5, 114.5, 130.2, 131.9, 164.6, 190.5.

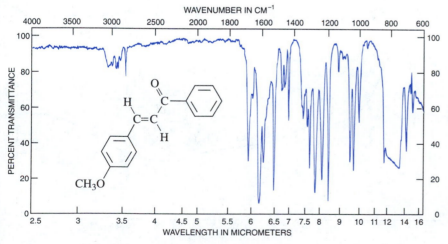

FIGURE 18-14   IR spectrum of *trans-p*-anisalacetophenone (in CCl$_4$ solution).

FIGURE 18-15   NMR data for *trans-p*-anisalacetophenone.

(a) ^{1}H NMR spectrum (90 MHz).

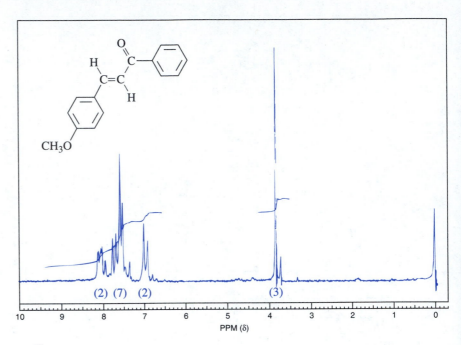

(b) ^{13}C NMR data.
Chemical shifts: δ 55.3, 114.4, 119.7, 127.6, 128.4, 128.5, 130.2, 132.5, 138.5, 144.6, 161.7, 190.4.

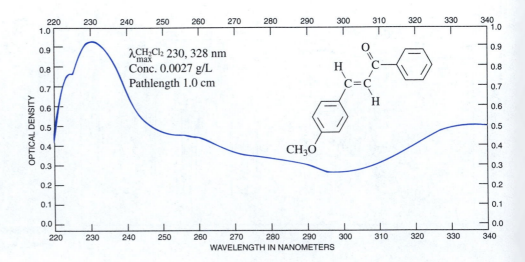

λ$_{max}^{CH_2Cl_2}$ 230, 328 nm
Conc. 0.0027 g/L
Pathlength 1.0 cm

FIGURE 18-16   UV spectrum of *trans-p*-anisalacetophenone.

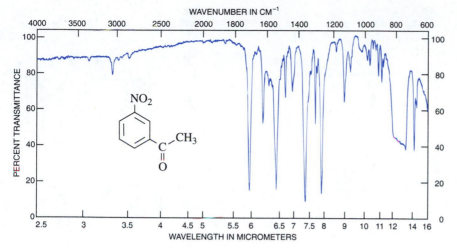

FIGURE 18-17   IR spectrum of *m*-nitroacetophenone (in CCl$_4$ solution).

(a) ^{1}H NMR spectrum (90 MHz).

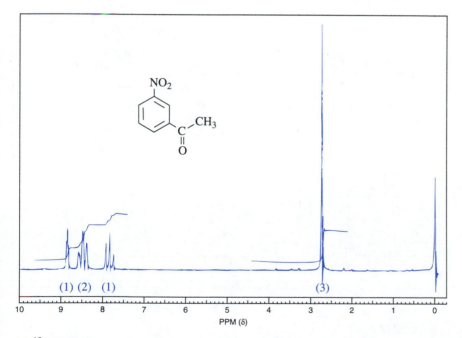

(b) ^{13}C NMR data.
Chemical shifts: δ 26.6, 122.9, 127.2, 130.1, 133.9, 138.5, 148.6, 195.6.

FIGURE 18-18   NMR data for *m*-nitroacetophenone.

FIGURE 18-19 IR spectrum of 3-(2-furyl)-1-(3-nitrophenyl)propenone (in CCl₄ solution).

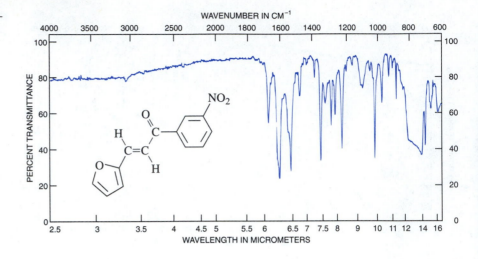

(a) ¹H NMR spectrum (90 MHz).

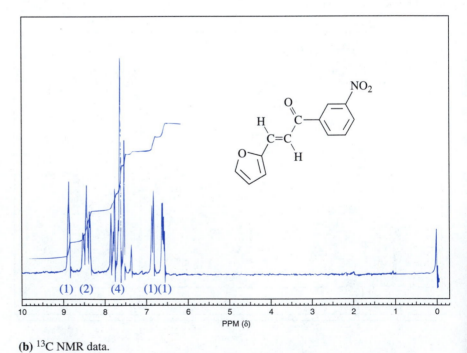

(b) ¹³C NMR data.

Chemical shifts: δ 113.0, 117.6, 117.8, 123.1, 126.9, 132.0, 133.9, 145.6, 148.4, 151.3, 187.1.

FIGURE 18-20 NMR data for 3-(2-furyl)-1-(3-nitrophenyl)propenone.

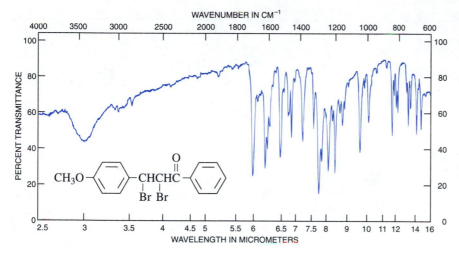

FIGURE 18-21 IR spectrum of *p*-anisalacetophenone dibromide (KBr pellet).

(a) ^{1}H NMR spectrum (90 MHz).

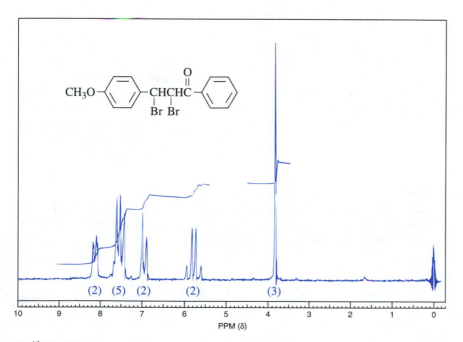

(b) ^{13}C NMR data.

Chemical shifts: δ 46.8, 49.9, 55.0, 113.8, 128.5, 128.6, 129.2, 129.9, 133.8, 134.1, 159.8, 190.9.

FIGURE 18-22 NMR data for *p*-anisalace-tophenone dibromide.

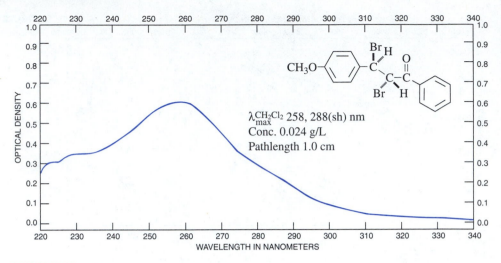

FIGURE 18-23   UV spectrum of *p*-anisalacetophenone dibromide.

18.2.3 ALKYLATION OF DIMETHYL MALONATE

The **aldol addition** (Eq. 18.13) and **condensation reactions** (Eqs. 18.13 and 18.14) are of great importance in organic chemistry because they are methods for generating new carbon-carbon bonds and because the products retain functional groups on which a variety of further synthetic transformations may be performed. In a general way, the crucial carbon-carbon bond-forming step (Eq. 18.13) in the process is reaction of a **nucleophile,** the **enolate ion,** with an **electrophile,** the carbonyl carbon atom. This general approach to forming carbon-carbon bonds may be extended to the reaction of enolate ions with other carbon electrophiles. The use of one such class of electrophiles, namely alkyl halides, R–X, is the subject of the experimental procedure contained in this subsection.

Alkylation of the enolate ions from aldehydes and ketones is plagued with problems, such as competing aldol reactions and polyalkylation (Eq. 18.19), that are difficult to overcome within the constraints of equipment and chemicals available in the introductory organic laboratory. However, such problems may be circumvented by employing a precursor to the enolate ion that is less prone to self-addition than is an aldehyde or ketone and by using a specialized alkylating agent with which polyalkylation is a disfavored process.

$$
\underset{\substack{\text{H} \qquad \text{H}}}{\overset{\displaystyle O}{\text{C–C–C}}} \xrightarrow[\substack{\text{R–X} \\ (X\ =\ \text{halogen})}]{\text{Base}} \underset{\substack{\text{H} \qquad \text{R}}}{\overset{\displaystyle O}{\text{C–C–C}}} \xrightarrow[\text{R–X}]{\text{Base}} \underset{\substack{\text{R} \qquad \text{R}}}{\overset{\displaystyle O}{\text{C–C–C}}} \qquad (18.19)
$$

The source of the enolate ion in our procedure is the *di*ester, dimethyl malonate (**37**). This carbonyl-containing compound is more acidic than monofunctional aldehydes, ketones, or esters owing to the presence of *two* carbonyl groups that assist in the delocalization of the negative charge in the conjugate base (**38**, Eq. 18.20). As a result, **37** has a pK_a of about 13, a value that is substantially below that of an aldehyde or ketone (see Sec. 18.2.2) or of a *mono*ester like methyl acetate, $CH_3CO_2CH_3$, whose pK_a is about 25. This means that treatment of diester **37** with bases as weak as alkoxide ions, RO^-, provides the enolate ion **38** essentially quantitatively (see below).

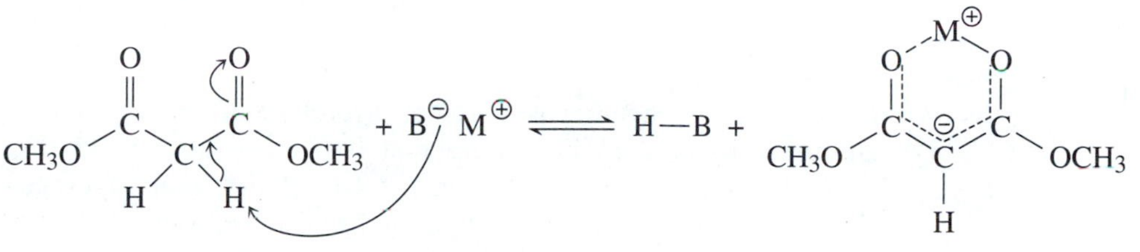

37
Dimethyl malonate

38

(18.20)

Although **38** might add to dimethyl malonate (**37**) and lead to products of self-condensation, this process is less important than the analogous reaction with aldehydes and ketones. The reason for this is that the π-bond of the carbonyl group of an ester is part of a delocalized system of π-electrons, which is a stabilizing feature that is lost upon addition of a nucleophile to the carbonyl carbon atom; this raises the **energy of activation,** E_a, for attack of the nucleophile according to Equation (18.21).

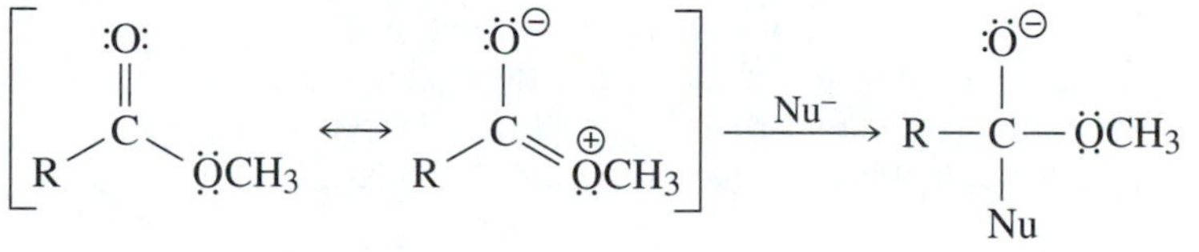

(18.21)

Ester resonance

The choice of the specific base for deprotonating dimethyl malonate (**37**) is important for two primary reasons. One of these is that the base must be strong enough to remove a proton from **37.** Given that the pK_a of this diester is approximately 13, bases such as hydroxide or alkoxide meet this criterion (see Ex. 8 at the end of this subsection). However, another requirement is that the base should *not* react at the carbonyl carbon atom of **37** by addition-elimination to give a *different* product (Eq. 18.21). This potential side-reaction, termed **nucleophilic acyl substitution,** excludes bases like hydroxide and ethoxide because they would substitute for methoxide in **37** and lead to hydrolysis and transesterification, respectively (Eq. 18.22, $B^- = HO^-$ and $C_2H_5O^-$, respectively). Transesterification would be less likely if the sterically bulky alkoxide, *tert*-butoxide, were utilized as the base, but this base is rather expensive. Methoxide ion is thus the preferred choice as the base for this reaction ion because the acyl substitution process simply regenerates dimethyl malonate (Eq. 18.22, $B^- = CH_3O^-$); the undesired reaction pathway is structurally "invisible"!

$$\underset{\text{CH}_3\text{OCCH}_2}{\overset{\text{O}}{\parallel}}\underset{\text{OCH}_3}{\overset{\ddot{\text{O}}:}{\underset{\parallel}{\text{C}}}} \quad \underset{\text{B}^-\ \text{M}^+}{\overset{}{\rightleftharpoons}} \quad \underset{\text{CH}_3\text{OCCH}_2}{\overset{\text{O}}{\parallel}}\underset{\underset{\text{B}}{|}}{\overset{\overset{:\ddot{\text{O}}^{\ominus}\ \text{M}^{\oplus}}{|}}{\underset{}{\text{C}}}}\text{OCH}_3 \quad \underset{-\text{MOCH}_3}{\overset{}{\rightleftharpoons}} \quad \underset{\text{CH}_3\text{OCCH}_2}{\overset{\text{O}}{\parallel}}\underset{\text{B}}{\overset{\overset{\text{O}}{\parallel}}{\text{C}}} \qquad \textbf{(18.22)}$$

The fact that hydroxide is not a suitable base because it would lead to hydrolysis of **37** is the basis for an important experimental precaution: All reagents, solvents, and apparatus employed in the alkylation of dimethyl malonate must be scrupulously *dry*. Water would react with methoxide ion to produce hydroxide ion, which in turn would promote hydrolysis of the diester **37.**

The alkylating agent used in the reaction is 1-bromohexane (**39,** Eq. 18.23). The overall reaction involves replacing one of the acidic hydrogen atoms of **37** with a new carbon-carbon bond to yield dimethyl heptane-1,1-dicarboxylate (**40**).

$$(\text{CH}_3\text{O}_2\text{C})_2\text{CH}_2 \ + \ \text{Br}-\text{CH}_2(\text{CH}_2)_4\text{CH}_3 \quad \xrightarrow[\text{CH}_3\text{OH}]{\text{NaOCH}_3} \quad \underset{\text{CH}_3\text{O}_2\text{C}}{\overset{\text{CH}_3\text{O}_2\text{C}}{\diagdown}}\text{C}\underset{\text{CH}_2(\text{CH}_2)_4\text{CH}_3}{\overset{\text{H}}{\diagup}} \qquad \textbf{(18.23)}$$

37	**39**	**40**
	1-Bromohexane	Dimethyl heptane-1,1-dicarboxylate

By-products may result from reactions other than the desired monoalkylation of the anion **38** (Eq. 18.20). For example, reaction of methoxide ion with 1-bromohexane (**39**) could produce hexyl methyl ether (**41,** Eq. 18.24) and/or 1-hexene (**42,** Eq. 18.25), by S_N2 and E2 processes, respectively. Fortunately, under the conditions used in this experiment, these side reactions are not competitive with that involving the formation of **38** (Eq. 18.20, $\text{B}^- = \text{CH}_3\text{O}^-$) and its subsequent reaction with **39,** according to Equation (18.23).

$$\text{CH}_3\text{O}^{\ominus} \ + \ \text{Br}-\text{CH}_2(\text{CH}_2)_4\text{CH}_3 \quad \longrightarrow \quad \text{CH}_3\text{O}-\text{CH}_2(\text{CH}_2)_4\text{CH}_3 \ + \ \text{Br}^{\ominus} \qquad \textbf{(18.24)}$$

39	**41**
	Hexyl methyl ether

$$\text{CH}_3\text{O}^{\ominus} \ + \ \text{Br}-\text{CH}_2-\overset{\overset{\text{H}}{|}}{\text{CH}}(\text{CH}_2)_3\text{CH}_3 \quad \longrightarrow \quad \text{H}_2\text{C}{=}\text{CH}(\text{CH}_2)_3\text{CH}_3 \ + \ \text{Br}^{\ominus} \qquad \textbf{(18.25)}$$

39	**42**
	1-Hexene

Another possible side reaction is further alkylation of the initial product **40** to afford the dialkylated malonic ester **43** (Eq. 18.26). Formation of this by-product is minimized by adding 1-bromohexane *to* the methanolic solution comprised of dimethyl malonate, the anion **38,** and sodium methoxide. This mode of addition maximizes the possibility that it will be **38,** rather than the corresponding enolate ion derived from **40,** that attacks 1-bromohexane. Moreover, this latter anion is sterically more hindered than is **38,** so its reaction with the alkylating agent is slower.

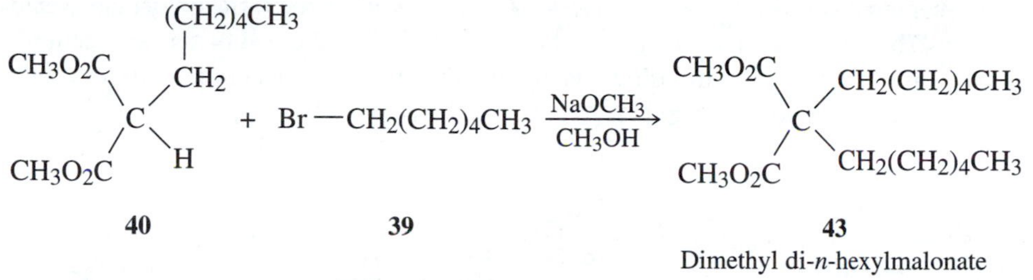

(18.26)

40   **39**   **43**

Dimethyl di-*n*-hexylmalonate

One additional possible side reaction bears noting. Enolate ion **38** is an **ambident ion;** that is, because of charge delocalization, **38** has *two* types of sites that may function as nucleophilic centers. Only reactions involving the nucleophilic *carbon* atom of this ion have been discussed thus far, but attack by an *oxygen* atom of **38** on **39** to produce **44** (Eq. 18.27), the *O*-alkylation product, must be considered. This possible complication is minimized under our reaction conditions by the choice of methanol as the solvent. Along with the sodium counter ion, the methanol molecules tend to cluster around the oxygen atom of **38** because this atom, being more electronegative than carbon, carries the majority of the negative charge in the delocalized enolate ion. Among other things, the phenomenon of clustering increases the steric hindrance about the oxygen atom and favors the desired nucleophilic attack by the carbon atom.

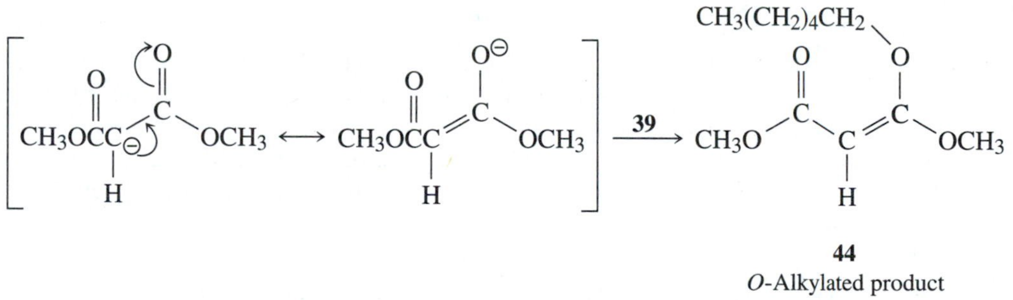

(18.27)

44

O-Alkylated product

Dimethyl heptane-1,1-carboxylate (**40**) is readily converted to the corresponding dicarboxylic acid **46** by hydrolysis under *basic* conditions, followed by acidification of the dibasic salt **45** that results (Eq. 18.28). A sequence for formation and isolation of **46** is provided in the Experimental Procedure.

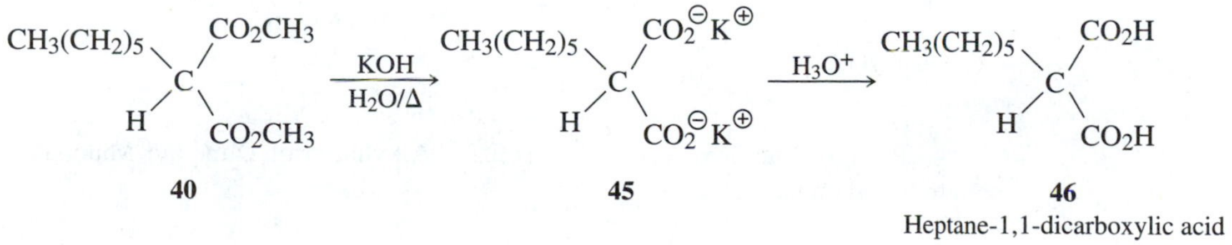

(18.28)

40   **45**   **46**

Heptane-1,1-dicarboxylic acid

Alternatively, hydrolysis of **40** under *acidic* conditions directly yields the diacid **46,** which undergoes *decarboxylation* to **47b** when heated to about 180 °C (Scheme 18.4). The decarboxylation reaction is believed to occur through a six-centered transition state to give **47a,** which is the enol form of octanoic acid (**47b**), the compound actually isolated (see Ex. 17 at the end of this subsection). This mode of decomposition is frequently observed in compounds in which there is a carbonyl group beta to a carboxylic acid function.

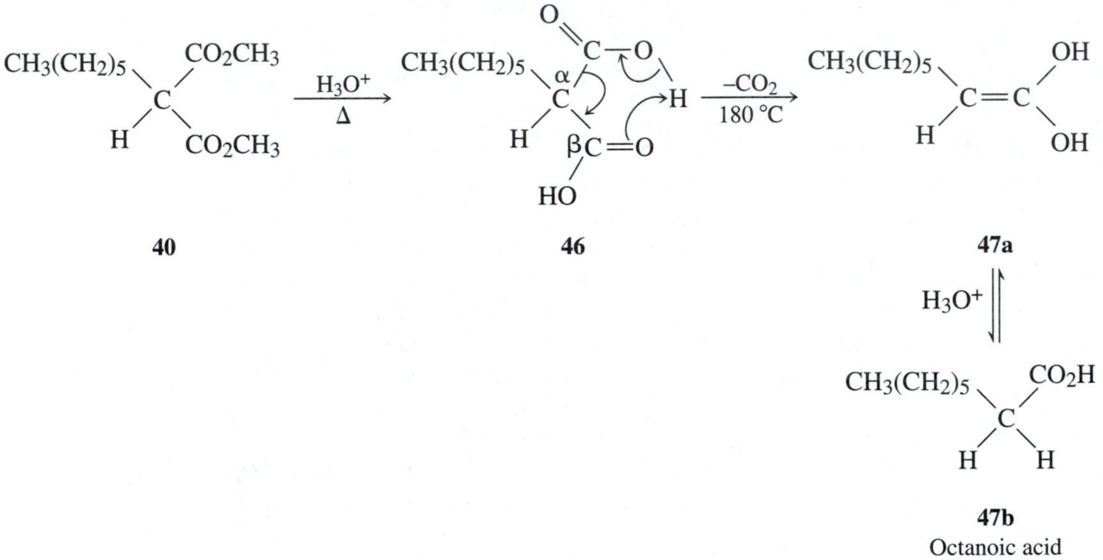

40   **46**   **47a**

47b
Octanoic acid

SCHEME 18.4

The overall reaction embodied in Equation (18.23) and Scheme 18.4 constitutes an example of the **malonic ester synthesis of substituted acetic acids** and represents a valuable way to make such compounds. Moreover, the preparation of mono- and dialkylated malonic esters themselves (Eqs. 18.23 and 18.26) is a key step in the synthesis of barbiturates (**48,** Eq. 18.29), a class of substances that serve as hypnotic drugs.

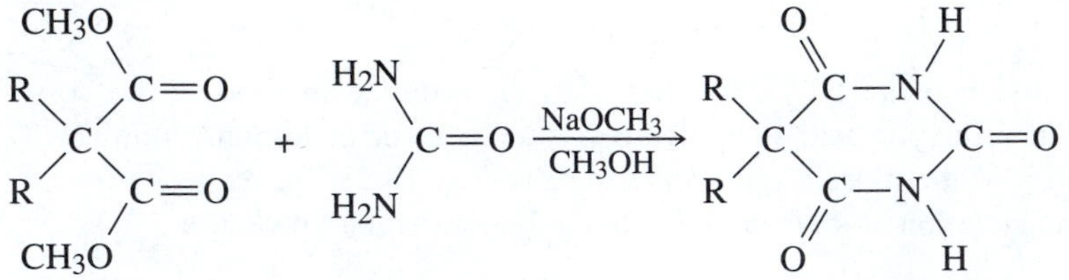

(18.29)

A dialkyl malonate   Urea   **48**
A barbiturate

Pre-Lab exercises for Section 18.2.3, "Alkylation of Dimethyl Malonate," are found on page PL. 101.

SAFETY ALERT

1. Sodium methoxide is a powdery solid that easily becomes airborne. Take care to avoid inhaling this chemical, as it causes irritation of the mucous membranes.

2. If sodium methoxide is to be prepared from the reaction of sodium and *anhydrous* methanol, use only *dry* containers, forceps, and other apparatus. Sodium metal reacts violently with water with the formation, and possible explosive combustion, of hydrogen gas.

3. The reaction between sodium metal and methanol is also very exothermic, and evolves hydrogen. *Strictly follow the directions given for portion-wise addition of the metal to the alcohol to keep the reaction under control.*

4. Wear rubber gloves if you need to handle metallic sodium, and minimize its exposure to atmospheric moisture.

5. Methanolic sodium methoxide and ethanolic potassium hydroxide are strongly caustic solutions. If they should come into contact with your skin, wash the affected area immediately with copious amounts of cold water.

6. Strong heating of mixtures containing undissolved solids can cause superheating that results in severe bumping. Heat such mixtures carefully and, if possible, stir them to minimize any bumping.

7. When performing a vacuum distillation, take care not to use any glassware that is cracked or otherwise damaged. Defective glassware may implode under reduced pressure, especially when heated, and this may result in sudden spillage of hot chemicals and danger of cuts from broken glass.

A.   PREPARATION OF DIMETHYL HEPTANE-1,1-DICARBOXYLATE

Note to Student.   The sodium methoxide used in this experiment is very hygroscopic; minimize its exposure to atmospheric moisture. The methanol used in this experiment must be anhydrous; dry it over 3-Å molecular sieves for 24 h before use. Distil the dimethyl malonate just prior to its use or dry it over 3-Å molecular sieves for 24 h.

Flame- or oven-dry a 50-mL round-bottom flask, fit it with a calcium chloride drying tube, and allow the apparatus to cool to room temperature. Equip the flask with several boiling stones or a magnetic stirring bar, if magnetic stirring is available, and transfer 15 mL of *anhydrous* methanol to it.

Prepare a solution of methanolic sodium methoxide in the flask by one of the following two methods. (Because handling of metallic sodium by a large class presents hazards, we recommend the use of solid sodium methoxide that has been preweighed in a dry vial.)

Method 1. Cool the methanol in an ice-water bath, and add 1.1 g of solid sodium methoxide *in 5 or 6 portions* to it. Because the heat of solution is high, the mixture should be chilled in an ice-water bath as the portions of methoxide are added with swirling.

Method 2. Place 5 mL of *dry* toluene in each of two *dry* 25-mL Erlenmeyer flasks, and weigh one of the two. The required amount of sodium is now to be added to the *weighed* beaker. To do this, obtain a pea-sized piece of metallic sodium by spearing it with a dry knife or spatula. Since the metal has probably been stored under mineral oil, remove any mineral oil from the piece by briefly swirling it in the *unweighed* beaker of toluene, blot the cleaned sodium with a paper towel, and immediately put it into the *weighed* beaker. Repeat the sequence until 0.5 g of cleaned sodium metal has been obtained. To prepare sodium methoxide, spear a piece of sodium with a knife or spatula, blot it briefly, and then drop it into the flask containing the anhydrous methanol. *Caution: No flames; hydrogen is evolved.* Replace the drying tube on the flask and wait until the initial vigorous gas evolution has subsided before adding the next piece of metallic sodium. Allow time for reaction of all of the metal before continuing.

After preparing the solution of sodium methoxide, replace the drying tube with a Claisen connecting tube and fit its two openings with a condenser and an addition funnel; the funnel should be on the arm of the connecting tube that leads directly to the flask. Alternatively, if you will add reagents with a syringe, put a serum cap rather than an addition funnel on the connecting tube. Protect both the condenser and the addition funnel with drying tubes. Rapidly add 2.6 g of dimethyl malonate to the reaction flask to prepare a methanolic solution of dimethyl sodiomalonate (**38,** M = Na); bring the solution to reflux. Next add 3.6 g of 1-bromohexane dropwise over a period of 5 min. Rinse the syringe or addition funnel with 2–3 mL of anhydrous methanol. Continue to reflux the reaction mixture until the solution is neutral, which should occur in less than 1 h. The pH can be tested by addition of 3 or 4 drops of the mixture to about 0.5 mL of water and then testing the solution with litmus or pHydrion paper. If neutrality, as evidenced by this test, has not been reached after 1 h at reflux, add sufficient glacial acetic acid to make the reaction mixture weakly acidic.✶

Allow the mixture to cool slightly, and arrange the flask for simple distillation, using a graduated cylinder as the collection vessel. Distil the methanol as rapidly as bumping caused by the precipitated sodium bromide will allow; if possible, continue stirring the mixture throughout the distillation to minimize bumping. After about 12 mL of distillate is collected, allow the pot residue to cool to room temperature.✶ Add 20 mL of water, and swirl the mixture until all salts have dissolved. Transfer the two-phase mixture to a separatory funnel and add 10 mL of diethyl ether to dissolve the organic products. Separate the layers and extract the aqueous phase two additional times with 10-mL portions of diethyl ether. Wash the combined ethereal extracts once with about 5 mL of saturated brine, and dry the solution over anhydrous sodium sulfate.✶

Decant the solution into a flask equipped for simple distillation, and remove the diethyl ether and any residual methanol by simple distillation (*no flames*), terminating the distillation when the head temperature reaches about 80 °C.✱ Place the residue in a round-bottom flask of appropriate size and equipped for short-path vacuum distillation (Fig. 2.15b), and distil the product under reduced pressure (see Sec. 2.9), using either a water aspirator or some other source of reduced pressure. The desired product has bp 129–130 °C (16 torr) and should be colorless. Weigh the product and calculate the yield. The residue in the stillpot is largely the dialkylated ester **43,** bp 160–163 °C (8 torr).

Finishing Touches   Flush the *methanol* recovered by distillation down the drain. Do the same with the *aqueous solution* that has been extracted with diethyl ether. Put the *diethyl ether* and any *pot residue* from the final distillation into a container for nonhalogenated organic liquids. Spread the *sodium sulfate* on a tray in the hood and then transfer it to the container for nonhazardous solids after the volatiles have evaporated. Put the *calcium chloride* in this same container.

B. PREPARATION OF HEPTANE-1,1-DICARBOXYLIC ACID

Combine 0.5 g of potassium hydroxide and 1 mL of 95% ethanol with each gram of dimethyl heptane-1,1-dicarboxylate that is to be hydrolyzed, and place this mixture in an appropriately sized flask equipped with a stirring bar, if magnetic stirring is available, or a few boiling stones and a reflux condenser. Heat the mixture at reflux for 0.5 h, and then allow it to cool to room temperature.✱ Acidify the solution to pH 2 by *cautious* addition of conc. HCl;✱ isolate the product by vacuum filtration.

The desired diacid should be colorless, mp 105–106 °C. If necessary, recrystallize it from toluene or ethyl acetate (*no flames*). After obtaining the melting point of the product, heat the capillary tube containing the sample at 150 °C for 10 min and then allow it to cool to room temperature, making note of any changes that occur in the sample.

Finishing Touches   Flush the *ethanol* and *methanol* that are recovered by distillation down the drain. Neutralize the acidic *aqueous solution* and pour it down the drain. Allow the residual volatiles to evaporate from the *sodium sulfate* in the hood before putting the drying agent in the container for nonhazardous wastes. Place the *diethyl ether* in the container for nonhalogenated organic liquids. Do the same with any *pot residue* from the distillation and with the *toluene* or *ethyl acetate* from the recrystallization of the diacid.

C. PREPARATION OF OCTANOIC ACID

Combine 3 mL of concentrated hydrochloric acid and 1.5 mL of water with each gram of dimethyl heptane-1,1-dicarboxylate that is to be hydrolyzed, and place this mixture in an appropriately sized flask equipped with a stirring bar, if magnetic stirring is available, or several boiling stones and a reflux condenser. Heat and stir or occasionally swirl the mixture at reflux until the solution becomes homogeneous,✶ and continue heating the solution for an additional hour.✶ Arrange the flask for simple distillation. Once the methanol, water, and hydrochloric acid have been removed by distillation, continue heating the stillpot until *gas* evolution ceases.✶ If an oil bath has been used for heating, it must ultimately be heated to about 180 °C to complete the decarboxylation. Isolate the crude product by a short path distillation under vacuum (see Fig. 2.15b), bp 129–130 °C (25 torr).

> **Finishing Touches**   Neutralize the *distillate* before pouring it down the drain. Put any *pot residues* from the distillations in the container for non-halogenated organic liquids

EXERCISES

1. Enolate ions are *more* nucleophilic than are enols. Explain why this is so.

2. What is the expected consequence of using methanolic sodium *ethoxide* rather than sodium *methoxide* in the alkylation reaction?

3. Write reactions showing the by-products expected if the alkylation procedure is conducted with reagents and/or solvents that are contaminated with water.

4. Why might the reaction mixture in the alkylation step *not* become neutral after the 1-h period of reflux?

5. What might the consequences be if the reaction mixture from the alkylation step were still basic during the work-up procedure? If acid is added to neutralize the reaction mixture, why is it important that the mixture not be made *strongly* acidic?

6. Explain why dimethyl malonate (**37**) is expected to be more acidic than its monoalkylation product **40**.

7. Explain why by-product **43** would be expected to have a higher boiling point than the desired product **40**.

8. Calculate the equilibrium concentration of dimethyl sodiomalonate (**38**) if 0.06 mol of dimethyl malonate (**37**) is added to 33 mL of 3 M sodium methoxide in methanol. The pK_a of **37** is about 13, and that of methanol is about 15.5.

9. Draw the structure of the *self-condensation* product expected to arise from the methoxide-promoted reaction of two molecules of dimethyl malonate (**37**). Predict whether this compound would be more or less acidic than **37** and explain your prediction.

10. In the alkylation step of the procedure, most of the methanol was removed by distillation *prior to* the addition of water and subsequent extraction with diethyl ether. Why is this preferred to an alternate work-up procedure in which water is first added to the reaction mixture, and this is followed by extraction with diethyl ether?

11. Why is base-promoted rather than acid-catalyzed hydrolysis the preferred method for preparing heptane-1,1-dicarboxylic acid (**46**) from the corresponding diester **40?**

12. The dipotassium salt (**45**) of heptane-1,1-dicarboxylic acid (**46**) does *not* decarboxylate when heated at 180 °C, whereas **46** does. Explain this difference in reactivity.

13. Explain the changes that occur when the melting point sample of heptane-1-dicarboxylic acid is heated to 150 °C and then cooled.

14. Why would calcium chloride *not* be an appropriate choice as the drying agent for the ethereal solution of heptane-1,1-dicarboxylic acid (**46**) obtained by base-promoted hydrolysis of **40?**

15. Why does sodiomalonate ion (**38**) react with 1-bromohexane (**39**) in preference to dimethyl malonate (**37**)?

16. Write a stepwise reaction mechanism for the acid-catalyzed tautomerization of **47a** to octanoic acid (**47b**).

SPECTRA OF STARTING MATERIALS AND PRODUCTS

The IR and NMR spectra of 1-bromohexane are presented in Figures 10.30 and 10.31, respectively.

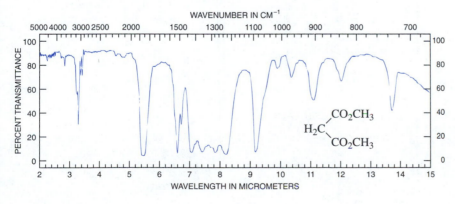

FIGURE 18-24   IR spectrum of dimethyl malonate.

FIGURE 18-25　NMR data for dimethyl malonate.

(a) ^1H NMR spectrum (60 MHz).

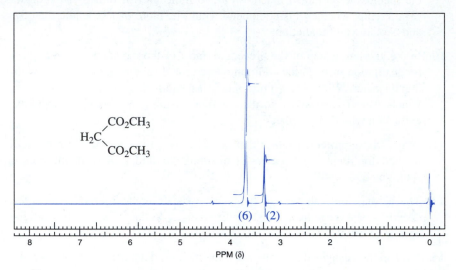

(b) ^{13}C NMR data.
　　Chemical shifts: δ 41.2, 52.4, 167.3.

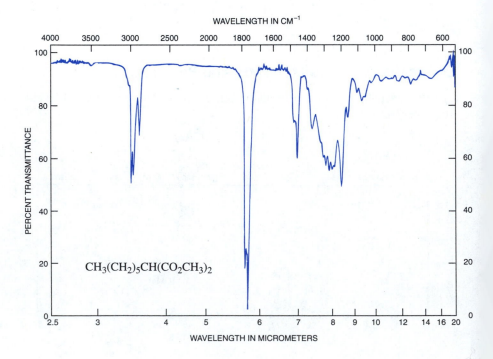

FIGURE 18-26　IR spectrum of dimethyl heptane-1,1-dicarboxylate.

(a) ^{1}H NMR spectrum (90 MHz).

FIGURE 18-27   NMR data for dimethyl
heptane-1,1-dicarboxylate.

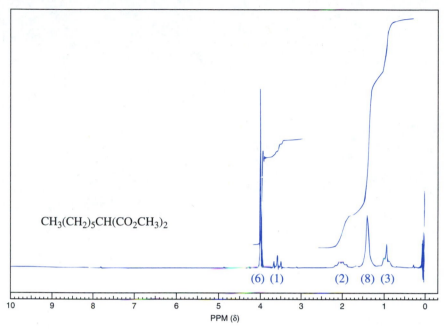

$CH_3(CH_2)_5CH(CO_2CH_3)_2$

(6) (1)   (2) (8) (3)

PPM (δ)

(b) ^{13}C NMR data.

Chemical shifts: δ 13.9, 22.4, 27.2, 28.7, 31.4, 41.0, 51.6, 52.3, 169.8.

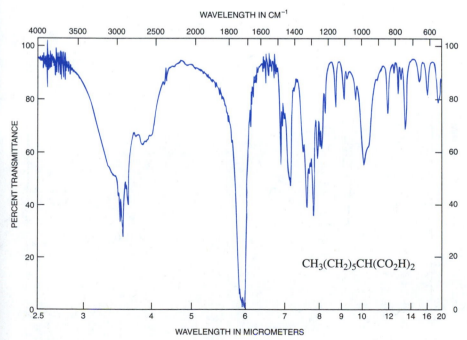

WAVELENGTH IN CM^{-1}

PERCENT TRANSMITTANCE

$CH_3(CH_2)_5CH(CO_2H)_2$

WAVELENGTH IN MICROMETERS

FIGURE 18-28   IR spectrum of heptane-
1,1-dicarboxylic acid (KBr pellet).

FIGURE 18-29 NMR data for heptane-1,1-dicarboxylic acid.

(a) ^{1}H NMR spectrum (90 MHz).

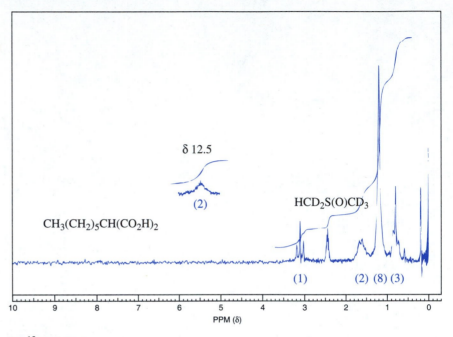

(b) ^{13}C NMR data.
Chemical shifts: δ 13.9, 22.0, 26.8, 28.4, 31.0, 40.6, 51.6, 170.8.

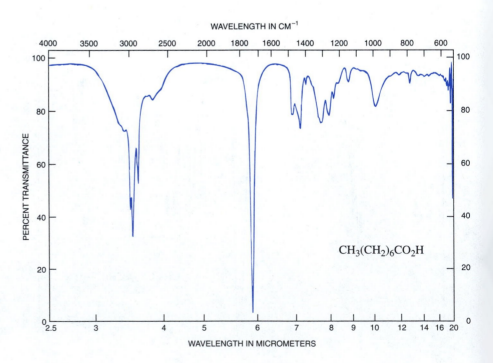

FIGURE 18-30 IR spectrum of octanoic acid.

(a) ^{1}H NMR spectrum (90 MHz).

FIGURE 18-31 NMR data for octanoic acid.

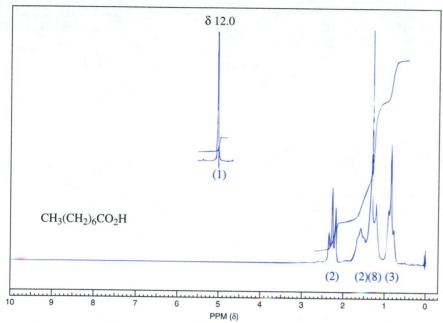

δ 12.0

(1)

$CH_3(CH_2)_6CO_2H$

(2)   (2)(8) (3)

10   9   8   7   6   5   4   3   2   1   0
PPM (δ)

(b) ^{13}C NMR data.

Chemical shifts: δ 14.1, 22.9, 25.0, 29.2, 29.3, 32.0, 34.3, 180.6.

18.3 CONJUGATE ADDITION TO AN α,β-UNSATURATED KETONE

The introduction to this chapter contained descriptions of two different methods by which a carbon atom alpha to a carbonyl function becomes nucleophilic by formation of an enol (Eq. 18.2) or an enolate (Eq. 18.3). The aldol and alkylation reactions of the preceding sections involved the latter option for promoting nucleophilic chemistry at the α-carbon atom. Moreover, it was pointed out in Section 18.2.2 that α,β-unsaturated carbonyl compounds undergo **1,4-,** or **conjugate, addition reactions** (Eq. 18.21) in addition to the reactions normally observed for nonconjugated carbon-carbon and carbon-oxygen double bonds. The procedure in this section involves an interesting example in which an enol serves as the nucleophile for conjugate addition to an α,β-unsaturated ketone and in which an **intramolecular aldol condensation** is acid-catalyzed rather than base-catalyzed.

The overall reaction is the conversion of 2-methylpropanal (**49**) and 3-buten-2-one (**50**) to 4,4-dimethyl-2-cyclohexen-1-one (**52**) by way of the conjugate addition product **51** (Eq. 18.30). Although it is possible to isolate **51,** this is not done in the present procedure. The conjugate addition to give **51** is an example of a **Michael addition,** whereas the combination of this addition followed by an aldol reaction to form the six-membered ring of **52** is called the **Robinson annulation.** The names honor the chemists who discovered the two types of reactions.

(18.30)

49	**50**	**51**	**52**
2-Methylpropanal (isobutyraldehyde)	3-Buten-2-one (methyl vinyl ketone)	2,2-Dimethylhexanal-5-one	4,4-Dimethyl-2-cyclohexen-1-one

A plausible overall mechanism for the reactions leading to **52** is shown in Scheme 18.5. 2-Methylpropanal (**49**) first undergoes acid-catalyzed tautomerization to its enol form **53**, in which the α-carbon atom is now nucleophilic. Protonation of the 3-buten-2-one (**50**) converts it into the delocalized cation **54**, which is more electrophilic than the unprotonated form. Because **53** is a weak nucleophile, relative to an enolate ion, the formation of **54** facilitates the next stage of the reaction, which is creating a new carbon-carbon bond between **49** and **50** to give the enol **55**. The highlighted atoms in structure **55** clearly show that the α-C–H bond of **49** has added in a conjugate, or 1,4-, manner to **50**. Acid-catalyzed tautomerization of **55** leads to the thermodynamically more stable keto form **51**.

Although **51** is isolable, it is not under our reaction conditions; rather it undergoes an acid-catalyzed **intramolecular aldol addition.** This process is initiated by the acid-catalyzed tautomerization of **51** to a different enol form, **56**. Activating the remaining carbonyl function by protonation sets the stage for cyclization to give the β-hydroxycarbonyl compound **57** after deprotonation. The fact that **51** undergoes efficient cyclization rather than intermolecular condensations with the enolic form of 2-methylpropanal or other potential nucleophiles is mainly attributable to *entropic* considerations that tend to favor *intra*molecular processes over *inter*molecular ones. In the final stage of the reaction, acid-catalyzed dehydration of **57** to give the desired **52** occurs; the mechanism of this reaction is left as an exercise (see Ex. 10 at the end of this section.)

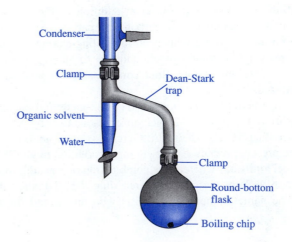

FIGURE 18-32 Dean-Stark trap for removal of water by azeotropic distillation.

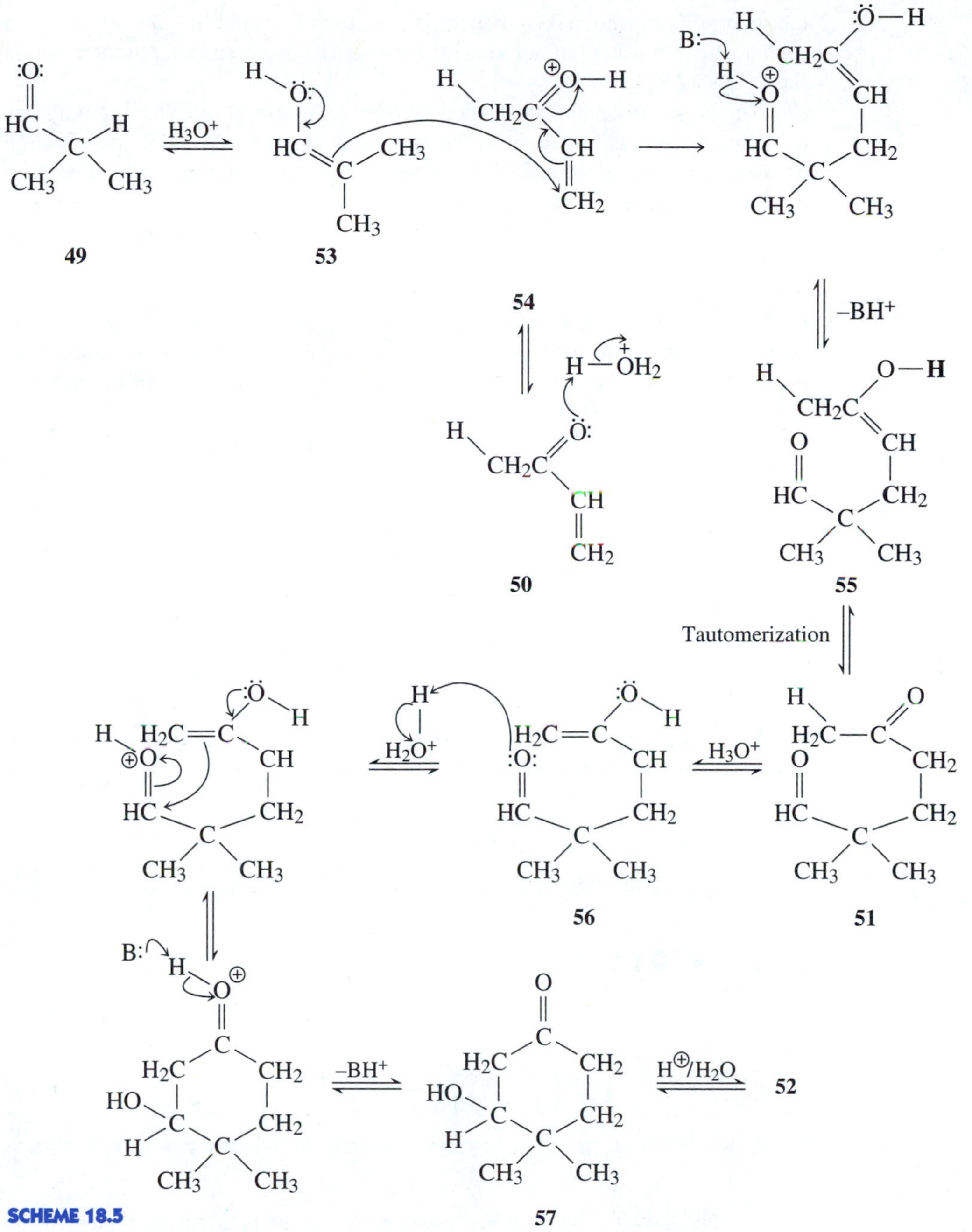

SCHEME 18.5

The dehydration of **57** to **52** is a reversible process, but it is driven to completion by removing the water from the reaction mixture. This is conveniently done by using toluene as the solvent for the reaction; water and toluene form an **azeotrope** (see Sec. 4.2.4), so azeotropic distillation allows the continuous separation of water as dehydration occurs. To minimize the amount of solvent that is required for distillations of this type, a Dean-Stark trap (Fig. 18.32) is commonly used. Although such traps are not usually available in the introductory

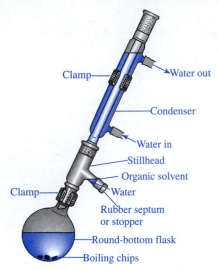

Water out

Condenser

Water in

Stillhead

Organic solvent

Clamp

Water

Rubber septum or stopper

Round-bottom flask

Boiling chips

FIGURE 18-33   *Apparatus for reaction of 2-methylpropanal (isobutyraldehyde) with 3-buten-2-one (methyl vinyl ketone).*

laboratory, its operational equivalent is possible by assembling the apparatus in such a way that water but not toluene is prevented from returning to the reaction flask (Fig. 18.33).

The choice of an aldehyde, such as 2-methylpropanal (**49**), having only *one* α-hydrogen atom is important to the success of this crossed aldol condensation because their self-addition is of no consequence. For example, although aldehyde **49** undergoes facile self-dimerization to **58** (Eq. 18.31), this process is *reversible* under the reaction conditions used. Moreover, unlike the aldol product **57**, **58** is incapable of dehydrating to give an α,β-unsaturated product and water, which could be removed to disrupt the equilibrium and lead to self-dimerization product. It is left to you to consider the possibility of self-condensation of 3-buten-2-one (**50**) and of the reaction of enol **53** with the carbon atom of the protonated carbonyl function in **54** (see Exs. 7 and 8 at the end of this section).

$$
2 \; (CH_3)_2C \underset{H}{\overset{O}{\underset{\|}{\overset{\|}{C}}}} H \; \overset{H_3O^+}{\rightleftharpoons} \; (CH_3)_2CH \underset{CH_3 \quad CH_3}{\overset{H \quad OH}{\underset{C}{\overset{C}{<}}}} {}^{CHO}
$$

(18.31)

49

58
2,2,4-Trimethyl-3-
hydroxypentanal

Pre-Lab exercises for Section 18.3, "Preparation of 4,4-Dimethyl-2-cyclohexene-1-one," are found on page PL. 103.

EXPERIMENTAL PROCEDURE

SAFETY ALERT

Volatile and flammable solvents are used in this experiment. Avoid the use of flames for heating purposes, if possible. If not, be certain that all joints in the apparatus are firmly mated in order to keep vapors from escaping and possibly igniting.

Equip a distillation head with a septum or rubber stopper firmly seated in the male joint that is normally used for connection to a condenser. To be certain that the septum or rubber stopper will not come out during the period of reflux, secure it to the joint with wire. Attach the distillation head to a 125-mL round-bottom flask containing 0.1 g of 2-naphthalenesulfonic acid. Tilt the flask at an angle of about 30° from the vertical so that when a condenser is attached, the condensate will drip into the sidearm of the distillation head (see Fig. 18.31). Add about 25 mL of toluene to the flask and swirl the resulting slurry to effect partial dissolution. To this mixture add a solution of 3.2 g of 3-buten-2-one (methyl vinyl ketone) and 4.2 g of 2-methylpropanal (isobutyraldehyde). Rinse the flask that contained this solution with a 15-mL portion of toluene and add the rinse to the reaction flask. Fit the distillation head with a reflux condenser, bring the solution to a brisk reflux, and continue reflux for at least 2.5 h.✷ As the reaction proceeds, water should collect in the plugged sidearm of the distillation head. Following reflux, allow the reaction mixture to cool and then carefully disconnect the apparatus, making certain that the collected water is not spilled back into the reaction flask. Pour the water and solvent present in the sidearm into a graduated cylinder to determine what percentage of the theoretical amount of water has been collected.✷ Transfer the reaction mixture to a separatory funnel and wash it with 30 mL of 0.6 *M* aqueous sodium bicarbonate.

Dry the organic layer over anhydrous sodium sulfate for at least 0.5 h,✷ and then filter or decant it into a 125-mL round-bottom flask.✷ Equip the flask for simple distillation and carefully (*caution:* foaming) remove the unchanged starting materials and solvent, collecting them as a single fraction boiling up to about 130 °C.✷

To complete isolating the product by distillation, remove the source of heat and dissemble the apparatus. Cool the pot residue to near room temperature; then transfer the pot residue to a 25-mL round-bottom flask with a micropipet. Rinse the original stillpot with 1–2 mL of diethyl ether and add the rinse to the smaller flask. Fit this flask for a short-path distillation (Fig. 2.15b). If the final stage of the distillation is to be performed at atmospheric pressure, add a fresh boiling stone to the stillpot, insulate the stillhead with cotton, glasswool, or some other equivalent material, and continue the distillation; 4,4-dimethyl-2-cyclohexen-1-one boils at about 200 °C (760 torr). Collect it as a single fraction boiling from 130 °C and higher; note that the head temperature may never reach the reported boiling point, owing to the high boiling temperature and the small amount of sample being distilled. Terminate heating when the dark pot residue becomes viscous and starts to evolve fumes. The distillate should be light to dark yellow in color.

Alternatively, and preferably, if the final stage of the distillation is to be done under vacuum, fit the stillpot for short-path distillation, as just described but add a capillary tube or a magnetic stir bar rather than a fresh boiling stone for ebullition (see Sec. 2.9). Attach the vacuum adapter to a water aspirator or some other source of a vacuum. Resume the distillation after a stable vacuum has been established in the apparatus. Terminate heating when the dark pot residue becomes viscous and only a few bubbles evolve from it. The boiling point of the product will depend on the quality of the vacuum that is available; a reported boiling point is 73–76 °C (14 torr), and estimates of boiling points of approximately 100 °C (40 torr) and 125 °C (100 torr) can be made with the aid of a nomograph (Fig. 2.18). A water aspirator should be capable of achieving a vacuum within the range of 40–100 torr if all joints in the apparatus are properly lubricated and tightly mated. The distillate should be colorless to slightly yellow.

The pleasant-smelling liquid obtained is contaminated with residual toluene (see Figs. 18.39 and 18.40), the amount of which will be dependent upon the efficiency with which the first stage of the distillation has been performed. However, toluene should constitute no more than 30% of the distillate. The exact amount can be determined by GLC or ^{1}H NMR spectroscopic analysis of the isolated product (see Ex. 4 at the end of this section).

4,4-Dimethyl-2-cyclohexen-1-one (**52**) forms a 2,4-DNP derivative, mp 140–142 °C; confirm the identity of your product by converting a 200–300 mg portion of it to this derivative (see Sec. 27.5.1A). Recrystallize the product from a minimum quantity of 95% ethanol if necessary.

Finishing Touches   Put the *toluene* obtained by azeotropic distillation and the *forerun* and *pot residue* from the distillation of the product into the container for nonhalogenated organic liquids. After residual solvent has evaporated from the *sodium sulfate* on a tray in the hood, place the drying agent in a container for nonhazardous waste.

EXERCISES

1. In the literature report used as the basis for this experiment, benzene was used as a solvent rather than toluene. For purposes of safety, toluene was substituted because benzene may be carcinogenic and toluene is not. The reaction appears to go faster in toluene than in benzene, as judged by the rate at which water is collected in the sidearm of the distillation head. How might you account for this rate acceleration?

2. Why is it important that water *not* return to the reaction flask during the reflux period?

3. Product **59** could potentially be produced by aldol condensation of **51** but is not. Provide a stepwise reaction mechanism for the possible formation of **59** in the presence of an acid catalyst, and explain why this alternate product is not observed.

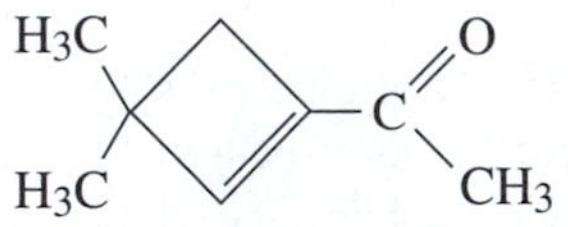

59

4. **a.** Figure 18.39 is the GLC trace of a sample of product obtained in this experiment. Calculate the ratio of toluene to **52** that is present, assuming that the response factor of the detector (see Sec. 6.4.4) is the same for the two compounds. Note that the GLC column used in this analysis separates substances on the basis of their relative boiling points.

   b. Figure 18.40 is the ^{1}H NMR spectrum of this same sample. Calculate the ratio of toluene to **52** that is present according to this method of analysis.

   c. Account for the discrepancy, if any, between the two ratios that you obtain.

   d. The distillate from which the samples for these analyses were drawn weighed a total of 2.8 g. Based on the ^{1}H NMR analysis of the mixture, calculate how many grams of the desired **52** were present in the distillate. Calculate the yield of the reaction, assuming that it was run on the scale described in the Experimental Procedure of this section.

5. Attempted acid-catalyzed reaction of propanal (propionaldehyde) with 3-buten-2-one (**50**) fails to give a good yield of the desired product, **60**. Explain why this might be.

$$
\underset{\substack{\text{Propanal}\\ \text{(propionaldehyde)}}}{CH_3CH_2\overset{\overset{\displaystyle O}{\|}}{C}H} \; + \; \underset{\textbf{50}}{CH_2{=}CH{-}\overset{\overset{\displaystyle O}{\|}}{C}CH_3} \;\xrightarrow{\;H_3O^+\;}\; \underset{\substack{\textbf{60}\\ \text{4-Methyl-2-cyclohexen-1-one}}}{CH_3{-}\!\!\!\bigcirc\!\!\!=\!O}
$$

6. a. Refer to Figure 18.38 and calculate log ε for the two maxima in the UV spectrum of 4,4-dimethyl-2-cyclohexen-1-one (**52**).

   b. What electronic excitation is responsible for each of these maxima?

7. Write a product of self-condensation of 3-buten-2-one (**50**), and explain why this process might be less favorable than the observed reaction between **49** and **50**.

8. Write the product expected to result from attack of enol **53** on the carbonyl carbon atom of **54**. Explain why formation of this product should be thermodynamically less favorable than formation of **51**.

9. Why is protonation of the enol **56** likely to be involved in the conversion of **56** to **57**?

10. Write a stepwise reaction mechanism for the acid-catalyzed dehydration of **57** to **52**.

SPECTRA OF STARTING MATERIALS AND PRODUCT

The IR spectrum and NMR data for 2-methylpropanal (isobutyraldehyde) are given in Figures 16.1 and 16.2, respectively.

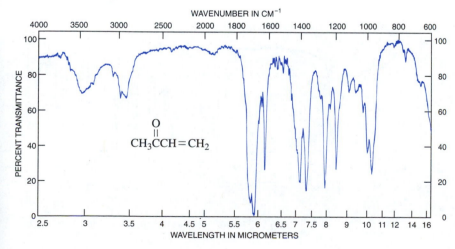

FIGURE 18-34 IR spectrum of 3-buten-2-one (methyl vinyl ketone)

FIGURE 18-35 NMR data for 3-buten-2-one (methyl vinyl ketone)

(a) ^{1}H NMR spectrum (90 MHz).

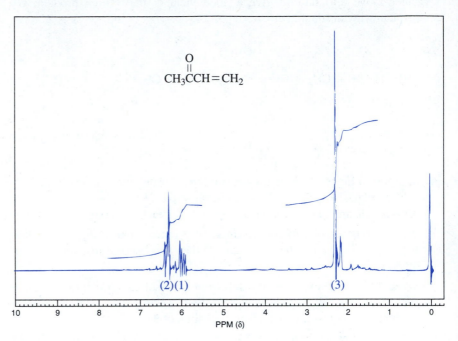

(b) ^{13}C NMR data.
Chemical shifts: δ 127.4, 136.3, 196.9.

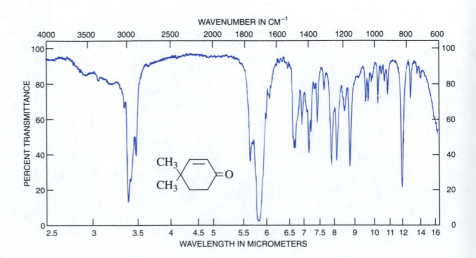

FIGURE 18-36 IR spectrum of 4,4-dimethyl-2-cyclohexen-1-one.

(a) ¹H NMR spectrum (90 MHz).

FIGURE 18-37   NMR data for 4,4-dimethyl-2-cyclohexen-1-one.

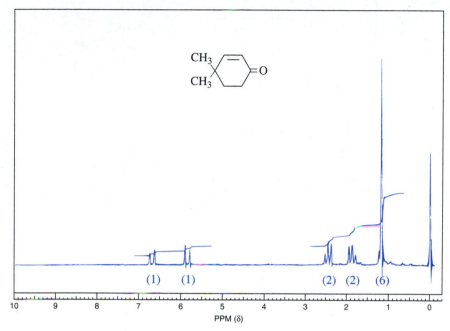

(b) ¹³C NMR data.

Chemical shifts: δ 27.7, 32.8, 34.4, 36.1, 126.8, 159.6, 199.0.

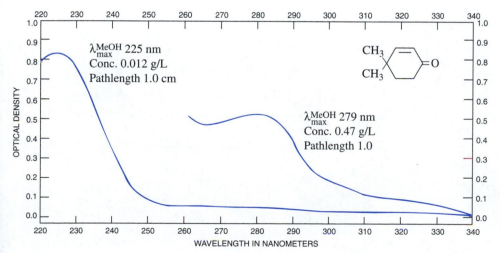

FIGURE 18-38   UV spectrum of 4,4-dimethyl-2-cyclohexen-1-one.

FIGURE 18-39 GLC trace of reaction mixture for Exercise 4a. Peak 1 is toluene; peak 2 is 4,4-dimethyl-2-cyclohexen-1-one.

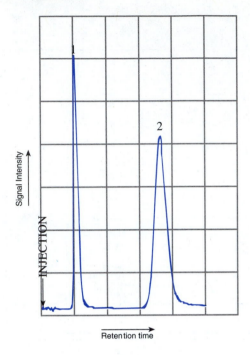

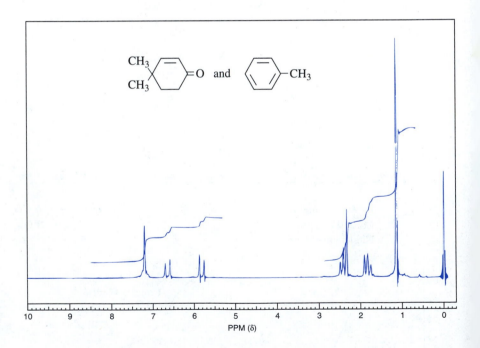

FIGURE 18-40 ^{1}H NMR spectrum (90 MHz) of 4,4-dimethyl-2-cyclohexen-1-one for Exercise 4b.

CHAPTER 19

ORGANOMETALLIC CHEMISTRY

Organometallic compounds are substances that contain carbon-metal bonds, and they may be generally represented by the structure **1**, in which the metal, **M,** may be Li, Na, Mg, Cu, Hg, Pd, or other transition elements. Organomagnesium compounds **1**, M = MgBr, which are commonly called **Grignard reagents,** were the first organometallic substances to be extensively studied, and they are still among the most important of all organometallic reagents. The polarization of the bond between the carbon atom and the electropositive metal in these reagents renders the carbon atom electron-rich, or Lewis basic, and the carbon atom bears a partial negative charge, δ^-, as shown. Thus, one of the characteristic properties of organometallic reagents **1** is that the carbon atom serves as a **nucleophile** in chemical reactions. In contrast, when a carbon atom is bonded to more electronegative elements such as the halogen atom in the alkyl halide **2** (X = Cl, Br, or I) or oxygen atom in the carbonyl compound **3**, it is electron-deficient or Lewis acidic and possesses a partial positive charge. Such carbon atoms then serve as **electrophiles** in chemical reactions.

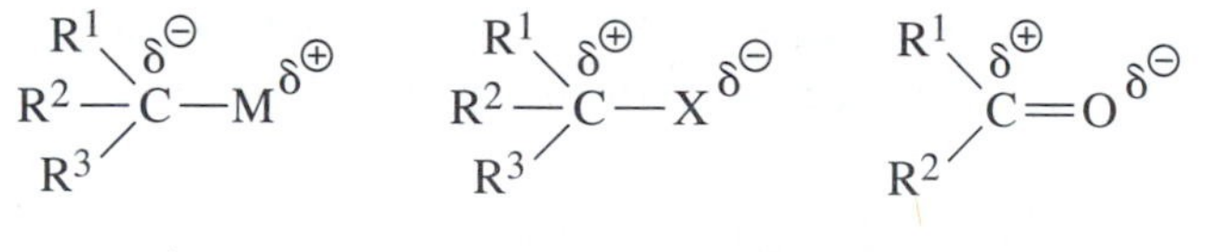

$$
\begin{array}{ccc}
\underset{R^3}{\overset{R^1}{\underset{|}{R^2-C}}}-M^{\delta\oplus} & \underset{R^3}{\overset{R^1}{\underset{|}{R^2-C}}}-X^{\delta\ominus} & \underset{R^2}{\overset{R^1}{C}}=O^{\delta\ominus}
\end{array}
$$

1	**2**	**3**
An organometallic reagent	An alkyl halide	A carbonyl compound

Because of their nucleophilic character, organometallic compounds are widely utilized as reagents in reactions that produce new carbon-carbon bonds. For example, the reactions of organometallic reagents, **1**, with carbon electrophiles such as alkyl halides, **2**, and carbonyl compounds, **3**, are illustrated by the general transformations shown in Equations (19.1) and (19.2). In each of these reactions, the nucleophilic carbon atom of one reactant becomes attached to the electrophilic carbon atom of the other reactant with the resulting formation of a new carbon-carbon bond. Thus, like many bond-forming processes, these reactions may be viewed in the simple context of *combinations of Lewis bases with Lewis acids.*

$$\overset{\delta\ominus}{R^1CH_2}-\overset{\delta\oplus}{M} \quad + \quad \overset{\delta\ominus}{X}-\overset{\delta\oplus}{CH_2R^2} \longrightarrow R^1CH_2-CH_2R^2 + MX \qquad \textbf{(19.1)}$$

An organometallic   An alkyl halide   An alkane
reagent

$$\overset{\delta\ominus}{R^1CH_2}-\overset{\delta\oplus}{M} \quad + \quad \overset{R^2}{\underset{R^3}{\diagdown}}\overset{\delta\oplus}{C}=\overset{\delta\ominus}{O} \longrightarrow R^1CH_2-\overset{R^2}{\underset{\underset{R^3}{|}}{\overset{|}{C}}}-O^\ominus M^\oplus \xrightarrow{H_3O^\oplus} R^1CH_2-\overset{R^2}{\underset{\underset{R^3}{|}}{\overset{|}{C}}}-OH \qquad \textbf{(19.2)}$$

An organometallic   A carbonyl   An alcohol
reagent   compound

The following discussions focus on the preparation and reactions of one important class of organometallic compounds, the **Grignard reagents, 1,** (M = Mg). However, many of the principles that are presented may be applied to the chemistry of other organometallic reagents. The experiments that may be performed in this chapter involve the preparation of two Grignard reagents from alkyl and aryl bromides, followed by their typical reactions with (1) an ester to produce a tertiary alcohol, (2) carbon dioxide to produce a carboxylic acid, and (3) an aldehyde to produce a secondary alcohol.

19.2   GRIGNARD REAGENTS: PREPARATION

19.2.1   REACTIONS OF ORGANIC HALIDES WITH MAGNESIUM METAL

Grignard reagents, R–MgX or Ar–MgX, are typically prepared by the reaction of an alkyl halide, R–X, or an aryl halide, Ar–X, with magnesium metal in an *anhydrous* ethereal solvent according to Equation (19.3); the reagent dissolves as it is formed. It is noteworthy that in this process *carbon is transformed from an electrophilic center in the starting material, R–X or Ar–X, into a nucleophilic center in the product, R–MgX or Ar–MgX.*

$$\overset{\delta\oplus}{R}-\overset{\delta\ominus}{X} \quad \text{or} \quad \overset{\delta\oplus}{Ar}-\overset{\delta\ominus}{X} \quad + \quad Mg^\circ \xrightarrow[\text{or THF (solvent)}]{\text{Dry ether}} \overset{\delta\ominus}{R}-\overset{\delta\oplus}{MgX} \quad \text{or} \quad \overset{\delta\ominus}{Ar}-\overset{\delta\oplus}{MgX} \qquad \textbf{(19.3)}$$

An alkyl   An aryl   An alkyl   An aryl
halide   halide   magnesium halide   magnesium halide

Although it is customary to represent the Grignard reagent by the formula R–MgX or Ar–MgX, the structure of the organometallic species in solution is rather complex and not well understood. For example, with alkyl magnesium halides, there is an equilibrium between RMgX, R_2Mg, and MgX_2 that depends on the solvent and the nature of the alkyl group. Moreover, the various organometallic species form aggregates in solution.

Grignard reagents are readily prepared from alkyl and aryl chlorides, bromides, and iodides; Grignard reagents derived from organofluorides are rarely used. The ease of formation of Grignard reagents from alkyl halides follows the order R–I > R–Br > R–Cl; aryl halides are less reactive than their alkyl counterparts, with aryl bromides and chlorides being comparable in reactivity. In the experiments in this chapter, the Grignard reagents derived from bromobenzene (**4**) and 1-bromobutane (**6**) are prepared according to Equations (19.4) and (19.5). The preparation of these reagents theoretically requires equivalent amounts of the organic halide and magnesium, but a slight excess of magnesium is normally used experimentally.

$$\text{Br} \quad + \text{Mg} \xrightarrow{\text{diethyl ether}} \text{MgBr} \qquad \textbf{(19.4)}$$

4	**5**
Bromobenzene	Phenylmagnesium bromide

$$CH_3CH_2CH_2CH_2-Br + Mg \xrightarrow{\text{diethyl ether}} CH_3CH_2CH_2CH_2-MgBr \qquad \textbf{(19.5)}$$

6	**7**
1-Bromobutane	*n*-Butylmagnesium bromide
(*n*-Butyl bromide)	

Although the formation of the Grignard reagent is quite general, it is important to recognize that hydroxyl (OH), amino (NH$_2$), and carbonyl groups must *not* be present in the alkyl or aryl halide, as they will react with the highly basic carbon atom of the Grignard reagent as it forms. For example, reaction of 5-bromo-1-pentanol with magnesium would not give the expected Grignard reagent, because it would be immediately converted to the corresponding bromomagnesium alkoxide by transfer of a proton from the acidic hydroxyl group, as illustrated in Equation (19.6). Moreover, because Grignard reagents react readily with all acidic protons, it is also necessary to prepare and use them under strictly *anhydrous* conditions (see also in Sec. 19.2.2).

$$H-OCH_2(CH_2)_3CH_2-Br + Mg° \xrightarrow{\text{dry Et}_2\text{O}} \left[H-OCH_2(CH_2)_3CH_2-MgBr \right]$$

5-Bromo-1-pentanol

$$\textbf{(19.6)}$$

$$\xrightarrow{\hspace{2cm}} BrMg-OCH_2(CH_2)_3CH_2-H$$

Use of an ethereal solvent is critical for the efficient preparation of the Grignard reagent because the basic oxygen atom of the ether complexes with the electropositive magnesium atom of R(Ar)–MgX to stabilize the organometallic species. The ethereal solvents most commonly used in this reaction are diethyl ether, $(C_2H_5)_2O$, and tetrahydrofuran (THF), $(CH_2)_4O$. Diethyl ether is often the solvent of choice because it is less expensive, may be purchased in anhydrous form, and is easily removed from the reaction mixture owing to its low boiling point (bp 36 °C, 760 torr). Tetrahydrofuran is a stronger Lewis base than diethyl ether, and it also has better solvating ability; it may be used when the Grignard reagent does not form readily in diethyl ether.

Care should be exercised when using anhydrous diethyl ether that has been stored in metal containers. If these containers have been opened so that the diethyl ether has been exposed to oxygen, **hydroperoxides,** CH_3–$CH(OOH)$–OCH_2CH_3, which are formed by a free-radical chain mechanism (see Chap. 9), may be present. These peroxides are *explosive,* and large volumes of anhydrous ether suspected to contain peroxides should *not* be evaporated to dryness. Moreover, emptied containers should be thoroughly rinsed with water to remove any peroxides before being discarded. Peroxides in anhydrous diethyl ether or tetrahydrofuran may be conveniently destroyed by distillation from alkali metals such as sodium or potassium metal or from lithium aluminum hydride, $LiAlH_4$. The presence of peroxides in diethyl ether and tetrahydrofuran may be detected by placing a drop on a piece of moistened starch-iodide test paper. If the paper turns dark violet, owing to formation of a starch-iodine complex, then the ether contains peroxides.

The magnesium metal used for the preparation of the Grignard reagent is normally in the form of turnings—thin shavings that have a high surface area relative to chunks of the metal. This type of magnesium is generally suitable for the preparation of most Grignard reagents. However, if the turnings have been repeatedly exposed to oxygen, their surface may be dull owing to a coating of magnesium oxide, which decreases the reactivity of the turnings. A highly activated form of magnesium having an extremely high surface area is prepared by reduction of magnesium chloride with potassium metal (Eq. 19.7) and this may be used in those cases for which the turnings are ineffective.

$$MgCl_2 + 2\,K° \longrightarrow Mg° + 2\,KCl \qquad \textbf{(19.7)}$$

Initiating the reaction between the organic halide and magnesium may sometimes be difficult, especially with unreactive halides. In these cases, the reaction may often be initiated by adding a small crystal of iodine, I_2, to the reaction mixture. The iodine facilitates the reaction either by activating the metal by removing some of its oxide coating, or by converting a small amount of an alkyl halide to the corresponding iodide, which is more reactive toward magnesium. Alternatively, a small amount of 1,2-dibromoethane can be added to the reaction mixture to activate the magnesium, presumably by removing the oxide coating. 1,2-Dibromoethane is very reactive toward magnesium metal giving ethylene and magnesium bromide (Eq. 19.8).

$$Br-CH_2CH_2-Br + Mg° \xrightarrow[\text{dry } (C_2H_5)_2O]{} CH_2{=}CH_2 + MgBr_2 \qquad \textbf{(19.8)}$$

1,2-Dibromoethane   Ethylene
                                                     (Ethene)

Formation of a Grignard reagent is an **exothermic** process. Since side reactions may occur if the reaction is allowed to proceed uncontrolled, experimental techniques have been devised to control the rate at which the reaction proceeds. One common tactic involves the slow, dropwise addition of the alkyl or aryl halide to a suspension of the magnesium metal in the dry ethereal solvent; the halide may be added either neat or as a solution in the ethereal solvent. Such *slow* addition maintains a low concentration of the halide, and the rate of the reaction and the evolution of heat may thus be easily regulated. Because it is ultimately transferred to the cooling water in the condenser, the heat generated in exothermic reactions is also removed by the vaporization and refluxing of the solvent. The rate of the reaction must be maintained within the capacity of the condenser to condense the refluxing solvent. Should it appear that the reaction is getting out of control, as evidenced by vapors escaping from the top of the condenser, the reaction mixture should be immediately cooled with an ice-water bath. Indeed, it is prudent always to have such a cooling bath prepared when performing an exothermic reaction.

19.2.2   SIDE REACTIONS

In the previous section, we noted that hydroxyl and amino groups must not be present on the alkyl or aryl halide because they would react with the Grignard reagent as it was generated. Other side reactions of Grignard reagents are also encountered. As the **conjugate bases** of the exceedingly weak organic acids R–H, Grignard reagents are *strong* bases that react rapidly with water according to Equation (19.9). The overall result is the destruction of the Grignard reagent and the formation of a hydrocarbon, RH, and a basic magnesium salt. This process can be utilized to standardize a solution of a Grignard reagent in cases where stoichiometry is important. For example, a known volume of a Grignard reagent may be added to an excess of a known volume of standardized mineral acid; the resulting solution is then back-titrated with standardized base.

$$R-MgX \ + \ H_2O \ \longrightarrow \ R-H \ + \ HO\text{-}Mg\text{-}X \qquad \text{(19.9)}$$

The reactivity of Grignard reagents with water (Eq. 19.9) dictates that all reagents, solvents, and apparatus used for their preparation must be thoroughly dry. Consequently, *anhydrous* ethereal solvents, which typically contain less than 0.01% water, must be used. Anhydrous diethyl ether rapidly absorbs atmospheric moisture, so if a container of diethyl ether has been opened a number of times over a period of several days, the ether will not be suitable. Only freshly opened cans should be used, and the cans should always be tightly sealed immediately after the necessary volume of ether has been removed. These operations will also minimize the formation of dangerous peroxides as mentioned previously.

Other side reactions that may occur during formation of a Grignard reagent are shown in Equations (19.10)–(19.12), but these can also be minimized by taking certain precautions. Reaction of the Grignard reagent with oxygen and carbon dioxide (Eqs. 19.10 and 19.11) is avoided simply by performing the reaction under an inert atmosphere, such as nitrogen (N_2) or argon. This precaution is routinely exercised in research laboratories. The coupling reaction (Eq. 19.12) is an example of a Wurtz-type reaction. Although this can be a useful process for preparing symmetrical hydrocarbons R–R, it is generally desirable to minimize this side reaction by using dilute solutions, thereby avoiding high localized concentrations of the alkyl halide. This is accomplished by efficient stirring and by slowly adding the halide to the suspension of magnesium in the ethereal solvent. Alkyl iodides are more prone to coupling than are the corresponding bromides and chlorides, so the latter, although less reactive, are preferable for preparing Grignard reagents.

$$RMgX + O_2 \longrightarrow R\text{-}O\text{-}O^{\ominus} \ \overset{\oplus}{M}gX \qquad (19.10)$$

$$RMgX + CO_2 \longrightarrow RCO_2^{\ominus} \ \overset{\oplus}{M}gX \qquad (19.11)$$

$$RMgX + RX \longrightarrow R\text{—}R + MgX_2 \qquad (19.12)$$

In the experiments that follow, phenylmagnesium bromide (**5**) and *n*-butyl-magnesium bromide (**7**) are prepared according to Equations (19.4) and (19.5), respectively. The use of an inert atmosphere to exclude the oxygen and moisture in the air is unnecessary, because diethyl ether, which is used as the solvent, has a very high vapor pressure that effectively excludes most of the air from the reaction vessel. The most important side reaction in these experiments thus involves the Wurtz-type coupling of the Grignard reagent with the organic halide. For example, during the preparation of phenylmagnesium bromide, small quantities of biphenyl (**8**) are formed according to Equation (19.13); however, the presence of **8** does not interfere with the subsequent reactions of phenylmag-nesium bromide. Although coupling may also occur during the preparation of *n*-butylmagnesium bromide, the *n*-octane that is produced is volatile and easily removed.

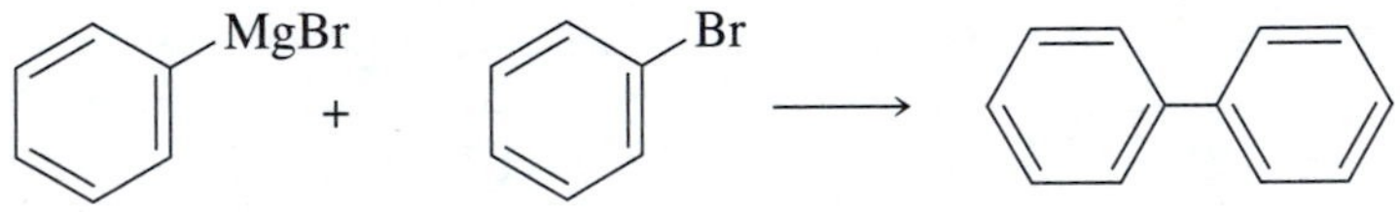

8
Biphenyl

(19.13)

19.3 GRIGNARD REAGENTS: REACTIONS

19.3.1 REACTION OF PHENYLMAGNESIUM BROMIDE WITH METHYL BENZOATE

The aryl carbon atom bearing the magnesium in phenylmagnesium bromide is nucleophilic and consequently reacts readily with electrophiles (see Sec. 19.1). In the experiment that follows, two equivalents of phenylmagnesium bromide (**5**, Eq. 19.4) react with an ester, methyl benzoate (**9**), in a highly exothermic reaction to give triphenylmethanol (**11**) via the sequence of reactions depicted in Scheme 19.1. In the first step of the reaction, nucleophilic attack of phenylmag-nesium bromide (**5**) on the electrophilic carbon atom of the ester group of **9** gives benzophenone (**10**) as an intermediate. Subsequent reaction of **10** with an additional mole of **5** then produces an alkoxide salt, which is converted to tri-phenylmethanol (**11**) upon protonation with acid. The reaction is worked up with acid rather than water alone in order to avoid precipitation of the basic magnesium salt, HOMgX.

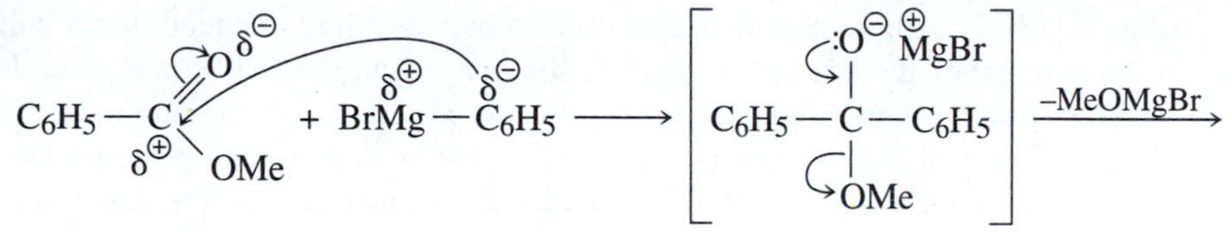

9   **5**
Methyl benzoate

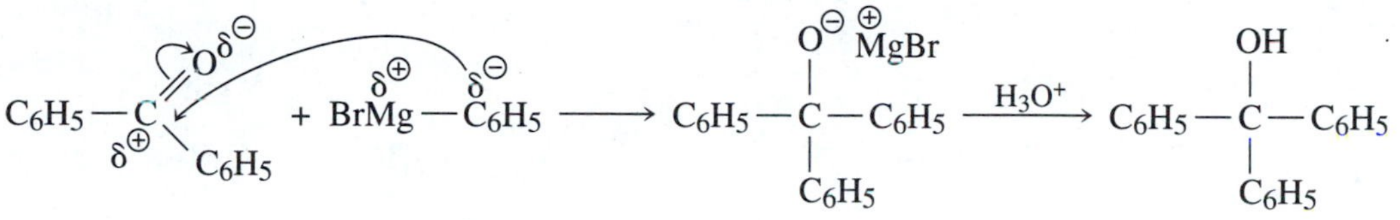

10   **5**   **11**
Benzophenone   Triphenylmethanol

SCHEME 19.1

The principal organic products present after the aqueous work-up are the desired triphenylmethanol (**11**), benzene from any unreacted **5**, and biphenyl (**8**). Fortunately, it is possible to separate **11** and **8** easily owing to their relative solubilities in nonpolar hydrocarbon solvents. Biphenyl is considerably more soluble in cyclohexane than triphenylmethanol, so recrystallization of the crude product mixture gives pure triphenylmethanol.

19.3.2   REACTION OF PHENYLMAGNESIUM BROMIDE WITH CARBON DIOXIDE

Nucleophilic addition of phenylmagnesium bromide (**5**, Eq. 19.4) to carbon dioxide produces an intermediate carboxylate salt, which may be converted into benzoic acid (**12**) by acidification of the reaction mixture, as shown in Equation (19.14). This process is accompanied by side reactions that lead to the formation of benzophenone (**10**) and/or triphenylmethanol (**11**), as shown in Equation (19.15). However, by controlling the reaction conditions, formation of these

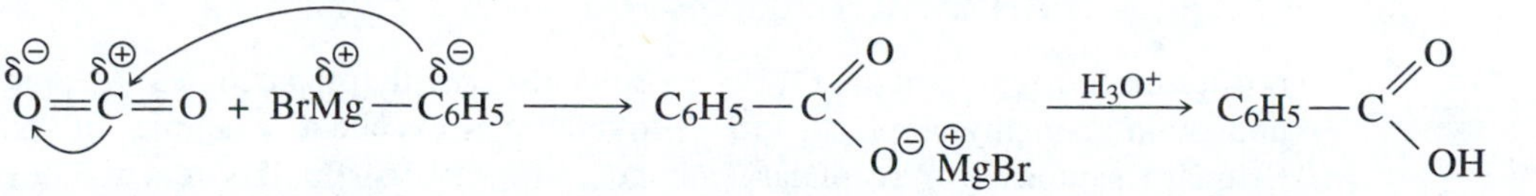

(19.14)

5   **12**
Benzoic acid

by-products is minimized. The bromomagnesium salt of benzoic acid is only slightly soluble in diethyl ether, so when the reaction is performed in this solvent, the salt precipitates from solution and cannot undergo further reaction. A large excess of carbon dioxide is also used to increase the likelihood that **5** will react with carbon dioxide rather than with the magnesium salt of benzoic acid. If the phenylmagnesium bromide is added slowly *to* the Dry Ice, carbon dioxide is always present in excess, thus favoring the desired reaction. Finally, when Dry Ice is used as the source of carbon dioxide, the temperature of the reaction is maintained at −78 °C until all of the Grignard reagent has been consumed. At this low temperature, the reaction of phenylmagnesium bromide with the bromomagnesium salt of benzoic acid is slow.

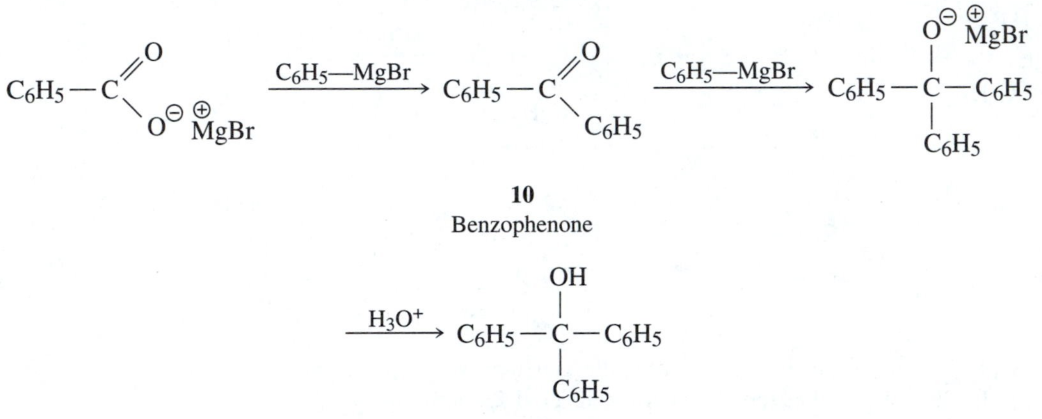

$$C_6H_5-C\underset{O^{\ominus}\ \overset{\oplus}{MgBr}}{\overset{O}{\Big\Vert}} \xrightarrow{C_6H_5-MgBr} C_6H_5-C\underset{C_6H_5}{\overset{O}{\Big\Vert}} \xrightarrow{C_6H_5-MgBr} C_6H_5-\underset{C_6H_5}{\overset{O^{\ominus}\ \overset{\oplus}{MgBr}}{\underset{\Big\vert}{\overset{\Big\vert}{C}}}}-C_6H_5$$

10
Benzophenone

(19.15)

$$\xrightarrow{H_3O^+} C_6H_5-\underset{C_6H_5}{\overset{OH}{\underset{\Big\vert}{\overset{\Big\vert}{C}}}}-C_6H_5$$

11
Triphenylmethanol

After the addition of the Grignard reagent to carbon dioxide is complete, the excess Dry Ice is allowed to evaporate, and the mixture is acidified. Extraction of the aqueous mixture with diethyl ether then gives a solution containing benzoic acid, benzophenone (**10**), triphenylmethanol (**11**), benzene, and biphenyl (**8**). The benzoic acid is readily separated from the neutral side products in the crude mixture by extracting the solution with dilute aqueous sodium hydroxide, whereby benzoic acid is converted to its water-soluble sodium salt; the neutral by-products remain in the organic layer. Cautious acidification of the aqueous layer regenerates the benzoic acid, which can be purified by recrystallization.

19.3.3 REACTION OF *n*-BUTYLMAGNESIUM BROMIDE WITH 2-METHYLPROPANAL

n-Butylmagnesium bromide (**7**, Eq. 19.5) is also readily prepared, and its reaction with 2-methylpropanal (**13**) provides an excellent example of the Grignard synthesis of secondary alcohols (Eq. 19.16). In this reaction, the Grignard reagent **7** adds to 2-methylpropanal to give the magnesium salt of 2-methyl-3-heptanol, and the alcohol **14** may be isolated after acidification of the mixture.

$$CH_3CH_2CH_2\overset{\delta\ominus}{C}H_2 - \overset{\delta\oplus}{M}gBr \ + \ \underset{\underset{H}{}}{\overset{\overset{\delta\ominus}{O}}{\overset{\|}{C}}} \overset{\delta\oplus}{-} CHCH_3 \ \longrightarrow \ CH_3(CH_2)_3 - \underset{CH_3}{\overset{\overset{O\ominus \ \overset{\oplus}{M}gBr}{|}}{CH}} - \underset{}{CHCH_3}$$

$$\underset{CH_3}{|}$$

7   **13**

n-Butylmagnesium bromide   2-Methylpropanal

(19.16)

$$\xrightarrow{H_3O^+} \ CH_3(CH_2)_3 - \underset{CH_3}{\overset{\overset{OH}{|}}{CH}} - CHCH_3$$

14

2-Methyl-3-heptanol

19.4 SPECIAL EXPERIMENTAL TECHNIQUES

Some of the side reactions that are encountered during the preparation of Grignard reagents (see Sec. 19.2) and during their reaction with electrophiles (see Sec. 19.3) can be controlled by prudent choice of solvent and reaction temperature. However, an experimental technique that can also be used to minimize undesired side reactions involves the **order of addition** of the Grignard reagent and the electrophilic reactant. For example, methyl benzoate and 2-methylpropanal, the electrophiles, are added *to* a solution of the appropriate Grignard reagent, the highly reactive nucleophile, to prepare triphenylmethanol (**11**) and 2-methyl-3-heptanol (**14**), respectively. This mode of addition is designated as **normal addition.** On the other hand, in the preparation of benzoic acid, the Grignard reagent, the *highly reactive* nucleophile, is added *to* the carbon dioxide, the electrophile. This procedure is called **inverse addition.** Whether an addition is performed according to the *normal* or *inverse* mode is dictated by the specific nature of the reactants and the potential side reactions that may be occur. For example, if carbon dioxide gas were bubbled into a solution of phenylmagnesium bromide (**5**) for the preparation of benzoic acid (normal addition), the by-products given in Equation (19.15) would become significant, perhaps predominant. There would always be an excess of the Grignard reagent **5** present relative to carbon dioxide, so further reaction between **5** and the bromomagnesium salt of benzoic acid would be favored. In the preparation of triphenylmethanol (**11**), the combination of the Grignard reagent with methyl benzoate could be done equally well in either the normal or inverse fashion, since two equivalents of **5** are necessary to complete the reaction, and the magnesium alkoxide salt of **11** is unreactive toward **5.** Inverse addition is less convenient in this case, however, because it requires transferring the Grignard reagent **5** from one flask to a dropping funnel and then addition of **5** to the reaction flask containing the methyl benzoate.

EXPERIMENTAL PROCEDURES

SAFETY ALERT

1. When flame-drying the glass apparatus, avoid excessive heating in the vicinity of the ring seals in the condenser and of the stopcock in the addition funnel, particularly if the stopcock is made of Teflon.

2. Diethyl ether is extremely flammable and volatile, and its vapors can easily travel several feet along the bench top or the floor and then be ignited. Consequently, be certain that there are no open flames in the *entire* laboratory when you are working with ether in this experiment.

3. Lubricate all ground-glass joints in the apparatus carefully, and mate them tightly to prevent escape of ether during the reaction.

4. The formation of the Grignard reagent is exothermic. Have an ice-water bath ready to use in the event it is necessary to moderate the rate of the reaction, as evidenced by an excessively rapid rate of reflux and the emission of vapors from the top of the condenser.

5. *Open containers of diethyl ether must not be kept at your laboratory bench.* Estimate the total volume of ether you will need, measure it in the hood into a container that is loosely stoppered before returning to your bench. Be certain to reclose the container from which you obtained the diethyl ether to prevent evaporation, absorption of moisture and oxygen, and accidental fires.

6. Do not store flasks containing diethyl ether in your laboratory drawer as vapors may build up and create a fire hazard. Rather, place the flasks, properly labeled, in the hood.

Pre-Lab exercises for Section 19.2, "Preparation of the Grignard Reagent," are found on page PL. 105.

A.   PREPARATION OF THE GRIGNARD REAGENT

Equip a 50-mL round-bottom flask with several boiling stones or with a magnetic stirring bar, if magnetic stirring is available, and with a Claisen adapter to which a reflux condenser and a dropping funnel, preferably with a ground-glass joint, are attached. Position the dropping funnel directly above the flask and the condenser on the sidearm of the adapter (Fig. 2.31b). Attach calcium chloride drying tubes to the top of the condenser and to the dropping funnel.

Be sure that no one in the laboratory is working with diethyl ether, and then dry the assembled apparatus using a flame (see Sec. 2.24). Allow the apparatus to cool to room temperature. Now verify that there are *NO flames* in the laboratory before continuing!

Working quickly, remove the dropping funnel from the Claisen adapter and add 0.5 g of magnesium turnings through the opening. Replace the funnel and add 2 mL of anhydrous diethyl ether to the round-bottom flask through the dropping funnel; close the stopcock. Prepare a solution of either 3.9 g of bromobenzene or 3.4 g of *n*-butyl bromide in 5 mL of anhydrous diethyl ether. Swirl the solution to achieve homogeneity, and add this solution to the dropping funnel. Be sure that water is running through the condenser.

Add a 0.5-mL portion of the halide-ether solution from the dropping funnel onto the magnesium turnings. If you are not using magnetic stirring, loosen the clamp to the flask and swirl the flask to mix the contents. If small bubbles form at the surface of the magnesium turnings or if the mixture becomes slightly cloudy or chalky, the reaction has started. The flask then becomes slightly warm, and the diethyl ether may begin to reflux. If the reaction has started, *disregard* the instructions in the next paragraph.

If the reaction does not start spontaneously, obtain one or two additional magnesium turnings and crush them thoroughly with a heavy spatula or the end of a clamp. Remove the dropping funnel just long enough to add these broken pieces of magnesium to the flask, and replace the funnel. The clean, unoxidized surfaces of magnesium that are exposed should aid in initiating the reaction. If the reaction still has not started after an additional 3 to 5 min, consult your instructor. The best remedy at this point is to warm the flask and add a crystal of iodine or 2 or 3 drops of 1,2-dibromoethane to the mixture.

Once the reaction has started, gently heat the reaction mixture so the solvent refluxes smoothly and add another 5-mL portion of anhydrous diethyl ether to the reaction mixture through the top of the condenser. This dilutes the mixture and minimizes the undesired coupling reaction. The rest of the halide-ether solution should now be added *dropwise* to the reaction mixture at a rate that is just fast enough to maintain a gentle reflux. If the halide is added too fast, the reaction may get out of control, and the yield of the Grignard reagent is reduced owing to increased coupling. If the reaction becomes too vigorous, cool the flask briefly with an ice-water bath and reduce the rate of addition. The addition should take about 3 to 5 min. If the spontaneous boiling of the mixture slows, increase the rate of addition slightly. If the rate of reflux does not increase, use a steam or hot-water bath or heating mantle to heat the mixture as necessary to maintain gentle reflux during the remainder of addition. At the end of the reaction, the solution normally has a tan to brown, chalky appearance, and most of the magnesium has disappeared, although residual bits of metal usually remain.

Use the Grignard reagent as soon as possible after its preparation, following one of the procedures given in Section 19.3. Phenylmagnesium bromide is used in Parts B and C, whereas *n*-butylmagnesium bromide is used in Part D.

Pre-Lab exercises for Section 19.3.1, "Preparation of Triphenylmethanol," are found on page PL. 107.

B. PREPARATION OF TRIPHENYLMETHANOL

Allow the reaction mixture for the preparation of phenylmagnesium bromide (Part A) to cool to room temperature. Dissolve 1.2 g of methyl benzoate in about 5 mL of anhydrous diethyl ether, and place this solution in the dropping or separatory funnel with the stopcock *closed.* Cool the reaction flask in an ice-water bath; then begin the *slow,* dropwise addition of the ethereal solution of methyl benzoate to the solution of phenylmagnesium bromide. This reaction is exothermic, so you should control the rate of reaction by adjusting the rate of addition *and* by occasional cooling of the reaction flask with the ice-water bath. The ring of condensate should be no more than one-third of the way up the reflux condenser. If you are not using magnetic stirring, swirl the flask from time to time during the addition to mix the reactants and to facilitate cooling. A white solid may form during the reaction and this is normal. After the addition is complete and the exothermic reaction subsides, you may complete the reaction in one of two ways. Consult with your instructor to determine whether you should (1) heat the reaction mixture at reflux for 30 min using a steam or oil bath or (2) stopper the flask after cooling the contents to room temperature and place it in the hood until the next laboratory period (no reflux required).✳

Place about 10 mL of cold 6 *M* sulfuric acid and about 5–10 g of crushed ice in a 100-mL beaker. Pour the reaction mixture gradually with stirring into the ice-acid mixture. Continue stirring until the heterogeneous mixture is completely free of undissolved solids. It may be necessary to add diethyl ether to dissolve all the organic material; the total volume of ether should be about 10–15 mL. Transfer the entire mixture to a 60-mL separatory funnel. Shake the funnel vigorously but carefully and vent the funnel often to relieve pressure; separate the aqueous layer.✳ Wash the organic layer with about 5 mL of 3 *M* sulfuric acid with two 5-mL portions of saturated aqueous sodium bicarbonate, and then with 5 mL of saturated sodium chloride solution.

Dry the organic layer with anhydrous sodium sulfate, and filter or decant the solution into a 50-mL round-bottom flask. Remove most of the diethyl ether by simple distillation (*no flames*); remove the last traces of solvent by connecting the *cool* stillpot to a water aspirator. After the crude solid residue has dried, determine its melting range, which may be wide.✳

Purify the product by dissolving it in a minimum amount of boiling cyclohexane (ca. 10 mL/g product). Perform this operation at the hood, or use a funnel that is attached to a vacuum source and inverted over the flask (see Fig. 2.37b). Once all the material is in solution, evaporate the solvent slowly until small crystals of triphenylmethanol start to form. Remove the flask from the heat and allow the crystallization to continue at room temperature and then in an ice-bath until no more crystals form. Isolate the colorless product by vacuum filtration. Determine the melting point and yield of the product. The reported melting point of triphenylmethanol is 160–163 °C.

Finishing Touches   Dilute the *combined aqueous layers and washes* with water, neutralize the solution if necessary and flush it down the drain with excess water. Place the *ether distillate* and the *cyclohexane mother liquor* in the container for nonhalogenated organic solvents. Spread the *calcium chloride* from the drying tube and the *sodium sulfate* on a tray in the hood and after the ether has evaporated place them and the filter paper in the container for nonhazardous solids.

Pre-Lab exercises for Section 19.3.2, "Preparation of Benzoic Acid," are found on page PL. 109.

C. PREPARATION OF BENZOIC ACID

In a 125-mL Erlenmeyer flask, place about 10 g of solid carbon dioxide (Dry Ice) that has been coarsely crushed and protected from moisture as much as possible. Add the phenylmagnesium bromide solution from Part A to the Dry Ice slowly with vigorous swirling; the mixture normally becomes rather viscous. After the addition is complete, allow the excess carbon dioxide to evaporate by letting the *unstoppered* flask stand in the hood, properly labeled with your name, until the next laboratory period.✳ You may expedite the removal of the excess CO_2 by shaking or swirling the flask while warming it *very slightly* in a warm-water bath or by adding small amounts of warm water to the reaction mixture with shaking or stirring. Because both of these methods may cause a sudden loss of CO_2 gas and splash the contents of the flask onto the floor or bench top, you should be careful using these methods. The flask should *never* be stoppered. Warming the flask may also cause the diethyl ether to evaporate, and you should replace the ether before going to the next step.

After the excess Dry Ice is gone, add 20 mL of diethyl ether and then treat the mixture with 10 mL of 3 *M* sulfuric acid that has been mixed with about 5 g of ice. Add the acid to the reaction mixture carefully to avoid excessive foaming. If the ether evaporates appreciably during this operation, more diethyl ether should be added so that the total volume of ether is about 20 mL. Swirl the mixture and transfer it to a small separatory funnel. Rinse the flask with a small portion of diethyl ether, and add the rinse to the separatory funnel. Shake the funnel cautiously, with venting, and separate the layers. Extract the aqueous layer two additional times with 5-mL portions of diethyl ether, and combine *all* the ethereal extracts.

Extract the combined ethereal solution with three 5-mL portions of a 1 *M* solution of sodium hydroxide, venting the funnel frequently during the extractions. Combine the alkaline extracts and slowly add 6 *M* hydrochloric acid until precipitation of the product is complete and the aqueous mixture is acidic to pHydrion paper. Isolate the solid benzoic acid by vacuum filtration.✳ After drying the product, determine its melting point and yield. If the melting point is not sharp, recrystallize the benzoic acid from water (see Sec. 3.1).

> **Finishing Touches**   Dilute the *combined aqueous layers and washes* with water, neutralize the solution if necessary and flush it down the drain with excess water. Place the *combined ethereal layers* in the container for nonhalogenated organic solvents. Spread the *calcium chloride* from the drying tube on a tray in the hood to allow the ether to evaporate; then place it and the filter paper in the container for nonhazardous solids.

Pre-Lab exercises for Section 19.3.3, "Preparation of 2-Methyl-3-heptanol," are found on page PL. 111.

D. PREPARATION OF 2-METHYL-3-HEPTANOL

Allow the reaction mixture of *n*-butylmagnesium bromide prepared in Part A to cool to room temperature. Fix a dropping or separatory funnel, which is preferably equipped with a ground glass joint, to the apparatus in which the Grignard reagent was prepared; close the stopcock. Dissolve 1.3 g of freshly distilled 2-methylpropanal in 5 mL of anhydrous diethyl ether, and place this solution in the funnel. Begin the dropwise addition of the ethereal solution of 2-methylpropanal *to* the solution of *n*-butylmagnesium bromide. Control the resulting exothermic reaction by adjusting the rate of addition so that the ring of condensate is no more than one-third of the way up the reflux condenser. Cool the reaction flask with a previously prepared ice-water bath if necessary. Either stir or swirl the reaction mixture occasionally during the course of the addition, which will require about 5 min. After completing the addition, allow the reaction mixture to stand for about 15 min.✳ The reaction mixture may be stored in the hood until the next laboratory period.

Place about 10 mL of ice-cold 6 *M* sulfuric acid and 5–10 g of crushed ice in a 100-mL beaker. Pour the reaction mixture slowly with stirring into the ice-acid mixture. After the addition is complete, transfer the cold mixture, which may contain some precipitate, to a small separatory funnel and shake it gently. The precipitate should dissolve. Separate the layers and extract the aqueous layer with two 5-mL portions of diethyl ether and add these extracts to the main ethereal layer. Transfer the combined ethereal extracts into the separatory funnel. Venting the funnel frequently to relieve pressure, wash the solution with 5 mL of saturated sodium bisulfite, with two 5-mL portions of saturated aqueous sodium bicarbonate, and finally with 5 mL of saturated sodium chloride solution. Dry the ethereal solution over anhydrous sodium sulfate.✳

Filter or carefully decant the dried organic solution into a 50-mL round-bottom flask equipped for simple distillation and remove the diethyl ether by simple distillation (*no flames*). Using a micropipet, transfer the residue to a 10-mL round-bottom flask equipped for short-path distillation (see Fig. 2.15b). Insulate the top of the distilling flask and the stillhead with glasswool wrapped with aluminum foil to ensure steady distillation of the high-boiling product. Heating the flask with a small Bunsen burner, distil the product. Although pure 2-methyl-3-heptanol boils at 165–168 °C (760 torr), it may be difficult to obtain an accurate boiling point owing to the relatively small amount of product. Collect all distillate that boils in the range 130–168 °C (760 torr). Determine the yield of product.

> **Finishing Touches** Dilute the *combined aqueous layers and washes* with water, neutralize the solution if necessary and flush it down the drain with excess water. Place the *ether distillate* in the container for nonhalogenated organic solvents. Spread the *calcium chloride* from the drying tube and the *sodium sulfate* on a tray in the hood, and then place them in the container for nonhazardous solids after the volatiles have evaporated.

EXERCISES

General Questions

1. Arrange the following compounds in order of increasing reactivity toward attack of the Grignard reagent at the carbonyl carbon atom: methyl benzoate, benzoic acid, benzaldehyde, acetophenone, benzoyl chloride. Explain the basis for your decision, making use of mechanisms where needed.

2. What is (are) the product(s) of reaction of each of the carbonyl-containing compounds in Exercise 1 with *excess* Grignard reagent, RMgBr?

3. How might primary, secondary, and tertiary alcohols be prepared from a Grignard reagent and a suitable carbonyl-containing compound? Write chemical reactions for these preparations using any starting materials you wish; indicate stoichiometry where important.

Questions for Part A

4. Why were you cautioned not to heat excessively in the vicinity of ring seals of the condenser when drying the apparatus for this experiment?

5. Ethanol is often present in technical diethyl ether. If this grade rather than anhydrous were used, what effect would the ethanol have on the formation of the Grignard reagent? Explain your reasoning.

6. Give a plausible, three-dimensional structure for the complex $RMgBr \cdot 2(C_2H_5)_2O$. How do you think the molecules of diethyl ether are bound to the Grignard reagent?

7. Which problems might be encountered if bromocyclohexane were used as an additive to help initiate the formation of phenylmagnesium bromide and *n*-butylmagnesium bromide. How does the use of 1,2-dibromoethane avoid such experimental difficulties?

8. Why is it unwise to allow the solution of the Grignard reagent to remain exposed to air, even if it is protected from moisture by drying tubes?

9. Why is it unwise to begin addition of the solution of methyl benzoate to the Grignard reagent before the latter has cooled to room temperature?

Questions for Part B

10. Why should anhydrous rather than technical diethyl ether be used to prepare the solution of methyl benzoate that is added to the Grignard reagent?

11. What is the solid that forms during the addition of the ester to the Grignard reagent?

12. Why does pressure develop when the separatory funnel containing aqueous sulfuric acid and the ethereal solution of organic products is shaken?

13. Comment on the use of steam distillation as a possible alternative procedure for purifying crude triphenylmethanol. Consider what possible starting materials, products, and by-products might be present, and indicate which of these should steam-distil and which should not. Would this method of purification yield pure triphenylmethanol? Give your reasoning.

Questions for Part C

14. Why does pressure develop when the separatory funnel containing aqueous sulfuric acid and the ethereal solution of organic products is shaken?

15. The yield of benzoic acid obtained when only enough acid is added to the aqueous solution of sodium benzoate to bring the pH to 7 is consistently lower than that obtained if the pH is brought below 5. Why is this?

16. What function does extracting the ethereal solution of organic products with aqueous base have in the purification of benzoic acid?

Questions for Part D

17. Given that commercially available aldehydes are usually produced by oxidative methods, what contaminant(s) might be present in the 2-methylpropanal that could lead to a decrease in yield?

18. Why is it unwise to begin addition of the solution of 2-methylpropanal to the Grignard reagent before the latter has cooled to room temperature?

19. What is the solid that forms upon reaction of *n*-butylmagnesium bromide with 2-methylpropanal?

20. Why does pressure develop when the separatory funnel containing aqueous sulfuric acid and the ethereal solution of organic products is shaken?

21. The work-up in this reaction calls for successive washes of an ethereal solution of the product with aqueous sodium bisulfite, sodium bicarbonate, and sodium chloride. What is the purpose of each of these steps?

22. The IR spectrum of 2-methyl-3-heptanol prepared by the procedure of Part D sometimes shows a band at 1720 cm^{-1}. Give a possible source for this absorption.

23. Explain why there are eight peaks in the ^{13}C NMR spectrum of 2-methyl-3-heptanol (Fig. 19.6b).

SPECTRA OF STARTING MATERIALS AND PRODUCTS

The IR, ^{1}H, and ^{13}C NMR spectra of bromobenzene are given in Figures 15.11 and 15.12(a) and (b), and the IR, ^{1}H, and ^{13}C NMR spectra of 1-bromobutane are given in Figures 14.3 and 14.4(a) and (b), respectively. The ^{13}C NMR spectrum of methyl benzoate is presented in Figure 8.53, and the IR, ^{1}H, and ^{13}C NMR spectra of benzoic acid are given in Figures 3.1 and 3.2(a) and (b) respectively.

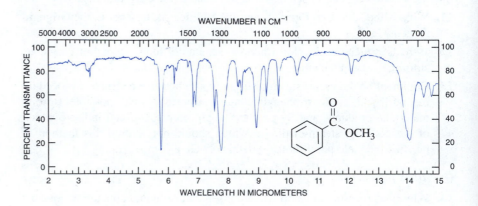

FIGURE 19-1   IR spectrum of methyl benzoate.

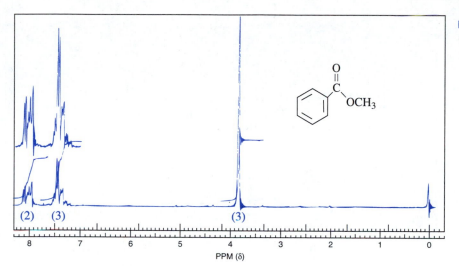

FIGURE 19-2 ^{1}H NMR spectrum (60 MHz) of methyl benzoate.

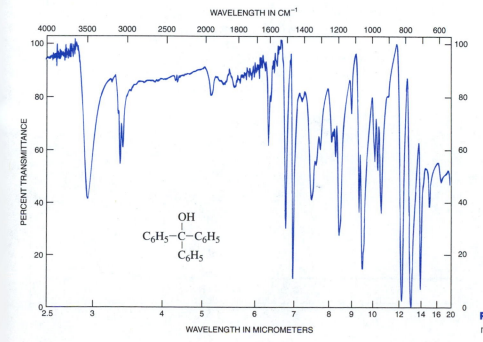

FIGURE 19-3 IR spectrum of triphenyl-methanol.

FIGURE 19-4 NMR data for triphenyl-methanol.

(a) ^{1}H NMR spectrum (60 MHz).

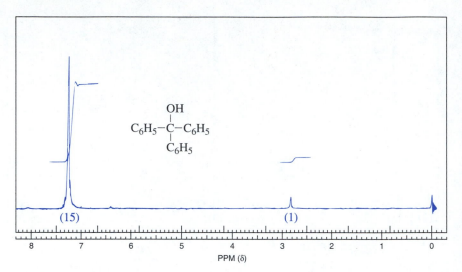

(b) ^{13}C NMR data.
Chemical shifts: δ 82.0, 127.2, 127.9, 146.9.

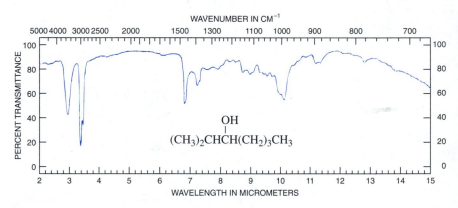

FIGURE 19-5 IR spectrum of 2-methyl-3-heptanol.

(a) ^{1}H NMR spectrum (60 MHz).

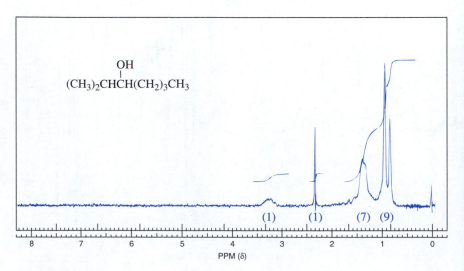

FIGURE 19-6 NMR data for 2-methyl-3-heptanol.

(b) ^{13}C NMR data.
Chemical shifts: δ 14.2, 17.5, 19.3, 23.1, 28.6, 33.8, 34.1, 76.6.

MULTISTEP ORGANIC SYNTHESIS

The synthesis of complex molecules from simpler starting materials is one of the most important aspects of modern organic chemistry. In the pharmaceutical and agricultural industries, large quantities of biologically active molecules, sometimes having very complex structures and numerous functional groups, must be prepared from commercially available compounds in the most efficient and cost-effective fashion. This often poses considerable challenges, and it is usually necessary to marshal the combined skills of expert synthetic and mechanistic organic chemists to solve the problems.

There are presently routes to biologically important molecules that require as many as 20 or 30 sequential reactions or steps; however the "arithmetic demon" haunts these multistep syntheses because the yield in each step is not quantitative. For example, even a five-step synthesis in which each step occurs in a respectable 80% yield gives an overall yield of only 33% [i.e., $(0.8)^5 \times 100 = 32.8\%$]. To offset this cumulative effect, the organic chemist must find the shortest and most efficient pathways to the targeted molecule. Each reaction must be carefully studied to minimize side reactions, and a thorough knowledge of the mechanism of each reaction is critical to success. Multistep syntheses that exceed 8–10 steps are economically feasible only for end products that are biologically highly active and cannot be obtained readily from natural sources.

The *optimal route* to a particular compound *is not always the shortest,* as is illustrated by the simple problem of preparing 2-chloropropane on an industrial scale, as shown in Scheme 20.1. Although 2-chloropropane is produced from propane in one step by direct chlorination, it must be separated from the 1-chloropropane formed concurrently; this is *not* a simple separation. However, if the mixture of 1- and 2-chloropropane is treated with a base, both isomers undergo dehydrochlorination to give propene, and subsequent addition of hydrogen chloride produces 2-chloropropane as the *sole* product. In this case a three-step process is the preferred way to obtain pure 2-chloropropane. As this example demonstrates, an important consideration in designing a synthesis is the ease of separating the desired product from unwanted by-products and purifying it. Other considerations include the availability and cost of starting materials, simplicity of equipment and instrumentation, energy costs, activity of catalysts, selectivity of reactions in polyfunctional molecules, and stereochemical control.

Because so many variables must be considered in planning the synthesis of a complex molecule, it is not surprising that organic chemists are investigating the possibility of using computers to design and analyze multistep syntheses. Computers can handle an enormous amount of information, and it will eventually be possible to design optimal sequences of reaction steps using sophisticated computer programs.

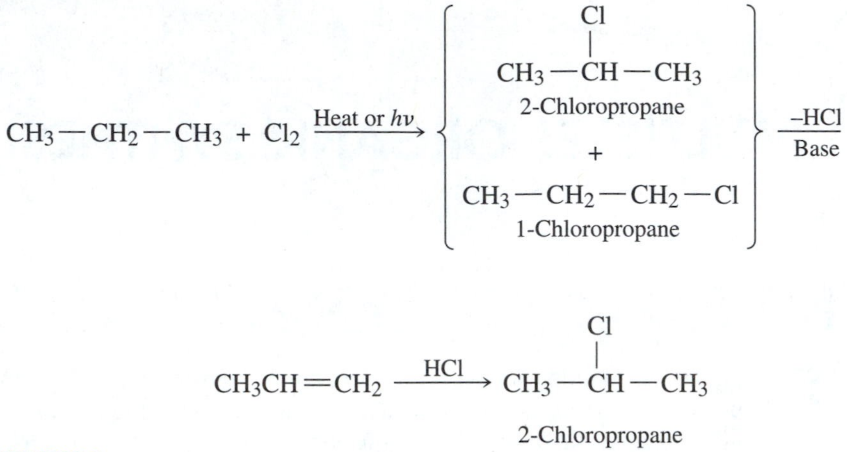

$$CH_3-CH_2-CH_3 \ + \ Cl_2 \ \xrightarrow{\text{Heat or } h\nu} \ \left\{ \begin{array}{c} \overset{\displaystyle Cl}{\underset{\displaystyle |}{}} \\ CH_3-CH-CH_3 \\ \text{2-Chloropropane} \\ + \\ CH_3-CH_2-CH_2-Cl \\ \text{1-Chloropropane} \end{array} \right\} \ \xrightarrow[\text{Base}]{-HCl}$$

$$CH_3CH{=}CH_2 \ \xrightarrow{\text{HCl}} \ CH_3-\overset{\displaystyle Cl}{\underset{\displaystyle |}{C}}H-CH_3$$
$$\text{2-Chloropropane}$$

SCHEME 20.1

In considering this fascinating prospect, however, it must be remembered that the information is provided to the computer by humans, specifically chemists, who will be vital to the success of such a project. Chemists still have at least one significant competitive advantage over computers—they are more creative!

Appreciation of and insight into some of the problems that may be encountered in planning a multistep synthesis is gained from preparing a compound that requires a fairly small number of steps. In this chapter, the synthesis of sulfanilamide (see Sec. 20.2) is used to illustrate the fundamental principles. There is also the possibility of devising additional synthetic sequences by combining some of the separate experiments described in other chapters in this book. Suggestions for two such sequences are given in Section 20.3. In these experiments, the product of one reaction is used in a subsequent step, and if you are to be successful, you *must* use good experimental technique. You may either isolate and purify the intermediates by distillation or recrystallization or you may use the crude material directly in the next step. This is the sort of choice the practicing synthetic organic chemist constantly faces. In general, extensive purification of each intermediate in a sequence is avoided *if* the impurities can eventually be removed *and* their presence does not interfere with the course of the desired reactions. However, even if the entire quantity of an intermediate is not purified, it is good scientific practice to purify a small sample of it for *complete* characterization by spectroscopic (see Chap. 8) and physical (bp, mp, etc.) methods.

20.2 SULFANILAMIDE: DISCOVERY AND SYNTHESIS OF THE FIRST ANTIBIOTIC

An important area of research in the pharmaceutical industry is the discovery and development of new orally active **antibiotics** to treat bacterial infections. Commonly used antibiotics fall into several important classes—including the **β-lactam antibiotics** such as ampicillin (**1**) and cephaclor (**2**), the **macrolide antibiotics** such as erythromycin A (**3**), the tetracycline antibiotics such as terramycin (**4**), the **aminocyclitol antibiotics** such as spectinomycin (**5**) and the **sulfa drugs** such as sulfamethoxazole (**6**) and sulfadiazine (**7**). All sulfa drugs are derivatives of *p*-aminobenzenesulfonamide, which is commonly called sulfanilamide (**8**), and these drugs were the first antibacterial agents available. Each antibiotic has a different profile of biological activity, and all of them are used in modern medicine to treat various infections.

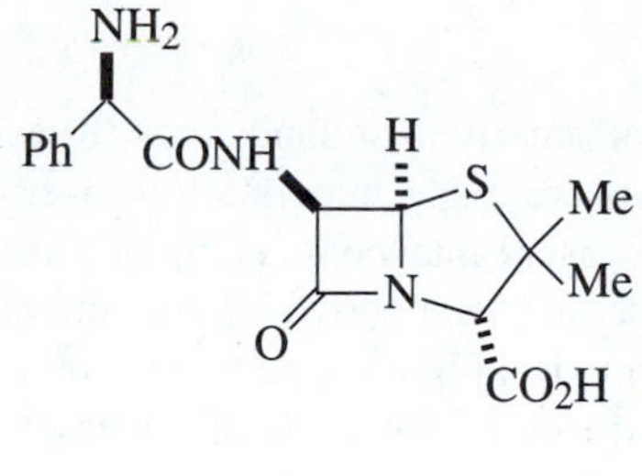

1
Ampicillin

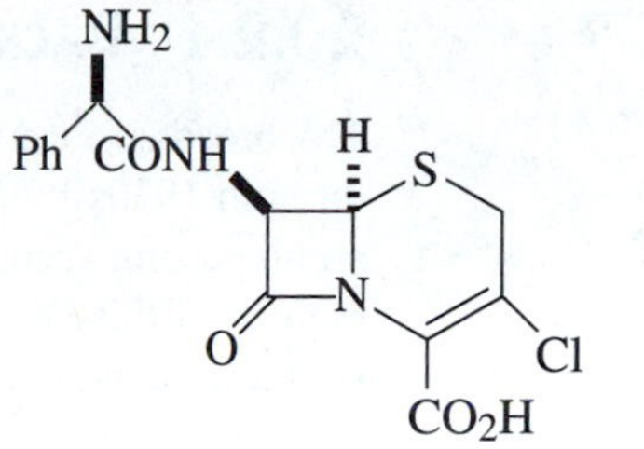

2
Cephaclor

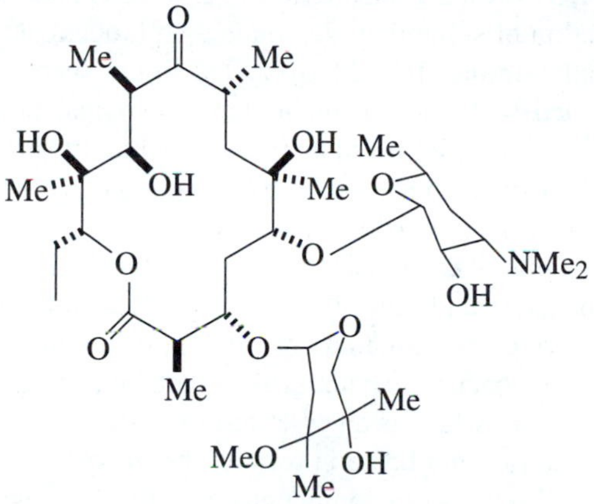

3
Erythromycin A

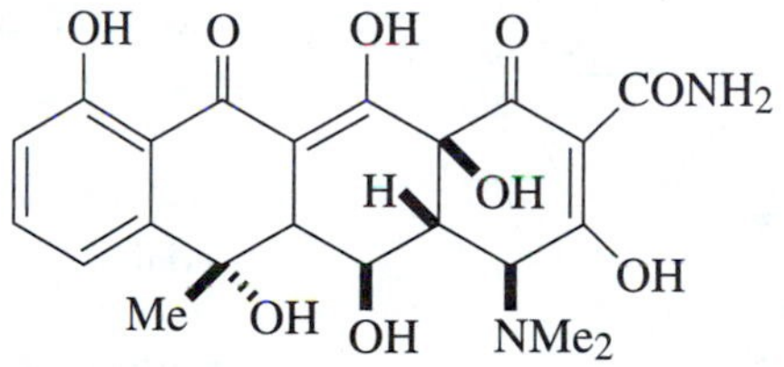

4
Terramycin

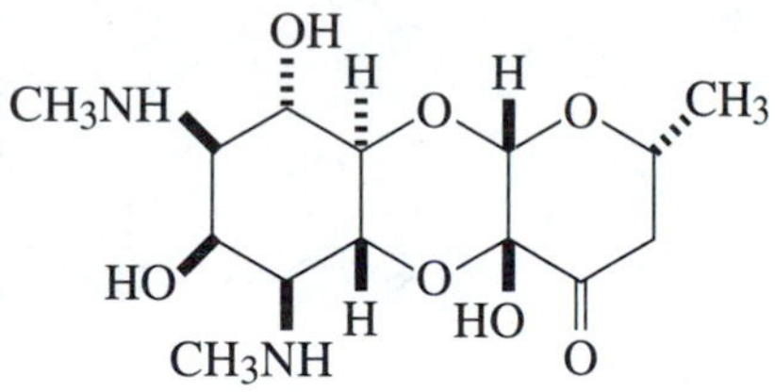

5
Spectinomycin

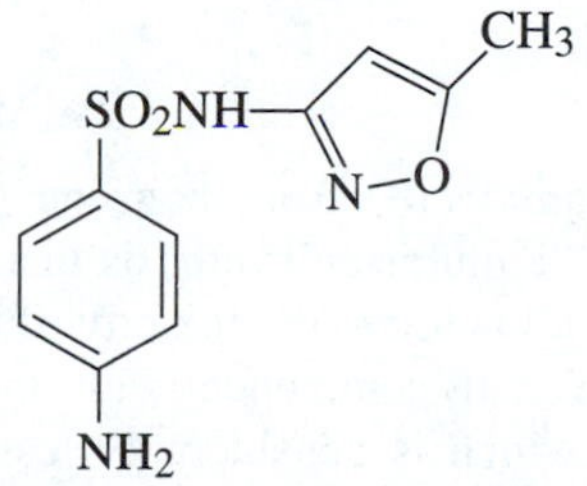

6
Sulfamethoxazole

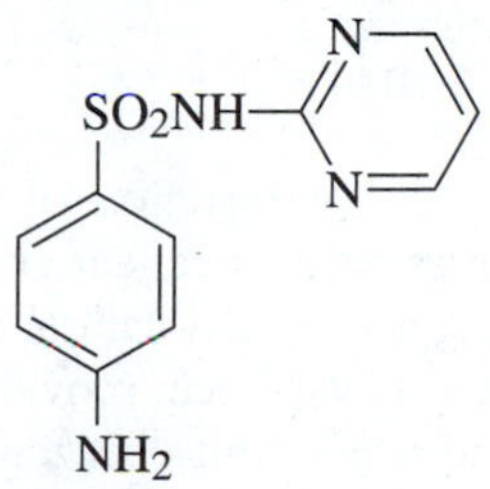

7
Sulfadiazine

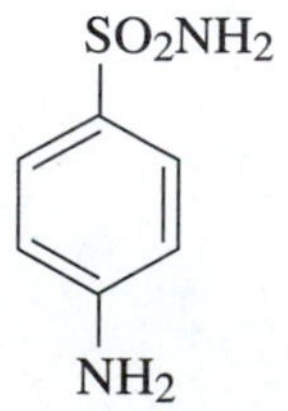

8
Sulfanilamide

20.2.1 BACKGROUND

The broad-spectrum antibacterial activity of sulfanilamide (**8**) was first revealed in the mid-1930s by serendipity, as are many important discoveries in science, and an interesting account of this discovery may be found in the Historical Highlight at the end of this chapter. Over one thousand derivatives of sulfanilamide have been synthesized and tested as potential antibiotics, and some of these, such as sulfamethoxazole (**6**), are still used today. The mode of action of the sulfa drugs is another interesting story because it provides some insights into strategies that might be generally exploited to design biologically active compounds.

Early in the development of sulfa drugs as antibiotics, it was found that *p*-aminobenzoic acid (PABA, **9**), which is also used in sunscreens and sunblocks, inhibits the antibacterial action of sulfanilamide. Since *p*-aminobenzoic acid and sulfanilamide are structurally similar, this discovery led to the speculation that the two compounds competed with each other in some biological process that was essential to bacterial growth. This speculation was eventually supported by experimentation. *p*-Aminobenzoic acid is used by bacteria in the synthesis of the essential enzyme cofactor folic acid (**10**). When sulfanilamide is present, it successfully competes with *p*-aminobenzoic acid for the active site in the enzyme that incorporates *p*-aminobenzoic acid into folic acid. By functioning as a competitive inhibitor of this enzyme, sulfanilamide blocks the biosynthesis of folic acid, and without folic acid, the bacteria cannot grow. On the other hand, animal cells cannot synthesize folic acid, which is an **essential vitamin,** and it therefore must be part of the diet. Because only bacteria rely on the biosynthesis of folic acid from *p*-aminobenzoic acid, the sulfa compounds are "ideal" drugs because they kill only the bacteria and not the animal host. Of course, they are not truly ideal, because some people have allergic reactions to sulfa drugs.

9
p-Aminobenzoic
acid (PABA)

10
Folic acid

The preparation of sulfanilamide (**8**) from benzene (**11**) (Scheme 20.2) serves as an excellent example of a multistep synthesis that produces a biologically active, non-natural molecule. Owing to the toxicity of benzene, the experimental sequence provided here will commence with the second step, the reduction of nitrobenzene (**12**), which is considerably less toxic. Each of the intermediate compounds in the synthesis may be either isolated or used directly in the subsequent step without further purification. A discussion of the reactions involved in each step follows.

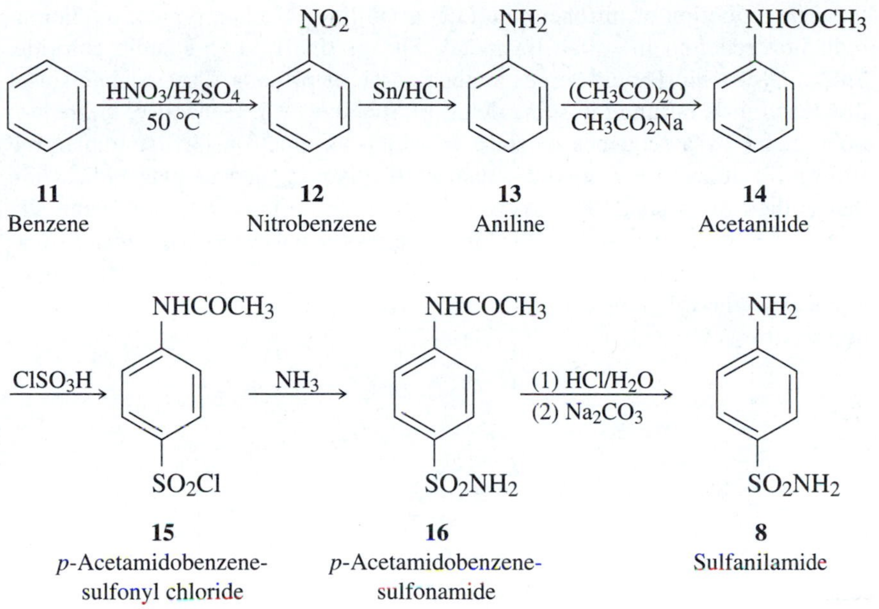

11 Benzene   **12** Nitrobenzene   **13** Aniline   **14** Acetanilide

15
p-Acetamidobenzene-
sulfonyl chloride

16
p-Acetamidobenzene-
sulfonamide

8
Sulfanilamide

SCHEME 20.2

20.2.2 REDUCTION OF AROMATIC NITRO COMPOUNDS: PREPARATION OF ANILINE (13)

Aromatic nitro compounds, which are readily prepared by electrophilic aromatic nitration (see Sec. 15.3), are **reduced** to the corresponding aromatic amines by a variety of methods (see Chap. 17). Chemical reduction is commonly used in the research laboratory, but the most important commercial method is catalytic hydrogenation. Although the precise mechanism of the reduction of nitro compounds is not fully understood, the isolation of various intermediates, some of which are shown in Equation (20.1), suggests that the reaction is stepwise. The symbol **[H]** over the arrows refers generally to a reduction without specifying the reagent. The product that is ultimately isolated from the reduction of an aromatic nitro compound can be controlled to a significant degree by the nature of the reducing agent. For example, the reduction of nitrobenzene (**12**) with zinc metal and ammonium chloride gives only *N*-phenylhydroxylamine (**18**), whereas the use of tin metal and hydrochloric acid gives only aniline (**13**) in excellent yield.

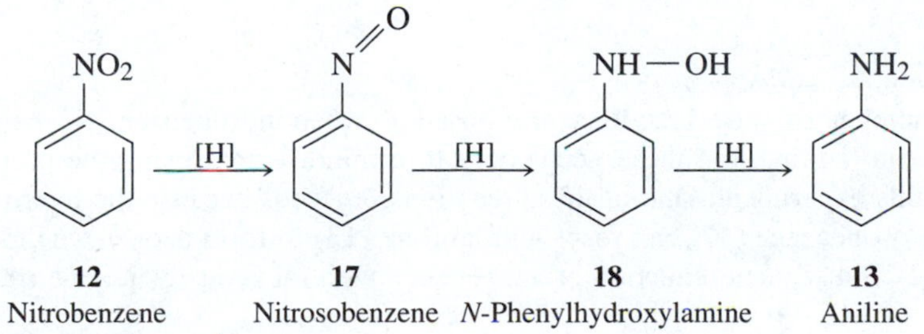

(20.1)

12
Nitrobenzene

17
Nitrosobenzene

18
N-Phenylhydroxylamine

13
Aniline

The reduction of nitrobenzene (**12**) to aniline (**13**) is a typical oxidation-reduction reaction in which tin metal, Sn°, is oxidized to stannic chloride, $SnCl_4$; hydrochloric acid serves as the source of protons. The mechanism of this reaction is outlined in some detail in Scheme 20.3. Generally, the reduction occurs by a sequence of steps in which an electron is first transferred from a tin atom to the organic substrate to give an intermediate radical ion that is then protonated. The oxygen atoms on the nitro group are eventually removed as water molecules. It is left as an exercise to write a *balanced* equation for the overall reaction and to provide a mechanism for the reduction of *N*-phenylhydroxylamine (**18**) into aniline (see Exercises 8 and 9 at the end of this section).

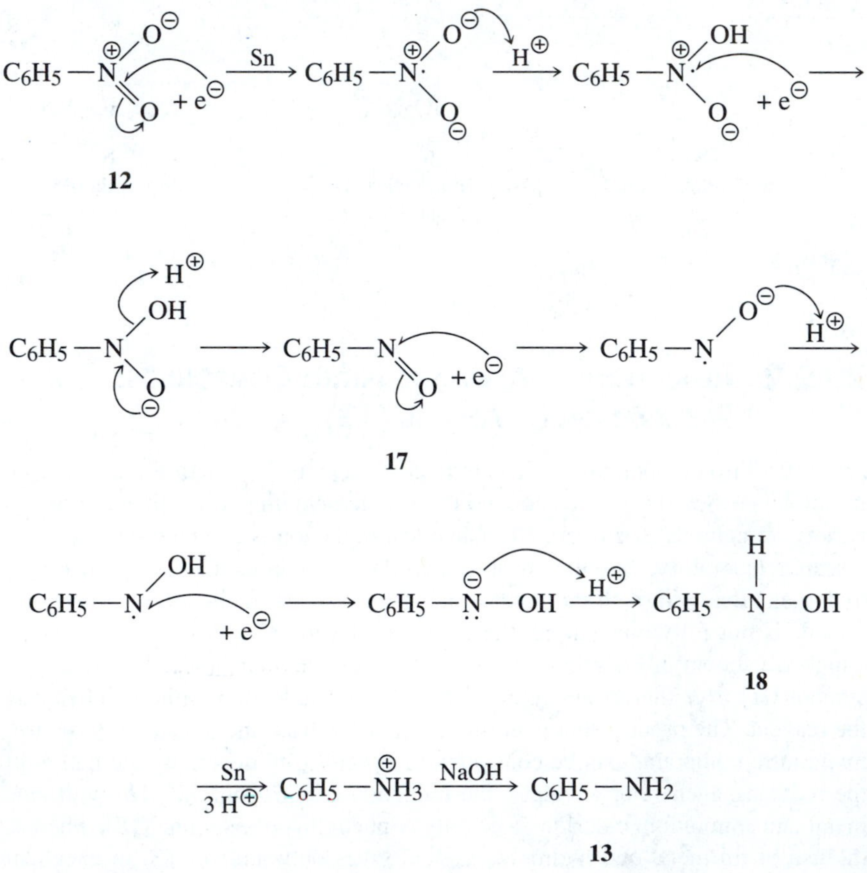

12

17

18

$$\xrightarrow[3\,H^{\oplus}]{Sn} \quad C_6H_5-\overset{\oplus}{N}H_3 \quad \xrightarrow{NaOH} \quad C_6H_5-NH_2$$

13

SCHEME 20.3

A number of side reactions are possible when nitrobenzene is being reduced into aniline, but these occur only to a minor extent using the procedure of this experiment. One of these reactions proceeds because the intermediate nitrosobenzene (**17**) can react with aniline (**13**) to form azobenzene (**19**) (Eq. 20.2). Subsequent reduction of azobenzene with tin gives hydrazobenzene

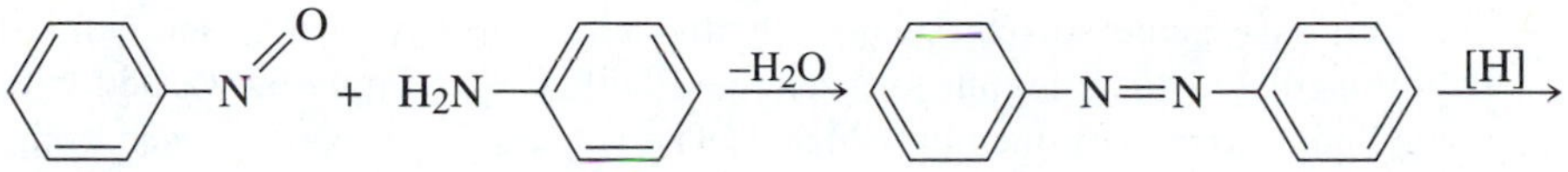

17   13   19
                                              Azobenzene

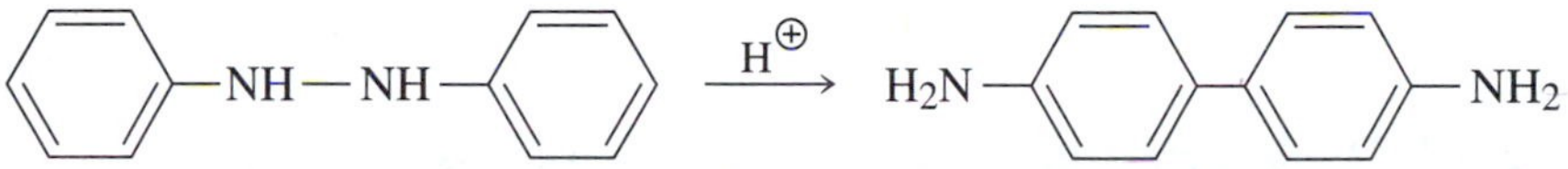

                                                                                    (20.2)

        20   21
  Hydrazobenzene   Benzidine

(**20**), which then can undergo an acid-catalyzed rearrangement known as the benzidine rearrangement to give benzidine (**21**). Another possible by-product in this reduction is *p*-aminophenol (**22**), which may be produced by the acid-catalyzed rearrangement of *N*-phenylhydroxylamine (**18**), as shown in Equation (20.3).

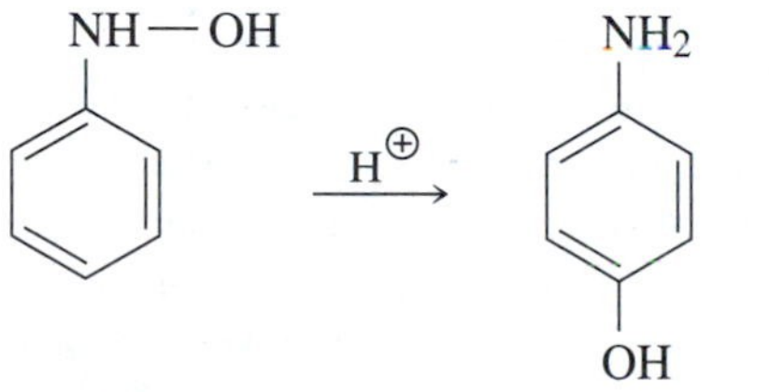

                                                                      (20.3)

        18   22
                                p-Aminophenol

   The procedure for isolating and purifying aniline in this experiment represents an excellent example of how the physical and chemical properties of an organic substance can be exploited to isolate a pure product *without* using chromatographic techniques. For example, after completing the procedure for the chemical reduction of nitrobenzene, it is necessary to remove aniline from its principal impurities, which are unchanged nitrobenzene (**12**), benzidine (**21**), and *p*-aminophenol (**22**). Steam distillation (see Sec. 2.10) is a good technique for separating volatile organic compounds from nonvolatile organic and inorganic substances. However, because the reaction mixture is acidic, it must be made basic prior to performing the steam distillation so aniline is present as the free base and the *p*-aminophenol is converted to its water-soluble sodium phenoxide salt. The aniline and nitrobenzene are then removed from the reaction mixture by steam distillation, and the nonvolatile salt of *p*-aminophenol together with the benzidine remains in the aqueous phase.
   Two methods may be used to separate aniline from nitrobenzene. In the first, the aqueous steam distillate is made acidic with hydrochloric acid to convert aniline to its water-soluble hydrochloride salt. Only nitrobenzene steam-distils from this mixture, and the anilinium salt will remain in the acidic aqueous residue. After the aqueous mixture is made basic, aniline is recovered by extrac-

tion into an organic solvent such as diethyl ether. Alternatively, the acidified steam distillate containing nitrobenzene and aniline can simply be washed with diethyl ether to remove the nitrobenzene. The aqueous layer is then made basic, and the aniline is isolated from the mixture by extraction with diethyl ether. Because this latter acid-base extraction procedure is operationally simpler, it will be used in this experiment.

20.2.3 PROTECTING GROUPS: PREPARATION OF ACETANILIDE (14)

The amino function on aniline is a highly electron-donating group that *activates* the aromatic ring toward certain electrophilic aromatic substitution reactions (see Chap. 15). However, if an aryl amino group is protonated or complexed with a Lewis acid, the resulting ammonium group *deactivates* the ring toward subsequent electrophilic aromatic substitution. Electrophilic reagents or other functional groups that may be present on the ring can also react directly with the amino group. Thus, because of its basic and nucleophilic properties, it is necessary to **protect** the amino group on aniline to ensure that the desired reaction, rather than a deleterious side reaction, will take place. In general, placing a **protecting group** on a reactive functional group alters the chemical properties of that group, thereby rendering it unreactive toward certain reagents and functional groups. The appropriate protecting group must be selected with care, however, because it must be stable to the reaction conditions employed in the various transformations, *and* it must be removable under conditions that do not affect other functional groups in the molecule.

In this present synthesis of sulfanilamide, a sulfonamido, $-SO_2NH_2$, group must be introduced onto the aromatic ring *para* to the nitrogen atom on aniline (see Secs. 20.2.3 and 20.2.4). A common method for converting an arene, Ar–H, into an aryl sulfonamide involves sequential chlorosulfonation, an electrophilic aromatic substitution reaction, followed by treatment of the intermediate aryl sulfonyl chloride (Ar–SO$_2$Cl) with ammonia to give the requisite sulfonamide, Ar–SO$_2$NH$_2$, (Eq. 20.4). However, if aniline were treated with chlorosulfonic

$$\text{Ar}\!-\!\text{H} \xrightarrow{\text{ClSO}_3\text{H}} \text{Ar}\!-\!\text{SO}_2\text{Cl} \xrightarrow{\text{NH}_3} \text{Ar}\!-\!\text{SO}_2\text{NH}_2 \qquad \textbf{(20.4)}$$

| An arene | An aryl sulfonyl chloride | An aryl sulfonamide |

acid, the amino group could react with either the SO$_3$ or hydrogen chloride, which are generated *in situ,* to provide ammonium salts such as **22** or **23;** these groups would then deactivate the ring toward electrophilic aromatic substitution. Moreover, a chlorosulfonyl group cannot be generated in the presence of an amino group since the chlorosulfonyl group of one molecule could react with the amino group of another, thereby forming a *polymeric* material containing sulfonamide linkages (see Exercise 18 at the end of this section).

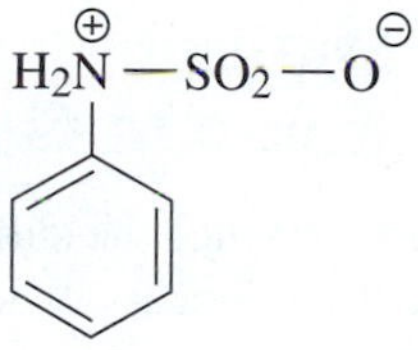

$$H_2\overset{\oplus}{N} - SO_2 - O^{\ominus} \qquad\qquad H_2\overset{\oplus}{N} - H$$

22 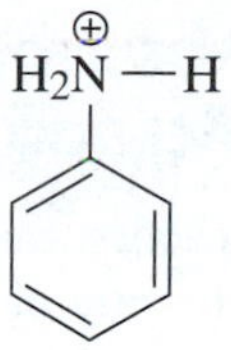   23

Since the presence of the free amino group of aniline would interfere with the preparation of the arylsulfonyl chloride, it is necessary to *protect* the amino group as an amide by acetylation. The free amine is regenerated by removing the acetyl group via controlled hydrolysis *after* the chlorosulfonyl moiety has been introduced onto the aromatic ring and converted into the sulfonamido group. The selective hydrolysis of the amido group in the presence of the sulfonamido group is possible because the sulfur atom of the sulfonamido group is tetrahedral and hence more sterically hindered toward nucleophilic attack by water than the trigonal carbonyl carbon of the amide function.

Aniline (**13**) is converted to acetanilide (**14**) by acetylation according to the sequence shown in Equation (20.5). In this reaction, aniline is first converted into its water-soluble hydrochloride salt (**24**). Acetic anhydride and sodium acetate are then added to give a mixture in which sodium acetate and aniline hydrochloride are in equilibrium with acetic acid and aniline. As free aniline is produced by this acid-base reaction, it rapidly undergoes acetylation by acetic anhydride to give acetanilide, which is isolated by filtration and purified by recrystallization from water. This method for acetylation of amines is general, and the yields are usually high.

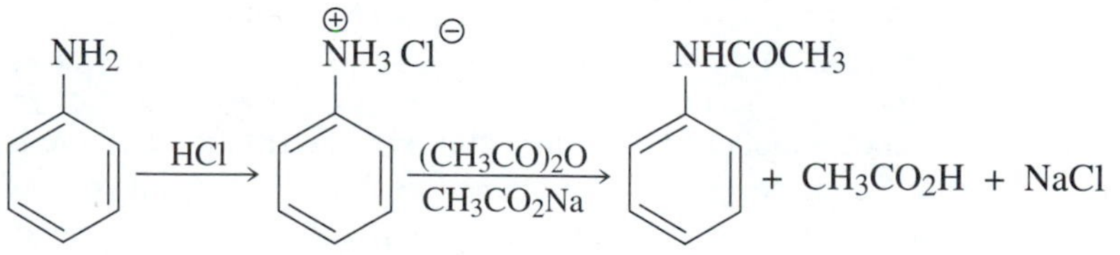

$$NH_2 \xrightarrow{\text{HCl}} \overset{\oplus}{NH_3}\,Cl^{\ominus} \xrightarrow[\text{CH}_3\text{CO}_2\text{Na}]{(CH_3CO)_2O} NHCOCH_3 + CH_3CO_2H + NaCl \quad \textbf{(20.5)}$$

13   24   14

Anilinium hydrochloride

A possible side reaction in this procedure is diacetylation, as illustrated in Equation (20.6). Although the acetylation of aniline in pure acetic anhydride gives substantial quantities of the diacetyl compound **25,** this side reaction is minimized in aqueous solution.

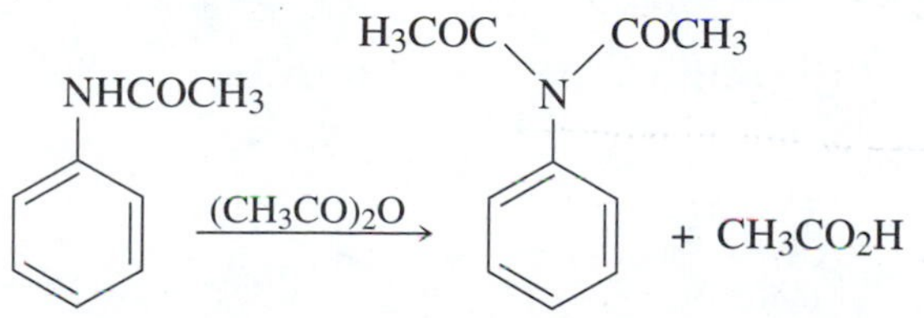

$$NHCOCH_3 \xrightarrow{(CH_3CO)_2O} \overset{H_3COC\diagdown\;\;\diagup COCH_3}{\underset{N}{\big|}} + CH_3CO_2H \quad \textbf{(20.6)}$$

14   25

20.2.4   CHLOROSULFONATION: SYNTHESIS OF *P*-ACETAMIDOBENZENESULFONYL CHLORIDE (15)

A simple one-step reaction can be used to introduce the sulfonyl group, –SO$_2$Cl, *para* to the *N*-acetyl group of acetanilide by a process called chlorosulfonation, which is shown in Equation (20.7). Although an electron-donating group can direct an incoming electrophile to either the *ortho* or *para* position, the acetamido group orients the incoming group predominantly *para,* presumably as a consequence of the steric bulk of the acetamido group; virtually none of the *ortho* isomer is observed. The reaction is known to proceed through the intermediate sulfonic acid (**26**), which is converted to the sulfonyl chloride (**15**) on further reaction with chlorosulfonic acid. At least two equivalents of chlorosulfonic

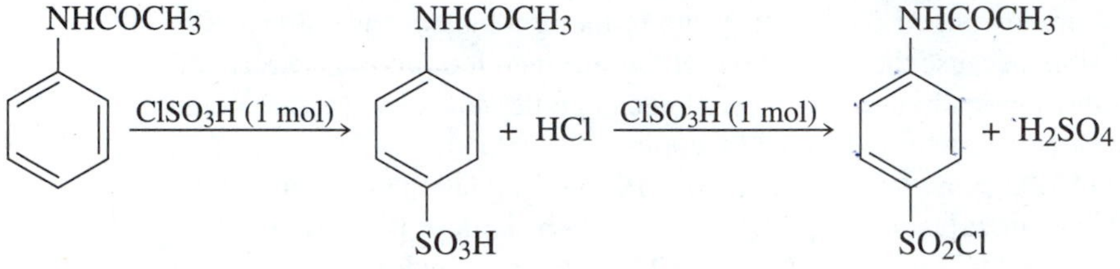

$$\text{(20.7)}$$

14

26
p-Acetamidobenzene-
sulfonic acid

15

acid per equivalent of acetanilide (**13**) are thus required for complete reaction. The attacking electrophile is probably SO$_3$, which is generated *in situ* from chlorosulfonic acid according to Equation (20.8).

$$ClSO_3H \rightleftharpoons SO_3 + HCl \qquad \text{(20.8)}$$

To isolate the *p*-acetamidobenzenesulfonyl chloride, the reaction mixture is poured into ice water, and the product is obtained as a white precipitate. The water rapidly hydrolyzes the excess chlorosulfonic acid (Eq. 20.9). In general, sulfonyl chlorides are much less reactive toward water than are carboxylic acid

$$ClSO_3H + H_2O \longrightarrow HCl + H_2SO_4 \qquad \text{(20.9)}$$

chlorides, but it is unwise to expose them to water for extended periods of time because they hydrolyze slowly to give the corresponding sulfonic acids according to Equation (20.10). Nevertheless, it is unnecessary to dry or purify the *p*-acetamidobenzenesulfonyl chloride (**15**) in this sequence; rather, **15** is simply treated with aqueous ammonia.

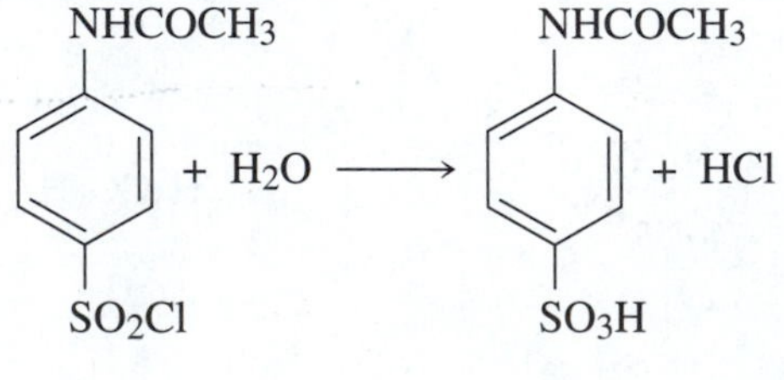

$$\text{(20.10)}$$

15

26

20.2.5 PREPARATION OF SULFANILAMIDE (8) (*P*-AMINOBENZENESULFONAMIDE)

The preparation of sulfanilamide (**8**) involves treatment of *p*-acetamidobenzene-sulfonyl chloride (**15**) with an excess of aqueous ammonia to give *p*-acetamidobenzenesulfonamide (**16**), followed by selective hydrolytic removal of the *N*-acetyl protecting group, as shown in Equation (20.11). In order to avoid hydrolysis of the sulfonyl chloride group (Eq. 20.10) before its reaction with ammonia, it is imperative to treat **16** with ammonia *immediately* after its isolation in the previous experiment. If the conversion of acetanilide (**13**) to *p*-acetamidobenzenesulfonamide (**16**) is not completed within a *single* laboratory period, the overall yield for the sequence is reduced significantly. Hydrolysis of the acetamido moiety of **16** may be effected using either aqueous acid or aqueous base without affecting the sulfonamido group, which hydrolyzes slowly, but an acid-catalyzed hydrolysis is performed in this experiment. Because the amine group will form a hydrochloride salt under the conditions of the hydrolysis, it is necessary to neutralize the solution with a base such as sodium carbonate in order to isolate sulfanilamide, which may be purified by recrystallization from water.

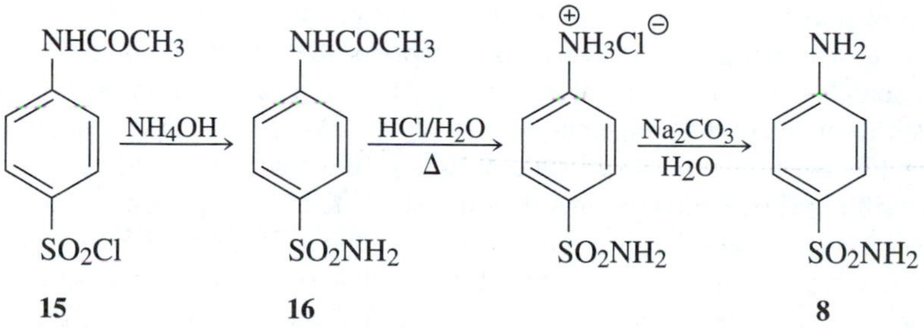

Pre-Lab exercises for Section 20.2, "The Synthesis of Sulfanilamide," are found on pages PL. 113–119.

EXPERIMENTAL PROCEDURES

The experimental procedures start with nitrobenzene and continue through to the preparation of sulfanilamide. Your instructor will indicate at which point to start and how far along the synthetic sequence to proceed. Quantities of reagents must be adjusted according to the amount of starting material that is available from the previous reaction, but you should not run the reactions on a *larger* scale than is indicated. If the amount of your starting material varies from that indicated, *do not change the reaction times*. You should obtain sufficient quantities of product from each reaction so that you can complete the sequence starting with the amount of nitrobenzene provided in the first step. Discuss any questions you may have with your instructor.

A. PREPARATION OF ANILINE

SAFETY ALERT

1. Nitrobenzene is toxic; avoid breathing its vapors or allowing it to come into contact with your skin. Wear rubber gloves when transferring this chemical.
2. Concentrated hydrochloric acid can cause burns if it comes in contact with the skin. Should contact occur, flood the affected area with water and rinse it thoroughly with dilute aqueous sodium bicarbonate solution. Wear rubber gloves when transferring this acid.
3. The solution of 12 *M* sodium hydroxide is highly caustic and can cause burns and loss of hair; avoid contact of this solution with your skin. Should contact occur, flood the affected area with water and rinse it thoroughly with dilute aqueous acetic acid solution. Wear rubber gloves when transferring this solution.

Place 13.1 g of finely divided tin powder in a 250-mL round-bottom flask and add 6.2 g of nitrobenzene. Add 28 mL of concentrated hydrochloric acid to the flask, insert a thermometer, and swirl the contents of the flask to mix the three-phase system. Monitor the temperature, and cool the flask as necessary in an ice-water bath to maintain the temperature below 60 °C. After swirling the mixture for about 15 min, the initial exothermic portion of the reaction should be complete. Attach a reflux condenser, and heat the contents of the flask either under reflux using an oil bath or on a steam bath for about 20 min with frequent swirling. During this time the color of the intermediate reduction product should disappear, and no droplets of nitrobenzene should be present in the condenser.

When the reflux period is complete, cool the acidic solution in an ice bath, and *slowly* add 50 mL of 12 *M* sodium hydroxide solution directly to the reaction mixture. Check the pH of the mixture to ensure that it is basic. Equip the flask for steam distillation (see Sec. 2.10), and steam-distil the mixture until the distillate emerging from the condenser no longer contains any visible quantity of organic material; about 35–40 mL should be collected. The distillate may still be slightly cloudy, but if there is no visible amount of oil in it, you may assume that the distillation is complete.* Add 4.5 mL of concentrated hydrochloric acid to the distillate, and transfer the mixture to a separatory funnel. Wash the aqueous layer twice with 15-mL portions of diethyl ether to remove any unreacted nitrobenzene. Transfer the *aqueous* layer to an Erlenmeyer flask; cool the contents of the flask in an ice-water bath and make the aqueous solution *basic* by slowly adding a minimum volume of 12 *M* sodium hydroxide solution. Saturate the aqueous solution with sodium chloride, cool the mixture to room temperature, and transfer it to a clean separatory funnel. Extract the crude aniline with two 15-mL portions of diethyl ether, using the first portion to rinse the flask in which the neutralization was done; separate the aqueous layer from the organic layer as thoroughly as possible each time. Dry the *combined* organic extracts over anhydrous sodium sulfate.* Transfer the solution by decantation into a round-bottom flask of appropriate size, and remove most of the diethyl ether by simple distillation (*no flames*). The last traces of diethyl ether may be removed by connecting the *cool* stillpot to a water aspirator (see Fig. 2.37b).

The aniline that remains in the stillpot may be used directly in the next step of the sequence. Alternatively, you may purify the aniline by shortpath distillation (see Fig. 2.15b). Wrap the stillhead with a layer of glasswool and then aluminum foil to minimize heat losses and facilitate the distillation of aniline, which boils at 183–185 °C (760 torr). Collect three fractions having boiling ranges (760 torr) of 35–90 °C, 90–180 °C, and 180–185 °C respectively. Pure aniline is colorless but may darken immediately following distillation, owing to air-oxidation. These oxidized products may be removed by redistillation from powdered zinc, but this purification step is *not* necessary for the present experiment.

water clear
irredescent diquid
practical terms pale
   yellow

Finishing Touches   Filter the *pot residue from the steam distillation* by vacuum filtration through a bed of a filter-aid and place the *tin salts* together with the *filter-aid* and *filter paper* in the container for nonhazardous solids. Neutralize the *filtrate* and the other *aqueous layers* with dilute hydrochloric acid and flush them down the drain with excess water. Place the combined *ethereal extracts and distillate* in a container for nonhalogenated organic solvents. After the diethyl ether evaporates from the sodium sulfate on a tray in the hood, place the drying agent in the container for nonhazardous solids.

collect, bottle, lable
distillation range -
have instructor look at
it right away so its
light colored

B. PREPARATION OF ACETANILIDE

SAFETY ALERT

Acetic anhydride is a lachrymator. Transfer this material in a hood and do not inhale its vapors.

x analine from teacher
                    make

Dissolve 3.7 g of aniline in 100 mL of 0.4 *N* hydrochloric acid in a 250-mL Erlenmeyer flask, and swirl the mixture to effect solution. Prepare a solution containing 6.0 g of sodium acetate trihydrate dissolved in 20 mL of water, and measure out 4.4 mL of acetic anhydride. Warm the solution containing the dissolved aniline hydrochloride to about 50 °C, and add the acetic anhydride in *one* portion. Swirl the flask vigorously and then add the solution of sodium acetate *immediately and in one portion*. Cool the reaction mixture to 5 °C in an ice-water bath and continue to swirl or stir the mixture, whereupon the crystalline product should separate.

Collect the acetanilide by vacuum filtration, wash it with a small portion of cold water, and dry.✻ Determine the melting point of the product and its yield. If impure or slightly colored acetanilide is obtained, recrystallize it from a minimum volume of hot water, using decolorizing carbon to give a colorless product (see Sec. 3.2.3). The melting point of pure acetanilide is 113–115 °C. *Be certain that your product is thoroughly dry before proceeding to the next step of the sequence.*

$M_1 V_1 = M_2 V_2$
            3,5 mL

must be dry
quickly
        1 minute at most

under 10°

tared beaker
dried to constant wt

Use handout

> **Finishing Touches**   Flush the *aqueous filtrate* and *mother liquor from crystallization* down the drain with a large excess of water.

C. PREPARATION OF *p*-ACETAMIDOBENZENESULFONYL CHLORIDE

Note to Student. *This entire experiment as well as the next one, up to the first stopping point, must be completed in a single laboratory period.*

SAFETY ALERT

1. If possible, perform this reaction in a hood. Otherwise, use the gas-removal apparatus described in Section 2.20 (see Fig. 2.35) to prevent escape of hydrogen chloride or oxides of sulfur (SO_2 or SO_3) into the laboratory.
2. Chlorosulfonic acid is highly corrosive and may cause serious burns if it is allowed to come in contact with your skin. Wear rubber gloves when handling or transferring this reagent, and exercise *extreme* care when working with it. Should any chlorosulfonic acid come in contact with your skin, flood the affected area *immediately* with cold water and then rinse it with 5% sodium bicarbonate solution.
3. Chlorosulfonic acid reacts *vigorously* with water. Open containers of chlorosulfonic acid will fume upon exposure to atmospheric moisture owing to the reaction of chlorosulfonic acid with water to give HCl and SO_3, both of which are noxious and corrosive gases. Measure and transfer this acid *only in the hood*. Several graduated cylinders should be kept in the hood for all students to use for measuring the volume of chlorosulfonic acid needed.
4. To destroy residual chlorosulfonic acid in graduated cylinders and other glassware that has contained it, add cracked ice to the glassware *in the hood* and let the glassware remain there until the ice has melted. Then rinse the apparatus with copious amounts of cold water.
5. Make sure your glassware is dry, lubricate the ground-glass joints of the apparatus carefully, and be certain that the joints are intimately mated. Otherwise, noxious gases will escape.

Equip a dry 50-mL round-bottom flask with a Claisen connecting tube. Place a vacuum adapter that is filled with calcium chloride granules on the sidearm of the Claisen tube, and connect the vacuum adapter to a vacuum trap (see Fig. 2.35). Since an airtight seal is required at all connections, carefully grease all joints. Place 2.7 g of *dry* acetanilide in the flask; if magnetic stirring is available, add a stirring bar to the flask. *Working at the hood,* measure 8.0 mL of chlorosulfonic acid into a 25-mL addition funnel or a small separatory funnel that is fitted with a standard-taper ground-glass joint; be certain that the stopcock of the funnel is firmly seated and closed prior to this transfer. *Use care in handling chlorosulfonic acid; see the Safety Alert section that precedes this experiment!* Stopper the funnel and place it on the straight arm of the Claisen tube so that the chlorosulfonic acid will drop directly onto the acetanilide contained in the reaction flask.

Cool the flask in a water bath maintained at 10–15 °C by adding a little ice; do *not* allow the temperature to go below 10 °C. Turn on the water fully, and regulate the aspirator vacuum using the release valve on the water trap so that there is a slight flow of air through the vacuum adapter. It is not necessary to have the maximum vacuum to achieve the desired result; however, if fumes evolve through the open end of the gas trap, adjust the vacuum until such emissions no longer occur. Fully open the stopcock of the funnel so that the chlorosulfonic acid is added as rapidly as possible to the flask containing the acetanilide; it may be necessary to lift the stopper on the funnel to equalize the pressure in the system if the flow of acid becomes erratic or slow. The reaction mixture darkens as the addition proceeds. When the addition has been completed, either swirl the mixture occasionally or stir it using a magnetic stirrer to facilitate dissolution of the acetanilide; maintain the temperature of the water bath below 20 °C. After most of the solid has dissolved and the initial reaction has subsided, remove the cooling bath and allow the reaction mixture to warm to room temperature with occasional swirling or stirring. To complete the reaction, heat the mixture on a steam bath or with a hot-water bath at about 70–80 °C with continued swirling or stirring until there is no longer any increase in the rate of gas evolution; about 10–20 min of heating will be required.

Cool the mixture to room temperature or slightly below using an ice-water bath. *Working at the hood,* place about 150 g of crushed ice in a 250-mL beaker and pour the reaction mixture slowly onto the ice with stirring using a glass rod. *Be careful and do not add the mixture too quickly to avoid splattering the chlorosulfonic acid.* Rinse the reaction flask with a little ice water and transfer this to the beaker; the remainder of the work-up procedure may be performed at the bench. The precipitate that forms in the beaker is crude *p*-acetamidobenzenesulfonyl chloride, which may be white to pink in color. If it is allowed to stand without stirring, a hard mass may form; any lumps that form should be broken up thoroughly with a stirring rod. Collect the crude material by vacuum filtration. Wash the solid with 15-mL portions of ice water *until the filtrate is neutral;* press the filter cake as dry as possible with a clean cork. The crude sulfonyl chloride should be used *immediately* in the next experiment to prepare sulfanilamide.

Purify a *small* sample of the product by dissolving it in a minimum amount of boiling dichloromethane in a test tube. Using a micropipet, remove the upper, aqueous layer as quickly as possible; be careful in separating the layers. Cool the organic layer and collect the colorless crystals of pure product by vacuum filtration and dry them in the air. The reported melting point of pure *p*-acetamidobenzenesulfonyl chloride is 149–150 °C.

Finishing Touches   Neutralize the combined *aqueous solutions* with sodium carbonate, and flush them down the drain with a large excess of water. Cover any spilled droplets of *chlorosulfonic acid* with sodium carbonate, and flush the resulting powder down the drain. Place the *calcium chloride* from the drying tube in the container for nonhazardous solids.

D. PREPARATION OF SULFANILAMIDE

Transfer all of the crude *p*-acetamidobenzenesulfonyl chloride obtained in the previous experiment to a 125-mL Erlenmeyer flask, and add 15 mL of concentrated (28%) ammonium hydroxide. A very rapid, exothermic reaction may occur if the crude *p*-acetamidobenzenesulfonyl chloride contains acidic contaminants that were not removed by the aqueous washings in the previous step. Use a stirring rod to break up any lumps of solid that may remain; the reaction mixture should be thick but homogeneous. Heat the mixture on a steam or hot water bath for about 30 min. Because vapors of ammonia evolve, it is best to heat this mixture in a hood; alternatively, invert a funnel over the flask and attach the funnel to a vacuum source such as a water aspirator pump (see Fig. 2.37b). You should notice that the mixture becomes more pasty as the sulfonyl chloride is transformed into the sulfonamide.*

Cool the reaction mixture in an ice-water bath and collect the product by vacuum filtration. Wash the crystals with cold water and air-dry them.* Determine the melting point of your product; the melting point of pure *p*-acetamidobenzenesulfonamide is 218–220 °C.

Weigh the dry *p*-acetamidobenzenesulfonamide and transfer it to a 50-mL round-bottom flask. Prepare a solution of dilute hydrochloric acid by mixing equal volumes of concentrated hydrochloric acid and water. Add to the round-bottom reaction flask an amount of dilute hydrochloric acid that is about twice the weight of the crude *p*-acetamidobenzenesulfonamide. Equip the flask for reflux, and gently heat the reaction mixture under reflux for 45 min; use magnetic stirring if available. Allow the mixture to cool to room temperature; if any solid, which is unreacted starting material, appears on cooling, heat the mixture again at reflux for an additional 15 min. Add an equal volume of water to the cooled solution, and transfer the resulting mixture to a 100-mL beaker. Neutralize the excess hydrochloric acid by adding *small* quantities (to minimize foaming) of solid sodium carbonate until the solution is slightly alkaline to pHydrion paper. A precipitate should form during neutralization; scratching at the liquid-air interface may be necessary to initiate crystallization. Cool the mixture in an ice-water bath to complete the precipitation of the product.*

Collect the crystals by vacuum filtration, wash them with a small amount of cold water, and allow them to air-dry.* Purify the crude product by recrystallization from the minimum volume of hot water. If necessary, decolorize the hot solution; you *must* remove the decolorizing carbon using the technique of hot filtration (see Fig. 2.27b) through a *preheated* funnel to avoid crystallization of the product in the funnel. Cool the aqueous solution of sulfanilamide in an ice-water bath. The product should separate as long, white needles which can be isolated by vacuum filtration.* Dry the crystals and determine the yield and the melting point, which is reported to be 164.5–166.5 °C.

Test the solubility of sulfanilamide in 1.5 *M* hydrochloric acid solution and in 1.5 *M* sodium hydroxide solution.

> **Finishing Touches** Neutralize the combined *aqueous solutions and filtrates* with 10% hydrochloric acid or sodium carbonate as required, and flush them down the drain with a large excess of water.

EXERCISES

General Questions

1. Calculate the overall yield of cortisone, an important drug, produced in a 33-step synthesis, assuming an average yield of 90% in each step.

2. Outline a possible synthesis of the sulfanilamide derivative, **28,** using benzene as the only source of an aromatic ring. Use any needed aliphatic or inorganic reagents.

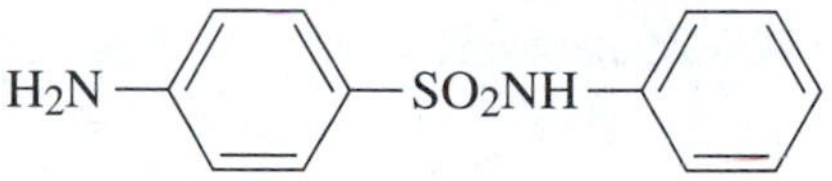

$$H_2N \text{—} \langle \text{benzene ring} \rangle \text{—} SO_2NH \text{—} \langle \text{benzene ring} \rangle$$

28

3. Two hypothetical sequences have been developed for converting an arbitrary compound A into E in the same overall yield.
   a. What is the overall yield for each sequence?
   b. As the production manager responsible for selecting the more economical of the two routes, what factor(s) would you consider in reaching a decision? Assume that no new capital investment would be required for either sequence.

$$A \xrightarrow{30\%} B \xrightarrow{49\%} C \xrightarrow{62\%} D \xrightarrow{57\%} E$$

$$A \xrightarrow{62\%} B \xrightarrow{57\%} C \xrightarrow{49\%} D \xrightarrow{30\%} E$$

4. A student proposed the following alternate sequence of reactions for preparing sulfanilamide from sulfanilic acid. Discuss this modified approach, commenting on any possible side reactions that might be encountered in either of the two steps. Propose a modification of this sequence that could be used to prepare sulfanilamide.

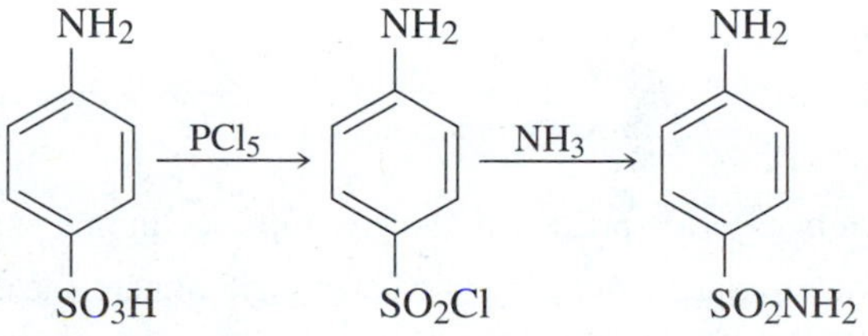

$$\underset{SO_3H}{\overset{NH_2}{\bigcirc}} \xrightarrow{PCl_5} \underset{SO_2Cl}{\overset{NH_2}{\bigcirc}} \xrightarrow{NH_3} \underset{SO_2NH_2}{\overset{NH_2}{\bigcirc}}$$

Questions for Part A

5. Outline in a flow diagram the procedure for the purification of aniline. Indicate the importance of each step in the procedure, and give reasons for doing the steam distillation with a basic solution. What is the purpose of performing the subsequent extractions from aqueous acid and then base. Write the equation(s) for the reactions that occur when base and acid are added.

6. What gas is evolved upon addition of concentrated hydrochloric acid to tin powder in this reduction?

7. Why is sodium chloride added to the steam distillate *after* it has been made basic but *before* extraction with diethyl ether?

8. Write the balanced half-reactions for the reduction of nitrobenzene (**12**) into aniline (**13**) and for the oxidation of tin metal into stannic ion, Sn^{+4}. Sum these equations and write the balanced equation for the overall process.

9. Write a mechanism for the reduction of *N*-phenylhydroxylamine (**18**) into aniline (**13**). Show all electron and proton transfer steps using curved arrows to symbolize the flow of electrons.

Questions for Part B

10. Why should the acetic anhydride not be allowed to stay in contact with the aqueous solution of the hydrochloric acid salt of aniline (anilinium hydrochloride, **24**) for an extended period of time before the solution of sodium acetate is added?

11. Why is aqueous sodium acetate preferred to aqueous sodium hydroxide for the conversion of anilinium hydrochloride (**24**) to aniline?

12. Why is aniline (**13**) soluble in aqueous hydrochloric acid whereas acetanilide (**14**) is not?

13. Give a stepwise reaction mechanism for the reaction of aniline (**13**) with acetic anhydride. Use curved arrows to symbolize the flow of electrons.

Questions for Part C

14. Explain why *p*-acetamidobenzenesulfonyl chloride (**15**) is much less susceptible to hydrolysis of the acid chloride function than is *p*-acetamidobenzoyl chloride (**29**).

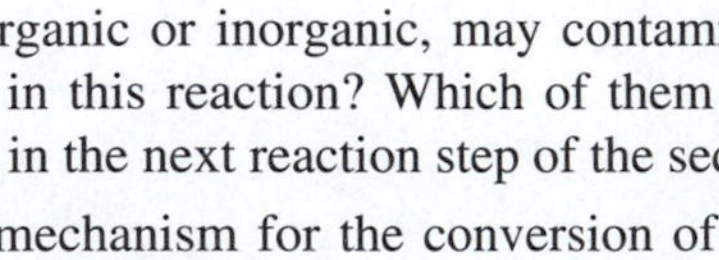

$$CH_3COHN \underset{}{\overset{}{\bigcirc}} COCl$$

29
p-Acetamidobenzoyl chloride

15. Why is calcium chloride present in the gas trap used in this experiment?

16. What materials, organic or inorganic, may contaminate the crude sulfonyl chloride prepared in this reaction? Which of them are likely to react with the ammonia used in the next reaction step of the sequence?

17. Write a stepwise mechanism for the conversion of acetanilide (**14**) into *p*-acetamidobenzenesulfonic acid; use curved arrows to symbolize the flow of electrons.

18. Write the structure of the material containing sulfonamide linkages that might be obtained upon polymerization of *p*-aminobenzenesulfonyl chloride (**30**).

$$H_2N-\!\!\!\!\!\raise2pt\hbox{⬡}\!\!\!\!\!-SO_2Cl$$

30
p-Aminobenzenesulfonyl chloride

Questions for Part D

19. In the preparation of sulfanilamide (**8**) from *p*-acetamidobenzenesulfonamide (**16**), only the acetamido group is hydrolyzed. Give an explanation for this difference in reactivity of the acetamido and sulfonamido groups toward aqueous acid.

20. Following hydrolysis of *p*-acetamidobenzenesulfonamide with aqueous acid, the reaction mixture is homogeneous, whereas *p*-acetamidobenzenesulfonamide is insoluble in aqueous acid. Explain the change in solubility that occurs as a result of the hydrolysis.

21. Explain the results obtained when the solubility of sulfanilamide was determined in 1.5 *M* hydrochloric acid and in 1.5 *M* sodium hydroxide. Write equations for any reaction(s) that occurred.

22. Why might there be a preference for using solid sodium carbonate instead of solid sodium hydroxide for basifying the acidic hydrolysis solution obtained in this experiment?

23. What would be observed if *p*-acetamidobenzenesulfonamide were subjected to vigorous hydrolysis conditions, such as concentrated hydrochloric acid and heat for a long period of time? Write an equation for the reaction that might occur.

24. What acids might be present in the crude sulfonyl chloride to cause the exothermic reaction with ammonia?

25. Why would you expect the yield of sulfanilamide to be lowered if the crude sulfonyl chloride was not combined with ammonia until the laboratory period following its preparation?

26. Calculate the overall yield of sulfanilamide obtained in the sequence of reactions that you performed.

SPECTRA OF STARTING MATERIALS AND PRODUCTS

The IR, ^{1}H, and ^{13}C NMR spectra of acetanilide are given in Figures 3.3 and 3.4(a) and (b), respectively.

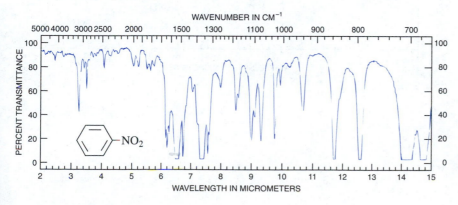

FIGURE 20-1 IR spectrum of nitrobenzene.

FIGURE 20-2 NMR data for nitrobenzene. **(a)** ^{1}H NMR spectrum (90 MHz).

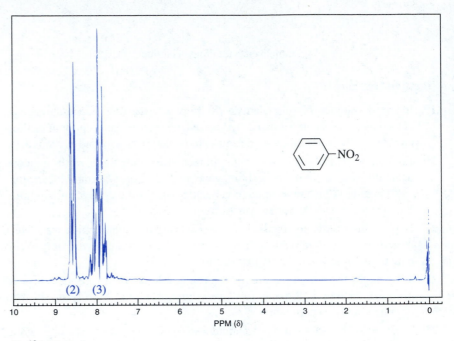

(b) ^{13}C NMR data.
Chemical shifts: δ 123.5, 129.5, 134.8, 148.3.

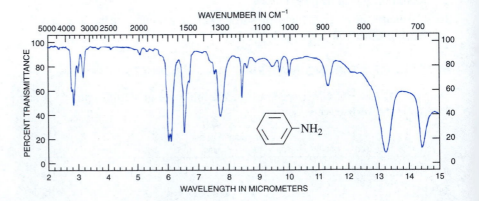

FIGURE 20-3 IR spectrum of aniline.

(a) ¹H NMR spectrum (60 MHz).

 FIGURE 20-4 NMR data for aniline.

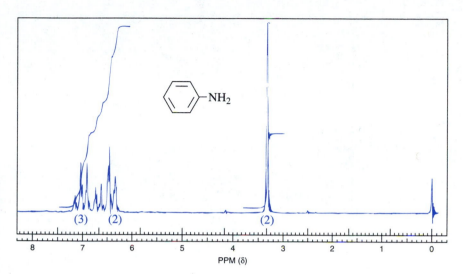

(b) ¹³C NMR data.

Chemical shifts: δ 115.1, 118.2, 129.2, 146.7.

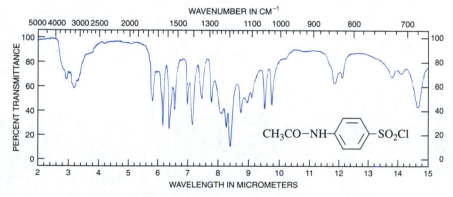

FIGURE 20-5 IR spectrum of *p*-acetamidobenzenesulfonyl chloride (in CCl$_4$ solution).

FIGURE 20-6 NMR data for *p*-acetamidobenzenesulfonyl chloride.

(a) ^{1}H NMR spectrum (90 MHz).

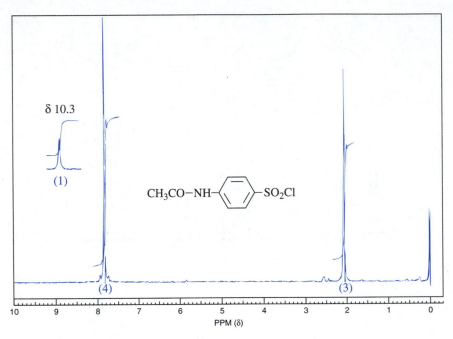

δ 10.3

(1)

$CH_3CO—NH—\langle\quad\rangle—SO_2Cl$

(4)

(3)

PPM (δ)

(b) ^{13}C NMR data.
Chemical shifts: δ 25.3, 119.7, 127.5, 141.6, 141.7, 170.3.

WAVELENGTH IN CM^{-1}

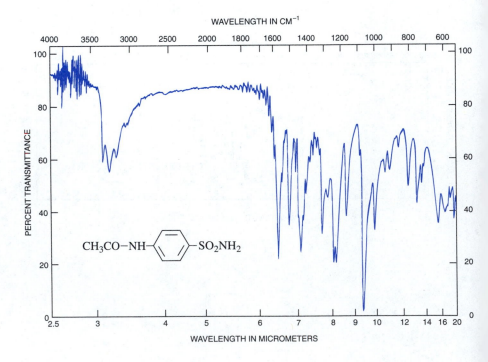

PERCENT TRANSMITTANCE

$CH_3CO—NH—\langle\quad\rangle—SO_2NH_2$

WAVELENGTH IN MICROMETERS

FIGURE 20-7 IR spectrum of *p*-acetamidobenzenesulfonamide.

(a) ^{1}H NMR spectrum (90 MHz).

FIGURE 20-8 NMR data for *p*-acetamidobenzenesulfonamide.

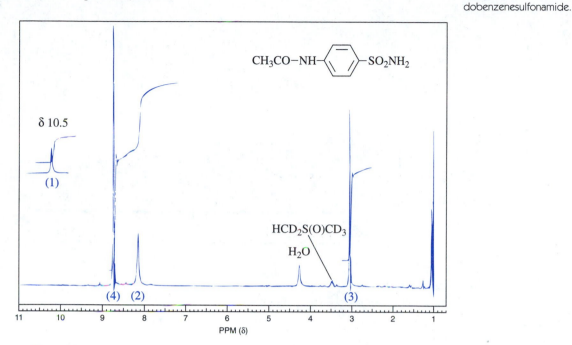

(b) ^{13}C NMR data.

Chemical shifts: δ 24.2, 118.5, 126.7, 138.1, 142.3, 169.0.

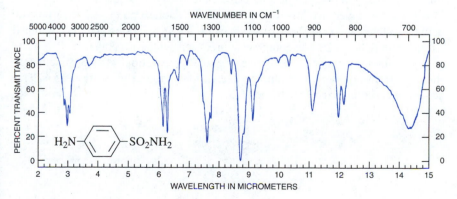

FIGURE 20-9 IR spectrum of sulfanilamide (KBr pellet).

FIGURE 20-10 NMR data for sulfanilamide.

(a) ^{1}H NMR spectrum (60 MHz).

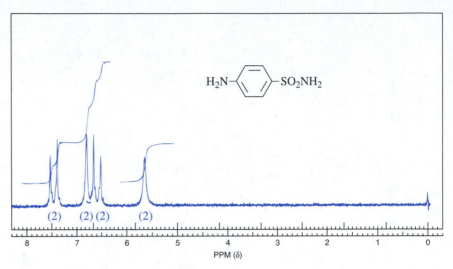

(2)   (2) (2)   (2)

PPM (δ)

(b) ^{13}C NMR data.
Chemical shifts: δ 113.1, 127.6, 130.2, 151.6.

REFERENCES

Amundsen, L. H. "Sulfanilamide and Related Chemotherapeutic Agents." *Journal of Chemical Education,* **1942,** *19,* 167.

Goodman, L. S.; Gilman, A. *The Pharmacological Basis of Therapeutics,* 7th ed. New York: Macmillan, 1985. Chap. 56, "The Sulfonamides," by L. Weinstein.

20.3 ADDITIONAL MULTISTEP SYNTHETIC SEQUENCES

In Section 12.3, the preparation of 4-cyclohexene-*cis*-1,2-dicarboxylic anhydride from maleic anhydride and 1,3-butadiene is described, as is the hydrolysis of this cycloadduct to 4-cyclohexene-*cis*-1,2-dicarboxylic acid. In Section 17.2, directions are given for hydrogenation of 4-cyclohexene-*cis*-1,2-dicarboxylic acid to cyclohexane-*cis*-1,2-dicarboxylic acid. These reactions may be combined as shown in Scheme 20.4 to comprise a three-step synthetic sequence.

A four-step synthetic sequence for the preparation of 1,3-dimethylan-thracene from *m*-xylene and phthalic anhydride is found in Section 15.4. This sequence is summarized in Scheme 20.5.

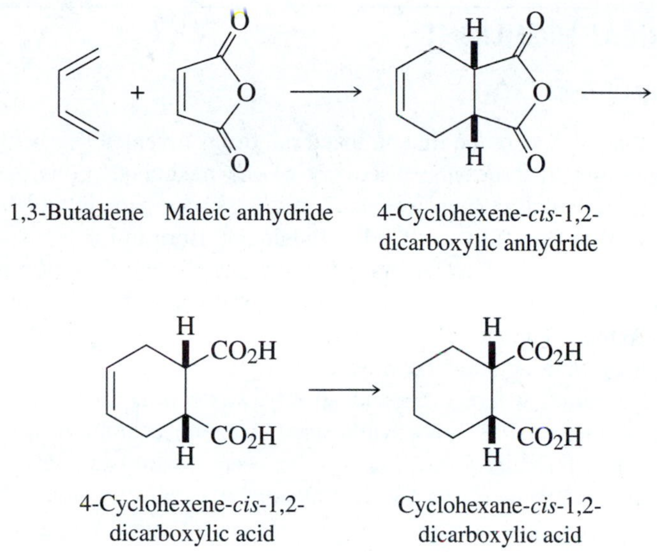

1,3-Butadiene   Maleic anhydride   4-Cyclohexene-*cis*-1,2-
dicarboxylic anhydride

4-Cyclohexene-*cis*-1,2-
dicarboxylic acid

Cyclohexane-*cis*-1,2-
dicarboxylic acid

SCHEME 20.4

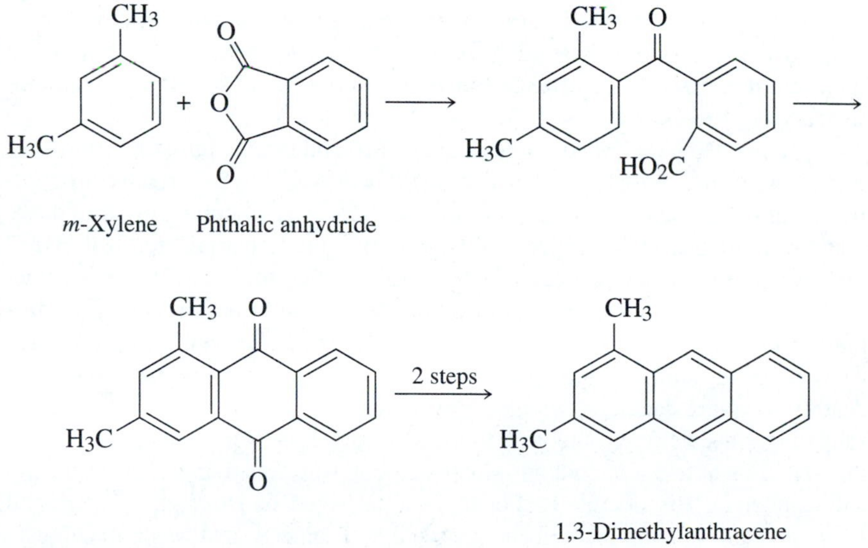

m-Xylene   Phthalic anhydride

2 steps

1,3-Dimethylanthracene

SCHEME 20.5

HISTORICAL HIGHLIGHT

Discovery of Sulfa Drugs

The sulfanilamides were the first antibacterial drugs invented by chemists, and a fascinating story underlies the discovery of the medicinal properties of these compounds. Although many scientists played important parts in this discovery, a key individual in the effort was Gerhard Domagk. Born in Lagow, Germany, in 1895, Domagk attended the University of Kiel intending to become a doctor, but World War I interrupted his medical studies. Following the Armistice in 1918, he reentered the University of Kiel and earned his medical degree in 1921. After a brief career in academia, he moved to I. G. Farbenindustrie (I.G.F.), the German dye cartel, where his responsibility was testing the pharmacological properties of the new dyes being synthesized by chemists at the firm.

At the time Domagk joined I.G.F., there were no antibacterials. This was a serious deficiency because bacteria were known to be the agents that caused pneumonia, meningitis, gonorrhea, and streptococcic and staphylococcic infections. The I.G.F. team of chemists and pharmacologists set out to find compounds that would kill these microbes without harming their animal or human hosts and they formulated a plan to determine whether certain dyes might be bacteriocidal. Their strategy evolved from the observation that particular dyes, specifically those containing a sulfonamido group, seemed to be particularly "fast," or tightly bound, to wool fabrics, indicating an affinity for the protein molecules comprising wool. Because bacteria are proteinaceous in nature, the researchers reasoned that the dyes might fasten to the bacteria in such a way as to inhibit or kill them selectively. As we shall see, this simple hypothesis was partially correct: The sulfonamido group was indeed essential, but the part of the molecule that made the substance function as a dye was irrelevant to its effectiveness as a bactericide.

One dye Domagk tested on laboratory mice and rabbits infected with streptococci was called Prontosil. This compound was found to be strongly disinfective against these bacteria and could be tolerated by the animals in large doses with no ill effects. This discovery of the bacteriocidal effect of Prontosil in animals was probably made in early 1932; I.G.F. applied for a patent in December of that year. Clinical tests on human patients apparently began soon after this, but the record is confused. Some accounts say that before any other tests had been made on humans, Domagk gave a dose of Prontosil in desperation to his deathly ill young daughter, who had developed a serious streptococcal infection following a needle prick; the girl then made a rapid recovery. Others report that the first clinical test was on a ten-month-old boy who was dying of staphylococcal septicemia. His doctor, R. Förster, was a friend of Domagk's superior at I.G.F. and through him learned about a red dye (Prontosil) that was miraculously effective in animals against streptococci. Since the baby was close to death anyway and feeling that there was nothing to lose if the dye was not effective against staphylococci, Förster gave the child two doses of the red dye; complete recovery rapidly followed.

Regardless of which of these two stories is correct, or whether both are true, it was widely recognized by the middle 1930s that the discovery of the bactericidal properties of Prontosil was a medical miracle, for which Domagk was awarded the Nobel Prize in Physiology or Medicine in 1939. There were other important developments in the years between 1933 and 1939, however. Domagk did not publish the results of his tests of Prontosil on animal infections until February 1935, more than two years after the work was done. Learning of his results, the Trefouels, a wife and husband team at the Pasteur Institute in Paris, were prompted to test the bactericidal properties of several compounds, all of which were "azo" dyes closely related in chemical structure to Prontosil. The feature common to their dyes was the sulfonamido portion, but the other parts of the molecules differed significantly. Remarkably, they found that the antibacterial properties of these dyes were virtually identical with those of Prontosil.

This finding led to an explanation of a puzzling fact about Prontosil: It was ineffective against bacteria *in vitro* but was strongly effective *in vivo*. Apparently, a metabolic process within animals was necessary to make the sulfonamide dyes antibacterial. The Trefouels reasoned that the dye is broken into two parts in animals and only the sulfonamido portion is effective as an antibacterial. To prove this, they synthesized the sulfonamido component of Prontosil, which was the known compound *p*-aminobenzenesulfonamide, or sulfanilamide, and found it to be as effective as Prontosil against bacterial infections. Comparison of the formulas of Prontosil and sulfanilamide makes it clear that cleavage of Prontosil at the azo double bond affords the skeleton of sulfanilamide. This cleavage occurs biochemically when Prontosil is injected or imbibed, and the sulfanilamide so produced is the actual antibacterial agent. The original hypothesis that sulfonamido dyes would be bacteriocidal was thus partly misconceived, in that only the sulfonamido part of the dye molecule kills microbes; the fact that it was a part of a dye molecule was incidental.

Prontosil   Sulfanilamide

Interestingly, the Trefouels' observations made the patent on Prontosil filed by I.G.F. useless. Sulfanilamide had been synthesized and patented many years before as a dye intermediate, but the patent had expired by the time the substance was found to be a potent bactericide. Moreover, their findings led to clinical trials of sulfanilamide in France, England, and the United States, all of which were highly successful. One case that gave great publicity to the new drug was the use of Prontosil to save the life of Franklin D. Roosevelt, Jr., son of the president. In 1936, young Roosevelt was dying from a streptococcic infection when his mother convinced a doctor to administer Prontosil, which saved his life.

By 1947 over 5,000 sulfonamides related to sulfanilamide had been prepared and tested for their efficacy as antibacterials. Although not all were effective, some were found to be better than sulfanilamide against certain infections.

Of the thousands of compounds prepared and tested, the active ones are almost always those in which the only variation in structure is a change in the group of atoms attached to the nitrogen atom of the sulfonamido moiety.

We mentioned that Domagk was awarded the Nobel Prize in 1939. This is not quite correct. He was selected for the prize in that year but did not actually receive it until many years later. When he received notice of the award in October 1939, Domagk sent a letter of acceptance, but a second letter declining it was received in Stockholm in November. The second letter was the result of pressure from the Nazis; Domagk was by this time in the hands of the Gestapo. In 1947, Domagk was finally able to visit Stockholm, deliver the Nobel lecture, and receive the medal and diploma—but not the prize money, which by then had reverted to the Nobel Foundation.

Relationship of Historical Highlight to Experiments

Many biologically important compounds commonly used as drugs must be prepared via multistep sequences of reactions. In such processes, the product of one reaction serves as the starting material for the next, so it is necessary to develop efficient procedures that afford products of sufficient purity that they may be used directly in the next step with minimal purification. In the series of experiments that are performed in this chapter, the antibiotic sulfanilamide is prepared from the readily available starting material nitrobenzene.

POLYMERS

Polymers are a class of molecules characterized by their high molecular weights, which range from the thousands to the hundreds of thousands, and by the presence of simple repeating structural units called **monomers.** Because of their large size, polymers are often referred to as **macromolecules.** A polymer comprised of a *single* recurring monomer, M, is termed a **homopolymer,** whereas one containing at least two structurally distinct monomeric units, M_1, M_2, distributed at random in the molecule is called a **copolymer.** These two classes of macromolecules are represented by **1** and **2,** respectively.

$$R—M—(M)_n—M—R \qquad R—M_1—(M_1—M_2)_n—M_2—R$$

1	**2**
Homopolymer	Copolymer

Polymers are found in nature and may also be produced by laboratory synthesis. Important examples of naturally occurring macromolecules, or **biopolymers,** are **proteins, polysaccharides, terpenes,** and **nucleic acids.** General representations of these substances are provided by structures **3–6,** respectively, in which the monomeric subunits of an α-amino acid, **3,** a pyranose, **4,** an isoprene (2-methyl-1,3-butadiene), **5,** and a ribonucleotide phosphate, **6,** are seen. **Synthetic,** or "manmade," **polymers** are represented by the myriad of **plastics, elastomers,** and **fibers** that are commonplace in contemporary society.

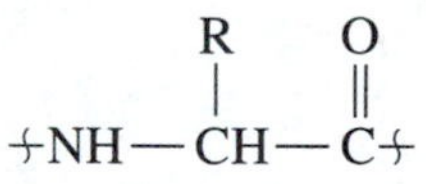

3	**4**	**5**	**6**
			(Base = purine or pyrimidine)

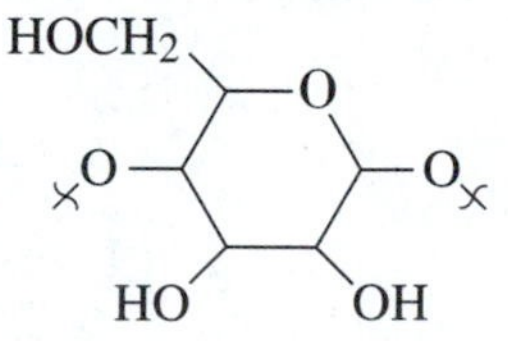

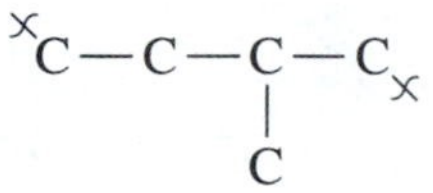

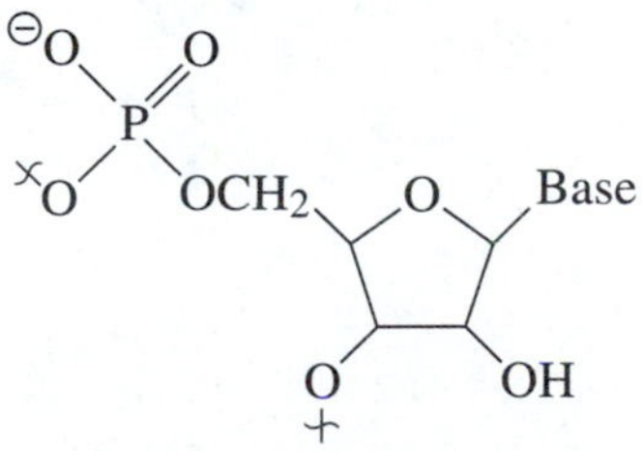

Two primary methods of polymerization are commonly used to convert monomers into synthetic polymers. In the older literature, these techniques are referred to as **addition** and **condensation polymerization,** but because of ambiguities in these terms, the preferred names now are **chain-reaction polymerization** and **step-growth polymerization,** respectively. As discussed in Sections 21.2 and 21.3, the major distinction between these two types of polymerization is the general mechanism by which the polymer forms.

The term addition polymerization arose because such polymers are produced by combining a large number of monomer molecules by addition reactions. For example, the self-addition of thousands of ethylene molecules yields polyethylene (**7,** Eq. 21.1), a homopolymer. Another homopolymer is polystyrene (**9**), which is formed by self-addition of styrene (**8**) (Eq. 21.2). It is appropriate to note that representation of the molecular formula of the polymer as essentially *n* times that of the monomer, as shown in these two equations, is only a slight over-simplification of the actual formula: Because *n* is such a large number, inclusion of the elemental compositions of the end-groups, R, that appear at the termini of the polymeric chain makes an insignificant change in the molecular formula in comparison with the rest of the molecule.

$$\text{n } CH_2 = CH_2 \xrightarrow[\text{and/or heat}]{\text{catalyst}} \{CH_2 - CH_2\}_n \qquad \textbf{(21.1)}$$

7

Ethylene   n = 10,000–30,000
(ethene)   Polyethylene
(polyethene)

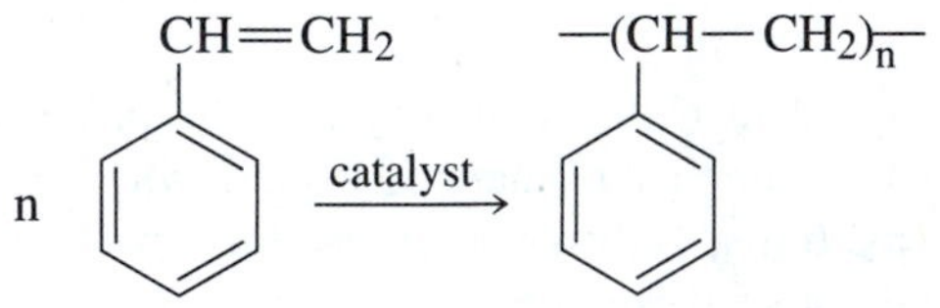

$$\textbf{(21.2)}$$

8   **9**

Styrene   Polystyrene
(ethenylbenzene)   (polyethenylbenzene)

Some other common addition polymers have trade names that do not indicate their structure. For example, Teflon is a homopolymer of tetrafluoroethylene (tetrafluoroethene, **10**), and Plexiglas™ is a polymer of methyl methacrylate (methyl 2-methylpropenoate, **11**).

$$CF_2 = CF_2 \qquad\qquad CH_2 = C \begin{array}{c} CO_2CH_3 \\ \diagdown \\ CH_3 \end{array}$$

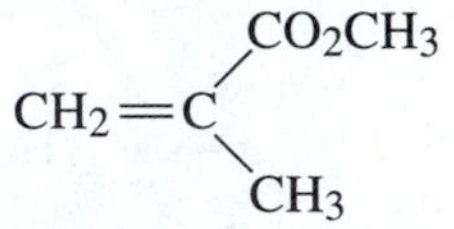

10   **11**

Tetrafluoroethylene   Methyl methacrylate
(tetrafluoroethene)   (methyl 2-methylpropenoate)

Copolymers are produced from a *mixture* of monomers, as was noted earlier. For example, saran (**14**), a widely used plastic film, is made by polymerizing a mixture of vinyl chloride (chloroethene, **12**) and 1,1-dichloroethene (vinylidene chloride, **13**), as depicted in Equation (21.3). Note that the abbreviated formula of **14** is not meant to imply that the two monomeric units appear as a sequence of two distinct blocks, each of which individually represents a homopolymer. Although such **block copolymers** can be produced by special techniques, copolymers usually have the two monomers distributed randomly along the chain, as in **15**.

$$\text{n CH}_2\!=\!\underset{\underset{\text{Cl}}{|}}{\text{CH}} + \text{m CH}_2\!=\!\text{CCl}_2 \longrightarrow \left(\!\!\text{CH}_2\!-\!\underset{\underset{\text{Cl}}{|}}{\text{CH}}\!\!\right)_{\!\!n}\!\!\left(\!\text{CH}_2\!-\!\text{CCl}_2\!\right)_{\!\!m} \quad \textbf{(21.3)}$$

12	**13**	**14**
Vinyl chloride (chloroethene)	Vinylidene chloride (1,1-dichloroethene)	Saran

$$\text{R-M}_1\text{-(M}_2\text{-M}_2\text{-M}_1\text{-M}_2\text{-M}_1\text{-M}_1\text{-M}_1)_n\text{-M}_2\text{-R)}$$

15

The following sections contain discussions of the mechanisms of chain-reaction and step-growth polymerization. The associated experimental procedures illustrate the preparation of examples of polymers derived from each type of polymerization.

21.2   CHAIN-REACTION POLYMERIZATION

As the name implies, **chain-reaction polymerization** is a *chain* reaction in which the initiator may be a cation, anion, or free radical. An example of a **cationic polymerization** is found in the polymerization of isobutylene (2-methylpropene) in the presence of protic or Lewis acid catalysts to give polyisobutylene (**16**, Eq. 21.4). The conversion of acrylonitrile (propenenitrile) to polyacrylonitrile (**17**) using sodium amide, a strong base, represents **anionic polymerization** (Eq. 21.5).

$$(\text{CH}_3)_2\text{C}\!=\!\text{CH}_2 \xrightarrow{\text{H}^+} (\text{CH}_3)_2\overset{\oplus}{\text{C}}\!-\!\text{CH}_2 \xrightarrow[\text{H}_2\text{C}=\text{C(CH}_3)_2]{} (\text{CH}_3)_3\text{C}\!-\!\text{CH}_2\overset{\oplus}{\text{C}}\overset{\text{CH}_3}{\underset{\text{CH}_3}{}}$$

Isobutylene (2-methylpropene)

$$\downarrow \text{n H}_2\text{C}\!=\!\text{C(CH}_3)_2 \qquad\qquad \textbf{(21.4)}$$

$$(\text{CH}_3)_3\text{C}\!-\!(\underset{\underset{\text{CH}_3}{|}}{\overset{\overset{\text{CH}_3}{|}}{\text{CH}_2\text{C}}})_n\!-\!\underset{\underset{\text{CH}_3}{|}}{\overset{\overset{\text{CH}_3}{|}}{\text{CH}_2\text{C}}}\!-\!\text{R}$$

16
Polyisobutylene

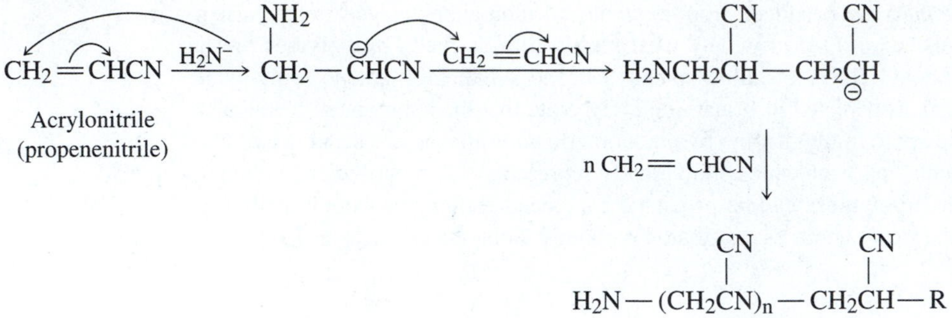

$$CH_2 = CHCN \xrightarrow{H_2N^-} CH_2 - \overset{NH_2}{\underset{|}{CHCN}} \xrightarrow{CH_2 = CHCN} H_2NCH_2\overset{CN}{\underset{|}{CH}} - CH_2\overset{CN}{\underset{|}{CH}}$$

Acrylonitrile
(propenenitrile)

$$n\,CH_2 = CHCN \downarrow$$

(21.5)

$$H_2N - (CH_2CN)_n - CH_2\overset{CN}{\underset{|}{CH}} - R$$

17

Polyacrylonitrile

Free-radical polymerization is a widely used method to induce chain-reaction polymerization, and its mechanistic course is parallel to that of the free-radical halogenation of hydrocarbons (see Sec. 9.2). The reaction is started by the thermal decomposition of an *initiator,* which in our experiment is *tert*-butyl peroxybenzoate (**18**); this compound produces the free radicals **19** and **20** when heated (Eq. 21.6). If

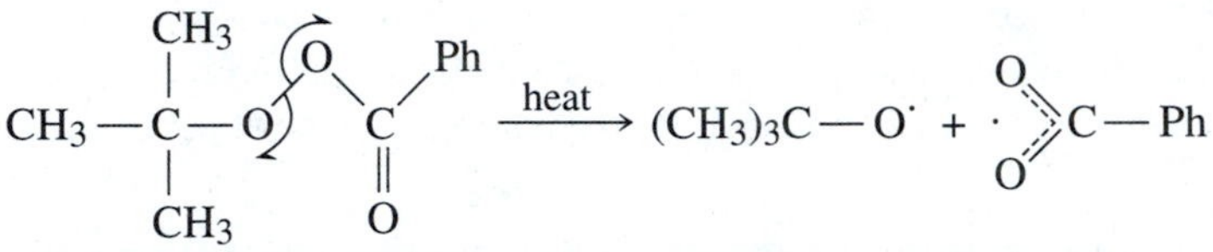

$$CH_3 - \overset{CH_3}{\underset{\underset{CH_3}{|}}{\overset{|}{C}}} - O - \overset{O}{\underset{\underset{O}{||}}{C}} - Ph \xrightarrow{heat} (CH_3)_3C - O^{\cdot} + {\cdot} \overset{O}{\underset{O}{}}C - Ph$$ (21.6)

18   **19**   **20**
tert-Butyl peroxybenzoate

In$^{\bullet}$ stands for one or both of these free radicals, the course of the polymerization may be illustrated as shown in Equations (21.7)-(21.10). Equation (21.7) indicates the function of the free radicals in **initiating** the polymerization. Equation (21.8) represents the **propagation** of the growing polymer chain. Equations (21.9) and (21.10) show possible **termination** processes. In Equation (21.9) the free-radical end of one growing polymer chain abstracts a hydrogen atom from the carbon atom next to the end of another polymer radical to produce the unsaturated and saturated polymer molecules **22** and **23,** respectively, in a process termed **disproportionation.** For the termination reaction illustrated by Equation (21.10), Rad$^{\bullet}$ may be one of the initiating radicals, In$^{\bullet}$, or another growing polymer chain.

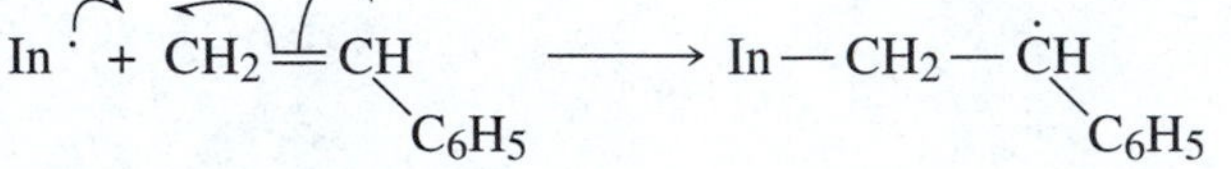

$$In^{\cdot} + CH_2 = \underset{C_6H_5}{CH} \longrightarrow In - CH_2 - \underset{C_6H_5}{\dot{C}H}$$ (21.7)

8

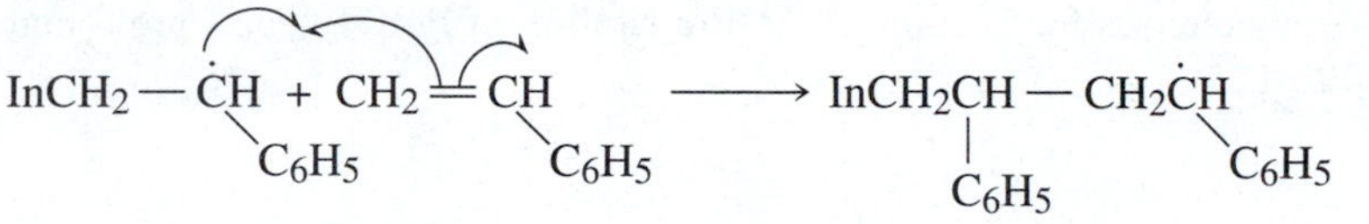

$$InCH_2 - \overset{\cdot}{C}H + CH_2 = CH \longrightarrow InCH_2CH - CH_2\overset{\cdot}{C}H \qquad \textbf{(21.8a)}$$
$$\overset{|}{C_6H_5} \qquad \overset{|}{C_6H_5} \qquad \qquad \overset{|}{C_6H_5} \qquad \overset{|}{C_6H_5}$$
$$\textbf{21}$$

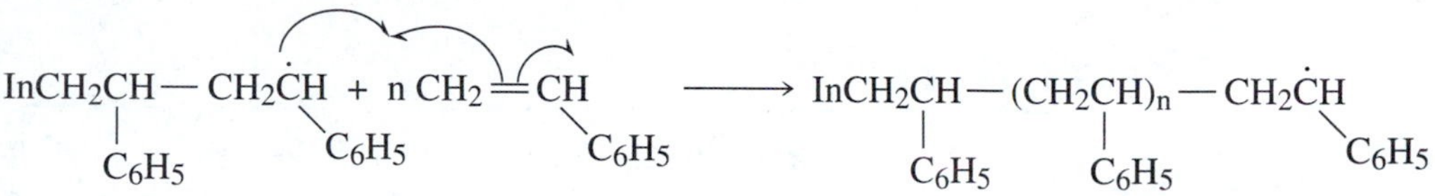

$$InCH_2CH - CH_2\overset{\cdot}{C}H + n\ CH_2 = CH \longrightarrow InCH_2CH - (CH_2CH)_n - CH_2\overset{\cdot}{C}H \qquad \textbf{(21.8b)}$$
$$\overset{|}{C_6H_5} \qquad \overset{|}{C_6H_5} \qquad \overset{|}{C_6H_5} \qquad \qquad \overset{|}{C_6H_5} \qquad \overset{|}{C_6H_5} \qquad \overset{|}{C_6H_5}$$
$$\textbf{21}$$

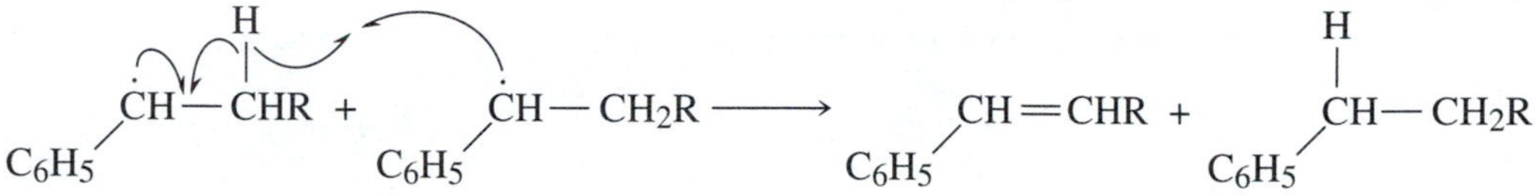

$$\overset{\cdot}{C}H - CHR + \overset{\cdot}{C}H - CH_2R \longrightarrow CH = CHR + \overset{\overset{H}{|}}{C}H - CH_2R \qquad \textbf{(21.9)}$$
$$\overset{|}{C_6H_5} \qquad \overset{|}{C_6H_5} \qquad \qquad \overset{|}{C_6H_5} \qquad \overset{|}{C_6H_5}$$
$$\qquad \qquad \qquad \qquad \qquad \textbf{22} \qquad \qquad \textbf{23}$$

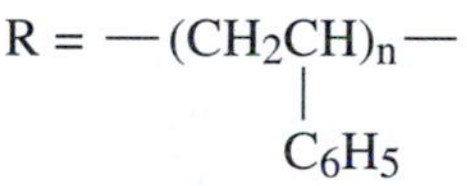

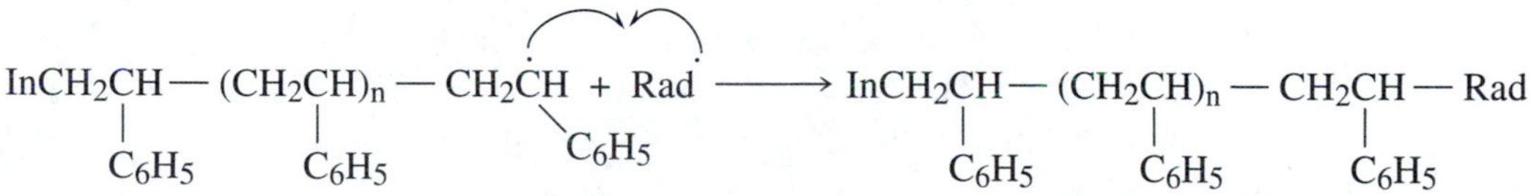

$$R = -(CH_2CH)_n -$$
$$\overset{|}{C_6H_5}$$

$$InCH_2CH - (CH_2CH)_n - CH_2\overset{\cdot}{C}H + Ra\overset{\cdot}{d} \longrightarrow InCH_2CH - (CH_2CH)_n - CH_2CH - Rad \qquad \textbf{(21.10)}$$
$$\overset{|}{C_6H_5} \qquad \overset{|}{C_6H_5} \qquad \overset{|}{C_6H_5} \qquad \qquad \overset{|}{C_6H_5} \qquad \overset{|}{C_6H_5} \qquad \overset{|}{C_6H_5}$$

The commercially available styrene used in our experiments contains *tert*-butyl-catechol (**24**, Eq. 21.11), a phenol that stabilizes styrene by functioning as a **radical scavenger.** The catechol does this by donating a hydrogen atom to reactive free radicals to convert them into nonradical products (Eq. 21.11, for example); the resulting phenoxy radical **25** that is produced is relatively unreactive as an initiator of a free-radical chain reaction (see Ex. 1 at the end of this section). The stabilizer is necessary to prevent premature polymerization of styrene during storage or shipment, because it is so readily polymerized by traces of substances such as atmospheric oxygen.

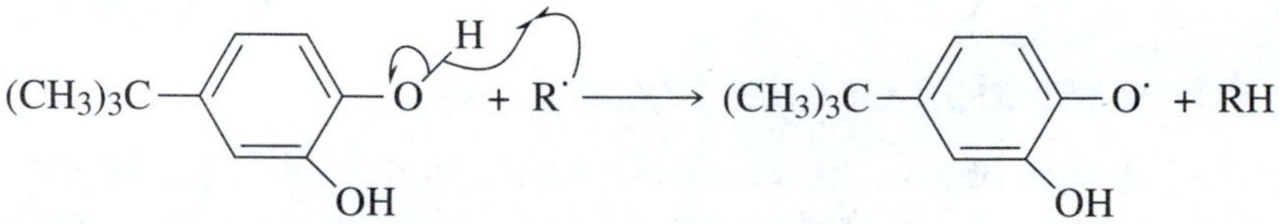

$$(CH_3)_3C - \overset{}{\underset{}{\bigcirc}} - O\overset{\curvearrow{H}}{} + R\overset{\cdot}{} \longrightarrow (CH_3)_3C - \overset{}{\underset{}{\bigcirc}} - O\overset{\cdot}{} + RH \qquad \textbf{(21.11)}$$
$$\overset{|}{OH} \qquad \qquad \qquad \qquad \overset{|}{OH}$$
$$\textbf{24} \qquad \qquad \qquad \qquad \textbf{25}$$
$$\textit{tert}\text{-Butylcatechol}$$

In the following procedure, directions are given for producing polystyrene in the form of an amorphous solid, a film, and a clear glass.

Pre-Lab exercises for Section 21.2, "Preparation of Polystyrene," are found on page PL. 121.

EXPERIMENTAL PROCEDURES

SAFETY ALERT

The free-radical initiator, *tert*-butyl peroxybenzoate, is a very safe material to use in this experiment, because it decomposes at a moderate rate when heated. Nonetheless, do not heat this catalyst excessively when performing the polymerization.

A. REMOVAL OF THE INHIBITOR FROM COMMERCIAL STYRENE

Place about 10 mL of commercial styrene in a small separatory funnel, and add 4 mL of 3 *M* sodium hydroxide and 15 mL of water. Shake the mixture thoroughly, allow the layers to separate, and withdraw the aqueous layer. Wash the organic layer with two 8-mL portions of water, carefully separating the aqueous layers after each wash. Dry the styrene by pouring it into a small Erlenmeyer flask containing a little anhydrous calcium chloride and swirling the flask. Allow the mixture to stand for 5–10 min, decant the liquid from the drying agent, and use the dried styrene in the following experiments.

Finishing Touches   Allow the volatiles to evaporate from the *calcium chloride* by placing it on a tray in the hood; then discard it in the container for nontoxic solids. Neutralize the *aqueous layers* if necessary before flushing them down the drain.

B. POLYMERIZATION OF PURE STYRENE

Place about 2–3 mL of pure, dry styrene in a small soft-glass test tube, and add 2 or 3 drops of *tert*-butyl peroxybenzoate (**18**). Clamp the test tube in a vertical position over a wire gauze, insert a thermometer so that its bulb is in the liquid, and heat the styrene and catalyst with a *small* burner flame. When the temperature reaches 140 °C, remove the flame temporarily. If boiling stops, replace the flame to maintain gentle boiling. The exothermicity of the polymerization increases the rate of formation of free radicals by thermal decomposition of the initiator, and this in turn increases the rate of polymerization. Thus be watchful for a rapid increase in the rate of boiling, and remove the flame if the refluxing liquid nears the top of the test tube.

After the onset of polymerization the temperature should rise to 180–190 °C, much above the boiling point of styrene. The viscosity of the liquid will increase rapidly during this time. As soon as the temperature begins to decrease, remove the thermometer and pour the polystyrene onto a watchglass. Do not touch the thermometer *before* the temperature decreases because movement of the thermometer in the boiling liquid might cause a sudden "bump," which could throw hot liquid out of the tube. Note the formation of fibers as the thermometer is pulled out of the polymer. The rate of solidification of the polystyrene depends on the amount of catalyst used, the temperature, and the length of time the mixture was heated.

C. SOLUTION POLYMERIZATION OF STYRENE

Place about 2 mL of pure, dry styrene and 5 mL of xylene in a 25-mL round-bottom flask and add 7 drops of *tert*-butyl peroxybenzoate from a micropipet. Connect a reflux condenser to the flask and heat the mixture under reflux with a small burner for 20 min. Cool the solution to room temperature; then pour about half of it into 25 mL of methanol. Collect the white precipitate of polystyrene that forms by decantation or, if decantation is not practical, by vacuum filtration. Resuspend the polystyrene in fresh methanol and stir it vigorously; collect the polystyrene by filtration and allow it to dry in the hood.

Pour the remaining half of the polystyrene solution onto a watch glass or the bottom of a large inverted beaker, and allow the solvent to evaporate. A clear film of polystyrene should form.

> **Finishing Touches**   Place the *filtrate* containing a mixture of xylene and methanol in the container for nonhalogenated organic liquids. Flush the *methanolic filtrate* obtained after resuspension of the polystyrene down the drain.

EXERCISES

1. *tert*-Butylcatechol (**24**) is capable of reacting with *two* equivalents of a radical, R•, to produce two moles of RH and a stable non-radical oxidation product of **24**. Propose a structure for this product and write a stepwise reaction mechanism for its formation. Use curved "fishhook" arrows to symbolize flow of electrons.

2. The use of phenols such as *tert*-butylcatechol (**24**) as free-radical scavengers is based on the fact that phenolic hydrogens are readily abstracted by radicals, producing relatively stable phenoxyl radicals, which interrupt chain

processes of oxidation and polymerization. Alcohols such as cyclohexanol, on the other hand, do *not* function as radical scavengers. Explain why the two types of molecules differ in their abilities to donate a hydrogen atom to a radical, R•.

3. Write an equation for the reaction involved in the removal of *tert*-butylcatechol from styrene by extraction with sodium hydroxide.

4. Why is the polymerization of styrene an exothermic reaction? Explain in terms of a calculation based on the following equation. Refer to a lecture textbook for the bond dissociation energies needed for the computation.

$$PhCH_2-H + CH_2=C \overset{Ph}{\underset{CH_3}{\big\langle}} \longrightarrow PhCH_2CH_2\overset{H}{\underset{CH_3}{C}}-Ph$$

5. Explain why polystyrene is soluble in xylene but insoluble in methanol.

6. What effect would using a smaller proportion of catalyst to styrene have on the average molecular weight of polystyrene?

7. In principle, radicals could add to styrene (**8**) at the carbon atom bearing the phenyl group rather than the other one, yet they do not. Explain the basis of this selectivity for the addition reaction.

SPECTRA OF STARTING MATERIALS AND PRODUCTS

The IR spectrum of polystyrene is provided as Figure 8.10.

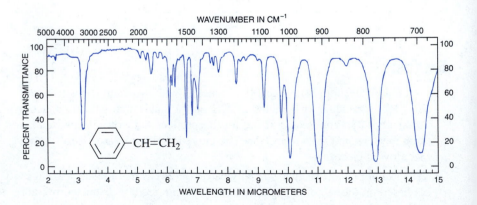

FIGURE 21.1   IR spectrum of styrene.

(a) ^{1}H NMR spectrum (90 MHz).

FIGURE 21.2 NMR data for styrene.

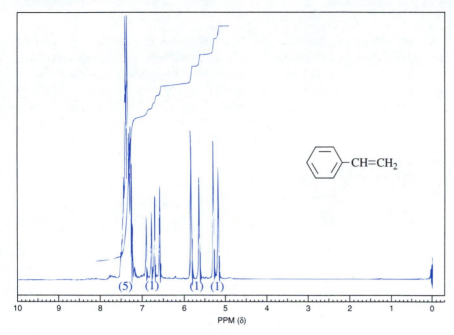

PPM (δ)

(b) ^{13}C NMR data.
Chemical shifts: δ 113.5, 126.2, 127.8, 128.5, 137.0, 137.7.

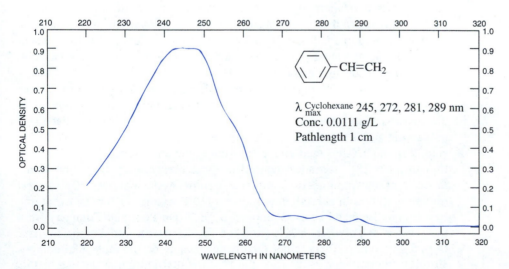

$\lambda_{max}^{Cyclohexane}$ 245, 272, 281, 289 nm
Conc. 0.0111 g/L
Pathlength 1 cm

OPTICAL DENSITY

WAVELENGTH IN NANOMETERS

FIGURE 21.3 UV spectrum of styrene.

21.3   STEP-GROWTH POLYMERIZATION

Step-growth polymerization typically involves the reaction between two different *difunctionalized* monomers. Both functionalities of each monomer react, and this leads to the formation of polymers. For example, a diacid such as terephthalic acid (**26**) can react with a diol such as ethylene glycol (**27**) in the presence of an acid catalyst to produce a polyester, as shown in Equation (21.12).

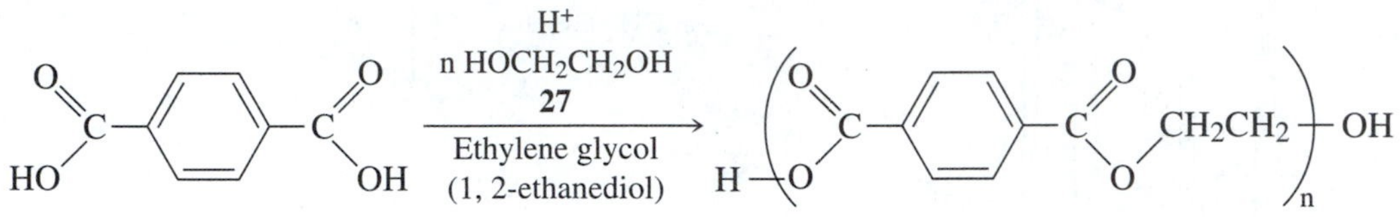

26
Terephthalic acid

A polyester

$$+ (2n - 1)\ H_2O$$

(21.12)

Chain growth is initiated by the acid-catalyzed reaction of a carboxyl group of the diacid with a hydroxy function of the diol to provide an ester and a molecule of water. The free carboxyl group or hydroxyl group of the resulting dimer then reacts with an appropriate functional group in another monomer or dimer, and the process is repeated in such *steps* until all of the monomers are converted into dimers, trimers, tetramers, and, eventually, polymers. The formation of new intermolecular bonds as the polymer grows involves elimination of a small molecule, water, which led to this type of polymerization being called **condensation polymerization** (see Sec. 21.1). Step-growth polymerization processes are much slower than chain-reaction processes. Because they also typically have higher activation energies, heating is often required to attain satisfactory rates of polymerization. Finally, condensation polymers generally have lower average molecular weights than polymers formed by chain-reaction polymerization.

Polyamides are one type of useful polymer that is produced by a step-growth process, and a variety of such polymers are preparable from various diacids and diamines. Nylon-6,6™ was the first commercially successful polyamide and is derived from the two monomers, hexanedioic acid (adipic acid, **28**) and 1,6-hexanediamine (hexamethylenediamine, **29**), as shown in Equation (21.13). Its trade name reflects the presence of six carbon atoms in each of the monomeric units that comprise the polymer. Of interest regarding the commercial importance of the nylons is the fact that the tremendous financial success enjoyed by E. I. Du Pont and Company from these types of polymers stems from the firm's patent on a method to draw the substance into fibers rather than from a patent on the molecular composition of the polymer itself! (See the Historical Highlight at the end of this chapter.)

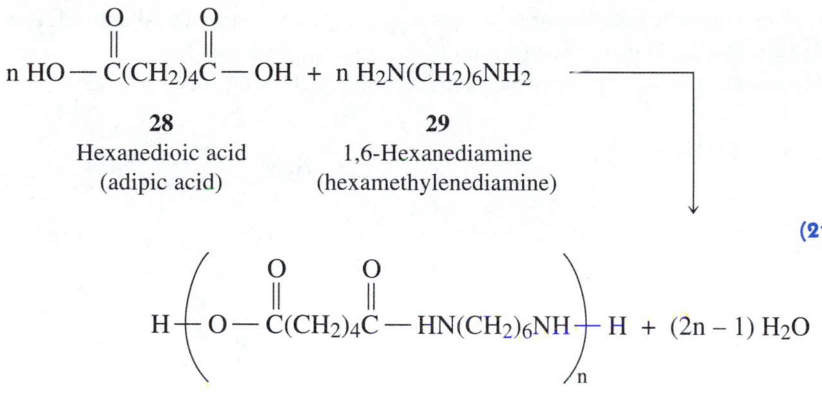

$$n \; HO - \overset{\overset{\displaystyle O}{\|}}{C}(CH_2)_4 \overset{\overset{\displaystyle O}{\|}}{C} - OH \; + \; n \; H_2N(CH_2)_6NH_2$$

28

Hexanedioic acid
(adipic acid)

29

1,6-Hexanediamine
(hexamethylenediamine)

(21.13)

$$H \!\!\left(\!\! O - \overset{\overset{\displaystyle O}{\|}}{C}(CH_2)_4\overset{\overset{\displaystyle O}{\|}}{C} - HN(CH_2)_6NH \!\!\right)_{\!\!n} \!\!\! H \; + \; (2n-1)\, H_2O$$

Nylon-6,6

In the typical industrial process for preparing polyamides, equimolar amounts of the diacid and diamine are mixed to give a salt, which is then heated to high temperature under vacuum to eliminate the water. The polymer so produced has a molecular weight of about 10,000 and a melting point of about 250 °C. Fibers may be spun from melted polymer, and if the fibers are stretched to several times their original length, they become very strong. This "cold-drawing" orients the polymer molecules parallel to one another so that hydrogen bonds form between C–O and N–H groups on adjacent polymer chains, as shown in **30,** greatly increasing the strength of the fibers. The strength of the fibers of silk, a well-known biopolymer involving protein molecules, is ascribed to the same stabilizing factor.

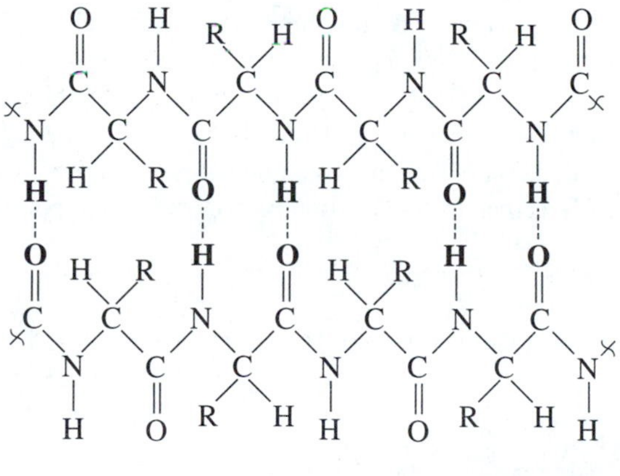

30

Three other industrially important polyamides are Qiana™, Nomex™, and Kevlar™. Qiana is produced from dodecanedioic acid, $HO_2C(CH_2)_{10}CO_2H$, and the diamine **31.** As a fiber, this polyamide feels much like silk. Nomex has earned considerable publicity as the insulator between the ceramic tiles and the aluminum surface of the space shuttles; its strength and especially its high melting point make it ideal for this purpose. Both Nomex and Kevlar belong to a family of "aramides," which are produced from aromatic diacyl chlorides and aromatic diamines. The monomers for Nomex are the *m*-disubstituted compounds **32** and

33, whereas those for Kevlar are the corresponding *para*-isomers **34** and **35.** It is interesting that Kevlar has been found to be superior to Nomex for use as automobile tire cord and in certain other applications, such as bullet-proof vests.

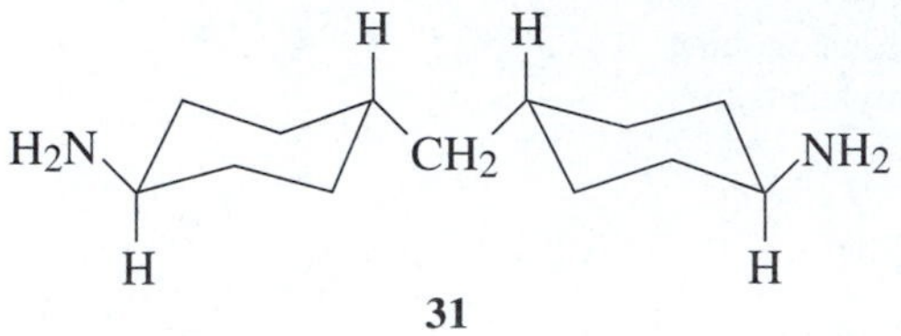

31

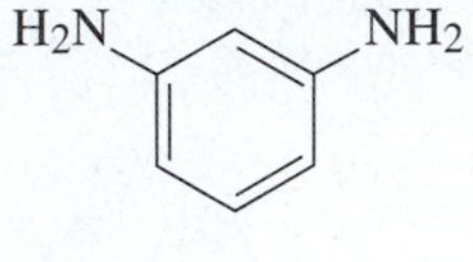

32
1,3-Diaminobenzene
(*m*-phenylenediamine)

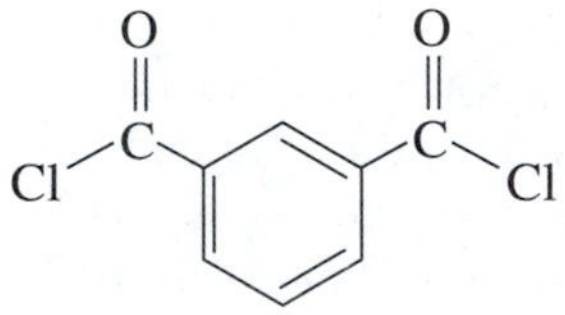

33
1,3-Benzenedicarbonyl chloride
(isophthaloyl dichloride)

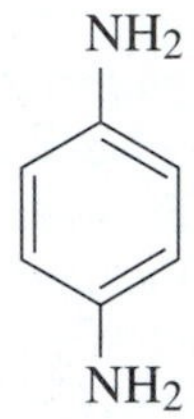

34
1,4-Diaminobenzene
(*p*-phenylenediamine)

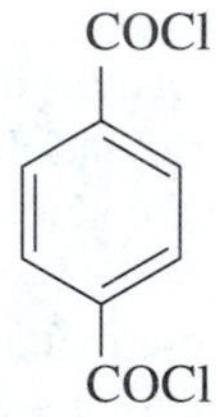

35
1,4-Benzenedicarbonyl chloride
(terephthaloyl dichloride)

The preparation of Nylon-6,10™ rather than Nylon-6,6 has been chosen for the present experiment to illustrate step-growth polymerization. This polyamide is commercially produced from decanedioic acid (sebacic acid, **36**) and 1,6-hexanediamine (**29**), as shown in Equation (21.14). To facilitate forming the polyamide under simple laboratory conditions, however, the diacyl dichloride of **36** is used because it is more reactive toward diamine **29.** This diacid dichloride is chosen because it is more stable toward hydrolysis than the corresponding six-carbon compound.

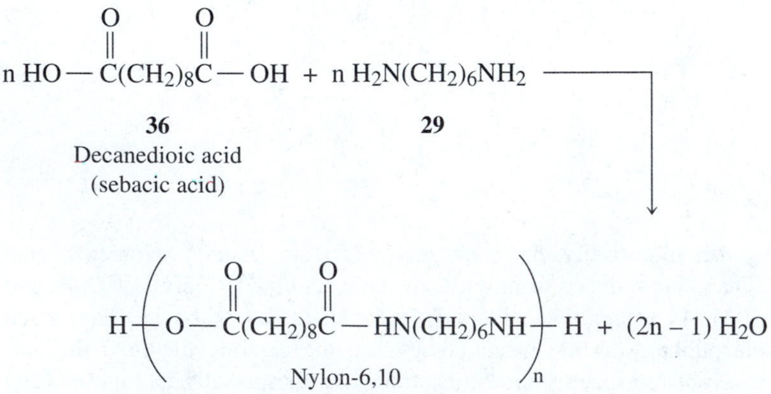

$$\text{n HO}-\overset{\overset{\displaystyle O}{\|}}{\text{C}}(\text{CH}_2)_8\overset{\overset{\displaystyle O}{\|}}{\text{C}}-\text{OH} \ + \ \text{n H}_2\text{N}(\text{CH}_2)_6\text{NH}_2 \longrightarrow$$

36   **29**
Decanedioic acid
(sebacic acid)

(21.14)

$$\text{H}\left(\!\!\text{O}-\overset{\overset{\displaystyle O}{\|}}{\text{C}}(\text{CH}_2)_8\overset{\overset{\displaystyle O}{\|}}{\text{C}}-\text{HN}(\text{CH}_2)_6\text{NH}\!\!\right)_{\!n}\!\!\text{H} \ + \ (2\text{n}-1)\,\text{H}_2\text{O}$$

Nylon-6,10

Using the diacyl dichloride means that the small molecule eliminated in this step-growth polymerization is hydrogen chloride rather than water (Eq. 21.15). Sodium carbonate is added to neutralize the acid formed to prevent consumption of the expensive diamine via an acid-base reaction. If the base were not added, an excess of diamine would be required for complete polymerization of the diacyl dichloride.

$$
\underset{\text{Cl}}{\overset{\text{O}}{\overset{\|}{\text{ClC(CH}_2)_{10}}}}\text{—CH}_2\text{C} \;\; + \;\; \text{H}_2\text{NCH}_2\text{—} \;\; \longrightarrow \;\; \underset{\text{NHCH}_2\text{—}}{\overset{\text{O}}{\overset{\|}{\text{ClC(CH}_2)_{10}}}}\text{—CH}_2\text{C} \;\; + \;\; \text{HCl} \qquad \textbf{(21.15)}
$$

The reactivity of the acid chloride toward nucleophilic acyl substitution (Eq. 21.15) allows this polymerization to be performed under mild conditions. When a solution of the dichloride in a water-immiscible solvent is brought into contact with an aqueous solution of the aliphatic diamine, a film of high-molecular-weight polymer forms immediately at the interface between the two solutions. The film is thin but strong, and can be pulled from the interface, where it is immediately and continuously replaced by further polymerization. In this way a long cord or rope of polyamide can be produced much as a magician pulls a string of silk handkerchiefs out of a top hat. When this experiment was first described by two Du Pont chemists, they characterized it as the "Nylon Rope Trick." It does seem to be almost magic that a polymer can attain an average molecular weight in the range 5000 to 20,000 in a fraction of a second!

To perform this experiment properly, the necessary equipment must be assembled so the polymer rope can be pulled from the reaction zone rapidly. A convenient way to do this is illustrated in Figure 21.4. A can, preferably with a

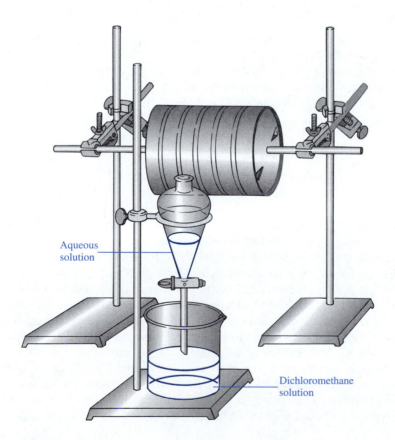

Aqueous
solution

Dichloromethane
solution

FIGURE 21.4 Apparatus for the "Nylon Rope Trick."

diameter of 10 cm or more, makes a good drum on which to wind the polymer. After puncturing the can in the center of each end, a wooden or metal rod is passed through the center holes to make an axle for the drum. The rod is supported horizontally by clamps attached to ring stands in the usual way.

The circumference of the drum should be measured and a reference point should be marked on the drum so that an estimate of the length of the nylon rope can be made by counting the revolutions made as the rope is produced. A length of 6 m or more can usually be obtained with the procedure described here.

Pre-Lab exercises for Section 21.3, "Preparation of Nylon-6,10," are found on page PL. 123.

EXPERIMENTAL PROCEDURE

SAFETY ALERT

1. If micropipets are used instead of syringes to measure the reactants, use a rubber bulb to draw up the liquid.
2. Until it has been washed free of solvent and reagents, do not handle the polymer rope with your bare hands any more than is necessary. Use rubber gloves, tongs, or forceps to manipulate it. If you touch the crude polymer, immediately wash your hands with soap and warm water.
3. If formic acid is used to form a film, do not let it get on your skin because it causes deep skin burns that are not immediately apparent. If the acid does accidentally come in contact with your skin, wash the affected area immediately with 5% sodium bicarbonate solution and then with copious amounts of water.

Note to student: Work in pairs on this experiment.

Measure 2 mL of decanedioyl dichloride into a 250-mL beaker using a syringe or pipet. *The size of the beaker is important:* In smaller beakers the polymer tends to stick to the walls, whereas in larger beakers poor "ropes" are obtained unless larger amounts of reagents are used. Dissolve the decanedioyl chloride in 100 mL of dichloromethane. Place 1.1 g of crystalline 1,6-hexanediamine, or 1.3 mL of a commercially available 80–95% aqueous solution, in a 125-mL separatory funnel, add 2.0 g of sodium carbonate, and 50 mL of water, gently shake the mixture to dissolve both substances. Arrange the drum on which the polymer is to be wound at a height such that the beaker containing the decanedioyl dichloride solution can be placed on the lab bench about 40 cm beneath and slightly in front of the drum.

Support the separatory funnel containing the other reagents so the lower tip of the funnel is centered no more than a centimeter above the surface of the dichloromethane solution of the decanedioyl dichloride. Open the stopcock of the separatory funnel slightly so the aqueous solution runs *slowly and gently* onto the surface of the organic solution. A film of polymer will form immediately at the interface of the two solutions. Use a long forceps or tongs to grasp the *center* of the polymer film and pull the rope that forms up to the front of the drum, loop it over the drum, and rotate the drum away from you to wind the rope onto the drum. For the first turn or two it may be necessary for you to use your fingers to secure the rope to the drum; if so, be sure to rinse your hands as soon as possible thereafter. Continue to rotate the drum and rapidly wind the nylon rope onto the drum until the reactants are consumed, remembering to count the revolutions of the drum as you wind.

Replace the beaker with a large dish or pan containing about 200 mL of 50% aqueous ethanol, and unwind the nylon rope into the wash solution. After stirring the mixture gently, decant the wash solution, and transfer the polymer to a filter on a Büchner funnel. Press the polymer as dry as possible, and then place it in your desk to dry until the next laboratory period. Dispose of the residual reaction mixture as described in **FINISHING TOUCHES**.

Examination of Dry Nylon-6,10.   You will probably encounter two surprises in this experiment. The first is the apparently enormous amount of nylon rope obtained from about 3 g of starting materials. The second surprise is the decrease in bulk of the polymer on drying. This is because the rope was really a delicate *tube* that appeared much larger when it was swollen with solvents. When the nylon is thoroughly dry, weigh it and calculate the yield.

Film Formation.   To produce a film of Nylon-6, 10, dissolve the dry polymer in about 10 times its weight of 90–100% formic acid (*Caution:* see the **SAFETY ALERT,** item 3) by stirring the mixture at *room temperature;* heating to achieve dissolution degrades the polymer. Spread the viscous solution on a glass plate to allow evaporation of the formic acid. Leave the plate *in a hood* until the next laboratory period for this evaporation.

Fiber Formation.   The dry polymer obtained in this experiment does not appear to have the properties expected of a nylon; it is fragile and of low density. However, fibers produced from it are much more dense, and have the appearance and strength more characteristic of typical polyamide. Form fibers by carefully melting the polymer in a metal spoon or spatula with gentle heating over a very small burner flame or an electric hot plate, and then drawing fibers from the melt with a small glass rod. If necessary, combine your polymer with that of several students to provide enough polymer to be melted and drawn successfully. Do not heat the polymer much above the melting temperature because it becomes discolored and charred.

Finishing Touches   After the rope has been drawn, stir the remaining reaction mixture thoroughly until no more polymer forms. Isolate any additional *polymer* that forms and, after thoroughly washing it with water, put it in the container for nonhazardous organic solids. Separate the *dichloromethane and aqueous layers* of the reaction mixture. Pour the *dichloromethane* into the container for halogenated organic liquids. Flush the *aqueous layer* and all *aqueous solutions* down the drain.

EXERCISES

1. Write an equation for the formation of the salt produced from one molecule of hexanedioic acid and two molecules of 1,6-hexanediamine.

2. Why is sodium carbonate used in the reaction?

3. Using full structural formulas, draw a typical portion of a Nylon-6,6 molecule; that is, expand a portion of the formula given in Equation (21.13). Show at least two hexanedioic acid units and two 1,6-hexanediamine units.

4. Draw formulas that illustrate the hydrogen bonding that may exist between two polyamide molecules after fibers have been "cold drawn."

5. Nylons undergo depolymerization when heated in aqueous acid. Propose a reaction mechanism that accounts for this fact, using curved arrows to symbolize flow of electrons.

6. Nylon-6 is produced from caprolactam by adding a small amount of aqueous base and then heating the mixture to about 270 °C.
   a. Draw a representative portion of the polyamide molecule.
   b. Suggest a mechanism for the polymerization, using curved arrows to symbolize flow of electrons, and decide whether it is of the chain-reaction or step-growth type.

Caprolactam

7. Why is Nylon-6,10 expected *not* to be optically active?

SPECTRA OF STARTING MATERIALS

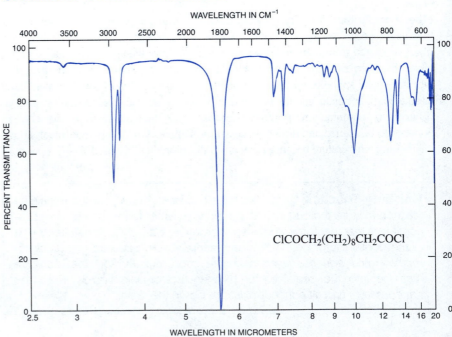

FIGURE 21.5 IR spectrum of decanedioyl dichloride.

(a) ^{1}H NMR spectrum (90 MHz).

FIGURE 21.6   NMR data for decanedioyl dichloride.

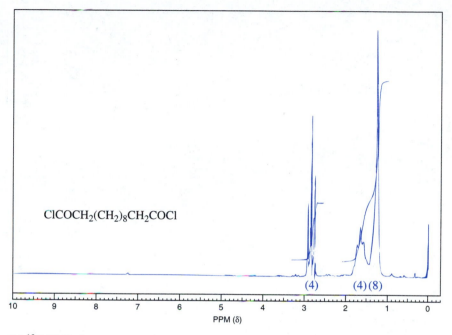

ClCOCH$_2$(CH$_2$)$_8$CH$_2$COCl

PPM (δ)

(b) ^{13}C NMR data.
   Chemical shifts: δ 24.7, 27.9, 28.4, 46.7, 173.2.

FIGURE 21.6   NMR data for decanedioyl dichloride.

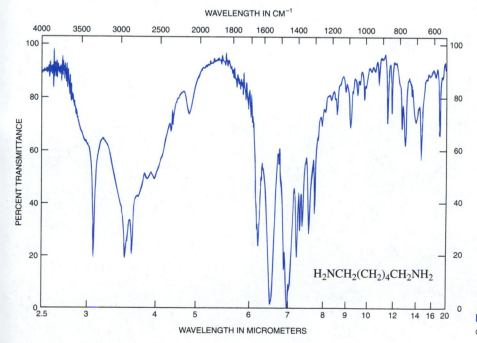

WAVELENGTH IN CM^{-1}

PERCENT TRANSMITTANCE

H$_2$NCH$_2$(CH$_2$)$_4$CH$_2$NH$_2$

WAVELENGTH IN MICROMETERS

FIGURE 21.7   IR spectrum of 1,6-hexanediamine.

FIGURE 21.8 NMR data for 1,6-hexanedi-amine.

(a) ^{1}H NMR spectrum (90 MHz).

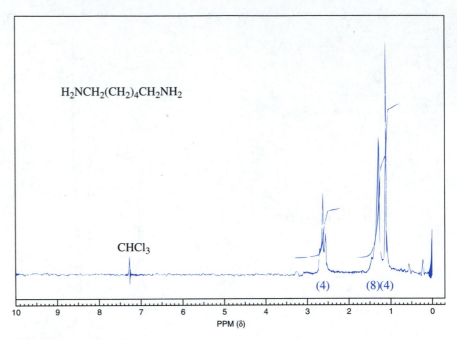

(b) ^{13}C NMR data.
Chemical shifts: δ 26.0, 33.0, 41.4.

HISTORICAL HIGHLIGHT:

Discovery of Polyethylene and Nylon

Polyethylene, or "polythene," as its British inventors called it, and the nylons are two types of polymers that have particularly interesting histories of discovery that bear repeating here. Both stories illustrate the role of serendipity in scientific achievements of great importance.

Polyethylene was discovered accidentally by British chemists at Imperial Chemicals Industries (I.C.I.) as an unexpected result of experiments on chemical reactions at very high pressures. In 1933, a reaction of benzaldehyde and ethylene at 170 °C and 1400 atmospheres gave no adducts involving the two reagents and was considered a complete failure. However, an observant chemist noticed a thin layer of "white waxy solid" on the walls of the reaction vessel used for the experiment. This was recognized as a polymer of ethylene, but additional experiments with ethylene alone to produce the same polymer only resulted in violent decompositions that destroyed the equipment.

Two years elapsed before better and stronger equipment was available for further experimentation. When ethylene was heated to 180 °C in this new equipment, the pressure in the apparatus dropped unexpectedly, so more ethylene was pumped in. Then, when the reaction vessel was opened, the I.C.I. chemists found a large amount of white powdery solid, which was the long-sought polyethylene. Because they knew that the polymerization could not account for all of the pressure drop that had been observed, they suspected a leak in one of the joints of the apparatus. This idea led to the proposal that the polymerization had been catalyzed by oxygen in the air that had leaked into the apparatus, and this hypothesis was confirmed by experiments in which air was intentionally included with the ethylene. Oxygen can act as a radical initiator and catalyze the polymerization by a chain-reaction mechanism analogous to Equations 21.7–21.10.

The polyethylene produced by the oxygen-catalyzed, high-pressure, high-temperature process developed by I.C.I. in the mid-1930s was ideal for many applications, including insulation of radar equipment, where it was used to great advantage by the Allies in World War II. Referring to the contribution radar made to naval operations, the British Commander-in-Chief said it enabled the Home Fleet to "find, fix, fight, and finish the Scharnhorst (the pride of Hitler's navy)."

The group of polymers called nylons was first produced in 1939 as a textile material for women's hose and other garments, but with the onset of World War II and the involvement of the U. S. by 1941, nylon was taken off the domestic market because it was found to be the best available material for military parachutes. The first nylon to be produced industrially was Nylon-6,6. The remarkable fact about the discovery of this polymer is not how it was first prepared from the two monomers, but how it was first prepared in a form suitable for a textile fiber. This depended on invention of the "cold-drawing process," and this technique was discovered almost completely by accident, as we shall see.

Wallace Hume Carothers was brought to Du Pont to direct its new basic chemical research program because his colleagues at the University of Illinois and Harvard University recommended him as the most brilliant organic chemist they knew. Carothers initiated a program aimed at understanding the composition of the high polymers of Nature such as cellulose, silk, and rubber, and of producing synthetic materials like them. By 1934 his group had contributed valuable fundamental knowledge in these areas, but Carothers had just about decided that their efforts to produce a synthetic fiber like silk was a failure.

It was a shrewd observation made during some "horseplay" among Carothers' chemists in the laboratory that turned this failure to compete with Nature into the enormous success ultimately advertised at the 1939 New York World's Fair as "Nylon, the Synthetic Silk Made from Coal, Air, and Water!" The Carothers group had already learned how to make Nylon-6,6, but even though this polyamide had a molecular structure similar to that of silk, they had "put it on the back shelf" without patent protection because the polymer did not have the tensile strength of silk, a necessary criterion for a good textile fiber. The group continued its research by investigating the polyester series, polymers which were more soluble, easier to handle, and thus simpler to work with in the laboratory. It was while working with one of these softer materials that Julian Hill noted that if he gathered a small ball of such a polymer on the end of a glass stirring rod and drew a thread out of the mass, the thread of polymer so produced became very silky in appearance. This attracted his attention and that of the others working with him, and it is reported that one day while Carothers was in downtown Wilmington, Hill and his cohorts tried to see how far they could stretch one of these samples. To do this, one chemist put a little ball of the polymer on a stirring rod and a second chemist touched a glass rod to the polymer ball and then ran down the hall to see how far he could stretch the thread of polymer. While doing this they noticed not only the silky appearance of the extended strands, but they also noticed the increased strength of the strands. They soon realized that this additional strength might be the result from some special orientation of the polymer molecules produced by the stretching procedure, as will be described below.

Because the polyesters they were working with at that time had melting points too low for use in textile products, a deficiency that has since been removed, the researchers returned to the polyamides (nylons) that had earlier been put aside. They soon found that these polymers, too, could be "cold-drawn" to increase their tensile strength so much that they could be made into excellent textiles. Filaments, gears, and other molded objects could also be made from the strong polymer produced by cold-drawing.

The alignment of the long polyamide molecules in a manner that produces extensive intermolecular hydrogen-bonding (**30,** Sec. 21.3) binds the individual polymer molecules together in much the same way that separate strands in a rope, when twisted together, form a cable. This association of linear polymer molecules through hydrogen bonding is responsible for the greatly increased strength of the nylon fibers. We believe that the same principle accounts for the strength of silk fibers; the natural polyamide molecules of silk are oriented in such a way that hydrogen bonds hold the individual molecules together. Interestingly, the silkworms accomplish the equivalent of "cold-drawing" as they extrude the viscous silk filaments to produce cocoons!

Relationship of Historical Highlight to Experiments

The experiments in this chapter represent examples of chain-reaction polymerization to produce polystyrene (Sec. 21.2) and step-growth polymerization (Sec. 21.3) to yield a nylon. The procedures of Section 21.2 provide an opportunity to observe formation of the same polymer in three physically different forms, whereas that of Section 21.3 illustrates how strong hydrogen-bound fibers can result from the "cold-drawing" technique patented by Du Pont.

CARBOHYDRATES

arbohydrates are one of the important classes of natural compounds that are found in living systems. The term carbohydrate owes its origin to the fact that many, but not all, members of this class have the general molecular formula $C_nH_{2n}O_n$ and thus were once considered to be "hydrates of carbon." The simplest carbohydrates are either polyhydroxy aldehydes, which have the general structure **1** and are referred to as **aldoses,** or polyhydroxy ketones, which have the general structure **2** and are called **ketoses.** Such simple carbohydrates are termed **monosaccharides.** Two monosaccharides may be combined with the loss of one molecule of water to form a **disaccharide. Polysaccharides** and other complex carbohydrates are then produced by the condensation of more monosaccharide subunits with the loss of a molecule of water for each additional monosaccharide. Depending upon its constitution, hydrolysis of a polysaccharide yields either a single monosaccharide or a mixture of monosaccharides. In this chapter, some of the fundamental chemical and physical properties of carbohydrates will be investigated using simple mono- and disaccharides; several classical qualitative tests to characterize and classify carbohydrates will also be performed.

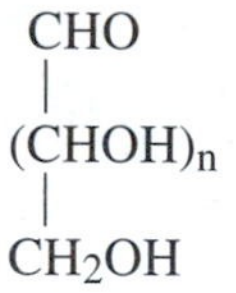

$$
\begin{array}{cc}
\text{CHO} & \text{CH}_2\text{OH} \\
| & | \\
(\text{CHOH})_n & \text{C}{=}\text{O} \\
| & | \\
\text{CH}_2\text{OH} & (\text{CHOH})_n \\
 & | \\
\mathbf{1} & \text{CH}_2\text{OH} \\
\text{An aldose} & \\
 & \mathbf{2} \\
 & \text{A ketose}
\end{array}
$$

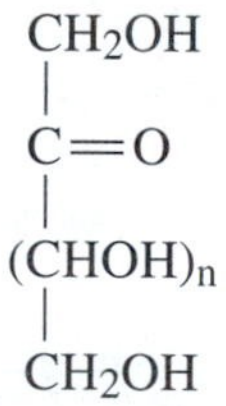

22.1 MONOSACCHARIDES: GENERAL PRINCIPLES

The monosaccharide D-glucose, whose chemistry is representative of all aldoses containing four or more carbon atoms, exists predominantly in the two **pyranose** forms **4** and **5,** which are six-membered **hemiacetals** formed from the acyclic form **3** by the reversible, intramolecular addition of the δ-hydroxyl group to the free aldehyde function (Eq. 22.1). In the cyclic forms **4** and **5,** the ring carbon that is derived from the carbonyl group is referred to as the **anomeric carbon.** The **specific rotation,** $[\alpha]_D^{25}$ (see Sec. 7.4), of α-D-(+)-glucose (**4**) is +112°, whereas that of the β-anomer **5** is +19°. When crystals of either pure **4** or pure **5** are dissolved in water, $[\alpha]_D^{25}$ changes to an equilibrium value +52.7°. This process is termed **mutarotation.** At equilibrium in water, the α- and β-forms are present in the ratio of 36:64; only about 0.03% of D-glucose is in the open form **3**.

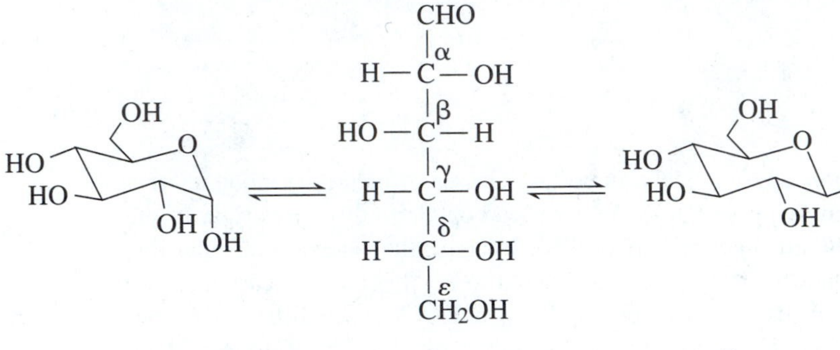

4	**3**	**5**
α-D-Glucopyranose	D-Glucose	β-D-Glucopyranose

(22.1)

Another common monosaccharide is D-fructose (**6**), a ketose. In aqueous solution, D-fructose also undergoes mutarotation to produce a complex equilibrium mixture of the acyclic form **6** (<1%), the five-membered hemiacetals **7** (31%) and **8** (9%), which are called **furanoses,** and the cyclic pyranoses **9** (57%) and **10** (3%); this mixture exhibits an $[\alpha]_D^{25} = -92°$ (Scheme 22.1).

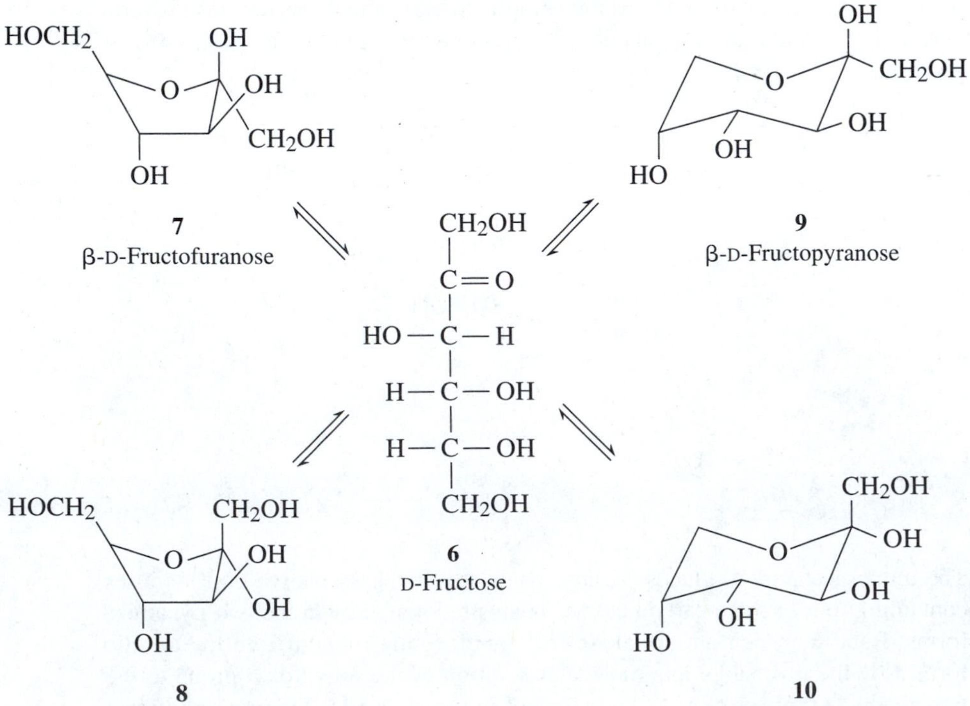

7		**9**
β-D-Fructofuranose		β-D-Fructopyranose

6
D-Fructose

8		**10**
α-D-Fructofuranose		α-D-Fructopyranose

SCHEME 22.1

The ready oxidation of the aldehyde group of an aldose to a carboxylic acid function forms the basis of a number of useful qualitative tests for classifying a carbohydrate as a **reducing sugar.** Ketoses also yield positive tests for reducing sugars even though no aldehyde group is present. The positive test for ketoses arises because the α-hydroxy keto group of the open form of a ketose undergoes base-catalyzed **tautomerization** to give an **enediol** that is protonated to provide a pair of epimeric aldehydes differing in configuration at C-2. These transformations are outlined in the form of partial structures in Equation (22.2). Since the tests for reducing sugars are performed under basic conditions that allow the equilibria of Equation (22.2) to be established, *all* known monosaccharides are reducing sugars.

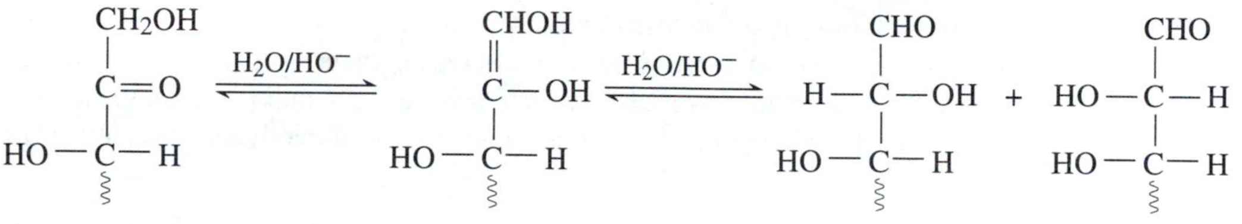

$$\underset{\text{A ketose}}{\begin{array}{c}CH_2OH \\ | \\ C=O \\ | \\ HO-C-H \end{array}} \overset{H_2O/HO^-}{\rightleftharpoons} \underset{\text{An enediol}}{\begin{array}{c}CHOH \\ || \\ C-OH \\ | \\ HO-C-H \end{array}} \overset{H_2O/HO^-}{\rightleftharpoons} \underset{\text{An aldose}}{\begin{array}{c}CHO \\ | \\ H-C-OH \\ | \\ HO-C-H \end{array}} + \underset{\text{An aldose}}{\begin{array}{c}CHO \\ | \\ HO-C-H \\ | \\ HO-C-H \end{array}} \qquad \textbf{(22.2)}$$

22.2 DISACCHARIDES: HYDROLYSIS OF SUCROSE

Disaccharides are *O*-glycosides in which the hydroxyl group attached to the anomeric carbon atom of one monosaccharide is replaced with a hydroxyl group of another monosaccharide, thereby forming a cyclic **acetal.** For example, maltose (**11**) is a disaccharide in which the anomeric hydroxyl group of one α-D-glucopyranose has been replaced with the C(4)-hydroxyl group of another D-glucopyranose subunit. Because the other D-glucopyranose ring of **11** is in equilibrium with the open-chain isomer having a free aldehyde function (Eq. 22.3), maltose is a reducing sugar. Indeed, any disaccharide in which *one* of the rings is a hemiacetal or hemiketal is a *reducing sugar* (see Sec. 22.1), since the cyclic hemiacetal or hemiketal moiety is in equilibrium with the open-chain isomer in which the aldehyde or the α-hydroxyketone function can be oxidized.

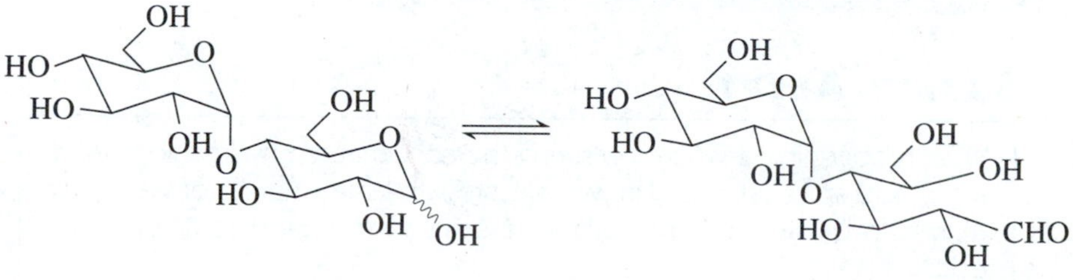

11
Maltose

(22.3)

12

Disaccharides in which *both* rings are in the **acetal** or **ketal** form are *not* reducing sugars because they *cannot* be in equilibrium with an aldehydo or keto form under neutral or basic conditions. For example, sucrose (**13**), a common foodstuff, is a **nonreducing** disaccharide because the glycosidic linkage between the two monosaccharide subunits is formed between the two anomeric carbon atoms, thereby incorporating the potential aldehyde and α-hydroxyketone functions into cyclic acetal and ketal groups. Thus sucrose, $[\alpha]_D^{25} = +66.5°$, does not undergo mutarotation under *neutral* or *alkaline* conditions. However, sucrose hydrolyzes under acidic conditions to give the reducing sugars D-glucose and D-fructose (Eq. 22.4). The 1:1 mixture of D-glucose and D-fructose that is obtained upon hydrolysis of sucrose is called **invert sugar** because the specific rotation of the mixture is dominated by the negative rotation of D-fructose. Fructose is the sweetest common sugar, being about twice as sweet as sucrose; consequently, invert sugar is sweeter than sucrose. The enzyme **invertase,** which bees use in making honey, accomplishes the same chemical result as does the acid-catalyzed hydrolysis of sucrose.

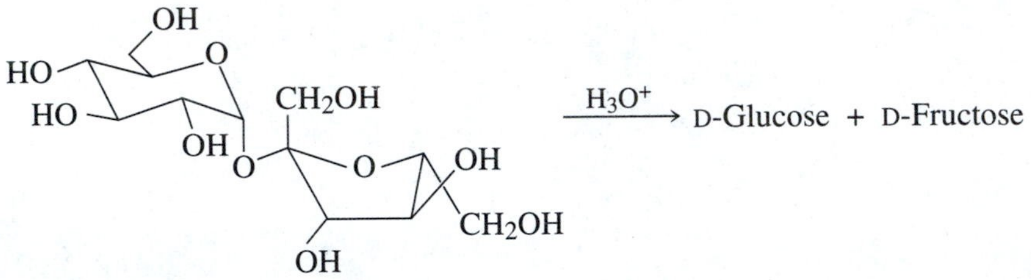

$$\text{13} \quad \xrightarrow{\text{H}_3\text{O}^+} \quad \text{D-Glucose} + \text{D-Fructose} \quad \textbf{(22.4)}$$

13
Sucrose

In the experiment in this section, you examine the acid-catalyzed hydrolysis of sucrose to give a mixture of D-glucose and D-fructose and monitor the consequent change in the specific rotation that occurs using a polarimeter.

Pre-Lab exercises for Section 22.2, "Hydrolysis of Sucrose," are found on page PL. 125.

EXPERIMENTAL PROCEDURE

SAFETY ALERT

Wear rubber gloves when measuring the concentrated hydrochloric acid. If any acid spills on your skin, wash it off with large volumes of water and then with dilute sodium bicarbonate solution to neutralize any residual acid.

Accurately weigh about 7.5 g of sucrose and place it in a 100-mL round-bottom flask. Add about 40 mL of water, swirl the contents of the flask to effect solution, and add about 0.5 mL of concentrated hydrochloric acid. Heat the solution at reflux for about 2 h.

While this solution is being heated, determine the specific rotation of sucrose according to the general directions that follow, together with any specific directions of your instructor. Carefully fill the sample tube for the polarimeter with water; *be certain that no air bubbles are trapped in it.* Place the sample tube in the polarimeter and determine the blank reading for the solvent. Record the blank rotation, the temperature, and the length of the tube. Empty and carefully dry the tube. Then accurately weigh 5–10 g of sucrose to the nearest 0.05 g, and transfer the sample quantitatively to a 50-mL volumetric flask. Fill the flask to within several milliliters of the volumetric mark, tightly stopper the flask, and shake the flask thoroughly. Carefully fill the flask to the mark and again shake the flask thoroughly. Transfer a portion of this solution to the polarimeter tube, measure the rotation and record this value.

At the end of the period of reflux, cool the reaction mixture to room temperature, and quantitatively transfer the solution to a 50-mL volumetric flask. Use small amounts of water to rinse the round-bottom flask, add the rinses to the flask, and fill the flask to the volumetric mark. Using the polarimeter as before, determine the specific rotation of the product mixture from the hydrolysis of sucrose. Compare this value with the specific rotation of sucrose determined earlier.

Finishing Touches   Neutralize the *aqueous solutions* with sodium carbonate and flush them down the drain.

EXERCISES

1. Explain the change in sign of the optical rotation that occurs when sucrose undergoes hydrolysis.

2. When determining the rotation, why is it important that no air bubbles be present in the polarimeter tube?

3. Calculate the specific rotation of *invert* sugar from the known rotations of the equilibrium mixtures of the anomers of D-glucose and D-fructose. How does this number compare with that determined experimentally?

4. The specific rotation of invert sugar and of a racemic mixture represent average rotations produced by the molecules in solution. Why is the rotation of the racemic mixture 0°, whereas that of invert sugar is not?

5. Write a stepwise reaction mechanism for the acid-catalyzed hydrolysis of sucrose to D-glucose and D-fructose.

6. Determine which of the structures in Scheme 22.1 are α-anomers and which are β-anomers.

7. Write a stepwise reaction mechanism for acid-catalyzed isomerization of **9** to **6,** followed by cyclization to produce a five-membered ring as in **7.**

22.3 CARBOHYDRATES: THEIR CHARACTERIZATION AND IDENTIFICATION

Determining the complete structure of an unknown monosaccharide was a formidable challenge to early organic chemists. Many carbohydrates, particularly when impure, have a tendency to form syrups rather than crystallize from solution, and this sometimes makes it difficult to obtain pure compounds for characterization and identification. The number and nature of functional groups present in carbohydrates further exacerbates the problem of assigning structures to carbohydrates.

Some of these difficulties are illustrated by considering the example of D-(+)-glucose, whose nature in solution is shown in Equation (22.1). The first step toward solving the structure of D-(+)-glucose was to identify D-(+)-glucose as a derivative of 2,3,4,5,6-pentahydroxyhexanal, **14.** Because D-(+)-glucose exists primarily as a cyclic hemiacetal at equilibrium, another issue involved determining the ring size. Namely, did D-(+)-glucose exist in the form of a *furanose,* **15,** or a *pyranose,* **16**? Finally, it was necessary to assign the relative and absolute configuration to each of the chiral centers (see Sec. 7.1), including the anomeric carbon atom (see Sec. 22.1). Despite these tremendous experimental challenges, the complete structures of many monosaccharides, some containing up to nine carbon atoms, are now known. It is amazing that much of this structural work was performed prior to the availability of the modern spectroscopic methods discussed in Chapter 8.

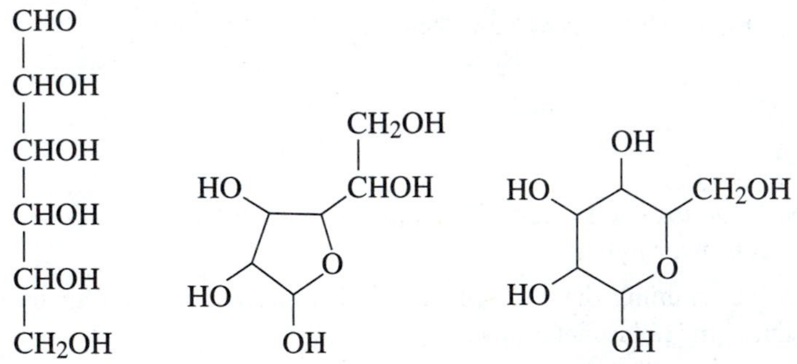

The structure elucidation of polysaccharides represents an even greater challenge. The individual monosaccharide subunits that constitute the unknown polysaccharide must first be identified. Then the ring size and position in the polysaccharide sequence must be elucidated for each monosaccharide. Finally, the nature of the glycosidic linkages that form the polysaccharide backbone must be defined. Toward this end, it is necessary to establish which hydroxyl group on one monosaccharide is involved in the formation of the acetal or ketal that forms the glycosidic bond to the adjoining monosaccharide; the stereochemistry at this anomeric center must also be determined.

22.3.1   CLASSIFICATION TESTS

A number of useful qualitative chemical tests have been invented to obtain information for unknown carbohydrates. Some of these tests are used to classify such molecules according to their structural type. In the experiments that follow, you use such tests to identify certain structural features that are found in mono- and polysaccharides. Information derived from these experiments can provide significant information that may be used to prove the structure of an unknown carbohydrate.

Tollens' Test.   Tollens' test is designed to distinguish between aldehydes and ketones. A complete discussion of this test together with an experimental procedure is included with the classification tests for aldehydes in Section 23.5.1B.

Benedict's Test.   This test is based upon the fact that cupric ion will oxidize aliphatic aldehydes, including α-hydroxyaldehydes, but not aromatic aldehydes. The reagent used in this test is a solution of cupric sulfate, sodium citrate, and sodium carbonate. The citrate ion forms a complex with the Cu(II) ion so that $Cu(OH)_2$ does not precipitate from the basic solution. A positive test for the presence of the aliphatic aldehyde group present in an aldose is evidenced by the formation of a red precipitate of cuprous oxide, Cu_2O (Eq. 22.5). A yellow precipitate is occasionally observed as a positive test. This yellow solid has apparently not been characterized, but its formation seems to depend on the amount of oxidizing agent present.

$$RCHO + 2\,Cu^{2+} + 5\,HO^- \xrightarrow{\text{citrate}} RCO_2^- + \underbrace{Cu_2O}_{\text{Brick red}} + 3\,H_2O \qquad (22.5)$$

Barfoed's Test for Monosaccharides.   Like Tollens' and Benedict's tests, this test depends on the *reducing* properties of the saccharides. However, because of the specific conditions employed for the test, it is possible to distinguish between monosaccharides and disaccharides. The test reagent consists of an aqueous solution of cupric acetate and acetic acid. Thus, in contrast to the previous tests, the reaction is carried out under *acidic* conditions. A positive test for monosaccharides is indicated by the formation of the brick-red precipitate of Cu_2O within *two* or *three* minutes. For reasons that are not understood, disaccharides require a longer time, providing the precipitate only after about ten minutes or more. Nonreducing sugars such as sucrose undergo slow hydrolysis under the aqueous acidic conditions of the test, and therefore they give a precipitate after *extended* time.

Pre-Lab exercises for Section 22.3.1, "Classification Tests for Carbohydrates," are found on page PL.127.

EXPERIMENTAL PROCEDURES

Obtain the carbohydrates for use in the following classification tests from your instructor. Perform each test on each carbohydrate and enter your observations and conclusions in your notebook.

Tollens' Test. *Before performing this test, read the* SAFETY ALERT *and* FINISHING TOUCHES *sections in 23.5.1B*. Perform Tollens' test according to the procedure outlined in Section 23.5.1B using about 0.1 g of the carbohydrate in a *clean* glass test tube and about 1 mL of Tollens' reagent. The formation of a silver mirror or a black precipitate constitutes a positive test.

Benedict's Test. Place about 0.1 g of the carbohydrate in a test tube and add 2 mL of water. Stir the mixture to effect solution and add 2–3 mL of the stock solution of Benedict's reagent. Heat the solution to boiling. The formation of a yellow to red precipitate is a positive test for aliphatic aldehydes and α-hydroxy aldehydes; the precipitate appears green when viewed in the blue solution of the reagent. For comparison, perform the test simultaneously on an unknown, on glucose, and on sucrose and record the results.

> **Finishing Touches**   Remove any precipitated *cuprous oxide* by vacuum filtration and place it in the container for nonhazardous solids. Flush the *filtrate* down the drain with an excess of water.

Barfoed's Test. Place 3 mL of Barfoed's reagent and 1 mL of a 1% solution of the carbohydrate in a test tube; place the test tube in a beaker of boiling water for *5 min.* Remove the test tube and cool it under running water. A red precipitate of cuprous oxide is a positive test. To see the precipitate, it may be necessary to view the tube against a dark background in good light. For comparison, run the test simultaneously on the unknown, on glucose, and on lactose, and record the results in your notebook.

> **Finishing Touches**   Remove any precipitated *cuprous oxide* by vacuum filtration and place it in the container for nonhazardous solids. Flush the *filtrate* down the drain with an excess of water.

22.3.2. FORMATION OF OSAZONES

Those carbohydrates that can exist in solution in an acyclic form with a free aldehyde or ketone function react with *three* equivalents of phenylhydrazine to form bright yellow crystalline derivatives called **phenylosazones,** together with aniline and ammonia as the other products of the reaction. For example, α- and β-D-glucopyranoses, **4** and **5,** are in dynamic equilibrium in aqueous solution with the ring-opened form **3,** which, because of its free aldehyde group, will react with phenylhydrazine to form glucosazone (**17**), as shown in Equation (22.6). Since phenylosazones may be readily identified by either their melting points or temperatures of decomposition, these derivatives may be used to characterize carbohydrates.

$$
\begin{array}{c}
\text{CHO} \\
|\\
\text{H—C—OH} \\
|\\
\text{HO—C—H} \\
|\\
\text{H—C—OH} \\
|\\
\text{H—C—OH} \\
|\\
\text{CH}_2\text{OH}
\end{array}
\quad + \ 3\ \text{C}_6\text{H}_5\text{NHNH}_2 \ \xrightarrow{\ \text{H}_3\text{O}^+\ }\quad
\begin{array}{c}
\text{CH}=\text{NNHC}_6\text{H}_5 \\
|\\
\text{C}=\text{NNHC}_6\text{H}_5 \\
|\\
\text{HO—C—H} \\
|\\
\text{H—C—OH} \\
|\\
\text{H—C—OH} \\
|\\
\text{CH}_2\text{OH}
\end{array}
\ +\ \text{C}_6\text{H}_5\text{NH}_2\ +\ \text{NH}_3
\qquad \textbf{(22.6)}
$$

3	**17**
D-Glucose	Phenylosazone of D-glucose

The accepted mechanism for the formation of phenylosazones is presented in Scheme 22.2. Following the initial formation of the phenylhydrazone **18** from the aldose, an internal oxidation-reduction occurs by tautomerization of *two* hydrogens from C-2 to the hydrazone moiety to give the ketone **19.** The newly formed carbonyl group then condenses with a second equivalent of phenylhydrazine to give **20,** which tautomerizes to **21.** After the 1,4-elimination of aniline to produce either **22** or **23,** a third equivalent of phenylhydrazine condenses with the imine group of **22** or **23** to yield the phenylosazone **24** and ammonia. Although it may appear that **24** should undergo further reaction by intramolecular oxidation-reduction between the secondary alcohol group at C-3 and the hydrazone group at C-2, the reaction stops at this point so that only two phenylhydrazine units are introduced. The formation of the intramolecular hydrogen bond shown in **24** limits the reaction to the first two carbons of the chain. Thus, when 1-methylphenylhydrazine [H$_2$NNMeC$_6$H$_5$] is used in place of phenylhydrazine, the reaction readily proceeds down the chain at least as far as C-5.

Pre-Lab exercises for Section 22.3.2, "Formation of Osazones," are found on page PL. 129.

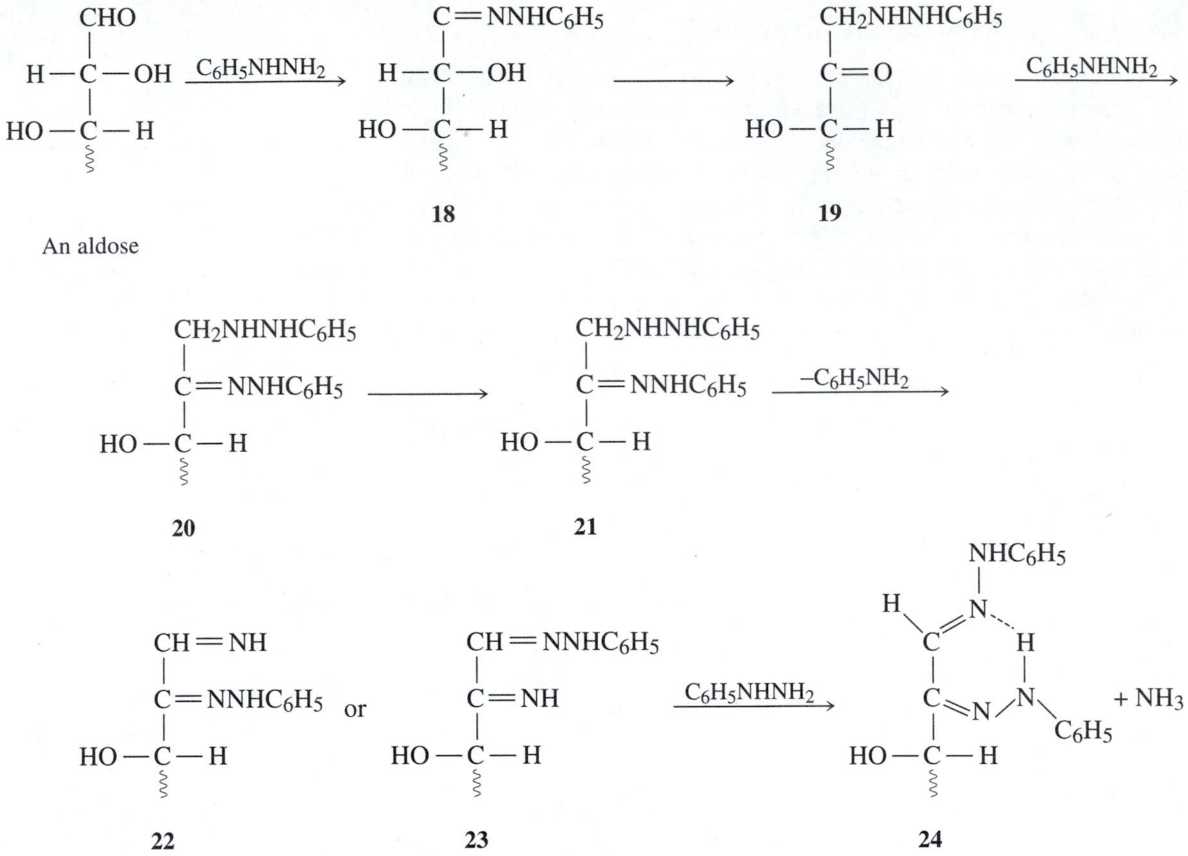

An aldose

SCHEME 22.2

EXPERIMENTAL PROCEDURE

SAFETY ALERT

Phenylhydrazine is toxic; avoid contact with it. If you spill some on your skin, rinse it off thoroughly with household bleach, dilute acetic acid, and then water.

Subject D-glucose, D-fructose, sucrose and the product of the hydrolysis of sucrose from Section 22.2 to the conditions required to form phenylosazones. For efficiency, perform all of these reactions simultaneously.

Heat a large beaker of water to boiling. In each of three test tubes, *separately* dissolve 0.2-g portions of D-glucose, D-fructose, and sucrose in 4 mL of water. In a fourth test tube place 4 mL of the solution that was used for rotation experiments for the hydrolysis of sucrose (see Sec. 22.2). Add 0.5 mL of saturated sodium bisulfite solution to each test tube to suppress oxidation of the phenylhydrazine during the reaction; this avoids contaminating the phenylosazone with tarry by-products. To each of the test tubes, add *either* 0.6 mL of glacial acetic acid, 0.6 g of sodium acetate, and 0.4 g of phenylhydrazine *or* 0.6 g of sodium acetate and 0.6 g of phenylhydrazine hydrochloride. Stir the solutions thoroughly, place the test tubes in the beaker of boiling water, and discontinue heating. Allow the test tubes to remain in the water bath for 30 min.

Remove the test tubes from the hot water bath and cool the contents to room temperature.✷ Collect any precipitates by vacuum filtration.✷ Recrystallize each of the phenylosazones formed from ethanol-water and determine their melting points. Make a mixture containing equal parts of the two phenylosazones obtained from the pure carbohydrates, and determine the melting point of this mixture. Because the melting points of phenylosazones may depend on the rate of heating, perform these determinations simultaneously. Make a mixture containing equal parts of the phenylosazone obtained from the hydrolysis of sucrose and each of the phenylosazones obtained from the pure carbohydrates and determine their melting points as well. Record your observations in your notebook, and use these results to answer Exercises 1–3 at the end of this subsection.

Finishing Touches Neutralize the *filtrates* obtained from isolating the crude phenylosazones with a small amount of sodium carbonate, and add 2 mL of laundry bleach (5.3% aqueous sodium hypochlorite) for every 1 mL of the filtrate. Heat the mixture at 45–50 °C for 2 h to oxidize and decompose any phenylhydrazine that is present. Cool the mixture, and then flush it down the drain with excess water. Flush the *filtrates* from recrystallization down the drain.

EXERCISES

1. Which of the pure carbohydrates formed phenylosazones?

2. What conclusions can be drawn from the results of the mixed melting-point determination on the phenylosazones formed from the pure carbohydrates?

3. What conclusions can be drawn from the results of the mixed melting-point determinations on the phenylosazones formed from the pure carbohydrates and that obtained from the hydrolysis product of sucrose?

4. D-Glucose and D-mannose give the same phenylosazone. Explain.

5. Draw the structure of the product that would be expected upon treatment of D-glucose with a large excess of 1-methylphenylhydrazine.

IDENTIFICATION OF ORGANIC COMPOUNDS

One of the greatest challenges to the chemist is identifying the substances that are obtained from chemical reactions or isolated from natural sources. Although structure elucidation may be difficult and time-consuming, the availability of modern spectroscopic techniques (see Chap. 8) in combination with chemical methods has greatly facilitated this aspect of experimental organic chemistry. Some of the techniques and approaches that are used to accomplish this goal are described in this chapter.

Systematic procedures for the identification of organic compounds were developed much later than those for inorganic compounds, ions, and elements. The first successful systematic scheme of qualitative organic analysis was developed by Professor Oliver Kamm and culminated in the publication of his textbook in 1922. This scheme, together with some more recent modifications, is termed **classical qualitative organic analysis,** and is the basis for most textbook discussions of the subject (see Ref. 1 at the end of Section 23.4, for example).

In recent years, the development of chromatographic methods of separation (see Chap. 6) and structural analysis by spectroscopic techniques (see Chap. 8) have revolutionized the laboratory practice of organic chemistry. Nevertheless, interest in classical qualitative organic analysis remains high because it is an effective and interesting means of understanding fundamental organic chemistry. Consequently, this chapter contains an outline of this classical scheme, and the experimental procedures associated with it, that is adequate for obtaining a good understanding of the approach. The scheme now is rarely used as the sole means of elucidating structures because a wealth of information is readily available from spectroscopic techniques. Thus, an introduction to identification and structure determination based on data from modern spectroscopic methods is included in Section 23.4. The advantages of the combined use of the classical and instrumental methods are also described in that section.

23.1 OVERVIEW OF ORGANIC ANALYSIS

An overview of a systematic procedure that can be used to identify an unknown pure organic compound serves to introduce the classical scheme for organic analysis. The first step is to ensure that the compound is pure, and this can be done in several ways. In the case of a liquid, gas-liquid chromatography (GLC, see Sec. 6.4) may be used to demonstrate that only one component is present; a pure compound produces only a single peak in a gas chromatogram, assuming that no decomposition occurs under the conditions of the analysis. If enough liquid is available, it can be distilled (see Sec. 4.2). A 1–2 °C boiling-point range implies that the compound is pure, although such a narrow range is also consistent with the distillation of an **azeotrope** (see Sec. 4.2.4). However, GLC provides better evidence for purity because of its ability to detect low levels of impurities. The purity of a solid substance can be ascertained from its melting point (see Sec. 3.3). A 1–2 °C melting-point range usually indicates a pure compound except in the rare instance of a **eutectic mixture** (see Sec. 3.3.3). Impure liquids can be purified by simple or fractional distillation (see Sec. 4.2) and by column or preparative GLC (see Secs. 6.2 and 6.4, respectively), whereas solids can be purified by recrystallization (see Sec. 3.2), by sublimation (see Sec. 2.17), or by column or thin-layer chromatography (see Secs. 6.2 and 6.3). The purification and identification of gases is not included in this text because of the difficulties in handling them and because they are not commonly encountered in organic chemistry.

Once purity is established, various physical properties are determined. The melting point of a solid or boiling point of a liquid is considered essential. Occasionally, the density and/or refractive index of a liquid may be useful, and for certain compounds, either liquid or solid, the specific rotation can be determined if a substance is optically active.

Establishing what elements other than carbon and hydrogen are present is critical for ultimately identifying the compound, and techniques for such **elemental analysis** are described later. Molecular weight, as determined by cryoscopic techniques or mass spectrometry, or percentage by weight composition of the elements present, also provides important data. The solubility of the unknown compound in water, in dilute acids and bases, or in various organic solvents may signal the presence or absence of various functional groups.

Perhaps the most important step in identifying an unknown substance is determining the functional group(s) that may be present, and IR and NMR spectroscopy (see Secs. 8.2 and 8.3, respectively) are now commonly used for this purpose. Before the development of these spectroscopic methods as routine experimental techniques, functional group determination involved performing qualitative chemical tests for each possible group. Although spectral analyses may not provide an *unequivocal* answer about the presence of certain functional groups, they at least permit narrowing the possibilities to a small number, so only one or two chemical tests are then needed to complete the identification of the functionalities present.

Final assignment of a structure to the "unknown" compound is achieved by one of several procedures. The classical method involves the chemical conversion of the substance into a *solid* derivative. The success of this technique depends on the availability of information about the unknown and its various derivatives. Of prime importance is knowledge of the melting or boiling points of possible candidates for the unknown as well as the melting points of solid derivatives. Many tabulations of organic compounds are available for this purpose, and some are provided in Tables 23.1–23.11. These tables are arranged by functional group. The compounds in them are listed in order of melting point or boiling point, and melting points of their derivatives are given.

The identification of the compound can also be completed by thoughtful analysis of the IR, ^{1}H, and ^{13}C NMR spectra of the compound. However, unequivocal proof of structure based *solely* upon NMR spectra is usually not possible unless the spectra are being used to compare the "unknown" with a number of known compounds; demonstrating that the IR and NMR spectra of an unknown and of a known compound are *identical* will suffice to prove that the substances are identical.

The following sections contain descriptions of the stepwise procedures that may be used to identify an unknown compound using classical methods alone or a combination of classical and spectroscopic methods. As a reminder, these procedures should *not* be performed unless the compound is pure. Since it is possible that you may be given a mixture of unknown compounds to identify, a procedure is provided in Section 23.3 for the separation of a mixture into its individual components so that each one can be identified.

23.2 CLASSICAL QUALITATIVE ANALYSIS PROCEDURE TO IDENTIFY A PURE COMPOUND

The *classical system of qualitative organic analysis* consists of the six steps that follow. The first four, which can be carried out in any order, should be completed before performing the qualitative tests for functional groups. The final step must always be the preparation of one or more solid derivatives.

1. Preliminary examination of physical and chemical characteristics
2. Determination of physical constants
3. Elemental analysis to determine the presence of elements other than carbon, hydrogen, and oxygen
4. Solubility tests in water, dilute acid, and dilute base
5. Functional group analysis using classification tests
6. Preparation of derivatives

It is a tribute to the effectiveness of the system that one can identify an unknown organic compound with certainty, in view of the fact that it may be one of several million known compounds. With the exception of a few general guidelines, there are no rigid directions to be followed. You must rely on good judgment and initiative in selecting a course of attack on the unknown, and it is particularly important to observe and consider each experimental result. Negative results may be as useful as positive ones in the quest to reveal the identity of an unknown.

23.2.1 THE PRELIMINARY EXAMINATION

If it is carried out intelligently, the **preliminary examination** may provide more information with less effort than any other part of the procedure. The simple observation that the unknown is a *crystalline* solid, for example, eliminates from consideration a major fraction of all organic compounds because many are liquids at room temperature. The **color** is also informative; most pure organic compounds are white or colorless. A brown color is often characteristic of small amounts of impurities; for example, aromatic amines and phenols quickly become discolored by the formation of trace amounts of highly colored air-oxidation products. Color in a pure organic compound is usually attributable to conjugated double bonds.

The **odor** of many organic compounds, particularly those of lower molecular weight, is highly distinctive. A conscious effort should be made to learn and recognize the odors that are characteristic of several classes of compounds such as the alcohols, esters, ketones, and aliphatic and aromatic hydrocarbons. The odors of certain compounds demand respect, even when they are encountered in small amounts and at considerable distance; for example, the unpleasant odors of thiols (mercaptans), isonitriles, and higher carboxylic acids and diamines cannot be described definitively, but they are recognizable once encountered. *Be extremely cautious* in smelling unknowns, because some compounds are not only disagreeable but also possibly toxic. Large amounts of organic vapors should *never* be inhaled.

The **ignition test** involves a procedure in which a drop or two of a liquid or about 50 mg of a solid is heated gently on a small spatula or crucible cover with a microburner flame. Whether a solid melts at low temperature or only upon heating more strongly is then noted. The flammability and the nature of any flame from the sample are also recorded. A yellow, sooty flame is indicative of an aromatic or a highly unsaturated aliphatic compound; a yellow but nonsooty flame is characteristic of aliphatic hydrocarbons. The oxygen content of a substance makes its flame more colorless or blue; a high oxygen content lowers or prevents flammability, as does halogen content. The unmistakable and unpleasant odor of sulfur dioxide indicates the presence of sulfur in the compound. If a white, nonvolatile residue is left after ignition, a drop of water is added and the resulting aqueous solution is tested with litmus or pHydrion paper; a metallic salt is indicated if the solution is alkaline.

23.2.2 PHYSICAL CONSTANTS

If the unknown is a solid, its **melting point** is measured by the capillary tube method (see Sec. 2.3) An observed melting-point range of more than 2–3 °C indicates that the sample should be recrystallized (see Sec. 3.1).

For an unknown that is a liquid, the **boiling point** is determined by the micro boiling-point technique (see Sec. 2.4). An indefinite or irreproducible boiling point or discoloration or inhomogeneity of the unknown requires that the sample be distilled (see Sec. 2.7); the boiling point is obtained during the distillation.

Other physical constants that *may* be of use for liquids are the **refractive index** and the **density.** Consult with your instructor about the advisability of making these measurements.

23.2.3 ELEMENTAL ANALYSIS

The technique of **elemental analysis** involves determining *which* elements may be present in a compound. The halogens, sulfur, oxygen, phosphorus, and nitrogen are the elements other than carbon and hydrogen that are most commonly found in organic molecules. Although there is no simple way to test for the presence of oxygen, it is fairly easy to determine the presence of the other heteroatoms, and the appropriate procedures are provided for the halogens, sulfur, and nitrogen.

The basis of these procedures is as follows. Because the bonding found in organic compounds is principally covalent, there are seldom direct methods analogous to those applicable to ionic inorganic compounds for determining the presence of the aforementioned elements. However, the covalent bonds between carbon and these heteroatoms may be broken by heating an organic compound with sodium metal. This process, called **sodium fusion,** results in the formation of inorganic ions involving these elements if they are present in the original compound; thus, the products are halide ions, X^-, from halogens, sulfide ion, S^{2-}, from sulfur, and cyanide ion, CN^-, from nitrogen. After the organic compound has been heated with sodium metal, the residue is *cautiously* treated with *distilled* water to destroy the excess sodium and to dissolve the inorganic ions that have been formed. The *fusion aqueous solution,* designated as **FAQS** can then be analyzed for the presence of halide ion, sulfide ion, and/or cyanide ion.

The presence of halide is determined by first acidifying a portion of the FAQS with dilute nitric acid and boiling the solution *in the hood* to remove any sulfide or cyanide ions as hydrogen sulfide or hydrogen cyanide, respectively; sulfide and cyanide must be removed because they interfere with the test for halogens. Silver nitrate solution is then added, and the formation of a precipitate of silver halide signals the presence of halide in the FAQS (Eq. 23.1).

$$ Ag^{\oplus} + X^{\ominus} \longrightarrow \underline{AgX} \text{ (solid)} \tag{23.1} $$

Silver halide

The color of the precipitate provides a *tentative* indication of which halogen is present: AgCl is *white* but turns *purple* on exposure to light, AgBr is *light yellow,* and AgI is *dark yellow. Definitive* identification is made by standard inorganic qualitative analysis procedures or by means of thin-layer chromatography (TLC, see Sec. 6.3).

Sulfur is detected by taking a second portion of the FAQS and carefully acidifying it. Any sulfide ion that is present will be converted to H_2S gas, which forms a dark precipitate of PbS when brought in contact with a strip of paper that has been saturated with lead acetate solution (Eq. 23.2).

$$ Pb(OAc)_2 + H_2S \longrightarrow \underline{PbS} \text{ (solid)} + 2\,HOAc \tag{23.2} $$

black

Lead acetate   Hydrogen sulfide   Lead sulfide   Acetic acid

The presence of nitrogen in the unknown sample is shown by the following qualitative test. A third portion of the FAQS is carefully acidified, and ferrous ion, Fe^{2+}, and ferric ion, Fe^{3+}, are added; this converts the cyanide ion into potassium ferric ferrocyanide, which precipitates as an intense blue solid called **Prussian blue** (Eq. 23.3).

$$ 4\,CN^- + Fe^{2+} \longrightarrow Fe(CN)_6^{4-} \xrightarrow[Fe^{3+}]{K^+ \text{ and}} \underline{KFeFe(CN)_6} \tag{23.3} $$

Potassium ferric ferrocyanide
(Prussian blue)

EXPERIMENTAL PROCEDURES

SAFETY ALERT

1. Sodium fusion involves heating sodium metal or a sodium-lead alloy to a high temperature and then adding the organic compound. Use extreme care when performing both the heating and addition.
2. Perform the sodium fusion in the hood if possible.
3. Use a Pyrex® test tube and check it for cracks or other imperfections before performing the sodium fusion.
4. If sodium metal is used, hydrolyze the residue *very* carefully as directed, because any excess metal reacts *vigorously* with alcohol or water.
5. Be careful when handling the test tube after the fusion is complete; remember that it may still be hot.
6. *Throughout this procedure, point the mouth of the test tube away from yourself and your neighbors;* the organic material may burst into flame when it contacts the hot metal, or it may react so violently that hot materials are splattered from the test tube.

A.   SODIUM FUSION

Sodium fusion may be executed using either sodium metal or a commercially available sodium-lead alloy, which contains nine parts lead and one part sodium. For safety reasons, we recommend using the sodium-lead alloy method because the alloy is easier to handle than is sodium metal and poses less potential danger during hydrolysis.

Sodium-Lead Alloy Method

Support a small Pyrex test tube in a vertical position using a clamp whose jaws have either an asbestos liner or none at all; the jaws must *not* be lined with materials such as rubber or plastic. Place a 0.5-g sample of sodium-lead alloy in the test tube and heat the alloy with a flame until it melts and vapors of sodium rise 1–2 cm in the test tube. *Do not* heat the test tube to redness. Then remove the flame and immediately add 2–3 drops of a liquid sample or about 10 mg of a solid sample to the hot alloy, being careful not to allow the sample to touch the sides of the hot test tube during the addition. If there is no visible reaction, heat the fusion mixture gently to initiate the reaction, then discontinue the heating and allow the reaction to subside. Next, heat the test tube to redness for a minute or two, and then *let it cool* to room temperature. Add 3 mL of *distilled water* to the cooled reaction mixture, and heat the mixture gently for a few minutes to complete hydrolysis. Decant or filter the solution. If the latter option is used, wash the filter paper with about 2 mL of water. Otherwise simply dilute the decanted solution with about 2 mL of water. Label the filtrate FAQS and use it in the appropriate tests for detecting sulfur, nitrogen, and the halogens.

> **Finishing Touches** Put the *lead pellet* that remains after hydrolysis of the reaction mixture in the container for heavy metals.

Sodium Metal Method

Support a small Pyrex test tube in a vertical position using a clamp whose jaws have either an asbestos liner or none at all; the jaws must *not* be lined with materials such as rubber or plastic. Place a clean cube of sodium metal about 3 to 4 mm on an edge in the tube and heat it gently with a microburner flame until the sodium melts and the vapors rise 1–2 cm in the test tube. Then remove the flame and add immediately 2 or 3 drops of a liquid or about 10 mg of a solid sample to the hot test tube, being careful not to allow the sample to touch the sides of the hot test tube during the addition. A brief flash of fire may be observed; this is normal. Reheat the bottom of the tube, remove the flame and add a second portion of the unknown organic compound in the same amount as for the first addition. Now heat the bottom of the tube until it is a dull red color; then remove the flame and allow the tube to cool to room temperature. Carefully add about 1 mL of 95% ethanol dropwise to the cooled reaction mixture to decompose the excess sodium; stir the contents of the tube with a stirring rod. After a few minutes, add another 1 mL of ethanol with stirring. After the reaction has subsided, apply gentle heat to boil the ethanol; either work at the hood or use an inverted funnel connected to a vacuum source (see Fig. 2.12c) over the mouth of the test tube to keep the vapors of ethanol from the room. When the ethanol has been removed, allow the tube to cool and add about 10 mL of *distilled water;* stir the mixture and pour it into a small beaker. Rinse the tube with an additional 5 mL of distilled water, and combine the rinse with the main solution. The total amount of water should be 15 to 20 mL. Boil the aqueous mixture briefly, filter it, and label the filtrate FAQS; use this filtrate in the tests for sulfur, nitrogen, and the halogens.

B. QUALITATIVE ANALYSIS FOR SULFUR, NITROGEN, AND HALOGENS

Halogens. Acidify about 2 mL of the FAQS by dropwise addition of 6 *M* nitric acid. Follow the acidification with litmus or pHydrion paper. Working at the hood, boil the solution gently for 2–3 min to expel any hydrogen sulfide or cyanide that may be present. Cool the solution and then add several drops of 0.3 *M* aqueous silver nitrate solution to it. A *heavy precipitate* of silver halide indicates the presence of chlorine, bromine, or iodine in the original organic compound. A faint turbidity of the solution should *not* be interpreted as a positive test.

Tentative identification of the particular halogen is made on the basis of color. *Definitive* identification must be made by standard inorganic qualitative procedures (see References 1–3 at the end of Section 23.4) or by means of TLC (see Sec. 6.2), using the following procedure.

Obtain a 3 × 10-cm strip of fluorescent silica gel chromatogram sheet. About 1 cm from one end, place four equivalently spaced spots as follows. Using a capillary, spot the FAQS at the left side of the sheet. Because this solution is likely to be relatively dilute in halide ion, several reapplications of it may be required. Allow the spot to dry following each application, a process that is hastened by blowing on the plate. Do not allow the spot to broaden to a diameter greater than about 2 mm. Next, and in order, spot samples of 1 M potassium chloride, 1 M potassium bromide, and 1 M potassium iodide. Develop the plate in a solvent mixture of acetone, 1-butanol, concentrated ammonium hydroxide, and water in the volume ratio of 13:4:2:1. Following development, allow the plate to air-dry, and, working at the hood, spray the plate lightly with an indicator spray prepared by dissolving 1 g of silver nitrate in 2 mL of water and adding this solution to 100 mL of methanol containing 0.1 g of fluorescein and 1 mL of concentrated ammonium hydroxide. Allow the now-yellow strip to dry, and then irradiate it for several minutes with a long-wavelength ultraviolet lamp (366 nm). Compare the spots formed from the test solution with those formed from the solutions of known halides. (Note that iodide gives two spots according to this procedure.)

Sulfur.   Acidify a 1–2 mL sample of the FAQS with acetic acid, and add a few drops of 0.15 M lead acetate solution. A black precipitate of PbS signifies the presence of sulfur in the original organic compound.

Nitrogen.   Determine the pH of a 1-mL sample of the FAQS with pHydrion E indicator paper. The pH should be about 13. If it is above 13, add a *small* drop of 3 M sulfuric acid to bring the pH down to about 13. If the pH of the fusion solution is below 13, add a *small* drop of 6 M NaOH to bring the pH up to about 13. Add 2 drops each of a saturated solution of ferrous ammonium sulfate and of 5 M potassium fluoride. Boil the mixture gently for about 30 sec, cool it, and add 2 drops of 5% ferric chloride solution. Then carefully add 3 M sulfuric acid dropwise to the mixture until the precipitate of iron hydroxide *just* dissolves. Avoid an excess of acid. At this point the appearance of the deep-blue color of potassium ferric ferrocyanide (Prussian blue) indicates the presence of nitrogen in the original organic compound. If the solution is green or blue-green, filter it; a blue color remaining on the filter paper is a weak but nonetheless positive test for nitrogen.

> **Finishing Touches**   Transfer precipitated *silver halides* to a container labeled for them so that the silver can be recovered. Put the precipitated *lead sulfide* in a container for heavy metals. Any FAQS that remains may be flushed down the drain, as can the *solution* from the test for nitrogen.

23.2.4   SOLUBILITY TESTS

The solubility of an organic compound in water, dilute acid, or dilute base can provide useful, but not definitive, information about the presence or absence of certain functional groups. Note that the assignment of an unknown to a formal solubility class is rather arbitrary because a large number of compounds exhibit *borderline* behavior. It is recommended that the solubility tests be done in the order presented here.

Water. Test the solubility of the unknown in water. For the present purposes, a compound is defined as soluble if it dissolves to the extent of 3 g in 100 mL of water, or more practically, 30 mg in 1 mL of water. As a general rule, few organic compounds exhibit appreciable water solubility as defined here.

Several structural features of the unknown can be deduced if it is water-soluble. It must be of low molecular weight and will usually contain no more than four to five carbon atoms, unless it is polyfunctional. It must contain a polar group that will hydrogen-bond with water, such as the hydroxy group of an alcohol or a carboxylic acid, the amino functionality of an amine, or the carbonyl group of aldehydes or ketones. Esters, amides, and nitriles dissolve to a lesser extent, and acid chlorides or anhydrides react with water rather than simply dissolving in it. On the other hand, alkanes, alkenes, alkynes, and alkyl halides are water-insoluble.

If the unknown is water-soluble as defined above, test its aqueous solution with litmus paper or pHydrion paper. If the solution is acidic, the unknown is likely to be a low-molecular-weight carboxylic acid such as acetic acid. If the solution is basic, a low molecular weight organic base such as diethylamine is possible. A neutral solution suggests the presence of a neutral polar compound such as an alcohol or a ketone.

The borderline for water solubility of *monofunctional* organic compounds is most commonly at or near the member of the homologous series containing five carbon atoms. Thus, butanoic acid is soluble, pentanoic acid is borderline, and hexanoic acid is insoluble in water; similarly, 1-butanol is soluble, 1-pentanol is borderline, and 1-hexanol is insoluble in this solvent. There are exceptions to this generalization, however. For example, if a molecule is spherical in shape, it can contain a larger number of carbon atoms and still remain soluble in water. A case in point is that 2-methyl-2-butanol (*tert*-pentyl alcohol) is water-soluble to the extent of 12.5 g/100 mL of water even though it contains five carbon atoms. The increased solubility is because the molecular surface areas of spherical molecules are less than those of nonspherical ones, and this decreases the **hydrophobicity** of the molecule.

Solubility in Aqueous Acid and Base. If an unknown is insoluble in water, test its solubility first in sodium hydroxide, then in sodium bicarbonate, and finally in hydrochloric acid. Solubility in one or more of these acids and bases is defined in terms of the compound being *more soluble in base or acid than in water* and reflects the presence of an acidic or basic functional group in the water-insoluble unknown compound.

In each of the following solubility tests, *shake the unknown with the test reagent at room temperature.* If it does not dissolve, warm the mixture for several minutes in a hot-water bath and continue shaking it. If the substance still appears not to dissolve, decant or filter the liquid from the undissolved sample and carefully *neutralize* the filtrate; observing a precipitate or turbidity is indicative of greater solubility in the aqueous acid or base than in water itself. The importance of bringing the filtrate to pH 7 arises because an unknown may show enhanced solubility in both acidic and basic solutions if it contains *both* basic and acidic functional groups. Assessing these possibilities necessitates performing *all* the solubility tests.

1. *Sodium Hydroxide.* If the compound is water-insoluble, test its solubility in 1.5 M NaOH solution. Carboxylic acids and sulfonic acids, which are strong acids, and phenols, which are weak acids, dissolve in sodium hydroxide because they are converted into their water-soluble sodium salts (Eq. 23.4). An unknown that is more soluble in NaOH than in water may be either a phenol or a carboxylic acid, and it must be tested for solubility in the weaker base, 0.6 M NaHCO$_3$, which may permit distinction between these two functional groups (see Part 2). If the unknown does not exhibit solubility in NaOH, its solubility in NaHCO$_3$ need not be tried; rather, it should next be tested for solubility in 1.5 M HCl.

$$\text{Carboxylic acids: } RCO_2H \xrightarrow{\text{NaOH}} RCO_2^{\ominus} Na^{\oplus} \qquad \textbf{(23.4a)}$$

Water-insoluble   *Water-soluble*

$$\text{Phenols: } \qquad ArOH \xrightarrow{\text{NaOH}} ArO^{\ominus} Na^{\oplus} \qquad \textbf{(23.4b)}$$

Water-insoluble   *Water-soluble*

2. *Sodium Bicarbonate.* An unknown that is soluble in dilute NaOH solution should also be tested for its solubility in 0.6 M NaHCO$_3$. If it is soluble, the *tentative* conclusion is that a carboxylic acid group is present, owing to the formation of the water-soluble sodium salt (Eq. 23.5a); phenols are normally not deprotonated in this medium (Eq. 23.5b). Dissolution should be accompanied by effervescence resulting from decomposition of the carbonic acid, H_2CO_3, formed from reaction of bicarbonate with the carboxylic acid, to carbon dioxide and water (Eq. 23.5a).

$$\text{Carboxylic acids: } RCO_2H \xrightarrow{\text{NaHCO}_3} RCO_2^{\ominus} Na^{\oplus} + H_2CO_3 \longrightarrow CO_2 + H_2O \qquad \textbf{(23.5a)}$$

Water-insoluble   *Water-soluble*

$$\text{Phenols: } ArOH \xrightarrow{\text{NaHCO}_3} \text{ no reaction} \qquad \textbf{(23.5b)}$$

Water-insoluble

Carboxylic acids are *usually* soluble in NaHCO$_3$ *and* in NaOH, whereas phenols usually dissolve *only* in NaOH. However, caution must be exercised in making *definitive* conclusions about the presence of a carboxylic acid or phenol based upon solubility in NaHCO$_3$. For example, a phenol containing one or more strong electron-withdrawing substituents, such as a nitro group, can be as acidic as a carboxylic acid and thus may form a water-soluble sodium salt by reaction with NaHCO$_3$. Similarly, the salt of a carboxylic acid of high molecular weight may *not* be completely soluble in the aqueous medium; nevertheless, evolution of CO$_2$ should still be observable when the acid comes in contact with the base.

3. *Hydrochloric Acid.* The solubility of the unknown in 1.5 M hydrochloric acid should be determined. If the unknown is soluble, the presence of an amino group in the compound is indicated because amines are organic bases that react with dilute acids to form ammonium salts that are usually water-soluble (Eq. 23.6). However, this solubility test does *not* permit the distinction between weak and strong organic bases.

$$RNH_2 \xrightarrow{\text{HCl}} RNH_3^{\oplus} Cl^{\ominus} \tag{23.6}$$

Water-insoluble　　　*Water-soluble*

4. *Concentrated Sulfuric Acid.* Many compounds that are too weakly basic or acidic to dissolve in dilute aqueous acid or base will dissolve in or react with concentrated H_2SO_4. Such solubility is often accompanied by the observation of a dark solution or the formation of a precipitate; any detectable reaction such as evolution of a gas or formation of precipitate is considered "solubility" in concentrated H_2SO_4. This behavior can usually be attributed to the presence of carbon-carbon π-bonds, oxygen, nitrogen, or sulfur in the unknown. The solubility is usually due to reaction of one of these functional groups with the concentrated acid, which results in the formation of a salt that is soluble in the reagent. For example, an alkene adds the elements of sulfuric acid to form an alkyl hydrogen sulfate (Eq. 23.7) that is soluble in the acid, and an oxygen-containing compound becomes protonated in concentrated acid to form a soluble oxonium salt (Eq. 23.8). Substances that exhibit this solubility behavior are termed "neutral" compounds.

$$R_2C{=}CR_2 + H_2SO_4 \longrightarrow \underset{\substack{| \quad\quad | \\ \text{H} \quad \text{OSO}_3\text{H}}}{R_2C{-}CR_2} \tag{23.7}$$

　　Alkene　　　Sulfuric acid　　　　　Alkyl sulfonate

$$R_2C{=}O \text{ or } ROR + H_2SO_4 \longrightarrow R_2C{=}\overset{\oplus}{O}H\ HSO_4^{\ominus} \text{ or } \underset{\substack{| \\ \text{H}}}{R{-}\overset{\oplus}{O}{-}R}\ HSO_4^{\ominus} \tag{23.8}$$

Aldehyde　　Ether　　　　　　　　　　　Oxonium ions
or ketone

　　Compounds that are soluble in dilute HCl or neutral water-insoluble compounds containing N or S need not be tested for their solubility in concentrated H_2SO_4 because they will invariably dissolve in or react with it. Note that solubility or insolubility of an unknown in concentrated H_2SO_4 does *not* yield a great deal of evidence for the presence or absence of any *specific* group, whereas the solubility of a compound in dilute HCl or NaOH or $NaHCO_3$ provides valuable information about the type of functional group present.

　　Assuming that a compound contains only *one* functional group, the following scheme classifies compounds according to their solubility in acid or base. This picture may be changed dramatically if a compound contains several polar functional groups that cause its solubility properties to be different from those of a monofunctional compound or if an acidic or basic compound contains one or more strongly electron-withdrawing groups.

Acidic Compounds Soluble in NaOH and NaHCO$_3$

    Carboxylic acids

    Phenols

Acidic Compounds Soluble in NaOH but not in NaHCO$_3$

    Phenols

Basic Compounds Soluble in HCl

    Amines

Neutral Compounds Soluble in Concentrated H$_2$SO$_4$

    Carbonyl compounds (aldehydes and ketones)

    Unsaturated compounds (alkenes and alkynes)

    Alcohols

    Esters

    Amides

    Nitriles

    Nitro compounds

    Ethers

Neutral Compounds Insoluble in Cold Concentrated H$_2$SO$_4$

    Alkyl halides

    Aryl halides

    Aromatic hydrocarbons

After determining the physical constants, elemental analysis, and solubility properties, provide a **preliminary report** of these findings to your instructor, who may give advice regarding the validity of these observations. This protocol serves to minimize unnecessary loss of time in finding the structure of the unknown.

23.2.5 CLASSIFICATION TESTS; IDENTIFICATION OF FUNCTIONAL GROUPS

The next step in identifying an unknown is to determine which functional groups are present. The classical scheme involves performing a number of chemical tests on a substance, each of which is specific for a type of functional group. These tests can normally be done quickly and are designed so that the observation of a color change or the formation of a precipitate indicates the presence of a particular functional group. The results of these tests usually allow the assignment of the unknown to a structural class such as alkene, aldehyde, ketone, or ester, for example. The following factors should be considered when performing qualitative classification tests for functional groups.

1. A compound may contain more than one functional group, so the complete series of tests must be performed unless you have been told that the compound is monofunctional.

2. Careful attention is required when the functional group tests are performed. Record *all* observations, such as the formation and color of any solid produced as a result of a test.

3. Some of the color-forming tests occur for several different functional groups, and although the *expected* colors are given in the experimental procedures, the *observed* color may be affected by the presence of other functional groups.

4. It is of utmost importance to perform a qualitative test on *both the unknown and a known compound that contains the group being tested.* Some functional groups may appear to give only a slightly positive test, and it is helpful to determine how a compound known to contain a given functional group behaves under the conditions of the test being performed. It is most efficient and reliable to do the tests on standards at the same time as on the unknown. In this manner, inconclusive positive tests may be interpreted correctly. Because aliphatic compounds are sometimes more reactive than aromatic ones, it is wise to perform a test on both of these types of standards along with the unknown.

5. The results obtained from the elemental analysis and the solubility tests can be used in deciding which functional group tests should be performed initially and which should not be done at all. The following examples illustrate the use of the preliminary work in making these decisions:

   a. A classification test for an amine should be applied first if a compound is found to be soluble in dilute hydrochloric acid and to contain nitrogen.

   b. The test for a phenol should be performed on an unknown that is soluble in dilute sodium hydroxide but insoluble in dilute sodium bicarbonate.

   c. The tests for alkyl or aryl halides should be omitted if the elemental analysis indicates the absence of halogen.

   d. The tests for amines, amides, nitriles, and nitro compounds need not be performed if nitrogen is absent, as shown by the elemental analysis.

6. A logical approach should be followed in deciding which tests are needed. The result obtained from one test, whether positive or negative, often has a bearing on which additional tests should be done. A random "hit-or-miss" approach is wasteful of time and often leads to erroneous results. Another error commonly made by beginners in qualitative organic analysis is to omit the tests for functional groups and immediately attempt preparation of a derivative. This tactic has a very low probability for success; for example, trying to make a derivative of a ketone is certain to fail if the unknown is actually an alcohol. The different classification tests should be performed until the nature of the functional group(s) present in the unknown is defined as completely as possible! This will minimize unproductive efforts at derivatization of the compound.

Qualitative tests for most of the common functional groups are presented in Section 23.5, which includes the following structural classes along with references to the sections in which the functional group is described.

Neutral Compounds

Alcohols, Section 23.5.5

Aldehydes, Section 23.5.1

Alkenes, Section 23.5.2

Alkyl halides, Section 23.5.3

Alkynes, Section 23.5.2

Amides, Section 23.5.12

Aromatic hydrocarbons, Section 23.5.4

Aryl halides, Section 23.5.4

Esters, Section 23.5.10

Ketones, Section 23.5.1

Nitriles, Section 23.5.11

Nitro compounds, Section 23.5.9

Acidic Compounds

Carboxylic acids, Section 23.5.7

Phenols, Section 23.5.6

Basic Compounds

Amines, Section 23.5.8

23.2.6   PREPARATION OF DERIVATIVES

It was mentioned earlier that the classical approach to structure elucidation usually involves converting an "unknown" liquid or solid into a second compound that is a solid, the latter being called a **derivative** of the first compound. It is better to prepare a derivative that is a solid rather than a liquid because solids can be obtained in pure form by recrystallization and because the melting point of a solid can be determined on a small quantity of material, whereas a larger amount of a liquid derivative would be required in order to determine its boiling point.

Two solid derivatives of an unknown compound should be prepared in order to double-check its identity. The melting points of the derivatives, along with the melting point or boiling point of the unknown compound, usually serve to identify the unknown completely. However, the success of this type of identification depends upon the availability of tables listing the melting points and/or boiling points of known compounds and the melting points of suitable derivatives. Tables containing this information for selected compounds are provided in Section 23.6, and more extensive listings of compounds are found in References 1, and 2 at the end of Section 23.4. These tabulations are by no means comprehensive, and many other compounds that have been identified on the basis of derivatives appear in the scientific literature (see Chap. 24).

Each of the classes of compounds listed in Section 23.2.5 are considered in the following manner in Section 23.5: **Classification tests** for the functional group characteristic of the type of compound are given, and these are followed by experimental procedures for preparing several different **solid derivatives.** References to the tables in Section 23.6, which list various specific compounds and their derivatives, are also made for each functional group.

The warning given near the end of Section 23.2.5 is repeated here: Do *not* proceed directly to the preparation of a derivative merely on the basis of a hunch about the class of compound to which your unknown belongs. Rather, make certain of the type of functional group present by obtaining one *or more* positive classification tests *before* attempting the preparation of any derivative.

The following example illustrates how preparation of derivatives is used to identify an unknown compound, for which the following experimental data are available.

Preliminary analysis	Colorless liquid with pleasant odor
Physical constant	bp 119–120 °C
Elemental analysis	No X, S, or N
Solubility tests	Slightly soluble in water; no increased solubility in dilute HCl, NaOH, or NaHCO$_3$; soluble in concentrated H$_2$SO$_4$.
Preliminary report	Observations confirmed by instructor
Classification tests	Negative test for aldehyde, alkene, alkyne; positive test for ketone, positive test for methyl ketone, negative tests for all other functional groups

The experimental error inherent in determining melting or boiling points means it is prudent to consider compounds melting or boiling within 5 °C of the observed melting or boiling point as possibilities for the unknown. Applying this principle to the entries in Table 23.15(a) for liquid ketones, provides the following list of possible substances for the unknown:

1. 1-methoxy-2-propanone (bp 115 °C, 760 torr), $CH_3OCH_2\overset{\displaystyle O}{\overset{\|}{C}}CH_3$

2. 4-methyl-2-pentanone (bp 117 °C, 760 torr), $(CH_3)_2CHCH_2\overset{\displaystyle O}{\overset{\|}{C}}CH_3$

3. 3-methyl-2-pentanone (bp 118 °C, 760 torr), $CH_3CH_2\overset{\displaystyle O}{\overset{\|}{C}}HCCH_3$

4. chloroacetone (bp 119 °C, 760 torr), $ClCH_2\overset{\displaystyle O}{\overset{\|}{C}}CH_3$

5. 3-penten-2-one (bp 122 °C, 760 torr), $CH_3CH=CH\overset{\displaystyle O}{\overset{\|}{C}}CH_3$

6. 2,4-dimethyl-3-pentanone (bp 124 °C, 760 torr), $(CH_3)_2CH\overset{\displaystyle O}{\overset{\|}{C}}CH(CH_3)_2$

From the experimental facts that were given, some of these compounds can be eliminated as possibilities for the following reasons: compound 4 because it contains a halogen, compound 5 because it contains a carbon-carbon double bond, and compound 6 because it is not a methyl ketone. Now suppose that two derivatives, the 2,4-dinitrophenylhydrazone (see Sec. 23.5.1A) and the semicarbazone (see Sec. 23.5.1F), were prepared from the unknown and found to melt at 93–95 °C and 69–71 °C, respectively. Further examination of Table 23.15(a) indicates that the derivatives of only one of the liquid ketones under consideration, namely 3-methyl-2-pentanone, melt at these temperatures. Hence, the identity of the unknown is deduced. Although it is possible that other ketones with similar boiling points *may* exist and may *not* be listed in Table 17.15(a), it is highly unlikely that any of these will give two derivatives with the same melting points as those of the derivatives obtained from 3-methyl-2-pentanone. This emphasizes the desirability of preparing *two* derivatives.

The preceding analysis indicates how positive *and* negative information can be utilized in determining the identity of an unknown substance. Of particular note is the importance of the functional group classification tests, which must be done carefully and thoroughly in order to exclude the possible presence of all groups other than the keto group in the unknown.

23.3 SEPARATING MIXTURES OF ORGANIC COMPOUNDS

The preceding section contains information identifying a *pure* organic compound. However, when a chemist is faced with the problem of identifying an organic compound, it is seldom pure; rather, it is often contaminated with by-products or starting materials if it has been synthesized in the laboratory. Modern methods of separation, particularly chromatographic techniques (see Chap. 6), make the isolation of a pure compound easier than it once was, but one must not lose sight of the importance of classical techniques of separation, which are treated in detail in Chapters 3–5.

The common basis of the procedures most often used to separate mixtures of organic compounds is the difference in **polarity** that exists or may be induced in the components of the mixture. This difference in polarity is exploited in nearly all the separation techniques, including distillation, recrystallization, extraction, and chromatography. The greatest differences in polarity, which make for the simplest separations, are those that exist between salts and nonpolar organic compounds. Whenever one or more of the components of a mixture can be converted to a salt, it can be separated easily and efficiently from the nonpolar components by distillation (see Chap. 4) or extraction (see Chap. 5).

In an introduction to the techniques of qualitative organic analysis, you may be given a *mixture* of unknown compounds, each of which is to be identified. Before this can be done, each component of the mixture must be obtained in pure form. The following general approach illustrates how this may be accomplished using the principles of separation based on differences in polarity and functionality.

GENERAL SCHEME FOR SEPARATING SIMPLE MIXTURES OF WATER-INSOLUBLE COMPOUNDS

A procedure that is adequate for separating mixtures of liquid or solid carboxylic acids, phenols, amines, and neutral compounds is outlined schematically in Figure 23.1. For this scheme to be applicable, each component of the mixture must have a little solubility in water and must not undergo appreciable hydrolysis by reaction with dilute acids or bases at room temperature. The procedure is based primarily on partitioning compounds of significantly different polarities between diethyl ether and water and separating these liquid layers in a separatory funnel. The underlying concept of this process involves extraction, the theory of which is discussed in Section 5.2.

Assuming that a mixture contains a carboxylic acid, a phenol, an amine, and a neutral compound, each of which is water-soluble, the separation is initiated by dissolving the mixture in a suitable organic solvent such as diethyl ether. The ethereal solution of the mixture is first extracted with sodium bicarbonate solution, which removes the carboxylic acid by converting it to its water-soluble sodium salt. This extraction is followed by one with sodium hydroxide solution, which removes the water-soluble sodium salt of the phenol. Finally, the ethereal solution of the mixture is treated with hydrochloric acid, which reacts with the amine, converting it into a water-soluble ammonium salt. The ethereal solution that remains after removal of the aqueous solution contains the neutral compound.

Note that each extraction is performed on the same ethereal solution that originally contained all the components of the mixture, and the *sequence* of extraction, namely, $NaHCO_3$, then NaOH, and finally HCl, is extremely important. The bases and acid each remove one type of organic compound from the

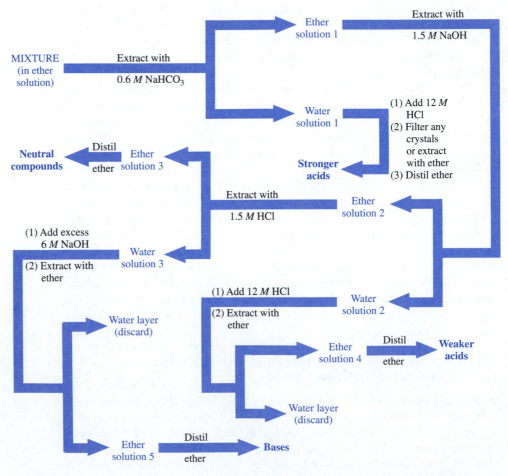

FIGURE 23.1 General scheme for separating a simple mixture of water-insoluble compounds.

mixture and leave the neutral compound in the other layer when the extractions have been completed. Each of the basic and acidic extracts is subsequently treated with acid or base to liberate the carboxylic acid, phenol, and amine from its salt; each of these compounds is then removed from the respective aqueous solutions by extraction with ether or by collecting the solid by vacuum filtration.

Many different layers and solutions are obtained in the experimental procedure that follows. This can lead to much confusion unless the flasks containing each liquid are clearly labeled as to the identity of their contents. Moreover, *it is prudent to retain the flasks containing all layers and solutions until it is certain that they are no longer needed.*

EXPERIMENTAL PROCEDURE

SAFETY ALERT

> Diethyl ether is removed from solutes after several extraction steps in this procedure. Exercise the usual precautions against fire, whether the ether is removed by distillation or by evaporation.

The following procedure, based on the scheme of Figure 23.1, may be used to separate the components of a mixture consisting of three or four compounds, 1–2 g of each, of the types described. It is important to understand that this generalized procedure may not give complete separation of all compounds of even the limited types for which it is intended. The separated products must be tested for purity by the usual methods of melting- or boiling-point determinations and, if possible, by GLC or TLC. Before attempting identification of the individual compounds by any of the classical or modern instrumental methods described later, the samples should be purified by recrystallization, distillation, or chromatography.

Stir or swirl the mixture (3–8 g) with 10 mL of diethyl ether at room temperature. If necessary, collect any solid material that does not dissolve by gravity filtration. Extract the ethereal solution with 4-mL portions of 0.6 M NaHCO$_3$ solution until the aqueous extract remains slightly basic. *Caution:* Vent the separatory funnel quickly after the first mixing because of possible build-up of pressure from carbon dioxide. This solution, **water solution 1** of Figure 23.1, should contain the sodium salt of any carboxylic acid present in the mixture; the ether layer, **ether solution 1,** should contain all other components of the mixture of unknowns. Regenerate the free organic acid by careful acidification of **water solution 1** to pH 2–3 with 12 M hydrochloric acid. If a solid acid separates, collect it by vacuum filtration. Otherwise, extract the aqueous acid solution with several 4-mL portions of diethyl ether, dry the ethereal solution over sodium sulfate, and remove the ether by distillation (*no flames*) or evaporation in the hood.

Extract **ether solution 1** with two 4-mL portions of 1.5 M NaOH solution to remove a phenol or any other weak acid from the solution and to give **ether solution 2.** Acidify the combined aqueous extracts, **water solution 2,** with 12 M hydrochloric acid, and extract the regenerated weak acid with several 4-mL portions of ether to yield **ether solution 4.** If **amphoteric** compounds were included in the unknown mixture, they would be carried through to the water layer separated from **ether solution 4.** Remove the ether from this solution as already described, and isolate the weak acid that remains as the residue.

Extract **ether solution 2** with one or more 4-mL portions of 1.5 *M* hydrochloric acid until the aqueous extract, **water solution 3,** remains acidic. The ether layer, designated as **ether solution 3** in Figure 23.1, should contain any neutral organic compounds; recover these by removing the ether as described previously.

Add sufficient 5 *M* NaOH solution to **water solution 3** to make the solution strongly basic; then extract it with several 4-mL portions of diethyl ether. The combined ether extracts, **ether solution 5,** should contain any organic base present in the original unknown mixture; isolate the base by removing the ether as described previously.

> **Finishing Touches**   Transfer any recovered *diethyl ether* in the container for nonhalogenated organic liquids. Neutralize all *basic or acidic aqueous solutions,* and flush them down the drain. Place the *filter paper* containing any solids that were not soluble in diethyl ether in the container for nonhazardous solids.

23.4 APPLICATIONS OF SPECTROSCOPIC METHODS TO QUALITATIVE ORGANIC ANALYSIS

A major limitation inherent to the classical system of qualitative organic analysis is that only *known* compounds can be identified. The research chemist is constantly faced with the task of identifying *new* substances. Although information about the type of compound can be derived from this classical approach, the complete identification of a new organic compound traditionally required a combination of degradation and synthesis in order to achieve a correlation with a known substance; this was usually a lengthy and laborious task. The advent of spectroscopic techniques changed this picture dramatically. New and unknown compounds may be identified quickly and with certainty by using a combination of spectroscopic methods such as those described in Chapter 8. A number of examples could be cited of structural elucidations that were completed in a matter of days or weeks for molecules of greater complexity than those of compounds that were the lifework of several of the great nineteenth-century and early twentieth-century organic chemists.

The modern approach to identifying organic compounds typically involves a combination of spectroscopic and classical methods. Ideally, students should be introduced to the application of spectroscopy in organic chemistry by using the instruments that produce the spectra, but this is not feasible in many instances because some of these instruments are very expensive. The next best alternative is to have access to the spectra of typical known compounds and then be provided with the spectra of "unknowns" to identify. This textbook contains nearly 300 IR, ^{1}H, and ^{13}C NMR, and UV spectra of starting materials and of the products obtained from various preparative experiments. A careful study of these spectra, aided by the material presented in Chapter 8 and by additional discussion by your instructor, should enable meaningful use of IR, NMR, and UV spectra in the identification of unknown organic compounds. Infrared and nuclear magnetic resonance spectroscopy are perhaps the most useful in this regard.

The spectral data serve to complement or supplement the "wet" qualitative classification tests and in many instances substitute for these tests entirely. For example, a strong IR absorption in the 1690–1760 cm^{-1} region is just as indicative of the presence of a carbonyl group as is formation of a 2,4-dinitrophenylhydrazone, and absorptions in the δ 6.0–8.5 region of ^1H NMR spectrum are a more reliable indication of the presence of an aromatic ring than a color test with $CHCl_3$ and $AlCl_3$.

It must be emphasized that spectroscopic analysis requires *careful interpretation of the data.* Although some problems can be solved quickly and uniquely by modern spectroscopy, others require the intelligent application of both the modern methods and the classical methods. The remainder of this section illustrates the use of spectroscopy in structure determination.

Because the primary use of IR spectroscopy is the identification of functional groups, the observation of certain IR absorptions provides information about the presence of particular functional groups in a compound. Conversely, the absence of certain peaks may be useful in *excluding* the possibility of specific functional groups. For example, the appearance of a strong IR band in the 1650–1760 cm^{-1} region is strong evidence that a substance contains the carbon-oxygen π-bond characteristic of an aldehyde, a ketone, a carboxylic acid, or a derivative of a carboxylic acid. On the other hand, the absence of such an absorption is interpreted to mean that the compound does not contain this functional group. Furthermore, a strong band in the 3500–3650 cm^{-1} region is indicative that a substance is an alcohol, a phenol, a carboxylic acid, a 1° or 2° amine or amide, whereas the complete absence of such an absorption excludes these functional groups from further consideration.

However, caution must be used in drawing conclusions based on the absence of expected absorption bands. As noted in Section 8.2, absorption of IR energy by a molecule requires a change in dipole moment as a result of the molecular vibration. This means that symmetrically substituted alkynes do *not* have an IR band for the stretching mode of the triple bond, and even disubstituted alkynes in which the two substituents are similar may have only a weak absorption in this region of the spectrum. For similar reasons, the absence of an absorption in the carbon-carbon double-bond stretching region does not necessarily exclude the presence of this functional group, because the nature of substitution at the double bond may render the functionality inactive in the IR spectrum. Finally, because certain structural features in a molecule cause small shifts in the expected absorptions of some functional groups, IR spectroscopy does not always provide a unique answer about the presence of a particular group. However, it is very useful in limiting the possibilities to a small number of functional groups, and the uncertainty may then be resolved by performing just a few qualitative tests in the laboratory.

As noted in Section 8.3, NMR spectra generally do not permit direct observation of specific functional groups, but they do provide indirect evidence regarding the presence or absence of some groups. The three major features of ^{1}H NMR spectra (see Sec. 8.3.3) that are useful in compound identification are the chemical shift, the splitting pattern, and the relative abundance of each type of hydrogen, as determined by the peak areas. The chemical shifts for carbon atoms in the ^{13}C NMR spectrum are also valuable for defining the nature of functional groups in unknown substances.

It may be desirable to obtain a UV spectrum of the compound, but this is normally done *only* after it has been determined that the molecule is likely to contain the structural features that lend themselves to analysis by this technique. Thus, UV spectroscopy (see Sec. 8.4) is mainly useful for studying conjugated systems such as 1,3-dienes, α,β-unsaturated carbonyl compounds, and aromatic compounds, and the IR and NMR spectra of the compound often provide clues about the presence of these structural features.

The research chemist usually obtains IR and NMR spectra of an unknown substance even before obtaining an elemental analysis or performing solubility tests because these types of spectra are easily and quickly measured on a small amount of sample. Only after analyzing them does the researcher undertake other experimental approaches for determining the structure. This process can save many hours of unnecessary laboratory work. Other useful information about a compound may be obtained from mass spectrometry, which provides the molecular weight of the compound and, if high-resolution data are available, its elemental composition. Elemental analysis gives the percentage composition by weight of the elements present and thus the empirical formula for the substance. If both the empirical formula and the molecular weight are known, this technique also provides the molecular formula of the compound.

An example that illustrates the value of the complementary application of classical qualitative analysis and spectral analysis for determining the structure of an unknown compound is given in the following paragraphs.

Example

The unknown is a liquid and the following information is available:

Preliminary analysis	Colorless liquid with pleasant odor
Physical constant	bp 143–145 °C
Elemental analysis	No X, S, or N
Solubility tests	Water-soluble to give a neutral solution
Preliminary report	Observations confirmed by instructor
Spectra	IR and ^{1}H NMR spectra are shown in Figure 23.2.
Classification tests	Positive test for ester, negative tests for all other functional groups

Analysis

Owing to its water-solubility, the compound most likely contains oxygen-bearing polar functional groups; it also probably contains a relatively small number of carbon atoms because compounds of more than five or six carbon atoms are usually water-insoluble. It is immediately evident from the IR spectrum shown in Figure 23.2 that the liquid contains an ester group because of the strong absorption peaks at about 1750 and 1230 cm^{-1}; the former peak is due to C–O stretching and the latter to the C–O–C bond (see Table 8.2). Although the 1750 cm^{-1} band could indicate a five-membered cyclic ketone, this and other possibilities are negated by the presence of numerous peaks in the ^{1}H NMR spectrum. The student confirmed the presence of an ester by obtaining a positive qualitative test for this group (see Sec. 23.5.10).

The ^{1}H NMR spectrum has four different resonances: a multiplet centered at δ 4.1, a second multiplet centered at δ 3.5, a singlet at δ 3.3, and another singlet at δ 2.0. Measurement of the height, in mm, of the integration steps above each absorption gives values of 16:16:24:24, respectively, which corresponds to a relative hydrogen abundance of 2:2:3:3 for the four resonances. This means that *at least* four chemically different carbon atoms must be present in the molecule. In fact, reference to the ^{13}C NMR data shows that *five* chemically distinct carbon atoms are contained in the unknown. The *relative* abundances of different types of protons are likely to correspond to the *absolute* ones, because the water-solubility of the unknown signals that the compound is likely to contain fewer than seven carbon atoms. Consequently, the two singlets at δ 2.0 and 3.3, with relative integrations of three, must represent two rather than some larger number of methyl groups, and the two groups have no nearest neighbors. The higher-field peak at δ 2.0 is characteristic of a methyl ketone or acetate (see Table 8.4), the

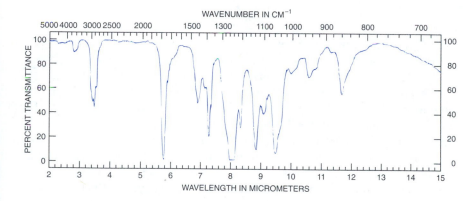

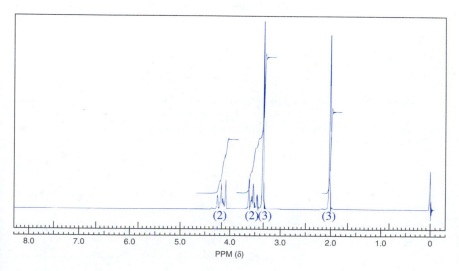

FIGURE 23.2 IR and ^{1}H NMR spectra for unknown compound in the Example.

latter being consistent with the deduction provided by the IR spectrum and the qualitative test that the unknown is an ester. The second methyl peak at δ 3.3 is shifted downfield, as would be expected if the methyl group were bonded to oxygen. This peak is too far upfield for a methyl ester, because methyl esters normally show the methyl absorption at about δ 3.7–4.1. However, it is within the δ 3.3–4.0 range frequently observed for aliphatic α-hydrogens of an alcohol or ether. Because neither spectrum shows any evidence for an –OH group, the compound probably contains a CH_3O– group, and the presence of an aliphatic ether is consistent with the C–O–C absorption observed at 1050 cm^{-1} in the IR spectrum. The multiplets centered at δ 4.1 and 3.5 each integrate for two hydrogens and show mutual spin-spin coupling. This pattern is diagnostic of chemically nonequivalent, adjacent methylene groups such as X–CH_2CH_2–Y. The downfield absorption for the CH_2 groups indicates that each is bonded to oxygen.

By reference to Table 23.13(a) for liquid esters, only a single structure, *2-methoxyethyl acetate* (**1,** bp 145 °C, 760 torr), appears to be consistent with all of the spectral observations.

$$CH_3O-CH_2CH_2-O-\overset{\overset{\textstyle O}{\|}}{C}-CH_3$$

1

This structural assignment was confirmed by hydrolyzing the ester and converting the resulting alcohol into an α-naphthyl urethane derivative (see Sec. 23.5.5D), which exhibited the correct melting point (see Table 23.3a).

Suppose that the elemental percentage composition by weight for the unknown, 50.83% C, 8.53% H, and 40.63% O, had also been provided. The empirical formula could then be determined as follows. A 100-g sample of the liquid would contain 50.83 g of oxygen, 8.53 g of hydrogen, and 40.63 g of oxygen. The weights of each element are first converted into moles; then the number of moles of each element is divided by the number of moles of that element present in the smallest amount. This provides a molar ratio, which is often a whole-number ratio; if it is not, the ratio must be multiplied by a number that gives whole numbers. In this example, oxygen is present in the smallest molar amount, so the number of moles of each other element is divided by 2.54. These calculations are summarized below.

Element	Weight (g)	Conversion to Moles	Conversion to Molar Ratio
C	50.83	$\dfrac{50.83\ g}{12.01\ g/mol} = 4.23\ mol$	$\dfrac{4.23}{2.54} = 1.67\ \text{or}\ 5$
H	8.53	$\dfrac{8.53\ g}{1.01\ g/mol} = 8.54\ mol$	$\dfrac{8.54}{2.54} = 3.33\ \text{or}\ 10$
O	40.63	$\dfrac{40.63\ g}{16.00\ g/mol} = 2.54\ mol$	$\dfrac{2.54}{2.54} = 1.00\ \text{or}\ 3$

This example of proving the structure of an unknown should provide an appreciation of the complementary aspects of spectral analysis and the "wet" classification scheme, including the use of physical properties. Although the spectra were not in themselves specifically definitive in the assignment of the structures of the unknown, they provided data regarding the possible presence of certain functional groups, which could be confirmed by performing the appropriate classification tests. Thus, the spectral analyses helped minimize the effort and time spent in performing numerous classification tests, most of which would have been negative. The final decision regarding the structure of the unknown came from preparing a solid derivative whose melting could be compared with that of the known compound. Of course, when solving the structure of an unknown, it is often possible to find the IR and NMR spectra of the suspected compounds in one of the catalogues of spectra given in the references in Chapter 8. If the reported spectra are *identical* to those obtained for the unknown, the identity of the unknown is confirmed.

REFERENCES

1. Shriner, R. L.; Fuson, R. C.; Curtin, D. Y.; Morrill, T. C. *The Systematic Identification of Organic Compounds,* 6th ed., John Wiley & Sons, New York, 1980.
2. *Handbook of Tables for Organic Compound Identification,* 3rd ed., Z. Rapoport, Ed., Chemical Rubber Company, Cleveland, OH, 1967. Provides physical properties and derivatives for more than 4000 compounds, arranged according to functional groups.

23.5 QUALITATIVE CLASSIFICATION TESTS AND PREPARATION OF DERIVATIVES

This section contains discussions about and experimental procedures for the qualitative classification tests and preparation of derivatives for compounds having most of the commonly encountered functional groups. References are also provided for the tables of compounds in Section 23.6, which list boiling points and melting points of many compounds and the melting points of suitable derivatives.

23.5.1 ALDEHYDES AND KETONES

Listings of aldehydes and ketones and their derivatives appear in Tables 23.4 and 23.15, respectively, of Section 23.6.

Classification Tests

A. 2,4-Dinitrophenylhydrazine. Arylhydrazines ($ArNHNH_2$), which include phenylhydrazine, *p*-nitrophenylhydrazine, and 2,4-dinitrophenylhydrazine, are commonly used to make crystalline derivatives of carbonyl compounds (see Sec. 18.1.1). The formation of a 2,4-dinitrophenylhydrazone **3** is represented by Equation (23.9). Since the products of such tests may be used as derivatives of the aldehyde or ketone, the arylhydrazines are valuable reagents for both classifying and forming derivatives; 2,4-dinitrophenylhydrazine is particularly useful in this regard.

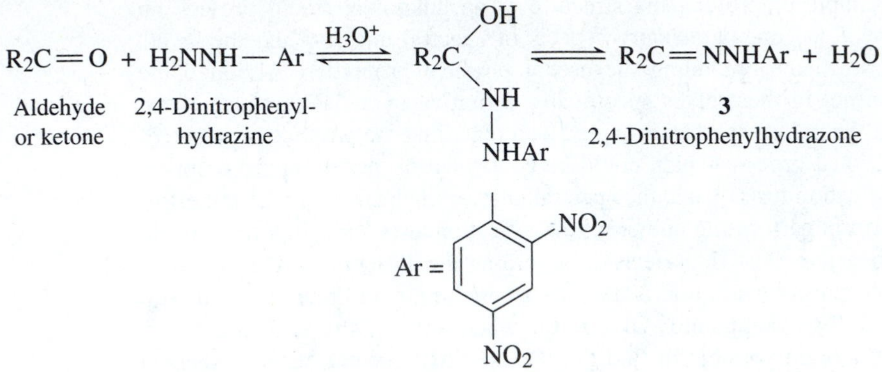

$$R_2C{=}O \ + \ H_2NNH{-}Ar \ \underset{}{\overset{H_3O^+}{\rightleftharpoons}} \ R_2C\overset{OH}{\underset{\underset{NHAr}{\overset{|}{NH}}}{\diagup}} \ \rightleftharpoons \ R_2C{=}NNHAr \ + \ H_2O \qquad \textbf{(23.9)}$$

Aldehyde   2,4-Dinitrophenyl-   **3**
or ketone   hydrazine   2,4-Dinitrophenylhydrazone

$$Ar = \underset{NO_2}{\overset{NO_2}{\bigcirc\!\!\!-\!\!\!-}}$$

The arylhydrazines will give a positive test for *either* an aldehyde *or* a ketone. Tollens' test given in Part B provides a method for distinguishing between these two types of compounds.

EXPERIMENTAL PROCEDURE

If the test reagent is not supplied, prepare it by dissolving 0.2 g of 2,4-dinitrophenylhydrazine in 1 mL of concentrated sulfuric acid. Add this solution, with stirring, to 1.5 mL of water and 5 mL of 95% ethanol. Stir the solution vigorously and then filter it to remove any undissolved solids.

Dissolve 1 or 2 drops of a liquid, or about 100 mg of a solid, in 2 mL of 95% ethanol, and add this solution to 2 mL of the 2,4-dinitrophenylhydrazine reagent. Shake the mixture vigorously; if a precipitate does not form immediately, let the solution stand for 15 min.

If more crystals are desired for a melting-point determination, dissolve 200–500 mg of the carbonyl compound in 20 mL of 95% ethanol, and add this solution to 15 mL of the reagent. Recrystallize the product from aqueous ethanol.

> **Finishing Touches**   Neutralize and then filter any excess *2,4-dinitrophenylhydrazine solution*. Put the filter cake in the container for nonhazardous solids. Flush the *filtrate*, as well as *filtrates* obtained from recrystallization, down the drain.

B. Tollens' Test. A method for distinguishing between aldehydes and ketones is Tollens' test. A positive test indicates the presence of an aldehyde function, whereas no reaction occurs with ketones. Tollens' reagent consists of silver-ammonia complex, $Ag(NH_3)_2^{\oplus}$, in an ammonia solution. This reagent oxidizes both aliphatic and aromatic aldehydes to the corresponding carboxylic acids; silver ion is reduced to elemental silver, which is deposited as a silver mirror on the glass wall of a *clean* test tube. Thus, the formation of the silver mirror or of a precipitate is considered a positive test. Equation (23.10) shows the reaction that occurs.

$$RCHO \; + \; 2\,Ag(NH_3)_2^{\oplus} + \; 2\,HO^{\ominus} \longrightarrow 2\,\underline{Ag} \; + \; RCO_2^{\ominus}NH_4^{\oplus} + \; H_2O \; + 3\,NH_3 \qquad \textbf{(23.10)}$$

Aldehyde $\qquad\qquad\qquad\qquad\qquad$ Metallic silver

Similar tests for aldehydes make use of Fehling's and Benedict's reagents, which contain complex tartrate and citrate salts, respectively, of cupric ion as the oxidizing agents. With these reagents a positive test is the formation of a brick-red precipitate of cuprous oxide (Cu_2O), which forms when Cu^{2+} is reduced to Cu^+ by the aldehyde. These two tests are more useful in distinguishing between aliphatic and aromatic aldehydes, because the aliphatic compounds give a faster reaction. They have been used widely to detect reducing sugars (see Chap. 22), whereas Tollens' test is used to distinguish between aldehydes, both aliphatic and aromatic, and ketones.

EXPERIMENTAL PROCEDURE

SAFETY ALERT

1. Avoid spilling Tollens' reagent on your skin. If you do come into contact with the solution, wash the affected area immediately with copious amounts of water.

2. *Do not store unused Tollens' reagent because it decomposes on standing and yields an explosive precipitate.* Follow the directions under **FINISHING TOUCHES** for proper disposal of the solution.

Tollens' reagent must be prepared at the time of use by combining two other solutions, *A* and *B,* which should be available as stock solutions. If not, make up solution *A* by dissolving 0.25 g of silver nitrate in 4.3 mL of distilled water and solution *B* by dissolving 0.3 g of potassium hydroxide in 4.2 mL of distilled water.

Prepare the test reagent according to the following directions. Add concentrated ammonium hydroxide solution dropwise to 3 mL of solution *A* until the initial brown precipitate begins to clear. The solution should be grayish and almost clear. Then add 3 mL of solution *B* to this mixture, and again add concentrated ammonium hydroxide dropwise until the solution is almost clear.

To carry out the test, add 0.5 mL of the reagent to 3 drops or 50–100 mg of the unknown compound; the formation of a silver mirror or black precipitate constitutes a positive test. The silver deposits in the form of a mirror only on a *clean* glass surface. A black precipitate, although not as aesthetically pleasing, still constitutes a positive test. If no reaction occurs at room temperature, warm the solution slightly in a beaker of warm water.

Finishing Touches   Remove the *silver mirrors* from the test tube with nitric acid, and pour the resulting *solution* into a beaker containing unused *Tollens' reagent, ammonium hydroxide,* and *sodium hydroxide.* Acidify this mixture with nitric acid to destroy the unreacted Tollens' reagent. Neutralize the solution with sodium carbonate and add saturated sodium chloride solution to precipitate silver chloride. Collect the silver chloride by vacuum filtration and place the *filter cake* in a container for recovered silver halides; flush the filtrate down the drain.

C. Chromic Acid Test.   Another method for distinguishing between aldehydes and ketones is the chromic acid test; the test reagent is prepared by dissolving chromic anhydride in sulfuric acid and used in acetone solution. This reagent oxidizes primary and secondary alcohols and all aldehydes, and a distinctive color change occurs (see Sec. 16.2), but it gives no visible reaction with tertiary alcohols and ketones under the test conditions. The reactions are shown in Equation (23.11).

$$RCH_2OH \xrightarrow[\text{(orange-red)}]{H_2CrO_4} RCHO \xrightarrow{H_2CrO_4} RCO_2H + \text{green solution} \quad \textbf{(23.11a)}$$
$$\text{or precipitate}$$

1° Alcohol

$$R_2CHOH \xrightarrow{H_2CrO_4} R_2CO + \text{green solution} \xrightarrow{H_2CrO_4} \text{no visible reaction} \quad \textbf{(23.11b)}$$
$$\text{or precipitate}$$

2° Alcohol

$$R_3COH \xrightarrow{H_2CrO_4} \text{no visible reaction} \quad \textbf{(23.11c)}$$

3° Alcohol

Thus, the chromic acid reagent gives a clear-cut distinction between primary and secondary alcohols and aldehydes on the one hand and tertiary alcohols and ketones on the other. Aldehydes may be distinguished from primary and secondary alcohols by means of Tollens', Benedict's, or Fehling's test, and lower-molecular-weight primary and secondary alcohols may be differentiated on the basis of their rates of reaction with concentrated hydrochloric acid containing zinc chloride—the Lucas reagent (see Sec. 23.5.5B).

EXPERIMENTAL PROCEDURE

SAFETY ALERT

1. The preparation of chromic acid requires diluting a paste of chromic anhydride and concentrated sulfuric acid with water. Be certain to add the water *slowly* to the acid and swirl the mixture to ensure continuous mixing. Swirling keeps the denser sulfuric acid from layering at the bottom of the flask and avoids possible splattering because of the heat generated when the two layers are suddenly mixed.

2. When preparing and handling solutions of chromic acid, wear rubber gloves to keep the acids from contacting your skin. Chromic acid causes unsightly stains on your hands for several days, and may cause severe chemical burns. Wash the affected area thoroughly with warm water if the oxidant comes in contact with the skin. Also rinse the area with 5% sodium bicarbonate solution.

If the chromic acid reagent is not available, prepare it as follows. Add 1 g of chromic anhydride (CrO_3) to 1 mL of concentrated sulfuric acid and stir the mixture until a smooth paste is obtained. Dilute the paste *cautiously* with 3 mL of distilled water, and stir this mixture until a clear orange solution is obtained.

Dissolve 1 drop of a liquid or about 10 mg of a solid alcohol or carbonyl compound in 1 mL of *reagent grade* acetone. Add 1 drop of the acidic chromic anhydride reagent to the acetone solution, and shake the tube to mix the contents. A positive oxidation reaction is indicated by disappearance of the orange color of the reagent and the formation of a green or blue-green precipitate or emulsion.

Primary and secondary alcohols and aliphatic aldehydes give a positive test within 5 sec. Aromatic aldehydes require 30 to 45 sec. Color changes occurring after about 1 min should not be interpreted as positive tests; other functional groups such as ethers and esters may slowly hydrolyze under the conditions of the test, releasing alcohols that in turn provide "false-positive" tests. Tertiary alcohols and ketones produce no visible change in several minutes. Phenols and aromatic amines give dark precipitates, as do aromatic aldehydes having hydroxyl or amino groups on the aromatic ring.

Finishing Touches Add sodium sulfite to the *aqueous solution* of chromium salts in order to destroy excess Cr^{VI}. Make the solution slightly basic, to form chromium hydroxide, and isolate this salt by vacuum filtration. Place the *filter paper* and the *filter cake* in the container for heavy metals; flush the *filtrate* down the drain.

D. Iodoform Test. When an aldehyde or ketone that has α-hydrogens is treated with a halogen in basic medium, halogenation occurs on the α-carbon (Eq. 23.12). This reaction involves the formation of an enolate ion (see Eq. 18.11),

$$R_2C \overset{\displaystyle |}{\underset{\displaystyle H}{-}} C \overset{\displaystyle O}{\underset{\displaystyle R}{\diagup}} \xrightarrow{\text{Base, } X_2} R_2C \overset{\displaystyle |}{\underset{\displaystyle X}{-}} C \overset{\displaystyle O}{\underset{\displaystyle R}{\diagup}} \tag{23.12}$$

Aldehyde or
ketone

which subsequently reacts with halogen to produce the substitution product. In the case of acetaldehyde or a methyl ketone, all three of the α-hydrogens of the methyl group are replaced by halogen to give **4** (Eq. 23.13), which then reacts with excess base to give products **5** and **6** (Eq. 23.14).

$$H \overset{\displaystyle H}{\underset{\displaystyle H}{-}} C \overset{\displaystyle O}{\underset{\displaystyle R}{\diagup}} \xrightarrow{\text{NaOH, } X_2} X \overset{\displaystyle X}{\underset{\displaystyle X}{-}} C \overset{\displaystyle O}{\underset{\displaystyle R}{\diagup}} \tag{23.13}$$

Aldehyde or
ketone   **4**

$$\underset{X_3C \qquad R}{\overset{\displaystyle O}{\overset{\displaystyle \|}{C}}} \xrightarrow{\text{NaOH}} CHX_3 + RCO_2^{\ominus} \, Na^{\oplus} \tag{23.14}$$

4   **5**   **6**
Haloform

Although chlorine, bromine, and iodine all react in this manner, the qualitative test for a methyl ketone utilizes iodine because it is safer to use and because the product is *iodoform,* CHI_3, a highly *insoluble* crystalline *yellow* solid that is readily observed and identified on the basis of its characteristic odor and melting point. It should be pointed out that a positive iodoform test involves two experimental observations: (1) the disappearance of the characteristic red-brown color of iodine as the test reagent is added to the compound being tested, *and* (2) the formation of a yellow precipitate of iodoform. The reason both observations are required is that *all* aldehydes and ketones containing α-hydrogens react with iodine in base to decolorize the test reagent. Only methyl ketones, after being trisubstituted with iodine on the methyl group, react with base to produce iodoform and the salt of a carboxylic acid.

A positive test also occurs for alcohols having a hydroxymethyl (–CHOHCH₃) functionality, as in ethanol and isopropyl alcohol, and for acetaldehyde, which is analogous to a methyl ketone in having three α-hydrogens. The reason that a hydroxymethyl compound gives a positive iodoform test is that the combination of iodine and sodium hydroxide produces sodium hypoiodite (Eq. 23.15) which, like sodium hypochlorite (Eq. 16.31, X = Cl), is a mild oxidant. Consequently, the test reagent is able to oxidize a hydroxymethyl function to a methyl ketone, and the methyl ketone then reacts with iodine in the presence of base to give iodoform. It is thus always necessary to consider this possibility before making a final decision regarding the functional group responsible for a positive test. The ambiguity in interpreting this test is resolved by determining whether a keto group is present or absent by using another qualitative test, such as that involving 2,4-dinitrophenylhydrazine (Eq. 23.9).

$$I_2 \ + \ 2\,NaOH \ \longrightarrow \ NaI \ + \ NaOI \ + \ H_2O \qquad \textbf{(23.15)}$$

The iodoform test is sometimes called the **hypoiodite test** because sodium hypoiodite is formed according to Equation 23.15. In more general terms, the reactions just described can be called the **haloform test** or the **sodium hypohalite test** when the specific halogen is not specified.

EXPERIMENTAL PROCEDURE

If the iodine reagent is not available as a stock solution, prepare it by dissolving 1 g of iodine in a solution of 2 g of potassium iodide in 8 mL of water. The potassium iodide is added to increase the solubility of iodine in water by formation of potassium triiodide, KI_3 (Eq. 23.16).

$$I_2 \ + \ KI \ \longrightarrow \ KI_3 \qquad \textbf{(23.16)}$$

If the substance is water-soluble, dissolve 2 to 3 drops of a liquid or an estimated 50 mg of a solid in 2 mL of water in a small test tube, add 2 mL of 3 M sodium hydroxide, and then slowly add 3 mL of iodine solution. In a positive test the brown color disappears, and yellow iodoform separates. If the substance tested is insoluble in water, dissolve it in 2 mL of dioxane, proceed as above, and dilute the mixture with 10 mL of water *after* addition of the iodine solution.

Iodoform is recognizable by its odor and yellow color and, more definitively, by its melting point, 118–119 °C. Isolate this product either by vacuum filtration of the test mixture or by the following sequence: Add 2 mL of dichloromethane to the mixture, shake the stoppered test tube to extract the iodoform into the small lower layer, withdraw the clear part of this layer with a Pasteur pipet, and evaporate it in a small test tube or on a watchglass with a steam bath in the hood. Whichever mode of isolation is used, recrystallize the crude solid from methanol-water.

Finishing Touches   Place the residual *aqueous solutions* in the container for halogenated liquids. Put the *iodoform* in the container for halogenated solids. Flush the *filtrate* down the drain.

Derivatives

Two of the most useful solid derivatives of aldehydes and ketones are the **2,4-dinitrophenylhydrazone** and the **semicarbazone**. The **oxime** is also sometimes useful, but it often forms as an oil rather than a solid.

E. 2,4-Dinitrophenylhydrazone. The qualitative test for aldehydes and ketones is described in Section 23.5.1A, and the solid that forms may be isolated and purified as indicated in that section.

F. Semicarbazone. Semicarbazide reacts with aldehydes and ketones to produce a derivative that is called a semicarbazone (Eq. 23.17). Because semicarbazide is unstable as the free base, it is usually stored in the form of its hydrochloric acid salt. In the procedure that follows, the free base is liberated from the salt by addition of sodium acetate.

$$
\underset{\substack{\text{An aldehyde}\\\text{or ketone}}}{\overset{\displaystyle O\atop\displaystyle\|}{\underset{R^1 \quad R^2}{C}}} \;+\; \underset{\text{Semicarbazide}}{H_2N-NHCNH_2} \xrightarrow{H^+} \underset{\text{A semicarbazone}}{\overset{\displaystyle O\atop\displaystyle\|\atop\displaystyle NNHCNH_2}{\underset{R^1 \quad R^2}{C}}} \;+\; H_2O \qquad (23.17)
$$

EXPERIMENTAL PROCEDURE

Dissolve 0.5 g of semicarbazide hydrochloride and 0.8 g of sodium acetate in 5 mL of water in a test tube, and then add about 0.5 mL of the carbonyl compound. Stopper and shake the tube vigorously, remove the stopper, and place the test tube in a beaker of boiling water. Discontinue heating the water, and allow the test tube to cool to room temperature in the beaker of water. Transfer the test tube to an ice-water bath, and if crystals have not formed, scratch the side of the tube with a glass rod at the interface between the liquid and air to induce crystallization. Recrystallize the semicarbazone from water or aqueous ethanol.

If the carbonyl compound is insoluble in water, dissolve it in 5 mL of ethanol. Add water until the solution becomes turbid, then add a little ethanol until the turbidity disappears. Add the semicarbazide hydrochloride and sodium acetate, and continue as above from this point.

> **Finishing Touches**   Flush *excess reagent* and *filtrates* down the drain.

G. Oximes. Hydroxylamine reacts with aldehydes or ketones to yield an oxime (Eq. 23.18). Hydroxylamine is usually stored as the hydrochloric acid salt because it is not stable as the free base; it is liberated from its salt by addition of base. The use of an oxime as a derivative has the limitation that such compounds frequently are *not* crystalline solids.

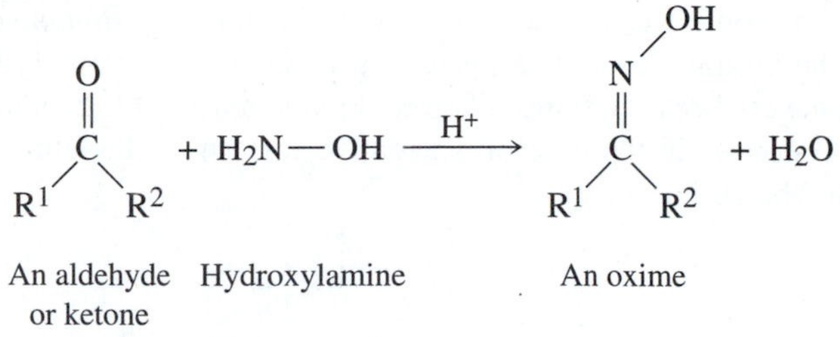

$$\underset{\substack{R^1 \\ \text{An aldehyde} \\ \text{or ketone}}}{\overset{O}{\underset{}{\parallel}}\underset{R^2}{C}} + \underset{\text{Hydroxylamine}}{H_2N-OH} \xrightarrow{H^+} \underset{\substack{R^1 \\ \text{An oxime}}}{\overset{\substack{OH \\ N \\ \parallel}}{\underset{R^2}{C}}} + H_2O \qquad \textbf{(23.18)}$$

EXPERIMENTAL PROCEDURE

Dissolve 0.5 g of hydroxylamine hydrochloride in 5 mL of water and 3 mL of 3 *M* sodium hydroxide solution, and add 0.5 g of the aldehyde or ketone. If the carbonyl compound is insoluble in water, add just enough ethanol to give a clear solution. Warm the mixture on a steam bath or boiling-water bath for 10 min; then cool it in an ice-water bath. If crystals do not form immediately, scratch the side of the tube with a glass rod at the air-liquid interface to induce crystallization. Recrystallize the oxime from water or aqueous ethanol.

In some cases the use of 3 mL of pyridine and 3 mL of absolute ethanol in place of the 3 mL of 3 *M* sodium hydroxide solution and 5 mL of water is more effective. A longer heating period is often necessary. After the heating is finished, pour the mixture into an evaporating dish and remove the solvent with a current of air in a hood. Grind the solid residue with 3–4 mL of cold water and filter the mixture. Recrystallize the oxime from water or aqueous ethanol.

Finishing Touches Flush *excess reagent* and *filtrates* down the drain.

23.5.2 ALKENES AND ALKYNES

A list of alkenes and alkynes is found in Table 23.6 of Section 23.6.

Classification Tests

Two common types of unsaturated compounds are alkenes and alkynes, characterized by the carbon-carbon double and triple bonds, respectively, as the functional group. The two common qualitative tests for unsaturation are the reaction of the compounds with *bromine in carbon tetrachloride* and with *potassium permanganate*. In both cases a positive test is denoted by decoloration of the reagent. There are no simple direct ways to prepare solid derivatives of unsaturated aliphatic compounds having no other functional groups.

A. Bromine in Carbon Tetrachloride. Bromine adds to the carbon-carbon double bond of alkenes to produce dibromoalkanes (Eq. 23.19) and reacts with alkynes to produce tetrabromoalkanes (Eq. 23.20). When this reaction occurs, molecular bromine is rapidly consumed, and its characteristic dark red-brown color disappears if bromine is not added in excess. The nearly *instantaneous* disappearance of the bromine color is a positive test for unsaturation. The test is ambiguous, however, because some alkenes do not react with bromine, and some react very slowly. In the case of a negative test, the potassium permanganate test should be performed.

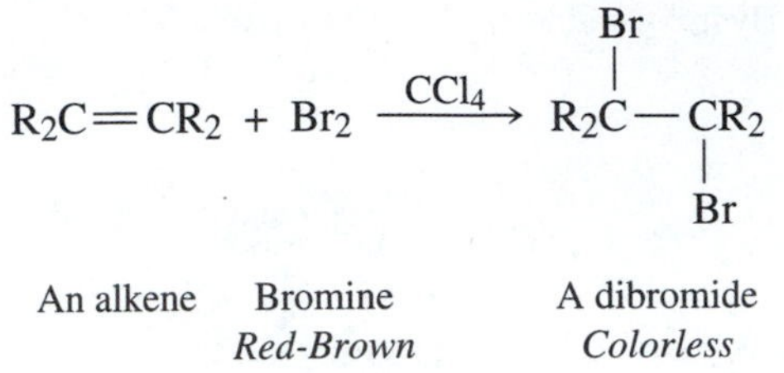

$$R_2C{=}CR_2 + Br_2 \xrightarrow{\text{CCl}_4} \overset{\displaystyle Br}{\underset{\displaystyle Br}{R_2C{-}CR_2}} \qquad \text{(23.19)}$$

An alkene   Bromine   A dibromide
Red-Brown   *Colorless*

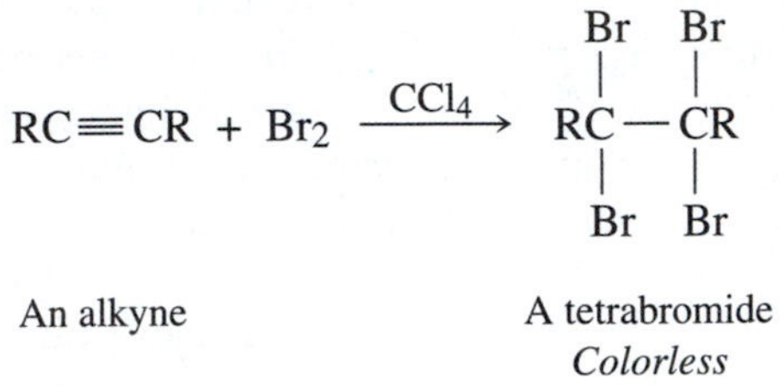

$$RC{\equiv}CR + Br_2 \xrightarrow{\text{CCl}_4} \overset{\displaystyle Br \quad Br}{\underset{\displaystyle Br \quad Br}{RC{-}CR}} \qquad \text{(23.20)}$$

An alkyne   A tetrabromide
Colorless

More details about the reaction of bromine in carbon tetrachloride with alkenes are found in Section 10.3.

EXPERIMENTAL PROCEDURE

SAFETY ALERT

Bromine is a hazardous chemical. Do not breathe its vapors or allow it to come into contact with your skin because it may cause serious chemical burns. Perform all operations involving the transfer of solutions of bromine at a hood; wear rubber gloves. If you get bromine on your skin, wash the area quickly with warm water and soak the area in 0.6 M sodium thiosulfate solution for up to 3 h if the burn is particularly serious.

Add 1 or 2 drops of the unknown to 1 or 2 mL of 0.1 M bromine in carbon tetrachloride solution. *Rapid* disappearance of the bromine color to give a colorless solution is a positive test for unsaturation.

Finishing Touches   Decolorize any *solutions* in which the color of *bromine* is visible by dropwise addition of cyclohexene; then discard the resulting mixtures in the container for halogenated organic liquids.

B. Potassium Permanganate. A second qualitative test for unsaturation, the Baeyer test, depends on the ability of potassium permanganate to oxidize the carbon-carbon double bond to give alkanediols (Eq. 23.21) or the carbon-carbon triple bond to yield carboxylic acids (Eq. 23.22). The permanganate is destroyed in the reaction, and a brown precipitate of MnO_2 is produced. The disappearance of the characteristic purple color of the permanganate ion is a positive test for unsaturation. However, care is needed in interpreting this test, because compounds containing certain other functional groups, such as aldehydes, also decolorize permanganate ion because they undergo oxidation.

$$R_2C{=}CR_2 \; + \; MnO_4^{\ominus} \; \xrightarrow{H_2O} \; R_2\overset{\overset{\textstyle OH}{|}}{C}{-}\overset{\overset{\textstyle OH}{|}}{C}R_2 \; + \; \underline{MnO_2} \qquad \textbf{(23.21)}$$

An alkene   Permanganate   A glycol Manganese dioxide
             Purple   *Colorless*   *Brown*

$$R^1C{\equiv}CR^2 \; + \; MnO_4^{\ominus} \; \xrightarrow{H_2O} \; R^1CO_2H \; + \; R^2CO_2H \; + \; \underline{MnO_2} \qquad \textbf{(23.22)}$$

An alkyne   Permanganate   Carboxylic acids Manganese dioxide
            Purple   *Colorless*   *Brown*

SAFETY ALERT

Wear rubber gloves to prevent solutions of potassium permanganate from contacting your skin. These solutions cause unsightly stains on your hands for several days. If this oxidant comes in contact with the skin, wash the affected area thoroughly with warm water.

Dissolve 1 or 2 drops of the unknown in 2 mL of 95% ethanol, and then add 0.1 *M* KMnO$_4$ solution dropwise; note the results. Count the number of drops added before the permanganate color persists. For a **blank determination,** count the number of drops of aqueous permanganate that can be added to 2 mL of 95% ethanol before the color persists. A significant difference in the number of drops required in the two cases is a positive test for unsaturation.

Finishing Touches   Filter all solutions to remove manganese dioxide. Place the *filter paper* and the *filter cake* containing manganese salts in the container for heavy metals. Flush all *filtrates* down the drain.

23.5.3 Alkyl Halides

A list of alkyl halides appears in Table 23.7 of Section 23.6.

Classification Tests

Qualitative tests for alkyl halides are useful in deciding whether the compound in question is a primary, secondary, or tertiary halide. In general it is quite difficult to prepare solid derivatives of alkyl halides, so this discussion is limited to two qualitative tests: (a) reaction with **alcoholic silver nitrate** solution, and (b) reaction with **sodium iodide in acetone.**

A. Alcoholic Silver Nitrate. If a compound is known to contain bromine, chlorine, or iodine, information concerning the environment of the halogen is obtained from reacting the substance with alcoholic silver nitrate. The overall reaction is shown in Equation (23.23).

$$\text{RX} + \text{AgNO}_3 \xrightarrow{\text{ethanol}} \underline{\text{AgX}} + \text{RONO}_2 \qquad \textbf{(23.23)}$$

| An alkyl | Silver | A silver | An alkyl |
| halide | nitrate | halide | nitrate |

Such a reaction mechanistically will be of the S_N1 type. As asserted in Section 14.1, tertiary halides are more reactive in an S_N1 reaction than are secondary halides, which are in turn more reactive than primary halides. Thus, the rate of precipitation of silver halide in this test is expected to correspond to the ease of ionization of the alkyl halide to a carbocation, that is, primary < secondary < tertiary. From a practical standpoint, these differences are best determined by testing authentic samples of primary, secondary, and tertiary halides with silver nitrate in separate test tubes and comparing the results.

Alkyl bromides and iodides react more rapidly than chlorides, and the latter may even require warming to produce a reaction in a reasonable time-period.

Aryl halides are *unreactive* toward the test reagent, as in general are any vinyl or alkynyl halides. Allylic and benzylic halides, even when primary, show reactivities as great as or greater than tertiary halides because of resonance stabilization of the resulting allyl or benzyl carbocations.

Experimental Procedure

Add 1 drop of the alkyl halide to 2 mL of a 0.1 *M* solution of silver nitrate in 95% ethanol. If no reaction is observed within 5 min at room temperature, warm the mixture in a beaker of boiling water and observe any change. Note the color of any precipitates: Silver chloride is white, but turns purple on exposure to light, silver bromide is pale yellow, and silver iodide is yellow. If there is any precipitate, add several drops of 1 *M* nitric acid solution to it, and note any changes; the silver halides are insoluble in acid. To determine expected reactivities, test known primary, secondary, and tertiary halides in this manner. If possible, use alkyl iodides, bromides, and chlorides so that differences in halogen reactivity can also be observed.

> **Finishing Touches** Neutralize and then filter all *solutions* and place the *filter cake* in a container for recovered silver halides. Flush the *filtrate* down the drain.

B. Sodium Iodide in Acetone. Another method for distinguishing between primary, secondary, and tertiary halides makes use of sodium iodide dissolved in acetone. This test complements the alcoholic silver nitrate test, and when these two tests are used together, it is possible to determine the gross structure of the attached alkyl group with reasonable accuracy.

The basis of this test is that both sodium chloride and sodium bromide are *not* very soluble in acetone, whereas sodium iodide is. Mechanistically, the reactions occurring are S_N2 substitutions in which iodide ion is the nucleophile (Eq. 23.24 and 23.25); the order of reactivity is primary > secondary > tertiary.

$$RCl + NaI \xrightarrow{\text{acetone}} RI + \underline{NaCl} \qquad \text{(23.24)}$$

$$RBr + NaI \xrightarrow{\text{acetone}} RI + \underline{NaBr} \qquad \text{(23.25)}$$

With the test reagent, *primary bromides* give a precipitate of sodium bromide in about 3 min at room temperature, whereas the *primary* and *secondary chlorides* must be heated to about 50 °C before reaction occurs. *Secondary* and *tertiary bromides* react at 50 °C, but the *tertiary chlorides* fail to react in a reasonable time. This test is necessarily limited to bromides and chlorides, of course.

EXPERIMENTAL PROCEDURE

Place 1 mL of the sodium iodide–acetone test solution in a test tube, and add 2 drops of the chloro or bromo compound. If the compound is a solid, dissolve about 50 mg of it in a minimum volume of acetone and add this solution *to* the reagent. Shake the test tube, and allow it to stand for 3 min at room temperature. Note whether a precipitate forms; if no change occurs after 3 min, warm the mixture in a beaker of water at 50 °C. After 6 min of heating, cool the solution to room temperature and note whether any precipitate forms. Occasionally a precipitate forms immediately after combination of the reagents; this represents a positive test *only* if the precipitate remains after the mixture is shaken and allowed to stand for 3 min.

Carry out this reaction with a series of primary, secondary, and tertiary chlorides and bromides. Note in all cases the differences in reactivity as evidenced by the rate of formation of sodium bromide or sodium chloride.

> **Finishing Touches**　Place all *solutions* and *solids* involved in this test in the container for halogenated liquids, unless instructed to do otherwise.

23.5.4　AROMATIC HYDROCARBONS AND ARYL HALIDES

Listings of aromatic hydrocarbons and aryl halides and their derivatives appear in Tables 23.10 and 23.11, respectively, of Section 23.6.

Classification Test

A. Chloroform and Aluminum Chloride.　This test for the presence of an aromatic ring should be performed only on compounds that have been shown to be *insoluble* in concentrated sulfuric acid (see Sec. 23.2.4). The test involves the reaction between an aromatic compound and chloroform in the presence of anhydrous aluminum chloride as a catalyst. The colors produced in this type of reaction are often quite characteristic for certain aromatic compounds, whereas aliphatic compounds give little or no color with this test. Some typical examples are tabulated below. Often these colors change with time and ultimately yield brown-colored solutions. Carbon tetrachloride may be used in place of chloroform; it yields similar colors.

Type of Compound	Color
Benzene and homologs	Orange to red
Aryl halides	Orange to red
Naphthalene	Blue
Biphenyl	Purple

The test is based upon a series of Friedel-Crafts alkylation reactions; for benzene, the ultimate product is triphenylmethane (Eq. 23.26).

$$3 \text{ C}_6\text{H}_6 + \text{CHCl}_3 \xrightarrow{\text{AlCl}_3} (\text{C}_6\text{H}_5)_3\text{CH} + 3 \text{ HCl} \qquad \textbf{(23.26)}$$

Benzene　Chloroform　　　Triphenylmethane

The colors arise because species such as triphenylmethyl cations, $(\text{C}_6\text{H}_5)_3\text{C}^+$, form and remain in the solution as AlCl_4^- salts; ions of this sort are highly colored owing to the extensive delocalization of charge that is possible throughout the three aromatic rings.

The test is significant if positive, but a negative test does *not* rule out an aromatic structure; some compounds are so unreactive that they do not readily undergo Friedel-Crafts reactions.

Positive tests for aryl halides are difficult to obtain directly, and some of the best evidence for their presence involves indirect methods. Elemental analysis indicates the presence of halogen. If *both* the silver nitrate and sodium iodide–acetone tests are negative, the compound is most likely a vinyl or an aromatic halide, both of which are very unreactive toward silver nitrate and sodium iodide. Distinction between a vinyl and an aromatic halide can be made by means of the aluminum chloride–chloroform test.

SAFETY ALERT

Anhydrous aluminum chloride reacts vigorously with water, even the moisture on your hands, producing fumes of hydrogen chloride that are highly corrosive if inhaled. Do not allow aluminum chloride to come in contact with your skin. If it does, flush the affected area with copious amounts of water. Because aluminum chloride is a powdery solid that can easily become airborne, weigh and transfer it into the reaction flask *in the hood*.

Heat about 100 mg of *anhydrous* aluminum chloride in a Pyrex test tube held almost horizontally until the material has sublimed to 3 or 4 cm above the bottom of the tube. Allow the tube to cool until it is almost comfortable to touch; then add about 20 mg of a solid, or 1 drop of a liquid down the side of the tube, followed by 2 or 3 drops of chloroform. The appearance of a bright color ranging from red to blue where the sample and chloroform come in contact with the aluminum chloride is a positive indication of an aromatic ring.

Finishing Touches   Cool the contents of the test tube in an ice-water bath and cautiously add a few milliliters of cold water to hydrolyze the aluminum salts. Flush the resulting *aqueous layer* down the drain and place the *organic layer* in the container for halogenated organic liquids.

Derivatives

Two types of derivatives are used to characterize aromatic hydrocarbons and aryl halides. These are prepared by nitration (Eq. 23.27) and side-chain oxidation (Eq. 23.28).

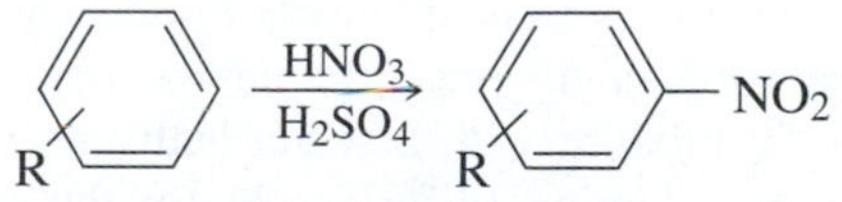

$$\underset{\text{An aromatic hydrocarbon}}{R-\!\!\!\bigcirc} \xrightarrow[\text{H}_2\text{SO}_4]{\text{HNO}_3} \underset{\text{A nitroaromatic}}{R-\!\!\!\bigcirc-\text{NO}_2} \qquad \textbf{(23.27)}$$

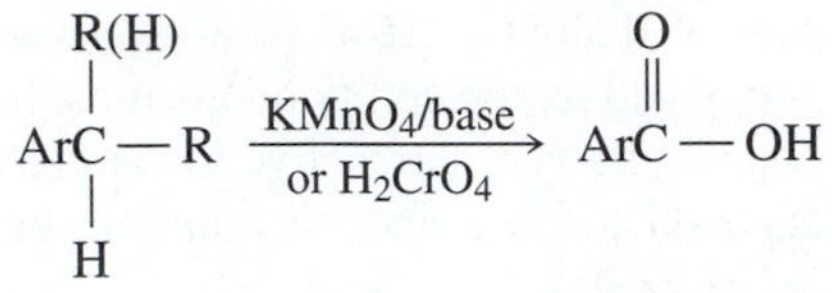

$$\underset{\substack{\text{An alkylated or} \\ \text{acylated aromatic hydrocarbon}}}{\overset{\overset{\displaystyle R(H)}{|}}{\underset{\underset{\displaystyle H}{|}}{ArC}}-R} \xrightarrow[\text{or } H_2CrO_4]{KMnO_4/base} \underset{\text{A benzoic acid}}{ArC}\overset{\overset{\displaystyle O}{\|}}{-}OH \qquad \textbf{(23.28)}$$

B. Nitration. Some of the best solid derivatives of aryl halides are mono- and dinitration products. Two general procedures can be used for nitration.

EXPERIMENTAL PROCEDURES

SAFETY ALERT

1. Use care whenever performing a nitration, whether it is with a known or an unknown substance, because many compounds react vigorously under typical nitration conditions.
2. Concentrated sulfuric and nitric acids are used in this procedure. These acids are very corrosive; handle them with care. If they contact your skin, immediately wash the affected area with copious amounts of water, followed by 5% sodium bicarbonate solution.

Method A. (Note: This method yields *m*-dinitrobenzene from benzene or nitrobenzene and the *p*-nitro derivative from chloro- or bromobenzene, benzyl chloride, or toluene; dinitro derivatives are obtained from phenol, acetanilide, naphthalene, and biphenyl.) Carry out the reaction in a large test tube or small Erlenmeyer flask. Add about 0.5 g of the compound to 2 mL of concentrated sulfuric acid. Add 2 mL of concentrated nitric acid dropwise to this mixture; stopper the container and shake the mixture after each addition. After the addition is complete, heat the mixture at 45 °C on a water bath for about 5 min. Then pour the reaction mixture onto 15 g of ice, and collect the precipitate by vacuum filtration. Recrystallize the solid from aqueous ethanol.

Method B. (Note: This method is the better one to use for nitrating halogenated benzenes, because dinitration occurs to give compounds that have higher melting points and are easier to purify than are the mononitration products obtained from Method *A*. The xylenes, mesitylene, and pseudocumene yield trinitro compounds using the present method.) Follow the procedure used for Method A, except use 2 mL of *fuming* nitric acid in place of the *concentrated* nitric acid and warm the mixture with a steam bath for 10 min. If little or no nitration occurs, substitute *fuming* sulfuric acid for the *concentrated* sulfuric acid. *Carry out this reaction in a hood.*

> **Finishing Touches**   Neutralize all *acidic solutions* before flushing them down the drain.

C. Side-Chain Oxidation.   A second method for characterizing aromatic hydrocarbons involves oxidation of an alkyl side chain to a carboxylic acid group, a reaction that is discussed in Section 16.4. Since carboxylic acids are often solids, they themselves serve as suitable derivatives.

EXPERIMENTAL PROCEDURES

SAFETY ALERT

1. Wear rubber gloves to keep the acids from contacting your skin when you prepare and handle solutions of potassium permanganate or chromic acid. These solutions cause unsightly stains on your skin for several days, and the chromic acid–sulfuric acid solution may cause severe chemical burns. Wash the affected area thoroughly with copious amounts of water if these oxidants come in contact with the skin. In the case of chromic acid, also rinse the area with 5% sodium bicarbonate solution.

2. The preparation of chromic acid requires diluting sulfuric acid with water. Be certain to add the acid *slowly* to the water and swirl the container to ensure continuous mixing. The dissolution of sulfuric acid in water generates heat, and when the acid is added to water, the heat is dispersed through warming of the water. Swirling prevents the layering of the denser sulfuric acid at the bottom of the flask and the attendant possibility that hot acid will be splattered by the steam generated when the two layers are suddenly mixed later by agitation.

1. Permanganate Method.   Add 0.5 g of the compound to a solution prepared from 80 mL of water and 2 g of potassium permanganate. Add 0.5 mL of 3 *M* sodium hydroxide solution, and heat the mixture at reflux until the purple color of permanganate ion has disappeared; this will normally take 0.5–3 h. At the end of the reflux period, cool the mixture and carefully acidify it with 3 *M* sulfuric acid. Now heat the mixture for an additional 30 min and cool it again; remove excess brown manganese dioxide, if any, by addition of sodium bisulfite solution. The bisulfite reduces the manganese dioxide to manganous ion, which is water-soluble. Collect the solid acid that remains by vacuum filtration. Recrystallize the acid from toluene or aqueous ethanol. If little or no solid acid is formed, this may be due to the fact that the acid is somewhat water-soluble. In this case, extract the aqueous layer with small portions of diethyl ether or dichloromethane, dry the organic extracts over sodium sulfate, decant the solution, and remove the organic solvent by means of a steam bath in the hood. Recrystallize the acid that remains. In this particular method, the presence of base during the oxidation often means that some silicic acid, derived from the glass of the reaction vessel, forms on acidification. Thus, purify the carboxylic acid by recrystallization prior to determination of the melting point.

Finishing Touches Place the *aqueous filtrate* containing manganese salts in the container for heavy metals. Flush all *ethanolic filtrates* down the drain. Put *filtrates* involving other organic solvents in the container for halogenated or nonhalogenated liquids, as appropriate.

2. Chromic Acid Method. Dissolve 1 g of sodium dichromate in 3 mL of water contained in a round-bottom flask, and add 0.5 g of the compound to be oxidized. Carefully add 1.5 mL of concentrated sulfuric acid to the mixture with mixing and cooling. Attach a reflux condenser to the flask, and heat the mixture gently until a reaction ensues; as soon as reaction begins, remove the flame and cool the mixture if necessary. After spontaneous boiling subsides, heat the mixture at reflux for 2 h. Pour the reaction mixture into 5 mL of water, and collect the precipitate by filtration. Transfer the solid into a flask, add 3 mL of 2 *M* sulfuric acid, and then warm the mixture on a steam cone with stirring. Cool the mixture, collect the precipitate, and wash it with about 3 mL of cold water. Dissolve the residue in 3 mL of 1.5 *M* sodium hydroxide solution and filter the solution. Add the filtrate, with stirring, to 6 mL of 2 *M* sulfuric acid. Collect the new precipitate, wash it with cold water, and recrystallize the carboxylic acid from either toluene or aqueous ethanol.

Finishing Touches Add sodium sulfite to the *aqueous filtrate* of chromium salts in order to destroy excess Cr^{VI}. Make the solution slightly basic, to form chromium hydroxide, and isolate this salt by vacuum filtration. Place the *filter paper* and the *filter cake* in the container for heavy metals; flush the *filtrate* down the drain. Neutralize all other *aqueous filtrates* before flushing them down the drain. Put any *filtrates* containing toluene in the container for nonhalogenated organic liquids.

23.5.5 ALCOHOLS

A list of alcohols and their derivatives appears in Table 23.3 of Section 23.6.

Classification Tests

The tests for the presence of a hydroxy group not only detect this functionality but may also indicate whether it is attached to a primary, secondary, or tertiary carbon atom.

A. Chromic Acid in Acetone. This test may be used to detect the presence of a hydroxy group, provided that it is shown previously that the molecule does *not* contain an aldehyde function. The reactions and experimental procedures for this test are given in Section 23.5.1(C). Chromic acid does *not* distinguish between primary and secondary alcohols because they *both* give a positive test; tertiary alcohols give a negative test.

B. Lucas Test. This test is used to distinguish among primary, secondary, and tertiary alcohols. The reagent is a mixture of concentrated hydrochloric acid and zinc chloride, which converts alcohols to the corresponding alkyl chlorides. With this reagent, primary alcohols give no appreciable reaction (Eq. 23.29), secondary alcohols react more rapidly (Eq. 23.30), and tertiary alcohols react very rapidly (Eq. 23.31). A positive test depends on the fact that the alcohol is soluble in the reagent, whereas the alkyl chloride is not; thus the formation of a second layer or an emulsion constitutes a positive test. The *solubility of the alcohol in the reagent places limitations on the utility of the test,* and in general only monofunctional alcohols with six or fewer carbon atoms, as well as polyfunctional alcohols, may be used successfully.

$$RCH_2-OH + HCl \xrightarrow{ZnCl_2} \text{No reaction} \qquad \textbf{(23.29)}$$

1° Alcohol

$$R_2CH-OH + HCl \xrightarrow[\text{slow}]{ZnCl_2} R_2CH-Cl + H_2O \qquad \textbf{(23.30)}$$

2° Alcohol $\qquad\qquad$ 2° Alkyl halide

$$R_3C-OH + HCl \xrightarrow[\text{fast}]{ZnCl_2} R_3C-Cl + H_2O \qquad \textbf{(23.31)}$$

3° Alcohol $\qquad\qquad$ 3° Alkyl halide

The similarity should be noted between this reaction and the nucleophilic displacement reactions between alcohols and hydrohalic acids as discussed in Sections 14.3 and 14.4. In the Lucas test, the presence of zinc chloride, a Lewis acid, greatly increases the reactivity of alcohols toward hydrochloric acid.

EXPERIMENTAL PROCEDURE

Add 5 mL of the hydrochloric acid–zinc chloride reagent (Lucas reagent) to about 0.5 mL of the compound in a test tube. Stopper the tube and shake it; allow the mixture to stand at room temperature. Try this test with known primary, secondary, and tertiary alcohols, and note the *time* required for the formation of an alkyl chloride, which appears either as a second layer or as an emulsion. Repeat the test with an unknown, and compare the result with the results from the knowns.

> **Finishing Touches** Use a micropipet to separate the layer of alkyl chloride, if formed, from the test reagent. Put the *alkyl chloride* in the container for halogenated liquids. Neutralize the *aqueous solutions* before flushing them down the drain.

C. Ceric Nitrate Test. Although this test is used primarily for detecting phenols, it can also be applied as a qualitative test for alcohols. Discussion about and experimental procedures for this test are given in Section 23.5.6.

Derivatives

Two common derivatives of alcohols are urethanes and the benzoate esters; the former are best for primary and secondary alcohols, whereas the latter are useful for all types of alcohols.

D. Urethanes. When an alcohol is allowed to react with an aryl substituted isocyanate, ArN=C=O, addition of the alcohol occurs to give a urethane (Eq. 23.32). Some commonly used isocyanates are α-naphthyl, *p*-nitrophenyl, and phenyl.

$$\text{ArN}=\text{C}=\text{O} + \text{ROH} \longrightarrow \underset{\underset{\text{A urethane}}{\text{ArNH}}}{\overset{\overset{\displaystyle O}{\parallel}}{\text{C}}}\diagdown\text{OR} \tag{23.32}$$

An aryl isocyanate   An alcohol   A urethane

A major side reaction involves hydrolysis of the isocyanate to an amine, which then reacts with more isocyanate to give a disubstituted urea (Eqs. 23.33 and 23.34). Because of their symmetry, ureas are high-melting, and their presence makes purification of the desired urethane difficult. Therefore, take precautions to ensure that the alcohol is *anhydrous* when using this procedure. The method works best for water-insoluble alcohols because they are more easily obtained in anhydrous form.

$$\text{ArN}=\text{C}=\text{O} + \text{H}_2\text{O} \longrightarrow \underset{\underset{\substack{\text{A carbamic acid}\\\text{(unstable)}}}{\text{ArNH}}}{\overset{\overset{\displaystyle O}{\parallel}}{\text{C}}}\diagdown\text{OH} \xrightarrow{-\text{CO}_2} \underset{\text{An arylamine}}{\text{ArNH}_2} \tag{23.33}$$

An aryl isocyanate

$$\text{ArN}=\text{C}=\text{O} + \text{ArNH}_2 \longrightarrow \underset{\underset{\text{A disubstituted urea}}{\text{ArNH}}}{\overset{\overset{\displaystyle O}{\parallel}}{\text{C}}}\diagdown\text{NHAr} \tag{23.34}$$

This type of derivative is also useful for phenols; the procedure given here has been generalized so that it can be used for both alcohols and phenols. Other derivatives of phenols are provided in Section 23.5.6.

EXPERIMENTAL PROCEDURE

SAFETY ALERT

> Aryl isocyanates such as those used in the preparation of these derivatives are toxic. Take normal precautions in handling them. Immediately wash with soap and warm water any areas of skin that come in contact with isocyanates.

Place 1 g of the *anhydrous* alcohol or phenol in a round-bottom flask, and add 0.5 mL of phenyl isocyanate or α-naphthyl isocyanate; be sure to recap the bottle of isocyanate tightly to minimize exposure of the reagent to atmospheric moisture. If you are preparing the derivative of a phenol, also add 2 or 3 drops of dry pyridine as a catalyst. Affix a calcium chloride drying tube to the flask. Warm the reaction mixture with a heating mantle or a steam or oil bath for 5 min. Cool the mixture in an ice-water bath, and scratch the mixture with a stirring rod at the air-liquid interface to induce crystallization. Recrystallize the crude derivative from petroleum ether. (*Note:* 1,3-Di(α-naphthyl)urea has mp 293 °C, and 1,3-diphenylurea (carbanilide) has mp 237 °C; if your product shows one or these melting points, repeat the preparation, taking greater care to maintain anhydrous conditions.)

> **Finishing Touches** Flush any *filtrates* that contain the starting isocyanate down the drain. Pour the *filtrates* from the recrystallization of the product into the container for nonhalogenated organic liquids. Put the *calcium chloride* in the container for nonhazardous solids.

E. 3,5-Dinitrobenzoates. The reaction between 3,5-dinitrobenzoyl chloride and an alcohol gives the corresponding ester (Eq. 23.35). This method is useful for primary, secondary, and tertiary alcohols, especially those that are water-soluble and thus are likely to contain traces of water.

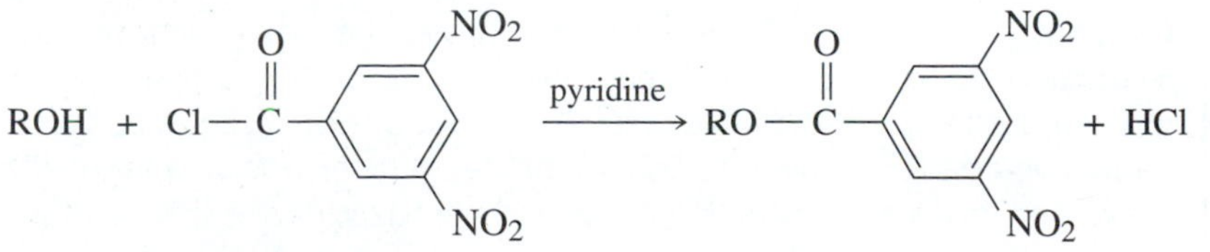

| An alcohol | 3,5-Dinitrobenzoyl chloride | | A 3,5-dinitrobenzoate | |

$$ROH \ + \ Cl-\overset{O}{\underset{}{C}}-\!\!\!\!\bigcirc\!\!\!\!\!\begin{smallmatrix}NO_2\\ \\NO_2\end{smallmatrix} \ \xrightarrow{\text{pyridine}} \ RO-\overset{O}{\underset{}{C}}-\!\!\!\!\bigcirc\!\!\!\!\!\begin{smallmatrix}NO_2\\ \\NO_2\end{smallmatrix} \ + \ HCl \qquad (23.35)$$

EXPERIMENTAL PROCEDURES

SAFETY ALERT

Thionyl chloride is a lachrymator. Measure this material in a hood and do not inhale its vapors.

Method A. (*Note:* 3,5-Dinitrobenzoyl chloride is reactive toward water; it should be used *immediately* after weighing. Take care to minimize its exposure to air and to keep the bottle tightly closed.) Mix 0.6 mL of the alcohol with about 0.2 g of 3,5-dinitrobenzoyl chloride and 0.2 mL of pyridine in a round-bottom flask bearing a reflux condenser. Boil the mixture gently for 30 min, although 15 min is sufficient for a primary alcohol. Cool the solution and add about 5 mL of 0.6 *M* aqueous sodium bicarbonate solution. Cool this solution in an ice-water bath, and collect the crude crystalline product. Recrystallize the product from aqueous ethanol, using a minimum volume of solvent.

Finishing Touches   Flush all *filtrates* down the drain.

Method B.   Perform this reaction in the hood, if possible, or use a gas trap such as that shown in Figure 2.35. Add 0.5 g of 3,5-dinitrobenzoic acid, 1.5 mL of thionyl chloride, and 1 drop of pyridine to a 50-mL round-bottom flask fitted with a reflux condenser. Heat the mixture at reflux until the acid has dissolved and then for an additional 10 min. The total reflux time should be about 30 min.

   Equip the flask for vacuum distillation. Cool the receiving flask with an ice-salt bath, and attach the vacuum adapter to an aspirator by means of a safety trap such as that shown in Figure 2.19. Evacuate the system and distil the excess thionyl chloride by heating with a steam bath. When the thionyl chloride has been removed, cautiously release the vacuum. To the residue in the stillpot, which is 3,5-dinitrobenzoyl chloride, add 1 mL of the alcohol and 0.3 mL of pyridine. Fit the flask with a reflux condenser bearing a calcium chloride drying tube. Proceed with the period of reflux according to the procedure of Method A.

Finishing Touches   Flush all *filtrates* down the drain with copious amounts of water. Discard the excess *thionyl chloride* by pouring it slowly down a drain *in the hood* while running water or by putting it in a container for recovered thionyl chloride. Spread the *calcium chloride* on a tray in the hood and discard it in the container for nonhazardous solids after the volatiles are removed.

23.5.6 PHENOLS

A listing of phenols and their derivatives appears in Table 23.18 of Section 23.6.

Classification Tests

Several tests can be used to detect the presence of a phenolic hydroxy group: (a) bromine water, (b) ceric ammonium nitrate reagent, and (c) ferric chloride solution. In addition to these, solubility tests give a preliminary indication of a phenol because phenols are soluble in 1.5 M sodium hydroxide solution, but generally are insoluble in 0.6 M sodium bicarbonate solution. Care must be exercised in interpreting solubility data, however, because phenols containing highly electronegative groups are stronger acids and *may* be soluble in 0.6 M sodium bicarbonate; examples are 2,4,6-tribromophenol and 2,4-dinitrophenol.

A. Bromine Water. Phenols usually are highly reactive toward electrophilic substitution and consequently are brominated readily by bromine water (Eq. 23.36, for example). The rate of bromination is much greater in water than in carbon tetrachloride solution. Water, being more polar than carbon tetrachloride, increases the ionization of bromine and thus enhances the rate of ionic bromination. Although hydrogen bromide is liberated, it is not observed when water is used as the solvent.

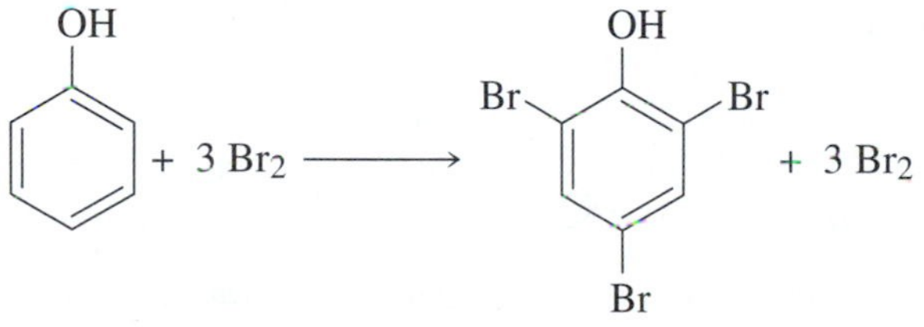

$$\text{Phenol} + 3\ Br_2 \longrightarrow \text{2,4,6-Tribromophenol} + 3\ Br_2 \qquad (23.36)$$

Phenols are so reactive toward electrophilic aromatic substitution that all unsubstituted positions *ortho* and *para* to the hydroxy group are brominated. The brominated compounds so formed are often solids and may be used as derivatives (see Sec. 23.5.6E). Aniline and substituted anilines are also very reactive toward bromine and react analogously; however, solubility tests are normally used to distinguish between anilines and phenols.

EXPERIMENTAL PROCEDURE

SAFETY ALERT

Bromine is a hazardous chemical. Do not breathe its vapors or allow it to contact your skin because it may cause serious chemical burns. Perform all operations involving the transfer of solutions of bromine in a hood; wear rubber gloves when handling this chemical. If you get bromine on your skin, wash the area immediately with warm water and soak the skin in 0.6 M sodium thiosulfate solution for up to 3 h if the burn is particularly serious.

Prepare a 1% aqueous solution of the unknown. If necessary, add dilute sodium hydroxide solution dropwise to effect solution of the phenol. Add a saturated solution of bromine in water dropwise to this solution; continue addition until the bromine color remains. Note how much bromine water is used and perform this experiment on phenol and aniline for purposes of comparison.

> **Finishing Touches**   Neutralize and then filter all *test solutions*. Flush the *filtrates* down the drain and place the *filter paper* in the container for halogenated organic compounds.

B. Ceric Nitrate Test.   Alcohols are capable of replacing nitrate ions in complex cerate anions, resulting in a change in color of the solution from yellow to red (Eq. 23.37). Phenols are oxidized by this reagent to give a brown to greenish-brown precipitate in aqueous solution; a red-to-brown solution is produced in 1,4-dioxane. Because of solubility problems, the alcohol or phenol should have no more than ten carbon atoms for this test. Aromatic amines are oxidized by the reagent to give a colored solution or precipitate; consequently the presence of this functionality must be excluded before concluding that a positive test signals an alcohol or phenol moiety.

$$(NH_4)_2^{2+}\,Ce(NO_3)_6^{\,2-} \;+\; ROH \longrightarrow (NH_4)_2^{2+}\,Ce(NO_3)_5^{\,2-} \;+\; HNO_3 \qquad \textbf{(23.37)}$$

OR
|

Ceric ammonium nitrate   An alcohol   An alkoxy ceric ammonium nitrate
Yellow   *Red*

EXPERIMENTAL PROCEDURE

Dissolve about 20 mg of a solid or 1 drop of a liquid unknown in 1–2 mL of water, and add 0.5 mL of the ceric ammonium nitrate reagent; shake the mixture and note the color. If the unknown is insoluble in water, dissolve it in 1 mL of 1,4-dioxane, and proceed as before.

> **Finishing Touches**   Flush the *test solutions* down the drain.

C. Ferric Chloride Test.   Most phenols and enols react with ferric chloride to give colored complexes. The colors vary, depending on the nature of the phenol or enol and also on the solvent, concentration, and time of observation. Some phenols not giving coloration in aqueous or alcoholic solution do so in chloroform solution, especially after addition of a drop of pyridine. The nature of the colored complexes is still uncertain; they may be ferric phenoxide salts that absorb visible light. The production of a color is typical of phenols and enols; however, many of them do *not* give colors, so a negative ferric chloride test must *not* be taken as significant without supporting information such as that available from the ceric ammonium nitrate and bromine water tests.

Dissolve 30 to 50 mg of the unknown compound in 1–2 mL of water, or a mixture of water and 95% ethanol if the compound is not water-soluble, and add several drops of a 0.2 M aqueous solution of ferric chloride. Most phenols produce red, blue, purple, or green coloration; enols give red, violet, or tan coloration.

Finishing Touches   Flush all *solutions* down the drain.

Derivatives

Two useful solid derivatives of phenols are α-naphthyl urethane and bromo compounds.

D. Urethanes.   The preparation of urethanes has already been discussed in Section 23.5.5D. Although either a phenyl- or naphthyl-substituted urethane could be prepared, the majority of the derivatives reported are the α-naphthyl urethanes, and they are generally the urethanes of choice.

E. Bromophenols.   The high reactivity of phenols toward electrophilic aromatic bromination has been discussed in Section 23.5.6A. This reaction can be used to prepare bromophenols, which may serve as solid derivatives.

SAFETY ALERT

Bromine is a hazardous chemical. Do not breathe its vapors or allow it to contact your skin because it may cause serious chemical burns. Perform all operations involving the transfer of solutions of bromine in a hood; wear rubber gloves when handling this chemical. If you get bromine on your skin, wash the area immediately with warm water and soak the skin in 0.6 M sodium thiosulfate solution for up to 3 h if the burn is particularly serious.

A stock solution for bromination in this preparation should be supplied; if not prepare it by dissolving 1 g of potassium bromide in 6 mL of water and adding 1 g of bromine. Dissolve 0.5 g of the phenolic compound in water or 95% ethanol, and add the bromine-containing solution to it *dropwise*. Continue the addition until the reaction mixture begins to develop a yellow color, indicating the presence of excess bromine. Let the mixture stand for about 5 min; if the yellow coloration begins to fade, add another drop or two of the solution of bromine. Add 5 mL of water to the mixture and then a few drops of 0.5 *M* sodium bisulfite solution to destroy the excess bromine. Shake the mixture vigorously, and remove the solid derivative by vacuum filtration. If necessary, neutralize the solution with concentrated hydrochloric acid to promote precipitation of the derivative. Purify the solid by recrystallization from 95% ethanol or aqueous ethanol.

> **Finishing Touches** Flush all *filtrates* down the drain.

23.5.7 Carboxylic Acids

A list of carboxylic acids and their derivatives appears in Table 23.12 of Section 23.6.

Classification Test

One of the best qualitative tests for the carboxylic acid group is solubility in basic solutions. Carboxylic acids are soluble both in 1.5 *M* sodium hydroxide solution *and* in 0.6 *M* sodium bicarbonate solution, from which they are regenerated by addition of hydrochloric acid. Solubility properties are discussed in Section 23.2.4.

The relatively high acidity of carboxylic acids enables ready determination of the **equivalent weight** or **neutralization equivalent** of the acid by titration with standard base. The equivalent weight of an acid is that weight, in grams, of acid that reacts with one equivalent of base. As an example, suppose that 0.1000 g of an unknown acid requires 16.90 mL of 0.1000 *N* sodium hydroxide solution to be titrated to a phenolphthalein end point. This means that 0.1000 g of the acid corresponds to (16.90 mL)(0.1000 equivalent/1000 mL) or 0.0016901 equivalent of the acid, or that one equivalent of the acid weighs 0.1000/0.00169 or 59.201 g. Thus the following expression applies:

$$equivalent\ weight = \frac{grams\ of\ acid}{(volume\ of\ base\ consumed\ in\ liters)(N)}$$

where *N* is the *normality* of the standard base.

Because each carboxylic acid function in a molecule is titrated with base, the equivalent weight corresponds to the molecular weight of the acid divided by *n,* where *n* is the number of acid functions present in the molecule. Thus, for the example given, the molecular weight is 59.2 for a single acid function, 118.4 for two, and 177.6 for three. If the molecular weight of an unknown compound is known, then the number of acid groups in the molecule is calculated by dividing the molecular weight by the equivalent weight. Hence, if the molecular weight of the unknown compound is 118 and its equivalent weight is 59.2, the unknown must have *two* titratable acid functions.

EXPERIMENTAL PROCEDURE

Accurately weigh about 0.2 g of the acid and dissolve it in 50 to 100 mL of water or 95% ethanol or a mixture of the two. It may be necessary to warm the mixture to dissolve the compound completely. Using phenolphthalein as the indicator, titrate the solution with a *standardized* sodium hydroxide solution having a concentration of about 0.1 *M*. From the data obtained, calculate the equivalent weight.

Finishing Touches   Flush the *aqueous solution* down the drain.

Derivatives

Three satisfactory solid derivatives of carboxylic acids are amides (Eq. 23.38), anilides (Eq. 23.39), and *p*-toluidides (Eq. 23.40). These derivatives are prepared by treatment of the corresponding acid chlorides with ammonia, aniline, or *p*-toluidine. The amides are generally less satisfactory derivatives than the other two because they tend to be more soluble in water and thus are harder to isolate. The acid chlorides are most conveniently prepared from the acid, or its salt, and thionyl chloride (Eq. 23.41).

$$RCOCl + 2\,NH_3 \xrightarrow{\text{cold}} RCONH_2 + NH_4Cl \qquad \textbf{(23.38)}$$

An acid
chloride   An amide

$$RCOCl + 2\,H_2NC_6H_5 \longrightarrow RCONHC_6H_5 + C_6H_5\overset{\oplus}{N}H_3\ \overset{\ominus}{Cl} \qquad \textbf{(23.39)}$$

Aniline   An anilide

$$RCOCl + 2\,H_2NAr \longrightarrow RCONHAr + Ar\overset{\oplus}{N}H_3\ \overset{\ominus}{Cl} \qquad \textbf{(23.40)}$$

p-Toluidine   A *p*-toluidide

$$Ar = -\!\!\left\langle \bigcirc \right\rangle\!\!-CH_3$$

$$\text{RCO}_2\text{H or RCO}_2^{\ominus} \text{ Na}^{\oplus} + \text{SOCl}_2 \longrightarrow \text{RCOCl} + \text{SO}_2 + \text{HCl (or NaCl)} \qquad \textbf{(23.41)}$$

| A carboxylic acid | A sodium carboxylate | Thionyl chloride | An acid chloride | Sulfur dioxide |

EXPERIMENTAL PROCEDURES

SAFETY ALERT

Thionyl chloride is a lachrymator. Measure out this material in a hood and do not inhale its vapors.

To prepare the acid chloride from either a carboxylic acid or its salt, place the acid or its sodium salt in a small round-bottom flask with 1 mL of thionyl chloride and 5 drops of *N,N*-dimethylformamide (DMF). Attach a calcium chloride drying tube *directly* to the flask, and clamp the flask into a 60–65 °C water bath *in the hood*. Bubbling or fuming usually begins shortly after the addition of the DMF. The reaction is sufficiently complete when the bubbling greatly slows, which typically occurs within 30 min. Use the mixture containing the acid chloride to make the amide, anilides, or *p*-toluidide according to the procedures that follow.

A. Amides. Use 1 g of the acid or its sodium salt to prepare the acid chloride. *At the hood* pour the mixture containing the acid chloride and unchanged thionyl chloride into 15 mL of *ice-cold*, concentrated ammonium hydroxide solution. Be very careful when performing this addition because the reaction is quite exothermic. Collect the precipitated amide derivative by vacuum filtration, and recrystallize the crude product from water or aqueous ethanol.

Finishing Touches Neutralize all *filtrates* and then flush them down the drain.

B. Anilides and *p*-Toluidides. Use 0.5 g of the carboxylic acid or its sodium salt to prepare the acid chloride. After the reflux period, cool the mixture to room temperature. Dissolve 0.5 g of either aniline or *p*-toluidine in 15 mL of cyclohexane; slight warming may be necessary to effect complete solution. Pour the acid chloride into the cyclohexane solution of the amine, and heat the resulting mixture on a steam bath for 2–3 min. A heavy white precipitate of the amine hydrochloride forms; isolate this precipitate by vacuum filtration and set it aside, but do *not* discard it. Transfer the filtrate to a separatory funnel and wash it sequentially with 3-mL portions of water, 1.5 *M* HCl, 1.5 *M* NaOH, and again with water. In some cases precipitation may occur in the organic layer during one or more of these washings. If this occurs, warm the solution gently with a warm-water bath to redissolve the precipitate.

Following the sequential washings, remove the cyclohexane from the organic layer either by distillation or by evaporation on a steam bath in the hood. Recrystallize the derivative of the carboxylic acid from aqueous ethanol.

If little residue remains after evaporating the cyclohexane, dissolve the precipitate that was removed earlier in about 10 mL of water. Stir the mixture and then remove any undissolved solid by vacuum filtration; any such solid is the desired derivative, which became entrapped in the precipitated amine hydrochloride. Combine this material with the residue obtained from cyclohexane.

Finishing Touches   Neutralize all *filtrates* and then flush them down the drain. Put any recovered *cyclohexane* into the container for nonhalogenated organic liquids.

23.5.8   AMINES

A list of amines and their derivatives is found in Table 23.9 of Section 23.6.

Classification Tests

Two common qualitative tests for amines are the Hinsberg test and the nitrous acid test. The nitrous acid test is not included here because the *N*-nitroso derivatives of some secondary amines are carcinogenic. The risk of producing an as-yet-unrecognized carcinogenic material in this test outweighs any possible benefit of a test that can be misleading and difficult to interpret. The modified sodium nitroprusside test has been included as an alternative.

A. Hinsberg Test.   The reaction between primary or secondary amines and benzenesulfonyl chloride (Eqs. 23.42 and 23.44, respectively) yields the corresponding substituted benzenesulfonamide. The reaction is performed in excess aqueous base; if the amine is primary, the sulfonamide, which has an acidic amido hydrogen, is converted by base (Eq. 23.43) to the potassium salt, which is normally soluble in the medium. With few exceptions, which are discussed in the next paragraph, primary amines react with benzenesulfonyl chloride to provide *homogeneous* reaction mixtures. Acidification of this solution regenerates the insoluble primary benzenesulfonamide. On the other hand, the benzenesulfonamides of secondary amines bear no acidic amido hydrogens and thus typically are insoluble in *both* acid and base. Therefore, secondary amines react to yield *heterogeneous* reaction mixtures, with formation of either an oily organic layer or a solid precipitate.

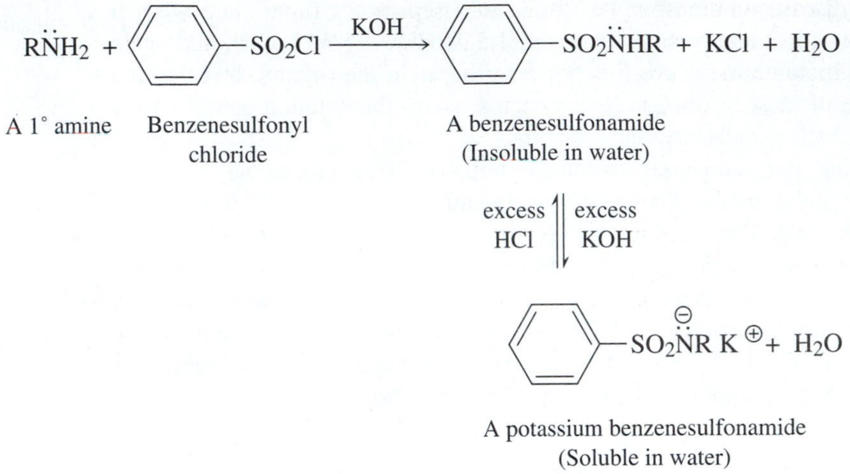

$$RNH_2 + \underset{\text{Benzenesulfonyl}}{\underset{\text{chloride}}{\overset{\text{A 1° amine}}{\bigcirc}}} -SO_2Cl \xrightarrow{\text{KOH}} \bigcirc -SO_2\overset{..}{N}HR + KCl + H_2O \qquad \textbf{(23.42)}$$

A 1° amine   Benzenesulfonyl
                 chloride

A benzenesulfonamide
(Insoluble in water)

excess ⇅ excess
HCl   KOH

$$\bigcirc -SO_2\overset{\ominus}{N}R \ K^{\oplus} + H_2O \qquad \textbf{(23.43)}$$

A potassium benzenesulfonamide
(Soluble in water)

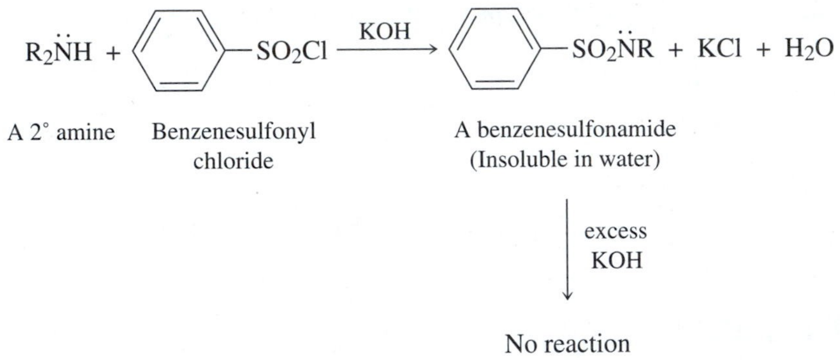

$$R_2\overset{..}{N}H + \bigcirc -SO_2Cl \xrightarrow{\text{KOH}} \bigcirc -SO_2\overset{..}{N}R + KCl + H_2O \qquad \textbf{(23.44)}$$

A 2° amine   Benzenesulfonyl
                 chloride

A benzenesulfonamide
(Insoluble in water)

excess
KOH

No reaction

The distinction between primary and secondary amines thus depends on the different solubility properties of their benzenesulfonamide derivatives. However, the potassium salts of *certain* primary sulfonamides are not completely soluble in basic solution. Examples usually involve primary amines of higher molecular weight and those having cyclic alkyl groups. To avoid confusion and possible misassignment of a primary amine as secondary, the basic solution is separated from the oil or solid and acidified. The formation of an oil or a precipitate indicates that the derivative is partially soluble and that the amine is primary. It is important not to overacidify the solution because this may precipitate certain side products that may form, resulting in an ambiguous test. The original oil or solid should be tested for solubility in water and acid to substantiate the test for a primary or a secondary amine.

Tertiary amines behave somewhat differently. Under the conditions of the Hinsberg test, the processes shown in Equation (23.45) typically provide for converting benzenesulfonyl chloride to potassium benzenesulfonate with recovery of the tertiary amine. Tertiary amines are nearly always insoluble in the aqueous potassium hydroxide solution, so the test mixture remains heterogeneous. Proof that the nonaqueous layer is a tertiary amine can generally be obtained simply by noting the relative densities of the oil layer and the aqueous test solution: Tertiary amines are less dense than the solution, whereas benzene-sulfonamides are generally more dense. Further support for the conclusion that the oil is a tertiary amine is available from testing its solubility in aqueous acid; solubility usually indicates a tertiary amine.

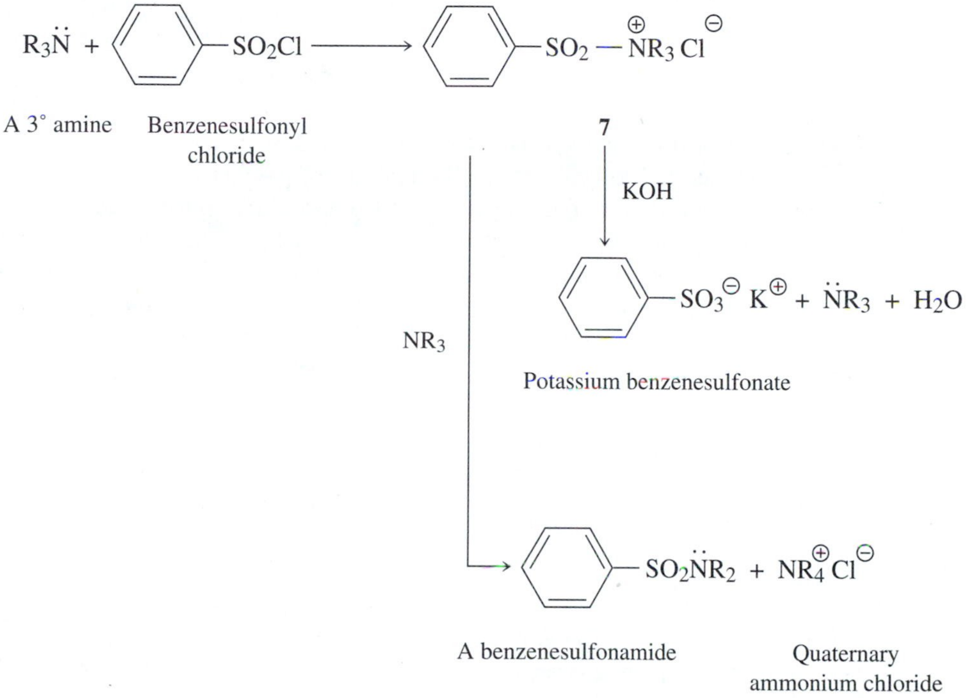

The procedure for the Hinsberg test must be followed as closely as possible because it is designed to minimize complications that arise because of side reactions of tertiary amines with benzenesulfonyl chloride. For example, **7**, the initial adduct between the chloride and the amine, is subject to an S_N2 reaction with another molecule of amine to produce the benzenesulfonamide of a *secondary* amine and a quaternary ammonium salt (Eq. 23.46). Although this process is normally unimportant, particularly when excess amine is avoided, the observation of *small* amounts of an insoluble product formed by this pathway may erroneously cause designation of a tertiary amine as secondary.

Formation of adduct **7** is usually less of a problem with tertiary *aryl*amines because of their lesser nucleophilicity and lower solubility as compared to *tri-alkyl*amines. Consequently, the competing hydrolysis of benzenesulfonyl chloride by hydroxide ion (Eq. 23.47) allows the recovery of most of the amine. However, this class of amines is often subject to other side reactions that produce a complex mixture of mainly insoluble products (Eq. 23.48). The ambiguity caused by side reactions of tertiary arylamines is minimized by keeping the reaction time short and the temperature low.

$$\text{C}_6\text{H}_5-\text{SO}_2\text{Cl} \xrightarrow{\text{KOH}} \text{C}_6\text{H}_5-\text{SO}_3^{\ominus}\ \text{K}^{\oplus} + \text{KCl} + \text{H}_2\text{O} \qquad \textbf{(23.47)}$$

$$\xrightarrow{\text{Ar}\ddot{\text{N}}\text{R}_2} \text{Complex mixture including} \quad \text{C}_6\text{H}_5-\text{SO}_2-\ddot{\text{N}}\text{RAr} \qquad \textbf{(23.48)}$$

A further complication with tertiary amines is that they often contain quantities of secondary amines as impurities. If it is not possible to obtain a reliable boiling point or if the amine is not carefully distilled, such contaminants may be present and lead to formation of small quantities of precipitate, making the test results ambiguous. To summarize the discussion, tertiary amines may produce small amounts of insoluble products if the concentration of the amine in the test solution is too high and if the reaction time is too long. If the directions of the procedure are followed and *care is taken not to interpret small amounts of insoluble product as a positive test for secondary amines,* the Hinsberg test may be used with confidence to designate an amine as primary, secondary, or tertiary.

EXPERIMENTAL PROCEDURE

SAFETY ALERT

Benzenesulfonyl chloride is a lachrymator. Measure out this material in a hood and do not unnecessarily inhale its vapors.

Mix 5 mL of 2 *M* aqueous potassium hydroxide, 0.2 mL (5 drops) or 0.2 g of the amine, and 0.7 mL (15 drops) of benzenesulfonyl chloride in a test tube. Stopper the tube and shake the mixture *vigorously,* with cooling if necessary, until the odor of benzenesulfonyl chloride is gone (*caution!*). In even the slowest case this should take no more than about 5 min. Test the solution to see that the mixture is still basic; if it is not, add sufficient 2 *M* potassium hydroxide solution dropwise until it is.

If the mixture forms two layers or a precipitate, note the relative densities, and separate the oil or solid by decantation or filtration. Test any oil for solubility in 0.6 *M* hydrochloric acid. The sulfonamide of a secondary amine is insoluble, whereas an amine is at least partially soluble. A solubility test indicating the presence of an amine could be caused by either a tertiary amine or secondary amines that for steric reasons react very slowly with benzenesulfonyl chloride.

Test any solid for solubility in water and in dilute acid. The potassium salt of a sulfonamide that is insoluble in base solution is usually soluble in water; acidifying the salt yields a sulfonamide that is insoluble in aqueous acid. A solid sulfonamide of a secondary amine is insoluble in both water and acid. Acidify the solution from the original reaction mixture to pH 4 as signalled by pHydrion paper or a few drops of Congo red indicator solution; the formation of a precipitate or oil indicates a primary amine.

If the original mixture from the Hinsberg test does not form two layers, a primary amine is indicated. Acidify the solution to pH 4; a sulfonamide of a primary amine will either separate as an oil or precipitate as a solid.

Finishing Touches   Neutralize all *test solutions.* Isolate any solids by filtration and any oils with the aid of a Pasteur pipet. Flush the *filtrates* down the drain and put any *filter cakes* or *oils* in the container for nonhalogenated organic compounds.

B. Sodium Nitroprusside Tests.   Two color tests to distinguish primary and secondary *aliphatic* amines have been available for many years, although they have not been widely used. More recently, these tests have been extended to primary, secondary, and tertiary *aromatic* amines. No attempt is made here to explain the complex color-forming reactions that occur. However, they most likely involve the reaction of the amine with either acetone, the Ramini test, or acetaldehyde, the Simon test, and the interaction of the resulting products with sodium nitroprusside to form colored complexes.

To apply these tests to an unknown amine, the *conventional* Ramini or Simon tests are first performed. These will give positive results in the cases of primary and secondary aliphatic amines. If these tests are negative and an aromatic amine is suspected, the *modified* versions of these tests are then performed. Reference to Figure 23.3 is helpful for interpreting the results of these tests.

	1° Aliphatic	2° Aliphatic	1° Aromatic	2° Aromatic	3° Aromatic
Ramini	Deep red	Deep red			
Simon	Pale yellow to red-brown	Deep blue			
Modified Ramini			Orange-red to red-brown	Orange-red to red-brown	Green
Modified Simon			Orange-red to red-brown	Purple	Usually green

FIGURE 23.3   Colors formed in the Ramini and the Simon tests

EXPERIMENTAL PROCEDURES

Conventional Tests. Prepare the sodium nitroprusside reagent for use in both the conventional Ramini and Simon tests by dissolving 0.4 g of sodium nitroprusside ($Na_2[Fe(NO)(CN)_5] \cdot 2\,H_2O$) in 10 mL of 50% aqueous methanol.

Ramini Test. To 1 mL of the sodium nitroprusside reagent, add 1 mL of water, 0.2 mL (5 drops) of acetone, and about 30 mg of an amine. In most cases the characteristic colors given in Figure 23.4 appear in a few seconds, although in some instances up to about 2 min may be necessary.

Simon Test. To 1 mL of the sodium nitroprusside reagent, add 1 mL of water, 0.2 mL (5 drops) of 2.5 M aqueous acetaldehyde solution, and about 30 mg of an amine. As in the Ramini test, color formation normally occurs in a few seconds, although occasionally times up to 2 min are necessary.

Modified Tests. Prepare the reagent for use in both the *modified* Ramini and Simon tests by dissolving 0.4 g of sodium nitroprusside in a solution containing 8 mL of dimethylsulfoxide and 2 mL of water.

Modified Ramini Test. To 1 mL of the *modified* sodium nitroprusside reagent, sequentially add 1 mL of saturated aqueous zinc chloride solution, 0.2 mL (5 drops) of acetone, and about 30 mg of an amine. Primary and secondary aromatic amines provide orange-red to red-brown colors within a period of a few seconds to 5 min. Tertiary aromatic amines give a color that changes from orange-red to green over a period of about 5 min.

Modified Simon Test. To 1 mL of the *modified* sodium nitroprusside reagent, sequentially add 1 mL of saturated aqueous zinc chloride solution, 0.2 mL (5 drops) of 2.5 M aqueous acetaldehyde solution, and about 30 mg of an amine. Primary aromatic amines give an orange-red to red-brown color within 5 min; secondary aromatic amines give a color changing from red to purple within 5 min; tertiary aromatic amines give a color that changes from orange-red to green over a period of 5 min.

> **Finishing Touches**   Flush the *test solutions* down the drain.

Derivatives

Suitable derivatives of primary and secondary amines are benzamides and benzenesulfonamides (Eq. 23.49 and 23.50, respectively).

$$R\ddot{N}H_2 \text{ or } R_2\ddot{N}H + C_6H_5COCl \xrightarrow{\text{pyridine}} C_6H_5\overset{\overset{\displaystyle O}{\|}}{C}-\ddot{N}HR \text{ or } C_6H_5\overset{\overset{\displaystyle O}{\|}}{C}-\ddot{N}R_2 \qquad \textbf{(23.49)}$$

1° Amine   2° Amine   Benzoyl chloride   A benzamide

$$R\overset{\cdot\cdot}{N}H_2 \text{ or } R_2\overset{\cdot\cdot}{N}H + C_6H_5SO_2Cl \longrightarrow C_6H_5SO_2 - \overset{\cdot\cdot}{N}HR \text{ or } C_6H_5SO_2 - \overset{\cdot\cdot}{N}R_2 \qquad \textbf{(23.50)}$$

1° Amine   2° Amine   Benzenesulfonyl
                 chloride
                                   A benzenesulfonamide

These transformations are satisfactory for derivatizing most primary and secondary amines, but tertiary amines do not undergo such reactions. However, such amines form salts that constitute solid derivatives. Two useful crystalline salts are those formed upon reaction with methyl iodide (methiodides) and picric acid (picrates), as shown in Equations (23.51) and (23.52), respectively.

$$R_3N: + CH_3I \longrightarrow R_3\overset{\oplus}{N}CH_3 \ I^{\ominus} \qquad \textbf{(23.51)}$$

3° Amine   Methyl
          iodine
                A quaternary
              ammonium iodide

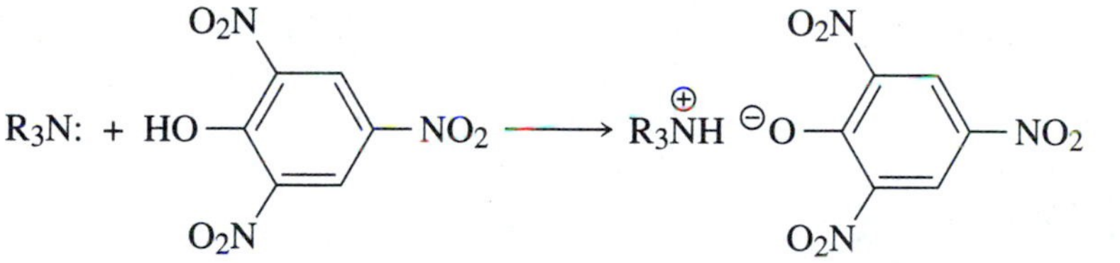

3° Amine   Picric acid   An ammonium picrate

A. Benzenesulfonamides.   The method for preparing the benzenesulfonamides has been discussed under the Hinsberg Test (see Sec. 23.5.8A). The derivatives may be prepared using that method, but sufficient amounts of material should be used so that the final product can be purified by recrystallization from 95% ethanol. If the derivative is obtained as an oil, it *may* crystallize by scratching at the liquid-liquid or air-liquid interface with a stirring rod. If the oil cannot be made to crystallize, separate it and dissolve it in a minimum quantity of hot ethanol and allow the solution to cool. Note that some amines do not give *solid* benzenesulfonamide derivatives.

SAFETY ALERT

Benzoyl chloride is a lachrymator. Transfer this material in a hood and do not inhale its vapors.

B. Benzamides.   In a 50-mL round-bottom flask dissolve 0.3 g of the amine in 3 mL of dry pyridine. *Slowly* add 0.3 mL of benzoyl chloride to this solution. Affix a calcium chloride drying tube to the flask and heat the reaction mixture to 60–70 °C for 30 min, using a water bath; then pour the mixture into 25 mL of water. If the solid derivative precipitates at this time, isolate it by vacuum filtration, and when it is nearly dry, dissolve it in 10 mL of diethyl ether. If no precipitate forms, extract the aqueous mixture twice with 5-mL portions of diethyl ether. Combine the extracts. Wash the ethereal solution sequentially with 5-mL portions of water, 1.5 *M* HCl, and 0.6 *M* sodium bicarbonate solution. Dry the ethereal layer over anhydrous sodium sulfate, filter or decant the dried solution, and remove the diethyl ether by simple distillation (*no flames*). Recrystallize the solid derivative from one of the following solvents: cyclohexane-hexane mixtures, cyclohexane-ethyl acetate mixtures, 95% ethanol, or aqueous ethanol.

Finishing Touches   Transfer any recovered *diethyl ether* into the container for nonhalogenated organic liquids. Flush all *aqueous* and *ethanolic solutions* down the drain, and pour any other *filtrates* containing organic solvents into the container for non-halogenated liquids.

C. Methiodides.   To prepare the methiodide derivative, mix 0.3 g of the amine with 0.3 mL of methyl iodide, and warm the test tube with a water bath for several minutes. Cool the test tube in an ice-water bath. If necessary, scratch at the air-liquid interface with a glass rod to induce crystallization. Isolate the product by vacuum filtration and purify it by recrystallization from absolute ethanol or methanol or from ethyl acetate.

Finishing Touches   Put the *filtrate* from the reaction mixture in the container for halogenated liquids and flush all other *filtrates* down the drain.

D. Picrates.   To prepare the picric acid derivative, mix 0.2 to 0.4 g of the compound with 5 mL of 95% ethanol. If dissolution is not complete, remove any excess solid by gravity filtration. Add 5 mL of a saturated solution of picric acid in 95% ethanol to the mixture, and heat the mixture to boiling. Allow the solution to cool slowly, and remove the yellow crystals of the picrate salt by vacuum filtration. Recrystallize the salt from 95% ethanol.

Finishing Touches   Flush the *filtrates* down the drain.

23.5.9 NITRO COMPOUNDS

A list of some nitro compounds and their derivatives appears in Table 23.17 of Section 23.6.

Classification Test

Ferrous Hydroxide Test. Organic compounds that are oxidizing agents will oxidize ferrous hydroxide (blue) to ferric hydroxide (brown). The most common organic compounds that function in this way are aliphatic and aromatic *nitro* compounds, which are reduced to amines by the reaction (Eq. 23.53). Other less common types of compounds that give the same test are nitroso compounds, hydroxylamines, alkyl nitrates, alkyl nitrites, and quinones.

$$RNO_2 \ + \ Fe(OH)_2 \ + \ 4\,H_2O \longrightarrow RNH_2 \ + \ 6\,Fe(OH)_3 \qquad \textbf{(23.53)}$$

A nitro   Ferrous hydroxide   An amine   Ferric hydroxide
compound   *Blue*   *Brown*

EXPERIMENTAL PROCEDURE

In a small test tube, mix about 20 mg of a solid or 1 drop of a liquid unknown with 1.5 mL of freshly prepared 5% ferrous ammonium sulfate solution. Add 1 drop of 3 *M* sulfuric acid and 1 mL of 2 *M* potassium hydroxide in *methanol*. Stopper the tube immediately and shake it. A positive test is indicated by the blue precipitate turning rust-brown within 1 min. A slight darkening or greenish coloration of the blue precipitate is *not* considered a positive test.

Finishing Touches   Flush the *solutions* down the drain.

Derivatives

Two different types of derivatives of nitro compounds may be prepared. Aromatic nitro compounds can be di- and trinitrated with nitric acid and sulfuric acid. Discussion of and procedures for nitration are given in Section 23.5.4.

The other method for preparation of a derivative may be utilized for both aliphatic and aromatic nitro compounds. This involves the reduction of the nitro compound to the corresponding primary amine (Eq. 23.54), followed by conversion of the amine to a benzamide or benzenesulfonamide, as described in Section 23.5.8. The reduction is most often carried out with tin and hydrochloric acid.

$$RNO_2 \ \text{ or } \ ArNO_2 \ \xrightarrow[\text{(2) NaOH}]{\text{(1) Sn, HCl}} \ R\ddot{N}H_2 \ \text{ or } \ Ar\ddot{N}H_2 \qquad \textbf{(23.54)}$$

A nitro compound   An amine

EXPERIMENTAL PROCEDURE

Perform the reduction of the nitro compound by combining 1 g of the compound and 2 g of granulated tin in a small flask. Attach a reflux condenser, and add 20 mL of 3 M hydrochloric acid in small portions. Shake the mixture after each addition. After addition is complete, warm the mixture for 10 min, using a steam bath. If the nitro compound is insoluble, add 5 mL of 95% ethanol to increase its solubility. Decant the warm, homogeneous solution into 10 mL of water, and add enough 12 M sodium hydroxide solution so that the tin hydroxide completely dissolves. Extract the basic solution with several 10-mL portions of diethyl ether. Dry the ethereal solution over potassium hydroxide pellets, decant the liquid, and remove the ether by simple distillation (*no flames*).

The residue contains the primary amine. Convert it to one of the derivatives described in Section 23.5.8.

> **Finishing Touches**   Transfer any recovered *diethyl ether* into the container for nonhalogenated organic liquids. Neutralize the *aqueous solution* before flushing it down the drain.

23.5.10   ESTERS

A list of esters appears in Table 23.13 of Section 23.6.

Classification Tests

Two tests for the ester functionality are the hydroxylamine test, a color reaction, and the saponification equivalent. The latter procedure can provide *quantitative* information regarding the number of ester groups present.

A. Hydroxylamine.   This test for an ester group involves the use of hydroxylamine and ferric chloride. The former converts the ester to a hydroxamic acid, which then complexes with Fe(III) to give a colored species (Eqs. 23.55 and 23.56).

$$R^1 \!-\! \overset{\overset{\displaystyle O}{\|}}{C} \!-\! OR^2 \; + \; H_2\ddot{N}OH \longrightarrow R^1 \!-\! \overset{\overset{\displaystyle O}{\|}}{C} \!-\! \ddot{N}HOH \; + \; R^2OH \qquad \textbf{(23.55)}$$

An ester   Hydroxylamine   A hydroxamic   An alcohol
                                          acid

$$R^1 \!-\! \overset{\overset{\displaystyle O}{\|}}{C} \!-\! \ddot{N}HOH \; + \; FeCl_3 \longrightarrow \left[R^1 \!-\! C \overset{\diagup O}{\underset{\diagdown \underset{H}{\overset{..}{N} - O}}{\diagdown}} \right]_3 \! Fe \; + \; 3\,HCl \qquad \textbf{(23.56)}$$

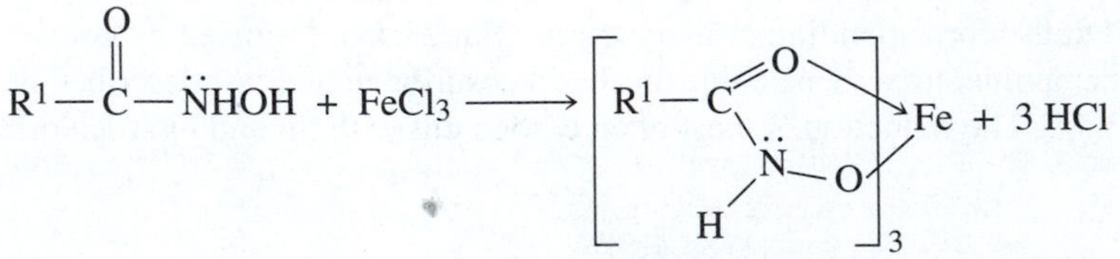

Iron complex
Colored

All carboxylic acid esters, including polyesters and lactones, give magenta colors that vary in intensity depending on other structural features of the molecule. Acid chlorides and anhydrides also give positive tests. Formic acid produces a red color, but other free acids give negative tests. Primary or secondary aliphatic nitro compounds give a positive test because ferric chloride reacts with the *aci* form, which is equivalent to the enol form of a ketone, that is present in basic solution. Most imides also provide positive tests. Some amides, but not all, produce light magenta coloration, whereas most nitriles give a negative test. A modification of the following procedure will yield a positive test for nitriles and amides, and the details of it are provided in Sections 23.5.11 and 23.5.12, respectively.

EXPERIMENTAL PROCEDURE

Before the final test is performed, it is necessary to run a preliminary test as a "blank" because some compounds produce a color in the absence of hydroxylamine hydrochloride and even though they do *not* contain an ester linkage.

Preliminary Test. Mix 1 mL of 95% ethanol and 50–100 mg of the compound to be tested, and add 1 mL of 1 *M* hydrochloric acid. Note the color that is produced when 1 drop of 0.6 *M* aqueous ferric chloride solution is added. If the color is orange, red, blue, or violet, the following test for the ester group does *not* apply and *cannot* be used.

Final Test. Mix 40–50 mg of the unknown, 1 mL of 0.5 *M* hydroxylamine hydrochloride in 95% ethanol, and 0.2 mL of 6 *M* sodium hydroxide. Heat the mixture to boiling, and after cooling it slightly, add 2 mL of hydrochloric acid. If the solution is cloudy, add about 2 mL more of 95% ethanol. Add 1 drop of 0.6 *M* ferric chloride, and observe any color. If needed, add more ferric chloride solution until the color persists. Compare the color obtained here with that from the preliminary test. If the color is burgundy or magenta, as compared to the yellow color in the preliminary experiment, the presence of an ester group is indicated.

Finishing Touches Neutralize the *aqueous solutions* before flushing them down the drain.

B. Saponification Equivalent. It is possible to carry out the hydrolysis of an ester with alkali in a *quantitative* manner so that the **saponification equivalent, SE,** results. This value is analogous to the equivalent weight of an acid (see Sec. 23.5.7) in that it is the molecular weight of the ester divided by the number of ester functions in the molecule. Therefore the SE is the number of grams of ester required to react with one gram-equivalent of alkali.

The SE of an ester is determined by hydrolyzing a weighed amount of the ester with standardized alkali and then titrating the excess alkali to a phenolphthalein end-point using standardized hydrochloric acid. The SE is then calculated as follows:

$$Saponification\ equivalent = \frac{grams\ of\ ester}{equivalents\ of\ alkali\ consumed}$$

$$= \frac{grams\ of\ ester}{(volume\ of\ alkali\ in\ liters)(N) - (volume\ of\ acid\ in\ liters)(N')}$$

where N is the normality of the standard base and N' is the normality of the standard acid.

EXPERIMENTAL PROCEDURE

Dissolve approximately 3 g of potassium hydroxide in 60 mL of 95% ethanol. Allow the small amount of insoluble material, if any, to settle to the bottom of the container, and by decantation fill a 50-mL buret with the clear supernatant. Measure 25.0 mL of the alcoholic solution into each of two round-bottom flasks. Quantitatively transfer an *accurately* weighed 0.3- to 0.4-g sample of pure, dry ester into one of the flasks; the other basic solution is used for a blank determination. Fit each flask with a reflux condenser.

Heat the solutions in both flasks under gentle reflux for 1 h. When the flasks have cooled, rinse each condenser with about 10 mL of distilled water, catching the rinse water in the flask. Add phenolphthalein, and separately titrate the solutions in each flask with *standardized* hydrochloric acid that is approximately 0.5 *M* in concentration.

The difference in the volumes of hydrochloric acid required to neutralize the base in the flask containing the sample and in the flask containing the blank corresponds to the amount of potassium hydroxide that reacted with the ester. The volume difference, in milliliters, multiplied by the molarity of the hydrochloric acid equals the number of *milli*moles of potassium hydroxide consumed. Using the titration data, calculate the saponification equivalent of the unknown ester.

The ester may not completely saponify in the allotted time, as evidenced by a *non*-homogeneous solution; if this is the case, heat the mixture under reflux for a longer period of time (2–4 h). Higher temperatures may be required in some cases; if so, diethylene glycol must be used as a solvent *in place of* the original 60 mL of 95% ethanol.

> **Finishing Touches** Flush the *neutralized solution* down the drain.

Derivatives

To characterize an ester completely, it is necessary to prepare solid derivatives of both the acid *and* the alcohol components. However, isolating both of these components in pure form so that suitable derivatives can be prepared may present problems. One way to do the isolation is to perform the ester hydrolysis in base (Eq. 23.57) in a high-boiling solvent. If the alcohol is low-boiling, it is distilled from the reaction mixture and characterized. The acid that remains in the mixture as a carboxylate salt is obtained upon acidification of the solution. Derivatives of alcohols and acids are discussed in Sections 23.5.5 and 23.5.7, respectively.

$$R^1CO_2R^2 + HO^{\ominus} \longrightarrow R^1CO_2^{\ominus} + R^2OH \xrightarrow{H_3O^+} R^1CO_2H \quad \textbf{(23.57)}$$

An ester   A carboxylate An alcohol   A carboxylic
                                  salt   acid

EXPERIMENTAL PROCEDURE

Mix 3 mL of diethylene glycol, 0.6 g (2 pellets) of potassium hydroxide, and 10 drops of water in a small round-bottom flask. Heat the mixture until the solution is homogeneous, and then cool it to room temperature. Add 1 mL of the ester and equip the flask with a few boiling stones and a reflux condenser. Reheat the mixture to boiling, with swirling; after the ester layer dissolves (3–5 min), allow the solution to cool. Equip the flask for simple distillation, and heat the solution strongly so that the alcohol distils; all but high-boiling alcohols can be removed by direct distillation. The distillate, which should be an alcohol in a fairly pure and dry state, is used directly for the preparation of a solid derivative (see Sec. 23.5.5).

The residue that remains after distillation contains the salt of the carboxylic acid. Add 10 mL of water to the residue and mix the two thoroughly. Acidify the resulting solution with 6 M sulfuric acid. Allow the mixture to stand, and collect any crystals by vacuum filtration. If crystals do not form, extract the aqueous acidic solution with small portions of diethyl ether or dichloromethane, dry the organic solution over anhydrous sodium sulfate, decant the liquid, and remove the solvent by distillation or by evaporation (in the hood). Use the residual acid to prepare a derivative (see Sec. 23.5.7).

> **Finishing Touches**   Flush the *pot residue* from the distillation of the alcohol down the drain. Neutralize the *aqueous filtrate* and also flush it down the drain. Transfer any *organic solvent* recovered by distillation to the appropriate container, that for nonhalogenated organic liquids in the case of *diethyl ether* and the one for halogenated organic liquids if *dichloromethane* was used.

23.5.11 NITRILES

A list of nitriles appears in Table 23.16 of Section 23.6.

Classification Test

A. Hydroxylamine.   A qualitative test used for nitriles is similar to the hydroxylamine test for esters (see Sec. 23.5.10A). Nitriles, as well as amides, typically give a colored solution on treatment with hydroxylamine and ferric chloride (Eq. 23.58).

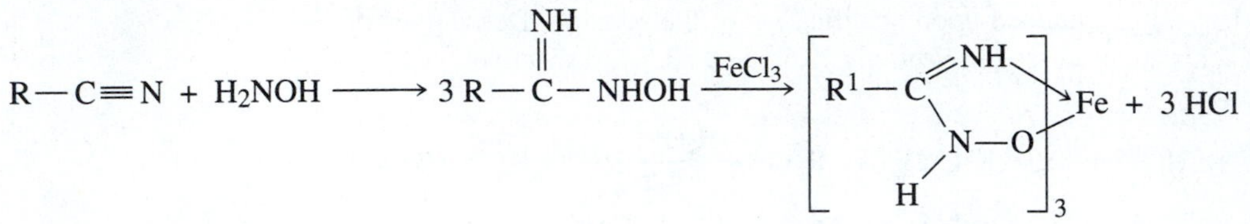

$$R-C\equiv N + H_2NOH \longrightarrow 3\ R-\overset{\overset{NH}{\|}}{C}-NHOH \xrightarrow{FeCl_3} \left[R^1-C{\overset{\diagup NH}{\diagdown N-O}} \right]_3 Fe + 3\ HCl \qquad \textbf{(23.58)}$$

A nitrile   Hydroxylamine   Iron complex
                                                              Colored

EXPERIMENTAL PROCEDURE

Prepare a mixture consisting of 2 mL of 1 *M* hydroxylamine hydrochloride in propylene glycol, 30–50 mg of the compound that has been dissolved in a minimum amount of propylene glycol, and 1 mL of 1 *M* potassium hydroxide. Heat the mixture to boiling for 2 min, and then cool it to room temperature; add 0.5–1.0 mL of a 0.5 *M alcoholic* ferric chloride solution. A red-to-violet color is a positive test. Yellow colors are negative, and brown colors and precipitates are neither positive nor negative.

Finishing Touches   Flush the *solutions* down the drain.

Derivatives

B. Hydrolysis.   On hydrolysis in either acidic or basic solution, nitriles are ultimately converted to the corresponding carboxylic acids (Eqs. 23.59 and 23.60). It is then possible to prepare a derivative of the acid using methods provided in Section 23.5.7.

Basic hydrolysis: $RC\equiv N + NaOH \xrightarrow{H_2O} RCO_2^{\ominus}\ Na^{\oplus} + {:}NH_3$ $\qquad$ **(23.59)**

$\quad$ A nitrile $\qquad\qquad\qquad\qquad$ A sodium carboxylate

Acidic hydrolysis: $RC\equiv N \xrightarrow[H_2SO_4]{H_2O} RCONH_2 \xrightarrow[H_2SO_4]{H_2O} RCO_2H + \overset{\oplus}{N}H_4$ $\quad$ **(23.60)**

$\quad$ A nitrile $\qquad\qquad$ An amide $\qquad\qquad$ A carboxylic
$\qquad\qquad\qquad\qquad\qquad\qquad\qquad\qquad\qquad$ acid

EXPERIMENTAL PROCEDURES

Basic Hydrolysis.   Mix 10 mL of 3 *M* sodium hydroxide solution and 1 g of the nitrile. Heat the mixture to boiling, and note the odor of ammonia, or hold a piece of moist pHydrion over the container and note the color change. After the mixture is homogeneous, cool and then acidify it to pHydrion paper. If the acid solidifies, collect the crystals by vacuum filtration. If it is a liquid, extract the acidic solution with small portions of diethyl ether, dry and decant the ethereal solution, and then remove the solvent by distillation (*no flames*). The residue is the acid. Prepare a suitable derivative of the acid (see Sec. 23.5.7).

> **Finishing Touches**   Neutralize all *aqueous solutions* before flushing them down the drain. Pour any *diethyl ether* isolated into the container for nonhalogenated organic liquids.

Acidic Hydrolysis.   Treat 1 g of the nitrile with 10 mL of concentrated sulfuric acid or concentrated hydrochloric acid, and warm the mixture to 50 °C for about 30 min. Dilute the mixture by addition of 20 mL of water. (*Caution:* add the mixture slowly *to* the water if sulfuric acid has been used.) Heat the mixture under gentle reflux for 30 min to 2 h and then allow it to cool. The acid usually forms a separate layer. If the acid solidifies upon cooling, collect it by vacuum filtration. If it is a liquid, extract the acidic mixture with small portions of diethyl ether, dry and decant the ethereal solution, and remove the solvent by distillation (*no flames*). Prepare suitable derivatives of the acid (see Sec. 23.5.7).

> **Finishing Touches**   Neutralize the *aqueous solution* before flushing it down the drain. Pour any *diethyl ether* isolated into the container for nonhalogenated organic liquids.

23.5.12 AMIDES

Classification Test

A list of amides appears in Table 23.8 of Section 23.6.

A. Hydroxylamine.   A qualitative test for an amide group (Eq. 23.61) is the same as that for an ester, as given in Section 23.5.10A. Follow the experimental procedure provided there. The colors observed with amides are the same as those with esters.

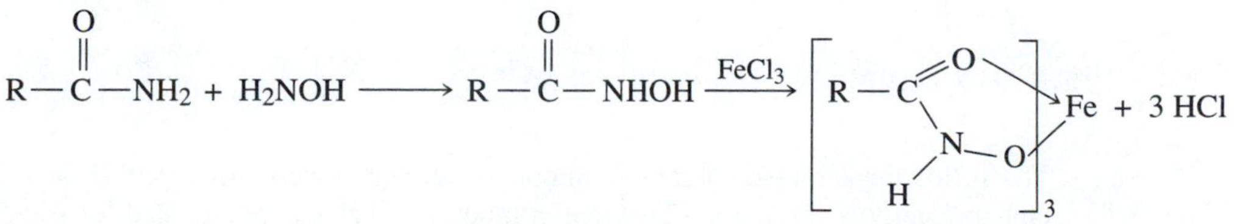

Iron complex
Colored

(23.61)

Derivatives

B. Hydrolysis.   Like nitriles (see Sec. 23.5.11), amides may be hydrolyzed under acidic or basic conditions, in this case giving an amine and a carboxylic acid (Eq. 23.62). In the instance of unsubstituted amides, ammonia is liberated. Substituted amides provide a substituted amine as a product. In those cases it is necessary to classify the amine as being primary or secondary and to prepare derivatives of both the acid (see Sec. 23.5.7) and the amine (see Sec. 23.5.8).

$$R^1CONR_2^2 \xrightarrow[\text{HO}^-]{\text{H}_3\text{O}^+ \text{ or}} R^1CO_2H + HNR_2^2 \tag{23.62}$$

An amide   A carboxylic An amine
(R = H, alkyl, aryl)   acid

EXPERIMENTAL PROCEDURE

Hydrolysis.   Carry out the procedure just as described for the hydrolysis of nitriles (see Sec. 23.5.11B). Distil the ammonia or volatile amine from the alkaline solution into a container of dilute hydrochloric acid. Neutralize this acidic solution, carry out the Hinsberg test (see Sec. 23.5.8A), and prepare a derivative of the amine (see Sec. 23.5.8). If the amine is *not* volatile, extract it from the aqueous layer with small portions of diethyl ether, dry the solution over potassium hydroxide pellets, decant it, and then remove the solvent by distillation to give the amine. After the amine is obtained, make the alkaline solution acidic and isolate the acid by either vacuum filtration or by extraction with diethyl ether. Characterize the acid by preparing a suitable solid derivative (see Sec. 23.5.7).

> **Finishing Touches**   Neutralize all *aqueous solutions* before flushing them down the drain. Pour any *diethyl ether* isolated into the container for nonhalogenated organic liquids.

23.6 TABLES OF COMPOUNDS AND DERIVATIVES

The following tables of organic compounds and their derivatives are arranged alphabetically according to classes of compounds. Tables are included for some functional groups for which derivatives are not described in Section 23.5

because you may be assigned unknowns that are to be identified solely on the basis of physical properties and spectroscopic methods.

The melting points (mp) and boiling points (bp), in °C, of the compounds listed represent the highest point in the range actually observed. Some compounds and derivatives may decompose at or below their melting or boiling points, and these are indicated by the designation *dec* beside the temperature. The abbreviation *di* beside some melting points indicates that the value is for the *di*substituted derivative. All boiling points are for a pressure of 760 torr.

Acid Anhydrides	Table 23.1
Acid Chlorides	Table 23.2
Alcohols	Table 23.3
Aldehydes	Table 23.4
Alkanes	Table 23.5
Alkenes and Alkynes	Table 23.6
Alkyl Halides	Table 23.7
Amides	Table 23.8
Amines	Table 23.9
Aromatic Hydrocarbons	Table 23.10
Aryl Halides	Table 23.11
Carboxylic Acids	Table 23.12
Esters	Table 23.13
Ethers	Table 23.14
Ketones	Table 23.15
Nitriles	Table 23.16
Nitro Compounds	Table 23.17
Phenols	Table 23.18

TABLE 23.1 Acid Anhydrides

Name of Compound	Boiling Point	Melting Point	Corresponding Carboxylic Acid	
			bp	mp
Acetic anhydride	140		118	
Propionic anhydride	167		141	
Isobutyric anhydride	182		155	
Butyric anhydride	198		164	
Dichloroacetic anhydride	216 *dec*		194	
cis-1,2-Cyclohexanedicarboxylic anhydride		34		192
Benzoic anhydride		42		122
Chloroacetic anhydride		46	189	63
Maleic anhydride		54		130
Glutaric anhydride		56		98
4-Methylbenzoic anhydride		95		179
Succinic anhydride		120		189
Phthalic anhydride		132		210
trans-1,2-Cyclohexanedicarboxylic anhydride		147		230
3-Nitrophthalic anhydride		162		218
4-Nitrobenzoic anhydride		189		241
d,l-Camphoric anhydride		225		187
Tetrachlorophthalic anhydride		256		250 *dec*
1,8-Naphthalic anhydride		274		270 *dec*
Tetrabromophthalic anhydride		280		266

TABLE 23.2 Acid Chlorides

Name of Compound	Boiling Point	Melting Point	Corresponding Carboxylic Acid	
			bp	mp
Acetyl chloride	52		118	
Oxalyl chloride	64			101
Methyl chloroformate	72			
Propionyl chloride	80		141	
Isobutyryl chloride	92		155	
Ethyl chloroformate	93			

(Continued)

TABLE 23.2   Acid Chlorides *(Continued)*

Name of Compound	Boiling Point	Melting Point	Corresponding Carboxylic Acid	
			bp	mp
Methacrylyl chloride	95		163	
Butyryl chloride	102		164	
Chloroacetyl chloride	108		189	63
Methoxyacetyl chloride	113		204	
Isovaleryl chloride	115		177	
Trichloroacetyl chloride	118		197	58
trans-Crotonyl chloride	126		189	72
Pentanoyl chloride	126		186	
Isobutyl chloroformate	129			
Hexanoyl chloride	153		205	
Fumaryl chloride	162			289 (200)
Cyclohexanecarboxylic acid chloride	184			31
Succinyl chloride	190 *dec*	20		189
Octanoyl chloride	196		239	16
Benzoyl chloride	197			122
Phenylacetyl chloride	210			76
Nonanoyl chloride	215		255	12
Glutaryl chloride	218			98
4-Chlorobenzoyl chloride	222			243
3-Chlorobenzoyl chloride	225			158
Phenoxyacetyl chloride	226			99
4-Methylbenzoyl chloride	226			179
2-Chlorobenzoyl chloride	238			142
Phthaloyl chloride	280	15		210
Adipoyl chloride	*dec*			154
Sebacoyl chloride	*dec*			134
trans-Cinnamoyl chloride	258	35		133
4-Bromobenzoyl chloride	247	42		251
Isophthaloyl chloride	276	44		348
2,4-Dinitrobenzoyl chloride		46		183
3,5-Dinitrobenzoyl chloride		69		207
4-Nitrobenzoyl chloride		75		241
Terephthaloyl chloride		84		300

TABLE 23.3   Alcohols   (a) Liquids

Name of Compound	Boiling Point	Melting Points of Derivatives	
		3,5-Dinitrobenzoate	α-Naphthylurethane
Methanol	65	108	124
Ethanol	78	93	79
2-Propanol	82	123	106
2-Methyl-2-propanol (*tert*-butyl alcohol)	83	142	101
3-Buten-2-ol	95	54	
2-Propen-1-ol (allyl alcohol)	97	50	108
1-Propanol	97	74	80
2-Butanol	100	76	97
2-Methyl-2-butanol	102	116	72
2-Methyl-1-propanol (isobutyl alcohol)	108	87	104
3-Buten-1-ol	113	59	
3-Methyl-2-butanol	114	76	109
3-Pentanol	116	101	95
1-Butanol	117	64	71
2-Pentanol	120	62	75
3,3-Dimethyl-2-butanol	120	107	
2,3-Dimethyl-2-butanol	120	111	101
3-Methyl-3-pentanol	123	96	84
2-Methyl-2-pentanol	123	72	104
2-Methoxyethanol	125		113
1-Chloro-2-propanol	127	77	
2-Methyl-3-pentanol	128	85	
2-Methyl-1-butanol	129	70	82
2-Chloroethanol	130	95	101
4-Methyl-2-pentanol	132	65	88
3-Methyl-1-butanol	132	61	68
2-Ethoxyethanol	135	75	67
3-Hexanol	136	97	72
2,2-Dimethyl-1-butanol	137	51	81
1-Pentanol	138	46	68
2-Hexanol	139	39	61
2,4-Dimethyl-3-pentanol	140		99
Cyclopentanol	141	115	118
4-Methyl-1-pentanol	153	72	58
4-Heptanol	156	64	80
1-Hexanol	157	60	62
2-Heptanol	159	49	54

(Continued)

TABLE 23.3 Alcohols (a) Liquids *(Continued)*

Name of Compound	Boiling Point	Melting Points of Derivatives	
		3,5-Dinitrobenzoate	α-Naphthylurethane
Cyclohexanol	161	112	129
3-Chloro-1-propanol	161	77	76
Furfuryl alcohol	172	81	130
1-Heptanol	177	47	62
2-Octanol	179	32	63
2-Ethyl-1-hexanol	185		60
1,2-Propanediol	187		
1-Octanol	195	62	67
1,2-Ethanediol	198	169	176
2-Nonanol	198	43	56
1-Linalool	199		53
1-Phenylethanol	202	95	106
Benzyl alcohol	206	113	134
1-Nonanol	213	52	66
1,3-Propanediol	215	178	164
2-Phenylethanol	220	108	119
Geraniol	230	63	48
1-Decanol	231	58	73
3-Phenyl-1-propanol	237	92	

TABLE 23.3 Alcohols (b) Solids

Name of Compound	Melting Point	Melting Points of Derivatives	
		3,5-Dinitrobenzoate	α-Naphthylurethane
Cinnamyl alcohol	33	121	114
α-Terpineol	36	79	152
1-Tetradecanol	39	67	82
Menthol	44	153	126
1-Hexadecanol	50	66	82
2,2-Dimethyl-1-propanol	52		100
4-Methylbenzyl alcohol	60	118	
1-Octadecanol	60	77	
Benzhydrol	68	141	139
4-Nitrobenzyl alcohol	93	157	
Benzoin	137		140
Lanosterol	140	201	
Cholesterol	148	195	176
Triphenylmethanol	161		

TABLE 23.4 Aldehydes (a) Liquids

Name of Compound	Boiling Point	Melting Points of Derivatives		
		Semi-carbazone	2,4-Dinitro-phenylhydrazone	Oxime
Ethanal (acetaldehyde)	20	169	168	47
Propanal (propionaldehyde)	48	89	150	40
Glyoxal	50	270	328	178
2-Propenal (acrolein)	52	171	165	
2-Methylpropanal (isobutyraldehyde)	64	126	187	oil
2-Methyl-2-propenal	68	198	206	
Butanal (*n*-butyraldehyde)	75	96	123	oil
Trimethylacetaldehyde	75	190	210	41
Chloroacetaldehyde	86	148		oil
3-Methylbutanal	93	107	123	49
Pentanal	103		98	52
2-Butenal	104	199	190	119
2-Ethylbutanal	117	99	129	
4-Methylpentanal	121	127	99	oil
Paraldehyde	125	169	168	47
Hexanal	131	106	104	51
5-Methylhexanal	144	117	117	
Heptanal	155	109	108	57
Furfural	162	202	230	92
Octanal	171	101	106	60
Benzaldehyde	179	222	237	35
Nonanal	185	100	100	64
Glutaraldehyde	189			178
Phenylethanal (phenylacetaldehyde)	194	153	121	100
Salicylaldehyde	197	231	252	63
3-Methylbenzaldehyde (*m*-tolualdehyde)	199	204	195	60
2-Methylbenzaldehyde (*o*-tolualdehyde)	200	209	194	49
4-Methylbenzaldehyde (*p*-tolualdehyde)	205	234	234	80
Decanal	207	102	104	69
2-Chlorobenzaldehyde	214	230	209	103
3-Chlorobenzaldehyde	214	228	256	70
3-Methoxybenzaldehyde	230	233		40
3-Bromobenzaldehyde	234	228		72
2-Ethoxybenzaldehyde	247	219		59
4-Methoxybenzaldehyde	248	210	254	64
Cinnamaldehyde	252	216	255	65

TABLE 23.4 Aldehydes (b) Solids

Name of Compound	Melting Point	Melting Points of Derivatives		
		Semi-carbazone	2,4-Dinitro-phenylhydrazone	Oxime
1-Naphthaldehyde	34	221		98
Phenylethanal (phenylacetaldehyde)	34	163	121	100
Piperonal	37	234	266	146
2-Methoxybenzaldehyde	39	215	254	92
4-Diethylaminobenzaldehyde	41	241		93
3,4-Dichlorobenzaldehyde	44		301	120
2-Nitrobenzaldehyde	44	256	265	102 (154)
3,4-Dimethoxybenzaldehyde	45	177	261	95
4-Chlorobenzaldehyde	48	233	254	110 (146)
2,3-Dimethoxybenzaldehyde	54	231		99
4-Bromobenzaldehyde	57	229	128 (257)	157 (111)
3-Nitrobenzaldehyde	58	246	293	122
2-Naphthaldehyde	60	245	270	156
3,5-Dichlorobenzaldehyde	65			112
2,6-Dichlorobenzaldehyde	71			150
2,4-Dimethoxybenzaldehyde	71			106
4-Aminobenzaldehyde	72	153		124
2-Chloro-4-nitrobenzaldehyde	74	234	247 *dec*	
2,4-Dichlorobenzaldehyde	74			137
4-Dimethylaminobenzaldehyde	74	222	325	185
3,4,5-Trimethoxybenzaldehyde	78	219		84
2-Chloro-5-nitrobenzaldehyde	79		277	176
4-Hydroxy-3-methoxybenzaldehyde	81	230	271 *dec*	122
3,5-Dibromosalicylaldehyde	85			220
Isophthalaldehyde	89			180
3-Hydroxybenzaldehyde	104	198	257 *dec*	90
5-Bromosalicylaldehyde	106	297 *dec*		126
4-Nitrobenzaldehyde	106	221	322	133
4-Hydroxybenzaldehyde	116	224	271	72 (112)
Terphthalaldehyde	116			200
2,4,6-Trimethoxybenzaldehyde	118			203
5-Nitrosalicylaldehyde	126			218
2,4-Dihydroxybenzaldehyde	136	260 *dec*	286	192
3,4-Dihydroxybenzaldehyde	154	230 *dec*	275 *dec*	157
3,5-Dihydroxybenzaldehyde	156	223		
Benzaldehyde-3-carboxylic acid	175	265		188 *dec*

* Values in parentheses are for different crystalline forms that may be isolated.

TABLE 23.5 Alkanes

Name of Compound	Boiling Point
Pentane	36
Cyclopentane	49
2,2-Dimethylbutane	50
2,3-Dimethylbutane	58
2-Methylpentane	60
3-Methylpentane	63
Hexane	69
Methylcyclopentane	72
2,2-Dimethylpentane	79
2,4-Dimethylpentane	80
Cyclohexane	81
3,3-Dimethylpentane	86
2,3-Dimethylpentane	89
2-Methylhexane	90
3-Methylhexane	92
3-Ethylpentane	93
Heptane	98
2,2,4-Trimethylpentane	99
Methylcyclohexane	101
2,5-Dimethylhexane	109
2-Methylheptane	118
Cycloheptane	119
trans-1,4-Dimethylcyclohexane	119
1,1-Dimethylcyclohexane	120
trans-1,2-Dimethylcyclohexane	123
Octane	126
2,2-Dimethylheptane	133
Cyclooctane	151
Nonane	151
Isopropylcyclohexane	154
1-Isopropyl-4-methylcyclohexane	169
Decane	174
trans-Decahydronaphthalene (*trans*-decalin)	187
Undecane	196

TABLE 23.6   Alkenes and Alkynes

Name of Compound	Boiling Point	Melting Point
1-Pentene	30	
2-Methyl-1-buten-3-yne	32	
2-Methyl-1,3-butadiene	34	
2-Methyl-2-butene	38	
1-Pentyne	40	
1,3-Pentadiene	41	
Cyclopentene	44	
4-Methyl-1-pentene	54	
3-Methyl-1-pentene	54	
2-Pentyne	56	
1,5-Hexadiene	59	
1-Hexene	63	
2-Ethyl-1-butene	65	
2,3-Dimethyl-1,3-butadiene	69	
1-Hexyne	71	
1,4-Hexadiene	72	
2,3-Dimethyl-2-butene	73	
2,4-Hexadiene	82	
3-Hexyne	82	
Cyclohexene	83	
2-Hexyne	84	
1-Heptene	94	
1-Heptyne	100	
2,4,4-Trimethyl-1-pentene	101	
4-Methylcyclohexene	103	
2,4,4-Trimethyl-2-pentene	105	
Cycloheptene	114	
1,3,5-Cycloheptatriene	116	
1-Octene	121	
1-Octyne	126	
4-Vinyl-1-cyclohexene	129	
2,5-Dimethyl-2,4-hexadiene	134	
1,3,5,7-Cyclooctatetraene	141	
Phenylethyne (phenylacetylene)	142	
Phenylethene (styrene)	145	
Cyclooctene	145	
1-Nonene	147	

(Continued)

TABLE 23.6 Alkenes and Alkynes *(Continued)*

Name of Compound	Boiling Point	Melting Point
1-Nonyne	151	
3-Phenylpropene (allylbenzene)	157	
d,l-Camphene	160	50
β-Pinene	164	
2-Phenylpropene	165	
Myrcene	166	
1-Phenylpropene	170	
Dicyclopentadiene	170	
1-Decene	171	
1-Decyne	174	
Limonene	178	
Indene	182	
1-Tetradecene	251	13
1,2-Diphenylethyne	298	62
trans-1,2-Diphenylethene (stilbene)		124

TABLE 23.7 Alkyl Halides (Haloalkanes)

Name of Compound	Boiling Point
(a) *Alkyl Chlorides*	
2-Chloropropane	36
1-Chloro-1-propene	37
Dichloromethane	41
3-Chloro-1-propene	45
1-Chloropropane	47
trans-1,2-Dichloroethene	48
2-Chloro-2-methylpropane	51
2-Chloro-1,3-butadiene	59
Chloroform	61
2-Chlorobutane	68
1-Chloro-2-methylpropane (isobutyl chloride)	68
1,1,1-Trichloroethane	74
Carbon tetrachloride	77
1-Chlorobutane	78
1-Chloro-2,2-dimethylpropane	84
1,2-Dichloroethane	84

(Continued)

TABLE 23.7 Alkyl Halides (Haloalkanes) *(Continued)*

Name of Compound	Boiling Point
2-Chloro-2-methylbutane	86
1,2-Dichloropropane	96
2-Chloropentane	97
3-Chloropentane	98
1-Chloro-3-methylbutane	101
1-Chloropentane	108
Chlorocyclopentane	115
3-Chloro-3-methylpentane	116
1,1,2,2-Tetrachloroethene	121
3-Chlorohexane	123
2-Chlorohexane	125
1,3-Dichloropropane	125
1-Chlorohexane	134
Chlorocyclohexane	143
1-Chloro-3-bromopropane	143
1-Chloroheptane	159
Benzyl chloride	179
1-Chlorooctane	180
2-Chloro-1-phenylethene	198
1-Chlorononane	203
α,α-Dichlorotoluene	205
m-Chlorobenzyl chloride	216
o-Chlorobenzyl chloride	217
p-Chlorobenzyl chloride	222
1-Chlorodecane	223
(b) *Alkyl Bromides*	
Bromoethane	38
2-Bromopropane	60
1-Bromopropane	71
3-Bromo-1-propene (allyl bromide)	71
2-Bromo-2-methylpropane (*tert*-butyl bromide)	73
2-Bromobutane	91
1-Bromo-2-methylpropane (isobutyl bromide)	93
Dibromomethane	99
1-Bromobutane	102
2-Bromopentane	117
3-Bromopentane	119
1-Bromo-3-methylbutane	120

(Continued)

TABLE 23.7   Alkyl Halides (Haloalkanes) *(Continued)*

Name of Compound	Boiling Point
3-Bromo-3-methylpentane	130
1-Bromopentane	130
1,2-Dibromoethane	132
Bromocyclopentane	137
1-Bromo-3-chloropropane	143
Bromoform	151
1-Bromohexane	155
Bromocyclohexane	165
1,3-Dibromopropane	168
1,4-Dibromobutane	198
1-Bromooctane	201
1-Bromononane	221
1-Bromodecane	241
(c) *Alkyl Iodides*	
Iodomethane	42
Iodoethane	72
2-Iodopropane	90
3-Iodo-1-propene (allyl iodide)	102
1-Iodopropane	103
2-Iodobutane	118
1-Iodo-2-methylpropane (isobutyl iodide)	120
1-Iodobutane	131
1-Iodopentane	157
Iodocyclopentane	167
Iodocyclohexane	179
1-Iodohexane	181
Diiodomethane	181
1-Iodoheptane	204
1-Iodooctane	225
Iodoform	119*

*Melting point

TABLE 23.8 Amides (a) Liquids

Name of Compound	Boiling Point
N,N-Dimethylformamide	153
N,N-Dimethylacetamide	165
N,N-Diethylformamide	178
N-Methylformamide	185
N,N-Diethylacetamide	186
Formamide	193 (195 *dec*)
N-Ethylformamide	199
N-Methyl-2-pyrrolidinone	202
N-Ethylacetamide	205
N-Methylformanilide	244
2-Pyrrolidinone (γ-butyrolactam)	250

TABLE 23.8 Amides (b) Solids

Name of Compound	Melting Point
N-Methylacetamide	31
δ-Valerolactam (2-piperidone)	39
Ethyl urethane	49
Formanilide	50
Methyl urethane	52
Phenyl urethane	53
N-Ethylacetanilide	54
Butyl urethane	54
Acetoacetamide	54
Propyl urethane	60
N-Benzylformamide	60
Pentananilide	63
Heptananilide	70
Decananilide	70
ε-Caprolactam	71
3-Butenamide	73
N,N-Diphenylformamide	73
Oleamide	76
N-Acetylacetamide	79
α-Chloropropionamide	80
Propionamide	81

(Continued)

TABLE 23.8 Amides   (b) Solids *(Continued)*

Name of Compound	Melting Point
N-Methylbenzamide	82
Acetamide	82
Acrylamide	85
Acetoacetanilide	86
3-Bromoacetanilide	87
2-Chloroacetanilide	88
N-Phenylmaleimide	91
Bromoactamide	91
2-Nitroacetanilide	92
Maleimide	93
Hexananilide	95
Iodoacetamide	95
Heptanamide	96
Butyranilide	96
m-Toluamide	97
Dichloroacetamide	98
Nonanamide	99
Hexanamide	100
N-Methylacetamide	102
Undecanamide	103
Isobutyranilide	105
Propionanilide	106
Hexadecanamide (palmitamide)	106
Pentanamide	106
Tetradecanamide (myristamide)	107
Decanamide	108
Heptadecanamide	108
Anthranilamide	109
Octadecanamide (stearamide)	109
Dodecanamide (lauramide)	110
Octanamide	110
β-Bromopropionamide	111
4-Aminobenzamide	114
Acetanilide	114
Butyramide	115
Methacrylamide	116
Chloroacetamide	118
Cyanoacetamide	120
Succinimide	126

(Continued)

TABLE 23.8 Amides (b) Solids *(Continued)*

Name of Compound	Melting Point
Isobutyramide	129
2-Methoxybenzamide	129
Benzamide	130
2-Ethoxybenzamide	130
Urea	133
3-Chlorobenzamide	134
Phenacetin	134
3-Methylbutanamide	136
Salicylanilide	136
2-Chlorobenzamide	142
Salicylamide	142
o-Toluamide	143
3-Nitrobenzamide	143
Cinnamamide	148
Trimethylacetamide (pivalamide)	155
2,5-Dichlorobenzamide	155
3-Bromobenzamide	155
Phenylacetamide	156
Succinic acid monoamide	157
N-(1-Naphthyl)acetamide	159
p-Toluamide	159
2-Bromobenzamide	161
4-Hydroxybenzamide	162
Benzanilide	163
3,4-Dimethoxybenzamide	164
4-Bromoacetanilide	167
4-Methoxybenzamide	167
4-Hydroxyacetanilide	169
3-Hydroxybenzamide	170
Malonamide (diamide)	170
N-Bromosuccinimide	173
2-Nitrobenzamide	176
4-Chloroacetanilide	179
4-Aminobenzamide	183
3,5-Dinitrobenzamide	183
2-Iodobenzamide	184
3-Iodobenzamide	186
4-Bromobenzamide	189

(Continued)

TABLE 23.8 Amides   (b) Solids *(Continued)*

Name of Compound	Melting Point
4-Nitrobenzamide	200
1-Naphthamide	202
2,6-Dichlorobenzamide	202
2,4-Dinitrobenzamide	203
4-Nitroacetanilide	215
4-Iodobenzamide	217
Hydantoin	218
Phthalamide (diamide)	220
2,4-Dihydroxybenzamide	222
Phthalimide	238
sym-Diphenylurea	240
Succinamide (diamide)	260 *dec*

TABLE 23.9 Amines   (a) Liquids

Name of Compound	Boiling Point	Melting Points of Derivatives			
		Benzamide	Benzene-sulfon-amide	Methio-dide	Picrate
Isopropylamine	33		26		
Ethylmethylamine	36				
tert-Butylamine	46	134			
n-Propylamine	49	84	36		
Diethylamine	56	42	42		
sec-Butylamine	63	76	70		
Isobutylamine	69	57	53		
n-Butylamine	77	42			
Diisopropylamine	84		94		
Pyrrolidine	89				
Triethylamine	89				173
2-Aminopentane	92				
Isopentylamine	96				
n-Pentylamine	104				
Piperidine	106	48	93		
Di-*n*-propylamine	110		51		
Ethylenediamine	116	244	168		

(Continued)

TABLE 23.9   Amines   (a) Liquids *(Continued)*

Name of Compound	Boiling Point	Melting Points of Derivatives			
		Benzamide	Benzene-sulfon-amide	Methio-dide	Picrate
Pyridine	116			117	167
2-Methylpyridine (2-picoline)	129			230	169
Morpholine	130	75	118		
n-Hexylamine	132	40	96		
Cyclohexylamine	134	149	89		
2-Dimethylaminoethyl alcohol	135				96
1,3-Diaminopropane	136	148	96		
Diisobutylamine	139		55		
2,6-Dimethylpyridine (2,6-lutidine)	143			233	168
3-Methylpyridine (3-picoline)	143				150
4-Methylpyridine (4-picoline)	146				167
n-Heptylamine	156				
Tri-*n*-propylamine	157				116
Di-*n*-butylamine	159				
1,4-Diaminobutane	159	177			
2-Aminoethanol	171				
2,4,6-Trimethylpyridine (2,4,6-collidine)	172				155
1,5-Diaminopentane	178	135	119		
n-Octylamine	180				
Benzyldimethylamine	181			179	
Benzylmethylamine	181				
Aniline	184	160	112		
Benzylamine	185	105	88		
1-Amino-1-phenylethane	187	120			
N, N-Dimethylaniline	193			228 *dec*	163
N-Methylaniline	196	63	79		
2-Amino-1-phenylethane	198	116	69		
2-Methylaniline (*o*-toluidine)	200	146	124		
n-Nonylamine	201	49			
3-Methylaniline (*m*-toluidine)	203	125	95		
N-Ethylaniline	205	60			
2-Chloroaniline	208	99	129		
4-Methylbenzylamine	208	137			
Tri-*n*-butylamine	211			186	105

(Continued)

TABLE 23.9   Amines   (a) Liquids (*Continued*)

		Melting Points of Derivatives			
Name of Compound	Boiling Point	Benzamide	Benzene-sulfon-amide	Methio-dide	Picrate
2,6-Dimethylaniline	215	168			
2,5-Dimethylaniline	215	140	138		
2,4-Dimethylaniline	216	192	130		
N,N-Diethylaniline	218			102	142
3,5-Dimethylaniline	220	144			
2,3-Dimethylaniline	221	189			
2-Methyoxyaniline (*o*-anisidine)	225	60	89		
4-Isopropylaniline	225	162			
2,4,6-Trimethylaniline	229	204	137		
3-Chloroaniline	230	119	121		
Quinoline	237			72* 133†	203
2-Chloro-6-methoxyaniline	246	135			
4-Ethoxyaniline	248	173	143		
3-Bromoaniline	251	120			
3-Methoxyaniline (*m*-anisidine)	251				
Dicyclohexylamine	255	153			
Tri-*n*-pentylamine	257				
Dibenzylamine	300	112	68		

*Hydrated.
†Anhydrous.

TABLE 23.9   Amines   (b) Solids

		Melting Points of Derivatives			
Name of Compound	Melting Point	Benzamide	Benzene-sulfon-amide	Methio-dide	Picrate
2-Bromoaniline	32	116			
3-Iodoaniline	33	157			
N-Benzylaniline	37	107	119		
2,6-Dichloroaniline	39		157		
1,6-Diaminohexane	42	155	154		

(*Continued*)

TABLE 23.9   Amines   (b) Solids *(Continued)*

Name of Compound	Boiling Point	Melting Points of Derivatives			
		Benzamide	Benzene-sulfon-amide	Methio-dide	Picrate
4-Methylaniline (*p*-toluidine)	44	158	120		
3,4-Dimethylaniline	49	185	118		
2,5-Dichloroaniline	50	120			
3,5-Dichloroaniline	50	147			
Indole	52	68	254		
Diphenylamine	53	180	124		
2-Aminopyridine	57	165			
4-Methoxyaniline (*p*-anisidine)	58	154	95		
4-Iodoaniline	62	222			
1,3-Diaminobenzene (*m*-phenylenediamine)	63	125 240	194		
2,4-Dichloroaniline	63	117	128		
4-Bromoaniline	66	204	134		
2-Nitroaniline	71	110	104		
4-Chloroaniline	72	192	122		
3,4-Dichloroaniline	72	144	130		
8-Hydroxyquinoline	75	240		143 *dec*	204
4-Methyl-3-nitroaniline	78	172	160		
2,4,6-Trichloroaniline	78	174	154		
2,4-Dibromoaniline	79	134			
3,4-Diaminotoluene	89	264	179		
Tribenzylamine	91			184	190
2-Methyl-3-nitroaniline	92	168			
2-Methyl-6-nitroaniline	97	167			
2,4-Diaminotoluene	99	224	192		
1,2-Diaminobenzene (*o*-phenylenediamine)	102	301	185		
2-Bromo-4-nitroaniline	105	160			
4-Aminoacetophenone	106	205	128		
2-Chloro-4-nitroaniline	107	161			
2-Methyl-5-nitroaniline	107	186	172		
3-Nitroaniline	114	157	136		
4-Methyl-2-nitroaniline	115	148	102		
4-Chloro-2-nitroaniline	116	133			
2,4,6-Tribromoaniline	122	198			
Triphenylamine	127				

(Continued)

TABLE 23.9 Amines (b) Solids

Name of Compound	Boiling Point	Melting Points of Derivatives			
		Benzamide	Benzene-sulfon-amide	Methio-dide	Picrate
2-Nitro-4-methoxyaniline	129	140			
2-Methyl-4-nitroaniline	130		158		
2-Methoxy-4-nitroaniline	139	150	181		
1,4-Diaminobenzene (p-phenylenediamine)	142	300	247		
4-Nitroaniline	147	199	139		
2-Aminobenzoic acid (anthranilic acid)	147	182	214		
4-Nitro-N-methylaniline	152	112	121		
4-Aminopyridine	159	202			
2-Hydroxyaniline	174	167	141		
3-Aminobenzoic acid	174	113			
2,4-Dinitroaniline	180	202			
4-Aminophenol	184	216	125		
4-Aminobenzoic acid	188	278	212		
2,4,6-Trinitroaniline	190	196	211		

TABLE 23.10 Aromatic Hydrocarbons

Name of Compound	Melting Point	Boiling Point	Nitration Product	
			Position	mp
Benzene	5	80	1,3	89
Toluene		111	2,4	70
Ethylbenzene		136	2,4,6	37
p-Xylene	13	138	2,3,5	139
m-Xylene		139	2,4,6	183
o-Xylene		142	4,5	118
Isopropylbenzene (cumene)		153	2,4,6	109
n-Propylbenzene		159	2,4	liquid
1,3,5-Trimethylbenzene		165	2,4	86
			2,4,6	235

(Continued)

TABLE 23.10 Aromatic Hydrocarbons *(Continued)*

Name of Compound	Melting Point	Boiling Point	Nitration Product Position	mp
tert-Butylbenzene		169	2,4	62
			2,4,6	124
1,2,4-Trimethylbenzene		169	3,5,6	185
1,2,3-Trimethylbenzene		176		
Indane		177	5	40
4-Isopropyltoluene		177	2,6	54
Indene		182		
1,2,3,5-Tetramethylbenzene		198	4,6	181(157)
1,3-Diisopropylbenzene		203	4,6	77
1,2,3,4-Tetramethylbenzene		205	5,6	176
1,3-Dimethyl-5-*tert*-butylbenzene		206	2,4,6	107(114)
1,2,3,4-Tetrahydronaphthalene (tetralin)		207	5,7	95
1,4-Diisopropylbenzene		210		
1-Phenylhexane		226		
Cyclohexylbenzene		236		
1-Methylnaphthalene		245	4	71
Diphenylmethane	27	264	2,2',4,4'	172
2-Methylnaphthalene	38	240	1	81
1,2-Diphenylethane	53	284	4,4'	180
			2,2',4,4'	169
Pentamethylbenzene	54	232	6	154
Biphenyl	69	254	4,4'	237
			2,2',4,4'	150
1,2,4,5-Tetramethylbenzene (durene)	80	198	3,6	205
Naphthalene				
Triphenylmethane	92	358	4,4',4"	206
Acenaphthene	96	278	5	101
Phenanthrene	101	340		
2,3-Dimethylnaphthalene	104	266		
2,6-Dimethylnaphthalene	111	262		
Fluorene	114	295	2	156
			2,7	199
trans-Stilbene	124			
1,4-Diphenyl-1,3-butadiene	152			
Anthracene	216			

TABLE 23.11 Aryl Halides

Name of Compound	Boiling Point	Melting Point	Nitration Product	
			Position	mp
Chlorobenzene	132		2,4	52
Bromobenzene	156		2,4	70
2-Chlorotoluene	159		3,5	63
3-Chlorotoluene	162		4,6	91
4-Chlorotoluene	162	7	2	38
1,3-Dichlorobenzene	173		4,6	103
1,2-Dichlorobenzene	180		4,5	110
2-Bromotoluene	182		3,5	82
3-Bromotoluene	184		4,6	103
4-Bromotoluene	184	28	2	47
Iodobenzene	188		4	171
2,6-Dichlorotoluene	199		3	50
2,4-Dichlorotoluene	200		3,5	104
3,4-Dichlorotoluene	201		2,6	91
3,5-Dichlorotoluene	201	26	2,6	99
3-Iodotoluene	204		4,6	108
1,2,4-Trichlorobenzene	213	17	5	56
1,2-Dibromobenzene	225	7	4,5	114
1-Chloronaphthalene	259		4,5	180
1-Bromonaphthalene	281	6	4	85
3,5-Dichlorotoluene	201	26	2,6	99
4-Bromotoluene	184	28	2	47
2,4,6-Trichlorotoluene		34	3	54
4-Iodotoluene	211	35		
1,2,3,4-Tetrachlorobenzene	275	46	5	64
			5,6	151
1,2,3,5-Tetrachlorobenzene	246	51	4	41
			4,6	162
1,4-Dichlorobenzene	173	53	2	54
			2,6	106
1,3,5-Trichlorobenzene	208	63	2	68
			2,4	131
1-Bromo-4-chlorobenzene	197	67	2	72
4-Chlorobiphenyl	293	77		
2,4,5-Trichlorotoluene		82	3	92
			3,6	227
1,4-Dibromobenzene	219	89	2,5	84
1,3,5-Tribomobenzene	271	120		
4,4'-Dichlorobiphenyl	315	149	2	102
			2,2'	138

TABLE 23.12   Carboxylic Acids   (a) Liquids

Name of Compound	Boiling Point	Melting Points of Derivatives		
		Anilide	*p*-Toluidide	Amide
Methanoic acid (formic acid)	101	50	53	
Ethanoic acid (acetic acid)	118	114	147	82
Propenoic acid	140	105	141	85
Propanoic acid (propionic acid)	141	106	126	81
2-Methylpropanoic acid (isobutyric acid)	155	105	107	129
2-Methylpropenoic acid	163			109
Butanoic acid	164	96	75	115
3-Butenoic acid	164	58		73
cis-2-Butenoic acid	169	102	132	102
3-Methylbutanoic acid (isovaleric acid)	177	110	107	137
3,3-Dimethylbutanoic acid	184	132	134	132
2-Chloropropanoic acid	186	92	124	80
Pentanoic acid (valeric acid)	186	63	74	106
Dichloroacetic acid	194	118	153	98
4-Methylpentanoic acid (isocaproic acid)	199	112	63	121
2-Bromopropanoic acid	205 *dec*	99	125	123
Hexanoic acid (caproic acid)	205	95	75	100
2-Bromobutanoic acid	217 *dec*	98	92	112
Heptanoic acid	223	70	81	96
2-Ethylhexanoic acid	228			102
Octanoic acid	239	57	70	110

TABLE 23.12   Carboxylic Acids   (b) Solids

Name of Compound	Melting Point	Boiling Point	Melting Points of Derivatives		
			Anilide	*p*-Toluidide	Amide
2-Bromopropanoic acid	26	205 *dec*	99	125	123
Undecanoic acid	29	284	71	89	103
Cyclohexanecarboxylic acid	31	233	146		186
Decanoic acid	32	270	70	78	108
2-Oxobutanoic acid	32				117
4-Oxopentanoic acid	34	245	102	108	108 *dec*
Pivalic acid (trimethylacetic acid)	36	164	130	120	157
Acetoacetic acid (3-oxobutanoic acid)	37	*dec*	86	95	54

(Continued)

TABLE 23.12 Carboxylic Acids (b) Solids (Continued)

Name of Compound	Melting Point	Boiling Point	Melting Points of Derivatives		
			Anilide	*p*-Toluidide	Amide
3-Chloropropanoic acid	42	204			101
2-Phenylbutanoic acid	42	270			85
Dodecanoic acid	44	299	78	87	100
Hydrocinnamic acid (3-phenylpropanoic acid)	48	280	98	135	105
Bromoacetic acid	50	208	131	91	91
4-Phenylbutanoic acid	52	290			84
Tetradecanoic acid	54		84	93	107
Trichloroacetic acid	58	198	97	113	141
5-Phenylpentanoic acid	60		90		109
3-Bromopropanoic acid	61				111
Hexadecanoic acid	62		91	98	106
Chloroacetic acid	63	189	137	162	118
cis-2-Methyl-2-butenoic acid	65	199	77	71	76
Cyanoacetic acid	66		198		120
Benzoylformic acid	66				91
Octadecanoic acid	70		96	102	109
trans-2-Butenoic acid (crotonic acid)	72	189	118	132	160
m-Methoxyphenylacetic acid	73				125
Phenylacetic acid	76		117	136	156
Glycolic acid (hydroxyacetic acid)	79		97	143	120
α-Methylcinnamic acid	81				128
Iodoacetic acid	83		144		95
p-Methoxyphenylacetic acid	87				189
o-Methylphenylacetic acid	90				161
3-Chloro-2-butenoic acid	94		124		101
o-Chlorophenylacetic acid	95		139	170	175
p-Methylphenylacetic acid	95				185
3,4-Dimethoxyphenylacetic acid	95				147
Glutaric acid (pentanedioic acid)	98		224	218	176
3-Phenoxypropionic acid	98				119
Phenoxyacetic acid	99		99		102
Citric acid hydrate	100		192	189	210 *dec*
2-Methoxybenzoic acid (*o*-anisic acid)	101	200	131		129
Malic acid (2-hydroxybutanedioic acid)	101		197	207	157
Oxalic acid dihydrate	101		254	268	419 *dec*

(Continued)

TABLE 23.12   Carboxylic Acids   (b) Solids *(Continued)*

Name of Compound	Melting Point	Boiling Point	Melting Points of Derivatives		
			Anilide	*p*-Toluidide	Amide
o-Toluic acid (2-methylbenzoic acid)	104		125	144	143
Heptanedioic acid	105			206	175
Nonanedioic acid	107		186	201	172
3-Methoxybenzoic acid (*m*-anisic acid)	110				136
Ethylmalonic acid	111		150		214
m-Toluic acid	112		126	118	95
2-Phenylbenzoic acid	113				177
2-Phenoxybenzoic acid	113				131
p-Bromophenylacetic acid	114				194
2-Acetylbenzoic acid	115				116
Methylsuccinic acid	115		200	164	225
β-Benzoylpropionic acid	116		150		146
2,6-Dimethylbenzoic acid	116				139
4-Isopropylbenzoic acid	117				133
Benzylmalonic acid	117 *dec*		217		225
2-Phenyl-3-hydroxypropanoic acid	117				169
Mandelic acid	118		152	172	132
m-Nitrophenylacetic acid	120				110
3-Furoic acid	121				169
Benzoic acid	122		160	158	130
Picric acid	123				
2,4-Dimethylbenzoic acid	127		141		180
2-Benzoylbenzoic acid	127		195		165
Dodecanedioic acid	128		191	165	185
Maleic acid (*cis*-butenedioic acid)	130		187	142	260
1-Naphthylacetic acid	132		155		180
2,5-Dimethylbenzoic acid	132		140		186
m-Chlorocinnamic acid	133		135	142	76
2-Furoic acid	133		124	108	143
trans-Cinnamic acid	133		153	168	148
Decanedioic acid (sebacic acid)	134		202	201	210
Malonic acid (propanedioic acid)	135		230	253	170
o-Acetylsalicylic acid	135		136		138
1,3-Acetonedicarboxylic acid	135 *dec*		155		
Pyridine-2-carboxylic acid	137		76	104	107
Phenylpropynoic acid	137		126	142	100

(Continued)

TABLE 23.12 Carboxylic Acids (b) Solids *(Continued)*

Name of Compound	Melting Point	Boiling Point	Melting Points of Derivatives		
			Anilide	*p*-Toluidide	Amide
Methylmalonic acid	138 *dec*		182	228	217
5-Chloro-2-nitrobenzoic acid	139		164		154
3-Nitrobenzoic acid	140		154	162	143
meso-Tartaric acid	140				187
2-Chloro-4-nitrobenzoic acid	141		168		172
o-Nitrophenylacetic acid	141				161
2,4-Dichlorophenoxyacetic acid	141				130
4-Chloro-2-nitrobenzoic acid	142				172
2-Naphthylacetic acid	142				200
2-Chlorobenzoic acid	142		118	131	142
Octanedioic acid	144		186	218	217
2,4,5-Trimethoxybenzoic acid	144		155		185
2,6-Dichlorobenzoic acid	144				202
o-Chlorophenoxyacetic acid	146		121		150
2-Nitrobenzoic acid	146		155		176
2-Aminobenzoic acid	147		131	151	109
Diphenylacetic acid	148		180	173	168
p-Hydroxyphenylacetic acid	148				175
2-Bromobenzoic acid	150		141		155
Benzilic acid	150		175	190	154
Citric acid (anhydrous)	153		192	189	210 *dec*
p-Nitrophenylacetic acid	153		198	210	198
2,5-Dichlorobenzoic acid	153				155
Hexanedioic acid (adipic acid)	154		241	239	220
3-Bromobenzoic acid	155		136		155
2,4,6-Trimethylbenzoic acid	155				188
p-Chlorophenoxyacetic acid	156		125		133
Hydroxymalonic acid	157 *dec*				198
3-Chlorobenzoic acid	158		123		134
Salicylic acid (2-hydroxybenzoic acid)	158		136	156	142
1-Naphthoic acid	162		163		205
2-Iodobenzoic acid	162		141		110
4-Nitrophthalic acid	164		192		200 *dec*
2,4-Dichlorobenzoic acid	164				194
3,4-Dinitrobenzoic acid	165		189		166
Propene-2,3-dicarboxylic acid	166				192 *di*

(Continued)

TABLE 23.12   Carboxylic Acids   (b) Solids *(Continued)*

Name of Compound	Melting Point	Boiling Point	Melting Points of Derivatives		
			Anilide	*p*-Toluidide	Amide
5-Bromosalicylic acid	165		222		232
2-Chloro-5-nitrobenzoic acid	165				178
1,2,3-Propane tricarboxylic acid	166		252		207 *dec*
3,4-Dimethylbenzoic acid	166		104		130
3-Methylsalicylic acid	166		83		112
3,5-Dimethylbenzoic acid	166				133
d-Tartaric acid	170		264 *dec*		196 *dec*
3,4,5-Trimethoxybenzoic acid	171				177
5-Chlorosalicylic acid	172				227
3-Aminobenzoic acid	174		140		111
3,5-Dinitrosalicylic acid hydrate	174				181
Acetylenedicarboxylic acid	179				249 *dec*
p-Toluic acid (4-methylbenzoic acid)	179	275	145	160	160
3,4-Dimethoxybenzoic acid	181		154		164
4-Chloro-3-nitrobenzoic acid	182		131		156
2,4-Dinitrobenzoic acid	183				203
4-Methoxybenzoic acid (*p*-anisic acid)	185		171	186	167
2-Naphthoic acid	185		171	192	192
3-Iodobenzoic acid	187				186
Coumarin-3-carboxylic acid	188		250		236
p-Nitrophenoxyacetic acid	188		170		158
4-Aminobenzoic acid	188				183
Succinic acid (butanedioic acid)	189		230	255	260 *dec*
Hippuric acid	190		208		183
Dimethylmalonic acid	193				269 *di*
4-Ethoxybenzoic acid	198		170		202
trans-m-Nitrocinnamic acid	199				196
3,4-Dihydroxybenzoic acid	200 *dec*		166		212
Fumaric acid	200		314		266
3-Hydroxybenzoic acid	200		157	163	170
2,5-Dihydroxybenzoic acid	204				218
d,l-Tartaric acid	204		236		226
2,3-Dihydroxybenzoic acid	204				175
3,5-Dinitrobenzoic acid	207		234	147	183
3,4-Dichlorobenzoic acid	209				168
Phthalic acid	210		253	201	220

(Continued)

TABLE 23.12 Carboxylic Acids (b) Solids *(Continued)*

Name of Compound	Melting Point	Boiling Point	Melting Points of Derivatives		
			Anilide	*p*-Toluidide	Amide
o-Chlorocinnamic acid	212		176		168
2,4-Dihydroxybenzoic acid	213		126		222
trans-o-Hydroxycinnamic acid	214 *dec*				209 *dec*
4-Hydroxybenzoic acid	215		197	204	162
3-Nitrophthalic acid	218		234	226	201 *dec*
4-Cyanobenzoic acid	219		179		223
4-Phenylbenzoic acid	226				223
Piperonylic acid	229				169
5-Nitrosalicylic acid	230		224		225
3-Chloro-2-nitrobenzoic acid	235		186		
trans-o-Nitrocinnamic acid	240				185
4-Nitrobenzoic acid	241		211	204	201
4-Chlorobenzoic acid	243		194		179
4-Dimethylaminobenzoic acid	245		183		206
4-Bromobenzoic acid	251		197		190

TABLE 23.13 Esters (a) Liquids

Name of Compound	Boiling Point
Methyl formate	31
Ethyl formate	54
Methyl acetate	57
Isopropyl formate	71
Vinyl acetate	73
Ethyl acetate	77
Methyl propionate	80
Methyl acrylate	80
Propyl formate	81
Isopropyl acetate	90
Methyl carbonate	91
Methyl isobutyrate	93
Isopropenyl acetate	94
tert-Butyl acetate	98

(Continued)

TABLE 23.13   Esters   (a) Liquids *(Continued)*

Name of Compound	Boiling Point
Methyl methacrylate	100
Ethyl propionate	100
Ethyl acrylate	101
Propyl acetate	102
Methyl butyrate	102
Allyl acetate	104
Ethyl isobutyrate	110
Isopropyl propionate	110
sec-Butyl acetate	112
Methyl isovalerate	117
Isobutyl acetate	117
Ethyl pivalate (ethyl trimethylacetate)	118
Methyl crotonate	119
Ethyl butyrate	122
Propyl propionate	123
Butyl acetate	126
Diethyl carbonate	127
Methyl valerate	127
Methyl methoxyacetate	130
Methyl chloroacetate	131
Ethyl isovalerate	135
Methyl pyruvate	137
Methyl α-hydroxyisobutyrate	137
Ethyl crotonate	138
3-Methylbutyl acetate (isoamyl acetate)	142
Methyl lactate	145
Ethyl chloroacetate	145
2-Methoxyethyl acetate	145
Ethyl valerate	146
Ethyl α-chloropropionate	146
Diisopropyl carbonate	147
Pentyl acetate	149
Methyl hexanoate	151
Cyclopentyl acetate	153
Ethyl lactate	154
Ethyl pyruvate	155
Ethyl dichloroacetate	158
Ethyl α-bromopropionate	162

(Continued)

TABLE 23.13   Esters   (a) Liquids *(Continued)*

Name of Compound	Boiling Point
Butyl butyrate	167
Ethyl hexanoate	168
Ethyl trichloroacetate	168
Methyl acetoacetate	170
Hexyl acetate	172
Methyl heptanoate	172
Cyclohexyl acetate	175
Furfuryl acetate	176
Ethyl β-bromopropionate	179
Ethyl acetoacetate	181
Methyl furoate	181
Dimethyl malonate	182
Diethyl oxalate	185
Ethyl δ-chlorobutyrate	186
Ethyl heptanoate	187
Ethylene glycol diacetate	190
Heptyl acetate	192
Methyl octanoate	193
Dimethyl succinate	196
Ethyl cyclohexanecarboxylate	196
Dimethyl methylsuccinate	196
Phenyl acetate	197
Diethyl malonate	199
Methyl benzoate	199
γ-Butyrolactone	204
Dimethyl maleate	204
Ethyl levulinate	206
γ-Valerolactone	207
Ethyl octanoate	208
Octyl acetate	210
Ethyl benzoate	212
Dimethyl glutarate	215
Methyl nonanoate	215
Benzyl acetate	217
Diethyl succinate	217
Diethyl fumarate	218
Methyl phenylacetate	220
Diethyl maleate	223

(Continued)

TABLE 23.13   Esters   (a) Liquids *(Continued)*

Name of Compound	Boiling Point
Methyl salicylate	224
Methyl decanoate	225
Ethyl phenylacetate	228
Propyl benzoate	231
Diethyl glutarate	234
Ethyl salicylate	234
Methyl β-phenylpropionate	238
Propylene carbonate	240
Diethyl adipate	245
Methyl undecylenate	248
Diethyl pimelate	255
Ethyl benzoylacetate	265
Dimethyl suberate	268
Ethyl cinnamate	271
Methyl 2-nitrobenzoate	275
Diethyl tartrate	280
Diethyl suberate	282
Dimethyl phthalate	284
Diethyl phthalate	290
Ethyl 3-aminobenzoate	294
Diethyl benzylmalonate	300
Methyl myristate	323
Diisobutyl phthalate	327
Dibutyl phthalate	340

TABLE 23.13   Esters   (b) Solids

Name of Compound	Melting Point
Dimethyl succinate	18
Methyl myristate	18
Diethyl tartarate	18
Benzyl benzoate	21
Methyl anthranilate (methyl 2-aminobenzoate)	24
Dimethyl sebacate	27
Bornyl acetate	29

(Continued)

TABLE 23.13 Esters (b) Solids *(Continued)*

Name of Compound	Melting Point
Methyl palmitate	30
Ethyl 2-nitrobenzoate	30
Methyl 4-toluate	33
Ethyl stearate	33
Ethyl 2-furoate	36
Methyl cinnamate	36
Ethylene carbonate	37
Ethyl mandelate	37
Dimethyl itaconate	39
Methyl stearate	39
Phenyl salicylate	42
Diethyl terephthalate	44
Ethyl 3-nitrobenzoate	47
Dimethyl tartrate	49
1-Naphthyl acetate	49
Methyl mandelate	53
Dimethyl oxalate	54
Ethyl 4-nitrobenzoate	56
Coumarin	67
Dimethyl isophthalate	68
Phenyl benzoate	69
Methyl 3-hydroxybenzoate	70
Diphenyl phthalate	74
Diphenyl carbonate	78
Methyl 3-nitrobenzoate	78
Methyl 4-bromobenzoate	83
Ethyl 4-aminobenzoate	90
Dimethyl *d,l*-tartarate	90
3-Carbethoxycoumarin	93
Methyl 4-nitrobenzoate	96
Propyl 4-hydroxybenzoate	96
Dimethyl fumarate	102
Cholesteryl acetate	114
Ethyl 4-hydroxybenzoate	116
Hydroquinone diacetate	124
Methyl 4-hydroxybenzoate	131
Ethyl 4-nitrocinnamate	137
Dimethyl terephthalate	141
Propyl gallate	150

TABLE 23.14 Ethers

Name of Compound	Boiling Point	Melting Point
Furan	31	
Diethyl ether	35	
Ethyl vinyl ether	36	
Methyl *n*-propyl ether	39	
Ethyl isopropyl ether	53	
tert-Butyl methyl ether	55	
Ethyl *n*-propyl ether	64	
Tetrahydrofuran	66	
Diisopropyl ether	68	
2-Methyltetrahydrofuran	79	
1,2-Dimethoxyethane	85	
3,4-Dihydropyran	86	
Tetrahydropyran	88	
Di-*n*-propyl ether	90	
n-Butyl vinyl ether	94	
1,4-Dioxane	101	
β-Chloroethyl ethyl ether	107	
1,2-Epoxy-3-chloropropane	117	
Diisobutyl ether	123	
Di-*n*-butyl ether	142	
Anisole	155	
Diethylene glycol dimethyl ether	162	
o-Methylanisole	171	
Ethoxybenzene	172	
p-Methylanisole	174	
m-Methylanisole	176	
2,2′-Dichloroethyl ether	178	
Di-*n*-pentyl ether	188	
3-Chloroanisole	194	
2-Chloroanisole	195	
4-Chloroanisole	200	
1,2-Dimethoxybenzene	207	
Butyl phenyl ether	210	
1,3-Dimethoxybenzene	217	
Di-*n*-hexyl ether	229	
Safrole	233	
4-Propenylanisole	235	
2-Nitroanisole	277	

(Continued)

TABLE 23.14   Ethers *(Continued)*

Name of Compound	Boiling Point	Melting Point
Dibenzyl ether	298	
Diphenyl ether	258	28
2-Ethoxynaphthalene	282	36
3-Nitroanisole	258	39
1,2,3-Trimethoxybenzene	241	47
4-Iodoanisole	240	52
1,3,5-Trimethoxybenzene	255	53
4-Nitroanisole	274	54
1,4-Dimethoxybenzene	213	56

TABLE 23.15   Ketones   (a) Liquids

Name of Compound	Boiling Point	Melting Points of Derivatives		
		Semi-carbazone	2,4-Dinitro-phenylhydrazone	Oxime
Acetone	56	190	128	59
3-Buten-2-one (methyl vinyl ketone)	81	141		
2-Butanone	82	136	117	
3-Butyn-2-one	86		181	
3-Methyl-2-butanone	94	114	120	
3-Methyl-3-buten-2-one	98	173	181	
Cyclobutanone	100		146	
3-Pentanone	102	139	156	
2-Pentanone	102	112	144	
1-Penten-3-one	103		129	
3,3-Dimethyl-2-butanone (pinacolone)	106	158	125	79
1-Methoxy-2-propanone	115		163	
4-Methyl-2-pentanone	117	135	95	
3-Methyl-2-pentanone	118	95	71	
Chloroacetone	119	150	125	
3-Penten-2-one	122	142	155	
2,4-Dimethyl-3-pentanone	124	160	88	
3-Hexanone	125	113	130	
4,4-Dimethyl-2-pentanone	125		100	

(Continued)

TABLE 23.15   Ketones   (a) Liquids *(Continued)*

Name of Compound	Boiling Point	Melting Points of Derivatives		
		Semi-carbazone	2,4-Dinitro-phenylhydrazone	Oxime
2-Hexanone	128	125	106	49
5-Hexen-2-one	130	102	108	
4-Methyl-3-penten-2-one (mesityl oxide)	130	164	206	49
Cyclopentanone	131	210	146	56
5-Methyl-3-hexanone	136	152		
2-Methyl-3-hexanone	136	119		
2,4-Pentanedione (acetylacetone)	139	209	209	149
4-Heptanone	144	132	75	
1-Hydroxy-2-propanone	146	196	129	
3-Heptanone	148	103		
2-Heptanone	151	123	89	
Cyclohexanone	156	167	162	91
2,3-Hexanedione	158			175
3,5-Dimethyl-4-heptanone	162	84		
2-Methylcyclohexanone	165	191	137	43
2,6-Dimethyl-4-heptanone	168	126	66	
4-Octanone	170	96	41	
4-Methylcyclohexanone	171	199		39
2-Octanone	173	123	58	
2,2,6-Trimethylcyclohexanone	179	209	141	
Ethyl acetoacetate	181	133	93	
5-Nonanone	186	90		
3-Nonanone	187	112		
2,5-Hexanedione	194	224	257	137
2-Nonanone	195	119	56	
Acetophenone	202	199	240	60
Menthone	209	189	146	59
2-Methylacetophenone	214	205	159	61
1,5,5-Trimethylcyclohexen-3-one	215	200		79
1-Phenyl-2-propanone	216	200	156	70
Propiophenone	220	174	191	54
3-Methylacetophenone	220	198	207	55
Isobutyrophenone	222	181	163	94
1-Phenyl-2-butanone	226	135		
3-Chloroacetophenone	228	232		88

(Continued)

TABLE 23.15   Ketones   (a) Liquids *(Continued)*

		Melting Points of Derivatives		
Name of Compound	Boiling Point	Semi-carbazone	2,4-Dinitro-phenylhydrazone	Oxime
2,4-Dimethylacetophenone	228	187		63
2-Chloroacetophenone	229	160		113
n-Butyrophenone	230	188	190	50
d-Carvone	230	163	191	73
4-Chloroacetophenone	232	204	231	95
3,5-Dimethylacetophenone	237			114
2-Methoxyacetophenone	239	183		83
3-Methoxyacetophenone	240	196		
n-Valerophenone	248	160	166	52
2,5-Dichloroacetophenone	251			130

TABLE 23.15   Ketones   (b) Solids

		Melting Points of Derivatives		
Name of Compound	Melting Point	Semi-carbazone	2,4-Dinitro-phenylhydrazone	Oxime
4-Methylacetophenone	28	205	260	88
2-Hydroxyacetophenone	28	210	212	118
2,6-Dimethyl-2,5-heptadien-4-one	28	221	118	48
2,4-Dichloroacetophenone	34	208		148
4-Chloropropiophenone	36	176	223	63
4-Methoxyacetophenone	38	198	220	87
2-Hydroxybenzophenone	39			143
2-Methoxybenzophenone	39		251	148
3-Bromopropiophenone	40	183		
4-Phenyl-3-buten-2-one (benzalacetone)	41	187	227	115
1-Indanone	42	233	258	146
4-Bromopropiophenone	46	171		91
Benzophenone	48	165	239	143
4-Bromoacetophenone	51	208	230	128
3,4-Dimethoxyacetophenone	51	218	207	140

(Continued)

TABLE 23.15   Ketones   (b) Solids *(Continued)*

Name of Compound	Melting Point	Melting Points of Derivatives		
		Semi-carbazone	2,4-Dinitro-phenylhydrazone	Oxime
Methyl 2-naphthyl ketone	54	235	262	149
4-Methyl benzophenone	57	121	200	154
Benzalacetophenone (chalcone)	58	170	245	140
α-Chloroacetophenone	59	156	214	89
Desoxybenzoin (benzyl phenyl ketone)	60	148	204	98
Benzoylacetone	61		151	
1,1-Diphenylacetone	61	170		165
4-Methoxybenzophenone	62		180	116
Cinnamalacetone	68	186	223	153
2,6-Dimethyl-1,4-benzoquinone	73			175
4-Chlorobenzophenone	78		185	105
1,4-Cyclohexanedione	79	231	240	188
3-Nitroacetophenone	80	257	228	132
4-Nitroacetophenone	81		258	174
4-Bromobenzophenone	82	350	230	116
9-Fluorenone	83		284	196
4,4′-Dimethylbenzophenone	95	143	219	163
Benzil	95	243	189	237
3-Hydroxyacetophenone	96	195	257	
1,3-Cyclohexanedione	104			156
4-Hydroxyacetophenone	109	199	261	145
3,4-Dihydroxyacetophenone	116			184 *dec*
1,4-Benzoquinone	116	243	231 *di*	240
1,4-Naphthoquinone	125	247	278	198
4-Hydroxybenzophenone	135	194	242	81
Benzoin	137	206 *dec*	245	152
2,4-Dihydroxyacetophenone	147	218	208	200 *dec*
4,4′-Dichlorobenzophenone	148		241	135
Camphor	178	248 *dec*	177	119
4,4′-Dihydroxybenzophenone	210		192	

TABLE 23.16 Nitriles

Name of Compound	Boiling Point	Melting Point
Acrylonitrile	77	
Acetonitrile	81	
Propanenitrile	97	
2-Methylpropanenitrile	108	
Butanenitrile	117	
4-Methylbutanenitrile	130	
Pentanenitrile	141	
4-Methylpentanenitrile	155	
Hexanenitrile	165	
3-Chloropropanenitrile	178	
5-Methylhexanenitrile	180	
Heptanenitrile	183	
Benzonitrile	190	
4-Chlorobutanenitrile	196	
2-Methylbenzonitrile (2-tolunitrile)	205	
Octanenitrile	206	
Ethyl cyanoacetate	207	
3-Methylbenzonitrile (3-tolunitrile)	212	
Nonanenitrile	224	
Phenylacetonitrile (benzyl cyanide)	234	
Decanenitrile	245	
1,3-Dicyanopropane	286	
Cinnamonitrile	256	20
4-Methylbenzonitrile (4-tolunitrile)	217	27
Malononitrile	219	30
4-Chlorobenzyl cyanide	267	30
1-Cyanonaphthalene	299	34
3-Bromobenzonitrile	225	38
3-Chlorobenzonitrile		41
2-Chlorobenzonitrile		43
4-Cyanobutanoic acid		45
3-Cyanopropanoic acid		48
2-Aminobenzonitrile		51
2-Bromobenzonitrile		53
2,4,6-Trimethylbenzonitrile		55
Succinonitrile		57
4-Methoxybenzonitrile		62
3,5-Dichlorobenzonitrile		65

(Continued)

TABLE 23.16   Nitriles *(Continued)*

Name of Compound	Boiling Point	Melting Point
Cyanoacetic acid		67
3,4-Dichlorobenzonitrile		72
Diphenylacetonitrile		75
4-Cyanopyridine		78
3-Cyanobenzaldehyde		80
2-Chloro-6-methylbenzonitrile		82
4-Aminobenzonitrile		86
4-Chlorobenzonitrile		96
2-Cyanophenol		98
2,4-Dinitrobenzonitrile		104
2-Nitrobenzonitrile		110
4-Bromobenzonitrile		112
4-Cyanophenol		113
p-Nitrophenylacetonitrile		116
3-Nitrobenzonitrile		118
2,5-Dichlorobenzonitrile		130
1,2-Dicyanobenzene		141
2,6-Dinitrobenzonitrile		145
4-Nitrobenzonitrile		147
1,3-Dicyanobenzene		162
2-Cyanobenzoic acid		187
3-Cyanobenzoic acid		217
4-Cyanobenzoic acid		219

TABLE 23.17   Nitro Compounds

Name of Compound	Boiling Point	Melting Point	Nitration Product	
			Position	mp
Nitromethane	101			
Nitroethane	115			
2-Nitropropane	120			
1-Nitropropane	131			
2-Nitrobutane	140			
1-Nitrobutane	153			
1-Nitropentane	173			
Nitrobenzene	211		1,3	90
2-Nitrotoluene	222		2,4	71

(Continued)

TABLE 23.17 Nitro Compounds *(Continued)*

Name of Compound	Boiling Point	Melting Point	Nitration Product Position	mp
1,3-Dimethyl-2-nitrobenzene	226	13	1,3,5	182
3-Nitrotoluene	233	16		
1,4-Dimethyl-2-nitrobenzene	241		1,2,4	139
1,3-Dimethyl-4-nitrobenzene	246	2	1,3,5	182
1,2-Dimethyl-3-nitrobenzene	248	15	1,2	82
1-Isopropyl-4-methyl-2-nitrobenzene	264		2,6	54
2-Nitroanisole	273	10	2,4,6	68
2-Methyl-2-nitropropane	127	26		
1,2-Dimethyl-4-nitrobenzene		31	1,2	82
2-Chloronitrobenzene	246	32	2,4	52
2,4-Dichloronitrobenzene	258	33		
2-Chloro-6-nitrotoluene	238	37	4,6	49
3-Nitroanisole		38	3,5	106
4-Chloro-2-nitrotoluene		38	2,6	77
2-Bromonitrobenzene		43	1,3	72
2,4,6-Trimethylnitrobenzene		44	1,3	86
3-Chloronitrobenzene		45		
4-Nitrotoluene	234	52	2,4	70
1-Chloro-2,4-dinitrobenzene		52	2,4,6	183
4-Nitroanisole		53	2,4	89
3-Bromonitrobenzene		56	3,4	59
1-Nitronaphthalene		57		
3,4-Dinitrotoluene		61		
2,6-Dinitrotoluene		66	2,4,6	82
2,4-Dinitrotoluene		70	2,4,6	82
3,5-Dimethylnitrobenzene		75		
4-Chloronitrobenzene	242	84	2,4	52
1,3-Dinitrobenzene		90		
1-Chloro-8-nitronaphthalene		94		
2,4-Dinitroanisole		95	2,4,6	68
4-Nitrobiphenyl		114	4,4′	240
1,2-Dinitrobenzene		118		
4-Bromonitrobenzene		126		
9-Nitroanthracene		146		
1,8-Dinitronaphthalene		170		
1,4-Dinitrobenzene		173		
1,5-Dinitronaphthalene		217		
4,4′-Dinitrobiphenyl		240		

TABLE 23.18 Phenols

Name of Compound	Melting Point	Boiling Point	Melting Points of Derivatives	
			α-Naphthyl-urethane	Bromo Derivative
2-Chlorophenol	7	176	120	76 *di*
Phenol	42	182	132	95 *tri*
2-Methylphenol (*o*-cresol)	31	191	142	56 *di*
2-Bromophenol	5	195	129	95 *tri*
Salicylaldehyde	2	197		
3-Methylphenol (*m*-cresol)	12	202	128	84 *tri*
4-Methylphenol (*p*-cresol)	35	202	146	198 *tetra*
2-Ethylphenol		207		
2,4-Dimethylphenol	27	212	135	
Methyl salicylate		224		
3-Methoxyphenol		243	129	104 *tri*
4-Allyl-2-methoxyphenol		255	122	118 *tetra*
2-Methoxy-4-propenylphenol		268	150	
2-Methoxyphenol	32	205	118	116 *tri*
3-Bromophenol	32	236	108	
3-Chlorophenol	33	214	158	
4-Methylphenol (*p*-cresol)	35	202	146	198 *tetra*
4-Methyl-2-nitrophenol	36			
2,4-Dibromophenol	36			95 *tri*
Phenol	42		132	95 *tri*
4-Chlorophenol	43		166	
2,4-Dichlorophenol	43			68 *mono*
2-Nitrophenol	45		113	117 *di*
4-Ethylphenol	47		128	
4-Chloro-2-methylphenol	49			
2,6-Dimethylphenol	49		176	79 *mono*
5-Methyl-2-isopropylphenol (thymol)	49		160	55 *mono*
4-Methoxyphenol	55			
2,5-Dichlorophenol	59			
3,4-Dimethylphenol	63		142	171 *tri*
4-Bromophenol	64		169	95 *tri*
4-Chloro-3-methylphenol	66		153	
2,6-Dichlorophenol	67			
3,5-Dimethylphenol	68			166 *tri*
3,5-Dichlorophenol	68			189 *tri*
3,4-Dichlorophenol	68			
2,4,5-Trichlorophenol	68			
2,4,6-Trichlorophenol	68			
2,4,6-Trimethylphenol	69			158 *di*

(Continued)

TABLE 23.18   Phenols *(Continued)*

Name of Compound	Melting Point	Boiling Point	Melting Points of Derivatives	
			α-Naphthyl-urethane	Bromo Derivative
2,6-Di-*tert*-butyl-4-methylphenol	70			
2,5-Dimethylphenol	75		173	178 *tri*
8-Hydroxyquinoline	76			
4-Hydroxy-3-methoxybenzaldehyde	81			
1-Naphthol	94		152	105 *di*
2,3,5-Trimethylphenol	96			
2-Methyl-4-nitrophenol	96			
3-Nitrophenol	97		167	91 *di*
4-*tert*-Butylphenol	100		110	50 *mono*
1,2-Dihydroxybenzene (catechol)	105		175	193 *tetra*
3,5-Dihydroxytoluene	106		160	104 *tri*
1,3-Dihydroxybenzene (resorcinol)	110			112 *tri*
2-Chloro-4-nitrophenol	111			
4-Nitrophenol	114		151	145 *di*
2,4-Dinitrophenol	114			118 *mono*
4-Hydroxybenzaldehyde	117			
1,3,5-Trihydroxybenzene	117			151 *tri*
2,3,5,6-Tetramethylphenol	118			118 *mono*
2,4,6-Trinitrophenol	122			
2-Naphthol	123		157	84 *mono*
3-Methyl-4-nitrophenol	129			
1,2,3-Trihydroxybenzene	133			158 *di*
2,4-Dihydroxyacetophenone	147			
Salicylic acid	158			
2,3-Dihydroxynaphthalene	160			
1,4-Dihydroxybenzene (hydroquinone)	171			186 *di*
3,5-Dinitrosalicylic acid	173			
2-Aminophenol	174			
1,4-Dihydroxynaphthalene	176		220	
4-Aminophenol	184			
2,7-Dihydroxynaphthalene	190			
Pentachlorophenol	190			
3-Hydroxybenzoic acid	200			
4-Hydroxybenzoic acid	215			
1,3,5-Trihydroxybenzene (phloroglucinol anhydrous)	217			151 *tri*
1,5-Dihydroxynaphthalene	265			
4,4′-Biphenol	274			

THE LITERATURE OF ORGANIC CHEMISTRY

This chapter is designed to assist those interested in obtaining additional information on different aspects of synthetic, physical, and theoretical organic chemistry. The chapter is not intended to be a comprehensive guide to the literature of organic chemistry. Rather, it should provide the reader with a general knowledge of the important literature sources of organic chemistry and of valuable lead references to initiate a specific search.

24.1 CLASSIFICATION OF THE LITERATURE

For common applications, most of the literature of organic chemistry can be divided into nine major categories: (1) primary research journals, (2) review journals, (3) encyclopedias and dictionaries, (4) abstract journals, (5) advanced textbooks, (6) monographs, (7) general multivolume references, (8) general reference works on synthetic procedures and techniques, and (9) catalogs of physical data. In this section, selected examples from each of these classes are given, along with brief explanatory notes. In Section 24.2, two examples are given to illustrate how the literature may be used to find information about a specific organic compound.

24.1.1 PRIMARY RESEARCH JOURNALS

Primary research journals are the ultimate source of most of the information in organic chemistry; these journals publish original research results in several formats. **Articles** are full papers that provide historical discussions together with a presentation of the important findings, conclusions, and experimental details for preparing new compounds. **Communications** are short articles, sometimes with brief experimental details, that are restricted to a single topic or important discovery; such articles must be especially timely and of general interest to the chemical community. Some journals also publish short papers called **Notes,** that contain limited discussions and experimental details. Some of the more useful journals for the practicing organic chemist and a brief description of their contents follow.

A. *Angewandte Chemie, International Edition in English.* Concurrent translation of the German journal *Angewandte Chemie.* This is an outstanding journal that publishes critical reviews of selected topics and communications covering all areas of chemistry.

B. *Justus Liebigs Annalen der Chemie.* Excellent journal with articles and notes in German and English covering all areas of organic chemistry; one of the two major German journals in organic chemistry.

C. *Biochemistry.* Articles and communications in biochemistry, but many of the contributions tend toward the biological side of organic chemistry.

D. *Bulletin of the Chemical Society of Japan.* Articles and notes in English covering all areas of chemistry.

E. *Canadian Journal of Chemistry.* Articles in English and French in all areas of chemistry.

F. *Chemische Berichte.* Excellent journal with articles and notes in German and English covering all areas of organic and organometallic chemistry; one of the two major German journals in organic chemistry.

G. *Chemistry Letters.* Communications in English covering all areas of organic chemistry.

H. *Helvetica Chimica Acta.* Articles and notes in English, French, or German covering all areas of organic chemistry

I. *Heterocycles.* Reviews, communications, and articles in all areas of heterocyclic chemistry.

J. *Journal of the American Chemical Society.* Articles and communications covering all areas of chemistry. Generally, the articles and communications appearing in this journal, which is one of the foremost chemical journals in the world, tend to have broader interest to chemists of various disciplines.

K. *Journal of the Chemical Society, Perkin Transactions.* Divided into two sections: I. Organic and bio-organic chemistry. II. Physical organic chemistry. Primarily articles but a few communications, in all areas of organic and bio-organic chemistry. Most communications are published in a separate journal titled *Chemical Communications.*

L. *Journal of Medicinal Chemistry.* Articles, communications, and notes in English covering the preparation of new organic compounds having biological activity.

M. *Journal of Organic Chemistry.* Articles, communications, and notes in all areas of organic chemistry; this is arguably the best journal in the world dedicated to publishing work covering the broad scope of organic chemistry and is one of the journals published by the American Chemical Society.

N. *Synlett.* Articles and communications in the general area of synthetic organic chemistry.

O. *Synthesis.* Reviews, articles, and communications in the general area of synthetic organic chemistry, mostly in English but some in German and French.

P. *Tetrahedron.* One of the best international journals dedicated to publishing articles, reviews and "symposia in print" in the general areas of organic and bio-organic chemistry; most articles are in English, but some are in German and French.

Q. *Tetrahedron: Asymmetry.* Reviews, articles, and communications in the specialized area of asymmetric synthesis and methods.

R. *Tetrahedron Letters.* Perhaps the best international journal dedicated to publishing brief two- or four-page communications in English, French, or German in all areas of organic and bio-organic chemistry.

24.1.2   REVIEW JOURNALS

Review journals publish longer articles covering selected topics in various areas of chemistry; these articles do not present new research, but they do reference the original work in the primary journals. Some review journals publish reviews in all areas of chemistry, while others cover only specific areas. Those journals that publish reviews in addition to original research are also listed.

A. *Accounts of Chemical Research.* Provides concise reviews of areas of active research in all areas of chemistry.

B. *Angewandte Chemie, International Edition in English.* (See A in Sec. 24.1.1.)

C. *Annual Reports on the Progress of Chemistry.* A series of timely reports in areas of general interest and importance to organic chemists; since 1968 this has been divided into sections, and Section B is devoted to organic chemistry. Supplemented by a series of *Specialist Periodical Reports,* which cover specific topics such as alkaloids, peptides, carbohydrates, nuclear magnetic resonance, and organometallic compounds.

D. *Chemical Reviews.* A review journal published eight times per year by the American Chemical Society that covers all areas of chemistry.

E. *Chemical Society Reviews.* A review journal with broad scope published quarterly by the Royal Chemical Society (London); succeeded *Quarterly Reviews.*

F. *Heterocycles.* (See I in Sec. 24.1.1)

G. *Journal of Chemical Education.* Contains some reviews written at a level that students and others unfamiliar with the subject can understand. New, tested experiments and modifications of old experiments suitable for organic laboratory courses are frequently published in the monthly issues.

H. *Synthesis.* (See O in Sec. 24.1.1.)

I. *Tetrahedron.* (See P in Sec. 24.1.1.)

J. *Index of Reviews in Organic Chemistry.* Published by the Chemical Society. Second cumulative edition in 1976; supplements in 1979 and 1981. A bibliography of published reviews, organized by topic; useful for locating a review on a particular subject.

24.1.3 ENCYCLOPEDIAS AND DICTIONARIES

Encyclopedias and dictionaries of organic chemistry provide specific information such as physical properties and uses of known compounds. Although all of these are useful, the most exhaustive compilation of such data is found in entry A.

A. *Beilstein's Handbüch der Organischen Chemie,* which was first published in 1881–1883 in two volumes, is perhaps the most complete reference work in any branch of science. In contrast to other collections of physical and other data, those included in the *Beilstein Handbüch* have been critically evaluated and checked for internal consistency by scientists. The fourth edition (1918–1938) contains data on and references for the 140,000 organic compounds known in 1909. This monumental work was originally edited by Friedrich Konrad Beilstein, and the first compilation is known as the **Hauptwerk,** abbreviated H. It is continued by a series of supplements called **Ergänzungswerke,** abbreviated E, the fifth of which is not yet complete. The main work, H, and the first four supplements, E I–E IV, are in German, but the fifth supplement, E V, is published in English. The first supplement (E I) covers the literature for the years 1910–1919, and the second supplement (E II) covers that for the years 1920–1929. Supplements three (E III) and four (E IV) cover the years 1930–1959, and the fifth supplement (E V) covers 1960–1979; Volumes 17–27 of Supplementary Series III and IV, covering heterocyclic compounds, are contained in a joint issue. Thus, the series of the *Beilstein* handbook have the following coverage:

H	Vol 1–27	up to 1909
E I	Vol 1–27	1910–1919
E II	Vol 1–27	1920–1929
E III	Vol 1–16	1930–1949
E III/IV	Vol 17–27	1930–1959
E IV	Vol 1–16	1950–1959
E V	Vol 17–22	1960–1979

The size and complexity of *Beilstein* can make it a daunting work to consult, so an understanding of how it is organized will help make it more approachable and usable. *Beilstein* is arranged by functional classes into a series of systems, and a compound is always treated in the same system, no matter in what supplement it may appear. This feature of the set greatly facilitates location of information; once a **system number** of a compound has been obtained, it is easy to trace it through the whole set, even when an index for a particular volume is not yet available. There are molecular-formula, **Formelregister,** and compound-name, **Sachregister,** indexes for the main set and the first two supplements. In addition, molecular-formula and compound-name indexes, which are called **Gesamtregister,** are now available for all volumes of the **Hauptwerk** and the **Ergänzungswerke** E I–E IV. Of these indexes, the molecular formula index provides the easiest access to *Beilstein* for those with little knowledge of German. Indeed, in order to facilitate the use of *Beilstein,* a **Centennial Index** in English is now being compiled that will allow access via both substance name and formula to all compounds in *Beilstein* for the main work, H, and the supplements E I–E IV, thereby covering the period 1779–1959. Publication of this collective index began in 1991. The first 10 volumes comprise the **Substance Name Index,** and later volumes will include the **Formula Index,** publication of which began in 1992.

In addition to the above indexes, the organic compound tables in the recent *CRC Handbooks of Chemistry and Physics and Lange's Handbook of Chemistry,* which are discussed later, and even catalogues from chemical companies such as Aldrich include references to *Beilstein.* There are several guides and a brief German-English dictionary designed specifically for use with *Beilstein.* The computer program SANDRA (Structure and Reference Analyzer) can also be used to search for a compound in *Beilstein.* SANDRA allows searching by the structure or substructure of a substance and predicts the system number of the requested substance, even if the substance is not yet included in *Beilstein.*

Two additional components complete the *Beilstein Information System.* The *Beilstein Online* database is available in a computerized format through the STN, Dialog, and BRS online vendors. *Beilstein Online* includes information in English from the printed Handbook, 1779–1959 as well as some unchecked data from 1960 onward. The Beilstein Institute also plans to include data from 1980–1989. The online file is indexed and thus searchable in many ways, including by Chemical Abstracts' Registry Number and other ways not available in the print format. While users may be able to search the online file without help, to make the fullest, most cost-effective use of the online file, it is best to consult a specialist for assistance. Beilstein's *Current Facts,* a computer-based information system that uses CD-ROM technology, completes the information system. Updated quarterly, it allows access to new data on previously known substances and provides basic data on new compounds.

B. *CRC Atlas of Spectral Data and Physical Constants for Organic Compounds,* Grasselli, J. G.; Ritchey, W. M., Eds. Chemical Rubber Co., Cleveland, OH, 1975. Six volumes. Gives data on some 21,000 organic compounds.

C. *CRC Handbook of Chemistry and Physics,* annual editions, CRC Press, Boca Raton, FL. Gives physical properties and Beilstein references for about 15,000 organic compounds.

D. *Dictionary of Organic Compounds,* 5th ed., Buckingham, J., Ed. Chapman and Hall, New York, 1982. This edition is in seven volumes. Volumes 1–5 contain the data for the compounds, Volume 6 is a name index with cross-references, and Volume 7 contains a molecular-formula index, a heteroatom index, and Chemical Abstracts Service registry number index. The set is updated in annual supplements, the first of which appeared in 1983.

E. *Handbook of Tables for Identification of Organic Compounds,* 3rd ed., Rapoport, Z., Ed. Chemical Rubber Co., Cleveland, OH, 1967. Gives physical properties and derivatives for more than 4000 compounds, arranged according to functional groups.

F. *Lange's Handbook of Chemistry,* 14th ed., Dean, J. A., Ed. McGraw-Hill, New York, 1992. Gives physical properties for about 6500 organic compounds.

G. *Merck Index of Chemicals and Drugs,* 11th ed., Merck and Co., Rahway, NJ, 1989. Gives a concise summary of the physical and biological properties of more than 10,000 compounds, with some literature references. Organization is alphabetical by name; synonyms and trade names; cross-index and index of Chemical Abstracts Service registry numbers.

24.1.4 ABSTRACT JOURNALS

Abstract journals provide concise summaries of original research articles in journals and listings of reviews and books. Although abstracts are the quickest way to locate chemical information, they are always incomplete and sometimes even misleading, so they should not be relied upon as the final source. A complete literature search should always include reference to the primary research publication.

A. *Chemical Abstracts.* This publication is one of the most important reference sources that indexes the original literature to provide "the key to the world's chemical literature." It began in 1907 and now contains abstracts for nearly half a million items each year, more than 550,000 in 1992, from some 20,000 sources. For material appearing after 1940, *Chemical Abstracts* provides the most complete coverage of the chemical literature. *Chemical Abstracts* is published weekly; at the end of each six months, indexes by author, general subject, chemical substance, patent and molecular formula appear. Prior to 1956, these indexes were cumulated every 10 years, but they are now cumulated at five-year intervals; the most recent covered the period 1987–1991.

Effective use of *Chemical Abstracts,* especially the chemical-substance indexes, requires an understanding of the nomenclature systems used in the past and currently. Current practice is summarized in the 1991 **Index Guide,** which is a listing of indexing terms and cross references; the *Guide* also has the indexing rules and procedures currently used. Supplementing the regular indexes are the **Ring Systems Handbook,** which contains information on the various organic ring compounds, and the **Registry Number Handbook,** which provides a guide to registry numbers. In *Chemical Abstracts,* a unique registry number is assigned to each chemical substance contained in the *Abstracts,* and these numbers are widely used throughout the chemical literature. Articles that describe the preparation or the reactions of a particular substance may be easily located by simply using the registry number of that substance.

It is now possible to search *Chemical Abstracts* using computerized databases provided by several different vendors. For example, Chemical Abstracts Service maintains a database called **CAS On-Line,** that covers the period 1967 to the present and is presently available in most college and university libraries as well as at most chemical companies. There are many advantages to using a computer to search *Chemical Abstracts.* For example, information frequently appears more quickly in the computer file than it does in the printed index. Computer searches using CAS On-Line may be tailored to find information based upon the author, the subject, the name or formula of a particular substance, and chemical structure or substructure. Another key feature of searching *Chemical Abstracts* by computer is that different terms and key words may be linked by Boolean operators to focus the search so that small sets of relevant information may be quickly found. Because there is a modest charge for using CAS On-Line, it is advisable to obtain training and assistance from a search specialist before beginning online searching. That assistance will improve the quality and the cost-effectiveness of the search.

In addition to *Chemical Abstracts,* similar databases provide access to other specialized types of literature, including medicine, patents, polymers, toxicology and pharmacy. There are also databases that provide the entire text of articles from some selected chemistry journals. The realm of online access, especially to scientific information, is growing quickly and changing rapidly. Contact the librarian or information specialist at your institution to find out about new developments in this area.

B. *Chemisches Zentralblatt* (1830–1970) is the second major abstract journal and is published in German. It predates *Chemical Abstracts* by more than 70 years, and prior to 1940 it was more complete and more reliable than *Chemical Abstracts.* For this reason, in an exhaustive search for an organic compound in the years before 1940, it is advisable to consult the collective formula indexes of *Chemisches Zentralblatt.*

C. *Chemical Titles,* published biweekly by Chemical Abstracts Service since 1961, provides lists of the titles of articles in more than 700 chemical journals. The unique value of this publication derives from the fact that not only the title but also every significant word in the title is listed in alphabetical order. Although it is not an abstract journal, it serves a similar function, and a title appears much sooner than an abstract.

D. *Science Citation Index* is a unique reference source because it enables one to search forward from a particular literature reference. For example, to find all of the applications of a reaction or method that was described in a specific paper, one simply looks up the appropriate literature reference for that article in *Science Citation Index,* and all papers published subsequently that referenced the original article in a footnote are listed.

24.1.5   ADVANCED TEXTBOOKS

Advanced textbooks in organic chemistry provide useful information for students who are interested in a more sophisticated or advanced treatment of the information found in typical undergraduate organic textbooks. Furthermore, material regarding a variety of interesting aspects of organic chemistry that are not covered in standard textbooks may be found. A selected listing includes:

A. Carey, F. A.; Sundberg, R. M. *Advanced Organic Chemistry; Part A: Structure and Mechanisms;* and *Part B: Reactions and Synthesis,* 3rd ed. Plenum, New York, 1990. An excellent survey of reactions and their applications.

B. Carruthers, W. *Some Modern Methods of Organic Synthesis,* 3rd. ed. Cambridge University Press, Cambridge, UK, 1986. Selected reactions used in synthetic organic chemistry are discussed in detail.

C. House, H. O. *Modern Synthetic Reactions,* 2nd ed. W. A. Benjamin, Menlo Park, CA, 1972. The general classes of reactions used for reduction, for oxidation, and for the formation of new carbon-carbon bonds are surveyed in terms of their scope, limitations, stereochemistry, and mechanisms.

D. Lowry, T. H.; Richardson, K. S. *Mechanism and Theory in Organic Chemistry,* 3rd ed. Harper and Row, New York, 1987. An excellent treatment of mechanistic and theoretical aspects of organic chemistry.

E. March, J. *Advanced Organic Chemistry: Reactions, Mechanisms, and Structures,* 4th ed., Wiley-Interscience, New York, 1992. Advanced text with broad coverage and many references to the original literature.

24.1.6 MONOGRAPHS

Numerous monographs contain reviews of selected topics in organic and related areas of chemistry. A partial listing follows:

A. Collman, J. P.; Hegedus, L. S.; Norton, J. R.; Finke, R. G. *Principles and Applications of Organotransition Metal Chemistry,* University Science Books, Mill Valley, CA, 1987. An excellent survey of organometallic chemistry

B. Eliel, E. L.; Allinger, N. L.; Angyal, S. J.; Morrison, G. A. *Conformational Analysis,* Interscience, New York, 1967. An important reference work covering all major aspects of conformational analysis.

C. Green, T. W.; Wuts, P. G. M. *Protective Groups in Organic Synthesis,* 2nd ed., Wiley-Interscience, New York, 1991. An extensive compilation of methods for protection and deprotection of various functional groups.

D. Larock, R. C. *Comprehensive Organic Transformations: A Guide to Functional Group Preparation,* VCH Publishers, New York, 1989. An exhaustive compilation of reactions of functional groups.

E. Paquette, L. A. *Principles of Modern Heterocyclic Chemistry,* W. A. Benjamin, New York, 1968. An introduction to the basic principles of heterocyclic chemistry.

F. Perrin, D. D.; Armarego, W. L. F.; Perrin, D. R. *Purification of Laboratory Chemicals,* 3rd Ed., Pergamon Press, Oxford, UK, 1988. Complete procedures for purifying organic solvents and reagents.

G. Turro, N. J. *Modern Molecular Photochemistry,* Benjamin/Cummings, Menlo Park, CA, 1978. A treatise on the applications of photochemistry in organic chemistry.

24.1.7 GENERAL, MULTIVOLUME REFERENCES

General, multivolume references provide a comprehensive review of various topical and general aspects of organic chemistry. A partial listing follows.

A. *Asymmetric Synthesis,* Morrison, J. D., Ed., Vol 1–5, Academic Press, New York, 1983–1985. Series dedicated to critical review of modern techniques of organic asymmetric synthesis.

B. *Comprehensive Heterocyclic Chemistry,* Katritzky, A. R.; Rees, C. W., Eds., Pergamon Press, Oxford, 1984. An eight-volume treatise covering the reactions, structure, synthesis, and uses of heterocyclic compounds.

C. *Comprehensive Organic Chemistry,* Barton, D. H. R.; Ollis, W. D., Eds., Pergamon Press, Oxford, UK, 1978. A six-volume treatise on the synthesis and reactions of organic compounds written by more than 100 authors, with over 20,000 references to the original literature.

D. *Comprehensive Organic Synthesis,* Trost, B. M.; Fleming, I., Eds., Pergamon Press, Oxford, UK, 1991. A nine-volume treatise with index of all aspects of synthetic organic chemistry including carbon-carbon bond formation, heteroatom manipulation, and oxidation and reduction.

E. *Comprehensive Organometallic Chemistry,* Wilkinson, G., Ed., Pergamon Press, Oxford, 1982. A nine-volume treatise covering the synthesis, reactions, and structure of organometallic compounds

F. *The Chemistry of the Functional Groups,* Patai, S., Senior Ed., Wiley-Interscience, New York. Series devoted to an in-depth review of each of the functional groups.

G. *Methoden der Organischen Chemie* (Houben-Weyl), 1952–present, Georg Thieme Verlag, Stuttgart. As of 1987 consisted of 20 volumes with supplements reviewing all classes of organic compounds; originally in German but more recent contributions are in English.

H. *Rodd's Chemistry of Carbon Compounds,* 2nd ed., Coffey, S., Ed., Elsevier, Amsterdam; 1964–present. A comprehensive survey of all classes of organic compounds, giving properties and syntheses for many individual compounds; consists of four volumes in 30 parts, with supplements to maintain currency.

24.1.8   REFERENCE WORKS ON SYNTHETIC PROCEDURES AND TECHNIQUES

It is often necessary to survey the scope and limitations of a reaction, a synthetic method, a reagent or a technique to determine whether it may be applied to solving a specific problem. To facilitate access to this information, a number of reference works are available that contain reviews of reactions, reagents, techniques and methods in organic chemistry; often these are serial in nature.

A. *Annual Reports in Organic Synthesis,* Academic Press, New York, 1970–present. An annual review of synthetically useful reactions.

B. *Compendium of Organic Synthetic Methods,* Vol 1–7, John Wiley & Sons, New York, 1971–1992. These volumes present in outline form possible interconversions among the major functional groups of organic compounds. References to the primary literature sources are given.

C. *Organic Functional Group Preparations,* 2nd ed., Sandler, W.; Karo, W., Eds., Vol 1, 1983; Vol 2, 1986, Academic Press, New York. A listing of different methods for preparing all functional groups.

D. *Organic Reactions,* by various contributors. John Wiley & Sons, New York, 1942–present, 42 volumes through 1992. Each volume contains from 5 to 12 chapters, each of which deals with an organic reaction of wide applicability. Typical experimental procedures are given in detail, and extensive tables of examples with references are given. Each volume contains a cumulative author and chapter-title index.

E. *Organic Syntheses,* John Wiley & Sons, New York, 1932–present, 71 volumes through 1992. Every ten volumes have been collected, indexed, and published as *Collective Volumes I–VI,* through Volume 59 as of 1988. Beginning with the *Collective Volume VII,* which contains Volumes 60–64, the collective volumes are compiled every five years. Detailed directions for synthesis of more than 1000 compounds are given. The procedures have all been thoroughly checked by independent investigators before publication. Many of the methods are general and can be applied to the synthesis of related compounds other than those described. The collective volumes contain indexes of formulas, names, types of reaction, types of compounds, purification of solvents and reagents, and illustrations of special apparatus. A cumulative index to the first five collective volumes was published in 1976, and an excellent reaction guide has been published that provides a compilation of all the reactions in *Collective Volumes I–VII* and the annual volumes 65–68.

F. *Synthetic Methods of Organic Chemistry,* Theilheimer, W.; Finch, A. F., Eds., Interscience, New York, Karger, Basel, 1948–present, 46 volumes through 1992. Emphasis is upon functional group interconversions and chemistry and general reactions that may be used to construct carbon-carbon and carbon-heteroatom bonds. The reactions are classified by symbols that are arranged systematically. There are cumulative indexes.

G. *Techniques of Organic Chemistry,* 3rd ed., Weissberger, A., Ed., Interscience, New York, 1959–present; some volumes have been revised. Title broadened in 1970 to *Techniques of Chemistry.* Examples of useful volumes are *Elucidation of Organic Structures,* Vol 4, parts 1–3; *Investigation of Rates and Mechanisms of Reactions,* Vol 6, parts 1 and 2; *Separation and Purification,* Vol 12; and *Thin Layer Chromatography,* Vol 14.

H. *Vogel's Elementary Practical Organic Chemistry,* 3rd ed. Vol 1; Smith, B. V.; Waldron, N. M., Eds., Longman, London, 1980. Good coverage of basic experimental techniques with a selection of contemporary laboratory preparations.

I. *Vogel's Textbook of Practical Organic Chemistry,* 4th ed., Furniss, B. S., et al., Eds., Longman, London, 1978. For the more experienced researcher; furnishes additional tips on technique and a wide range of preparations.

J. Buehler, C. A.; Pearson, D. E. *Survey of Organic Syntheses.* Wiley-Interscience, New York, Vol 1, 1970; Vol 2, 1972. This extensive two-volume work covers the principal methods of synthesizing the main types of organic compounds. The limitations of the reactions, the preferred reagents, the newer solvents, and experimental conditions are considered.

K. Fieser, M.; Fieser; L. F. *Reagents for Organic Synthesis.* Wiley-Interscience, New York, 1967-present, 16 volumes as of 1992. These volumes list some 8000 reagents and solvents, described in terms of methods of preparation or source, purification, and use in typical reactions. Ample references to original literature are given.

L. Hilgetag, G; Martini, A. *Preparative Organic Chemistry,* John Wiley & Sons, New York, 1972. An extensive compilation of reactions and methods for interconversion of functional groups and forming carbon-carbon bonds.

M. Wagner, R. B.; Zook, H. D. *Synthetic Organic Chemistry,* John Wiley and Sons, New York, 1953. An indispensible compilation of synthetic methods that may be used to prepare organic molecules having one or more functional groups.

24.1.9 CATALOGS OF PHYSICAL DATA

A problem commonly encountered in organic chemistry is identifying a compound that has been obtained from a chemical reaction or isolated from a natural source or the environment is a commonly encountered problem in organic chemistry. The spectroscopic characteristics of a sample provide important clues to the identity of the substance. Collections of spectral data of known compounds provide an important source of such information that may be used either to identify known compounds or to assist in the determination of the structure of an unknown substance. Some useful compilations of spectral data are available.

A. ^{1}H NMR Spectra

Aldrich Library of NMR Spectra, 2nd ed., Pouchert, C. J., Ed. Aldrich Chemical Company, Milwaukee, WI, 1983; two volumes. A collection of about 37,000 spectra.

High Resolution NMR Spectra Catalog, compiled by the staff of Varian Associates, Palo Alto, CA, 1962–1963; two volumes. Hydrogen NMR spectra of 587 representative organic molecules are collected, and the peaks are assigned to the hydrogen nuclei responsible for the absorptions.

Nuclear Magnetic Resonance Spectra, Sadtler Research Laboratories, Philadelphia. Hydrogen NMR spectra of more than 55,000 compounds have been published as of 1992, with 1000 being added annually. Assignment of peaks are made as in the Varian spectra, and integration of the signals is shown on many of the spectra.

B. ^{13}C NMR Spectra

Breitmaier, E.; Haas, G.; Voelter, W. *Atlas of Carbon-13 NMR Data.*; two volumes plus index. Heyden, Philadelphia, 1979. Tabular data on 3017 compounds, with chemical shifts for ^{13}C given and ^{1}H–^{13}C multiplicities indicated.

Carbon-13 NMR Nuclear Magnetic Resonance Spectra, Sadtler Research Laboratories, Philadelphia, beginning in 1976. By 1992, 32,000 proton-decoupled spectra had been published.

C. IR Spectra

Sadtler Standard Spectra, Midget Edition, Sadtler Research Laboratories, Philadelphia. As of 1992, the prism series had 83,000 spectra and the grating series had 83,000 spectra.

Aldrich Library of Infrared Spectra, 2nd ed., Pouchert, C. J., Ed. Aldrich Chemical Company, Milwaukee, WI, 1975. A collection of about 10,000 spectra.

D. UV Spectra

Sadtler Standard Spectra: Ultraviolet Spectra, Sadtler Research Laboratories, Philadelphia. Through 1984, contains 75,000 spectra.

24.2 USE OF THE LITERATURE OF ORGANIC CHEMISTRY

The literature outline given in this chapter may be used in a variety of ways, according to the aims and needs of different courses and the library facilities available. Even in those cases where the pressure of time and/or lack of facilities preclude the use of literature beyond the pages of this textbook itself, this chapter is still valuable to the student who decides to go further in the study of organic chemistry.

In many organic laboratory courses, instructors make part of the experimentation open-ended—encouraging the students to plan and carry out experiments with some independence. This is a desirable objective, but there is an element of risk if the experimental procedure has not been carefully checked; *procedures found in the literature are not always easily reproduced!* One of the better sources of different experiments is *Organic Syntheses;* although the experiments are typically reported on a large scale, they may often be easily scaled down. The experiments that occasionally appear in the *Journal of Chemical Education* also deserve mention.

For more experience in identifying unknown organic compounds, the reference books cited at the end of the different sections in Chapter 8 are most useful. The catalogs of spectra listed in Section 24.1.9 represent a vast reservoir of spectra that may be used for instructional purposes, although many molecules give IR and ^{1}H NMR spectra that are not easily interpreted by beginners.

The synthesis of known and unknown compounds is a task commonly encountered by the research organic chemist. In the course of this work, it is often necessary to determine whether a particular compound has been previously prepared; if it has been, then the questions of when, how, and by whom arise. If the compound has not been reported, then the organic chemist searches for similar compounds that might have been prepared, because the same methods might apply to the synthesis of the substance of interest. The examples that follow will provide some guidelines to performing literature searches to solve such problems and the exercises at the end pose many types of problems that are routinely encountered in research.

24.2.1 WHAT IS MUSTARD GAS?

"Mustard gas" is one of the names that has been applied to the compound ClCH$_2$CH$_2$–S–CH$_2$CH$_2$Cl. Find the answers to the following questions: (1) Has this compound been synthesized or isolated? (2) By whom? (3) When? (4) Where can the most recent information on this compound be found?

First, write the molecular formula as C$_4$H$_8$Cl$_2$S, and look in *Beilstein's* General Formelregister, Zweites Ergänzungswerk. On page 65 is the entry "β,β′-Dichlordiäthylsulfid, Senfgas **1**, 349, I 174, II 348, 940," and just below it the entry "α,α′-Dichlordiäthylsulfid (**1**) II 685." The first entry is recognized as that pertaining to the subject compound. The references are to page 349 in Volume 1 of the main work, page 175 in Volume 1 of the first supplement, and to pages 348 and 940 in Volume 1 of the second supplement. Although the third supplement was published after the general index, by noting the system number of the subject compound, which is 23, one may locate it on page 1382 of the third supplement using the index in the appropriate volume.

Referring to page 349 in Volume 1 of the main work, one finds "β,β'-Dichlordiäthylsulfid $C_4H_8Cl_2S$ = $(CH_2Cl \cdot CH_2)_2S$. B. Aus Thiodiglykol S $(CH_2 \cdot CH_2 \cdot OH)_2$ und PCl_3 (V. Meyer, B. **19**, 3260)." This translates: "B. = Bildung, Formation. From thiodiglycol and PCl_3 (V. Meyer, *Berichte,* **19**, 3260)." Looking up the reference in the *Berichte der Deutschen Chemischen Gesellschaft* (Vol. 19, p. 3260, published in 1886) reveals that Victor Meyer first prepared **1** in two steps as illustrated in Equations (24.1) and (24.2).

$$\text{(24.1)}$$

$$\text{(24.2)}$$

Thiodiglykol   Mustard gas
1

The entry in the main work of *Beilstein* (Vol. 1, p. 349) gives in six lines the boiling point and solubility properties of **1** and one chemical reaction. The final two words are *Sehr giftig,* "very poisonous"—a terse commentary on a material that was much feared as a lethal military weapon in World War II, but was never used. Apparently the compound does not adversely affect everyone who is exposed to it, because Meyer noted in his *Berichte* article that his laboratory assistant developed skin eruptions and eye inflammation after preparing the compound, but he suffered no ill effects even though he took no precautions in handling it. In the first supplement on page 175, 15 lines are devoted to "β,β'-diäthylsulfid," and in the second supplement (p. 348), five and one-half pages appear, indicating the increased interest in this compound during the period 1920–1929.

A search of *Chemical Abstracts* may be begun by using the *Chemical Abstracts* Collective Formula Index for 1920–1946. Thus, under the molecular formula $C_4H_8Cl_2S$ is found the entry "(See also Sulfide, bis(chloroethyl)) Sulfide, 1-chloroethyl 2-chloroethyl, **25**:2114[8]," wherein the reference is to volume 25, page 2114. Since "Sulfide, 1-chloroethyl 2-chloroethyl" is not the compound of interest, we look in the *Chemical Abstracts* Decennial Subject Index, 1917–1926, for the entry "Sulfide bis (β-chloroethyl)," which is followed by the names *mustard gas, yperite* and by six general references and two columns of more specific references beginning with "absorption by skin, mechanism of, **14**: 300[4]" and ending with "toxicity and skin-irritant effects of, **15**: 1943[6]."

The *Chemical Abstracts* Decennial Subject Indexes can be used for more recent decades, but *one must be on guard for changes in nomenclature.* Consequently, it is usually advantageous to use formula indexes first. For example, upon examining the January–June 1972 Formula Index, under $C_4H_8Cl_2S$ we find "Ethane, 1,1'-thiobis[2-chloro-" as the name for our compound, with seven references to abstracts. Under this name in the *Chemical Substance Index* for the same period, we find the same seven references, but with specific subject headings such as "DNA, cross-linking induced by, 95344w."

24.2.2 IS A COMPOUND OF INTEREST KNOWN?

A second example of a search illustrates how *Beilstein* may be used to find compounds reported after 1929, the last year covered by the Formula Index of Volume 29. The preparation of compound 3-(2-furyl)-1-(3-nitrophenyl)propenone (**2**) is described in Section 18.2. If one looks for this compound in the *Beilstein* Formelregister (Vol. 29), under the molecular formula $C_{13}H_9NO_4$, no name fitting the structure is found, indicating that this compound was not reported prior to 1929. However, the analogous compound without the nitro group is listed under the formula $C_{13}H_{10}O_2$ in the Formelregister, with the notation *17* 353; II 377, which means that it is described in Vol. 17 of the Hauptwerk on p. 353 and in the Zweites Ergänzungswerk on p. 377. Looking up these entries, one finds that the system number for this compound is 2467. In Vol. 17 of the combined Drittes und Viertes Ergänzungswerk (E III/IV), one then finds that compounds having System Number 2467 are in Part 6 of Vol. 17. In the formula index of Part 6 under $C_{13}H_9NO_4$ one finds the name of the compound sought, followed by the page number 5262, where a description of the preparation and properties of this compound is given. Alternatively, the Gesamtregister Formelregister for Volumes 17 and 18 directs one to the same page in Volume 17.

2

As mentioned earlier, *Chemical Abstracts* Collective Formula Indexes may conveniently be used for locating compounds appearing in the literature after 1929. The 1920–1946 Index has no entry for the compound located in *Beilstein* as described earlier, but the 1947–1956 Index lists under the formula $C_{13}H_9NO_4$ the name "Acrylophenone, 3-(2-furyl)-3′-nitro", which may, *or may not*, be recognized as another name for the compound. Both *Beilstein* E III/IV, *17* and *Chemical Abst*racts, *43,* 3429g (1949) refer to the primary source, the *Journal of the American Chemical Society, 71,* 612 (1949), where D. L. Turner describes the first preparation of this compound.

This example illustrates the advantage of using the formula indexes, of *Beilstein* and/or *Chemical Abstracts* for locating an organic compound in the literature. Several different names may be used for the same compound, but the formula will be more distinctive, and one can be on the lookout for different ways of naming the compound.

EXERCISES

1. Find the melting points of the following crystalline derivatives, none of which are listed in the tables of Chap. 23: (a) 2,4-dinitrophenylhydrazone of trichloroacetaldehyde; (b) semicarbazone of 3-methylcyclohexanone; (c) 3,5-dinitrobenzoate of 1,3-dichloro-2-propanol; (d) amide of 2-methyl-3-phenylpropanoic acid; (e) benzamide of 4-fluoroaniline.

2. Locate an article or a chapter on each of the following types of organic reactions: (a) the aldol reaction; (b) the Wittig reaction; (c) reactions of diazoacetic esters with unsaturated compounds; (d) hydration of alkenes and alkynes through hydroboration; (e) metalation with organolithium compounds; (f) reactions of lithium dialkylcuprates; (g) asymmetric synthesis of amino acids.

3. Give a literature reference for a practical synthetic procedure for each of the following compounds and state the yield that may be expected: (a) 1,2-dibromocyclohexane; (b) α-tetralone; (c) 3-chlorocyclopentene; (d) 2-carboethoxycyclopentanone; (e) norcarane; (f) tropylium fluoborate; (g) 1-methyl-2-tetralone; (h) adamantane; (i) buckminsterfullerene (buckyball).

4. Locate descriptions of procedures for the preparation or purification of the following reagents and solvents used in organic syntheses: (a) Raney nickel catalysts; (b) sodium borohydride; (c) dimethyl sulfoxide; (d) sodium amide; (e) diazomethane.

5. Find IR spectra for the following compounds: (a) N-cyclohexylbenzamide; (b) 4,5-dihydroxy-2-nitrobenzaldehyde; (c) benzyl acetate; (d) diisopropyl ether; (e) 3,6-diphenyl-2-cyclohexen-1-one; (f) 4-amino-1-butanol.

6. Find ^{1}H and ^{13}C NMR spectra of the following compounds: (a) benzyl acetate; (b) diisopropyl ether; (c) 4-amino-1-butanol; (d) 1-propanol; (e) indan.

7. N-Mesityl-N'-phenylformamidine (3) was first synthesized between 1950 and 1960. Find the primary research article in which this compound is described, and write an equation for the reaction used to prepare it.

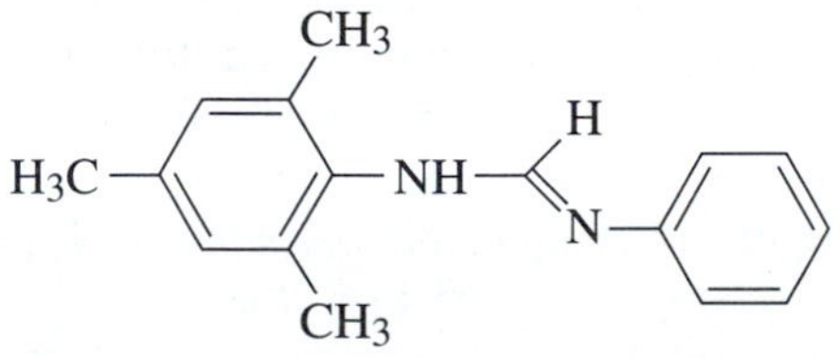

3

8. N-Phenyl-N'-p-tolylformamidine (4) (the German name for this compound is the same as in English except that the final "e" is omitted) is reported in Beilstein to have a melting point of 86 °C. If you check the first reference given in Beilstein, however, you will find the surprising fact that the same chemist who reported this pure compound to have a melting point of 86 °C had described it as melting at 103.5–104.5 °C two years previously. The discrepancy between these reports was not explained until the ambiguity was reexamined during the period 1947–1956. Find the article that solved this mystery.

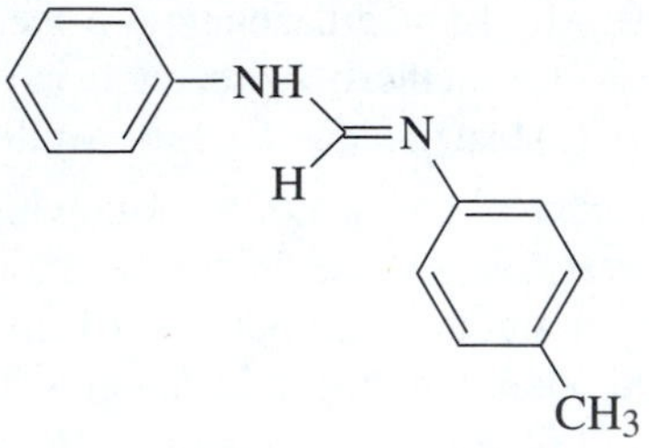

4

9. The *N*-benzoyl derivative of α-phenylethanamine (**5**) was first described in the form of the optically active (–)-isomer in 1905. Using the formula index of *Beilstein* and given the German name "benzosäure-1-α-phenyläthylamid," find the first reference to this compound in a primary research journal. If you can read the German, give (a) the method of preparation by writing the equation; (b) the melting point of the pure compound and the recrystallization solvent; and (c) the $[\alpha]_D$. Find an article published in English between 1910 and 1920 for a description of the *N*-benzoyl derivative of **5**.

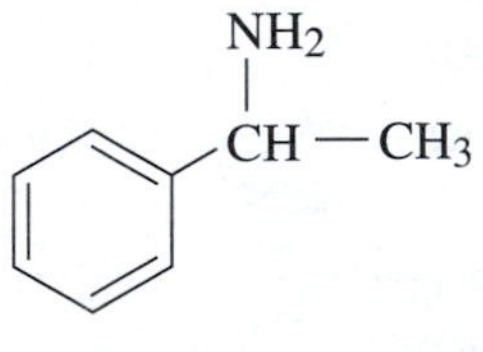

5

10. The name used in the formula indexes of *Chemical Abstracts* for the compound described in Exercise 9 is "benzamide, *N*-methylbenzyl" or "benzamide, *N*-α-methylbenzyl." (a) Find a second reference to the (-)-isomer that was published between 1930 and 1940, and compare the physical constants given there with the earlier data. (b) Find a reference to a paper published in a Czechoslovakian journal between 1950 and 1960 giving data on the racemic form of the compound. (c) Find a reference to data on the (+)-isomer of the compound in a paper published between 1960 and 1970.

11. Determine whether or not each of the following compounds whose names are provided below has ever been synthesized, and if it has, give the reference to the first appearance of its synthesis in the literature.

(a) Vitamin A	(f) Penicillin G	(k) Morphine
(b) Strychnine	(g) Prostaglandin E_2	(l) Vitamin B_{12}
(c) Testosterone	(h) Lysergic acid diethylamide	(m) Cocaine
(d) Cephalosporin	(i) Erythromycin A	(n) Taxol
(e) Reserpine	(j) Vinblastine	(o) Vernolepin

12. Determine whether or not each of the compounds **(a)**–**(l)** has ever been synthesized; if it has, give the reference to the first appearance of its synthesis in the literature.

(a) CH₂CH₂NMe₂

OCH₃
OCH₃

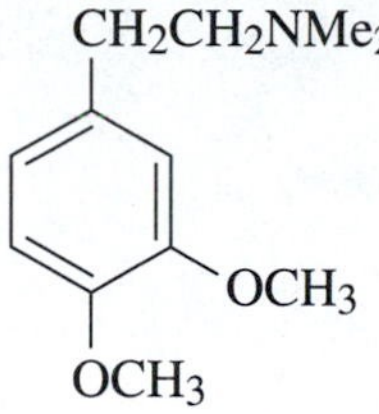

(b) H₃C, CH₃
C=C=C=C
H₃C, CH₃

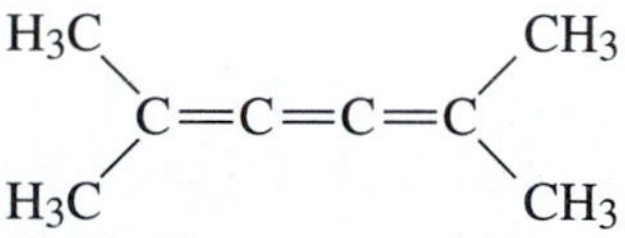

(c)

(d)

(e)

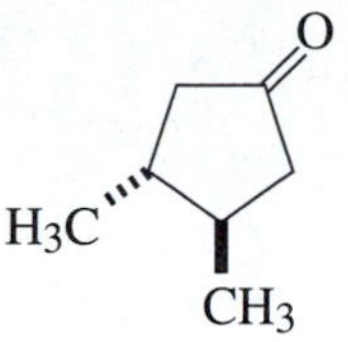

H₃C
CH₃

(f) H

H

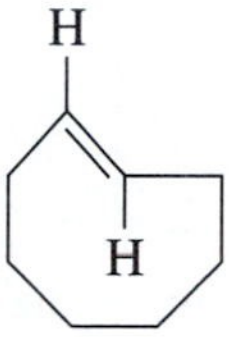

(g)

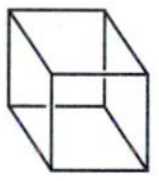

(h)

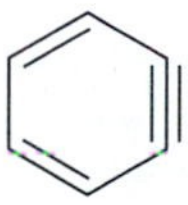

(i) H₃CO, OCH₃
C=C
H₃CO, OCH₃

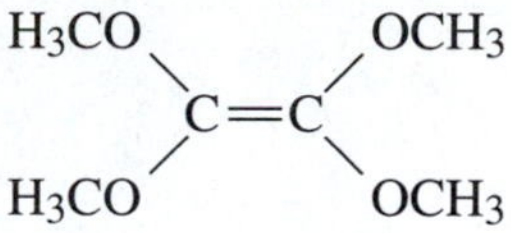

(j)

(k) Cl, O, Cl

Cl, O, Cl

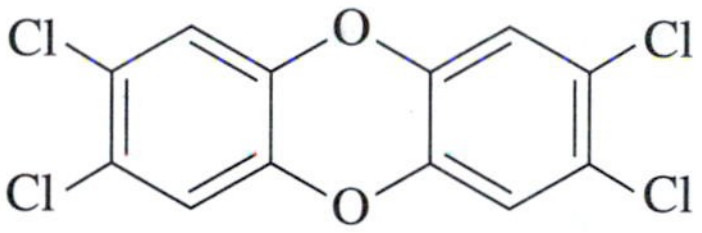

(l)

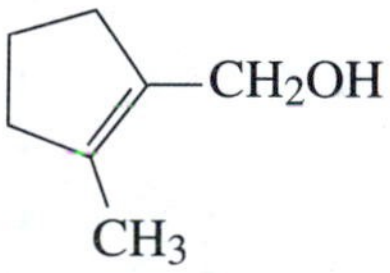

—CH₂OH

CH₃

13. Using *Science Citation Index,* list five research papers by complete title and journal citation that have cited the review by Martin, S. F. *Tetrahedron* **1980,** *36,* 419.

Index*

*__Boldface__ page numbers indicate pages on which the structures of compounds appear.

Approximate ^{1}H and ^{13}C NMR Shifts

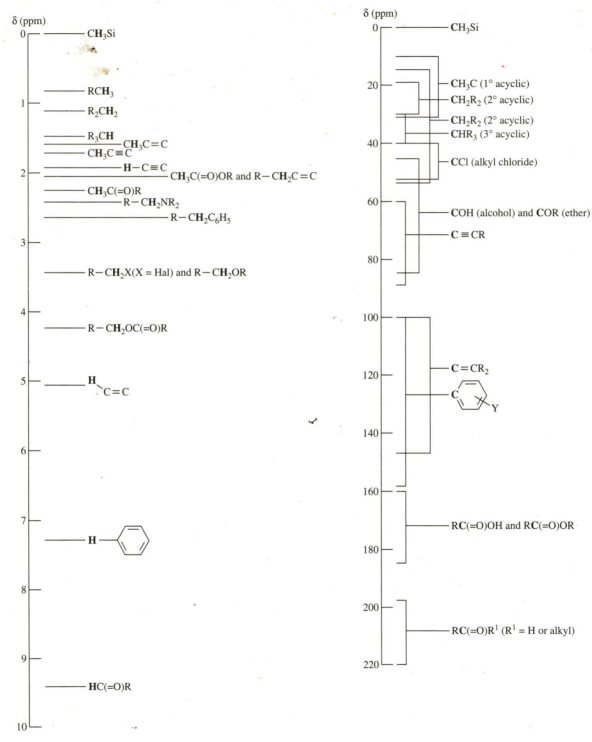

δ (ppm)

0 — CH$_3$Si

1 — RCH$_3$

— R$_2$CH$_2$

— R$_3$CH
— CH$_3$C=C
— CH$_3$C≡C
2 — H−C≡C
— CH$_3$C(=O)OR and R−CH$_2$C=C
— CH$_3$C(=O)R
— R−CH$_2$NR$_2$
— R−CH$_2$C$_6$H$_5$

3 — R−CH$_2$X(X = Hal) and R−CH$_2$OR

4 — R−CH$_2$OC(=O)R

5 — $\overset{H}{\underset{}{}}$C=C

6

7 — H−⬡

8

9

— HC(=O)R

10

δ (ppm)

0 — CH$_3$Si

20 — CH$_3$C (1° acyclic)
— CH$_2$R$_2$ (2° acyclic)
— CH$_2$R$_2$ (2° acyclic)
40 — CHR$_3$ (3° acyclic)
— CCl (alkyl chloride)

60 — COH (alcohol) and COR (ether)
— C≡CR

80

100

120 — C=CR$_2$
— C⬡Y
140

160
— RC(=O)OH and RC(=O)OR
180

200
— RC(=O)R^1 (R^1 = H or alkyl)
220